CAMBRIDGE STUDIES IN ADVANCED MATHEMATICS 203

HOMOLOGICAL METHODS IN BANACH SPACE THEORY

Many researchers in geometric functional analysis are unaware of algebraic aspects of the subject and the advances they have permitted in the last half century. This book, written by two world experts on homological methods in Banach space theory, gives functional analysts a new perspective on their field and new tools to tackle its problems. All techniques and constructions from homological algebra and category theory are introduced from scratch and illustrated with concrete examples at varying levels of sophistication. These techniques are then used to present both important classical results and powerful advances from recent years. Finally, the authors apply them to solve many old and new problems in the theory of (quasi-) Banach spaces and outline new lines of research. Containing a lot of material unavailable elsewhere in the literature, this book is the definitive resource for functional analysts who want to know what homological algebra can do for them.

Félix Cabello Sánchez is Professor of Mathematics at the Universidad de Extremadura. He is co-author of the monograph *Separably Injective Banach Spaces* (2016).

Jesús M. F. Castillo is Professor of Mathematics at the Universidad de Extremadura. He is co-author of the monographs *Three-Space Problems in Banach Space Theory* (1997) and *Separably Injective Banach Spaces* (2016).

CAMBRIDGE STUDIES IN ADVANCED MATHEMATICS

All the titles listed below can be obtained from good booksellers or from Cambridge University Press. For a complete series listing, visit www.cambridge.org/mathematics.

Already Published
165 P. Gille & T. Szamuely *Central Simple Algebras and Galois Cohomology (2nd Edition)*
166 D. Li & H. Queffelec *Introduction to Banach Spaces, I*
167 D. Li & H. Queffelec *Introduction to Banach Spaces, II*
168 J. Carlson, S. Müller-Stach & C. Peters *Period Mappings and Period Domains (2nd Edition)*
169 J. M. Landsberg *Geometry and Complexity Theory*
170 J. S. Milne *Algebraic Groups*
171 J. Gough & J. Kupsch *Quantum Fields and Processes*
172 T. Ceccherini-Silberstein, F. Scarabotti & F. Tolli *Discrete Harmonic Analysis*
173 P. Garrett *Modern Analysis of Automorphic Forms by Example, I*
174 P. Garrett *Modern Analysis of Automorphic Forms by Example, II*
175 G. Navarro *Character Theory and the McKay Conjecture*
176 P. Fleig, H. P. A. Gustafsson, A. Kleinschmidt & D. Persson *Eisenstein Series and Automorphic Representations*
177 E. Peterson *Formal Geometry and Bordism Operators*
178 A. Ogus *Lectures on Logarithmic Algebraic Geometry*
179 N. Nikolski *Hardy Spaces*
180 D.-C. Cisinski *Higher Categories and Homotopical Algebra*
181 A. Agrachev, D. Barilari & U. Boscain *A Comprehensive Introduction to Sub-Riemannian Geometry*
182 N. Nikolski *Toeplitz Matrices and Operators*
183 A. Yekutieli *Derived Categories*
184 C. Demeter *Fourier Restriction, Decoupling and Applications*
185 D. Barnes & C. Roitzheim *Foundations of Stable Homotopy Theory*
186 V. Vasyunin & A. Volberg *The Bellman Function Technique in Harmonic Analysis*
187 M. Geck & G. Malle *The Character Theory of Finite Groups of Lie Type*
188 B. Richter *Category Theory for Homotopy Theory*
189 R. Willett & G. Yu *Higher Index Theory*
190 A. Bobrowski *Generators of Markov Chains*
191 D. Cao, S. Peng & S. Yan *Singularly Perturbed Methods for Nonlinear Elliptic Problems*
192 E. Kowalski *An Introduction to Probabilistic Number Theory*
193 V. Gorin *Lectures on Random Lozenge Tilings*
194 E. Riehl & D. Verity *Elements of ∞-Category Theory*
195 H. Krause *Homological Theory of Representations*
196 F. Durand & D. Perrin *Dimension Groups and Dynamical Systems*
197 A. Sheffer *Polynomial Methods and Incidence Theory*
198 T. Dobson, A. Malnič & D. Marušič *Symmetry in Graphs*
199 K. S. Kedlaya *p-adic Differential Equations*
200 R. L. Frank, A. Laptev & T. Weidl *Schrödinger Operators:Eigenvalues and Lieb–Thirring Inequalities*
201 J. van Neerven *Functional Analysis*
202 A. Schmeding *An Introduction to Infinite-Dimensional Differential Geometry*
203 F. Cabello Sánchez & J.M.F. Castillo *Homological Methods in Banach Space Theory*
204 G. P. Paternain, M. Salo & G. Uhlmann *Geometric Inverse Problems*

Homological Methods in Banach Space Theory

FÉLIX CABELLO SÁNCHEZ
Universidad de Extremadura

JESÚS M. F. CASTILLO
Universidad de Extremadura

CAMBRIDGE
UNIVERSITY PRESS

CAMBRIDGE
UNIVERSITY PRESS

Shaftesbury Road, Cambridge CB2 8EA, United Kingdom

One Liberty Plaza, 20th Floor, New York, NY 10006, USA

477 Williamstown Road, Port Melbourne, VIC 3207, Australia

314–321, 3rd Floor, Plot 3, Splendor Forum, Jasola District Centre,
New Delhi – 110025, India

103 Penang Road, #05–06/07, Visioncrest Commercial, Singapore 238467

Cambridge University Press is part of Cambridge University Press & Assessment,
a department of the University of Cambridge.

We share the University's mission to contribute to society through the pursuit of education, learning and research at the highest international levels of excellence.

www.cambridge.org
Information on this title: www.cambridge.org/9781108478588

DOI: 10.1017/9781108778312

First published 2023

A catalogue record for this publication is available from the British Library.

ISBN 978-1-108-47858-8 Hardback

Contents

Preface

This is a book about homological techniques applied to Banach and quasi-Banach space theory. It has been our aim to show how much this modern branch of functional analysis has to offer analysts. To do so, we will present the basic elements and techniques of homological algebra in the concrete categories of (quasi-) Banach spaces and will show how to use them and how they work to approach and solve classical problems in analysis. Years of working collaboration with many mathematicians convinced us of a simple truth: people do not understand diagrams. This is the cause of the seasickness so customary when sailing the homological sea. There are good reasons for that; mainly, that the foundational ideas, basic principles and visual techniques that make the theory powerful are couched in categorical jargon and not in the language of functional analysis. Thus considerable effort is required simply to decode the information.

Accordingly, it has been one of our main concerns to present all homological tools without losing contact with classical (quasi-) Banach space theory: all categorical constructions are introduced to explain some concrete (quasi-) Banach space constructions appearing in nature. Our guide in this line was settled by Olek Pełczyński: examples always go first. In fact, probably the first paper in Banach space theory with clear homological content is Pełczyński's monograph [377]. Semadeni's book [430], which followed, was already written in an entirely categorical mood. To discover what happened after that, just turn back the page. The role of 3-space problems for Hilbert spaces and local convexity cannot go unmentioned. From these came the seminal works of Enflo, Lindenstrauss and Pisier [167], Ribe [401] and Kalton and Peck [280], where exact sequences and quasilinear maps came to light. After that, categorical methods in (quasi-) Banach space theory emerged at a steady pace: pullback/pushout constructions, splitting, extension/lifting of operator connections and, finally, homology sequences.

There is at least one serious omission in the topics covered in this book: Kalton centralizers, which are everything but a special extremely nice type of quasilinear map. On one hand because centralizers are always related to some underlying Banach algebra and thus appear in a Banach module context (though, to be blunt, Banach spaces not carrying a natural module structure over some familiar Banach algebra are, well, expendable). On the other hand, centralizers live their own lives: they enjoy a homology of their own, there are specific estimates and averaging techniques for them and, moreover, centralizers, Banach space structures and interpolation methods form a whole together. Endorsements for these assertions can be found in the articles of Cwikel, Milman and Rochberg [140; 141; 140] and in Pisier's article [392] about the meaning of Kalton's work in the wondrous Kalton Selecta Birkhäuser volumes.

Writing this book took us 12 years, from the day when the picture of Atenea was taken by the Obluda until the day when the picture of Ariel, Celia and Pipo was taken by Gema. During this time, parts of the theory were discovered and lectured on in many seminars and conferences, and a lot of people shared ideas and their vision of the topics with us, and to all of them must go our gratitude. A special mention is for the members of the research group Banext of the Institute of Mathematics Imuex of the University of Extremadura, Ricardo García, Yolanda Moreno and Jesús Suárez de la Fuente, for a non-trivial and perhaps unbounded quantity of reasons, mathematical and not. To Manuel González and Pier Luigi Papini, for many friendly conversations, many mathematical conversations and a lot of proofreading. Antonio Avilés, Gilles Godefroy, Piotr Koszmider, Jordi López-Abad and Grzegorz Plebanek read parts of the book and suggested a good number of improvements. Whether we followed them or not, it is entirely our responsibility. Financial support came from projects MTM2007-6994-C02-02, MTM2010-20190-C02-01, MTM2013–C2-1-P, MTM2016-76958-C2-1-P and PID2019-103961GB-C21 of the Ministerio de Ciencia e Innovación, España, and from Projects IB16056 and IB20038 de la Junta de Extremadura. The second author benefited from a 2007 visit to the Institute of Mathematics of the Polish Academy of Sciences supported by EC FP6 Marie Curie ToK programme SPADE2. The diligence of our editor at Cambridge in checking our work nipped many of our errors and oversights in the bud.

Let us conclude with a special mention for Aleksander Pełczyński and Nigel Kalton: each in their own way a one-of-a-kind mathematician and both friends whom we continue to miss and for whose work we maintain an endless admiration.

Y
X
Z

Preliminaries

Anyone familiar with ℓ_p spaces can follow a healthy 50 percent of this book; if familiar with L_p spaces, the percentage rises to 75 percent. All the rest can be found in the text. Anyway, since once a man indulges himself in murder…, a reasonable list of prerequisites that could help smooth reading would be some acquaintance with classical Banach space theory; lack of fear when local convexity disappears; a certain bias towards abstraction; calm when non-linear objects show off and some fondness for exotic spaces. Better yet, perhaps, would be if, when in doubt, the reader is reminded about:

Sets and Functions

If S is a set, $\mathbf{1}_S$ denotes the identity on S, while 1_S stands for the characteristic function of S. If $S = \{s\}$ is a singleton, we write 1_s instead of $1_{\{s\}}$. We write $|S|$ for the cardinality of S. Sometimes 2 denotes the set $\{0, 1\}$. The power set of S is denoted $\mathcal{P}(S)$ or 2^S, and if $|S| = \aleph$ then $|2^S| = 2^{\aleph}$. We will write $\mathcal{P}_n(S) = \{a \in 2^S : |a| = n\}$ and $\mathrm{fin}(S) = \bigcup_n \mathcal{P}_n(S)$ for the family of all finite subsets of S, while $\mathcal{P}_\infty(S)$ denotes the family of all infinite subsets of S. The axiomatic system in which we work is ZFC, the usual Zermelo–Fraenkel axioms for set theory, including the axiom of choice. CH is the continuum hypothesis ($2^{\aleph_0} = \aleph_1$), and GCH is the generalised continuum hypothesis ($2^{\aleph} = \aleph^+$ for all infinite cardinals $\aleph$). Wherever additional axioms are assumed for some statement, the axioms appear in square brackets before the corresponding statement. Given a function $f: A \longrightarrow B$, its domain is $\operatorname{dom} f = A$, and its codomain is $\operatorname{codom} f = B$. Given functions $f: A \longrightarrow B$ and $g: C \longrightarrow D$, we write $f \times g: A \times C \longrightarrow B \times D$ for the function $(f \times g)(a, c) = (fa, gc)$. If $f: A \longrightarrow B$ and $g: A \longrightarrow D$, then $(f, g): A \longrightarrow B \times D$ is the function $(f, g)(a) = (fa, ga)$. If $f: A \longrightarrow B$ and $g: C \longrightarrow B$ then $f \oplus g: A \times C \longrightarrow B$ is the function $(f \oplus g)(a, c) = fa + gc$

when this makes sense. Two functions $f, g\colon S \longrightarrow \mathbb{R}^+$ will be called equivalent if there are constants $C, c > 0$ such that $cg(s) \le f(s) \le Cg(s)$ for all $s \in S$.

Boolean Algebras

An algebra of sets is a non-empty subfamily of the family $\mathcal{P}(X)$ of all subsets of a set X which contains the sets $\varnothing$ and X and is closed under finite unions, finite intersections and taking complements. A Boolean algebra is a set $\mathfrak{B}$ endowed with two abstract binary operations of 'union' $\vee$ and 'intersection' $\wedge$, an operator of 'complementation' $A \mapsto A^c$ and two distinguished elements 0 and 1 which satisfy the same laws of the union, intersection and complements as algebras of sets, with 0 in place of the empty set $\varnothing$ and 1 in place of the ambient set X. More precisely, it is required that union and intersection be commutative and associative, that each is distributive with respect to the other and that the absorption laws $A \vee (A \wedge B) = A$ and $A \wedge (A \vee B) = A$ and the complementation laws $A \wedge A^c = 0$ and $A \vee A^c = 1$ are satisfied for every $A, B \in \mathfrak{B}$. The simplest Boolean algebras are algebras of sets, and in fact, every Boolean algebra is isomorphic to some algebra of sets. A Boolean homomorphism $f\colon \mathfrak{A} \longrightarrow \mathfrak{B}$ is a function that preserves the Boolean operations and the distinguished elements.

The space of ultrafilters on $\mathfrak{A}$, denoted by $\mathrm{ult}(\mathfrak{A})$, is the set of Boolean homomorphisms $\mathfrak{A} \longrightarrow \mathbf{2}$, where $\mathbf{2} = \{0, 1\}$ has the obvious Boolean structure. Clearly, $\mathrm{ult}(\mathfrak{A})$ is a closed subset of $2^{\mathfrak{A}}$, hence it is a compact space, which is called the Stone space of $\mathfrak{A}$.

Ordinals and Cardinals

Ordinals and cardinals are generalisations of natural numbers. Cardinals represent equivalence classes of sets under the equivalence relation: there is a bijective map between them. Ordinals represent equivalence classes of well-ordered sets (a well order on a set is an order such that any subset has a first element) under the equivalence relation that there exists a bijective order-preserving map. The axiom of choice in the equivalent form of the well ordering principle (every set admits a well order) means that every cardinal is an ordinal. One can identify cardinals with some ordinals as follows: the cardinal κ corresponds to the ordinal β of a set of cardinal κ that cannot be bijected with any set having ordinal $\alpha < \beta$. Every set S can be bijected with a unique cardinal $|S|$, called the cardinality of S: the cardinal (and ordinal) of any set with n elements is just n, the cardinal of $\mathbb{N}$ is called $\aleph_0$ and its ordinal (in its standard order) is ω. The set $\mathbb{N} \cup \{\bullet\}$ endowed with the well order in which $n < \bullet$ for all n has cardinal $\aleph_0$, but its ordinal is different from ω and is usually called $\omega + 1$. The cardinal of $\mathbb{R}$ is called $\mathfrak{c}$. Even while ignoring what they actually are, ordinals can be constructed *inductively* with three rules:

- 0 is an ordinal.
- Given an ordinal α, there is an ordinal $\alpha + 1$ or α^+ called its *successor*.
- Given any *set* of ordinals, there is an ordinal that is the supremum of the set.

Ordinals that are not successors are called *limit ordinals*. If one replaces the third rule by 'any *countable set* of ordinals has a supremum', one will never abandon the realm of countable sets. Thus 0, $1 = 0^+$, $2 = 1^+$ and $3 = 2^+$ are ordinals, as well as $\omega = \sup\{n : n \in \mathbb{N}\}$, $\omega + 1 = \omega^+$, etc. Sets of ordinals are well ordered, and thus ordinals can be used to perform inductive arguments even on uncountable sets. Precisely, to accept that a statement $P(\alpha)$ is valid for all ordinals α, it is enough to prove that $P(0)$ is true and then that if $P(\alpha)$ is true for every ordinal $\alpha < \beta$ then $P(\beta)$ is also true. Analogously, to define a function F on all ordinals, it is enough to describe how to determine $F(\beta)$ once $F(\alpha)$ is known for all $\alpha < \beta$. Given a cardinal κ, there exists a minimum cardinal κ^+ greater than κ. The first infinite cardinal is clearly $\aleph_0$. The cardinal $\aleph_0^+$ is denoted $\aleph_1$, and this notation continues by declaring $\aleph_{n+1} = \aleph_n^+$. The cardinal $\aleph_\omega$ is the supremum of the $\aleph_n$ for $n < \omega$, and so on. Given a cardinal κ, 2^κ is the cardinality of the product $\{0, 1\}^\kappa$. It is always the case that $\kappa < 2^\kappa$. The cardinality of the continuum $\mathbb{R}$ is $\mathfrak{c} = 2^{\aleph_0}$. The cofinality of a limit ordinal α is the least cardinal λ for which there is a subset of α of cardinality λ whose supremum is α. Thus, $\aleph_\omega$ has cofinality $\aleph_0$. The cofinality of $\mathfrak{c}$ is strictly greater than $\aleph_0$.

Compact Spaces

Each compact space has an associated important cardinal, its *weight*, and an important ordinal, its *height*. The *weight* is the smallest cardinal of a base of open sets. To define the height, we explain the derivation process: the derived set S' of a topological space S is the subset of its accumulation points. Given a compact K, its αth derived space K^α is defined by transfinite induction $K^0 = K$, $K^{\alpha+1} = (K^\alpha)'$ and $K^\beta = \bigcap_{\alpha<\beta} K^\alpha$ for a limit ordinal β. The *height* is the smallest ordinal (if it exists) α such that the αth derived set K^α is empty. A compact space satisfying $K' = K$ is called *perfect*. A compact is said to be scattered if each subset admits a relatively isolated point. When K is scattered, its height exists and is always a successor ordinal. Ordinals can be viewed as compact spaces when endowed with the order topology, defined as follows: a fundamental system of neighbourhoods of α is formed by the sets $(\gamma, \alpha]$ for $\gamma < \alpha$. Limit ordinals are accumulation points, while successor ordinals are isolated points. The ordinal β has weight $|\beta|$, while the ordinals ω^N have height $N+1$ and ω^ω has height $\omega+1$. A compact space is said to be zero-dimensional or totally disconnected if its clopen (closed and open) subsets form a base of the topology, or, equivalently, if, for every two different points $x, y \in K$, there exists a clopen set A such that $x \in A$, $y \notin A$. Scattered compacta are totally disconnected, while the Cantor set is totally disconnected and perfect. The topological structure of zero-dimensional compacta is completely described by the algebraic structure of their family of clopen sets through Stone duality: if K is a compact space, then its clopen sets $\mathfrak{cl}(K)$ form a Boolean algebra whose Stone space is naturally homeomorphic to K itself when K is zero-dimensional. See Note 4.6.1 for more on this duality.

Quasinormed Spaces and Operators

We will work with quasi and p-normed spaces (see Section 1.1), which can be either real or complex. If there is no need to specify whether the ground field is $\mathbb{R}$ or $\mathbb{C}$ then it will simply be $\mathbb{K}$. A map $f\colon X \longrightarrow Y$ acting between linear spaces is called *homogeneous* if $f(\lambda x) = \lambda f(x)$. It is called *positively homogeneous* if $f(\lambda x) = |\lambda| f(x)$ for every $\lambda \in \mathbb{K}$ and every $x \in X$. A map $f\colon X \longrightarrow Y$ acting between quasinormed spaces is bounded if there is a constant K such that $\|f(x)\| \leq K\|x\|$ for all $x \in X$. The least possible K for which this holds is denoted by $\|f\|$ and provides a quasinorm on the space $\mathsf{B}(X, Y)$ of bounded homogeneous maps $X \longrightarrow Y$. Linear continuous maps are called operators, and the space of all operators $X \longrightarrow Y$ is denoted by $\mathfrak{L}(X, Y)$, or just $\mathfrak{L}(X)$ when $Y = X$. A (linear) isomorphism is an operator admitting an inverse operator. We write $X \simeq Y$ when the spaces X, Y are isomorphic, i.e., there exists an isomorphism between them. We say that they are C-isomorphic if there is an isomorphism $u\colon X \longrightarrow Y$ such that $\|u\|\|u^{-1}\| \leq C$. The Banach–Mazur distance between X and Y is defined as $d(X, Y) = \inf\{\|u\|\|u^{-1}\|\colon u$ is an isomorphism between X and $Y\}$. An isometry is an operator $u\colon X \longrightarrow Y$ such that $\|u(x)\| = \|x\|$ for all $x \in X$. Isometries are not assumed to be surjective! However, we will say that X and Y are isometric, $X \approx Y$, if there is a surjective isometry between X and Y. An ε-isometry, $\varepsilon \in [0, 1)$, is an operator u satisfying $(1+\varepsilon)^{-1}\|x\| \leq \|u(x)\| \leq (1+\varepsilon)\|x\|$ for all x in the domain of u. The range or image of an operator $u\colon X \longrightarrow Y$ is denoted by $u[X]$, and we say that u is an embedding when it is an isomorphism between X and $u[X]$, in which case we also say that $u[X]$ is a copy of X in Y. We say that Y contains a copy of X if there is an embedding $u\colon X \longrightarrow Y$. If $u\colon X \longrightarrow Y$ is an ε-isometry then u^{-1} is an ε-isometry from $u[X]$ to X. A quotient map is a surjective operator that is open. A quotient map $u\colon X \longrightarrow Y$ is called isometric if $\|y\| = \inf_{y=u(x)} \|x\|$ for all $y \in Y$. If Y is a closed subspace of X then the quotient quasinorm on the quotient space X/Y is $\|x + Y\| = \inf_{y\in Y} \|x + y\|$. Let $\jmath\colon Y \longrightarrow X$ be an embedding, and let $\tau\colon Y \longrightarrow E$ be an operator. An extension of τ is an operator $T\colon X \longrightarrow E$ such that $T\jmath = \tau$. A λ-extension of τ is an extension T such that $\|T\| \leq \lambda\|\tau\|$. Given a p-normed space X, we will denote the set of all its finite-dimensional subspaces by $\mathscr{F}(X)$. Sometimes we will use $\mathscr{F}^{(p)}$ to represent the class of all finite-dimensional p-Banach spaces, with $\mathscr{F}^{(1)}$ shortened to $\mathscr{F}$. We will denote the set of all its separable subspaces by $\mathscr{S}(X)$. Sometimes we will use $\mathscr{S}^{(p)}$ to represent the class of all separable p-Banach spaces, with $\mathscr{S}^{(1)}$ shortened to $\mathscr{S}$.

In this book we will use an unusual notation: by the dimension dim(X) of a quasi-Banach space X, we shall not mean the dimension of its underlying vector space but rather the cardinal of a smallest subset spanning a dense subspace.

Classical Spaces

Most of the time we work with quite honest spaces. Given a set I and $p \in (0, \infty)$, we write $\ell_p(I)$ for the space of all functions $f\colon I \longrightarrow \mathbb{K}$ such that $\|f\|_p = \left(\sum_{i\in I} |f(i)|^p\right)^{1/p} < \infty$. This is a quasi-Banach space with the obvious quasinorm and is a Banach space when $p \geq 1$. The space of all bounded functions $f\colon I \rightarrow \mathbb{K}$ endowed with the supremum norm is denoted by $\ell_\infty(I)$, and $c_0(I)$ is the closed subspace spanned by the characteristic functions of the singletons of I. The isometry type of $\ell_p(I)$ depends only on $\aleph = |I|$, and sometimes we write $\ell_p(\aleph)$ with the obvious meaning. Similar conventions apply to $c_0(\aleph)$. When $I = \mathbb{N}$ or $\aleph = \aleph_0$, we just write ℓ_p and c_0, while c denotes the subspace of ℓ_∞ formed by all convergent sequences. Clearly, $c_0 \simeq c$. Given a compact Hausdorff space K, we write $C(K)$ for the Banach space of all continous functions $f\colon K \longrightarrow \mathbb{K}$, with the sup norm. A $\mathscr{C}$-space is a Banach space isometric to $C(K)$ for some (often unspecified) compact K. Given a sequence (F_n) of finite-dimensional p-Banach spaces that are dense in $\mathscr{F}^{(p)}$ in the Banach–Mazur distance one can form the by now classical spaces $C_\infty^{(p)} = \ell_\infty(\mathbb{N}, F_n)$, $C_r^{(p)} = \ell_r(\mathbb{N}, F_n)$ and $C_0^{(p)} = c_0(\mathbb{N}, F_n)$. Let T be a compact operator on a fixed, separable Hilbert space H. The singular numbers of T are the eigenvalues of $|T| = (T^*T)^{1/2}$, arranged in decreasing order and counting multiplicity. The Schatten class S_p consists of those operators on H whose sequence of singular numbers $(s_n(T))$ belongs to ℓ_p. It is a quasi-Banach space endowed with the quasinorm $\|T\|_p = |(s_n(T))|_p$. Each $T \in S^p$ can be represented as $T = \sum_n s_n(T)\, x_n \otimes y_n$ for some orthonormal sequences $(x_n), (y_n)$ in H. The Schatten class S_p can be considered as the non-commutative version of ℓ_p. The terms *function space* and *sequence space* have very specific meanings in this book. The definition of function space is in 1.1.4, while that of sequence space is in 3.2.4. This otherwise coherent notation unfortunately introduces some points that may perplex the casual reader: $C[0,1]$ is not a function space, and ℓ_∞ is not a sequence space.

Approximation Properties

Let X be a quasi-Banach space. A sequence $(x_n)_{n\geq 1}$ is a (Schauder) basis of X if for every $x \in X$ there is a unique sequence of scalars $(c_n)_{n\geq 1}$ such that $x = \sum_{n=1}^{\infty} c_n x_n$. If all these series are unconditionally convergent, then the basis is called *unconditional*, and there is a constant C, called the *unconditionality constant* of (x_n), such that $\left\|\sum_{1\leq i\leq k} t_i x_i\right\| \leq C\left\|\sum_{1\leq i\leq k} c_i x_i\right\|$ provided $|t_i| \leq c_i$ for every i. In this case an equivalent quasinorm with unconditionality constant 1 can be defined as $\left|\left|\left|\sum_{i=1}^{\infty} c_i x_i\right|\right|\right| = \sup_{|t_i|\leq|c_i|}\left\|\sum_{i=1}^{\infty} c_i x_i\right\|$. If (x_n) is a normalised 1-unconditional basis of X, then X can be seen as a sequence space in the obvious way. A sequence (x_n) of elements of a quasi-Banach space is called a basic sequence if it is a basis of its closed linear span. The basic constant of a basic sequence (x_n) is the smallest $K > 0$ such that $\|\sum_{n=1}^{N} \lambda_n x_n\| \leq K\|\sum_{n=1}^{M} \lambda_n x_n\|$ for all $N \leq M$.

Basic sequence criterion *If (x_n) is a normalised basic sequence with constant K in a p-Banach space and $\sum \|x_n - y_n\|^p < (2K)^{-p}$ then (y_n) is a basic sequence.*

The meaning of *unconditional basic sequence* should be obvious. A Banach space with unconditional basis contains c_0, or ℓ_1, or a reflexive subspace, and thus the same is true for a Banach space containing an unconditional basic sequence. Quasi-Banach spaces with unconditional bases have (many) infinite-dimensional complemented subspaces: if (x_n) is an unconditional basis of X and $I \subset \mathbb{N}$, then $X(I) = \left\{x \in X : x = \sum_{i\in I} c_i x_i\right\}$ is a complemented subpace of X. The next best thing to a basis is a finite-dimensional decomposition (FDD): a sequence of finite-dimensional subspaces $(X_n)_{n\geq 1}$ is a FDD if each $x \in X$ has a unique expansion of the form $x = \sum_{n=1}^{\infty} x_n$, with $x_n \in X_n$ for all n. If these series converge unconditionally then the FDD is said to be an UFDD.

Definition 0.0.1 A quasi-Banach space X is said to have the λ-approximation property (λ-AP) if for each finite-dimensional subspace $F \subset X$ and each $\varepsilon > 0$ there is a finite-rank operator $T \in \mathfrak{L}(X)$ such that $\|T\| \leq \lambda$ and $\|f - Tf\| \leq \varepsilon\|f\|$ for each $f \in F$. The space is said to have the bounded approximation property (BAP) if it enjoys the λ-AP for some λ.

One could alternatively require that the operator fix F, that is, $Tf = f$ for $f \in F$, transferring the error to the norm of the operator so that we can just ask for $\|T\| \leq \lambda + \varepsilon$. When X is separable, the BAP (resp. the λ-AP) is equivalent to the existence of a sequence of finite-rank operators $T_n \in \mathfrak{L}(X)$ converging pointwise to the identity (resp. with $\sup_n \|T_n\| \leq \lambda$). A Banach space X is said to have the uniform approximation property (UAP) when it has the λ-AP and there exists a 'control function' $f\colon \mathbb{N} \longrightarrow \mathbb{N}$ such that, given F and $\lambda' > \lambda$, we can choose T such that $\mathrm{rk}(T) \leq f(\dim F)$ and $Tf = f$ for all $f \in F$, with $\|T\| \leq \lambda'$. This means exactly that every ultrapower of X has the BAP. A sequence (E_n) of spaces is said to have the joint-UAP if all the spaces E_n have the λ-UAP with the same control function. This occurs if and only if $\ell_\infty(\mathbb{N}, E_n)$ has the λ-UAP. Since X^{**} is complemented in some ultrapower of X, when X has the UAP, then all even duals have the UAP. And since the BAP passes from the dual to the space, when X has the UAP, all its duals have the UAP. See either [227, p. 60] or [83, Section 7] for details.

Selected Operator Ideals

Many interesting classes of operators between Banach spaces are *operator ideals*, classes $\mathfrak{A}$ such that (a) finite-rank operators are in $\mathfrak{A}$, (b) $\mathfrak{A} + \mathfrak{A} \subset \mathfrak{A}$ and (c) The composition of any operator with an operator in $\mathfrak{A}$ is in $\mathfrak{A}$. Following Pietsch's traditional notation, we use "fraktur" types for operator ideals. The fundamental operator ideals that appear often in these pages are:

- The ideal $\mathfrak{L}$ of all operators.
- The ideal $\mathfrak{F}$ of finite-rank operators: an operator $T\colon X \longrightarrow Y$ is in $\mathfrak{F}$ if $T[X]$ is finite-dimensional; finite-rank operators can be represented as $T = \sum_{n\leq N} x_n^* \otimes y_n$, where $x_i \in X^*$, $y_i \in Y$ and $N \in \mathbb{N}$.
- The ideal $\mathfrak{N}_p$ of p-nuclear operators, $0 < p \leq 1$: those admitting a representation $T = \sum_{n=1}^{\infty} x_n^* \otimes y_n$, with $\sum_n \|x_n^*\|^p \|y_n\|^p < \infty$.
- The ideal $\mathfrak{K}$ of compact operators: those transforming the unit ball of X into a relatively compact subset of Y.
- The ideal $\mathfrak{V}$ of completely continuous operators: those transforming the unit ball of X into a relatively compact subset of Y.
- The ideal $\mathfrak{W}$ of weakly compact operators between Banach spaces: those transforming the unit ball of X into a weakly compact subset of Y.
- The ideal $\mathfrak{S}$ of strictly singular operators: those whose restriction to an infinite-dimensional closed subspace is never an isomorphism.

0.0.2 *An operator $T\colon X \longrightarrow Y$ acting between Banach spaces is strictly singular if and only if, for every $\varepsilon > 0$, every infinite-dimensional subspace $A \subset X$ contains an infinite-dimensional subspace $B \subset A$ such that $\|T|_B\| \leq \varepsilon$.*

The result fails for quasi-Banach spaces, as explained in Note 9.4.4.

- The ideal $\mathfrak{P}_p$ of p-summing operators: $T : X \longrightarrow Y$ is p-summing if it transforms weakly p-summable sequences (sequences (x_n) such that $(\langle x^*, x_n \rangle)_n \in \ell_p$ for all $x^* \in X^*$) into absolutely p-summable sequences (sequences (y_n) such that $(\|y_n\|)_n \in \ell_p$).

The Grothendieck–Pietsch domination / factorisation theorem establishes that T is p-summing if and only if there is a factorisation

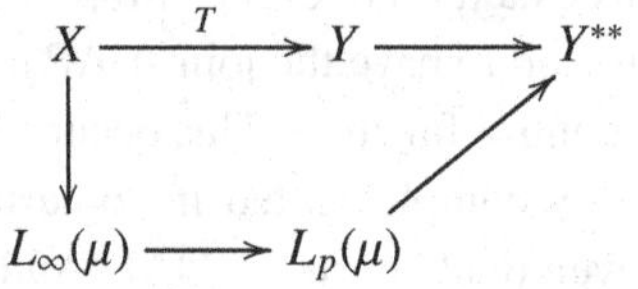

for some measure μ on B_{X^*} so that $L_\infty(\mu) \longrightarrow L_p(\mu)$ is the canonical inclusion. When $p = 2$, the right upwards arrow goes $L_2(\mu) \longrightarrow Y$. This means that 2-summing operators factorise through Hilbert spaces (operators that factorise through a Hilbert space are sometimes called 2-factorable) and that they extend anywhere [153, 4.15]. Grothendieck's theorem [153, Theorem 3.7] establishes that every operator between an $\mathscr{L}_1$ and an $\mathscr{L}_2$ space is 2-summing.

1

Complemented Subspaces of Banach Spaces

1.1 Banach and Quasi-Banach Spaces

A quasinorm on a real or complex vector space X is a map $\|\cdot\|\colon X \longrightarrow \mathbb{R}^+$ that satisfies

- $\|x\| = 0 \implies x = 0$,
- $\|\lambda x\| = |\lambda|\|x\|$,
- $\|x + y\| \le \Delta(\|x\| + \|y\|)$,

for all $x, y \in X$, all scalars λ and some constant $\Delta \ge 1$, called the *modulus of concavity of the quasinorm*. If $\Delta = 1$, then $\|\cdot\|$ is a norm. A quasinorm induces a linear topology on the underlying space: the coarsest linear topology for which the unit ball $B_X = \{x \in X : \|x\| \le 1\}$ is a neighbourhood of the origin. A quasinormed space is a vector space equipped with a quasinorm; when the space is moreover complete, it is called a *quasi-Banach space*. If the quasinorm is a norm then it will be called a *normed space* and, if it is complete, a *Banach space*. An operator is a linear continuous map. Topologies induced by quasinorms are so uncomplicated that a linear map $u\colon X \longrightarrow Y$ acting between quasinormed spaces is continuous if and only if it is bounded, in the sense that

$$\|u\| = \sup_{\|x\|\le 1} \|u(x)\| < \infty. \tag{1.1}$$

The space of all operators from X to Y is denoted by $\mathfrak{L}(X, Y)$ and is a quasinormed space endowed with the quasinorm (1.1). It is complete (or p-normed, see later) when Y is. The space $\mathfrak{L}(X, \mathbb{K})$ is the *dual* space of X, usually denoted by X^*, and is always a Banach space. We say that X has *separating dual* if, for every non-zero $x \in X$, there is $x^* \in X^*$ such that $\langle x^*, x\rangle \neq 0$. The space is said to have *trivial dual* if $X^* = 0$.

Why do we bother the reader with quasinorms instead of using just plain norms? Because one must face right from the start the awful truth: it is not possible to make a serious study of twisted sums of Banach spaces without considering quasinorms. That is simply the way things are. It will be explained at length through the book, but a motivating example can be given now: if everything one knows about a space Z is that it contains a Banach subspace Y such that Z/Y is a Banach space, it may simply happen that Z is not locally convex, but, even if it is, there is no way to specify a particular norm on Z, while it could be feasible to describe a quasinorm on it. Moreover, the necessity to distinguish quasinorms from the far more popular norms is because $\Delta > 1$ and has, in practice, side effects: the unit ball is no longer convex, and thus the Hahn–Banach theorem ceases to work, up to the point that a quasi-Banach space might have trivial dual or that a quasinorm is not necessarily continuous with respect to its own topology. In spite of these facts, there is no need to panic: the open mapping, closed graph and Banach–Steinhaus theorems work perfectly well on quasi-Banach spaces. Moreover, a simple remedy to correct the possible discontinuity of the quasinorm is to work with p-norms, $0 < p \leq 1$, which are quasinorms satisfying the additional inequality

- $\|x + y\|^p \leq \|x\|^p + \|y\|^p$.

Obviously p-norms are q-norms for $0 < q < p$. A p-norm satisfies the inequality $\big|\|x\|^p - \|y\|^p\big| \leq \|x - y\|^p$, and this makes it continuous. Even better than that, quasinorms can be judiciously replaced by p-norms:

1.1.1 Aoki–Rolewicz Theorem *Each quasinorm is equivalent to some p-norm.*

Indeed, if $\|\cdot\|$ has modulus of concavity Δ and $2^{1/p-1} = \Delta$ then

$$\|x\|_{(p)} = \inf\left\{\left(\sum_{1\leq i\leq n} \|x_i\|^p\right)^{1/p} : x = \sum_{1\leq i\leq n} x_i\right\} \tag{1.2}$$

defines a p-norm on X such that $\|\cdot\|_{(p)} \leq \|\cdot\| \leq 2\Delta\|\cdot\|_{(p)}$. This is somehow optimal since an elementary computation reveals that the modulus of concavity of a p-norm is at most $2^{1/p-1}$. There is moreover an effective way to detect when a quasinorm is equivalent to a p-norm for a pre-established p:

Lemma 1.1.2 *Let $0 < p \leq 1$. A quasinormed space X is isomorphic to a p-normed space if and only if there is a constant C such that, for finitely many $x_i \in X$, one has*

$$\left\|\sum_i x_i\right\| \leq C\left(\sum_i \|x_i\|^p\right)^{1/p}. \tag{1.3}$$

Proof It is clear that X is isomorphic to a p-normed space precisely when there is a p-norm $|\cdot|$ on X such that $|\cdot| \leq \|\cdot\| \leq C|\cdot|$. If this holds, then given finitely many $x_1, \ldots, x_n \in X$, one has

$$\Big\| \sum_{1\leq i\leq n} x_i \Big\| \leq C \Big| \sum_{1\leq i\leq n} x_i \Big| \leq C \Big(\sum_{1\leq i\leq n} |x_i|^p \Big)^{1/p} \leq C \Big(\sum_{1\leq i\leq n} \|x_i\|^p \Big)^{1/p}.$$

As for the converse, if (1.3) holds then the functional (1.2) is a p-norm satisfying $\|\cdot\|_{(p)} \leq \|\cdot\| \leq C\|\cdot\|_{(p)}$. □

A p-norm defines an invariant metric by the formula $d(x,y) = \|x-y\|^p$, and thus the Aoki–Rolewicz theorem implies that quasinormed spaces are metrisable. The absolute summability criterion for completeness of a p-norm is:

1.1.3 *Let X be a p-normed space. Then X is complete if and only if every sequence (x_n) such that $\sum_{n\geq 1} \|x_n\|^p < \infty$ is summable in X.*

The Completion

The Aoki–Rolewicz theorem places quasinormed spaces among metric linear spaces; thus quasinormed spaces can be completed. Quasinorms, however, can be rather nasty functions for which there can be no natural extension to the completion. Thus, perhaps the simplest way to construct a completion $\kappa\colon X \longrightarrow \widehat{X}$ of a quasinormed space X is to first put an equivalent p-norm to then use the induced metric to construct a completion via equivalent classes of Cauchy sequences. The operations and the p-norm extend by uniform continuity. If for some reason we want to keep the original quasinorm on X, we can do so. Completions, like many other universal constructions in this book, come with a universal property: every operator u from X into a complete space Y factors through the inclusion $\kappa\colon X \longrightarrow \widehat{X}$ as $u = \widehat{u}\kappa$ with $\|\widehat{u}\| = \|u\|$ in the form

$$\begin{array}{ccc} X & \xrightarrow{\ \kappa\ } & \widehat{X} \\ & {\scriptstyle u}\searrow \quad \swarrow{\scriptstyle \widehat{u}} & \\ & Y & \end{array} \tag{1.4}$$

The p-Banach Envelope

The proof of Lemma 1.1.2 suggests the following construction. Given a quasinormed space X and $p \in (0,1]$, the formula (1.2) defines a positively homogeneous and p-subadditive function such that $\|\cdot\|_{(p)} \leq \|\cdot\|$. If $N_{(p)} = \{x \in X \colon \|x\|_{(p)} = 0\}$, then $\|\cdot\|_{(p)}$ becomes a genuine p-norm on $X/N_{(p)}$. The completion of $X/N_{(p)}$ with the natural extension of $\|\cdot\|_{(p)}$ is a p-Banach space called the *p-Banach envelope* of X and is denoted $X_{(p)}$. The universal property that corresponds to p-Banach envelopes is that every operator $u: X \longrightarrow Y$ from X into a p-Banach space Y factors as (the unlabeled arrow is the obvious map)

$$\begin{array}{ccc} X & \longrightarrow & X_{(p)} \\ & {\scriptstyle u}\searrow \quad \swarrow{\scriptstyle \tilde{u}} & \\ & Y & \end{array}$$

and $\|\tilde{u}\| = \|u\|$. The 1-Banach envelope is called the *Banach envelope*. It should be clear that the dual of a quasinormed space and that of its Banach envelope coincide so that $X_{(1)} = 0$ if and only if X has trivial dual. A good account of Banach envelopes is [289].

Some Fundamental Examples

Given a σ-finite measure space (S, μ), we write $L_0(\mu)$ for the space of real or complex measurable functions on S modulo almost everywhere equality. The space $L_0(\mu)$ comes with the topology of convergence in measure on sets of finite measure, for which a typical neighbourhood of zero is

$$\{f \in L_0(\mu) : \mu\{s \in A : |f(s)| > \varepsilon\} < \varepsilon\},$$

where $\mu(A) < \infty$ and $\varepsilon > 0$. This space, mostly used as an ambient space, is not locally bounded (unless S consists of a finite number of atoms) nor locally convex (unless μ is purely atomic); see 1.8.2.

1.1.4 By a (quasinormed) function space, we will mean a linear subspace $X \subset L_0(\mu)$ equipped with a quasinorm $\|\cdot\|$ such that

- if A has finite measure, then the characteristic function 1_A belongs to X,
- if $g \in X$ and $f \in L_0(\mu)$ is such that $|f| \le g$, then $f \in X$ and $\|f\| \le \|g\|$,
- the inclusion of X into $L_0(\mu)$ is continuous.

If X is complete, we call it a quasi-Banach function space.

The simplest such spaces are the Lebesgue spaces $L_p(\mu)$ of p-integrable functions quasinormed, when $0 < p < \infty$, with

$$\|f\|_p = \left(\int_S |f|^p d\mu \right)^{1/p}$$

and the space $L_\infty(\mu)$ of essentially bounded measurable functions endowed with the essential supremum norm. These are Banach spaces for $p \ge 1$ and quasi-Banach (actually p-Banach) spaces when $0 < p < 1$. When μ is a Lebesgue measure on $[0, 1]$, we omit it, and we just write L_p. The Hardy classes

H_p, closely related to the spaces L_p, are thoroughly studied in Duren's classic [163]. The space H_p is the space of analytic functions $f\colon \mathbb{D} \longrightarrow \mathbb{C}$ such that $\|f\|_p = \sup_{0<r<1} M(f,r) < \infty$, where

$$M(f,r) = \left(\frac{1}{2\pi}\int_0^{2\pi} |f(re^{it})|^p dt\right)^{1/p}.$$

If $f \in H_p$, then the limit $f^*(e^{it}) = \lim_{r\to 1} f(re^{it})$ exists for almost all $t \in (0, 2\pi]$ and $M(f,r)$ is increasing on $r \in (0,1)$. The mapping $f \longmapsto f^*$ defines an isometry between H_p and the closed subspace of $L_p(\mathbb{T})$ spanned by the exponentials e^{int} for $n = 0, 1, 2, \ldots$

The sequence spaces ℓ_p and the Schatten classes S_p have separating dual for all p, and the same is true for H_p since an obvious application of Cauchy's integral formula shows that for every $z \in \mathbb{D}$ and every $f \in H_p$, we have

$$|f(z)| \leq 2^{1/p}(1-|z|)^{-p}\|f\|_p. \tag{1.5}$$

When a quasi-Banach space X has non-trivial dual, as is the case if X is a normed space, then $\mathfrak{L}(X,Y) \neq 0$ for all quasinormed spaces (or topological vector spaces) Y. This is no longer true when X is a quasi-Banach space:

1.1.5 *If $0 < p < 1$, then $\mathfrak{L}(L_p, Y) = 0$ for all q-normed spaces Y with $q > p$.*

The proof can be made easy: let Y be a q-normed space, $0 < p < q \leq 1$, and let $u\colon L_p \longrightarrow Y$ be an operator. If f is a norm 1 element of L_p then, by the intermediate value theorem, there is $a \in (0,1)$ such that

$$\int_0^a |f|^p = \int_a^1 |f|^p = \frac{1}{2}.$$

Set $g = 1_{[0,a)}f$, $h = 1_{[a,1]}f$. Then $\|g\| = \|h\| = 2^{-1/p}$ and $f = g + h$, so

$$\|uf\| = \|ug + uh\| \leq (\|ug\|^q + \|uh\|^q)^{1/q} \leq \|u\|(2^{-q/p} + 2^{-q/p}) = 2^{1-q/p}\|u\|.$$

As f is arbitrary, this implies that $\|u\| \leq 2^{1-q/p}\|u\|$, which is possible only if $\|u\| = 0$. The same argument works replacing L_p by any vector valued $L_p(X)$, no matter the quasi-Banach space X one considers.

1.2 Complemented Subspaces

A projection P on a quasinormed space X is an idempotent of $\mathfrak{L}(X)$, i.e., an operator $P\colon X \longrightarrow X$ such that $P^2 = P$. A closed subspace $Y \subset X$ is complemented if there exists a projection on X whose range is Y, usually called a projection onto Y. It is clear that a projection of X onto Y is just an extension of the identity $\mathbf{1}_Y$ to X.

Lemma 1.2.1 *A closed subspace Y of a quasi-Banach space X is complemented if and only if there exists a closed subspace V of X such that $Y \cap V = 0$ and $Y + V = X$.*

Proof If a projection P as above exists then $\ker P$ is closed, and letting $V = \ker P$, we have $X = Y \oplus V$. Conversely, if such a V exists, every $x \in X$ can be decomposed as $x = y + v$ with $y \in Y$ and $v \in V$ in a unique way, and we can define a mapping $P\colon X \longrightarrow X$ taking $y = P(x)$. This map is linear and idempotent and satisfies $P[X] = Y$. It is continuous since it has a closed graph. □

The subspace V in the lemma is sometimes called a *complement* of Y, which explains the terminology. It is clear that all complements of a given subspace are isomorphic as they are isomorphic to the quotient space X/Y. A quantified version of the notion of complemented subspace can be defined by means of the quasinorm of the projection: we say that Y is λ-complemented in X if there is a projection of X onto Y whose quasinorm is at most λ. The relative projection constant of Y in X is defined as $\lambda(Y, X) = \inf\{\|P\| : P \text{ is a projection of } X \text{ onto } Y\}$. While the finite-dimensional subspaces of a Banach space are all complemented, by the Hahn–Banach theorem, that is not always the case for quasi-Banach spaces, and indeed 1.1.5 shows that no finite-dimensional subspace of L_p can be complemented when $0 < p < 1$.

Pełczyński's Decomposition Method and Minimality

Pełczyński's decomposition method appears in [376] to prove that all infinite-dimensional complemented subspaces of ℓ_p for $p \in [1, \infty)$ or c_0 are isomorphic to the whole space. We could not resist giving its proof.

Proposition 1.2.2 *Let X and Y be quasi-Banach spaces, each of them isomorphic to a complemented subspace of the other. Assume that either both X and Y are isomorphic to their squares or that Y is isomorphic to $\ell_p(Y)$ for some $0 < p < \infty$ or to $c_0(Y)$. Then, $X \simeq Y$.*

Proof The first case is as follows: let V be a complement of X in Y. Then

$$Y \simeq X \times V \simeq X \times X \times V \simeq Y \times X$$

and, for the same reason, $X \simeq X \times Y$. The second case is similar. We will give the proof for ℓ_p; the case of c_0 is analogous. First note that $Y \simeq \ell_p(Y) \simeq Y^2$.

Thus, as before, $X \simeq Y \times X$. Let V be a complement of X in Y such that $Y \simeq X \times V$. Then,

$$X \simeq Y \times X \simeq \ell_p(Y) \times X \simeq \ell_p(X \times V) \times X \simeq \ell_p(X) \times \ell_p(V) \simeq \ell_p(Y) \simeq Y. \quad \square$$

An example of Gowers [195] shows that the two hypotheses $X \simeq X^2$ and $Y \simeq Y^2$ cannot be simultaneously dropped. Now, a gliding hump argument shows that if X is one of the spaces c_0 or ℓ_p for $0 < p < \infty$, then every infinite-dimensional closed subspace of X contains a further subspace spanned by a basic sequence equivalent to a sequence of blocks of the canonical basis, and thus is isomorphic to the whole space. If, moreover, $1 \leq p < \infty$, that subspace can be taken as complemented. This shows that a complemented subspace of ℓ_p with $1 \leq p < \infty$ or c_0 is either finite-dimensional or isomorphic to the whole space. The result is also true for $p = \infty$, but for quite different reasons (see the discussion after 1.6.3). It is also true for $0 < p < 1$ because each *complemented* subspace (but not any subspace, see later) of ℓ_p contains a further complemented copy of ℓ_p; cf. [443; 444]. The non-separable versions (see the later proposition) also hold: the case $p = 1$ was obtained in [304] (see also [411, Corollary, p. 29] and [175]), the case $c_0(I)$ in [199], the case $0 < p < 1$ in [372] (notice that here the non-separable version holds, while the separable version does not: see the comment after Corollary 2.7.4) and the case $1 < p < \infty$ in [406].

Proposition 1.2.3 *Each complemented subspace of either $\ell_p(I)$, $0 < p < \infty$ or $c_0(I)$ is isomorphic to some $\ell_p(J)$ or $c_0(J)$. Each complemented subspace of ℓ_∞ is either finite-dimensional or isomorphic to ℓ_∞.*

There exist complemented subspaces of $\ell_\infty(I)$ that are not isomorphic to any $\ell_\infty(J)$ [413]; see also [22].

Definition 1.2.4 A quasi-Banach space X is said to be minimal (resp. complementably minimal) if each of its closed infinite-dimensional subspaces contains a copy of X (resp. a copy of X complemented in X). We say that X is prime if all its infinite-dimensional complemented subspaces are isomorphic to X and primary if $X = A \oplus B$ implies that either $A \simeq X$ or $B \simeq X$.

Proposition 1.2.5 *If X is a minimal Banach space then there is $C \geq 1$ such that every closed infinite-dimensional subspace of X contains a subspace that is C-isomorphic to X.*

Proof Throughout this proof, *subspace* means closed infinite-dimensional subspace. If no such C exists, fix $f \in \mathbb{R}^{\mathbb{N}}$ and obtain a decreasing sequence $Y_1 \supset Y_2 \supset \cdots$ such that X does not $f(n)$-embed into Y_n. Pick $(y_n)_{n\geq 1}$ a basic sequence in X with $y_n \in Y_n$, and let $E_n = \overline{[y_n, y_{n+1}, \dots]}$. There is a subspace $Z \subset E_1$ and a $C > 1$ such that Z is C-isomorphic to X and for which we may assume $\dim E_1/Z = \infty$. Set $H_n = Z \cap E_n$, and choose $H_n \subset F_n \subset E_n$ such that $k = \dim E_n/F_n = \dim Z/H_n \leq n$. Thus, for some k-dimensional $A \subset E_n$ and $B \subset Z$, we have that E_n is (at most) $(1 + \sqrt{n})$-isomorphic to $A \oplus_1 F_n$ and $\overline{[H_n + B]}$ is (at most) $(1 + \sqrt{n})$-isomorphic to $B \oplus_1 H_n$. Since A and B are (at most) n-isomorphic, it follows that $A \oplus_1 H_n$ is at most $n(1+\sqrt{n})^2$-isomorphic to Z, hence $Cn(1+\sqrt{n})^2$-isomorphic to X. And since $\overline{[H_n + B]} \subset E_n$, it turns out that X does actually $Cn(1 + \sqrt{n})^2$-embed in E_n, in flagrant contradiction of the assumption that X does not $f(n)$-embed into Y_n for, say $f(n) = n^2(1+\sqrt{n})^2$. □

Thus, ℓ_p spaces, $0 < p < \infty$, and c_0 are prime. It is an open problem whether L_p spaces are prime for $0 < p < 1$: Kalton proved in [250] that they are primary and that there exists, up to isomorphisms, at most one complemented subspace of L_p other than L_p itself [253]; see also [256, Section 3]. The list of minimal or complementably minimal spaces is not long. All ℓ_p spaces, $0 < p < \infty$, and c_0 are minimal, and complementably minimal when $1 \leq p < \infty$ or c_0. The spaces ℓ_p for $p \in (0, 1)$ are not complementably minimal [397]. Schlumprecht's arbitrarily distortable space [426; 13] and its dual [87] are complementably minimal, as are its superreflexive variations [87] and their duals. Tsirelson's space T, or its dual, are not complementably minimal [88, pp. 54–59], but its p-convexified version T_p, $1 < p < \infty$, is complementably minimal [88]. It is obvious that every minimal space must be separable. A bit less clear is that minimal spaces must be subspaces of spaces with unconditional basis: by the Gowers dichotomy [196], a Banach space X contains either a subspace with unconditional basis or an H.I. subspace, in which case it cannot be isomorphic to any proper subspace. Since a space with unconditional basis contains either c_0, ℓ_1 or a reflexive subspace [334, Theorem 1.c.12 (a)], a minimal space must be either reflexive or a subspace of c_0 or ℓ_1. Thus, a complementably minimal space must be c_0, ℓ_1 or reflexive.

1.3 Uncomplemented Subspaces

It is a basic fact in functional analysis that each closed subspace of a Hilbert space has a complement, namely the *orthogonal* complement. A classical result by Lindenstrauss and Tzafriri [333] provides the converse: a Banach space all of whose subspaces are complemented is isomorphic to a Hilbert space. Thus, each non-Hilbert infinite-dimensional Banach space contains uncomplemented subspaces. The introduction of the nicely written paper

[396] contains an historical account of the first discovered uncomplemented subspaces. The minimality of the spaces ℓ_p allows us to more or less easily locate some uncomplemented subspaces of ℓ_p: all subspaces not isomorphic to ℓ_p itself. For instance, if $Q: \ell_p \longrightarrow X$ is a quotient map onto an infinite-dimensional space X not isomorphic to ℓ_p then $\ker Q$ is uncomplemented in ℓ_p, since any complement of $\ker Q$ should be isomorphic to X and also complemented, which is impossible. Still, further insight into properties of ℓ_p spaces is needed to obtain such quotients. But it can be done. Worse yet, uncomplemented subspaces of ℓ_p exist and can even be isomorphic to ℓ_p:

Proposition 1.3.1 *For each $p \in (0, \infty)$ different from 2, the space ℓ_p contains an uncomplemented subspace isomorphic to ℓ_p.*

This result is local in nature, in the sense that it depends upon proving that for every $0 < p < \infty$, $p \neq 2$, there is a sequence of subspaces $E_n \subset \ell_p^n$ such that

- $\ell_p(\mathbb{N}, E_n)$ is isomorphic to ℓ_p,
- $\lambda(E_n, \ell_p^n) \longrightarrow \infty$ as $n \longrightarrow \infty$.

The uncomplemented copy of ℓ_p inside ℓ_p is thus $\ell_p(\mathbb{N}, E_n)$ inside $\ell_p(\mathbb{N}, \ell_p^n)$. Let us tell the story as p decreases. The case $2 < p < \infty$ is solved by Rosenthal in [417] finding subspaces E_n that are badly complemented but still uniformly isomorphic to the ℓ_p space of the corresponding dimension. Bennett, Dor, Goodman, Johnson and Newman [37] solved the case $1 < p < 2$ locating badly complemented subspaces of ℓ_p^n uniformly isomorphic to the corresponding Euclidean space. Then $\ell_p(\mathbb{N}, E_n) \simeq \ell_p(\mathbb{N}, \ell_2^k) \simeq \ell_p$, by Pełczyński's decomposition method (see later). In fact, Rosenthal had previously settled the case $1 < p < \frac{4}{3}$ in [409] using harmonic ideas of Rudin. The case $p = 1$, by far the most difficult and least understood of all, resisted until Bourgain's paper [48]; see also Section 2.2. Curiously enough, the result is almost trivial for $0 < p < 1$, since in this case one can take E_n of dimension 1! Indeed, let E_n be the line spanned by $s_n = \sum_{i=1}^n e_i$ in ℓ_p^n. Any projection of ℓ_p^n onto E_n has the form $P(x) = f(x)s_n$, where f is a linear functional on ℓ_p^n such that $f(s_n) = 1$. Clearly, $\|P\| = \|f\|\|s_n\|$, and the minimum is attained when $f(x) = n^{-1} \sum_{i=1}^n x(i)$, which gives $\|P\| = n^{1/p-1}$. A more sophisticated argument of Stiles [443, Theorem 2.3] produces a subspace X of ℓ_p isometric to ℓ_p without infinite-dimensional subspaces complemented in ℓ_p. The case of c_0 stands apart: every copy of c_0 inside c_0 is complemented by Sobczyk's theorem (Section 1.7), and quotients of c_0 contain c_0 complemented [334]:

Proposition 1.3.2 *Every quotient of c_0 is isomorphic to a subspace of c_0.*

If $(F_n)_n$ is a sequence of finite-dimensional Banach spaces increasingly badly complemented in ℓ_∞^n then $c_0(\mathbb{N}, F_n)$ cannot be isomorphic to c_0. Deciding the converse is an open question: if $c_0(\mathbb{N}, F_n) \simeq c_0$, must the F_n be uniformly isomorphic to $\ell_\infty^{\dim F_n}$?

The space c_0 is not complemented in its bidual ℓ_∞. Phillips proved it in [386] for c and Sobczyck in [439] for c_0 itself:

Proposition 1.3.3 *c_0 is not complemented in ℓ_∞.*

Proof A weak*-null sequence of extensions of the coordinate functionals must have weak null restrictions to c_0. The Schur property of ℓ_1 makes them norm null, and that is impossible. □

Indecomposable and H.I. Spaces

While the only complemented subspaces in a prime space are copies of the space itself, there is a much more extreme way of having very few complemented subspaces: not having complemented subspaces at all.

Definition 1.3.4 A Banach space X is said to be indecomposable if, whenever $X = A \oplus B$, either A or B is finite-dimensional. A Banach space is said to be *hereditarily indecomposable* (H.I.) if every subspace is indecomposable.

H.I. Banach spaces exist [197], can be uniformly convex [169] or $\mathscr{L}_\infty$-spaces [17] and solve Banach's unconditional basis problem: does every Banach space contain an unconditional basic sequence? No: H.I. spaces do not. Actually, the Gowers dichotomy theorem [196] states that every Banach space contains either an H.I. subspace or an unconditional basic sequence. Thus, H.I. spaces are deeply entwined with the structure of general Banach spaces and can no longer be regarded as an anecdotal pathology. An H.I. space cannot be isomorphic to any proper subspace: actually, every operator from an H.I. space to any of its proper subspaces must be strictly singular [197, §4 theorem and corollary]. Indecomposable spaces isomorphic to their hyperplanes (which also exist [198]) must be prime: indeed, their infinite-dimensional complemented subspaces are finite-codimensional, and a space isomorphic to its hyperplanes is also isomorphic to its finite-codimensional subspaces. Indecomposable spaces can be $C(K)$-spaces [297], large [299] and also arbitrarily large [301], while H.I. spaces must be subspaces of ℓ_∞ [396], hence they have at most the dimension of the continuum. Abandon Banach spaces and you will find stranger things: rigid spaces (whose only endomorphisms are the scalar multiples of the identity; see [404; 449] for surprising news), spaces without basic sequences (see Section 9.4.4) and so on.

1.4 Local Properties and Techniques

The so-called local theory studies the structure of (quasi-) Banach spaces by means of their finite-dimensional subspaces. Very often the asymptotic behaviour of quantitative information is the key. Of paramount importance is the:

1.4.1 Principle of Local Reflexivity *Let X be a Banach space, F a finite-dimensional subspace of X^{**}, G a finite-dimensional subspace of X^* and $\varepsilon > 0$. Then there is an operator $T\colon F \to X$ such that*

- $\|T\|\|T^{-1}\| \leq 1 + \varepsilon$,
- $Tx = x$ *for every* $x \in F \cap X$,
- $\langle x^{**}, x^* \rangle = \langle x^*, Tx^{**} \rangle$ *for every* $x^* \in G$ *and every* $x^{**} \in F$.

The reader is referred to [227, § 9, p. 53] or [5, § 12.2] for proofs that have been refined over the years.

The $\mathscr{L}_p$-Spaces and Related Classes

Definition 1.4.2 Let $1 \leq p \leq \infty$ and $1 \leq \lambda < \infty$. An infinite-dimensional Banach space X is said to be an $\mathscr{L}_{p,\lambda}$-space if every finite-dimensional subspace of X is contained in another finite-dimensional subspace of X whose Banach–Mazur distance to the corresponding ℓ_p^n is at most λ. A space X is said to be an $\mathscr{L}_p$-space if it is an $\mathscr{L}_{p,\lambda}$-space for some $\lambda \geq 1$.

$\mathscr{L}_p$-spaces can be considered the local version of $L_p(\mu)$-spaces and $\mathscr{L}_\infty$-spaces the local version of $C(K)$-spaces. An infinite-dimensional Banach space is an $\mathscr{L}_{\infty,1+}$-space if and only if its dual is isometric to $L_1(\mu)$ for some measure μ. The latter are usually called *Lindenstrauss spaces* and include, among other interesting classes – see Note 8.8.1 – all $C(K)$ spaces. If the reader tries to define $\mathscr{L}_p$-space for $0 < p < 1$ in the most standard form, there will be a traumatic moment when they discover that it is not even known whether L_p satisfies the definition. There is, however, a satisfactory notion of $\mathscr{L}_p$-space, due to Kalton, that works fine for $0 < p < 1$, but it has to wait until Chapter 5.

Type and Cotype

Our basic source for the study of type and cotype of Banach spaces is the masterpiece of Diestel, Jarchow and Tonge [153, Chapter 11]. In what follows we will work in the wider context of quasi-Banach spaces.

Definition 1.4.3 A quasi-Banach space X is said to have type p for $0 < p \leq 2$ if there is a constant T such that for every finite sequence $(x_i)_{1\leq i\leq n}$ of points of X we have

$$\left(\int_0^1 \Big\| \sum_{1\leq i\leq n} r_i(t)x_i \Big\|^p dt\right)^{1/p} \leq T\left(\sum_{1\leq i\leq n} \|x_i\|^p\right)^{1/p}, \tag{1.6}$$

where $(r_i)_{i\geq 1}$ is the Rademacher sequence.

If $0 < p \leq 1$, this is just the randomised version of (1.3), which was used to characterise the p-normability of X. We are not really interested in type $p < 1$ because a quasi-Banach space has type $0 < p < 1$ if and only if it has an equivalent p-norm [252, Theorem 4.2]. Incidentally, the L_p quasinorm on the left-hand side of (1.6) can be replaced by any other L_q quasinorm with $q \in (0, \infty)$. This fact is due to Kahane when X is a Banach space [153, 11.1] and to Kalton in general [252, Theorem 2.1].

Proposition 1.4.4 *A quasi-Banach space having type $p > 1$ is locally convex.*

The proof is based on the behaviour of the following sequences, defined for any quasi-Banach space X:

$$a_n(X) = \sup_{\|x_i\|\leq 1} \Big\| \sum_{1\leq i\leq n} x_i \Big\|, \qquad b_n(X) = \sup_{\|x_i\|\leq 1} \inf_{\varepsilon_i=\pm 1} \Big\| \sum_{1\leq i\leq n} \varepsilon_i x_i \Big\|. \tag{1.7}$$

It is clear that both $(a_n)_n$ and $(b_n)_n$ are increasing and submultiplicative, that is, $a_{nm} \leq a_n a_m$ and $b_{nm} \leq b_n b_m$ for all $n, m \in \mathbb{N}$. The proof follows by simply assembling the three parts of the next result:

Lemma 1.4.5

- *If X has type $p > 1$, then $n^{-1}b_n \longrightarrow 0$.*
- *If $n^{-1}b_n \longrightarrow 0$, then $(n^{-1}a_n)_n$ is bounded.*
- *If $(n^{-1}a_n)_n$ is bounded, then X is locally convex.*

Proof The first part is obvious, since for $(x_i)_{1\leq i\leq n}$ in X and $p \in (0, \infty)$, one has

$$\inf_{\varepsilon_i \pm 1} \Big\| \sum_{1\leq i\leq n} \varepsilon_i x_i \Big\| \leq \left(\int_0^1 \Big\| \sum_{1\leq i\leq n} r_i(t)x_i \Big\|^p dt\right)^{1/p},$$

and so, if X has type p then $b_n(X) \leq Tn^{1/p}$. Let us check the third point. With no loss of generality, we may assume that X is r-normed for some $0 < r \leq 1$. Now, if $n^{-1}a_n \leq R$, then the ball of radius R contains the set of means

$$\left\{\frac{x_1 + \cdots + x_n}{n} : \|x_i\| \leq 1, n \in \mathbb{N}\right\}$$

whose closure is exactly the closed convex hull of B_X. The same argument shows that X is locally convex if and only if $(n^{-1}a_n)_n$ has a bounded subsequence. The second point accounts for the lion's share of the proof. We keep assuming that X is r-normed. Pick $2n$ points $(x_i)_{1\le i\le 2n}$ from B_X and $\varepsilon_i = \pm 1$ such that $\|\sum_{1\le i\le 2n}\varepsilon_i x_i\| \le b_{2n}$. Rearranging, we may assume that $\varepsilon_i = 1$ for $i \le k \le n$ and $\varepsilon_i = -1$ for $k < i \le 2n$, that is, $\|x_1 + \cdots + x_k - x_{k+1} - \cdots - x_{2n}\| \le b_{2n}$. Writing $\sum_{1\le i\le 2n} x_i = 2\sum_{1\le i\le k} x_i - \sum_{1\le i\le 2n}\varepsilon_i x_i$ and using the r-subadditivity of the quasinorm, we obtain

$$\Big\|\sum_{i=1}^{2n} x_i\Big\|^r = 2^r\Big\|\sum_{i=1}^{k} x_i\Big\|^r + b_{2n}^r \le 2^r a_n^r + b_{2n}^r \quad\Longrightarrow\quad a_{2n}^r \le 2^r a_n^r + b_{2n}^r.$$

As (b_n) is increasing and submultiplicative and $n^{-1}b_n \longrightarrow 0$, there is some $s > 1$ such that $n^{-s}b_n$ is bounded, say, by C, and dividing by $(2n)^r$, we get

$$\left(\frac{a_{2n}}{2n}\right)^r \le \left(\frac{a_n}{n}\right)^r + \left(\frac{C2^s n^s}{2n}\right)^r = \left(\frac{a_n}{n}\right)^r + C^r(2n)^{r(s-1)}.$$

If, for $n \ge 0$, we put $\alpha_n = 2^{-n}a_{2^n}$, then the preceding inequality becomes $\alpha_{n+1}^r - \alpha_n^r \le C^r 2^{r(s-1)(n+2)}$, and since $\sum_n 2^{r(s-1)(n+2)} < \infty$, we obtain that (α_n^r) is bounded, which is enough. □

Banach spaces for which $n^{-1}b_n \longrightarrow 0$ are traditionally called *B-convex* and Banach spaces having type $p > 1$ are traditionally called spaces having non-trivial type. A very deep result [153, 13.10 Theorem plus 13.16 Theorem] states:

1.4.6 *Let X be an infinite-dimensional Banach space. The following are equivalent:*

(i) *X is B-convex,*
(ii) *X does not contain ℓ_1^n uniformly,*
(iii) *X has non-trivial type.*

We have established that quasi-Banach spaces satisfying (iii) are Banach spaces satisfying (i). A still deeper result by Pisier shows that for each infinite-dimensional B-convex Banach space, there is a constant C so that, for every n, there are operators $I: \ell_2^n \longrightarrow X$ and $P: X \longrightarrow \ell_2^n$ such that PI is the identity on ℓ_2^n and $\|I\|\|P\| \le C$ [153, 19.3 Theorem]; in particular, X contains ℓ_2^n uniformly complemented. All this produces the following dichotomy [153, 13.3 and 19.3]:

1.4.7 *A Banach space either contains ℓ_1^n uniformly or contains ℓ_2^n uniformly complemented.*

Definition 1.4.8 A (quasi-) Banach space X is said to have cotype q , where $2 \leq q < \infty$, if there is a constant C such that

$$\left(\sum_{1\leq i\leq n} \|x_i\|^q\right)^{1/q} \leq C\left(\int_0^1 \Big\|\sum_{1\leq i\leq n} r_i(t)x_i\Big\|^q dt\right)^{1/q} \tag{1.8}$$

for every $x_1, \ldots, x_n \in X$, where $(r_i)_{i\geq 1}$ is the Rademacher sequence.

The spaces L_p have type $\min(p, 2)$ and cotype $\max(2, p)$ for $0 < p < \infty$. It is relatively easy to prove that the dual of a type p space has cotype p^*, where p^* is given by $1 = 1/p + 1/p^*$ [153, 11.10 Proposition]. The converse is true for B-convex spaces [153, 13.17 Proposition] and false in general (consider the case of ℓ_1, which has cotype 2). Kwapien's theorem [153, 12.19 and 12.20] establishes:

1.4.9 Kwapien's theorem *A (quasi-) Banach space having type* 2 *and cotype* 2 *is isomorphic to a Hilbert space.*

The Maurey–Pisier *Great* Theorem (well, one of them) states:

1.4.10 Maurey–Pisier theorem *Every Banach space X contains almost isometric copies of $\ell^n_{p(X)}$ and $\ell^n_{q(X)}$, where $p(X) = \sup\{p\colon X \text{ has type } p\}$ and $q(X) = \inf\{q\colon X \text{ has cotype } q\}$.*

See [362, § 13] for a reasonably accessible proof. Spaces with $p(X) = 2 = q(X)$ have been called *near Hilbert* and will be encountered later. Kwapień's result is contained in Maurey's extension theorem [153, 12.22]: *every operator from a subspace of a type* 2 *space to a cotype* 2 *space can be extended to an operator on the whole space that still factorises through a Hilbert space.* The following definition should then come as no surprise:

1.4.11 Maurey extension property A Banach space X is said to have the Maurey extension property (MEP) if every Hilbert valued operator defined on a subspace of X can be extended to X.

Type 2 spaces have MEP, and thus Hilbert subspaces of type 2 spaces are complemented.

Ultraproducts

The Banach space ultraproduct construction originates in model theory and has been, and continues to be, the main channel of communication between logic and Banach space theory. Even emancipated from model theory, ultraproducts of Banach spaces have a surprisingly large number of applications, ranging from the local theory to the Lipschitz and uniform classification of Banach spaces. For a detailed study of this construction at the elementary level needed here, we refer the reader to Heinrich's survey paper [211] or Sims' notes [434]. A more complete exposition with the necessary model-theoretic background is [212]. Ultraproducts of quasi-Banach spaces have never had a comparable prestige; nevertheless, they are even more useful, for the same reasons as in the case of Banach spaces and because they provide an operative substitute for the bidual (which may well be trivial now). The ultraproduct construction is based on the notion of convergence along an ultrafilter, which we pause to explain.

Let I be a set and $\mathcal{F}$ a filter on I (a family of subsets that does not contain the empty set, is closed under finite intersections and such that if $A \in \mathcal{F}$ and $A \subset B$ then $B \in \mathcal{F}$). Given a topological space S, a function $f\colon I \longrightarrow S$, a filter $\mathcal{F}$ on I and $s \in S$, we write $\lim_{\mathcal{F}(i)} f(i) = s$ and say that f converges to s along $\mathcal{F}$ if $f^{-1}[V] \in \mathcal{F}$ for every neighbourhood V of s. The definition makes sense even if f is *only* defined on some $A \in \mathcal{F}$: just consider the limit along the family $\{B \in \mathcal{F}\colon B \subset A\}$, which is a filter on A. Keep this fact in mind since it will be used without further mention. An ultrafilter on I is a maximal filter with respect to inclusion. Not exactly trivial, but nonetheless graspable, is the very elegant characterisation of ultrafilters as those filters $\mathcal{F}$ such that for every partition of I into two (or finitely many) subsets, exactly one of them belongs to $\mathcal{F}$. The only ultrafilters that one will ever see explicitly are the principal, or fixed, ultrafilters: an ultrafilter $\mathcal{U}$ is fixed if it contains a finite set (hence a singleton) in which case there is $i \in I$ such that $\mathcal{U} = \{A \subset I\colon i \in A\}$. Otherwise, $\mathcal{U}$ is called free. An ultrafilter $\mathcal{U}$ is said to be *countably incomplete* if there is a decreasing sequence of elements of $\mathcal{U}$ with empty intersection. This happens if and only if there is a strictly positive function $f\colon I \longrightarrow (0, \infty)$ such that $f(i) \longrightarrow 0$ along $\mathcal{U}$. Every free ultrafilter on $\mathbb{N}$ is countably incomplete. The simplest way to produce a free ultrafilter is to start with a filter containing no finite subsets (for instance, the cofinite subsets of I when this is infinite) and use Zorn's lemma to refine it to an ultrafilter, or use the following variation that will appear over and over: when I carries a partial *directed* order, i.e. such that for every $i, j \in I$, there exists $k \in I$ such that $i, j \leq k$, the sets $\{j \in I\colon i \leq j\}$ generate the so-called order filter, and every ultrafilter containing it is free unless I has a maximal element. The key point is that every function with values on a compact (Hausdorff) space has a (unique) limit along any ultrafilter, and this is actually a topological version of the ultrafilter characterisation alluded to earlier.

To avoid unnecessary complications, we will work with families of p-Banach (instead of arbitrary quasi-Banach) spaces: this yields the continuity of the quasinorms and, more importantly, a uniform bound for the moduli of concavity. So, let $(X_i)_{i\in I}$ be a family of p-Banach spaces indexed by I, and let $\mathcal{U}$ be an ultrafilter on I. The space of bounded families $\ell_\infty(I, X_i)$ is a p-Banach space, and $c_0^{\mathcal{U}}(I, X_i) = \{(x_i) \in \ell_\infty(X_i) : \lim_{\mathcal{U}(i)} \|x_i\| = 0\}$ is a closed subspace of $\ell_\infty(I, X_i)$. The ultraproduct of the spaces $(X_i)_{i\in I}$ following $\mathcal{U}$ is defined as the quotient $[X_i]_{\mathcal{U}} = \ell_\infty(I, X_i)/c_0^{\mathcal{U}}(I, X_i)$. We denote by $[(x_i)]$ the element of $[X_i]_{\mathcal{U}}$ which has the family (x_i) as a representative. Using the continuity of p-norms, it is not difficult to show that $\|[(x_i)]\| = \lim_{\mathcal{U}(i)} \|x_i\|$. It is clear that two bounded families $(x_i)_{i\in I}$ and $(y_i)_{i\in I}$ define the same element of $[X_i]_{\mathcal{U}}$ if the set $\{i \in I : x_i = y_i\}$ belongs to $\mathcal{U}$. As we remarked before (claiming falsely that the fact will be used without further mention), to define an element of $[X_i]_{\mathcal{U}}$, one just needs a bounded family $(x_i)_{i\in A}$ defined *only* on some subset $A \in \mathcal{U}$. When $X_i = X$ for all $i \in I$, we denote the ultraproduct by $X_{\mathcal{U}}$ and call it the ultrapower of X following $\mathcal{U}$. The diagonal mapping $X \longrightarrow X_{\mathcal{U}}$ sending each $x \in X$ to $[(x)]$ is an isometric embedding, and so each p-Banach space is isometric to a subspace of its ultrapowers. If $T_i : X_i \longrightarrow Y_i$ is a uniformly bounded family of operators, where X_i and Y_i are all p-Banach spaces, the ultraproduct operator $[T_i]_{\mathcal{U}} : [X_i]_{\mathcal{U}} \longrightarrow [Y_i]_{\mathcal{U}}$ is given by $[T_i]_{\mathcal{U}}[(x_i)] = [T_i(x_i)]$. Quite clearly, $\|[T_i]_{\mathcal{U}}\| = \lim_{\mathcal{U}(i)} \|T_i\|$.

Definition 1.4.12 An ultrasummand is a quasi-Banach space complemented in all its ultrapowers through the diagonal embedding.

The following result confirms the intuition that the unit ball of an ultrasummand enjoys a kind of 'compactness' (after all, a projection of $X_{\mathcal{U}}$ onto X must select one point of X from each bounded family of points).

Proposition 1.4.13 *Let X be a Banach space. The following are equivalent!*

(i) *X is an ultrasummand,*
(ii) *X is complemented in its bidual,*
(iii) *X is a complemented subspace of some dual space.*

Proof We prove the implications (ii) $\Longrightarrow$ (iii) $\Longrightarrow$ (i) $\Longrightarrow$ (ii). The first one is trivial. To prove (iii) $\Longrightarrow$ (i), we assume that Y is a Banach space whose dual contains X and that $P : Y^* \longrightarrow X$ is a bounded projection. Now, if $\mathcal{U}$ is an ultrafilter on I and $X_{\mathcal{U}}$ is the corresponding ultrapower, we can define a projection $L : X_{\mathcal{U}} \longrightarrow X$ as $L[(x_i)] = \lim_{\mathcal{U}(i)} x_i$, where the limit is taken in the weak* topology of Y^*. The proof of (iii) $\Longrightarrow$ (i) relies on the principle of local reflexivity, which is responsible for embedding X^{**} as a very well-placed subspace of a suitable ultrapower of X: indeed, consider the order in $\mathscr{F}(X^{**}) \times \mathscr{F}(X^*) \times (0, \infty)$ given by $(F, G, \varepsilon) \leq (F', G', \varepsilon')$ if $F \subset F', G \subset G'$

and $\varepsilon' \le \varepsilon$. Let $\mathfrak{U}$ be a free ultrafilter refining the order filter on $\mathscr{F}(X^{**}) \times \mathscr{F}(X^*) \times (0, \infty)$. Given $F \in \mathscr{F}(X^{**}), G \in \mathscr{F}(X^*)$ and $\varepsilon > 0$, we consider the operator $T_{(F,G,\varepsilon)}: F \longrightarrow X$ provided by the principle of local reflexivity. We define a map $\Delta: X^{**} \longrightarrow X_{\mathfrak{U}}$ by letting $\Delta(x^{**}) = [(x_{(F,G,\varepsilon)})]$, where $x_{(F,G,\varepsilon)} = T_{(F,G,\varepsilon)}(x^{**})$ if $x^{**} \in F$, and 0 otherwise. Clearly, Δ is a linear isometry of X^{**} into $X_{\mathfrak{U}}$. Note that $x_{(F,G,\varepsilon)}$ and $T_{(F,G,\varepsilon)}(x^{**})$ agree 'eventually', and so the linearity of Δ is not a problem due to our choice of $\mathfrak{U}$. The isometric copy thus obtained is moreover 1-complemented via the operator $\nabla: X_{\mathfrak{U}} \longrightarrow X^{**}$ sending $[(x_{(F,G,\varepsilon)})]$ to the weak* limit of $(x_{(F,G,\varepsilon)})$ along $\mathfrak{U}$. Clearly, ∇ is a well-defined, contractive operator and $\nabla\Delta = \mathbf{1}_{X^{**}}$, since the family $x_{(F,G,\varepsilon)}$ converges to x^{**} in the weak* topology along $\mathfrak{U}$ because each $x^* \in X^*$ eventually falls in G. □

Thus, reflexive and $L_1(\mu)$-spaces are ultrasummands. Since c_0 is not complemented in ℓ_∞, it cannot be an ultrasummand, and the same happens to any space containing c_0 complemented. To present typical non-locally convex ultrasummands, note that the only property of a dual Banach space which is needed to carry out the proof of the implication (iii) $\Longrightarrow$ (i) in the preceding result is that the weak* topology is a linear topology weaker than the norm topology and makes the unit ball compact. Quasi-Banach spaces admitting a weaker-than-the-quasinorm topology, making the unit ball compact, are termed *pseudoduals*. One has

1.4.14 *The spaces ℓ_p, H_p and S_p are pseudoduals for all $0 < p < \infty$ and, therefore, ultrasummands.*

We only sketch the proof. The case ℓ_p is clear since the topology of pointwise convergence makes its unit ball compact. As for the Hardy classes H_p, the estimate (1.5) is exactly what we need to invoke Montel's theorem on normal families to conclude that the ball of H_p is compact under the topology of pointwise convergence on points of the open disc. The case of the Schatten classes S_p is because the quasinorm of S_p is lower semicontinuous with respect to the weak operator topology of $\mathfrak{L}(H)$ [157, Corollary 2.3]. Nevertheless, L_p emphatically refuses to be an ultrasummand for $0 < p < 1$ (see Note 1.8.3 for details).

1.5 The Dunford–Pettis, Grothendieck, Pełczyński and Rosenthal Properties

Although the borders between global and local properties are somewhat permeable, we now discuss some 'global' properties important in the study of Banach spaces mainly because of their connections with the structure of $\mathscr{C}$-spaces. A terse exposition can be found in [22, Appendix A1].

Definition 1.5.1 A Banach space X is said to have

- the Dunford–Pettis property (DPP) if every weakly compact operator defined on X sends weakly convergent sequences to convergent sequences,
- Pełczyński's property (V) if every operator defined on X is either weakly compact or an isomorphism on a copy of c_0,
- Rosenthal's property (V) – the same as before, replacing c_0 by ℓ_∞,
- Grothendieck's property if every operator from X to a separable Banach space is weakly compact.

Dunford and Pettis themselves established that $L_1(\mu)$ and $C(K)$ spaces have DPP. The extension to $\mathscr{L}_\infty$-spaces and the name are due to Grothendieck; see [151]. It is clear that a complemented subspace of a space with the DPP has DPP and that an infinite-dimensional space with DPP cannot be reflexive since, otherwise, weakly convergent sequences must be convergent, which makes the unit ball compact. General background about the Dunford–Pettis property can be found in [151] and [102, Chapter 6]. Pełczyński proved in [376, Theorem 5] that $\mathscr{C}$-spaces have property (V). This property clearly passes to quotients and, since Johnson and Zippin showed in [234] that every separable Lindenstrauss space is a quotient of $C[0, 1]$ (plus the obvious fact that each separable subspace of a Lindenstrauss space is contained in a separable Lindenstrauss subspace), all Lindenstrauss spaces have property (V). However, not all $\mathscr{L}_\infty$-spaces have property (V) (see Section 10.5 for accessible counterexamples). The combination of property (V) and DPP shows that *every infinite-dimensional complemented subspace of a Lindenstrauss space contains* c_0. Reflexive spaces enjoy Grothendieck's property for obvious reasons, but so do injective spaces [152, VII, Theorem 15]. The information we need is that $\mathscr{L}_\infty$-spaces with Grothendieck's property do not contain separable complemented subspaces [22, Proposition 2.8]. Ultrasummands of type $\mathscr{L}_\infty$ or, equivalently, injective Banach spaces [22, Proposition 1.5] even have Rosenthal's property (V) [22, Proposition 2.8].

1.6 $C(K)$-Spaces and Their Complemented Subspaces

We focus on the following idea: the homeomorphism type of a compact space is independent of the particular realisation of it one encounters. The web formed of the seven compacta appearing in this book that we describe now has been woven with that silk:

1. ω^N is the only countable compact whose Nth derived set is one point.
2. ω^ω is the only countable compact whose ωth derived set is one point.
3. The Cantor set Δ is the only compact that is totally disconnected perfect and metrisable, regardless of whether it appears as $\{0,1\}^{\mathbb{N}}$, as $\{-1,0,1\}^{\mathbb{N}}$ or wearing other clothes.
4. The unit interval $[0,1]$.
5. $\beta\mathbb{N}$ is the Stone–Čech compactification of the discrete space $\mathbb{N}$, i.e. the only compact space containing a dense copy of $\mathbb{N}$ such that every bounded function on $\mathbb{N}$ extends to a continuous function on $\beta\mathbb{N}$. Thus, $\ell_\infty = C(\beta\mathbb{N})$. $\beta\mathbb{N}$ can be obtained as the Stone space of $\mathcal{P}(\mathbb{N})$, aka the space of ultrafilters on $\mathbb{N}$.
6. $\mathbb{N}^*$ is $\beta\mathbb{N}\setminus\mathbb{N}$. Two continuous functions on $\beta\mathbb{N}$ coincide on $\mathbb{N}^*$ if and only if their difference converges to 0 on $\mathbb{N}$, and thus $C(\mathbb{N}^*)$ can be identified with ℓ_∞/c_0. Under CH, $\mathbb{N}^*$ is the only totally disconnected Hausdorff F-space without isolated points of weight $\mathfrak{c}$ and such that every non-empty G_δ subset has a non-empty interior by Parovičenko's *other* theorem, see [453, Chapter 3, p. 80–83]. Parovičenko's *first* theorem [45] asserts that $\mathbb{N}^*$ maps continuously onto each compact space of weight $\aleph_1$ or less.
7. The unit ball B_X^* of the dual of a Banach space X is endowed with the weak* topology.

The passage from the compact K to the corresponding $C(K)$-space tears the cobweb apart:

1.6.1 Classification of separable $\mathscr{C}$-spaces *Let K be a metrisable compact space. If K is uncountable then $C(K)$ is isomorphic to $C[0,1]$. If K is countable then $C(K)$ is isometric to $C(\alpha)$ for some countable ordinal α.*

The first assertion is Milutin's theorem [364], which in particular means that $C[0,1] \simeq C\Delta) \simeq C(B_X^*)$ for all separable X. No proof for Milutin's theorem is perhaps clearer than the one presented in [5, §4.4]. The second assertion is a famous theorem of Mazurkiewicz and Sierpiński [360, Théorème 1], see also [430, Theorem 8.6.10]. Moreover, so far as separability is involved, $C(\Delta)$ is the guy to deal with:

Lemma 1.6.2 *Let K be any compact metric space. Then there exist positive contractive operators $S\colon C(K) \longrightarrow C(\Delta)$ and $R\colon C(\Delta) \longrightarrow C(K)$ such that $R1_\Delta = 1_K, S1_K = 1_\Delta$ and $RS = \mathbf{1}_{C(K)}$.*

The problem of identifying the complemented subspaces of $\mathscr{C}$-spaces was hermetically open until Plebanek and Salguero [393] betrayed our trust by finding a complemented subspace of a non-separable $C(K)$ space that is not isomorphic to a $C(K)$-space while this book was in print.

1.6.3 *This is (most of) what is currently known:*

(a) *Every complemented subspace of ℓ_∞ is isomorphic to ℓ_∞.*

(b) *A subspace of $c_0(I)$ is complemented if and only if it is isomorphic to $c_0(J)$ for some J, and if and only if it is an $\mathscr{L}_\infty$-space.*

(c) *Every complemented subspace of $C(\omega^\omega)$ is isomorphic to either c_0 or $C(\omega^\omega)$.*

(d) *A complemented subspace of $C[0,1]$ with non-separable dual must be isomorphic to $C[0,1]$.*

Assertion (a) results from a combination of the DPP and Rosenthal's property (V) of ℓ_∞: together they yield that complemented subspaces of ℓ_∞ contain ℓ_∞, necessarily complemented, and therefore, by Pełczyński's decomposition method, they must be isomorphic to ℓ_∞. Incidentally, this is the case $p = \infty$ in Proposition 1.2.3. The first part of (b) for countable I is consequence of the DPP and Pełczyński's property of c_0 plus Pełczyński's decomposition method ('only if'). The 'if' part is a particular case of Sobczyk's theorem (the main topic of Section 1.7). The general non-separable version is due to Suárez Granero [199]. The second part of (b) is due to Godefroy, Kalton and Lancien [190, Remark 5.4], who proved that an $\mathscr{L}_\infty$-subspace of $c_0(I)$ is isomorphic to some $c_0(J)$. Part (c) is due to Benyamini [40]. The proof of (d) is built on a beautiful lemma of Rosenthal [415]: *an operator $T\colon C[0,1] \longrightarrow X$ for which $T^*[X^*]$ is non-separable fixes a copy of* $C[0,1]$. Therefore, let X be a complemented subspace of $C[0,1]$, and let $P\colon C[0,1] \longrightarrow X$ be a projection. If X has a non-separable dual then $P^*[X^*]$ is non-separable, and thus P fixes a copy of $C[0,1]$. Therefore, X contains a copy of $C[0,1]$. Now we isolate a useful result of Pełczyński [380, Theorem 1]

1.6.4 *A subspace of a separable $C(K)$ that contains a copy of $C(K)$ must also contain a complemented copy of $C(K)$.*

The proof of (d) can be completed as follows: if X is complemented in, and contains a copy of, $C[0,1]$, apply Pełczyński's decomposition method using c_0-vector sums. □

1.7 Sobczyk's Theorem and Its Derivatives

It is a straightforward consequence of the Hahn–Banach theorem that when Y is a subspace of a Banach space X, every operator $u\colon Y \longrightarrow \ell_\infty$ can be extended, with the same norm to X. In particular, but in the end equivalently, ℓ_∞ is 1-complemented in any Banach space containing it. The following fundamental result of Sobczyk establishes that c_0 has a similar *homological* property for separable spaces, although doubling the norm of the extension. What follows is not Sobczyk's original proof but the beautiful proof of Veech [451], one of the masterpieces in THE BOOK, which Diestel will be annotating now.

1.7.1 Sobczyk's theorem *If X is a separable Banach space and Y is a subspace of X, then every operator $u\colon Y \longrightarrow c_0$ has a 2-extension to X.*

Proof Assume $\|u\| = 1$ and write $u(y) = (\langle y_n^*, y\rangle)_{n\geq 1}$, where $(y_n^*)_{n\geq 1}$ is a weak*-null sequence in the unit ball of Y^*. The strategy is to find a weak*-null sequence of extensions of these functionals. For each n, let $x_n^* \in X^*$ be a Hahn–Banach extension of y_n^*. Let D be the set of weak*-accumulation points of the sequence $(x_n^*)_n$, and recall that the dual ball of a separable space is weak*-metrisable by some metric d. The ridiculously simple observation 'a sequence such that every subsequence contains a further subsequence converging to zero is itself convergent to zero' yields $\lim_n d(x_n^*, D) = 0$. Choose $f_n \in D$ such that $d(x_n^*, f_n) \leq d(x_n^*, D) + \frac{1}{n}$. The sequence $(x_n^* - f_n)_n$ is weak*-null and extends (y_n^*) since an accumulation point of (x_n^*) must vanish on Y. Thus, the mapping $U\colon X \longrightarrow c_0$ given by $U(x) = (\langle x_n^* - f_n, x\rangle)_{n\geq 1}$ is an extension of u and, quite clearly, $\|U\| \leq 2$. □

The bound 2 cannot be improved because the norm of any projection of c onto c_0 is at least 2, as can be seen just considering the 'obvious projection' with kernel $[1_{\mathbb{N}}]$. The notion looming over Sobczyk's theorem is isolated in the next definition.

Definition 1.7.2 A Banach space E is separably injective if, for every subspace Y of a separable space X, every operator $u\colon Y \longrightarrow E$ has an extension $U\colon X \longrightarrow E$. If the extension can be achieved with $\|U\| \leq \lambda\|u\|$ then E is said to be λ-separably injective.

It is easy, though not entirely trivial, to see that each separably injective space must be λ-separably injective for some $\lambda \geq 1$ [22, Proposition 1.6]. The theory of separably injective spaces is surprisingly rich with examples and applications, as can be deduced from the mere existence of [22]. The theory of separably injective spaces concerns non-separable Banach spaces, in view of the outstanding

1.7.3 Zippin's theorem *c_0 is the only separable separably injective Banach space, up to isomorphism.*

Despite its pristine formulation, the proof requires a clever combination of Banach space machinery and delicate computations. Following Wittgenstein's mandate, we remain silent on the issue and refer to Zippin [466].

En Route to Non-separable Versions of Sobczyk's Theorem

The only well-known aspect of Sobczyk's result is the scalar separable case. To handle large non-separable spaces, we introduce a further batch of properties.

Definition 1.7.4 A Banach space is weakly compactly generated (WCG) if it contains a weakly compact subset with dense linear span.

Reflexive spaces and separable spaces are the two basic types of WCG spaces. The space $c_0(I)$ is WCG for all I since the inclusion $\ell_2(I) \longrightarrow c_0(I)$ has dense range, and so is $L_1(\mu)$ for finite μ since the inclusion $L_2(\mu) \longrightarrow L_1(\mu)$ has dense range. Let us present an accurate formulation for the idea that WCG spaces have many complemented subspaces.

Definition 1.7.5 A Banach space X has the separable complementation property (SCP) if every separable subspace of X is contained in a complemented separable subspace of X.

All WCG spaces have SCP [148, Chapter IV, Lemma 2.4], but there are many more, as can be seen in [396].

Definition 1.7.6 A projectional resolution of the identity (PRI) on a Banach space X is a system of projections $(P_\alpha)_{\omega\leq\alpha\leq\mu}$, where $\mu = \dim(X)$ such that

- $\|P_\alpha\| = 1$,
- $P_\alpha P_\beta = P_{\min(\alpha,\beta)}$,
- $\dim P_\alpha[X] \leq |\alpha|$,
- $\overline{\bigcup_{\beta<\alpha} P_\beta[X]} = P_\alpha[X]$,
- $P_\mu[X] = \mathbf{1}_X$.

It is not simple to prove, but every WCG space admits a PRI [148, VI, Theorem 2.5]; however, it is simple that a Banach space with dimension $\aleph_1$ and a PRI has the SCP. Being obvious that Sobczyk's theorem works in SCP spaces, it works in WCG spaces with dimension $\aleph_1$. A straight projectional-resolution-free proof can be given:

Proposition 1.7.7 *c_0 is complemented in every WCG superspace (or, more generally, in a Banach space with the property that the closure of a countable set is metrisable in the weak topology).*

Proof Since weakly compact sets in a separable space are metrisable, separable sets in a WCG space are metrisable in the weak topology too. Veech's proof applies. □

We now consider special types of $\mathscr{C}$-spaces with PRI.

Definition 1.7.8 A compact space K is said to be

- Eberlein if it is homeomorphic to a weakly compact set of a Banach space,
- Corson if it is homeomorphic to a compact subset of some $\Sigma(I)$, the subspace of all countably supported elements of $[0, 1]^I$,
- Valdivia if there exist an I and an embedding of $\varphi\colon K \longrightarrow [0, 1]^I$ such that $\varphi[K] \cap \Sigma(I)$ is dense in $\varphi[K]$.

It would be very hard, if not impossible, to show the implications Eberlein $\Longrightarrow$ Corson $\Longrightarrow$ Valdivia more clearly than [148, VI. Theorem 7.2], and the same is true for [148, VI. Lemma 7.4]: K is an Eberlein compact if and only if $C(K)$ is WCG. Focusing now on the largest class of Valvidia compacta, we have the following decomposition [148, VI. Lemma 7.5]:

Lemma 1.7.9 *Let $K \subset [0, 1]^I$ be a Valdivia compact for which we assume that $K \cap \Sigma(I)$ is dense in K. Assume that $K \cap \Sigma(I)$ contains a dense subset of cardinality μ. Then there exists an increasing family $(I_\alpha)_{\omega\leq\alpha\leq\mu}$ of subsets of I such that, for every $\omega \leq \alpha < \mu$,*

(a) $|I_\alpha| \leq \alpha$,
(b) $I_\alpha = \bigcup_{\beta<\alpha} I_{\beta+1}$,
(c) $I_\mu = I$.

Moreover, if we define $R_L\colon [0, 1]^I \to [0, 1]^I$ by $R_L(x) = 1_L x$ when $L \subset I$, then $K_\alpha = R_{I_\alpha}[K] \subset K$ is a Valdivia compact.

Taking the embeddings $R^\circ_{I_\alpha}\colon C(K_\alpha) \longrightarrow C(K), R^\circ_{I_\alpha}(f) = f \circ R_{I_\alpha}$, associated to the family $(I_\alpha)_\alpha$ leads to [148, VI. Theorem 7.6 and Remark 7.7, p. 256]:

Proposition 1.7.10 *If K is a Valdivia compact, then $C(K)$ has a PRI $(P_\alpha)_\alpha$ such that $P_\alpha[C(K)]$ is isometric to $C(K_\alpha)$, with K_α again a Valdivia compact.*

And thus we are ready to discuss:

Sobczyk's Theorem for $c_0(I)$

Sobczyk's theorem steadily percolates out from the separability reservoir, first to copies of c_0 in WCG spaces. Then comes the smashing surprise that Sobczyk's original proof 'remains valid for $c_0(I)$', with the meaning that copies of $c_0(I)$ inside spaces X such that $X/c_0(I)$ is separable must be complemented. See [440] but also [79] for a lively discussion and full details. This leads to the question, first formulated by Yost, of whether copies of $c_0(I)$ must be complemented in WCG spaces (or beyond). Our first example of set theoretic considerations bursting into a seemingly remote domain appears while trying to answer this question: Yost's problem has a positive solution if $|I| < \aleph_\omega$, while there exists an Eberlein compact K such that $C(K)$ admits an uncomplemented copy of some $c_0(I)$ with $|I| = \aleph_\omega$ (the latter result shows that the former is optimal). We present now the positive part of the answer and postpone the negative part to Proposition 2.2.15.

Definition 1.7.11 A Banach space X is said to be K-Sobczyk if every κ-isomorphic copy of $c_0(I)$ inside X is $K\kappa$-complemented.

For instance, separable Banach spaces are 2-Sobczyk. We now begin our practice of using 'long' decompositions based on ordinals. To this end, let us explain the meaning of $x = \sum_{\alpha<\mu} x(\alpha)$: form the function $\widehat{x}\colon [0,\mu) \longrightarrow X$ inductively defined with $\widehat{x}(0) = 0$, then

$$\begin{cases} \widehat{x}(\beta) = \widehat{x}(\alpha) + x(\alpha+1), & \text{if } \beta = \alpha+1 \text{ is a successor ordinal,} \\ \widehat{x}(\beta) = x(\beta) + \lim_{\alpha\to\beta} \widehat{x}(\alpha), & \text{if } \beta \text{ is a limit ordinal,} \end{cases}$$

and finally asking for $x = \lim_{\beta\to\mu} \widehat{x}(\beta)$. Let X be a Banach space and μ an ordinal. We say that a family $(X_\alpha)_{\alpha<\mu}$ of subspaces, indexed by the ordinals below μ, is a decomposition of X if every point $x \in X$ can be written in a unique way as $x = \sum_{\alpha<\mu} x(\alpha)$, where $x(\alpha) \in X_\alpha$. A cardinal μ is said to be *regular* when the union of fewer than μ sets of cardinality smaller than μ has cardinality smaller than μ.

Lemma 1.7.12 *Let μ be an uncountable regular cardinal, and let X be a Banach space which admits a decomposition $(X_\alpha)_{\alpha<\mu}$. Assume that for some constants $K, M < \infty$ and each $\alpha < \mu$, we have*

(a) $\dim(X_\alpha) < \mu$,
(b) *is a projection* $P_\alpha : X \longrightarrow \overline{[X_\beta : \beta < \alpha]}$ *such that* $\|P_\alpha\| \le M$,
(c) $\overline{[X_\beta : \beta < \alpha]}$ *is K-Sobczyk.*

Then X is $2MK$-Sobczyk.

Proof For $x \in X$, denote $\operatorname{supp}(x) = \{\alpha < \mu : x(\alpha) \neq 0\}$ and observe that $|\operatorname{supp}(x)| \leq \aleph_0$. Since $\operatorname{cf}(\mu) = \mu > \aleph_0$, there exists $\beta < \mu$ such that $x(\alpha) = 0$ for all $\beta < \alpha < \mu$. Let Y be a κ-isomorphic copy of $c_0(I)$ in X. For each $i \in I$, let $y_i \in X$ be the isomorphic image of $e_i \in c_0(I)$. Let us dispose of the case where $|I| < \mu$. Since each y_i has countable support, there is some $\nu < \mu$ such that $\overline{[X_\beta : \beta < \nu]}$ contains every y_i, hence the whole of Y; the result follows using the M-projection P_ν provided by (b) followed by any K-projection $\overline{[X_\beta : \beta < \nu]} \longrightarrow Y$, whose existence is guaranteed by (c). For the remainder of the proof, we assume $|I| = \mu$. As the only relevant property of I is its cardinality, we treat it also as a cardinal (namely μ), and we identify $I = \{i_\alpha : \alpha < \mu\}$. We maintain the different names, I and μ, mostly for notational (and psychological) reasons.

Claim 1 For every $\alpha < \mu$, we have $|\{i \in I : y_i(\alpha) \neq 0\}| < \mu$.

Proof of Claim 1 Let $J = \{i \in I : y_i(\alpha) \neq 0\}$. Let ν be the smallest cardinal of a subset spanning a weak*-dense subspace of X_α^*. Since $\nu \leq \dim(X_\alpha) < \mu$ (because of (a)), the vectors $\{y_j\}_{j \in J}$ can be separated from 0 using no more than ν functionals of X_α^*. So, the unit basis $\{e_j\}_{j \in J}$ of $c_0(J)$ can be separated from 0 using ν functionals of $\ell_1(J)$, which cannot be if $|J| = \mu > \nu$. □

Since μ is regular, for each $\alpha < \mu$, one has

$$|\{i \in I : \operatorname{supp}(y_i) \cap [0, \alpha) \neq \emptyset\}| < \mu. \tag{1.9}$$

This allows us to introduce a correspondence between points of I, certain subsets of I and limit ordinals below μ that 'stabilises supports' as follows. Given $\beta, \gamma \leq \mu$, we write $I[\beta, \gamma) = \{i \in I : \operatorname{supp} y_i \subset [\beta, \gamma)\}$.

Claim 2 Given $\beta < \mu$ and $j \in I[\beta, \mu)$, there exist a limit ordinal $\rho = \rho(\beta, j)$ with $\beta < \rho < \mu$ and a subset $J(\beta, j) \subset I[\beta, \mu)$ containing j, with $|J(\beta, j)| < \mu$, such that

- for every $i \in J(\beta, j)$, we have $\operatorname{supp} y_i \subset [\beta, \rho)$,
- if $i \in I[\beta, \mu) \setminus J(\beta, j)$, then $\operatorname{supp} y_i \cap [\beta, \rho) = \emptyset$.

Proof of Claim 2 Take any limit ordinal $\alpha_1 \in [\beta, \mu)$ such that $[\beta, \alpha_1)$ contains $\operatorname{supp} y_j$. By (1.9), the set $\{i \in I : \operatorname{supp}(y_i) \cap [\beta, \alpha_1) \neq \emptyset\}$ has cardinality smaller than μ. As μ is regular, we can find another limit ordinal $\alpha_2 \in (\alpha_1, \mu)$ such that

$$i \in I[\beta, \mu) \text{ and } \operatorname{supp} y_i \cap [\beta, \alpha_1) \neq \emptyset \implies \operatorname{supp} y_i \subset [\beta, \alpha_2).$$

Iterating the argument, we obtain a strictly increasing sequence of limit ordinals $(\alpha_k)_{k \geq 1}$ with $\alpha_k < \mu$ for all k in such a way that

$$i \in I[\beta, \mu) \text{ and } \operatorname{supp} y_i \cap [\beta, \alpha_k) \neq \emptyset \implies \operatorname{supp} y_i \subset [\beta, \alpha_{k+1}).$$

We set the limit ordinal $\rho = \sup_{k\geq 1} \alpha_k$. Clearly, $\rho < \mu$, and the set

$$J = \{i \in I[\beta, \mu)\colon\ \operatorname{supp} y_i \cap [\beta, \rho) \neq \emptyset\} = \bigcup_{k\geq 1} \{i \in I[\beta, \mu)\colon\ \operatorname{supp} y_i \cap [0, \alpha_k) \neq \emptyset\}$$

has the required properties. □

The main piece of the proof is the following:

Claim 3 There exists an increasing family of ordinals $(\rho_\alpha)_{\alpha<\mu}$ such that

- $\rho_\alpha < \mu$ is a limit ordinal for all $\alpha > 0$ and $\rho_0 = 0$,
- the sets $I_\alpha = I[\rho_\alpha, \rho_{\alpha+1})$ form a partition of I, with $|I_\alpha| < \mu$ for all $\alpha < \mu$.

Proof of Claim 3 Note that μ, being an infinite cardinal, cannot be a successor ordinal, so $\alpha + 1 < \mu$ for $\alpha < \mu$. The proof is by transfinite induction on α. For the initial step we set $\beta = 0$ and $j = i_0$ in Claim 2. Then $I_0 = J(0, i_0)$, $\rho_0 = 0$ and $\rho_1 = \rho(0, i_0)$. Note that $i_0 \in I_0$ and that ρ_1 is a limit ordinal. Let us perform the inductive step: to this end, assume that for some $\eta < \mu$, we have already obtained an increasing family of limit ordinals $(\rho_{\alpha+1})_{\alpha<\eta}$ such that

($\dagger_\eta$) $|I_\alpha| < \mu$ for all $\alpha < \eta$ and $i_\beta \in \bigcup_{\alpha\leq\beta} I_\alpha$ for all $\beta < \eta$,
($\ddagger_\eta$) if $i \notin \bigcup_{\alpha<\eta} I_\alpha$, then $\operatorname{supp} y_i \bigcap \bigcup_{\alpha<\eta} [\rho_\alpha, \rho_{\alpha+1}) = \emptyset$,

and let us focus on i_η. If i_η is already in $\bigcup_{\alpha<\eta} I_\alpha$, there is nothing to do but wait: set $\rho_\eta = \rho_{\eta+1} = \sup_{\alpha<\eta} \rho_{\alpha+1}$. Needless to say, in this case, we have $I_\eta = \emptyset$, but we are at peace with that. Otherwise, if i_η is not yet in $\bigcup_{\alpha<\eta} I_\alpha$, we distinguish two cases, as is by now customary:

- If $\eta = \gamma + 1$ is a successor, use Claim 2 with $j = i_\eta$ and $\beta = \rho_\eta = \rho_{\gamma+1}$ and set $\rho_{\eta+1} = \rho(\rho_\eta, \eta)$. Note that I_η corresponds to the output set $J(\rho_\eta, \eta)$.
- If η is a limit ordinal, set $\beta = \sup_{\alpha<\eta} \rho_\alpha$ and $j = i_\eta$ in Claim 2 and $\rho_\eta = \beta$, $\rho_{\eta+1} = \rho(\beta, j)$. This yields $I_\eta = J(\beta, j)$, and everything works fine.

This finishes the induction process. Iterating the construction until $\eta = \mu$ proves the claim. □

Finally, we use Claim 3 to conclude the proof. It is clear that $\overline{[y_i\colon i \in I_\alpha]}$ is a subspace of $\overline{[X_\beta\colon \rho_\alpha \leq \beta < \rho_{\alpha+1}]}$ κ-isomorphic to $c_0(I_\alpha)$, so hypothesis (c) provides a projection $R_\alpha\colon \overline{[X_\beta\colon \rho_\alpha \leq \beta < \rho_{\alpha+1}]} \longrightarrow \overline{[y_i\colon i \in I_\alpha]}$ with norm at most $K\kappa$. Since

$$(P_{\rho_{\alpha+1}} - P_{\rho_\alpha})[X] = \overline{[X_\beta\colon \rho_\alpha \leq \beta < \rho_{\alpha+1}]}$$

and $\lim_{\alpha\to\mu} \|(P_{\rho_{\alpha+1}} - P_{\rho_\alpha})x\| = 0$ for each $x \in X$, it is possible to define a projection $Q\colon X \longrightarrow Y$ taking

$$Q(x) = \sum_{\alpha<\mu} R_\alpha(P_{\rho_{\alpha+1}} - P_{\rho_\alpha})(x).$$

Quite clearly, $\|Q\| \leq 2MK\kappa$. □

We are thus ready to provide a lavish solution to Yost's problem.

Proposition 1.7.13 *Let $m < \omega$. If K is a Valdivia compact of weight at most $\aleph_m$ then $C(K)$ is 2^{m+1}-Sobczyk.*

Proof Note that the dimension of $C(K)$ equals the weight of K. The proof proceeds by induction on m. Before we begin, recall that Proposition 1.7.10 asserts that $C(K)$ spaces with K a Valdivia compact are overt examples of spaces with a PRI decomposition *and* that the spaces in the decomposition can be chosen $C(S)$-spaces with S Valdivia compact again. Now trust us, just sit at the peak of the induction roller coaster and let yourself go down with it: when $m = 0$, separable ($C(K)$ or not) spaces are 2-Sobczyk; when $m = 1$, Lemma 1.7.12 shows that Banach spaces with a PRI and dimension $\aleph_1$ ($C(K)$ or not) are 4-Sobczyk. This includes $C(K)$ spaces with K a Valdivia compact. And from that point on, recall that we only need to consider $C(K)$-spaces with K Valdivia and apply Lemma 1.7.12. □

A class of compacta $\mathcal{K}$ produces two classes of Banach spaces: one by the simple method of isolating those Banach spaces X such that $B_X^* \in \mathcal{K}$, the other by generation – an element of $\mathcal{K}$ spans a dense subspace of X. Sometimes the match is perfect, as in the case of Eberlein compacta, but not always; that explains the formulation of the next result.

Proposition 1.7.14 *Let X be a Banach space such that B_X^* is a Valdivia compact. If $m < \omega$, then every M-isomorphic copy of $c_0(\aleph_m)$ in X is $2^{m+1}M$-complemented.*

Proof We show that when K is a Valdivia compact, any copy of $c_0(\aleph_m)$ inside $C(K)$ is contained in a complemented subspace $C(K_m)$ of $C(K)$ such that K_m is a Valdivia compact with $\dim C(K_m) = \aleph_m$ in order to then apply Proposition 1.7.13 to get a 2^{m+1} projection. To do that, just let yourself go down the cardinal slide: let I be a set of cardinality $\aleph_m$ and $K \subset [0,1]^I$ a Valdivia compact. Let ω_m be the first ordinal with cardinal $|\omega_m| = \aleph_m$, let $\mu = \dim C(K)$, and pick a subset $H = \{h_i \colon i < \omega_m\}$ of $K \cap \Sigma(I)$ that norms $c_0(I)$. Use Lemma 1.7.9 to obtain the increasing family $\{I_\alpha : \omega \leq \alpha \leq \mu\}$ and then Proposition 1.7.10 to get a PRI $(P_\alpha)_{\omega\leq\alpha\leq\mu}$ on $C(K)$ for which $P_\alpha[C(K)] = C(K_\alpha)$ with each $K_\alpha \subset K$ a Valdivia compact. By construction, P_{ω_m} is an isometry on $c_0(I)$, and $\dim C(K_m) = \aleph_m$. □

In due course, Proposition 2.2.15 will show that the preceding result is optimal. On the other hand, despite that ordinal spaces $[0, \alpha]$ are not Valdivia compacta for $\alpha \geq \omega_2$ [147, II. Proposition 2], the decomposition method of Lemma 1.7.12 still works, providing:

Proposition 1.7.15 *The space $C[0, \omega_2]$ is 2^3-Sobczyk.*

Proof The space $C[0, \omega_2]$ admits a PRI $(P_\alpha)_{\omega \leq \alpha < \omega_m}$ given by

$$P_\alpha(f)(\beta) = \begin{cases} f(\beta) & \text{if } \beta \leq \alpha \\ f(\alpha) & \text{if } \beta \geq \alpha \end{cases}$$

(details can be seen in [148, p. 259]). Taking into account that the range of P_α is isometric to $C[0, \alpha]$ for $\alpha < \omega_2$, and that $[0, \alpha]$ is a Valdivia compact, we are ready for induction: Proposition 1.7.13 yields that it is 4-Sobczyk, and the induction Lemma 1.7.12 applies to obtain that $C[0, \omega_2]$ is 8-Sobczyk. □

It could be interesting to determine relations between I and α so that copies of $c_0(I)$ are complemented in $C[0, \alpha]$. A remarkable example in [302, Theorem 2.7] isolates a compact scattered space K of height 3 and Lindelöf (every open cover contains a countable subcover) in its weak topology that contains an uncomplemented copy of $c_0(\aleph_1)$.

1.8 Notes and Remarks

1.8.1 Topological Stuff

The reader can skip this section now and eventually return when some non-Hausdorff space pops up. Because non-Hausdorff topologies will pop up. Indeed, there are places in this book where non-Hausdorff linear topologies are unavoidable. Let X be a linear space. A mapping $\varrho \colon X \longrightarrow \mathbb{R}_+$ is said to be a semi-quasinorm if it is positively homogeneous (that is, $\varrho(\lambda x) = |\lambda|\varrho(x)$ for every $x \in X$ and every scalar λ) and there is a constant Δ such that $\varrho(x + y) \leq \Delta(\varrho(x) + \varrho(y))$ for all $x, y \in X$. If, moreover, one has $\varrho(x + y)^p \leq \varrho(x)^p + \varrho(y)^p$ for all $x, y \in X$ then we say that ϱ is a semi-p-norm, or just a seminorm when $p = 1$. Yes, right: this is nothing different from a quasinorm, just omitting the requirement that if an element has 'size' zero, it has to be zero. Let us agree that a semi-quasinormed space is a linear space X endowed with a semi-quasinorm ϱ. In a semi-quasinormed space, we can form the linear topology for which the sets $\{x \in X \colon \varrho(x) \leq \varepsilon\}$ are a fundamental system of neighbourhoods of zero as in the quasinormed case. Needless to say, such a topology is Hausdorff precisely when ϱ is a quasinorm. There is also a uniform structure whose (basic) neighbourhoods of the diagonal are the sets $\{(x, y) \in X \times X : \varrho(y - x) \leq \varepsilon\}$. It turns out that X is complete if and only if every Cauchy sequence converges. This can be taken as the definition, if one prefers. Of course, by a Cauchy sequence, we mean a sequence (x_n) such that for every $\varepsilon > 0$, there is $k \in \mathbb{N}$ such that $\varrho(x_n - x_m) < \varepsilon$ for all $n, m \geq k$. And (x_n) converges to x if $\varrho(x - x_n) \to 0$ as $n \to \infty$. Note that if (x_n) converges to x and $\varrho(x' - x) = 0$, then (x_n) converges to x' too. By the very definition, $\ker \varrho = \{x \in X : \varrho(x) = 0\}$ is a closed subspace of X and $\varrho(x)$ essentially depends only on the class of x in $X/\ker \varrho$, because when $y \in \ker \varrho$, $\Delta^{-1}\varrho(x) \leq \varrho(x + y) = \Delta\varrho(x)$. When ϱ is a semi-p-norm, we actually have $\varrho(x + y) = \varrho(x)$. In any case, ϱ induces a quasinorm $\varrho[x] = \inf_{\varrho(y)=0} \varrho(x + y)$ on the quotient space $X/\ker \varrho$. Since any linear projection $X \longrightarrow \ker \varrho$ is continuous, $\ker \varrho$ is complemented in X, and X is linearly isomorphic to $\ker \varrho \times X/\ker \varrho$ endowed with the product topology, corresponding to the functional $(y, [x]) \longmapsto \max(\varrho(y), \varrho[x]) = \varrho[x]$. No open mapping theorem exists for non-Hausdorff spaces: consider the formal identity $X \longrightarrow Y$, where X is your favourite Banach space and Y is the same space with the trivial seminorm. The two basic examples of semi-quasinormed spaces to keep in mind are:

- The quotient of a quasinormed space X by a possibly non-closed subspace Y endowed with the quotient semi-quasinorm

$$\|x + Y\| = \inf_{y \in Y} \|x + y\|.$$

The class of x in X/Y is zero if and only if $x \in Y$. However, we have $\|x+Y\| = 0$ if and only if x belongs to the closure of Y in X. It is clear that if X is complete (a quasi-Banach space), then so is X/Y, no matter if it is Hausdorff or not.

- The space $\mathsf{Q}(X, Y)$ of homogeneous maps $\Phi : X \longrightarrow Y$ acting between two quasinormed spaces X, Y such that

$$Q(\Phi) = \sup_{x,y \neq 0} \frac{\|\Phi(x + y) - \Phi(x) - \Phi(y)\|}{\|x\| + \|y\|} < \infty.$$

It is clear that $\Phi \in \mathsf{Q}(X, Y) \longmapsto Q(\Phi) \in \mathbb{R}_+$ is a semi-quasinorm whose modulus of concavity does not exceed that of Y and that $Q(\Phi) = 0$ if and only if Φ is linear (maybe unbounded). These maps will have their moments of glory in Chapter 3 and after that throughout the book.

A subset B of a topological vector space X is said to be bounded if it is absorbed by all neighbourhoods of zero; i.e. for every neighbourhood V of zero, there is $\lambda > 0$ such that $B \subset \lambda V$. A topological vector space is said to be locally bounded if it has a bounded neighbourhood of the origin. Semi-quasinormed spaces are locally bounded because the unit ball $B_X = \{x \in X : \varrho(x) \leq 1\}$ is a bounded neighbourhood of zero. Conversely, if B is a bounded symmetric neighbourhood of zero in X, the functional $\|x\| = \inf\{t > 0: x \in tB\}$ is a semi-quasinorm giving the topology of X.

1.8.2 Orlicz, Young, Fenchel and L_0 Too

The terminology in this section may differ from that used in other, more respectable texts. An Orlicz function is a continuous, increasing function $\varphi: [0, \infty) \longrightarrow [0, \infty)$ vanishing only at zero *and* satisfying the 'Δ_2-condition': there is a constant C such that $\varphi(2t) \leq C\varphi(t)$ for all $t \geq 0$. We do not require Orlicz functions to be convex or that $\varphi(t) \to \infty$ as $t \to \infty$, conditions which are basically equivalent to the fact that the associated Orlicz space is a Banach space (see, for instance, Lindenstrauss-Tzafriri [334, pp. 137]). If (S, μ) is a measure space then the associated Orlicz space $L_\varphi(\mu)$ is the space of those measurable functions $f: S \longrightarrow \mathbb{K}$ such that

$$|f|_\varphi = \int_S \varphi(|f(s)|)d\mu(s) < \infty.$$

Although $|\cdot|_\varphi$ need not be homogeneous or subadditive, it defines a linear topology on $L_\varphi(\mu)$ for which the sets $\{f: |f|_\varphi < \varepsilon\}$ are a neighbourhood base at zero. The condition $\lim_{\lambda\to 0} \sup_t \frac{\varphi(\lambda t)}{\varphi(t)} = 0$ guarantees that $L_\varphi(\mu)$ is a locally bounded space, in which case, the functional $\|f\|_\varphi = \inf\{r > 0: |r^{-1}f|_\varphi \leq 1\}$ is a quasinorm, called the *Luxemburg quasinorm*. When μ is the counting measure on $\mathbb{N}$, we obtain the so-called Orlicz sequence spaces.

Let V be a finite-dimensional linear space, possibly of dimension 1. A Young function $\Phi: V \longrightarrow \mathbb{R}^+$ is an even convex function such that $\Phi(tv) \to \infty$ for each $v \in V$ as $|t| \to \infty$. We do not require that Φ vanish only at zero. A family $\Phi_k: V_k \longrightarrow \mathbb{R}^+$ of Young functions defines a *modular* sequence space

$$h((\Phi_k)_k) = \left\{v \in \prod_{k\geq 1} V_k: \sum_{k=1}^{\infty} \Phi_k(tv_k) < \infty \text{ for all } t > 0\right\},$$

equipped with the Luxemburg norm

$$\|v\|_{(\Phi_k)_k} = \inf\left\{t > 0: \sum_{k=1}^{\infty} \Phi_k(v_k/t) \leq 1\right\}.$$

If $V_k = \mathbb{K}^n$ for some n and all k and all the Φ_k agree with some Young function $\Phi: \mathbb{K}^n \longrightarrow \mathbb{R}_+$, then the modular sequence space $h(\Phi)$ is called a Fenchel-Orlicz space; if, moreover, $n = 1$, then $h(\Phi)$ agrees with the small Orlicz space as defined in [334, bottom of p. 137]. There is a considerable overlap between Orlicz sequence spaces and modular spaces in the locally convex zone. These subtleties will be necessary only in Chapter 8: note that, according to our fussy definitions, c_0 is a modular sequence space, but not an Orlicz space.

The space L_0 of all measurable functions on the unit interval deserves a special mention. The topology of convergence in measure is metrisable: set

$$|h|_0 = \int_0^1 \frac{|h|}{1+|h|}dt$$

so that the formula $d(f,g) = |f-g|_0$ defines a complete (invariant) metric on L_0. Thus, L_0 is an Orlicz function space in the wide sense adopted earlier. However, L_0 is not locally bounded. More yet:

Proposition *Each operator from L_0 to a quasinormed space is zero.*

Proof The key is that $|f|_0 \leq \lambda(\text{supp}(f))$ regardless of the values assumed by f. Let $u\colon L_0 \longrightarrow Y$ be an operator, where Y is a p-normed space. Take $\delta > 0$ such that $\|u(f)\| \leq 1$ for $|f|_0 \leq \delta$. Divide $[0,1]$ into n subintervals $I_1, \ldots, I_n$ of measure less than δ. Pick any $f \in L_0$. Then $f = \sum_{i=1}^n f_i$, where $f_i = 1_{I_i} f$. For each scalar c, we have

$$\|u(cf)\|^p \leq \sum_{1\leq i\leq n} \|u(cf_i)\|^p \leq n,$$

and since n is fixed and c arbitrary, we see that $\|u(f)\| = 0$. □

1.8.3 Ultrapowers of L_p When $0 < p < 1$

We emphatically concluded Section 1.4 with the assertion that L_p is not an ultrasummand if $0 < p < 1$. The following result shows that L_p is uncomplemented even in its 'countable' ultrapowers. The proof, a re-elaboration of a quip of Kalton [255, Proof of Lemma 8.1], will be eased by recalling that, given a family $(S_i)_{i\in I}$ of sets and an ultrafilter $\mathfrak{U}$ on I, the *set theoretic ultraproduct* $\langle S_i\rangle_{\mathfrak{U}}$ is the set $\prod_i S_i$ factored by the equivalence relation $(s_i) = (t_i) \iff \{i \in I : s_i = t_i\} \in \mathfrak{U}$. The class of (s_i) in $\langle S_i\rangle_{\mathfrak{U}}$ will be denoted $\langle (s_i)\rangle$.

Proposition *Let $\mathfrak{U}$ be a free ultrafilter on $\mathbb{N}$. Then $\mathfrak{L}((L_p)_{\mathfrak{U}}, Y) = 0$ for every separable quasi-Banach space Y.*

Proof Let us treat the Lebesgue measure on $[0,1]$ as a probability and, accordingly, the elements of L_p as random variables. Our first observation is that if $f \in L_p$ is simple then there is a sequence of 'Rademacher-like' functions (r_n) mutually independent and independent with f such that $\lambda\{t\colon r_n(t) = \pm 1\} = \frac{1}{2}$ for all $n \in \mathbb{N}$, where λ is Lebesgue measure on the unit interval: just write $f = \sum_k a_k 1_{A_k}$ with A_k a partition of $[0,1]$ and work on each A_k separately. Take finitely many non-zero scalars c_n. Applying Khintchine's inequality [153, p. 10] on each A_k, we get

$$\Big\|\sum_n c_n r_n f\Big\|_p \le \|f\|_p\|(c_n)\|_2. \tag{1.10}$$

Now, let $\mathcal{U}$ be a free ultrafilter on a countable set I and form the ultrapower $(L_p)_{\mathcal{U}}$. For notational reasons, it's better to keep using I instead of $\mathbb{N}$ for the index set of the ultrafilter. Take a normalised $f \in (L_p)_{\mathcal{U}}$ and a representative (f_i) with f_i simple and $\|f_i\|_p = 1$ for all $i \in I$. For each i we select a 'Rademacher' sequence $(r_i^n)_{n\ge1}$ of mutually independent functions which are, moreover, independent with f_i and $\lambda\{t\colon r_i^n(t) = \pm1\} = \frac{1}{2}$ for all n and i. Let $n\colon I \longrightarrow \mathbb{N}$ be any function. Consider the class of $(r_i^{n(i)})$ in the ultrapower $(L_\infty)_{\mathcal{U}}$ and write $[(r_i^{n(i)})]f = [(r_i^{n(i)} f_i)]$. The class of $(r_i^{n(i)})$ in $(L_\infty)_{\mathcal{U}}$ depends only on the class of $(n(i))$ in the set theoreric ultrapower $\langle\mathbb{N}\rangle_{\mathcal{U}}$. Thus, if $\alpha = \langle(n(i))\rangle$, then $r_\alpha = [(r_i^{n(i)})]$ is correctly defined as an element of $(L_\infty)_{\mathcal{U}}$, and so is the product $r_\alpha f = [(r_i^{n(i)} f_i)]$ for every $f \in (L_p)_{\mathcal{U}}$.

Claim 1 If $T\colon (L_p)_{\mathcal{U}} \longrightarrow Y$ is an operator and f and $\{r_\alpha\colon \alpha \in \langle\mathbb{N}\rangle_{\mathcal{U}}\}$ are as before, then there is $y \in Y$ such that for every $\varepsilon > 0$, the set

$$\{\alpha \in \langle\mathbb{N}\rangle_{\mathcal{U}} : \|y - T(r_\alpha f)\| \le \varepsilon\}$$

has the cardinality of the continuum.

Proof of Claim 1 Assume Y is q-normed for the remainder of the proof and write it as the union of countably many balls of radius 1. Since $\langle\mathbb{N}\rangle_{\mathcal{U}}$ has the cardinality of the continuum (use an almost disjoint (Definition 2.2.9) family of size $\mathfrak{c}$), some of its members must contain $T(r_\alpha f)$ for 'continuum many' αs. Take that ball and write it as the union of countably many balls of radius $\frac{1}{2}$, and so on. Continuing in this way, we get a sequence of closed balls $(B_n)_{n\ge1}$ in Y such that $B_{n+1} \subset B_n$ for every $n \in \mathbb{N}$; B_n has radius 2^{-n}, and for every n, the cardinality of the set of those $\alpha \in \langle\mathbb{N}\rangle_{\mathcal{U}}$ for which $T(r_\alpha f) \in B_n$ is the continuum. The intersection of the balls yields the point we were looking for. □

Claim 2 The point in Claim 1 has to be zero.

Proof of Claim 2 Let $y \in Y$ a point for which the conclusion of Claim 1 is true. Take a sequence $(\alpha_n)_n$ of different indices in $\langle\mathbb{N}\rangle_{\mathcal{U}}$ such that $\|y - T(r_{\alpha_n} f)\|_p^q \le 2^{-n}$. Then, for every integer n, we have

$$\Big\|ny - \sum_{k=1}^n T(r_{\alpha_k} f)\Big\|_p^q \le \sum_{k=1}^n \frac{1}{2^k} \le 1 \quad\Longrightarrow\quad \overbrace{\Big\|\sum_{k=1}^n T(r_{\alpha_k} f)\Big\|_p^q}^{(\dagger)} \ge n^q\|y\|^q - 1\,.$$

But if we apply (1.10) 'coordinatewise' to $\sum_{k=1}^{n} r_{\alpha_k} f$ with $c_k = 1$ for $1 \leq k \leq n$, we obtain

$$\left\| \sum_{k=1}^{n} r_{\alpha_k} f \right\|_p \leq n^{1/2} \quad \Longrightarrow \quad \left\| \sum_{k=1}^{n} T(r_{\alpha_k} f) \right\|_p^q \leq \|T\|^q n^{q/2},$$

which is compatible with (†) only if $\|y\| = 0$. This proves Claim 2. □

To conclude the proof, pick any normalised $f \in (L_p)_{\mathfrak{U}}$ and construct the corresponding family (r_α) as before. For each $\varepsilon > 0$, there is some α such that $\|T(r_\alpha f)\| \leq \varepsilon$. Note that, by independence, $\|(1 + r_\alpha)f\|^p = 2^{p-1}\|f\|^p$, that is, $\|(1 + r_\alpha)f\| = 2^{1-1/p}\|f\|$. Recalling that Y is q-normed, we have

$$\|Tf\|^q \leq \|T(1 + r_\alpha)f\|^q + \|T(r_\alpha f)\|^q \leq \varepsilon^q + \|T\|^q 2^{q-q/p}\|f\|^q.$$

But ε is arbitrary, and so $\|T\| \leq 2^{1-1/p}\|T\|$; that is, $T = 0$. □

This result trivially implies that every compact operator on L_p with values in a quasi-Banach space is zero if $0 < p < 1$, which is a result of Pallaschke. We will not explore this line of research here since anyone interested in operators on non-locally convex spaces should begin with Chapters 7 and 8 of [283] or with [166]. The conclusion we prefer to draw instead is that the proof of the proposition depends only on that fact that the cardinality of the set theoretic ultrapower $\langle \mathbb{N} \rangle_{\mathfrak{U}}$ is greater than the dimension of the target space Y; it actually works when $\dim(Y) < \mathfrak{c}$. On the other hand, $\langle \mathbb{N} \rangle_{\mathfrak{U}}$ can be as large as we want:

Lemma *For every cardinal $\aleph$, there is an ultrafilter $\mathfrak{U}$ on some index set such that $|\langle \mathbb{N} \rangle_{\mathfrak{U}}| \geq \aleph$.*

Proof Let A be any set and $I = \text{fin}(A)$ the set of all non-empty finite subsets of A ordered by inclusion. Let $\mathfrak{U}$ be any ultrafilter refining the order filter of I. Let $\langle F \rangle_{\mathfrak{U}}$ denote the set theoretic ultraproduct of the family $\{F : F \in \text{fin}(A)\}$ following $\mathfrak{U}$, that is, the 'elements' of $\langle F \rangle_{\mathfrak{U}}$ are (classes of) families (a_F) of the product space $\prod_{\text{fin}(A)} F$, where a_F belongs to F for every F and $\langle (a_F) \rangle = \langle (b_F) \rangle$ if the set of those F for which $a_F = b_F$ belongs to $\mathfrak{U}$. There is an obvious embedding of A into $\langle F \rangle_{\mathfrak{U}}$. Get the idea? Thus, $|A| \leq |\langle F \rangle_{\mathfrak{U}}| \leq |\langle \mathbb{N} \rangle_{\mathfrak{U}}|$. □

Corollary *If Y is an ultrasummand, then $\mathfrak{L}(L_p, Y) = 0$ for $0 < p < 1$.*

Proof Let $T : L_p \longrightarrow Y$ be an operator, and let $\mathfrak{U}$ be an ultrafilter on some index set I for which the cardinality of $\langle \mathbb{N} \rangle_{\mathfrak{U}}$ is strictly greater than $\dim(Y)$. Let $P : Y_{\mathfrak{U}} \longrightarrow Y$ be a projection through the diagonal embedding. The hypotheses imply that the composition $PT_{\mathfrak{U}} : (L_p)_{\mathfrak{U}} \longrightarrow Y$ is zero, and so is T. □

1.8.4 Sobczyk's Theorem Strikes Back

According to Veech's proof, Sobczyk's theorem is the statement that every weak*-null sequence on a subspace of a separable Banach space can be extended to a weak*-null sequence on the whole space. The norm of the elements in the sequence of extended functionals, however, doubles. This fact makes it apparently impossible to produce a proof à la Hahn–Banach obtaining a suitable extension to one more dimension and then iterating the argument. Such a proof is almost possible, nonetheless, and it was obtained in passing by Kalton [273, Section 5]. Following, pretty badly, Behrends [35], we call an ordered space as *groundless* when any decreasing sequence of elements has a lower bound. The groundless set we need is the space $\mathcal{P}^*_\infty(\mathbb{N})$ of all infinite subsets of the integers modulo finite sets endowed with the order $[A] \leq [B]$ if $B\backslash A$ is finite. It is not obvious that it is obvious that $\mathcal{P}^*_\infty(\mathbb{N})$ is groundless: if $(A_n)_{n\geq 1}$ is a sequence of infinite subsets of $\mathbb{N}$ such that $[A_{n+1}] \leq [A_n]$ for all n, then $A = \{k_n : n \in \mathbb{N}\}$, where $k_n \in \bigcap_{i\leq n} A_i$ is infinite and $[A] \leq [A_n]$ for all n. A mapping $f : P \longrightarrow Q$, acting between ordered sets, is order preserving if $y \leq x$ implies $f(y) \leq f(x)$. We say that a point $x \in P$ is stationary if $f(y) = f(x)$ for all $y \leq x$.

Behrends' lemma *Let $f : P \longrightarrow Q$ be an order-preserving map, where P is groundless. Then f has a stationary point in the following cases:*

- *Q is a subset of $\mathbb{R}^\mathbb{N}$.*
- *$Q = \mathcal{K}(M)$ is the set of all compact subspaces of a metric space M ordered by inclusion.*

Proof We first show the result for $Q = \mathbb{R}$: since P is groundless, there is $p_0 \in P$ such that f must be bounded on $\{y \in P : y \leq p_0\}$. If $m = \inf\{f(y) : y \leq p_0\} = \inf f(y_n)$ and $y \leq y_n$ for all n, then $f(a) = f(b)$ for all $a, b \leq y$. The result for $\mathbb{R}^\mathbb{N}$ follows by diagonalisation. The second case also follows by taking into account that $\mathcal{K}(M)$ is 'countably determined' in the sense that there is an order-preserving, injective mapping $g : \mathcal{K}(M) \longrightarrow [-1, 1]^\mathbb{N}$. Indeed, let $(h_n)_{n\in\mathbb{N}}$ be a dense sequence in the unit ball of the real-valued $C(M)$ and set $g(K) = (\sup_{s\in K} h_n(s))_{n\in\mathbb{N}}$. □

Let X be a Banach space, and let K_n be a sequence of weak*-compact convex subsets of the unit ball of X^*. If A is an infinite subset of $\mathbb{N}$, we set

$$K_A = \bigcap_{n\in\mathbb{N}} \overline{\bigcup_{i\in A, i\geq n} K_i}. \tag{1.11}$$

This definition is formally identical to the definition of an upper limit and, clearly, K_A depends only on $[A]$. One thus has an order-preserving function

$$K_{\bullet}\colon \mathcal{P}^*_\infty(\mathbb{N}) \longrightarrow \mathcal{K}(B^*_X).$$

Lemma *Assume that K_A is weak*-metrisable.*

(a) *If $x^* \in K_A$ then x^* is the weak*-limit of a sequence $(x^*_i)_{i\in A}$ with $x^*_i \in K_i$ for all $i \in A$ if and only if, for every infinite subset $B \subset A$, one has $x^* \in K_B$.*
(b) *There is an infinite subset $B \subset A$ with the property that $K_M = K_B$ for every $M \subset B$. For this B, the set K_B is convex.*

Proof The 'only if' implication in (a) is clear. For the converse, given $x^* \in K_A$, we can arrange a decreasing sequence of weak* open sets $(V_k)_{k\geq 0}$ such that $x^* = \bigcap_{k\geq 1}(V_k \cap K_A)$, with $V_0 = X^*$. The hypothesis implies that for each k the set $\{i \in A\colon K_i \cap V_k \neq \emptyset\}$ is cofinite in A. Pick an increasing sequence $(j(k))_k$ of indices in A such that $K_i \cap V_k \neq \emptyset$ for $i \geq j(k)$ and define a sequence $(x_i)_{i\in A}$ taking $x^*_i \in K_i \cap V_k$ if $j(k) \leq i < j(k+1)$. Clearly, $x^*_i \longrightarrow x^*$ weak* as i increases in A. The first part of (b) is an obvious application of Behrend's lemma to the map $K_{\bullet}\colon \mathcal{P}^*_\infty(A) \longrightarrow \mathcal{K}(K_A)$ defined as $B \longrightarrow K_B$. The convexity of K_B follows from part (a). □

It is plain that if the ambient compact $K_{\mathbb{N}}$ is metrisable then each infinite set contains a stationary subset, from which a kind of à-la-Hahn–Banach-better-than-the-original Sobczyk-like theorem follows:

Corollary *Let Y be a closed subspace of a real Banach space X such that X/Y is separable, and let $\tau\colon Y \longrightarrow c_0$ be an operator. If, for every $x \in X$, there is a λ-extension $T_x\colon Y + [x] \longrightarrow c_0$, then there is a λ-extension $T\colon X \longrightarrow c_0$.*

Proof Assume $\|\tau\| = 1$, and write $\tau = (\tau_n)$ as a sequence of functionals. Let K_n be the set of all extensions of τ_n with norm at most λ, and consider the family of compacta K_A defined by (1.11). The separability of X/Y makes the bounded subsets of $Y^\perp = \{x^* \in X^*\colon x^*|_Y = 0\}$ weak*-metrisable, as well as K_A for every $\mathcal{P}^*_\infty(\mathbb{N})$. Assume that no weak*-null extension of (τ_n) with norm at most λ exists. Applying the first part of the previous lemma with $A = \mathbb{N}$, we get an infinite subset $B \subset \mathbb{N}$ such that $0 \notin K_B$. Without loss of generality, we can assume this B is the stationary set appearing in (b), and so K_B is convex. We can thus separate K_B from 0 by an element $x \in X$; say $\langle x^*, x\rangle \geq \varepsilon$ for some $\varepsilon > 0$ and all $x^* \in K_B$. The hypothesis yields a λ-extension $T_x\colon Y + [x] \longrightarrow c_0$. If $(f_n)_{n\geq 1}$ is the corresponding sequence of functionals, we have $f_n \in K_n$ for all n since $\|f_n\| \leq \lambda$ and $\langle f_n, x\rangle \to 0$ as $n \to \infty$. If f is a weak*-accumulation point of the subsequence $(f_n)_{n\in B}$, then $f \in K_B$ and $\langle f, x\rangle = 0$, which contradicts the choice of x. □

The previous approach only really makes sense for $\lambda < 2$, because for $\lambda = 2$, we already have Sobczyk's theorem. For $\lambda = 2$, it would provide a proof for Sobczyk's theorem if it were true that c_0-valued operators extend to one more dimension, doubling the norm. But proving that is actually as hard as Sobczyk's theorem! A nice loophole would exist were it obvious, clear or at least true that hyperplanes of Banach spaces are 2-complemented, but this is false; see comments and examples before Definition 2.1.6. Two more remarkable results about c_0 complementation deserve mention [394, Theorem 3]: *for every $n \in \mathbb{N}$, there exists a $6(n+1)$-Sobczyk space that is not n-Sobczyk*, from which it follows [394, Corollary 4] that there exists a Banach space X admitting a countable chain of subspaces $(Y_n)_{n\geq 0}$ such that $X = \overline{\bigcup_n Y_n}$, $Y_0 = c_0$, Y_n is complemented in Y_{n+1} but c_0 is not complemented in X – even if every copy of c_0 in X contains a subspace isomorphic to c_0 and complemented in X [182].

Sources

General references on quasi-Banach spaces are [283; 269; 408]. The result 1.1.5 is from Day [144], but the proof we present is in the spirit of [303, §15.9.9]. Proposition 1.2.5 is from [84]. The proof of Proposition 1.3.3 is a clever insight in Phillips' proof taken from an exercise in Bourbaki [47, 55, Exercice 16]. Two long-standing problems have been whether complemented subspaces of spaces with unconditional basis have unconditional basis and whether every Banach space contains an unconditional basic sequence. The first one is still open [83, Problem 1.8], while the second was solved by Gowers and Maurey [197], leading, with the aid of W. B. Johnson, to the discovery of H.I. spaces, thoroughly studied by Argyros and his group. Precisely, that H.I. spaces are subspaces of ℓ_∞ is due to Argyros, although a proof can be traced back to Plichko-Yost [396], and the proof presented in the text has been taken from [2]. The principle of local reflexivity is the wondrous creation of Lindenstrauss and Rosenthal [331]. Besides WCG spaces, the list of known spaces with SCP includes weakly sequentially complete Banach lattices [187], Banach spaces with the commuting bounded approximation property [83], duals of Asplund spaces [224, p. 38]; see also [207, Theorem 3.42], spaces of continuous functions on any ordinal [244, Theorem 1.6] and Plichko spaces [243]. Just in case the definition has momentarily slipped our minds, recall that a Banach space X is called Plichko if there is a dense subset $A \subset X$ and norming subset $B \subset X^*$ such that for every $x^* \in B$, the set $\{x \in A : \langle x^*, x\rangle \neq 0\}$ is countable. There also exist $\mathscr{C}$-spaces with SCP that are not Plichko [309]. Proposition 1.7.7 is from Yost [460, corollary], who invented the Veech topological spaces: those in which every separable subset is metrisable. The paper [396] contains a lot of additional information about the SCP property. Proposition 1.7.13 has been taken from the Argyros versus Spain paper [16], where much more results and examples can be found. The results and ideas in 1.8.4 are from Kalton [273], although we followed a different path through Behrends' lemma, which has been obviously taken from [35, Section 2], even if Behrends generously attributes the idea to Hagler and Johnson (see [152, p. 231]).

2

The Language of Homology

In this chapter we introduce the basic elements of the homological language and translate the statements about complemented and uncomplemented subspaces presented in Chapter 1 into this language. The homological language has a few advantages over the classical one:

- It allows us to present all available information about the problem in question at a glance. To give an example, assume we want to prove the following statement: *if Y is a subspace of Z such that Z/Y is reflexive and Y is complemented in Y^{**}, then Z is also complemented in Z^{**}*. Try to do it. Done? Good. Now, the homological way. All the information appears displayed in the diagram:

$$\begin{array}{ccccccccc} & & 0 & & 0 & & 0 & & \\ & & \downarrow & & \downarrow & & \downarrow & & \\ 0 & \longrightarrow & Y & \xrightarrow{\jmath} & Z & \longrightarrow & R & \longrightarrow & 0 \\ & & \downarrow & & \downarrow & & \| & & \\ 0 & \longrightarrow & Y^{**} & \xrightarrow{\jmath^{**}} & Z^{**} & \longrightarrow & R^{**} & \longrightarrow & 0 \\ & & \downarrow{\scriptstyle Q_Y} & & \downarrow{\scriptstyle Q_Z} & & \downarrow & & \\ 0 & \longrightarrow & Y^{**}/Y & = & Z^{**}/Z & \longrightarrow & 0 & & \\ & & \downarrow & & \downarrow & & & & \\ & & 0 & & 0 & & & & \end{array}$$

  As for the proof (accept many terms here to be explained later), that Y is complemented in Y^{**} means that the left vertical sequence splits, so Q_Y admits a linear continuous section s. This obliges the middle vertical sequence to split since $\jmath^{**}s$ is a linear continuous section for Q_Z, thus Z

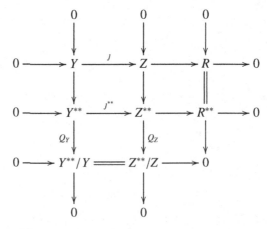

is complemented in Z^{**}. Simple? As it should be. Why is it so simple? Answering this brings us to the main feature of the homological approach compared to the classical language:

- Diagrams encode a large amount of information in a simple way. Consequently, once the reader becomes familiar with the language, complicated things can be said in simple forms, usually simpler than in the classical language. Even at this early stage, an example can be given: unlike the classical language, the homological language treats subspaces and quotients symmetrically. For instance, saying 'each subspace of X is complemented' in classical terms with quotients requires some thought, while in the homological language, 'every exact sequence $0 \longrightarrow \cdot \longrightarrow X \longrightarrow \cdot \longrightarrow 0$ whose middle term is X splits' cannot be simpler or clearer.

Thus, the general strategy for tackling a problem the homological way is:

- **Draw diagrams** Formulating the problem using the homological language means stating the problem as the possibility of constructing a more or less complex diagram. Keep in mind the unspoken rule that diagrams must start and end in 0. No loose ends allowed. See Note 2.15.2 for details.
- **Simplify diagrams** Find a way to simplify diagrams. Simplifying means different things: usually, making the diagram 'split' into elementary diagrams. Techniques of homological algebra allow us to understand diagrams, show us how to manipulate them and determine when they split.
- **Interpret diagrams** Find out the meaning of the simplified diagrams inside Banach space theory.

Let us show off this strategy in action:

Claim Given a subspace Y of a Banach space Z, there is a natural isomorphism $(Z^{**}/Y^{**})/(Z/Y) \simeq (Z^{**}/Z)/(Y^{**}/Y)$.

Of course, there is a classical way to do that (please, be our guest!). To use homological language, one begins by observing that the first line in the data 'given a subspace Y of a Banach space Z' is pick the exact sequence $0 \longrightarrow Y \longrightarrow Z \longrightarrow Z/Y \longrightarrow 0$. Since the question involves biduals, observe also that biduals form an exact sequence $0 \longrightarrow Y^{**} \longrightarrow Z^{**} \longrightarrow (Z/Y)^{**} \longrightarrow 0$. This, and the very meaning of 'exact sequence', already yields $(Z/Y)^{**} \simeq Z^{**}/Y^{**}$. To get the isomorphism we are looking for, form a commutative diagram

$$\begin{array}{ccccccccc} 0 & \longrightarrow & Y & \longrightarrow & Z & \longrightarrow & Z/Y & \longrightarrow & 0 \\ & & \downarrow & & \downarrow & & \downarrow & & \\ 0 & \longrightarrow & Y^{**} & \longrightarrow & Z^{**} & \longrightarrow & (Z/Y)^{**} & \longrightarrow & 0 \end{array}$$

where the descending arrows are the natural inclusion maps, and complete the diagram so that no loose ends remain. One way, then, is to use brute force to check that the sequence of quotients is also exact. The other is to appeal to a far more general result known as the Snake lemma (Note 2.15.2). Whichever way, one arrives at the diagram

$$\begin{array}{ccccccccc}
 & & 0 & & 0 & & 0 & & \\
 & & \downarrow & & \downarrow & & \downarrow & & \\
0 & \longrightarrow & Y & \longrightarrow & Z & \longrightarrow & Z/Y & \longrightarrow & 0 \\
 & & \downarrow & & \downarrow & & \downarrow & & \\
0 & \longrightarrow & Y^{**} & \longrightarrow & Z^{**} & \longrightarrow & (Z/Y)^{**} & \longrightarrow & 0 \\
 & & \downarrow & & \downarrow & & \downarrow & & \\
0 & \longrightarrow & Y^{**}/Y & \longrightarrow & Z^{**}/Z & \longrightarrow & \dfrac{(Z/Y)^{**}}{(Z/Y)} & \longrightarrow & 0 \\
 & & \downarrow & & \downarrow & & \downarrow & & \\
 & & 0 & & 0 & & 0 & &
\end{array} \qquad (2.1)$$

That this is a correct diagram we can be certain: it begins and ends with 0, as it should. Now, this diagram – more precisely, its right bottom corner – immediately provides the isomorphism claimed.

2.1 Exact Sequences of Quasi-Banach Spaces

The study of exact sequences of (quasi-) Banach spaces is the main theme of the book. The idea from now on is to consider the structure formed by a subspace and the correspondent quotient considered as a whole, and the same applies to a quotient operator and its kernel. Observe the asymmetry of these assertions: we spoke of the quotient by a subspace and the kernel of a quotient operator. To make everything symmetric, we should treat subspaces and quotients on equal terms by focusing on the operators rather than on the spaces themselves:

Definition 2.1.1 A short exact sequence of quasi-Banach spaces is a diagram composed by quasi-Banach spaces and operators

$$0 \longrightarrow Y \xrightarrow{\jmath} Z \xrightarrow{\rho} X \longrightarrow 0 \qquad (\mathrm{z})$$

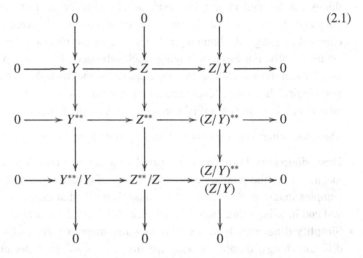

in which the kernel of each arrow coincides with the image of the preceding one. The middle space Z is usually called a *twisted sum* of Y and X.

The most natural (and, to some extent, the only) example of a short exact sequence is to start with an embedding $\jmath\colon Y \longrightarrow Z$ and form the diagram

$$0 \longrightarrow Y \xrightarrow{\ \jmath\ } Z \xrightarrow{\text{quotient}} Z/\jmath[Y] \longrightarrow 0$$

or to start with a quotient operator $\rho\colon Z \longrightarrow X$ and form the diagram

$$0 \longrightarrow \ker\rho \xrightarrow{\text{inclusion}} Z \xrightarrow{\ \rho\ } X \longrightarrow 0$$

In general, given a short exact sequence (z), exactness at Y means that $\jmath$ is injective; exactness at X means that ρ is onto and, by the open mapping theorem, open – i.e. a quotient map. The exactness at Z implies that $\jmath$ has closed range since $\jmath[Y] = \ker\rho$, and the open mapping theorem yields that $\jmath$ is an isomorphic embedding. Thus, any short exact sequence like (z) can be placed in a commutative diagram

$$\begin{array}{ccccccccc} 0 & \longrightarrow & Y & \xrightarrow{\ \jmath\ } & Z & \xrightarrow{\ \rho\ } & X & \longrightarrow & 0 \\ & & \downarrow & & \| & & \uparrow & & \\ 0 & \longrightarrow & \jmath[Y] & \longrightarrow & Z & \longrightarrow & Z/\jmath[Y] & \longrightarrow & 0 \end{array} \tag{2.2}$$

in which the vertical arrows are isomorphisms and the lower row is a 'natural' exact sequence. Summing up, the meaning of an exact sequence (z) is Y is (isomorphic to) a closed subspace of Z in such a way that the corresponding quotient is (isomorphic to) X. The sequence (z) is said to be *isometrically exact* if the embedding is an isometry and ρ is an isometric quotient in the sense that it maps the open unit ball of Z onto that of X.

Equality Notions for Short Exact Sequences

When must two short exact sequences be considered 'equal'? Although this is a matter of perspective, it is plain that any reasonable notion of equality for exact sequences must involve the operators appearing in the sequence and not only the spaces.

The following definition is the classical one in homology, in which the subspace and quotient space are fixed:

Definition 2.1.2 Two exact sequences $0 \longrightarrow Y \longrightarrow Z \longrightarrow X \longrightarrow 0$ and $0 \longrightarrow Y \longrightarrow Z' \longrightarrow X \longrightarrow 0$ are said to be equivalent if there exists an operator β making the following diagram commutative:

$$\begin{array}{ccccccccc} 0 & \longrightarrow & Y & \longrightarrow & Z & \longrightarrow & X & \longrightarrow & 0 \\ & & \| & & \downarrow{\scriptstyle\beta} & & \| & & \\ 0 & \longrightarrow & Y & \longrightarrow & Z' & \longrightarrow & X & \longrightarrow & 0 \end{array}$$

The following elementary lemma implies that the operator β must be an isomorphism, which somehow matches our expectations that equivalence of exact sequences is a true equivalence relation:

2.1.3 The 3-lemma *Given a commutative diagram of vector spaces and linear maps with exact rows*

$$\begin{array}{ccccccccc} 0 & \longrightarrow & Y & \longrightarrow & Z & \longrightarrow & X & \longrightarrow & 0 \\ & & {\scriptstyle\alpha}\downarrow & & \downarrow{\scriptstyle\beta} & & \downarrow{\scriptstyle\gamma} & & \\ 0 & \longrightarrow & Y' & \longrightarrow & Z' & \longrightarrow & X' & \longrightarrow & 0 \end{array}$$

if α and γ are injective / surjective / bijective, then so is β.

The proof is just chasing diagrams since no topology is involved. And, speaking about trivial matters, the simplest exact sequence in sight is the direct product exact sequence $0 \longrightarrow Y \xrightarrow{\imath_1} Y \times X \xrightarrow{\pi_2} X \longrightarrow 0$, in which $\imath_1(y) = (y, 0)$ and $\pi_2(y, x) = x$.

Definition 2.1.4 An exact sequence is said to be trivial when it is equivalent to the direct product sequence.

We write $\mathsf{z} \sim \mathsf{z}'$ to mean that the sequences z and z' are equivalent, and we denote by $[\mathsf{z}]$ the *class* of all sequences that are equivalent to z. The *set* of equivalence classes of exact sequences $0 \longrightarrow Y \longrightarrow \cdot \longrightarrow X \longrightarrow 0$, also called *extensions of X by Y*, will be denoted $\operatorname{Ext}(X, Y)$. The study of the assignment $X, Y \rightsquigarrow \operatorname{Ext}(X, Y)$ is the central topic of Chapter 4. There we will show, among other things, that $\operatorname{Ext}(\cdot, \cdot)$ is a functor and that $\operatorname{Ext}(X, Y)$ admits a natural vector space structure whose zero is the class of trivial sequences. Thus, $\operatorname{Ext}(X, Y) = 0$ means that all exact sequences $0 \longrightarrow Y \longrightarrow \cdot \longrightarrow X \longrightarrow 0$ are trivial. The trivial character of a sequence can be detected by looking at either the embedding or the quotient:

Lemma 2.1.5 *Given an exact sequence* $0 \longrightarrow Y \xrightarrow{\jmath} Z \xrightarrow{\rho} X \longrightarrow 0$, *the following are equivalent:*

(i) *The sequence is trivial.*
(ii) *The embedding $\jmath$ admits a left inverse in $\mathfrak{L}(Z, Y)$; i.e. there exists an operator $P\colon Z \longrightarrow Y$ such that $P\jmath = \mathbf{1}_Y$.*

(iii) *The quotient map* ρ *admits a right inverse in* $\mathfrak{L}(X, Z)$*; i.e. there exists an operator* $s\colon X \longrightarrow Z$ *such that* $\rho s = \mathbf{1}_X$.

Proof Assume (i) and simply stare at the diagram

$$\begin{array}{ccccccccc}
0 & \longrightarrow & Y & \xrightarrow{\ \jmath\ } & Z & \xrightarrow{\ \rho\ } & X & \longrightarrow & 0\\
 & & \| & & \big\downarrow{\scriptstyle\beta} & & \| & & \\
0 & \longrightarrow & Y & \underset{\imath_1}{\overset{\pi_1}{\rightleftarrows}} & Y\times X & \underset{\pi_2}{\overset{\imath_2}{\leftrightarrows}} & X & \longrightarrow & 0
\end{array}$$

until it becomes crystal clear that $\pi_1\beta$ is a left inverse of $\jmath$, while $\beta^{-1}\imath_2$ is a section of ρ. Hence, (i) implies (ii) and (iii). If (ii) holds and P is a left inverse of $\jmath$ then the restriction of ρ to $\ker P$ is an isomorphism onto X (it is injective since $\rho(x) = 0$ implies $x \in Y$, and since $x \in \ker P$, then $x = \rho(x) = 0$; it is surjective since $\rho(x - Px) = \rho(x)$), whose inverse is clearly a section of ρ. This shows (iii). That (iii) implies (ii) is by far the simplest implication: if s is a right inverse for ρ then $\mathbf{1}_Z - s\rho$ is a left inverse of $\jmath$. We finally prove that (iii) implies (i). If s is a section of ρ and $P = \mathbf{1}_Z - s\rho$ then the map $\beta\colon Z \longrightarrow Y \times X$ given by $\beta(x) = (P(x), \rho(x))$ is an isomorphism (its inverse is $(y, z) \mapsto \jmath(y) + s(z)$), making the following diagram commutative:

$$\begin{array}{ccccccccc}
0 & \longrightarrow & Y & \xrightarrow{\ \jmath\ } & Z & \xrightarrow{\ \rho\ } & X & \longrightarrow & 0\\
 & & \| & & \big\downarrow{\scriptstyle\beta} & & \| & & \\
0 & \longrightarrow & Y & \longrightarrow & Y\times X & \longrightarrow & X & \longrightarrow & 0
\end{array}$$

□

Left inverses are called *retractions* in the language of categories and *projections* in the language of Banach spaces. Right inverses are called *sections* in the language of categories and *liftings* in the language of Banach spaces. Retractions and sections appear in pairs: f is a retraction of $g \iff g$ is a section of f. Both sections and retractions are called *splitting morphisms* in algebraic jargon: if (ii) holds, then P splits Z as the direct sum $\jmath[Y] \oplus \ker P$; if (iii) holds, then $Z = s[X] \oplus \ker\rho$. For this reason, trivial sequences are said to *split*. The splitting of a sequence with embedding $\jmath\colon Y \longrightarrow Z$ means in classical terms that $\jmath[Y]$ is a complemented subspace of Z: if $P\jmath = \mathbf{1}_Y$, then $\jmath P$ is a projection of Z onto $\jmath[Y]$; we often say that P is a projection along $\jmath$. Conversely, if R is a projection of Z onto $\jmath[Y]$, then $\jmath^{-1}R$ is left inverse of $\jmath$.

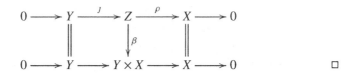

It is important to realise that when an exact sequence splits, the middle space is isomorphic to the direct product of the other two, but the converse is not true: pick any non-trivial sequence $0 \longrightarrow Y \longrightarrow Z \longrightarrow X \longrightarrow 0$. 'Multiplying' by $Z \times X$ on the left one gets the sequence

$$0 \longrightarrow (Z \times X) \times Y \xrightarrow{\mathbf{1}_{Z\times X} \times \jmath} (Z \times X) \times Z \xrightarrow{0 \oplus \rho} X \longrightarrow 0$$

which does not split (look at the quotient side). Proceeding analogously on the right side, we get the sequence $0 \longrightarrow Z \times X \times Y \longrightarrow Z \times X \times Z \times (Z \times Y) \longrightarrow X \times (Z \times Y) \longrightarrow 0$. Now, assuming that all the three spaces are isomorphic to their squares, this last sequence has the form $0 \longrightarrow X \times Y \times Z \longrightarrow (X \times Y \times Z)^2 \longrightarrow X \times Y \times Z \longrightarrow 0$ and is non-trivial. See Proposition 2.2.5 for less artificial examples.

This is a good place to discuss the norms of sections and projections in trivial isometric sequences of (quasi-) Banach spaces. It is clear that if $s \in \mathfrak{L}(X, Z)$ is a section for the quotient map then $P = \mathbf{1}_Z - s\rho$ is a projection onto $\jmath[Y]$, and we have $\|P\| \leq 1 + \|s\|$ when Z is a Banach space and $\|P\| \leq \Delta_Z(1 + \|s\|)$ in general. And, conversely, if $P \in \mathfrak{L}(Z)$ is a projection onto $\jmath[Y]$ then $\mathbf{1}_Z - P$ vanishes on $\ker \rho$ and induces an operator $s\colon X \longrightarrow Z$, which is a section of ρ. As before, $\|s\| \leq 1 + \|P\|$ in Banach spaces, and $\|s\| \leq \Delta_Z(1 + \|P\|)$ in general. The simplest sequence of Banach spaces $0 \longrightarrow \mathbb{K} \longrightarrow Z \longrightarrow X \longrightarrow 0$ admits a norm 1 projection, while it is not true, in general, that $0 \longrightarrow Y \longrightarrow Z \longrightarrow \mathbb{K} \longrightarrow 0$ admits a norm 1 section: that happens if and only if the norm 1 functional f used as a quotient map attains its norm on B_Z. And this happens for every $f \in Z^*$ if and only if Z is reflexive, by a famous theorem of James. Of course, norm $1 + \varepsilon$ sections exist for all $\varepsilon > 0$, and thus hyperplanes are always $(2+\varepsilon)$-complemented. A precise calculus of the norm of projections onto hyperplanes of ℓ_1 and c_0 has been given in [46] with surprising results: every hyperplane of c_0 admits a projection with norm strictly less than 2, and there are hyperplanes in ℓ_1 for which a projection of norm 2 does not exist; an explicit example is the kernel of the functional given by $(1/2, 2/3, \ldots, n/(n+1), \ldots) \in \ell_\infty$ [202, p. 199]. In 2.14.9 we will encounter a special situation in which 1-sections exist.

Returning to the main topic, the standard notion of isomorphism of quasi-Banach spaces translates naturally to exact sequences as follows:

Definition 2.1.6 Two exact sequences $0 \longrightarrow Y \longrightarrow Z \longrightarrow X \longrightarrow 0$ and $0 \longrightarrow Y' \longrightarrow Z' \longrightarrow X' \longrightarrow 0$ are said to be isomorphic if there exist isomorphisms α, β and γ making the following diagram commutative:

$$\begin{array}{ccccccccc} 0 & \longrightarrow & Y & \longrightarrow & Z & \longrightarrow & X & \longrightarrow & 0 \\ & & \downarrow{\scriptstyle \alpha} & & \downarrow{\scriptstyle \beta} & & \downarrow{\scriptstyle \gamma} & & \\ 0 & \longrightarrow & Y' & \longrightarrow & Z' & \longrightarrow & X' & \longrightarrow & 0 \end{array}$$

Diagram (2.2) says that every short exact sequence is isomorphic to a natural one. The notion of isomorphism for sequences contains most of what it is expected to have: the overall meaning that the twisted sum spaces in isomorphic sequences are isomorphic via isomorphisms that somehow safeguard the positions of subspaces and quotients. It is clear that equivalent exact sequences are isomorphic. The converse is true for trivial sequences:

Lemma 2.1.7 *An exact sequence is isomorphic to the direct product sequence if and only if it splits.*

Proof As we said before, simply stare at the diagram

$$\begin{array}{ccccccccc} 0 & \longrightarrow & Y & \stackrel{\jmath}{\longrightarrow} & Z & \stackrel{\rho}{\longrightarrow} & X & \longrightarrow & 0 \\ & & \downarrow{\scriptstyle \alpha} & & \downarrow{\scriptstyle \beta} & & \downarrow{\scriptstyle \gamma} & & \\ 0 & \longrightarrow & Y' & \underset{\imath_1}{\overset{\pi_1}{\rightleftarrows}} & Y' \times X' & \underset{\pi_2}{\overset{\imath_2}{\leftrightarrows}} & X' & \longrightarrow & 0 \end{array}$$

until it becomes crystal clear that $\alpha^{-1}\pi_1\beta$ is a retraction of $\jmath$ while $\beta^{-1}\imath_2\gamma$ is a section of ρ. □

The lemma might fuel a hope (or suspicion) that isomorphic and equivalent sequences coincide (when that is possible). But they do not: for an easy example of two isomorphic non-equivalent sequences, consider a quasi-Banach space Z and a subspace Y and form the corresponding sequence $0 \longrightarrow Y \longrightarrow Z \stackrel{\rho}{\longrightarrow} Z/Y \longrightarrow 0$ that we may call z. Pick a scalar $c \neq 0, 1$ and consider the sequence

$$0 \longrightarrow Y \stackrel{c}{\longrightarrow} Z \stackrel{\rho}{\longrightarrow} Z/Y \longrightarrow 0, \qquad (\mathsf{c})$$

where c denotes multiplication by c. The diagram

$$\begin{array}{ccccccccc} 0 & \longrightarrow & Y & \stackrel{\text{inclusion}}{\longrightarrow} & Z & \stackrel{\rho}{\longrightarrow} & Z/Y & \longrightarrow & 0 \qquad (\mathsf{z}) \\ & & \uparrow{\scriptstyle c1_Y} & & \| & & \| & & \\ 0 & \longrightarrow & Y & \stackrel{c}{\longrightarrow} & Z & \stackrel{\rho}{\longrightarrow} & Z/Y & \longrightarrow & 0 \qquad (\mathsf{c}) \end{array}$$

shows that the two sequences are isomorphic. They are, however, equivalent if and only if z is trivial: indeed, assume that $u \in \mathfrak{L}(Z)$ makes the following diagram commutative:

$$\begin{array}{ccccccccc} 0 & \longrightarrow & Y & \stackrel{\text{inclusion}}{\longrightarrow} & Z & \stackrel{\rho}{\longrightarrow} & Z/Y & \longrightarrow & 0 \qquad (\mathsf{z}) \\ & & \| & & \uparrow{\scriptstyle u} & & \| & & \\ 0 & \longrightarrow & Y & \stackrel{c}{\longrightarrow} & Z & \stackrel{\rho}{\longrightarrow} & Z/Y & \longrightarrow & 0 \qquad (\mathsf{c}) \end{array}$$

The operator $\mathbf{1}_Z - cu$ vanishes on Y, so it induces an operator $S : Z/Y \longrightarrow Z$, defined by $S(z+Y) = z - cu(z)$. Composing with ρ, we obtain $\rho(S(z+Y)) = \rho(z - cu(z)) = \rho(z) - c\rho(z) = (1-c)(z+Y)$, which shows that $(1-c)^{-1}S$ is a section of ρ. When the operators (α, β, γ) appearing in the definition of isomorphic sequences are isometries, the sequences are said to be *isometric*. When $Y' = Y$ and $X' = X$, and both α and γ are scalar multiples of the identity, the exact sequences are said to be *projectively equivalent*. The preceding example shows that this notion is strictly weaker than usual equivalence. It is clear that isometric or projectively equivalent sequences are isomorphic. There is a topologised version of the 3-lemma:

Lemma 2.1.8 *Assume one has a commutative diagram of quasi-Banach spaces and operators with exact rows*

$$\begin{array}{ccccccccc} 0 & \longrightarrow & Y & \xrightarrow{\jmath} & Z & \xrightarrow{\rho} & X & \longrightarrow & 0 \\ & & \downarrow{\scriptstyle \alpha} & & \downarrow{\scriptstyle \beta} & & \downarrow{\scriptstyle \gamma} & & \\ 0 & \longrightarrow & Y' & \xrightarrow{\jmath'} & Z' & \xrightarrow{\rho'} & X' & \longrightarrow & 0 \end{array}$$

(a) *If α and γ have dense range then β has dense range.*
(b) *If α and γ are isomorphic embeddings then so is β.*

Proof We may assume that Z' carries a p-norm for some $0 < p < 1$, that Y' is a subspace of Z' and that $X' = Z'/Y'$ carries the quotient quasinorm. (a) Pick $z' \in Z'$ and $\varepsilon > 0$. Set $x' = \rho'(z')$ and take $x \in X$ such that $\|x' - \gamma(x)\| < \varepsilon$. Now choose $z \in Z$ such that $x = \rho(z)$. As $\|\rho'(\beta(z) - z')\| < \varepsilon$, there is $y' \in Y'$ such that $\|y' + \beta(z) - z'\| < \varepsilon$, and since α has dense range, we may assume that $y' = \alpha(y)$ for some $y \in Y$. Clearly, $\|\beta(\jmath(y) + z) - z'\| < \varepsilon$. To prove (b), it suffices to check that if (z_n) is a sequence in Z and $\beta(z_n) \longrightarrow 0$ in Z', then $z_n \longrightarrow 0$ in Z. We have $\rho'(\beta(z_n)) = \gamma(\rho(z_n)) \longrightarrow 0$, hence $\rho(z_n) \longrightarrow 0$ in X, and we can write $z_n = \jmath(y_n) + \tilde{z}_n$, where $y_n \in Y$ and $\tilde{z}_n$ converges to zero in Z. Since $\beta(z_n) = \beta(\jmath(y_n) + \tilde{z}_n) = \jmath'(\alpha(y_n)) + \beta(\tilde{z}_n)$ and $\beta(\tilde{z}_n) \longrightarrow 0$, $\jmath'(\alpha(y_n)) \longrightarrow 0$, and thus y_n and z_n converge to zero. □

2.2 Basic Examples of Exact Sequences

In general, an exact sequence $0 \longrightarrow A \longrightarrow B \longrightarrow C \longrightarrow 0$ cannot split when B enjoys a certain property preserved by isomorphisms (say, the DPP or 'being an $\mathscr{L}_1$-space') that passes to complemented subspaces while either A or C does not have that property. A variation of the same argument is that if B has some hereditary property and C contains a subspace failing to have it, then the sequence cannot split.

Folklore on Exact Sequences Involving $C(K)$-Spaces

Let S and T be compact spaces. Every continuous mapping $\varphi\colon S \longrightarrow T$ induces an operator (well, a homomorphism of Banach algebras) $\varphi^\circ : C(T) \longrightarrow C(S)$. Clearly, $\|\varphi^\circ\| = 1$. There are two cases in which φ° can appear in a short exact sequence:

- φ is injective (hence a homeomorphism onto its range) $\Longleftrightarrow$ φ° is surjective $\Longleftrightarrow$ φ° is an isometric quotient.

All this is Tiezte's extension theorem. Assuming that S is a closed subset of T, we can interpret φ° as plain restriction, call it r and obtain the sequence

$$0 \longrightarrow \ker r \longrightarrow C(T) \xrightarrow{\ r\ } C(S) \longrightarrow 0 \tag{2.3}$$

in which the nature of the subspace cannot be clearer: it is the ideal of those functions on T vanishing on S, and thus it is naturally isometric to $C_0(T\backslash S)$. When $S = T'$ is the subset of accumulation points of T, we have $C_0(T\backslash S) = c_0(I)$, where I is the (discrete) set of isolated points of T. Bounded linear sections of r are called extension operators. They always exist if S is metrisable, and equivalently, if $C(S)$ is separable: this is the content of the Borsuk–Dugundji theorem, which will be generalised in Theorem 2.14.5.

- φ is surjective $\Longleftrightarrow$ φ° is injective $\Longleftrightarrow$ φ° is an isometry.

This is obvious and yields the exact sequence

$$0 \longrightarrow C(T) \xrightarrow{\ \varphi^\circ\ } C(S) \longrightarrow \cdot \longrightarrow 0 \tag{2.4}$$

A projection along φ° is called an averaging operator: the reason is that $C(T)$ sits in $C(S)$ as the subspace of those functions that remain constant on the fibers of φ – the sets $\varphi^{-1}(t)$ for $t \in T$. Thus, if P is a projection of $C(S)$ onto $\varphi^\circ[C(T)]$, then $P(f)$ must be a kind of average of f. No general criterion is known for the splitting of (2.4). The nature of the quotient space is also unclear, and many natural questions about it remain unanswered: perhaps the most glaring one is whether it is isomorphic to a $\mathscr{C}$-space, or even to a Lindenstrauss space. Even so, one can easily compute norms in the quotient space thanks to [377, Corollary 9.10]:

Lemma 2.2.1 *Let $\varphi\colon S \longrightarrow T$ be a surjection between compact spaces. Then the norm in $C(S)/\varphi^\circ[C(T)]$ is given by (real case)*

$$\left\|f + \varphi^\circ[C(T)]\right\| = \sup_{\varphi(s)=\varphi(s')} \frac{|f(s) - f(s')|}{2}.$$

A generalised form for this estimate [42, Proposition 1.18] will be useful later. Recall that the oscillation of a function $f\colon K \longrightarrow \mathbb{R}$ at a point $s \in K$ is defined by $\operatorname{osc} f(s) = \inf_V \sup_{r,t\in V} (f(r) - f(t))$, where V runs over the neighbourhoods of s in K. The oscillation of f on K is $\operatorname{osc}_K(f) = \sup_{s\in K} \operatorname{osc} f(s)$.

Lemma 2.2.2 *Let K be a compact space and $f\colon K \longrightarrow \mathbb{R}$ a bounded function. Then*

$$\operatorname{dist}(f, C(K)) = \|f + C(K)\| = \tfrac{1}{2}\operatorname{osc}_K(f).$$

Proof Only one of the inequalities needs a proof. Define

$$f_{\mathrm{lsc}}(s) = \sup_V \inf_{t\in V} f(t) = \max\left(f(s), \liminf_{t\to s} f(t)\right),$$

$$f^{\mathrm{usc}}(s) = \inf_V \sup_{t\in V} f(t) = \min\left(f(s), \limsup_{t\to s} f(t)\right),$$

where V runs over the neighbourhoods of s. Clearly, f^{usc} is upper semicontinuous, f_{lsc} is lower semicontinuous and $f^{\mathrm{usc}} \le f \le f_{\mathrm{lsc}}$. If $\delta = \frac{1}{2}\operatorname{osc}_K f$, it is clear that $f_{\mathrm{lsc}} - \delta \le f^{\mathrm{usc}} + \delta$. The Hahn–Tong separation theorem [430, 6.4.4. theorem] gives us a continuous function h satisfying $f_{\mathrm{lsc}} - \delta \le h \le f^{\mathrm{usc}} + \delta$. Hence,

$$f - \delta \le f_{\mathrm{lsc}} - \delta \le h \le f^{\mathrm{usc}} + \delta \le f + \delta \quad\Longrightarrow\quad \|f - h\| \le \delta. \qquad \square$$

It is important to realise that many Banach algebras are $\mathscr{C}$-spaces in disguise: a commutative, unital Banach algebra A is isometrically isomorphic to the algebra of all continuous functions on some compact space if and only if it is a C^*-algebra (complex case) or for every $f, g \in A$ we have $2\|fg\| \le \|f^2 + g^2\|$ (real case). The complex case is the celebrated Gelfand–Naimark theorem; its real companion is due to Albiac and Kalton [5, Theorem 4.2.1]. In both cases we recover the underlying compact space as the set of unital homomorphisms $A \longrightarrow \mathbb{K}$ with the relative weak* topology. The result applies, for instance, to the spaces $L_\infty(\mu)$ and $\ell_\infty(I)$ and their unital (and self-adjoint in the complex case) subalgebras, to the ultraproducts of families of $\mathscr{C}$-spaces and to many others that we will meet along the way. This implies that if $u\colon A \longrightarrow B$ is a unital homomorphism and A and B satisfy the corresponding condition, then there are compact spaces S and T isometric isomorphisms $\alpha\colon A \longrightarrow C(T)$,

$\beta\colon B \longrightarrow C(S)$ and continuous mapping $\varphi\colon S \longrightarrow T$ forming a commutative diagram which one should keep in mind for subsequent examples:

$$\begin{array}{ccc} A & \xrightarrow{\ u\ } & B \\ \downarrow{\scriptstyle\alpha} & & \downarrow{\scriptstyle\beta} \\ C(T) & \xrightarrow{\ \varphi^\circ\ } & C(S) \end{array}$$

The Foiaş–Singer Sequence and Its Variations

This construction appears in [176, Theorem 6]. Let $\Delta = \{0,1\}^{\mathbb{N}}$ be the Cantor set equipped with the product topology and the lexicographic order. We denote by Δ_0 the countable and dense subset of those $t \in \Delta$ having finitely many 1s. Let $D = D(\Delta;\Delta_0)$ be the space of all functions $\Delta \longrightarrow \mathbb{R}$ that are continuous at every $t \notin \Delta_0$ and left continuous with right limits at every $t \in \Delta_0$. It is *really* easy to prove that the sup norm makes $D(\Delta)$ into a Banach space containing $C(\Delta)$ and that the quotient $D/C(\Delta)$ is isometric to c_0. Indeed, if $J\colon D \longrightarrow \ell_\infty(\Delta_0)$ denotes the 'jump' function $J(f)(q) = \frac{1}{2}(f(q^+) - f(q))$, where $f(q^+)$ is the right limit of f at q, then J maps D onto $c_0(\Delta_0)$ and $\operatorname{dist}(f, C(\Delta)) = \|Jf\|_\infty$ for every $f \in D$, as an obvious application of Lemma 2.2.2. Thus one has an exact sequence

$$0 \longrightarrow C(\Delta) \xrightarrow{\ \imath\ } D \xrightarrow{\ J\ } c_0 \longrightarrow 0 \tag{2.5}$$

Besides, the space D is a unital subalgebra of $\ell_\infty(\Delta)$, and since $\imath$ is a homomorphism, it follows that D is a $\mathscr{C}$-space and that the Foiaş–Singer sequence (2.5) has the form (2.4).

Lemma 2.2.3 *Let (f_q) be any sequence in D such that $J(f_q) = e_q$ for every $q \in \Delta_0$. Then, given $\lambda^1,\dots,\lambda^n \in \mathbb{R}; q^1,\dots,q^n \in \Delta_0$ and $\varepsilon > 0$, there exist $q \in \Delta_0\backslash\{q^1,\dots,q^n\}$ and $\lambda = \pm 1$ such that*

$$\left\|\lambda f_q + \sum_{k=1}^{n} \lambda^k f_{q^k}\right\|_D \geq 1 + \left\|\sum_{k=1}^{n} \lambda^k f_{q^k}\right\|_D - \varepsilon.$$

Proof With no serious loss of generality we may assume that there is q in $\Delta_0\backslash\{q^1,\dots,q^n\}$ such that

$$\sum_{k=1}^{n} \lambda^k f_{q^k}(q) > \left\|\sum_{k=1}^{n} \lambda^k f_{q^k}\right\|_D - \varepsilon.$$

But $\sum_{k=1}^n \lambda^k f_{q^k}$ is continuous at q, and so $\sum_{k=1}^n \lambda^k f_{q^k}(q^+) = \sum_{k=1}^n \lambda^k f_{q^k}(q)$. Now, since $Jf_q = e_q$, if $f_q(q) \geq -1$, then $f_q(q^+) \geq 1$, hence

$$\left\| f_q + \sum_{k=1}^n \lambda^k f_{q^k} \right\|_D \geq \left(f_q + \sum_{k=1}^n \lambda^k f_{q^k} \right)(q^+) > 1 + \left\| \sum_{k=1}^n \lambda^k f_{q^k} \right\|_D - \varepsilon,$$

as required. And if $f_q(q) < -1$, then

$$\left\| -f_q + \sum_{k=1}^n \lambda^k f_{q^k} \right\|_D \geq \left(-f_q(q) + \sum_{k=1}^n \lambda^k f_{q^k}(q) \right) > 1 + \left\| \sum_{k=1}^n \lambda^k f_{q^k} \right\|_D - \varepsilon.$$

This is enough to make the sequence (2.5) non-trivial. □

Other arguments can be given [1, Remark (ii); 42, Example 1.20, p. 24; or 73, Lemma 2.2]. A variation of this construction, working now in $[0, 1]$, was presented by Aharoni and Lindenstrauss in [1]. Fix a countable subset $N \subset [0, 1]$ and form the space $D([0, 1]; N)$ of all real, bounded functions on $[0, 1]$ which are continuous except at points of N, where they are left-continuous and have right limits. Again, $C[0, 1]$ is a closed subspace of $D([0, 1]; N)$ and $D([0, 1]; N)/C[0, 1]$ is isometric to $c_0(N)$ via $Jf = \frac{1}{2}(f(q_n^+) - f(q_n))$. All this gives an exact sequence $0 \longrightarrow C[0, 1] \longrightarrow D([0, 1]; N) \longrightarrow c_0 \longrightarrow 0$ whose splitting depends on the location of N inside $[0, 1]$. Precisely:

Lemma 2.2.4 *If N is dense in $[0, 1]$ then no lifting of (e_n) is weakly Cauchy.*

Proof The assumption $J(f_n) = e_n$ means that $f_n(q_n^+) - f_n(q_n) = 2$ for all n. Let us assume (f_n) is weakly Cauchy and hence bounded. We first note that if I is any non-empty open interval in $(0, 1)$, $\alpha \in \mathbb{R}$ and $m \in \mathbb{N}$, there exists $n > m$ and a non-empty open interval J with $\overline{J} \subset I$ such that for some β with $|\beta - \alpha| \geq 1$, we have $|f_n(t) - \beta| \leq \frac{1}{4}$ for $t \in J$. Indeed, we just pick $n > m$ such that $q_m \in I$ and then let β be either $f_n(q_n)$ or $f_n(q_n^+)$. The interval J can then be chosen using the left- or right-hand limit condition. Now we can use this inductively to create a subsequence (f_{n_k}) of (f_n), a sequence of non-empty intervals (I_k) with $\overline{I}_{k+1} \subset I_k$ and a sequence of reals (α_k) with $|\alpha_{k+1} - \alpha_k| \geq 1$ such that $|f_{n_k}(t) - \alpha_k| \leq \frac{1}{4}$ for $t \in I_k$. If we pick $t_0 \in \bigcap_{k=1}^\infty I_k$ (which is non-empty by compactness), it is clear that $|f_{n_k}(t_0) - f_{n_{k+1}}(t_0)| \geq 1/2$ for all k, and this yields a contradiction. □

We now identify the isometry type of the spaces $D(\Delta; \Delta_0)$ and $D([0, 1]; N)$.

Proposition 2.2.5 *$D(\Delta; \Delta_0)$ is isometric to $C(\Delta)$ and so is $D([0, 1]; N)$ when N is a countable dense subset of $[0, 1]$.*

Proof We give the proof for $D([0, 1]; N)$. The case of $D(\Delta; \Delta_0)$ is easier. We know that the Banach algebra $A = D([0, 1]; N)$ is isometrically isomorphic to a $C(K)$ for some compact space K. Three obvious properties of A force this K to be homeomorphic to Δ.

- K is totally disconnected $\iff$ A is generated by its idempotents, which is clear since if $s, t \in N$, $s < t$, then $1_{(s,t]} \in A$.
- K does not have isolated points $\iff$ A does not have minimal idempotents, which is clear since each idempotent in A is the sum of two idempotents.
- K is metrisable $\iff$ A is separable, which is obvious. □

The Foiaş–Singer and Aharoni–Lindenstrauss constructions produce uncomplemented copies of $C(\Delta)$ and of $C[0,1]$ inside $C(\Delta)$ with quotient c_0. A variation of Amir and Lindenstrauss [10] considers uncountable sets of jumps, such as the set $\mathbb{I} \subset [0,1]$ of irrationals, thus obtaining a non-trivial sequence $0 \longrightarrow C[0,1] \longrightarrow D([0,1];\mathbb{I}) \longrightarrow c_0(\mathbb{I}) \longrightarrow 0$.

We now consider countable compacta. To produce a 'countable' version of the Foiaş–Singer sequence, we need the following 'reordering':

Lemma 2.2.6 *Any countable compact space K can be embedded in $\mathbb{R}$ in such a way that for each $n \in \mathbb{N}$ and each x in the nth derived set $K^{(n)}$ and each $\varepsilon > 0$, both the left neighbourhood $(x - \varepsilon, x)$ and the right neighbourhood $(x, x + \varepsilon)$ contain points of $K^{(n-1)}$.*

Proof By 1.6.1, we can think of K as a countable ordinal. The proof proceeds easily by transfinite induction: the case $\alpha = 1$ is clear, as well as the inductive step for $\alpha + 1$ assuming it is true for α. We thus prove the inductive step for $\alpha = \lim \alpha_n$. The induction hypothesis is that each interval $(\alpha_n, \alpha_{n+1}]$ admits an embedding into $\mathbb{R}$ with the required properties, so embed it into $(-1)^n(\frac{1}{n+1}, \frac{1}{n})$ and send α to the origin. □

Let K be a countable compact which we assume is embedded in the line as in the lemma. The simple plan now is to define $D(K;K')$ as the space of all functions $K \longrightarrow \mathbb{R}$ that are left continuous and possess right limits at every point of K'. The space $D(K;K')/C(K)$ is isometric to $c_0(K')$; via the jump function $J(f) = \frac{1}{2}(f(t^+) - f(t))_{t \in K'}$. As before, we have an exact sequence

$$0 \longrightarrow C(K) \xrightarrow{\text{inclusion}} D(K;K') \xrightarrow{J} c_0(K') \longrightarrow 0 \tag{2.6}$$

We denote the unit basis in $c_0(K')$ by $(e_t)_{t \in K'}$.

Lemma 2.2.7 *Let $\{f_x : x \in K'\}$ be any collection of functions in $D(K;K')$ for which $Jf_x = e_x$. Fix $\delta > 0$ and $n \in \mathbb{N}$ and choose a point $x_n \in K^{(n)}$. Then there exist $t \in K$ and distinct points $x_i \in K^{(i)}$ for $1 \le i < n$ and signs $\varepsilon_i = \pm 1$ for $1 \le i \le n$ such that $\varepsilon_i f_{x_i}(t) > 1 - \delta$ for $1 \le i \le n$.*

Proof We will write f_j instead of f_{x_j}. We are told that $f_n(x_n^+) - f_n(x_n) = 2$, which obviously implies that either $f_n(x_n^+) \geq 1$ or $f_n(x_n) \leq -1$. In the former case, put $\varepsilon_n = 1$, in the latter, $\varepsilon_n = -1$. In either case, x_n has a one-sided neighbourhood $O_n \subset K$ on which $\varepsilon_n f_n > 1 - \delta$. Thanks to our embedding of K into $\mathbb{R}$, we may choose a point $x_{n-1} \in K^{(n)} \cap O_n$. The same argument then gives some $\varepsilon_{n-1} = \pm 1$ and a one-sided neighbourhood O_{n-1} of x_{n-1} contained in O_n on which $\varepsilon_{n-1} f_{n-1} > 1 - \delta$. Repeat the process until exhaustion. □

Corollary 2.2.8 *If K is a countable compact such that $K^{(n)} \neq \emptyset$, then any linear section of J in* (2.6) *has norm at least n.*

Proof If $L\colon c_0(K') \longrightarrow D(K; K')$ is a section of J, then the preceding lemma applies to the family $f_x = L(e_x)$. Taking $x_n \in K^{(n)}$ and $s = \sum_{1\leq i\leq n} \varepsilon_i e_{x_i}$, it is clear that $\|s\| = 1$, while the norm of $L(s) = \sum_{1\leq i\leq n} \varepsilon_i f_{x_i}$ is at least $(1 - \delta)n$. □

The reader is invited to ponder the choices $K = \omega^N$ for $1 \leq N < \omega$.

Exact Sequences Involving $c_0(I)$

We present three essentially different non-trivial exact sequences of the type $0 \longrightarrow c_0(I) \longrightarrow \cdot \longrightarrow c_0(J) \longrightarrow 0$:

- The Nakamura–Kakutani sequences 2.2.10, mutated into the Johnson–Lindenstrauss sequences of Diagram (2.38) in Section 2.12.
- The Ciesielski–Pol sequence of 2.2.11.
- The Bell–Marciszewski construction in Proposition 2.2.15.

The first two types provide non-WCG (hence non-trivial) twisted sums of two $c_0(I)$ spaces, which is already weird; the last one provides non-trivial WCG twisted sums of two $c_0(I)$ spaces, which is weirder still.

Definition 2.2.9 A family $\mathcal{M}$ of infinite subsets of $\mathbb{N}$ is called almost disjoint if the intersection of any two elements of $\mathcal{M}$ is finite.

The existence of such families of cardinal $\mathfrak{c}$ was first observed by Sierpiński [433]. A good example could be enumerating the nodes of the dyadic tree: $\mathcal{M}$ will be the set of branches such that each α will be the set of naturals assigned to the nodes in the branch α. Or else, pick an enumeration of all rationals and identify each irrational α with the set of natural numbers corresponding to a sequence of rationals converging to it. Nakamura and Kakutani [369] observe that the images of the characteristic functions of an almost disjoint family generate an isometric copy of $c_0(\mathcal{M})$ in ℓ_∞/c_0.

2.2.10 The Nakamura–Kakutani sequences *An uncountable almost disjoint family* $\mathcal{M}$ *generates a non-trivial exact sequence*

$$0 \longrightarrow c_0 \longrightarrow C_0(\wedge_{\mathcal{M}}) \xrightarrow{\ \rho\ } c_0(\mathcal{M}) \longrightarrow 0$$

where $C_0(\wedge_{\mathcal{M}}) = \overline{[\{1_n : n \in \mathbb{N}\} \cup \{1_\alpha : \alpha \in \mathcal{M}\}]} \subset \ell_\infty$ *is not WCG.*

The sequence cannot split because the points of ℓ_∞, and so those of its subspace $C_0(\wedge_{\mathcal{M}})$, can be separated by a countable family of functionals; since those of $c_0(\mathcal{M})$ cannot, the conclusion is that no injective linear map from $c_0(\mathcal{M})$ to ℓ_∞ exists, let alone a bounded linear section of the quotient ρ. This also shows that $C_0(\wedge_{\mathcal{M}})$ cannot be WGC; see Proposition 1.7.7. To understand the nature of the space $C_0(\wedge_{\mathcal{M}})$ and justify the seemingly eccentric notation, just observe that $C_0(\wedge_{\mathcal{M}})$ is a subring of ℓ_∞ but does not have a unit. It follows that it can be represented as the ring of all continuous functions vanishing at infinity on some locally compact space $\wedge_{\mathcal{M}}$, which we now describe. The space $\wedge_{\mathcal{M}}$ has two classes of points: those corresponding to points of $\mathbb{N}$, which we declare isolated, and those corresponding to the elements of $\mathcal{M}$; a typical neighbourhood of A must contain A together with almost all the 'elements' of A. This space is locally compact, but not compact. The one point compactification of $\wedge_{\mathcal{M}}$ will be denoted by $\Delta_{\mathcal{M}}$. It is a fairly run-of-the-mill scattered compactum of height 3. The space $C(\Delta_{\mathcal{M}})$ can be viewed as the unitisation of $C_0(\wedge_{\mathcal{M}})$ in ℓ_∞ and can be placed in the obvious exact sequence $0 \longrightarrow c_0 \longrightarrow C(\Delta_{\mathcal{M}}) \longrightarrow c(\mathcal{M}) \longrightarrow 0$, which is of type (2.3). When $\mathcal{M}$ is the family of branches of the dyadic tree, in which case $|\mathcal{M}| = \mathfrak{c}$, the space $C(\Delta_{\mathcal{M}})$ could well be called the Johnson–Lindenstrauss space for reasons that will become clear in Section 2.12. Analogous constructions can be carried out for larger cardinals using the fact that given an infinite set I, there exists a family $\mathcal{M}$ of infinite subsets of I such that $|A \cap B| < |I|$ for each $A, B \in \mathcal{M}$ and with $|\mathcal{M}| > |I|$; see [314].

The Ciesielski–Pol Space

Ciesielski and Pol obtained the thereafter so-called *Ciesielski–Pol* compacta, namely height 3 compact spaces K such that (a) both $K \setminus K'$ and K' are uncountable and (b) for every $\alpha \in K' \setminus \{\infty\}$ there is an infinite countable set $C_\alpha \subset K \setminus K'$ such that (b.1) $C_\alpha \cup \{\alpha\}$ is clopen, and (b.2) every uncountable subset of $K \setminus K'$ contains some C_α. If CP is a Ciesielski–Pol compact, the space $C(\mathsf{CP})$ is far from being WCG, as 2.2.11 shows. The intricate construction of Ciesielski–Pol compacta together with the proof of 2.2.11 probably cannot be better described than in [148, VI, Example 8.8], so we do not even try:

2.2.11 Ciesielski–Pol sequence *Let* CP *be a Ciesielski–Pol compact. There is a non-trivial exact sequence* $0 \longrightarrow c_0(\mathfrak{c}) \longrightarrow C(\mathsf{CP}) \longrightarrow c_0(\mathfrak{c}) \longrightarrow 0$ *such that no injective operator* $C(\mathsf{CP}) \longrightarrow c_0(I)$ *exists for any set* I.

A WCG Non-trivial Twisted Sum of $c_0(\Gamma)$

The discussion leading to Definition 1.7.11 suggests Yost's question: must every copy of $c_0(I)$ in a WCG space be complemented? The previous examples also lead to, does a non-trivial WCG twisted sum of two spaces $c_0(I)$ exist? The answer to this second question is yes, and thus the answer to the first question is no. We will deal with the two preceding questions together, adding a cardinal delicacy: that $C(\aleph_\omega)$ is not K-Sobczyk for no $K > 0$, which moreover shows that Proposition 1.7.14 is optimal.

Given a set S, the closed subset $\sigma_n(2^S) = \{a \in 2^S : |a| \leq n\}$ of 2^S is an Eberlein compact and of height $n+1$ when S is infinite. It is an Eberlein compact because $\sigma_n(2^S)$ is a weakly compact subset of $c_0(S)$ under the identification $a \leftrightarrow 1_a$, and it is of finite height because the set of its isolated points is $\{a \in 2^S : |a| = n\}$, thus its derived set is $\sigma_n(2^S)' = \sigma_{n-1}(2^S)$. Obviously the compact $\sigma_n(2^S)$ is scattered. A cornerstone result here is from Godefroy, Kalton and Lancien [190, Theorem 4.8, plus comment on p. 800]:

2.2.12 *For a compact space K of weight strictly lesser than $\aleph_\omega$, we have $C(K) \simeq c_0(I)$ if and only if K is an Eberlein compact of finite height.*

It therefore follows that $C(\sigma_n(2^S)) \simeq c_0(I)$ for some I. One can rummage around to see what else is in this pocket: the natural exact sequence $0 \longrightarrow c_0(I) \longrightarrow C(\sigma_n(2^S)) \longrightarrow C(\sigma_{n-1}(2^S)) \longrightarrow 0$ splits since Granero [199] showed that every copy of $c_0(I)$ inside of $c_0(J)$ is complemented. Marciszewski [353, Proposition 3.1] provides a nice improvement:

Lemma 2.2.13 *Let $K \subset \sigma_n(2^S)$ be compact and let $r\colon C(\sigma_n(2^S)) \longrightarrow C(K)$ be the restriction operator. The following sequence splits:*

$$0 \longrightarrow \ker r \longrightarrow C(\sigma_n(2^S)) \longrightarrow C(K) \longrightarrow 0$$

Thus, if K is a compact that can be embedded in some $\sigma_n(2^S)$ for some S and some $n \in \mathbb{N}$, then $C(K) \simeq c_0(I)$ for some I. The converse also holds [353, Theorem 1.1]:

Lemma 2.2.14 *A compact K can be embedded in some $\sigma_n(2^S)$ for some S and some $n \in \mathbb{N}$ if and only if $C(K) \simeq c_0(I)$ for some I.*

Proof If $C(K) \simeq c_0(I)$ then K is an Eberlein compact that, by 2.2.12, must be of finite height, hence scattered. Some combinatorial work [353, Lemma 2.2], see also [36, Lemma 1.1] yields the existence of a family $\mathcal{F}$ of clopen subsets of K such that (i) given two points of K, there is some element of $\mathcal{F}$ containing exactly one of them, and (ii) there is $n \in \mathbb{N}$ such that intersection of any n elements of $\mathcal{F}$ is empty. With this family $\mathcal{F} = \{U(s): s \in S\}$ in hand, we can form the embedding $e: K \longrightarrow \sigma_n(2^S)$ given by $e(k)(s) = 1_{U(s)}(k)$. □

Proposition 2.2.15 *There is an Eberlein compact* BM *of weight* $\aleph_\omega$ *and height* 3 *such that* $C(\mathsf{BM})$ *contains an uncomplemented copy of* $c_0(\aleph_\omega)$ *for which there is a non-trivial exact sequence* $0 \longrightarrow c_0(\aleph_\omega) \longrightarrow C(\mathsf{BM}) \longrightarrow c_0(\aleph_\omega) \longrightarrow 0$.

Proof Let $S = \bigcup_n \mathcal{P}_n(\omega_n)$. Consider the set

$$\mathcal{A}_n = \{\emptyset\} \cup \{\{b\}: b \in \mathcal{P}_n(\omega_n) \cup \{A \subset \mathcal{P}_n(\omega_n): |A| = n+1 \ \wedge \ |\cup_{a\in A} a| = n+1\}$$

and $\mathcal{A} = \bigcup_n \mathcal{A}_n$. Identifying $a \in \mathcal{A}$ with 1_a in 2^S, let $\mathsf{BM} = \{1_a: a \in \mathcal{A}\}$. This is a compact space of weight $\aleph_\omega$ and height 3 since

$$\mathsf{BM} = \{\emptyset\} \cup \bigcup_{n=1}^{\infty} \Big(\underbrace{\{1_{\{a\}}: a \in \mathcal{P}_n(\omega_n)\}}_{\mathsf{BM}'} \cup \underbrace{\{1_A: A \subset \mathcal{P}_n(\omega_n): |A| = |\cup A| = n+1\}}_{\text{isolated points}} \Big)$$

with $\{\emptyset\} = \mathsf{BM}''$. If we set $K_n = \{1_a: a \in \mathcal{A}_n\}$ then for each point $a \in \mathcal{P}_n(\omega_n)$, it turns out that $1_{\{a\}}$ is the unique accumulation point of $\{1_A \in K_n: a \in A\}$, and that is why BM' corresponds to $\bigcup_n \mathcal{P}_n(\omega_n)$. Each K_n is an Eberlein compact, and thus, BM is an Eberlein compact. Since the isolated points of BM form a discrete subset I and $C(\mathsf{BM}') \simeq c_0(J)$ because of its height 3, the natural exact sequence $0 \longrightarrow C_0(I) \longrightarrow C(\mathsf{BM}) \longrightarrow C(\mathsf{BM}') \longrightarrow 0$ becomes

$$0 \longrightarrow c_0(I) \longrightarrow C(\mathsf{BM}) \longrightarrow c_0(J) \longrightarrow 0$$

It remains to show that the sequence does not split or, equivalently, that $C(\mathsf{BM})$ is not isomorphic to some $c_0(I)$. It is then enough to show that BM cannot be embedded into any $\sigma_n(2^T)$. To do that, we show that K_{2^n} cannot be embedded into any $\sigma_n(2^T)$. The combinatorial core of the argument is two lemmata: the almost obvious but dismaying

Lemma 2.2.16 *Let B be a set and $n \in \mathbb{N}$. There is a family $\{V_a: a \in \sigma_n(2^B)\}$ of open subsets such that $a \in V_a$ for every a and the intersection of any $2^n + 1$ elements of the family is empty.*

Proof Pick for each a the clopen neighbourhood $V_a = \{b: a \subset b\}$ of a. □

plus a disguised form of the diagonal argument that no surjective map from a set into its power set exists, no less dismaying nevertheless:

Lemma 2.2.17 *Let $n \in \mathbb{N} \cup \{0\}$ and let $\varphi: \mathcal{P}_n(\omega_n) \longrightarrow \operatorname{fin}(\omega_n)$ be a map. There is $f \in \mathcal{P}_{n+1}(\omega_n)$ such that $a \notin \varphi(f \setminus \{a\})$ for every $a \in f$.*

Proof The proof proceeds by induction:

(0) If $n = 0$ then $\{a \in 2^\omega : |a| = 0\} = \{\emptyset\}$. Pick $k \in \omega \setminus \varphi(\emptyset)$ and set $b = \{k\}$.

(n) Assume that the lemma holds for n and let $\varphi\colon \mathcal{P}_{n+1}(\omega_{n+1}) \longrightarrow \mathrm{fin}(\omega_{n+1})$. Since $\left|\bigcup_{b\in\mathcal{P}_{n+1}(\omega_n)} \{\, \varphi(b) \,\}\right| \le \omega_n$, there is $\beta \in \omega_{n+1} \setminus \left(\bigcup_{b\in\mathcal{P}_{n+1}(\omega_n)} \cup\, \omega_n\right)$, and we can define an auxiliary map $\psi\colon \mathcal{P}_n(\omega_n) \longrightarrow \mathrm{fin}(\omega_n)$ by $\psi(b) = \varphi(b \cap \{\beta\} \cap \omega_n$. The induction hypothesis provides some $c \in \mathcal{P}_{n+1}(\omega_{n+1})$ such that $a \notin \psi(c \setminus \{a\})$ for every $a \in c$. Then $c \cup \{\beta\}$ is the desired element f that works for φ. □

Let's go for the proof of Proposition 2.2.15. If K_{2^n} embeds into some $\sigma_n(2^T)$ then the first lemma provides a family of open neighbourhoods of its points such that every $2^n + 1$ elements have empty intersection. So, it is enough to prove that every family $\{V_a : a \in K_n\}$ of open neighbourhoods in K_n has $n + 1$ elements whose intersection is non-empty. Given $b \in \mathcal{P}_n(\omega_n)$, pick the clopen neighbourhood of $1_{\{b\}}$ given by $U_b = \{1_A \in K_n : b \in A\}$. Since $1_{\{b\}}$ is the only accumulation point, every neighbourhood of $1_{\{b\}}$ contains a set of the form $W_b = U_b \setminus F_b$, where F_b is a finite subset of $U_b \setminus \{1_{\{b\}}\}$. It is then enough to show that there exist $n + 1$ elements in $\{W_b : b \in \mathcal{P}_n(\omega_n)\}$ whose intersection is non-empty.

Claim For every $1_A \in U_b \setminus \{1_{\{b\}}\}$, we have $A = \mathcal{P}_n(b \cup \{\alpha\})$ for some $\alpha \in \omega_n \setminus b$.

This is in the definition: if $A \subset \mathcal{P}_n(\omega_n)$ satisfies the conditions $|A| = n + 1$ and $|\cup A| = n + 1$ then it has the form $A = \mathcal{P}_n(B)$ for some $B \in \mathcal{P}_{n+1}(\omega_n)$ of the form $B = b \cup \{\alpha\}$. We can thus define the function $\varphi\colon \mathcal{P}_n(\omega_n) \longrightarrow \mathrm{fin}(\omega_n)$ given by $\varphi(b) = \{\alpha \in \omega_n : 1_{\mathcal{P}_n(b\cup\{\alpha\})} \in F_b\}$, to which the second lemma applies to yield $f \in \mathcal{P}_{n+1}(\omega_n)$ such that for every $\alpha \in f$, we have $\alpha \notin \varphi(f \setminus \{\alpha\})$. This means that if $f = \{\alpha_1, \ldots, \alpha_n + 1\}$ then

$$1_{\mathcal{P}_n(f)} \in \bigcap_{j=1}^{n+1} W_{f\setminus\{\alpha_j\}}.$$

□

A less glittering example appears in [16]:

2.2.18 *Let* $\alpha = \lim \alpha_n$ *with* $\alpha_0 = \aleph_1$, $\alpha_{n+1} = 2^{\alpha_n}$. *The one-point compactification* ACGJM *of* $\cup_n \sigma_n(2^{\alpha_n})$ *is an Eberlein compact, and* $C(\mathsf{ACGJM})$ *contains an uncomplemented copy of some* $c_0(I)$

The argument is by showing that $\|P\| \ge 1 + \frac{n}{2}$ for any projection P in the natural sequence $0 \longrightarrow \ker r_n \longrightarrow C(\sigma_n(2^{\alpha_n})) \longrightarrow C(\sigma_{n-1}(2^{\alpha_n})) \longrightarrow 0$. Such projections exist by Lemma 2.2.13 or, under GCH, by Proposition 1.7.13 since, in this case, the compact $\sigma_n(2^{\alpha_n})$ has weight $\alpha_n = \aleph_n$. The space ACGJM is the one-point compactification of $\bigcup_n \sigma_n(2^{\alpha_n})$ so that the c_0-sum

$$0 \longrightarrow c_0(\mathbb{N}, \ker r_n) \longrightarrow c_0(\mathbb{N}, C(\sigma_n(2^{\alpha_n})) \longrightarrow c_0(\mathbb{N}, C(\sigma_{n-1}(2^{\alpha_n})) \longrightarrow 0$$

becomes $0 \longrightarrow c_0(I) \longrightarrow C(\mathsf{ACGJM}) \longrightarrow C(\mathsf{ACGJM}) \longrightarrow 0$. The space $C(\mathsf{ACGJM})$ cannot be isomorphic to any $c_0(J)$.

Ultraproduct Sequences

Given an ultrafilter $\mathcal{U}$ on a set I, the ultrapower $X_{\mathcal{U}}$ of a space X is the quotient space in the exact sequence $0 \longrightarrow c_0^{\mathcal{U}}(I, X) \longrightarrow \ell_\infty(I, X) \longrightarrow X_{\mathcal{U}} \longrightarrow 0$ (see Section 1.4). From now on the quotient map in this ultraproduct sequence will be denoted $[\cdot]_{\mathcal{U}}$ – when it is necessary to make $\mathcal{U}$ explicit – and $[\cdot]$ otherwise. A standard argument, see [102], shows that given $0 \longrightarrow Y \longrightarrow Z \longrightarrow X \longrightarrow 0$, the natural sequence of ultrapowers $0 \longrightarrow Y_{\mathcal{U}} \longrightarrow Z_{\mathcal{U}} \longrightarrow X_{\mathcal{U}} \longrightarrow 0$ is exact. No criterion is known to decide when the ultrapower sequence splits.

The Pełczyński–Lusky Sequence

Let X be a quasi-Banach space, and let (X_n) be an increasing sequence of subspaces whose union is dense in X. Let $c(\mathbb{N}, X_n)$ denote the space of those converging sequences (x_n) such that $x_n \in X_n$ for every $n \in \mathbb{N}$, equipped with the sup quasinorm. There is an obvious exact sequence

$$0 \longrightarrow c_0(\mathbb{N}, X_n) \longrightarrow c(\mathbb{N}, X_n) \xrightarrow{\lim} X \longrightarrow 0. \tag{2.7}$$

Proposition 2.2.19 *Let X be a separable p-Banach space, and let (X_n) be an increasing sequence of finite-dimensional subspaces whose union is dense in X. The Pełczyński–Lusky sequence* (2.7) *splits if and only if X has the BAP.*

Proof The space $c(\mathbb{N}, X_n)$ has the BAP since the sequence of finite-rank operators $T_k((x_n)) = (x_1, x_2, \ldots, x_{k-1}x_k, x_k, x_k, \ldots)$ converges pointwise to the identity. Thus, if the sequence splits, X, as a complemented subspace of $c(\mathbb{N}, X_n)$, would have the BAP. Now assume that X has the λ-AP and that there is an increasing sequence of integers $n(k)$ and a sequence of finite-rank operators (T_k) with $T_0 = 0$ and such that $T_k[X] \subset X_{n(k)}$ with $\|T_k\| \leq \lambda^+$ and $T_{k+1}(x) = x$ for $x \in X_{n(k)}$. A linear continuous section for the limit map is provided by the map $s \colon X \longrightarrow c(\mathbb{N}, X_n)$ given by $s(x)(n) = T_{k-1}(x)$ for $n(k) \leq n < n(k+1)$. □

The following key application is a formal adaptation of [384, Lemma 1.2]:

Lemma 2.2.20 *Every separable quasi-Banach space with the BAP is isomorphic to a complemented subspace of a space with a* 1*-FDD.*

Proof Assume X is a separable p-Banach space with the λ-AP. Then there is a sequence of finite-rank operators $(f_n)_{n\geq 1}$ converging pointwise to $\mathbf{1}_X$, with $\|f_n\| \leq \lambda$ for all n. Assuming $f_1 = 0$, we define $a_n = f_{n+1} - f_n$ so that $\|a_n\| \leq 2^{1/p}\lambda$ and $x = \sum_{n=1}^{\infty} a_n(x)$ for all $x \in X$; i.e. the sequence (a_n) is a finite-dimensional expansion of the identity in X. Set $Y_n = a_n[X]$, and consider the vector space

$$\Sigma(Y_n) = \left\{ (y_n) \in \prod_n Y_n : \sum_{n\geq 1} y_n \text{ converges in } X \right\}$$

p-normed by $\|(y_n)\| = \sup_k \|\sum_{n\leq k} y_n\|$. The following facts are nearly trivial:

- $\Sigma(Y_n)$ has a 1-FDD.
- The sum operator $s\colon \Sigma(Y_n) \longrightarrow X$ given by $s((y_n)_n) = \sum_{n\geq 1} y_n$ is contractive.
- The operator $a\colon X \longrightarrow \Sigma(Y_n)$ defined by $a(x) = (a_n x)_n$ is a right inverse of s, with $\|a\| \leq \lambda$. □

If $(Y_n)_{n\geq 1}$ is a chain of subspaces of X with dense union and $Y_1 = 0$ then $\Sigma(Y_n)$ is isomorphic to $c(\mathbb{N}, Y_n)$ via telescoping the series. Thus, Lemma 2.2.20 is roughly equivalent to Proposition 2.2.19.

The Bourgain ℓ_1-Sequence

We now revisit Bourgain's embedding $\ell_1 \longrightarrow \ell_1$ mentioned in Proposition 1.3.1. In proving [48, Theorem 7], Bourgain shows that there is some constant $C > 0$ such that for every $\varepsilon > 0$ and every sufficiently large $n \in \mathbb{N}$, there is $N(n)$ and an n-dimensional subspace E_n of $\ell_1^{N(n)}$ which is C-isomorphic to ℓ_1^n and such that every projection $P\colon \ell_1^{N(n)} \longrightarrow E_n$ has $\|P\| \geq C^{-1}(\log\log n)^{1-\varepsilon}$. Form the sequences $0 \longrightarrow E_n \longrightarrow \ell_1^{N(n)} \longrightarrow \ell_1^{N(n)}/E_n \longrightarrow 0$ and then their adjoints $0 \longrightarrow E_n^{\perp} \longrightarrow \ell_\infty^{N(n)} \longrightarrow E_n^* \longrightarrow 0$ and observe that each E_n^* is C-isomorphic to ℓ_∞^n. Amalgamating, we obtain the exact sequence

$$0 \longrightarrow c_0(\mathbb{N}, E_n^{\perp}) \longrightarrow c_0(\mathbb{N}, \ell_\infty^{N(n)}) \longrightarrow c_0(\mathbb{N}, E_n^*) \longrightarrow 0, \quad (2.8)$$

and its adjoint is

$$0 \longrightarrow \ell_1(\mathbb{N}, E_n) \longrightarrow \ell_1(\mathbb{N}, \ell_1^{N(n)}) \longrightarrow \ell_1(\mathbb{N}, \ell_1^{N(n)}/E_n) \longrightarrow 0.$$

Neither of these splits because if $P\colon \ell_1(\mathbb{N}, \ell_1^{N(n)}) \longrightarrow \ell_1(\mathbb{N}, E_n)$ is a projection, the restriction to the nth coordinate yields a projection $P_n\colon \ell_1^{N(n)} \longrightarrow E_n$ with $\|P_n\| \leq \|P\|$. Thus, $\ell_1(\mathbb{N}, E_n)$ provides an uncomplemented copy of ℓ_1 inside ℓ_1.

If we call $\mathcal{B} = c_0(\mathbb{N}, E_n^\perp)$ then, since $c_0(\mathbb{N}, \ell_\infty^{N(n)}) = c_0$ and $c_0(\mathbb{N}, E_n^*) \simeq c_0$, the sequence (2.8) becomes

$$0 \longrightarrow \mathcal{B} \longrightarrow c_0 \longrightarrow c_0 \longrightarrow 0 \tag{2.9}$$

and its adjoint

$$0 \longrightarrow \ell_1 \longrightarrow \ell_1 \longrightarrow \mathcal{B}^* \longrightarrow 0. \tag{2.10}$$

The space $\mathcal{B}$ cannot be an $\mathscr{L}_\infty$-space since $\mathscr{L}_\infty$-subspaces of c_0 are complemented by 1.6.3 (b). The space $\mathcal{B}^*$ cannot therefore be an $\mathscr{L}_1$-space.

2.3 Topologically Exact Sequences

In categories where no open mapping theorem exists (say, normed spaces), the exactness of a sequence

$$0 \longrightarrow Y \xrightarrow{\ \jmath\ } Z \xrightarrow{\ \rho\ } X \longrightarrow 0 \tag{2.11}$$

no longer means that $\jmath$ embeds Y as a subspace of Z or that ρ is a quotient map: consider the (very) short sequences where either $\jmath$ or ρ is the formal identity $\ell_p^0 \longrightarrow \ell_q^0$ with $0 < p < q \leq \infty$ and draw your own conclusions. Here ℓ_r^0 is the quasinormed space of finitely supported sequences with the restriction of the quasinorm of ℓ_r. Thus, what we used to get for free must now be courteously requested. A map $f\colon A \longrightarrow B$ acting between topological spaces is said to be relatively open if, whenever $U \subset A$ is an open set, $f[U]$ is open in (the relative topology of) $f[A]$.

Definition 2.3.1 An exact sequence of topological vector spaces is topologically exact if its arrows are relatively open operators.

This notion retains the meaning that Y is (embedded by $\jmath$) a closed subspace of Z in such a way that the corresponding quotient is (isomorphic to) X. In categories where operators with closed range are relatively open, such as quasi-Banach spaces, exactness implies topological exactness. A 3-space problem has the form: given a topologically exact sequence (2.11) in which Y, X have a certain property, does Z have it? Any 3-space problem implicitly carries a category, or at least a class of spaces, where the action takes place, which until further notice will be the category of topological vector spaces and operators. The first thing to know in the 3-space business is:

2.3.2 Roelcke's lemma *Let Z be a linear space and $Y \subset Z$ a linear subspace. Comparable linear topologies on Z that induce the same topologies on Y and X/Y agree.*

Proof The comparability assumption is necessary even if the linear topologies on a vector space form a lattice. Let $\mathcal{T} \leq \mathcal{T}'$ be linear topologies on Z that induce the same topologies on Y and X/Y. We denote by $\mathcal{O}$ and $\mathcal{O}'$ the respective filters of neighbourhoods at the origin of Z. Pick $U \in \mathcal{O}$ and then $U_1 \in \mathcal{O}$ such that $U_1 \pm U_1 \subset U$. As $U_1 \cap Y$ is a neighbourhood of zero in Y for the restriction of $\mathcal{T}'$, we can pick $V \in \mathcal{O}'$ such that $(V \pm V) \cap Y \subset U_1 \cap Y$. Let $\pi \colon Z \longrightarrow Z/Y$ be the natural quotient map, and take $W \in \mathcal{O}'$ such that $W \subset V$ and $\pi[W] \subset \pi[U_1 \cap V]$, that is, $W + Y \subset (U_1 \cap V) + Y$. Hence,

$$W \subset (U_1 + Y) \cap (V + Y) \subset (U_1 \cap V) + ((V - V) \cap Y) \subset U_1 + U_1 \subset U. \quad \square$$

The reader is invited to freely interpret Roelcke's lemma and uncover some of its many consequences. The following result gathers together some interesting 3-space properties:

Lemma 2.3.3 *The Hausdorff character, metrisability, local boundedness and completeness are* 3*-space properties.*

Proof Let $0 \longrightarrow Y \longrightarrow Z \xrightarrow{\pi} X \longrightarrow 0$ be a topologically exact sequence and assume $Y = \ker \pi$. In what follows, $\mathcal{O}_Y, \mathcal{O}_Z$ and $\mathcal{O}_X$ denote the filters of neighbourhoods of 0 in those spaces.
Hausdorff character Pick any non-zero $z \in Z$. If $\pi(z) \neq 0$ then there is $V \in \mathcal{O}_X$ such that $\pi(z) \notin V$ and every $U \in \mathcal{O}_Z$ such that $\pi[U] \subset V$ separates z from the origin in Z. If $\pi(z) = 0$ then z belongs to Y, and since Y is Hausdorff, there is $U \in \mathcal{O}_Z$ such that $z \notin U \cap Y$, which is enough.
Metrisability We use the Birkhoff–Kakutani theorem: a topological vector space is metrisable (by a translation-invariant metric) if and only if it is Hausdorff and there is a countable base of neighbourhoods of 0; see the argument leading to the corollary in [283, p. 5] for an elegant proof. Take countably many sets $U_n \in \mathcal{O}_Z$ such that $\{U_n \cap Y\}$ and $\{\pi[U_n]\}$ are bases for $\mathcal{O}_Y$ and $\mathcal{O}_X$, respectively. If $\mathcal{T}$ is the least linear topology on X containing the sets U_n, then $\mathcal{T}$ is metrisable, and the sequence $0 \longrightarrow Y \longrightarrow Z \longrightarrow X \longrightarrow 0$ remains topologically exact when Z carries $\mathcal{T}$. By Roelcke's lemma, $\mathcal{T}$ must be the original topology of Z.
Completeness Let $\mathcal{F}$ be a Cauchy filter on Z. Then $\pi[\mathcal{F}]$ is a Cauchy filter and converges in X. Applying a translation, if necessary, we may and do assume that $\pi[\mathcal{F}]$ converges to zero. Put $\mathcal{G} = \{F + U \colon F \in \mathcal{F}, U \in \mathcal{O}_Z\}$. This is another Cauchy filter on Z. Moreover, for each $U \in \mathcal{O}_X$, there is $F \in \mathcal{F}$ such that $\pi[F] \subset \pi[U]$, whence $F \subset Y + U$. Consequently, $\mathcal{G} \cap Y$ is a filter, hence a Cauchy filter on Y, and converges, say, to $y \in Y$. Thus, y is adherent to $\mathcal{G}$, which implies that $\mathcal{G}$ and $\mathcal{F}$ converge to y in Z.

Local boundedness Take a balanced $U \in \mathcal{O}_Z$ such that $(U + U) \cap Y$ and $\pi[U]$ are bounded. Let us verify that U is bounded. Given a balanced $W \in \mathcal{O}_Z$, there are $m, k \in \mathbb{N}$ such that $(U + U) \cap Y \subset mW$ and $\pi[U] \subset k\pi[U \cap W]$. Hence,

$$U \subset k(U + U) \cap Y + kW \subset mkW + kW \subset mk(W + W). \qquad \square$$

Since quasi-Banach = Hausdorff + locally bounded + complete, we get:

Proposition 2.3.4 *To be a quasi-Banach space is a 3-space property.*

However, to be a Banach space is not a 3-space property, as shown by Ribe's counterexample in Section 3.2. A careful study of the 3-space problem for local convexity can be found in Section 3.4. Let us close this section with a result that will be needed in due course. Let $\alpha\colon A \longrightarrow B$ be an operator acting between quasinormed spaces. If $\widehat{A}$ and $\widehat{B}$ are completions that we will consider to contain the corresponding spaces, it is obvious that there exists a unique operator $\widehat{\alpha}$ making the following diagram commute:

$$\begin{array}{ccc} A & \xrightarrow{\ \alpha\ } & B \\ {\scriptstyle \text{inclusion}}\downarrow & & \downarrow{\scriptstyle \text{inclusion}} \\ \widehat{A} & \xrightarrow{\ \widehat{\alpha}\ } & \widehat{B} \end{array}$$

Now, if $0 \longrightarrow Y \stackrel{\jmath}{\longrightarrow} Z \stackrel{\rho}{\longrightarrow} X \longrightarrow 0$ is an exact sequence of quasinormed spaces and operators, we have a commutative diagram (the vertical arrows are the corresponding inclusions)

$$\begin{array}{ccccccccc} 0 & \longrightarrow & Y & \xrightarrow{\ \jmath\ } & Z & \xrightarrow{\ \rho\ } & X & \longrightarrow & 0 \\ & & \downarrow & & \downarrow & & \downarrow & & \\ 0 & \longrightarrow & \widehat{Y} & \xrightarrow{\ \widehat{\jmath}\ } & \widehat{Z} & \xrightarrow{\ \widehat{\rho}\ } & \widehat{X} & \longrightarrow & 0 \end{array}$$

2.3.5 Completion of an exact sequence *If the upper row in the preceding diagram is topologically exact then the lower row is exact.*

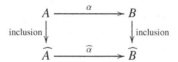

Proof Three things have to be proved: that $\widehat{\jmath}$ is injective, that $\ker\widehat{\rho} = \widehat{\jmath}[\widehat{Y}]$ and that $\widehat{\rho}$ is surjective. We may assume that $\widehat{Z}$ is a p-Banach space containing Z as a dense subspace; that Y and X carry the induced p-norms and that the p-norms of $\widehat{Y}$ and $\widehat{X}$ extend those of Y and X, respectively. Assume $\widehat{\jmath}(y) = 0$ for some $y \in \widehat{Y}$. Pick a sequence $y_n \longrightarrow y$, with $y_n \in Y$. Then $\jmath(y_n) \longrightarrow 0$ in X and so $y_n \longrightarrow 0$ in Y since $\jmath$ is an embedding. Let $\widehat{\rho}(z) = 0$, and pick a sequence $z_n \longrightarrow z$ with $z_n \in Z$. Then $\rho(z_n) \longrightarrow 0$ in X, and since ρ is open, we can pick a sequence (y_n) in Y for which $z_n - y_n \longrightarrow 0$ in Z. Since (z_n) is Cauchy, (y_n) is Cauchy in Y as well and thus convergent to some point of $\widehat{Y}$ which agrees with z. This shows that $\ker\widehat{\rho} \subset \widehat{\jmath}[\widehat{Y}]$ and that the reverse containment is trivial. To check that $\widehat{\rho}$ is surjective, pick $x \in \widehat{X}$ and write $x = \sum_{n=1}^{\infty} x_n$ with $\sum_{n=1}^{\infty} \|x_n\|^p < \infty$ and $x_n \in X$. Then choose $z_n \in Z$ such that $\rho(z_n) = x_n$ and $\|z_n\| \leq C\|x_n\|$ for some C independent on n. Then $\sum_n z_n$ converges to some $z \in \widehat{Z}$ whose image under $\widehat{\rho}$ is x. □

2.4 Categorical Constructions for Absolute Beginners

Anyone reading or even just flipping through this book will surely know what categories and functors are. And for those who do not, there are much better places than this book to learn such things, say [350; 27; 402]. So, rather than annoying anyone with definitions, let us present the rather short list of categories appearing onstage, in a speaking part, in this book.

A domestic atlas of categories

Name	Objects	Arrows (Morphisms)
A	Boolean algebras	Boolean homomorphisms
B	Banach spaces	operators
$\mathbf{B}_1$	Banach spaces	contractive operators
K	compact spaces	continuous maps
$\mathbf{K}_0$	Stone compacta	continuous maps
$p\mathbf{B}$	p-Banach spaces	operators
$p\mathbf{B}_1$	p-Banach spaces	contractive operators
Q	quasi-Banach spaces	operators
$\mathbf{Q}_1$	quasi-Banach spaces	contractive operators
sQ	semi-quasi-Banach spaces	operators
$\mathbf{s}(p\mathbf{B})$	semi-p-Banach spaces	operators
S	sets	mappings
V	vector spaces	linear maps

Kernel and Cokernel

A widespread slogan in category theory is ‘its the arrows that really matter’. Accordingly, one should define everything by means of arrows. For instance, kernel. In its categorical definition, the kernel of an arrow f is an arrow k such that $fk = 0$ and with the universal property that whenever $fg = 0$, the arrow g factorises through k. Thus, the kernel of an operator $f: A \longrightarrow B$ is the inclusion of the subspace $\ker f = \{x \in A : f(x) = 0\}$ into A. Composing on the left, one obtains the categorical definition of the cokernel of f: an arrow c such that $cf = 0$ and with the universal property that whenever $gf = 0$, the arrow g factorises through c. To identify the cokernel of an operator $f: A \longrightarrow B$, just observe that if $g: B \longrightarrow C$ is an operator in $\mathbf{Q}$ such that $gf = 0$ then g vanishes on $f[A]$, hence on $\overline{f[A]}$, and therefore it factors through the natural quotient map $\pi: B \longrightarrow B/\overline{f[A]}$. Since $\pi f = 0$, it is clear that the cokernel of 'coker f' in $\mathbf{Q}$ is precisely $\pi: B \longrightarrow B/\overline{f[A]}$. We relapse into bad habits and write 'coker f' for the space $B/\overline{f[A]}$. The notion of a cokernel is most useful for relatively open operators (equivalently, with closed range); these include embeddings, whose cokernel is the corresponding quotient, and quotient maps, whose cokernel is zero. Every operator $f: A \longrightarrow B$ factors as

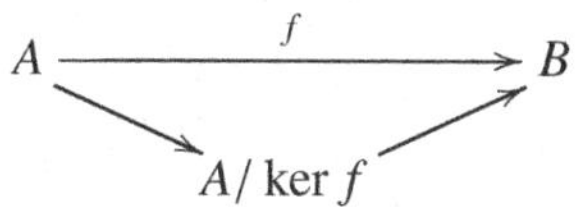

The left descending arrow is always a quotient map. The right ascending one is always injective and an embedding if and only if f has closed range, in which case, we can 'expand' it to a complete diagram with exact horizontal row

$$0 \longrightarrow \ker f \xrightarrow{\text{inclusion}} A \xrightarrow{\quad f \quad} B \xrightarrow{\text{quotient}} \operatorname{coker} f \longrightarrow 0$$
$$\searrow{\scriptstyle\text{quotient}} \quad A/\ker f \quad \nearrow{\scriptstyle\text{embedding}}$$

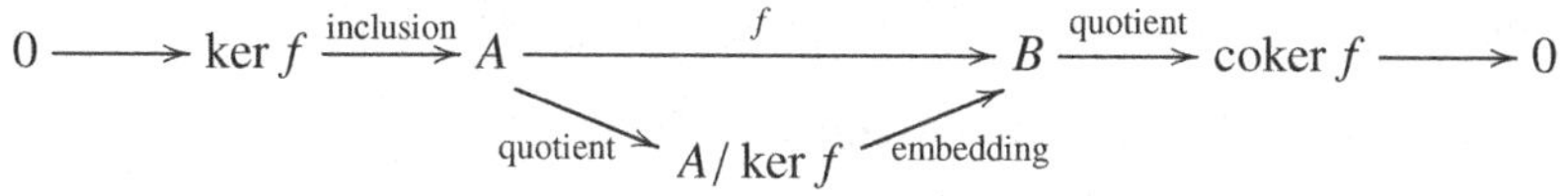

A non-closed range operator and its cokernel have a problematic relationship.

Product and Direct Sum

Elementary products of quasi-Banach spaces have already appeared in Chapter 1. Anyway, just to fix ideas, the product $A \times B$ of two quasi-Banach spaces A, B is the vector space product endowed with the $\| \cdot \|_\infty$ norm – or any equivalent quasinorm if one is prone to ignoring estimates. The product has the universal property that if $\pi_A: A \times B \longrightarrow A$ and $\pi_B: A \times B \longrightarrow B$ are the canonical projections then for every pair of operators $a: \diamond \longrightarrow A$ and $b: \diamond \longrightarrow B$, there exists a unique operator $c: \diamond \longrightarrow A \times B$ such that $a = \pi_A c$ and $b = \pi_B c$, and, moreover, such that $\|c\| \le \max\{\|a\|, \|b\|\}$ if we rightly set $\| \cdot \|_\infty$. The direct sum $A \oplus B$ of two quasi-Banach spaces A, B is (again) the vector space $A \times B$ (endowed with any equivalent quasinorm if estimates are to be ignored) and enjoys the universal property that if $\imath_A: A \longrightarrow A \times B$ and $\imath_B: B \longrightarrow A \times B$ are the canonical injections then for every pair of operators $a: A \longrightarrow \diamond$ and $b: B \longrightarrow \diamond$, there exists a unique operator $c: A \times B \longrightarrow \diamond$ such that

$a = c\iota_A$ and $b = c\iota_B$. If A, B are p-Banach spaces and $A \times B$ is given the p-norm $\|\cdot\|_p$, which we denote $A \oplus_p B$, then the estimate $\|c\| \leq \max\{\|a\|, \|b\|\}$ holds. However, no quasinorm yields that same estimate for arbitrary quasi-Banach spaces. There is a categorical way to say all this, including the caveat about estimates. The product of two objects in a category is an object $\prod$ of the category and two arrows $\bullet \longleftarrow \prod \longrightarrow \bullet$ with the universal property with respect to this diagram: for any other object $\diamond$ yielding a similar diagram $\bullet \longleftarrow \diamond \longrightarrow \bullet$, there is a unique arrow $\diamond \longrightarrow \prod$ making a commutative diagram

$$\begin{array}{ccccc} \bullet & \longleftarrow & \prod & \longrightarrow & \bullet \\ & \nwarrow & \uparrow & \nearrow & \\ & & \diamond & & \end{array}$$

The (categorically speaking) dual notion of coproduct $\coprod$ or direct sum is defined by the diagram

$$\begin{array}{ccccc} \bullet & \longrightarrow & \coprod & \longleftarrow & \bullet \\ & \searrow & \downarrow & \swarrow & \\ & & \diamond & & \end{array}$$

Summing up, this is how things are: given two objects A, B,

(a) the product in $\mathbf{Q}, p\mathbf{B}$ and in $\mathbf{B}_1, \mathbf{Q}_1$ and $p\mathbf{B}_1$ is $A \times B$,
(b) the direct sum in $\mathbf{Q}$ and $p\mathbf{B}$ is $A \oplus B$,
(c) the direct sum in $p\mathbf{B}_1$ is $A \oplus_p B$,
(d) no direct sum exists in $\mathbf{Q}_1$.

Moving beyond, a *universal construction* is a compressed way of speaking about a correspondence that assigns to a certain family of (quasi-) Banach spaces another (quasi-) Banach space in a canonical way (not a mere witticism). The categorical term for universal construction is *limit*, with its prefix *colimit* to isolate the corresponding construction obtained reversing arrows. Note 2.15.1 contains a We are Groot presentation of categorical limits.

2.5 Pullback and Pushout

Let us now consider the diagrams

$$\begin{array}{ccc} \cdot & \longrightarrow & \cdot \\ \downarrow & & \\ \cdot & & \end{array} \qquad\qquad \begin{array}{ccc} \cdot & \longrightarrow & \cdot \\ & & \uparrow \\ & & \cdot \end{array}$$

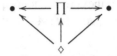

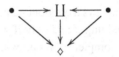

The pushout is the limit of the first (two arrows with the same domain), which means that given a diagram in the category of (quasi-) Banach spaces

$$Y \xrightarrow{\alpha} A$$
$$\beta \downarrow$$
$$B$$

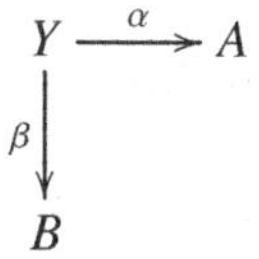

the pushout is a (quasi-) Banach space PO and two operators $\overline{\alpha} : B \longrightarrow$ PO and $\overline{\beta} : A \longrightarrow$ PO making the diagram

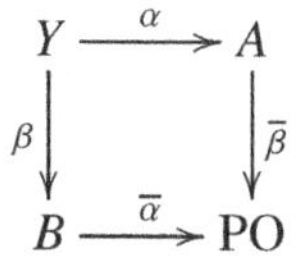

$$\begin{array}{ccc} Y & \xrightarrow{\alpha} & A \\ \beta \downarrow & & \downarrow \overline{\beta} \\ B & \xrightarrow{\overline{\alpha}} & \text{PO} \end{array} \qquad (2.12)$$

commute and with the universal property that given any other (quasi-) Banach space C and operators $\beta' : A \longrightarrow C$ and $\alpha' : B \longrightarrow C$ such that $\beta'\alpha = \alpha'\beta$, there is a unique operator $\gamma : \text{PO} \longrightarrow C$ such that $\alpha' = \gamma\overline{\alpha}$ and $\beta' = \gamma\overline{\beta}$, i.e. making the following diagram commute:

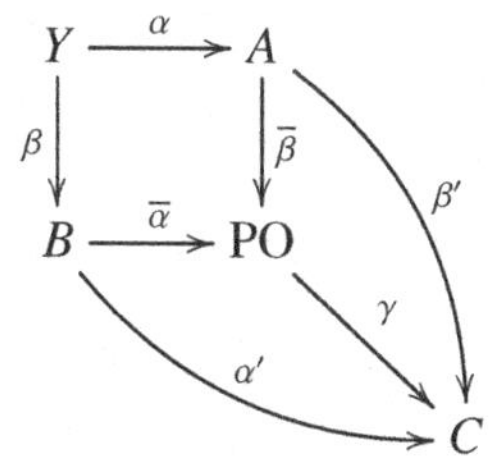

$$(2.13)$$

Pushouts exist in (quasi-) Banach spaces: if $\Delta = \{(\alpha y, -\beta y) : y \in Y\}$ then

$$\text{PO} = \text{PO}(\alpha, \beta) = (A \oplus B)/\overline{\Delta}.$$

The map $\overline{\alpha}$ is the composition of the inclusion of B into $A \oplus B$ and the natural quotient map $A \oplus B \longrightarrow (A \oplus B)/\overline{\Delta}$, such that $\overline{\alpha}(b) = (0, b) + \overline{\Delta}$ and, analogously, $\overline{\beta}(a) = (a, 0) + \overline{\Delta}$. All this makes a commutative diagram: $\overline{\beta}\alpha = \overline{\alpha}\beta$. Moreover, if $\beta' : A \longrightarrow C$ and $\alpha' : B \longrightarrow C$ are operators such that $\beta'\alpha = \alpha'\beta$ then there is a unique operator $\gamma : \text{PO} \longrightarrow C$ given by $\gamma((a, b) + \overline{\Delta}) = \beta'(a) + \alpha'(b)$ such that $\alpha' = \gamma\overline{\alpha}$ and $\beta' = \gamma\overline{\beta}$. Pushouts are unique, up to isomorphisms in the ambient category. It is simple to check that

2.5.1 *If α is an embedding, then Δ is closed and $\overline{\alpha}$ is an embedding.*

Indeed, the operator $(\alpha, -\beta)\colon Y \longrightarrow A \oplus B$ is an embedding and its range $(\alpha, -\beta)[Y] = \Delta$ is closed. The second part is trivial. Assume $A \oplus B$ carries the sum quasinorm. Then since Δ is closed, letting $c = \min\left((\|\beta\|\,\|\alpha^{-1}\|)^{-1}, 1\right)$,

$$\|\overline{\alpha}(b)\| = \|(0,b) + \Delta\| = \inf_{y\in Y} \|(0,b) + (\alpha(y), -\beta(y))\| = \inf_{y\in Y} \|(\alpha(y), b - \beta(y))\|$$

$$= \inf_{y\in Y} \|\alpha(y)\| + \|b - \beta(y)\| \geq c \inf_{y\in Y} \|\beta(y)\| + \|b - \beta(y)\| \geq \frac{c}{\Delta_B}\|b\|, \quad (2.14)$$

where Δ_B is the modulus of concavity of B. If we now move to the category of p-Banach spaces and perform the pushout via the right direct sum $A \oplus_p B$ then the maps in Diagram (2.12) enjoy additional metric properties:

Lemma 2.5.2

(a) $\|\gamma\| \leq \max\left(\|\alpha'\|, \|\beta'\|\right)$.
(b) $\max\left(\|\overline{\alpha}\|, \|\overline{\beta}\|\right) \leq 1$.
(c) *If α is an isometry and $\|\beta\| \leq 1$ then $\overline{\alpha}$ is an isometry.*
(d) *If $\|\beta\| \leq 1$ and α is an isomorphism then $\overline{\alpha}$ is an isomorphism and*

$$\|(\overline{\alpha})^{-1}\| \leq \max\{1, \|\alpha^{-1}\|\}.$$

Proof (a) is a direct consequence of the p-Banach structure involved:

$$\|\gamma((a,b) + \overline{\Delta})\|^p = \|\beta' a + \alpha' b\|^p \leq \|\beta' a\|^p + \|\alpha' b\|^p \leq \max\left(\|\beta'\|, \|\alpha'\|\right)^p \|(a,b)\|_p^p.$$

(b) is clear. To prove (c), keep in mind that Δ is closed. If $\|\beta\| \leq 1$ then

$$\|\overline{\alpha}(b)\|^p = \|(0,b) + \Delta\|^p = \inf_{y\in Y} \|\alpha y\|^p + \|b - \beta y\|^p \geq \inf_{y\in Y} \|\beta y\|^p + \|b - \beta y\|^p \geq \|b\|^p.$$

To prove (d), we first check that $\overline{\alpha}$ is onto. Pick $(a,b) \in A \oplus_p B$. Take $y \in Y$ such that $a = \alpha(y)$ and then set $b' = b + \beta(y)$. Clearly, $\overline{\alpha}(b') = (0,b') + \Delta = (a,b) + \Delta$ since $(a,b) - (0,b') = (\alpha(y), -\beta(y))$. To get a lower bound for $\|\overline{\alpha}(b)\|$, we can use the string of inequalities (2.14) and the p-subadditivity of the quasinorms to obtain

$$\|\overline{\alpha}(b)\| \geq \min\left(1, \frac{1}{\|\beta\|\|\alpha^{-1}\|}\right)\|b\| \geq \min\left(1, \frac{1}{\|\alpha^{-1}\|}\right)\|b\|,$$

which is exactly what the estimate in (d) says. □

The pullback is the colimit of the other diagram at the beginning of this section

(two arrows with the same codomain), which means that the pullback of a diagram of (quasi-) Banach spaces

$$\begin{array}{ccc} A & \xrightarrow{\alpha} & X \\ & & \uparrow{\scriptstyle \beta} \\ & & B \end{array}$$

is a (quasi-) Banach space PB and operators $\underline{\alpha}$: PB $\longrightarrow B$ and $\underline{\beta}$: PB $\longrightarrow A$, making a commutative diagram

$$\begin{array}{ccc} A & \xrightarrow{\alpha} & X \\ {\scriptstyle \underline{\beta}}\uparrow & & \uparrow{\scriptstyle \beta} \\ \mathrm{PB} & \xrightarrow{\underline{\alpha}} & B \end{array} \tag{2.15}$$

and with the universal property that given any other (quasi-) Banach space C and operators $\alpha' : C \longrightarrow B$ and $\beta' : C \longrightarrow A$ such that $\alpha\beta' = \alpha'\beta$, there is a unique operator $\gamma: C \longrightarrow$ PB such that $\beta' = \underline{\beta}\gamma$ and $\alpha' = \underline{\alpha}\gamma$; i.e. making a commutative diagram

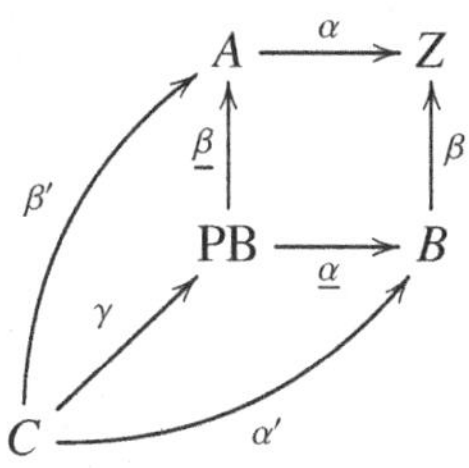

Pullbacks exist in (quasi-) Banach spaces:

$$\mathrm{PB} = \mathrm{PB}(\alpha, \beta) = \{(a, b) \in A \oplus_\infty B : \alpha(a) = \beta(b)\},$$

with operators $\underline{\alpha}(a, b) = b$ and $\underline{\beta}(a, b) = a$. All properties are immediate, $\gamma(c) = (\beta'(c), \alpha'(c))$, and we have the additional estimate $\|\gamma\| \leq \max(\|\alpha'\|, \|\beta'\|)$ for free. It is simple to check that when α is onto, $\underline{\alpha}$ is onto. Do it.

2.6 Pushout and Exact Sequences

Suppose we are given an exact sequence $0 \longrightarrow Y \xrightarrow{\jmath} Z \xrightarrow{\rho} X \longrightarrow 0$ and an operator $\tau: Y \longrightarrow B$. Consider the pushout of the pair $(\jmath, \tau)$ and draw the corresponding arrows:

$$\begin{CD} 0 @>>> Y @>\jmath>> Z @>\rho>> X @>>> 0\\ @. @V{\tau}VV @VV{\overline{\tau}}V \\ @. B @>{\overline{\jmath}}>> \mathrm{PO} \end{CD}$$

By Lemma 2.5.2(a), $\overline{\jmath}$ is an embedding. Now, the quotient operator ρ and the null operator $0\colon B \longrightarrow X$ satisfy $\rho\jmath = 0\tau = 0$, and thus the universal property of the pushout gives a unique operator $\overline{\rho}\colon \mathrm{PO} \longrightarrow X$, making a commutative diagram:

$$\begin{CD} 0 @>>> Y @>\jmath>> Z @>\rho>> X @>>> 0\\ @. @V{\tau}VV @VV{\overline{\tau}}V @|\\ 0 @>>> B @>{\overline{\jmath}}>> \mathrm{PO} @>{\overline{\rho}}>> X @>>> 0 \end{CD} \tag{2.16}$$

To make it explicit, $\overline{\rho}((x,b) + \Delta) = \rho(x)$. It is easy to check that the lower sequence in the preceding diagram is exact since $\ker\overline{\rho} = \{(x,b) + \Delta\colon \rho(x) = 0\} = \{(y,b)+\Delta\colon y \in Y\} = \{(0,b)+\Delta\colon b \in B\} = \overline{\jmath}\,[B]$. As $\jmath$ is an embedding, the operator $\overline{\jmath}$ is injective, and $\overline{\rho}$ is surjective since $\overline{\rho\tau} = \rho$, so the lower row in 2.16 is a short exact sequence, from now on referred to as the *pushout sequence*. Actually, the universal property of the pushout makes Diagram (2.16) work as the definition of pushout:

2.6.1 *Given a commutative diagram with exact rows*

$$\begin{CD} 0 @>>> Y @>\jmath>> Z @>\rho>> X @>>> 0\\ @. @V{\tau}VV @VV{T}V @|\\ 0 @>>> B @>{\imath}>> Z' @>{\pi}>> X @>>> 0 \end{CD} \tag{2.17}$$

the lower sequence is equivalent to the pushout sequence (2.16).

Proof Indeed, the universal property of the pushout implies that since $\imath\tau = T\jmath$, there must be an operator $\gamma\colon \mathrm{PO} \longrightarrow Z$ such that $\imath = \gamma\overline{\jmath}$ and $T = \gamma\overline{\tau}$; this makes the left bottom square commutative in the diagram

$$\begin{CD} 0 @>>> Y @>\jmath>> Z @>\rho>> X @>>> 0\\ @. @V{\tau}VV @VV{\overline{\tau}}V @|\\ 0 @>>> B @>{\overline{\jmath}}>> \mathrm{PO} @>{\overline{\rho}}>> X @>>> 0\\ @. @| @VV{\gamma}V @|\\ 0 @>>> B @>{\imath}>> Z' @>{\pi}>> X @>>> 0 \end{CD}$$

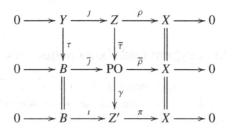

To check the commutativity on the right, observe that $\pi\gamma\overline{\tau} = \pi T = \rho$ and $\pi\gamma\overline{\jmath} = 0$. Since the arrow $\overline{\rho}$ appearing in the definition of *pushout* is unique, $\pi\gamma = \overline{\rho}$. □

For this reason, we usually refer to a diagram like (2.17) as a pushout diagram. It is implicit in the preceding argument that

2.6.2 *Making pushout preserves the equivalence of extensions. Precisely, if* z *and* z' *are exact sequences,* τ *is an operator and* τ z *denotes the pushout sequence, then* $[\mathrm{z}] = [\mathrm{z}'] \implies [\tau\,\mathrm{z}] = [\tau\,\mathrm{z}']$.

The following result is the key piece that connects the pushout construction with operator extension properties:

Lemma 2.6.3 *The lower sequence in a pushout diagram*

$$\begin{array}{ccccccccc} 0 & \longrightarrow & Y & \xrightarrow{\jmath} & Z & \xrightarrow{\rho} & X & \longrightarrow & 0 \\ & & \downarrow{\scriptstyle\tau} & & \downarrow{\scriptstyle\overline{\tau}} & & \| & & \\ 0 & \longrightarrow & B & \xrightarrow{\overline{\jmath}} & Z' & \xrightarrow{\overline{\rho}} & X & \longrightarrow & 0 \end{array} \tag{2.18}$$

splits if and only if there is an operator $T\colon Z \longrightarrow B$ *such that* $T\jmath = \tau$.

Proof One implication is trivial: if the lower sequence splits, composing $\overline{\tau}$ with any left inverse of $\overline{\jmath}$, one obtains the required 'extension' of τ. We provide two (two?) proofs for the converse. First, assume that the left square is the straight pushout diagram and $Z' = \mathrm{PO}$: if $T\colon Z \longrightarrow B$ satisfies $\tau = T\jmath$ then by applying the universal property of PO to the operators $T, \mathbf{1}_B$, we obtain $\gamma\colon Z' \longrightarrow B$ such that $\gamma\overline{\jmath} = \mathbf{1}_B$, so the lower row splits. The general case follows from this in view of 2.6.1. But even ignoring these facts, the proof is easy: if an extension operator T exists, there is an operator $s\colon X \longrightarrow Z'$ such that $\overline{\tau} - \overline{\jmath}T = s\rho$ since $(\overline{\tau} - \overline{\jmath}T)\jmath = 0$. This s is a linear continuous selection for $\overline{\rho}$ since $\overline{\rho}s\rho = \overline{\rho}\overline{\tau} - \overline{\rho}\overline{\jmath}\,T = \rho$, hence $\overline{\rho}s = \mathbf{1}_X$. □

Diagram (2.16) describes the pushout space PO as an enlargement of B whose main characteristic is that PO enlarges B in the same way that X enlarges Y, i.e. $\mathrm{PO}/B = Z/Y$. From the point of view of the operators, PO provides an enlargement of B that enables us to extend $\tau\colon Y \longrightarrow B$ to the whole of Z. The next question to ponder is then: where does one encounter pushouts? Or worse: are there pushout sequences at all? Yes, everywhere! Indeed, all exact sequences 'are' pushout sequences. That is the content of the next section.

2.7 Projective Presentations: the Universal Property of ℓ_p

The spaces $\ell_p(I)$ enjoy, for $0 < p \leq 1$, two very special properties among all p-Banach spaces. One of them is that they are *projective*, which means that whenever $\pi\colon Z \longrightarrow X$ is a quotient map between two p-Banach spaces, every operator $\tau\colon \ell_p(I) \longrightarrow X$ can be lifted to Z, that is, there exists an operator $T\colon \ell_p(I) \longrightarrow Z$ such that $\pi T = \tau$. In other words, there is a commutative diagram

$$\begin{array}{ccc} Z & \xrightarrow{\ \pi\ } & X \\ & {}^{T}\nwarrow & \uparrow{\scriptstyle\tau} \\ & & \ell_p(I) \end{array}$$

To obtain T, it is enough to pick a bounded family of elements z_i such that $\pi(z_i) = \tau(e_i)$ and define $T(e_i) = z_i$. This property admits the following reformulation: every exact sequence of p-Banach spaces

$$0 \longrightarrow Y \longrightarrow Z \longrightarrow \ell_p(I) \longrightarrow 0$$

splits; indeed, every quotient map $Z \longrightarrow \ell_p(I)$ admits a linear continuous section, namely any lifting of the identity. By Proposition 1.2.3, one has:

2.7.1 *$\ell_p(I)$ are the only projective p-Banach spaces,* $0 < p \leq 1$.

The other is that every p-Banach space X is a quotient of some $\ell_p(\alpha)$. A quotient map $\pi\colon \ell_p(\alpha) \longrightarrow X$ can be defined explicitly as follows: take $(x_i)_{i\in\alpha}$ a set of size α that is dense in the unit ball of X and set $\pi((\lambda_i)_{i\in\alpha}) = \sum_i \lambda_i x_i$. It is very easy to see that π is an operator onto X; the associated exact sequence $0 \longrightarrow \ker\pi \longrightarrow \ell_p(\alpha) \xrightarrow{\pi} X \longrightarrow$ is called a projective presentation of X in the category of p-Banach spaces. Whether a space is projective depends upon the category we consider it in: the space ℓ_1 is projective in the category of Banach spaces but not in the category of p-Banach spaces $0 < p < 1$ because there is a quotient map $\ell_p \longrightarrow \ell_1$ for which no linear continuous section is possible since ℓ_1 is not a subspace of ℓ_p. In general, an exact sequence $0 \longrightarrow \kappa \longrightarrow \mathcal{P} \longrightarrow X \longrightarrow 0$ in a category in which the object $\mathcal{P}$ is projective is called a *projective presentation* of X. When every object of a category admits a projective presentation, we sometimes say that the category has *enough projectives*. Therefore, the category of p-Banach spaces has enough projectives.

2.7.2 *Every exact sequence of p-Banach spaces* $0 \longrightarrow Y \xrightarrow{\jmath} Z \xrightarrow{\rho} X \longrightarrow 0$ *is (equivalent to) a pushout of any projective presentation of* X.

Indeed, if $\pi\colon \ell_p(I) \longrightarrow X$ is a projective presentation then π can be lifted to Z, so let $T\colon \ell_p(I) \longrightarrow Z$ be an operator such that $\rho T = \pi$. The restriction of T to $\ker\pi$ takes values in $\jmath[Y]$ since $\pi T(k) = \pi(k) = 0$ for $k \in \ker\pi$. Hence, if $\tau\colon \ker\pi \longrightarrow Y$ is given by $\tau(k) = \jmath^{-1}T(k)$, we have a commutative diagram

$$\begin{array}{ccccccccc} 0 & \longrightarrow & \ker\pi & \longrightarrow & \ell_p(I) & \xrightarrow{\pi} & X & \longrightarrow & 0 \\ & & \downarrow{\scriptstyle\tau} & & \downarrow{\scriptstyle T} & & \| & & \\ 0 & \longrightarrow & Y & \xrightarrow{\jmath} & Z & \xrightarrow{\rho} & X & \longrightarrow & 0 \end{array} \tag{2.19}$$

which, according to 2.6.1, is a pushout diagram. Therefore, exact sequences $0 \longrightarrow Y \longrightarrow Z \longrightarrow X \longrightarrow 0$ and operators $\ker\pi \longrightarrow Y$ are, roughly speaking, equivalent objects. Simple examples show that projective presentations are not unique, not even isomorphic: if X is a separable p-Banach space and J is uncountable, the two sequences

$$\begin{array}{ccccccccc} 0 & \longrightarrow & \ker\pi & \longrightarrow & \ell_p & \xrightarrow{\pi} & X & \longrightarrow & 0 \\ & & & & & & \| & & \\ 0 & \longrightarrow & \ell_p(J)\times\ker\pi & \longrightarrow & \ell_p(J)\times\ell_p & \xrightarrow{0\oplus\pi} & X & \longrightarrow & 0 \end{array}$$

define non-isomorphic projective presentations of X. Classical Banach space theory already noticed the phenomenon that all projective presentations of a separable Banach space are 'essentially the same' and provided the following ad hoc explanation: if X is a separable Banach space not isomorphic to ℓ_1, the kernels of any two quotient maps from ℓ_1 to X are isomorphic. But that is just a part of the picture:

Proposition 2.7.3 *Let $\pi\colon \ell_p(I) \longrightarrow X$ and $\pi'\colon \ell_p(J) \longrightarrow X$ be two quotient maps. Then there is a commutative diagram*

$$\begin{array}{ccccccccc} 0 & \longrightarrow & \ker\pi\times\ell_p(J) & \longrightarrow & \ell_p(I)\times\ell_p(J) & \xrightarrow{\pi\oplus 0} & X & \longrightarrow & 0 \\ & & {\scriptstyle\alpha}\downarrow & & {\scriptstyle\beta}\downarrow & & \| & & \\ 0 & \longrightarrow & \ell_p(I)\times\ker\pi' & \longrightarrow & \ell_p(I)\times\ell_p(J) & \xrightarrow{0\oplus\pi'} & X & \longrightarrow & 0 \end{array}$$

in which α and β are isomorphisms. In particular, $\ker\pi\times\ell_p(J) \simeq \ker\pi'\times\ell_p(I)$, *and the rows are isomorphic sequences.*

Proof Consider the quotient operator $Q\colon \ell_p(I) \times \ell_p(J) \longrightarrow X$ given by $Q(x,y) = \pi x - \pi' y$ whose kernel is $\{(x,y)\colon \pi x = \pi' y\}$. The map $\ker Q \longrightarrow \ell_p(J)$ given by $(x,y) \longmapsto y$ is surjective, and thus it admits a linear bounded section $\ell_p(J) \longrightarrow \ker Q$ given by $y \longmapsto (sy, y)$. We define an isomorphism $u\colon \ker\pi \times \ell_p(J) \longrightarrow \ker Q$ as $u(x,y) = (x + sy, y)$. It is well defined since $\pi(x + sy) = \pi sy = \pi' y$. It is obviously injective since $(x + sy, y) = (0,0)$ implies

$(x, y) = (0, 0)$. And it is surjective since $(x, y) = u(x - sy, y)$, and if $(x, y) \in \ker Q$ then $\pi x = \pi' y$ and thus $x - sy \in \ker \pi$ since $\pi(x - sy) = \pi x - \pi' y = 0$. Analogously, there is an isomorphism $v : \ell_p(I) \times \ker \pi' \longrightarrow \ker Q$ given by $(a, b) \longmapsto (a, s'a + b)$, where $x \longmapsto (x, s'x)$ is a linear continuous section for the map $\ker Q \longrightarrow \ell_p(J)$ given by $(x, y) \longmapsto x$. Then, define $\alpha = v^{-1}u$. The map $\beta(x, y) = (x + sy, y - s'(x + sy))$ is an automorphism of $\ell_p(I) \times \ell_p(J)$ whose inverse is $(a, b) \longmapsto (a - s(s'a + b), s'a + b)$. □

What you have just seen is an example of the use of the diagonal principles we will present in Section 2.11. From Proposition 2.7.3 we derive the result from Banach space folklore mentioned earlier. For this reason, we will use the notation $\kappa_p(X)$ to denote the kernel of any quotient map $\ell_p(I) \longrightarrow X$.

Corollary 2.7.4 *Let X be a separable p-Banach space, and let π, π' be two quotient maps $\ell_p \longrightarrow X$. Then $\ker \pi \times \ell_p$ and $\ker \pi' \times \ell_p$ are isomorphic. If $p = 1$ and X is not isomorphic to ℓ_1, then $\ker \pi$ and $\ker \pi'$ are isomorphic.*

Proof The first part is contained in the preceding proposition. Infinite-dimensional subspaces of ℓ_1 contain complemented copies of ℓ_1, thus we have $\ker \pi \simeq \ell_1 \times A \simeq \ell_1 \times \ell_1 \times A \simeq \ell_1 \times \ker \pi$; analogously, $\ker \pi' \simeq \ell_1 \times \ker \pi'$. □

This raises the apparently open question of whether subspaces of ℓ_p contain complemented copies of ℓ_p (the different issue of whether a subspace of ℓ_p contains a copy of ℓ_p complemented *in* ℓ_p is treated in [443; 444; 397; 21; 261]). We conclude the section by connecting projective presentations of the subspace and the quotient space in an exact sequence.

Lemma 2.7.5 *Given an exact sequence $0 \longrightarrow Y \xrightarrow{\jmath} Z \xrightarrow{\rho} X \longrightarrow 0$ of p-Banach spaces and projective presentations of Y and X, there exists a projective presentation of Z forming a commutative diagram*

$$\begin{array}{ccccccccc}
 & & 0 & & 0 & & 0 & & \\
 & & \downarrow & & \downarrow & & \downarrow & & \\
0 & \longrightarrow & \ker \pi_Y & \longrightarrow & \ker \pi_Z & \longrightarrow & \ker \pi_X & \longrightarrow & 0 \\
 & & \downarrow & & \downarrow & & \downarrow & & \\
0 & \longrightarrow & \mathcal{P}_Y & \longrightarrow & \mathcal{P}_Z & \longrightarrow & \mathcal{P}_X & \longrightarrow & 0 \\
 & & \downarrow{\scriptstyle \pi_Y} & & \downarrow{\scriptstyle \pi_Z} & & \downarrow{\scriptstyle \pi_X} & & \\
0 & \longrightarrow & Y & \xrightarrow{\jmath} & Z & \xrightarrow{\rho} & X & \longrightarrow & 0 \\
 & & \downarrow & & \downarrow & & \downarrow & & \\
 & & 0 & & 0 & & 0 & &
\end{array}$$

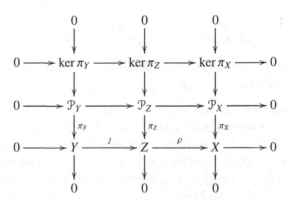

Proof Let $Q\colon \mathcal{P}_X \longrightarrow Z$ be a lifting of π_X, set $\mathcal{P}_Z = \mathcal{P}_Y \times \mathcal{P}_X$ and define $\pi_Z\colon \mathcal{P}_Z \longrightarrow Z$ by $\pi_Z(a,b) = \jmath\pi_Y(a) + Q(b)$. □

2.8 Pullbacks and Exact Sequences

Consider an exact sequence $0 \longrightarrow Y \stackrel{\jmath}{\longrightarrow} Z \stackrel{\rho}{\longrightarrow} X \longrightarrow 0$ and an operator $\tau\colon A \longrightarrow X$. Let us form the pullback diagram

$$\begin{array}{ccccccccc} 0 & \longrightarrow & Y & \stackrel{\jmath}{\longrightarrow} & Z & \stackrel{\rho}{\longrightarrow} & X & \longrightarrow & 0 \\ & & & & \uparrow{\scriptstyle \underline{\tau}} & & \uparrow{\scriptstyle \tau} & & \\ & & & & \mathrm{PB} & \stackrel{\underline{\rho}}{\longrightarrow} & A & & \end{array}$$

Recalling that $\underline{\rho}$ is onto and setting $\underline{\jmath}(y) = (0, \jmath(y))$, it is easily seen that the following diagram is commutative:

$$\begin{array}{ccccccccc} 0 & \longrightarrow & Y & \stackrel{\jmath}{\longrightarrow} & Z & \stackrel{\rho}{\longrightarrow} & X & \longrightarrow & 0 \\ & & \| & & \uparrow{\scriptstyle \underline{\tau}} & & \uparrow{\scriptstyle \tau} & & \\ 0 & \longrightarrow & Y & \stackrel{\underline{\jmath}}{\longrightarrow} & \mathrm{PB} & \stackrel{\underline{\rho}}{\longrightarrow} & A & \longrightarrow & 0 \end{array} \tag{2.20}$$

The lower sequence is exact and shall be referred to as the *pullback sequence*. Actually, the universal property of the pullback makes this diagram work as the definition of the pullback, which is why we will usually refer to the diagram in 2.8.1 as a pullback diagram:

2.8.1 *Given a commutative diagram*

$$\begin{array}{ccccccccc} 0 & \longrightarrow & Y & \longrightarrow & Z & \longrightarrow & X & \longrightarrow & 0 \\ & & \| & & \uparrow & & \uparrow & & \\ 0 & \longrightarrow & Y & \longrightarrow & Z' & \longrightarrow & A & \longrightarrow & 0 \end{array} \tag{2.21}$$

the lower exact sequence is equivalent to the pullback sequence (2.20).

The proof is entirely dual of that of 2.6.1 and, thus, it is implicit that

2.8.2 *Taking pullbacks preserves the equivalence of extensions. Precisely, if* z *and* z' *are exact sequences,* τ *is an operator and* $\mathsf{z}\,\tau$ *denotes the pullback sequence, then* $[\mathsf{z}] = [\mathsf{z}'] \Longrightarrow [\mathsf{z}\,\tau] = [\mathsf{z}'\,\tau]$.

The following result is the key piece that connects pullback properties with operator lifting properties.

Lemma 2.8.3 *The lower sequence in the pullback diagram* (2.20) *splits if and only if there is an operator* $T: A \longrightarrow Z$ *such that* $\rho T = \tau$.

Proof If $s: A \longrightarrow \mathrm{PB}$ is a section of $\underline{\rho}$, then $T = \underline{\tau} s$ is a lifting of τ. And if $T: A \longrightarrow Z$ is a lifting of τ, then $s(a) = (T(a), a)$ is a section of $\underline{\rho}$. □

Where does one encounter pullback diagrams? Are there pullback sequences at all? Yes: everywhere! Indeed, all exact sequences *are* pullback sequences, and that is the content of the next section.

2.9 Injective Presentations: the Universal Property of ℓ_∞

This section is dual, almost word-by-word, to Section 2.7, except for 2.9.1. Apart from that, if one goes to Section 2.7, fixes $p = 1$, changes 'quotient' to 'embedding', reverses arrows in diagrams, etc.... one gets this section. We stress the correspondence by reproducing the presentation as closely as possible. The spaces $\ell_\infty(I)$ enjoy two very special properties among all Banach spaces. First, they are *injective*: a p-Banach space X is said to be injective if all operators $A \longrightarrow X$ can be extended to any p-Banach superspace. If norm one operators can be extended to norm λ operators then X is called λ-injective.

2.9.1 *If* $0 < p < 1$, *the only injective space in* $\mathbf{pB}$ *is* 0.

Wow, that's a surprise, right? The proof follows from the corollary in Note 1.8.3: injective spaces Y in $\mathbf{pB}$ must obviously be ultrasummands, but if some non-zero $y \in Y$ exists then the operator $c \in \mathbb{K} \longmapsto cy \in Y$ cannot extend to L_p. Thus, this section is about Banach spaces only.

The space $\ell_\infty(I)$ is 1-injective among Banach spaces, but there are many others. Most of what is known about injective Banach spaces has been collected in [22]. Injectivity admits the formulation: every exact sequence of Banach spaces $0 \longrightarrow \ell_\infty(I) \longrightarrow \cdot \longrightarrow \cdot \longrightarrow 0$ splits; think of an extension of the identity of $\ell_\infty(I)$. Moreover, if I is a dense set of a Banach space Y then Y is isometric to a closed subspace of $\ell_\infty(I)$. An embedding $\jmath: Y \longrightarrow \ell_\infty(I)$ can be defined explicitly as follows: take for each $i \in I$ a norm one functional i^* such that $\langle i^*, i\rangle = \|i\|$ and set $\jmath(y) = \langle i^*, y\rangle_{i\in I}$. The associated exact sequence

$$0 \longrightarrow Y \longrightarrow \ell_\infty(I) \longrightarrow \ell_\infty(I)/Y \longrightarrow 0$$

or, more generally, any exact sequence $0 \longrightarrow Y \longrightarrow \mathcal{I} \longrightarrow \mathcal{I}/Y \longrightarrow 0$ in which $\mathcal{I}$ is an injective Banach space, is called an *injective presentation* of Y. The cokernel space $\mathcal{I}/Y$ is sometimes denoted $c\kappa(Y)$. A category in which every object admits an injective presentation is said to have *enough injectives*. Since the spaces $\ell_\infty(I)$ are injective, the category of Banach spaces admits enough injectives. A category may have injective objects, but not enough: the category of separable Banach spaces has injective objects (c_0, thanks to Sobczyk's theorem), but they are not enough by Zippin's theorem (c_0 is the only separable separably injective space). Exact sequences $0 \longrightarrow Y \longrightarrow Z \longrightarrow X \longrightarrow 0$ and injective presentations of Y can be connected to form commutative diagrams

$$\begin{array}{ccccccccc} 0 & \longrightarrow & Y & \longrightarrow & \ell_\infty(I) & \longrightarrow & \ell_\infty(I)/Y & \longrightarrow & 0 \\ & & \| & & \uparrow & & \uparrow & & \\ 0 & \longrightarrow & Y & \longrightarrow & Z & \longrightarrow & X & \longrightarrow & 0 \end{array} \tag{2.22}$$

In this way, exact sequences $0 \longrightarrow Y \longrightarrow Z \longrightarrow X \longrightarrow 0$ and operators $X \longrightarrow \ell_\infty(I)/Y$ are, roughly speaking, equivalent objects. There are many non-(isomorphically) equivalent injective presentations, but the result dual to Proposition 2.7.3 works and can be proved cleanly either with the Diagonal principle 2.11.7 or through an ad hoc dual rewriting of the proof of Proposition 2.7.3. However we go, we get

Proposition 2.9.2 *Let $\imath\colon Y \longrightarrow \ell_\infty(I)$ and $\jmath\colon Y \longrightarrow \ell_\infty(J)$ be two emdeddings. Then there is a commutative diagram*

$$\begin{array}{ccccccccc} 0 & \longrightarrow & Y & \longrightarrow & \ell_\infty(I)\times\ell_\infty(J) & \longrightarrow & (\ell_\infty(I)/Y)\times\ell_\infty(J) & \longrightarrow & 0 \\ & & \| & & \uparrow{\scriptstyle\beta} & & \uparrow{\scriptstyle\gamma} & & \\ 0 & \longrightarrow & Y & \longrightarrow & \ell_\infty(I)\times\ell_\infty(J) & \longrightarrow & \ell_\infty(I)\times(\ell_\infty(J)/Y) & \longrightarrow & 0 \end{array}$$

in which β and γ are isomorphisms. In particular, $(\ell_\infty(I)/Y)\times\ell_\infty(J) \simeq \ell_\infty(I)\times(\ell_\infty(J)/Y)$, and the rows are isomorphic sequences.

Corollary 2.9.3 *If $\imath$ and $\jmath$ are embeddings of a separable space Y into ℓ_∞, then $\ell_\infty/\imath[Y]$ and $\ell_\infty/\jmath[Y]$ are isomorphic.*

Proof The quotient $\ell_\infty/\imath[Y]$ must contain ℓ_∞ by Rosenthal's property (V) and therefore $\ell_\infty/\imath[Y] \simeq \ell_\infty/\imath[Y]\times\ell_\infty$. Analogously, $\ell_\infty/\jmath[Y] \simeq \ell_\infty/\jmath[Y]\times\ell_\infty$. □

The result applies to all subspaces Y of ℓ_∞ such that ℓ_∞/Y contains ℓ_∞. See 7.2.2 for further developments. Injective presentations of the subspace and the quotient of an exact sequence can be connected:

Lemma 2.9.4 *Given an exact sequence $0 \longrightarrow Y \stackrel{\jmath}{\longrightarrow} Z \stackrel{\rho}{\longrightarrow} X \longrightarrow 0$ of Banach spaces and injective presentations of Y and X, there exists an injective presentation of Z forming a commutative diagram*

$$\begin{array}{ccccccccc} & & 0 & & 0 & & 0 & & \\ & & \downarrow & & \downarrow & & \downarrow & & \\ 0 & \longrightarrow & Y & \xrightarrow{\jmath} & Z & \xrightarrow{\rho} & X & \longrightarrow & 0 \\ & & \downarrow{\scriptstyle \imath_Y} & & \downarrow{\scriptstyle \imath_Z} & & \downarrow{\scriptstyle \imath_X} & & \\ 0 & \longrightarrow & \mathfrak{J}_Y & \longrightarrow & \mathfrak{J}_Z & \longrightarrow & \mathfrak{J}_X & \longrightarrow & 0 \\ & & \downarrow & & \downarrow & & \downarrow & & \\ 0 & \longrightarrow & \mathfrak{J}_Y/Y & \longrightarrow & \mathfrak{J}_Z/Z & \longrightarrow & \mathfrak{J}_X/X & \longrightarrow & 0 \\ & & \downarrow & & \downarrow & & \downarrow & & \\ & & 0 & & 0 & & 0 & & \end{array}$$

Proof Indeed, let $I \colon Z \longrightarrow \mathfrak{J}_Y$ be an extension of $\imath_Y$, set $\mathfrak{J}_Z = \mathfrak{J}_Y \times \mathfrak{J}_X$ and define $\imath_Z \colon Z \longrightarrow \mathfrak{J}_Z$ as $\imath_Z(x) = I(x) + \imath_X(\rho x)$. □

2.10 All about That Pullback/Pushout Diagram

Once we are aware of their existence, there are a few essential things to know about pullback/pushouts.

First thing: how to recognise them. Imagine that reading and writing diagrams is reading and writing Japanese. The two basic ideograms to learn are *pullback* and *pushout*: we must learn to recognise that

$$\begin{array}{ccccccccc} 0 & \longrightarrow & \cdot & \longrightarrow & \cdot & \longrightarrow & \cdot & \longrightarrow & 0 \\ & & \| & & \uparrow & & \uparrow & & \\ 0 & \longrightarrow & \cdot & \longrightarrow & \cdot & \longrightarrow & \cdot & \longrightarrow & 0 \end{array}$$

is *always* a pullback diagram (Section 2.8), while

$$\begin{array}{ccccccccc} 0 & \longrightarrow & \cdot & \longrightarrow & \cdot & \longrightarrow & \cdot & \longrightarrow & 0 \\ & & \downarrow & & \downarrow & & \| & & \\ 0 & \longrightarrow & \cdot & \longrightarrow & \cdot & \longrightarrow & \cdot & \longrightarrow & 0 \end{array}$$

is *always* a pushout diagram (Section 2.6). This pictorial recognition of pullback/pushouts makes evident that taking the pushout along α and *then* along α' is the same as taking pushout along $\alpha'\alpha$ and that taking pullback along γ and *then* along γ' is the same as taking pullback along $\gamma\gamma'$.

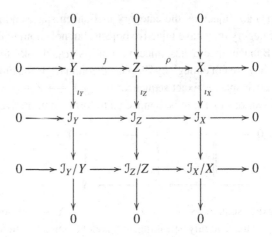

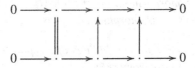

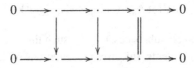

And taking a pushout along α and then a pullback along γ? Keep reading.

Second thing: they commute. Namely, given an exact sequence $0 \longrightarrow Y \longrightarrow Z \longrightarrow X \longrightarrow 0$ and two operators $\alpha\colon Y \longrightarrow Y'$ and $\gamma\colon X' \longrightarrow X$, first taking pushout along α and then the pullback along γ,

$$\begin{array}{ccccccccc}
0 & \longrightarrow & Y & \xrightarrow{\jmath} & Z & \xrightarrow{\rho} & X & \longrightarrow & 0\\
 & & \downarrow{\scriptstyle \alpha} & & \downarrow & & \| & & \\
0 & \longrightarrow & Y' & \xrightarrow{\jmath_o} & \mathrm{PO} & \xrightarrow{\rho_o} & X & \longrightarrow & 0\\
 & & \| & & \downarrow{\scriptstyle \underline{\gamma}} & & \uparrow{\scriptstyle \gamma} & & \\
0 & \longrightarrow & Y' & \xrightarrow{\jmath_{bo}} & \mathrm{PB(PO)} & \xrightarrow{\rho_{bo}} & X' & \longrightarrow & 0
\end{array} \tag{2.23}$$

or first taking the pullback along γ and then the pushout along α,

$$\begin{array}{ccccccccc}
0 & \longrightarrow & Y & \xrightarrow{\jmath} & Z & \xrightarrow{\rho} & X & \longrightarrow & 0\\
 & & \| & & \uparrow & & \uparrow{\scriptstyle \gamma} & & \\
0 & \longrightarrow & Y & \xrightarrow{\jmath_b} & \mathrm{PB} & \xrightarrow{\rho_b} & X' & \longrightarrow & 0\\
 & & \downarrow{\scriptstyle \alpha} & & \downarrow & & \| & & \\
0 & \longrightarrow & Y' & \xrightarrow{\jmath_{ob}} & \mathrm{PO(PB)} & \xrightarrow{\rho_{ob}} & X' & \longrightarrow & 0
\end{array} \tag{2.24}$$

produces equivalent sequences. Indeed, that the final resulting sequences are equivalent means that there is an operator $T\colon \mathrm{PO(PB)} \longrightarrow \mathrm{PB(PO)}$ making the following diagram commute:

$$\begin{array}{ccccccccc}
0 & \longrightarrow & Y' & \xrightarrow{\jmath_{ob}} & \mathrm{PO(PB)} & \xrightarrow{\rho_{ob}} & X' & \longrightarrow & 0\\
 & & \| & & \downarrow{\scriptstyle T} & & \| & & \\
0 & \longrightarrow & Y' & \xrightarrow{\jmath_{bo}} & \mathrm{PB(PO)} & \xrightarrow{\rho_{bo}} & X' & \longrightarrow & 0
\end{array} \tag{2.25}$$

There are two ways to get such a map: one relying on the fact that PO(PB) is a pushout and the other relying on the fact that PB(PO) is a pullback. We show the second case. Thus, consider the pullback square

$$\begin{array}{ccc}
\mathrm{PO} & \xrightarrow{\rho_o} & X\\
{\scriptstyle \underline{\gamma}}\uparrow & & \uparrow{\scriptstyle \gamma}\\
\mathrm{PB(PO)} & \xrightarrow{\rho_{bo}} & X'
\end{array} \tag{2.26}$$

Let us form another commutative square;

$$\begin{array}{ccc} \mathrm{PO} & \xrightarrow{\rho_{\mathrm{o}}} & X \\ {\scriptstyle\delta}\big\uparrow & & \big\uparrow{\scriptstyle\gamma} \\ \mathrm{PO(PB)} & \xrightarrow{\rho_{\mathrm{bo}}} & X' \end{array} \tag{2.27}$$

in which the arrow δ is obtained from the universal property of the pushout square

$$\begin{array}{ccc} Y & \xrightarrow{\jmath_{\mathrm{b}}} & \mathrm{PB} \\ {\scriptstyle\alpha}\big\downarrow & & \big\downarrow{\scriptstyle\underline{\alpha}} \\ Y' & \xrightarrow{\jmath_{\mathrm{ob}}} & \mathrm{PO(PB)} \end{array}$$

in combination with the fact that the square obtained by juxtaposition of the upper left squares of Diagrams 2.24 and 2.25, namely

$$\begin{array}{ccc} Y & \xrightarrow{\jmath_{\mathrm{b}}} & \mathrm{PB} \\ {\scriptstyle\alpha}\big\downarrow & & \big\downarrow{\scriptstyle\overline{\alpha}\,\overline{\gamma}} \\ Y' & \xrightarrow{\jmath_{\mathrm{o}}} & \mathrm{PO} \end{array}$$

is also commutative. Thus, there is a unique operator $\delta\colon \mathrm{PO(PB)} \longrightarrow \mathrm{PO}$ such that $\delta\underline{\alpha} = \overline{\alpha}\,\overline{\gamma}$ and $\jmath_{\mathrm{ob}} = \jmath_{\mathrm{o}}$.

Finally, the commutativity of Diagram (2.27) and the universal property of (2.26) immediately yield the existence of an operator $T\colon \mathrm{PO(PB)} \longrightarrow \mathrm{PB(PO)}$ such that $\rho_{\mathrm{bo}}T = \rho_{\mathrm{ob}}$ and $\underline{\gamma}T = \delta$. The first of those equalities is the commutativity of the right square in Diagram (2.25). Let us prove the commutativity of the left square, i.e. $T\jmath_{\mathrm{ob}} = \jmath_{\mathrm{bo}}$: since $\rho_{\mathrm{bo}}T = \rho_{\mathrm{bo}}$, it is clear that $\rho_{\mathrm{bo}}T\jmath_{\mathrm{ob}} = \rho_{\mathrm{ob}}\jmath_{\mathrm{ob}} = 0$, and therefore some operator $u\colon Y' \longrightarrow Y'$ must exist such that $T\jmath_{\mathrm{ob}} = \jmath_{\mathrm{bo}}u$. But since $\jmath_{\mathrm{o}}u = \underline{\gamma}\jmath_{\mathrm{bo}}u = \underline{\gamma}T\jmath_{\mathrm{ob}} = \delta\jmath_{\mathrm{ob}} = \jmath_{\mathrm{o}}$, it follows that u is the identity on Y', and this concludes the proof.

Third thing: they mix. They mix in a single diagram – this one:

$$\begin{array}{ccccccccc} 0 & \longrightarrow & Y & \xrightarrow{\jmath} & Z & \xrightarrow{\rho} & X & \longrightarrow & 0 \\ & & {\scriptstyle\alpha}\big\downarrow & & \big\downarrow{\scriptstyle\beta} & & \big\downarrow{\scriptstyle\gamma} & & \\ 0 & \longrightarrow & Y' & \xrightarrow{\jmath'} & Z' & \xrightarrow{\rho'} & X' & \longrightarrow & 0 \end{array} \tag{2.28}$$

since it can be artfully decomposed as

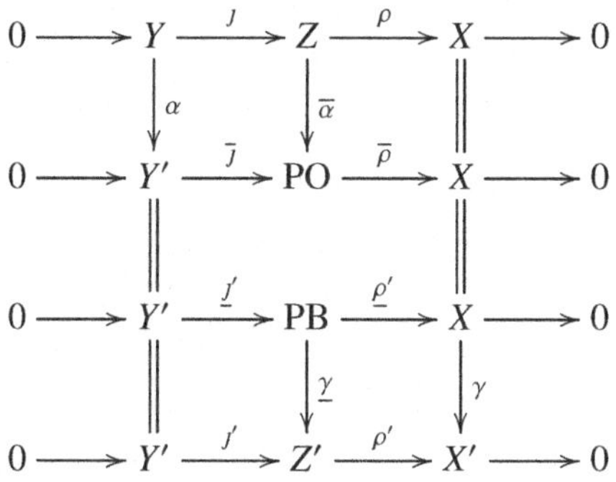

$$\begin{array}{ccccccccc}
0 & \longrightarrow & Y & \xrightarrow{\jmath} & Z & \xrightarrow{\rho} & X & \longrightarrow & 0 \\
 & & \downarrow{\scriptstyle\alpha} & & \downarrow{\scriptstyle\overline{\alpha}} & & \| & & \\
0 & \longrightarrow & Y' & \xrightarrow{\overline{\jmath}} & \mathrm{PO} & \xrightarrow{\overline{\rho}} & X & \longrightarrow & 0 \\
 & & \| & & & & \| & & \\
0 & \longrightarrow & Y' & \xrightarrow{\underline{\jmath}'} & \mathrm{PB} & \xrightarrow{\underline{\rho}'} & X & \longrightarrow & 0 \\
 & & \| & & \downarrow{\scriptstyle\underline{\gamma}} & & \downarrow{\scriptstyle\gamma} & & \\
0 & \longrightarrow & Y' & \xrightarrow{\jmath'} & Z' & \xrightarrow{\rho'} & X' & \longrightarrow & 0
\end{array} \tag{2.29}$$

where the two middle sequences are equivalent, as we show now (the argument is quite similar to the one used earlier). The commutative diagram

$$\begin{array}{ccc}
Z & \xrightarrow{\rho} & X \\
\downarrow{\scriptstyle\beta} & & \downarrow{\scriptstyle\gamma} \\
Z' & \xrightarrow{\rho'} & X'
\end{array}$$

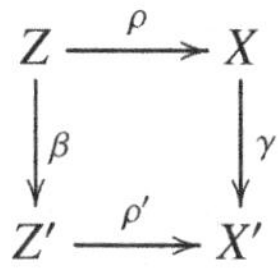

yields an operator $\tau\colon Z \longrightarrow \mathrm{PB}$ forming a commutative diagram

$$\begin{array}{cccc}
Z & & & \\
 & \searrow^{\tau} & & \\
 & & \mathrm{PB} & \xrightarrow{\underline{\rho}_1} & X \\
 & & \downarrow{\scriptstyle\underline{\gamma}} & & \downarrow{\scriptstyle\gamma} \\
 & & Z' & \xrightarrow{\rho'} & X'
\end{array}
\qquad (Z \xrightarrow{\rho} X,\ Z \xrightarrow{\beta} Z')$$

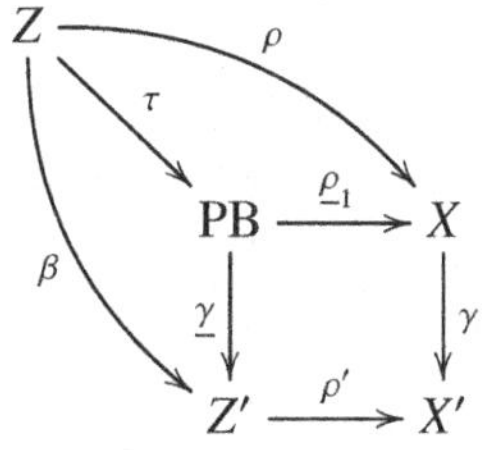

which therefore means that it also makes a commutative square

$$\begin{array}{ccc}
Y & \xrightarrow{\jmath} & Z \\
\downarrow{\scriptstyle\alpha} & & \downarrow{\scriptstyle\tau} \\
Y' & \xrightarrow{\underline{\jmath}'} & \mathrm{PB}
\end{array}$$

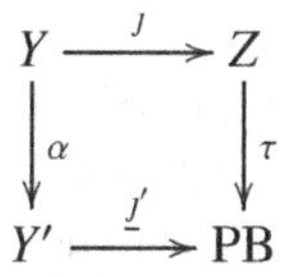

and in turn there is an operator $T : \mathrm{PO} \longrightarrow \mathrm{PB}$ forming a commutative diagram

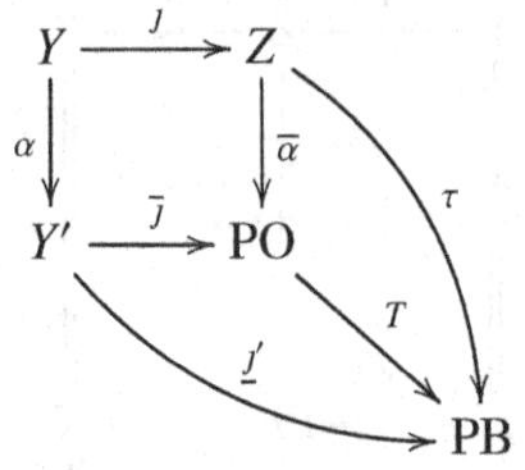

which, inserted correctly in (2.29), yields the commutative diagram

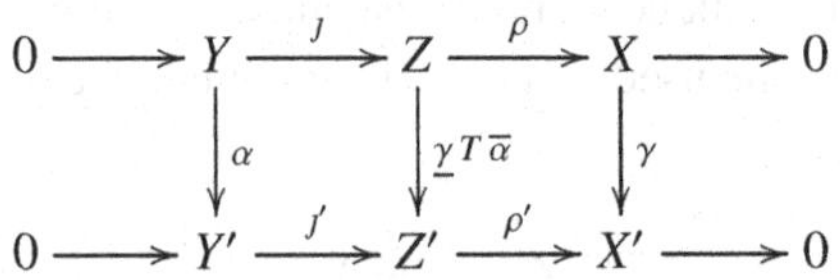

Only one task remains (for the sceptics): check that $\beta = \underline{\gamma} T \overline{\alpha}$. Really? Yes, $\underline{\gamma} T \overline{\alpha} = \underline{\gamma} \tau = \beta$.

Perhaps the best example one can give of Diagram (2.29) comes from considering an exact sequence of Banach spaces and combining it with projective and injective presentations of the quotient and subspace, respectively, namely

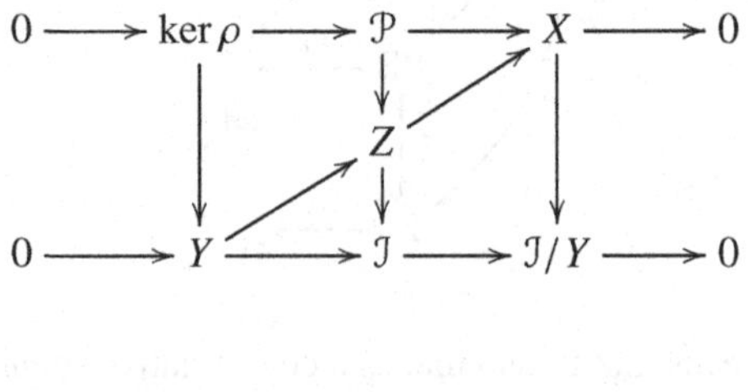

(2.30)

in which it is clear that the middle diagonal sequence is both a pushout of the upper sequence and a pullback of the lower one.

Fourth thing: they can be completed. It is time to show the value of completing diagrams. The natural context in which diagrams can be completed is when the involved operators are either quotient maps or embeddings, which are the two main cases of closed-range operators. Start with a pullback Diagram (2.20) whose upwards operator is a quotient map to obtain

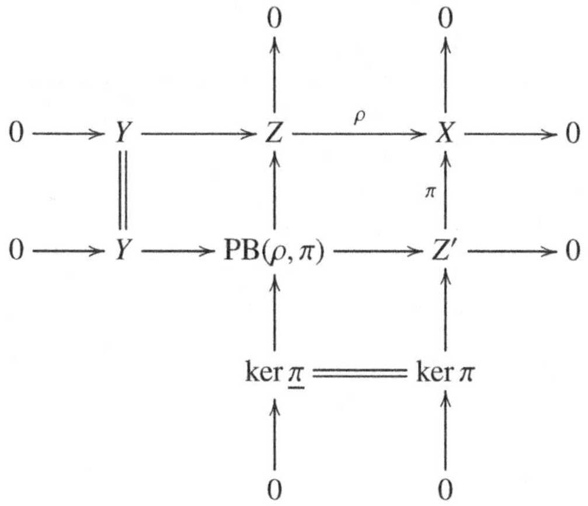

(2.31)

When the operator is an embedding, the completed diagram is

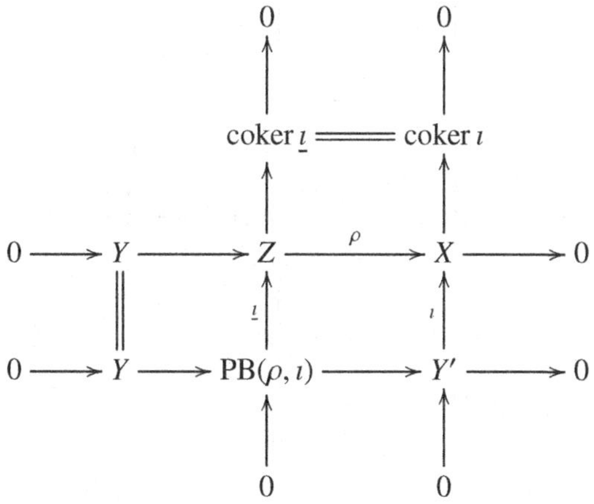

(2.32)

in which $\mathrm{PB}(\rho, \imath)$ is naturally isomorphic to $\rho^{-1}[\imath[Y']]$. The completion of a pushout Diagram (2.16) when the left downwards operator is a quotient map yields

$$
\begin{array}{ccccccccc}
 & & 0 & & 0 & & & & \\
 & & \downarrow & & \downarrow & & & & \\
 & & \ker\pi & = & \ker\bar{\pi} & & & & \\
 & & \downarrow & & \downarrow & & & & \\
0 & \longrightarrow & Y & \xrightarrow{\jmath} & Z & \longrightarrow & X & \longrightarrow & 0 \\
 & & \downarrow{\scriptstyle \pi} & & \downarrow{\scriptstyle \bar{\pi}} & & \| & & \\
0 & \longrightarrow & X' & \longrightarrow & \mathrm{PO}(\pi, \jmath) & \longrightarrow & X & \longrightarrow & 0 \\
 & & \downarrow & & \downarrow & & & & \\
 & & 0 & & 0 & & & &
\end{array}
\tag{2.33}
$$

and when it is an embedding, we get

$$
\begin{array}{ccccccccc}
 & & 0 & & 0 & & & & \\
 & & \downarrow & & \downarrow & & & & \\
0 & \longrightarrow & Y & \xrightarrow{\jmath} & Z & \longrightarrow & X & \longrightarrow & 0 \\
 & & \downarrow{\scriptstyle \imath} & & \downarrow{\scriptstyle \bar{\imath}} & & \| & & \\
0 & \longrightarrow & Z' & \longrightarrow & \mathrm{PO}(\imath, \jmath) & \longrightarrow & X & \longrightarrow & 0 \\
 & & \downarrow & & \downarrow & & & & \\
 & & \operatorname{coker}\imath & = & \operatorname{coker}\bar{\imath} & & & & \\
 & & \downarrow & & \downarrow & & & & \\
 & & 0 & & 0 & & & &
\end{array}
\tag{2.34}
$$

This makes a total of $2 + 2 = 3$ (!) diagrams, since (2.32) and (2.33) are exactly the same … when rotated. *This* diagram could then be written as

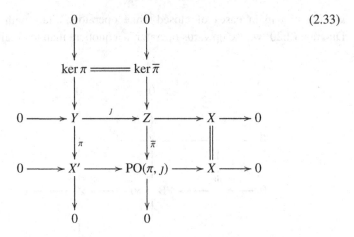

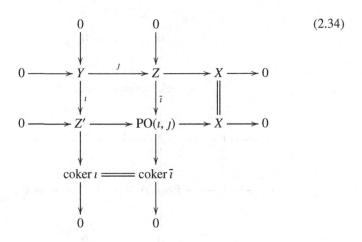

$$\begin{array}{ccccccccc} & & 0 & & 0 & & & & \\ & & \downarrow & & \downarrow & & & & \\ & & \ker\pi & = & \ker\pi & & & & \\ & & \downarrow & & \downarrow & & & & \\ 0 & \longrightarrow & \mathrm{PB}(\rho,\imath) & \xrightarrow{\jmath} & Z & \longrightarrow & X & \longrightarrow & 0 \\ & & \downarrow{\scriptstyle\pi} & & \downarrow{\scriptstyle\rho} & & \| & & \\ 0 & \longrightarrow & Y & \xrightarrow{\imath} & \mathrm{PO}(\pi,\jmath) & \longrightarrow & X & \longrightarrow & 0 \\ & & \downarrow & & \downarrow & & & & \\ & & 0 & & 0 & & & & \end{array} \tag{2.35}$$

Summing the situation up, the four fundamental diagrams

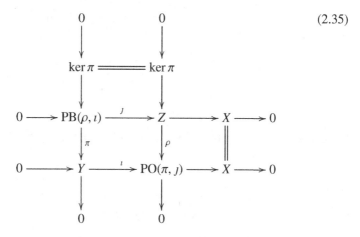

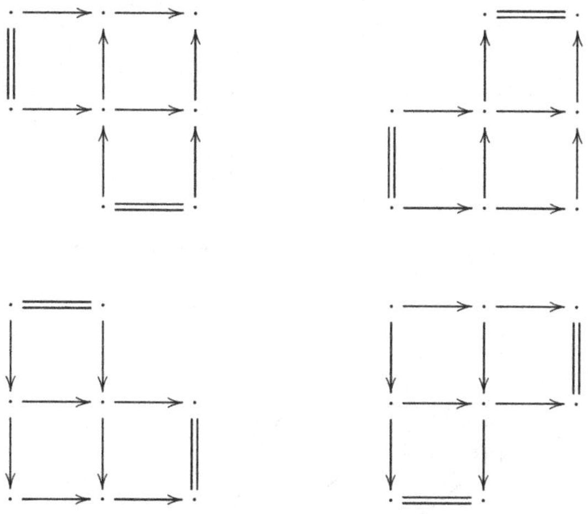

are actually three, since those in the positions $(0,0)$ and $(1,1)$ are the same.

Diagonals

Concealed in the diagrams are other 'diagonal' exact sequences. While at first glance they may seem to be the oompa loompas of homology, they are in fact essential both for the understanding of pairs of exact sequences and for the construction of counterexamples. Since the reader should start to think in terms of equivalence classes, we will from now on write $[z]$ instead of z whenever z

can be replaced by any other equivalent sequence to obtain equivalent results. Thus, given a pullback diagram

$$\begin{array}{ccccccccc} 0 & \longrightarrow & Y & \xrightarrow{\jmath} & Z & \xrightarrow{\rho} & X & \longrightarrow & 0 \\ & & \| & & \uparrow{\scriptstyle \underline{\tau}} & & \uparrow{\scriptstyle \tau} & & \\ 0 & \longrightarrow & Y & \xrightarrow{\imath} & \mathrm{PB} & \xrightarrow{\underline{\rho}} & X' & \longrightarrow & 0 \end{array} \qquad [\mathrm{z}]$$

the very definition of pullback space generates a *diagonal pullback sequence*:

$$0 \longrightarrow \mathrm{PB} \longrightarrow Z \times X' \xrightarrow{\rho \oplus (-\tau)} X \longrightarrow 0$$

(the unnamed arrow is plain inclusion).

2.10.1 *The diagonal pullback sequence is the pushout sequence* $[\imath\,\mathrm{z}]$.

The truth of the assertion is witnessed by the diagram

$$\begin{array}{ccccccccc} 0 & \longrightarrow & Y & \xrightarrow{\jmath} & Z & \xrightarrow{\rho} & X & \longrightarrow & 0 \\ & & \downarrow{\scriptstyle \imath} & & \downarrow{\scriptstyle \bar{\imath}} & & \| & & \\ 0 & \longrightarrow & \mathrm{PB} & \xrightarrow{\bar{\jmath}} & \mathrm{PO} & \longrightarrow & X & \longrightarrow & 0 \\ & & \| & & \downarrow{\scriptstyle w} & & \| & & \\ 0 & \longrightarrow & \mathrm{PB} & \longrightarrow & Z \times X' & \xrightarrow{\rho \oplus (-\tau)} & X & \longrightarrow & 0 \end{array} \qquad \begin{array}{c} [\mathrm{z}] \\ \\ [\imath\,\mathrm{z}] \\ \\ \\ \end{array}$$

in which the arrow w has been obtained from the universal property of the pushout

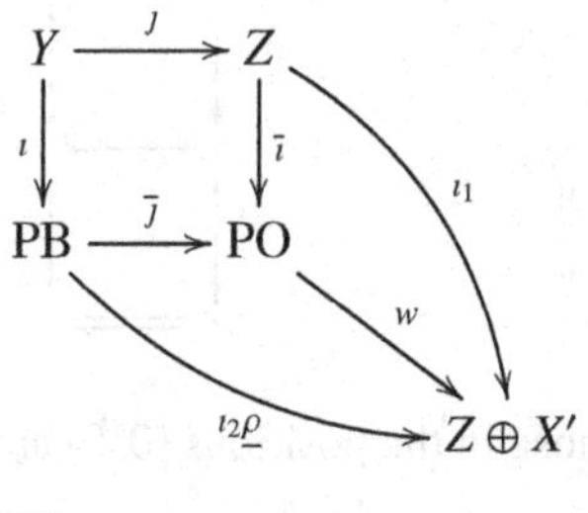

Given a pushout diagram

$$\begin{array}{ccccccccc} 0 & \longrightarrow & Y & \xrightarrow{\jmath} & Z & \xrightarrow{\rho} & X & \longrightarrow & 0 \\ & & \downarrow{\scriptstyle \tau} & & \downarrow{\scriptstyle \overline{\tau}} & & \| & & \\ 0 & \longrightarrow & Y' & \xrightarrow{\bar{\jmath}} & \mathrm{PO} & \xrightarrow{\bar{\rho}} & X & \longrightarrow & 0 \end{array}$$

the very nature of the pushout space yields a *diagonal pushout sequence*,

$$0 \longrightarrow Y \xrightarrow{(\tau, \jmath)} Y' \oplus Z \xrightarrow{\bar{\jmath} \oplus (-\overline{\tau})} \mathrm{PO} \longrightarrow 0$$

2.10.2 *The diagonal pushout sequence is the pullback sequence* $[\mathsf{z}\,\overline{\rho}]$.

That can be seen dualising what was done in the pullback case. We hope the reader will have fun with it. Summing up again, the four (three) pullback/pushout diagrams contain in them the diagonals

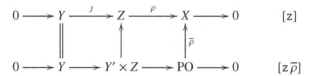

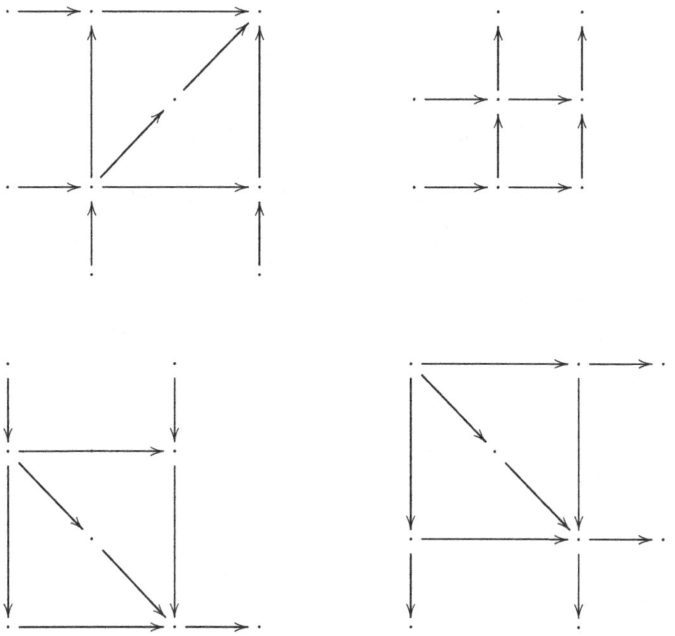

When do these diagonal sequences split? Good question. Observe that no matter where they are placed, those diagonals are plain pullbacks/pushouts: indeed, the pullback diagonal $0 \longrightarrow \mathrm{PB} \longrightarrow Z \times X' \longrightarrow X \longrightarrow 0$ is the pushout sequence $[\imath\,\mathsf{z}]$ (see 2.10.1) and the diagonal pushout sequence is the lower sequence in the pullback diagram (see your own diagram)

$$\begin{array}{ccccccccccc} 0 & \longrightarrow & Y & \xrightarrow{\jmath} & Z & \xrightarrow{\rho} & X & \longrightarrow & 0 & \qquad & [\mathsf{z}] \\ & & \| & & \uparrow & & \uparrow{\scriptstyle \overline{\rho}} & & & & \\ 0 & \longrightarrow & Y & \longrightarrow & Y' \times Z & \longrightarrow & \mathrm{PO} & \longrightarrow & 0 & \qquad & [\mathsf{z}\,\overline{\rho}] \end{array}$$

2.10.3 *The diagonal pullback sequence splits if and only if* $\imath\colon Y \longrightarrow \mathrm{PB}$ *can be extended to* Z. *The diagonal pushout sequence splits if and only if* $\overline{\rho}\colon \mathrm{PO} \longrightarrow X$ *can be lifted to* Z.

Ok, this is, admittedly, better than nothing, but not terribly informative for the reader hungry for something more substantial. But wait: the splitting of the diagonal pullback sequence obviously implies that $\mathrm{PB} \times X \simeq Z \times X'$, while the splitting of the diagonal pushout sequence implies $Y \times \mathrm{PO} \simeq Y' \times Z$. And a little bit of *this* (when pullback/pushout sequences split) in combination with some additional little bit of *that* (the pullback/pushout spaces are isomorphic to some product) leads to the main course served up next: diagonal principles.

2.11 Diagonal and Parallel Principles

We have, in fact, already encountered diagonal and parallel principles earlier. The sceptical reader is invited to look again to Propositions 2.7.3 and 2.9.2. Thus our aim here is to give (homological) shape to a real-life phenomenon and thus bring understanding. The *diagonal* and *parallel* principles we present now govern the behaviour of pairs of exact sequences and work in categories where pullback, pushout, finite products and exact sequences exist:

- The parallel principles are concerned with the following problem: assume we have an exact sequence $[\mathrm{z}]$ that is pullback (resp. pushout) of another $[\mathrm{z}']$; when can we conclude that $[\mathrm{z}']$ is also a pullback (resp. pushout) of $[\mathrm{z}]$?
- The diagonal principles are concerned with what occurs if we have success in the previous situation: if one of the exact sequences $[\mathrm{z}]$ and $[\mathrm{z}']$ is a pullback (resp. pushout) of the other then ... what?

The starting point is to get a criterion to detect when an exact sequence $[\mathrm{z}']$ is a pushout (or pullback) of another sequence $[\mathrm{z}]$.

Proposition 2.11.1 *Given two exact sequences*

$$\begin{array}{ccccccccccc} 0 & \longrightarrow & Y & \xrightarrow{\jmath} & Z & \xrightarrow{\rho} & X & \longrightarrow & 0 & \qquad & (\mathrm{z}) \\ & & & & & & \| & & & & \\ 0 & \longrightarrow & Y' & \xrightarrow{\jmath'} & Z' & \xrightarrow{\rho'} & X & \longrightarrow & 0 & \qquad & (\mathrm{z}') \end{array}$$

$[\mathrm{z}']$ *is a pushout of* $[\mathrm{z}]$ *if and only if* $[\mathrm{z}'\rho] = 0$.

Proof The necessity is clear: since $\mathbf{1}_Z$ is a lifting of $\rho\colon Z \longrightarrow X$, we have $[\mathrm{z}\rho] = 0$ and, therefore, $[\mathrm{z}'] = [\tau\,\mathrm{z}] \implies [\mathrm{z}'\rho] = [(\tau \mathrm{z})\rho)] = [\tau(\mathrm{z}\rho)] = 0$, by the commutativity of pullbacks and pushouts. The sufficiency is clear too:

$[\mathrm{z}'\rho] = 0$ means that ρ can be lifted to a map $L \in \mathfrak{L}(Z, Z')$, which yields a commutative diagram

$$\begin{CD} 0 @>>> Y @>\jmath>> Z @>\rho>> X @>>> 0\\ @. @VVV @VV L V @| \\ 0 @>>> Y' @>\jmath'>> Z' @>\rho'>> X @>>> 0 \end{CD}$$

because $\rho' L\jmath = \rho\jmath = 0$ means that $L\jmath$ takes values in $\jmath'[Y'] = \ker \rho'$. □

Definition 2.11.2 Two exact sequences

$$\begin{CD} 0 @>>> Y @>\jmath>> Z @>\rho>> X @>>> 0 \qquad (\mathrm{z})\\ @. @. @. @| \\ 0 @>>> Y' @>\jmath'>> Z' @>\rho'>> X @>>> 0 \qquad (\mathrm{z}') \end{CD}$$

will be called semi-equivalent (*on the quotient end*, if one needs to specify) if each is a pushout of the other.

It is clear now that z and z′ are semi-equivalent if and only if $[\mathrm{z}\rho'] = 0$ and $[\mathrm{z}'\rho] = 0$. The pullback version is analogous:

Proposition 2.11.3 *Given two exact sequences*

$$\begin{CD} 0 @>>> Y @>\jmath>> Z @>\rho>> X @>>> 0 \qquad (\mathrm{z})\\ @. @| @. @. \\ 0 @>>> Y' @>\jmath'>> Z' @>\rho'>> X' @>>> 0 \qquad (\mathrm{z}') \end{CD}$$

$[\mathrm{z}']$ *is a pullback of* $[\mathrm{z}]$ *if and only if* $[\jmath\,\mathrm{z}'] = 0$.

Proof The necessity is clear: since $\mathbf{1}_Z$ is an extension of $\jmath$, we have $[\jmath\,\mathrm{z}] = 0$, and thus $[\mathrm{z}'] = [\mathrm{z}\,\tau] \Longrightarrow [\jmath\,\mathrm{z}'] = [(\jmath\,\mathrm{z})\tau] = 0$. The sufficiency is clear too: $[\jmath\,\mathrm{z}'] = 0$ means that $\jmath$ can be extended to an operator $J \colon Z' \longrightarrow Z$, which yields a commutative diagram

$$\begin{CD} 0 @>>> Y @>\jmath>> Z @>\rho>> X @>>> 0\\ @. @| @AA J A @AAA \\ 0 @>>> Y' @>\jmath'>> Z' @>\rho'>> X' @>>> 0 \end{CD}$$

because $\rho J\jmath' = \rho\jmath = 0$ forces ρJ to factorise through ρ'. □

Definition 2.11.4 Two exact sequences

$$\begin{CD} 0 @>>> Y @>\jmath>> Z @>\rho>> X @>>> 0 \qquad (\mathrm{z})\\ @. @| @. @. \\ 0 @>>> Y' @>\jmath'>> Z' @>\rho'>> X' @>>> 0 \qquad (\mathrm{z}') \end{CD}$$

will be called semi-equivalent (*on the subspace end*, if necessary) if one is a pullback of the other, and vice versa.

The sequences (z) and (z′) are thus semi-equivalent if and only if $[\jmath\, z'] = 0$ and $[\jmath'\, z] = 0$. In what follows, we will call two exact sequences semi-equivalent if they are semi-equivalent on the appropriate end. Only when necessary will we specify the end. With these tools in hand, we get

2.11.5 Parallel lines principle *In any of the three diagrams*

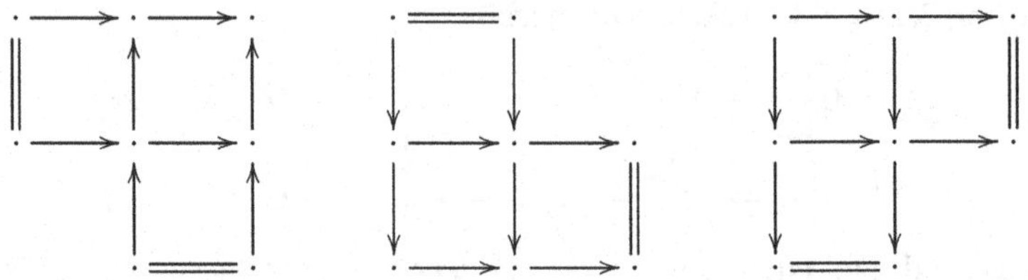

the two vertical sequences are semi-equivalent if and only if the two horizontal sequences are semi-equivalent.

Proof In all cases, the announced semi-equivalence corresponds to the splitting of the diagonal sequence. □

Of course, one can give standard proofs for each of these results, though each is over-long, gnarled and different to the others. That's why we prefer the homological approach. We now present the diagonal principles, furthering our understanding of semi-equivalence. Some more names will be of great help here. Given an exact sequence

$$0 \longrightarrow Y \xrightarrow{\ \jmath\ } Z \xrightarrow{\ \rho\ } X \longrightarrow 0 \qquad \text{(z)}$$

and a quasi-Banach space E, $E \times \mathrm{z}$ will denote the sequence obtained by multiplying on the left by E:

$$0 \longrightarrow E \times Y \xrightarrow{1_E \times \jmath} E \times Z \xrightarrow{0 \oplus \rho} X \longrightarrow 0$$

The sequence obtained multiplying by E on the right will be denoted $\mathrm{z} \times E$:

$$0 \longrightarrow Y \xrightarrow{(\jmath, 0)} Z \times E \xrightarrow{\rho \times 1_E} X \times E \longrightarrow 0$$

2.11.6 Diagonal principle: projective case *If the sequences*

$$\begin{array}{ccccccccc} 0 & \longrightarrow & Y & \xrightarrow{\ \jmath\ } & Z & \xrightarrow{\ \rho\ } & X & \longrightarrow & 0 & \qquad \text{(z)} \\ & & & & & & \| & & & \\ 0 & \longrightarrow & Y' & \xrightarrow{\ \jmath'\ } & Z' & \xrightarrow{\ \rho'\ } & X & \longrightarrow & 0 & \qquad \text{(z}'\text{)} \end{array}$$

are semi-equivalent then the sequences $Z' \times \mathrm{z}$ *and* $Z \times \mathrm{z}'$ *are isomorphic.*

Proof The hypothesis means that there are operators α, β such that $[\alpha\, \mathrm{z}] = [\mathrm{z}']$ and $[\beta\, \mathrm{z}'] = [\mathrm{z}]$, and therefore the diagonal pushout sequence

$$0 \longrightarrow Y \xrightarrow{(\jmath,\alpha)} Z \times Y' \longrightarrow Z' \longrightarrow 0$$

splits, which yields an isomorphism $\phi\colon Z \times Y' \longrightarrow Z' \times Y$ such that $\phi(\jmath, \alpha)(y) = (0, y)$. Back in the fast lane, we notice that since $[\alpha\, \mathrm{z}] = [\mathrm{z}']$ and $[\jmath\, \mathrm{z}] = 0$, $[\mathrm{Z} \times \mathrm{z}'] = [(\jmath, \alpha)\, \mathrm{z}]$ according with the agreed names, and therefore

$$[\phi\, (\mathrm{Z} \times \mathrm{z}')] = [\phi\, (\jmath, \alpha)\, \mathrm{z}] = [\mathrm{Z}' \times \mathrm{z}]$$

thus there is an operator $\overline{\phi}$ making the following diagram commute:

$$\begin{array}{ccccccccc} 0 & \longrightarrow & Z \times Y' & \longrightarrow & Z \times Z' & \longrightarrow & X & \longrightarrow & 0 \\ & & {\scriptstyle\phi}\downarrow & & \downarrow{\scriptstyle\overline{\phi}} & & \| & & \\ 0 & \longrightarrow & Z' \times Y & \longrightarrow & Z' \times Z & \longrightarrow & X & \longrightarrow & 0 \end{array}$$

The operator $\overline{\phi}$ is an isomorphism since ϕ is an isomorphism. □

Proposition 2.7.3 is an easy victim of this principle. There is a dual (pullback, injective, left end) version which anyone can perform by simple dualisation of this one. Let us present a classically knitted proof:

2.11.7 Diagonal principle: injective case *If the sequences*

$$\begin{array}{ccccccccccc} 0 & \longrightarrow & Y & \xrightarrow{\jmath} & Z & \xrightarrow{\rho} & X & \longrightarrow & 0 & & (\mathrm{z}) \\ & & \| & & & & & & & & \\ 0 & \longrightarrow & Y & \xrightarrow{\jmath'} & Z' & \xrightarrow{\rho'} & X' & \longrightarrow & 0 & & (\mathrm{z}') \end{array}$$

are semi-equivalent then $(\mathrm{z} \times \mathrm{Z}')$ *and* $(\mathrm{z}' \times \mathrm{Z})$ *are isomorphic sequences.*

Proof The hypothesis yields operators $I\colon Z' \longrightarrow Z$ and $J\colon Z \longrightarrow Z'$ such that $I\jmath = \jmath'$ and $J\jmath' = \jmath$. The maps $\tau, \tau' \in \mathfrak{L}(\mathrm{Z} \times \mathrm{Z}')$ given by

$$\tau(z, z') = (z, z' + J(z)),$$
$$\tau'(z, z') = (z - I(z'), z')$$

are both isomorphisms. So, $T = \tau'\tau$ is the isomorphism we are looking for since $T(z, z') = \tau'\tau(z, z') = \tau'(z, z' + J(z)) = (z - I(z' + J(z)), z' + J(z))$ and thus $T(\jmath'(y), 0) = (\jmath'(y) - I(0 + J(\jmath'(y)), 0 + J(\jmath'(y)) = (0, \jmath'(y))$. □

2.12 Homological Constructions Appearing in Nature

We have seen that all exact sequences of p-Banach spaces are pushouts of a projective presentation of the quotient space and that all exact sequences of Banach spaces are pullbacks of an injective presentation of the subspace. We now record a few more entries in the directory of natural situations in which one encounters pushouts and pullbacks and other homological or categorical constructions.

The Natural Embedding of X into $C(B_X^*)$

Here is an everyday example of functor and natural transformation (notions to be defined in Chapter 4). Recall that B_X^* denotes the unit ball of the dual of X with the weak* topology, which is a compact space by the Banach–Alaoglu theorem. There is a natural isometry $\delta_X\colon X \longrightarrow C(B_X^*)$ given by $\delta_X(x)(x^*) = \langle x^*, x\rangle$. We have

Lemma 2.12.1 *Every $\mathscr{C}$-valued operator defined on a Banach space X admits a 1-extension through the embedding $\delta_X\colon X \longrightarrow C(B_X^*)$.*

Proof Assume without loss of generality that $\tau\colon X \longrightarrow C(K)$ is a contractive operator. Let $\delta_K\colon K \longrightarrow C(K)^*$ be the canonical embedding, in which $\delta_K(k) = \delta_k$ is the evaluation functional at k. It is clear that this map is continuous when $C(K)^*$ carries the weak* topology. The sought-after extension $T\colon C(B_X^*) \longrightarrow C(K)$ is $T(f)(k) = f(\tau^*(\delta_k))$. The operator T is well defined since $T(f)$, being the composition of three continuous maps, is a continuous function. The linearity of T and the bound $\|T\| \leq 1$ are clear. That T extends τ through δ_X is clear as well:

$$(T\delta_X(x))(k) = \delta_X(x)\,(\tau^*(\delta_k)) = \langle \tau^*\delta_k, x\rangle = \langle \delta_k, \tau(x)\rangle = \tau(x)(k). \qquad \square$$

Regarding complementation, δ_X is the 'best' embedding that X can have into a $\mathscr{C}$-space:

2.12.2 *A Banach space X is complemented in a $\mathscr{C}$-space if and only if it is complemented in $C(B_X^*)$ through the natural embedding.*

Proof Let K be a compact space, and let $\jmath\colon X \longrightarrow C(K)$ and $P\colon C(K) \longrightarrow X$ be operators with $P\jmath = \mathbf{1}_X$. Let $J\colon C(B_X^*) \longrightarrow C(K)$ be an extension of $\jmath$ through δ_X, with $\|J\| = \|\jmath\|$. Then $Q = \delta_X PJ$ is a projection of $C(B_X^*)$ onto $\delta_X[X]$ and, clearly, $\|Q\| \leq \|\jmath\|\,\|P\|$. The other implication is obvious. $\square$

Interpolation Theory

We will have to wait until 10.8 for a fine-brush painting of the interlacing connections between complex interpolation and twisted sums. The purpose of that section is to derive the construction of the fundamental Kalton–Peck Z_p spaces from complex interpolation theory. However, we can take a broad-brush approach now. The reader can consider this and the next section as trailers for forthcoming films, including some spoilers.

As far as we currently know, most interpolation methods for pairs of Banach spaces follow the following schema. One starts with a pair (X_0, X_1) of Banach spaces that one assumes are linear and continuously embedded into some Banach space Σ. Then, there is a Banach space $\mathcal{H}$ and an operator $\Phi\colon \mathcal{H} \longrightarrow \Sigma$, which we will call an *interpolator on* $\mathcal{H}$, such that, for every linear operator $t\colon \Sigma \longrightarrow \Sigma$ acting continuously sending $X_0 \longrightarrow X_0$ and $X_1 \longrightarrow X_1$, there is an operator $T\colon \mathcal{H} \longrightarrow \mathcal{H}$ such that $t \circ \Phi = \Phi \circ T$. We denote by X_Φ the space $\Phi(\mathcal{H})$ endowed with the quotient norm $\|x\|_\Phi = \inf\{\|f\|_{\mathcal{H}} : f \in \mathcal{H}, \Phi f = x\}$, which is a Banach space. Given two interpolators Ψ, Φ on $\mathcal{H}$, consider the map $(\Psi, \Phi)\colon \mathcal{H} \longrightarrow \Sigma \times \Sigma$, and let $X_{\Psi,\Phi}$ denote the space $(\Psi, \Phi)[\mathcal{H}]$ endowed with the quotient norm. One thus has the following pushout diagram:

$$\begin{array}{ccccccccc}
 & & 0 & & 0 & & & & \\
 & & \downarrow & & \downarrow & & & & \\
 & & \ker\Psi \cap \ker\Phi & = & \ker(\Psi,\Phi) & & & & \\
 & & \downarrow & & \downarrow & & & & \\
0 & \longrightarrow & \ker\Phi & \longrightarrow & \mathcal{H} & \xrightarrow{\Phi} & X_\Phi & \longrightarrow & 0 \\
 & & \downarrow{\scriptstyle\Psi} & & \downarrow{\scriptstyle(\Psi,\Phi)} & & \| & & \\
0 & \longrightarrow & \Psi[\ker\Phi] & \xrightarrow{\imath} & X_{\Psi,\Phi} & \xrightarrow{\rho} & X_\Phi & \longrightarrow & 0 \\
 & & \downarrow & & \downarrow & & & & \\
 & & 0 & & 0 & & & &
\end{array} \tag{2.36}$$

where $\Psi[\ker\Phi]$ is endowed with the obvious quotient norm. The maps $\imath, \rho$ are defined by $\imath(\Psi g) = (\Psi g, 0)$ and $\rho(\Psi f, \Phi f) = \Phi f$.

The complex interpolation method, as described in Section 10.8, as well as the K and J real methods [82] and, in general, the unifying method of Cwikel, Kalton, Milman, and Rochberg [139], can be fit into this schema.

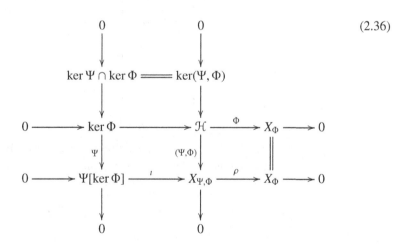

The 3-Space Problem for Dual Spaces

In [452], Vogt posed a quite natural problem: must an exact sequence

$$0 \longrightarrow A^* \longrightarrow X \longrightarrow B^* \longrightarrow 0$$

be the dual sequence of another exact sequence? Must X be a dual space? The answer is no. We present now a no-frills counterexample, while Section 10.5 displays a much more natural and elaborated example.

Proposition 2.12.3 *Let X be a Banach space such that X^{**}/X is an ultrasummand with the RNP. Then X is a complemented subspace of a twisted sum of two dual spaces.*

Proof Let $\delta\colon X \longrightarrow X^{**}$ be the canonical inclusion, and let $\pi\colon \ell_1(I) \longrightarrow X^{**}$ be a quotient map. Form the complete pullback diagram

$$\begin{CD}
@. @. 0 @. 0 \\
@. @. @AAA @AAA \\
@. @. X^{**}/X @= X^{**}/X \\
@. @. @AAA @AAA \\
0 @>>> \kappa(X^{**}) @>>> \ell_1(I) @>\pi>> X^{**} @>>> 0\\
@. @| @AAA @AA\delta A \\
0 @>>> \kappa(X^{**}) @>>> \mathrm{PB} @>>> X @>>> 0\\
@. @AAA @AAA @AAA\\
@. 0 @. 0 @. 0
\end{CD}$$

We need from the reader a leap of faith here and belief that the space PB is complemented in its bidual (the justification comes in Lemma 10.4.1). Form the diagonal exact sequence

$$0 \longrightarrow \mathrm{PB} \longrightarrow \ell_1(I) \times X \longrightarrow X^{**} \longrightarrow 0$$

Multiplying by a complement V of PB in PB^{**}, we get the sequence

$$0 \longrightarrow \mathrm{PB}^{**} \longrightarrow V \times \ell_1(I) \times X \longrightarrow X^{**} \longrightarrow 0 \qquad \square$$

To obtain specific examples, just set $X = \mathsf{JT}_*$, the natural predual of the James–Tree space JT. It is well known (see, e.g., [102]) that JT_* is uncomplemented in its bidual JT^* and that $\mathsf{JT}^*/\mathsf{JT}_*$ is a non-separable Hilbert space. All this yields the non-trivial exact sequence

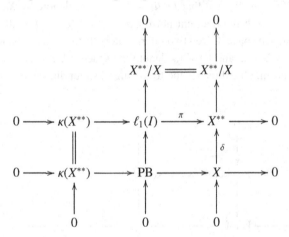

$$0 \longrightarrow \mathrm{PB}^{**} \longrightarrow (\mathrm{PB}^{**}/\mathrm{PB}) \times \ell_1(I) \times \mathsf{JT}_* \longrightarrow \ell_2(I) \longrightarrow 0$$

whose middle space cannot be an ultrasummand, let alone a dual space.

The more elaborate counterexample in Section 10.5 alluded to earlier will consist of an exact sequence of Banach spaces $0 \longrightarrow R \longrightarrow \diamond \longrightarrow U \longrightarrow 0$ in which R is reflexive, U is an ultrasummand and $\diamond$ is not an ultrasummand. The other two possible configurations of Banach spaces lead to ultrasummands:

2.12.4 *In any of the sequences of Banach spaces* $0 \longrightarrow R \longrightarrow U \longrightarrow \diamond \longrightarrow 0$ *or* $0 \longrightarrow U \longrightarrow \diamond \longrightarrow R \longrightarrow 0$, *the space* $\diamond$ *is an ultrasummand.*

Proof In the first situation, we can assume that R is a subspace of U, hence of U^{**}. Observe the commutative diagram

$$\begin{array}{ccccccccc}
0 & \longrightarrow & R & \longrightarrow & U & \xrightarrow{\ \pi\ } & \diamond & \longrightarrow & 0 \\
 & & \| & & \downarrow{\scriptstyle \delta_U} & & \downarrow{\scriptstyle \delta_\diamond} & & \\
0 & \longrightarrow & R & \longrightarrow & U^{**} & \xrightarrow{\ \pi^{**}\ } & \diamond^{**} & \longrightarrow & 0
\end{array}$$

If P is a projection along δ_U, then $P|_R = \mathbf{1}_R$, and so it induces an operator $\overline{P}\colon \diamond^{**} \longrightarrow \diamond$. This operator is a projection along $\delta_\diamond$ because $\overline{P}\delta_\diamond\pi = \overline{P}\pi^{**}\delta_U = \pi P\delta_U = \pi$ implies $\overline{P}\delta_\diamond = \mathbf{1}_\diamond$ since π is surjective. The second situation was already treated in the introduction to this chapter. □

The 3-Space Problem for the Dunford–Pettis Property

The DPP is not a 3-space property [102]. Moreover, a careful using of the pullback construction establishes that counterexamples are almost ubiquitous.

Proposition 2.12.5 *Every Banach space is a complemented subspace of a twisted sum of two Banach spaces with the Dunford–Pettis property.*

Proof Let X be any Banach space. Consider an embedding $\jmath\colon X \longrightarrow C(K)$; fix, then, a quotient $\pi\colon \ell_1(I) \longrightarrow C(K)$ and draw the pullback diagram:

$$\begin{array}{ccccccccc}
0 & \longrightarrow & \ker\pi & \longrightarrow & \ell_1(I) & \longrightarrow & C(K) & \longrightarrow & 0 \\
 & & \| & & \uparrow & & \uparrow{\scriptstyle \jmath} & & \\
0 & \longrightarrow & \ker\pi & \longrightarrow & \mathrm{PB} & \longrightarrow & X & \longrightarrow & 0
\end{array}$$

The diagonal pullback sequence $0 \longrightarrow \mathrm{PB} \longrightarrow \ell_1(I) \times X \longrightarrow C(K) \longrightarrow 0$ proves the assertion: the space PB, as any space with the Schur property does, has the DPP as well as any $C(K)$ space. □

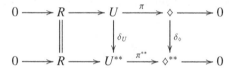

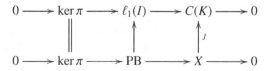

The Johnson–Lindenstrauss Spaces

Let $\mathcal{M}$ be an almost disjoint family of subsets of $\mathbb{N}$. The Nakamura–Kakutani sequence 2.2.10 it generates is the perfect example of a pullback sequence. Indeed, the inclusion $C_0(\wedge_{\mathcal{M}}) \longrightarrow \ell_\infty$ induces an isometry $\imath\colon c_0(\mathcal{M}) \longrightarrow \ell_\infty/c_0$ that generates the pullback diagram

$$\begin{array}{ccccccccc} 0 & \longrightarrow & c_0 & \longrightarrow & \ell_\infty & \longrightarrow & \ell_\infty/c_0 & \longrightarrow & 0 \\ & & \| & & \uparrow & & \uparrow{\scriptstyle \imath} & & \\ 0 & \longrightarrow & c_0 & \longrightarrow & C_0(\wedge_{\mathcal{M}}) & \longrightarrow & c_0(\mathcal{M}) & \longrightarrow & 0 \end{array} \tag{2.37}$$

Now pick the size $\mathfrak{c}$ family $\mathcal{D}$ of branches of the dyadic tree. Let $\imath_p\colon \ell_p(\mathcal{D}) \longrightarrow c_0(\mathcal{D})$ denote the canonical inclusion and form the pullback diagram

$$\begin{array}{ccccccccc} 0 & \longrightarrow & c_0 & \longrightarrow & \ell_\infty & \longrightarrow & \ell_\infty/c_0 & \longrightarrow & 0 \\ & & \| & & \uparrow & & \uparrow{\scriptstyle \imath} & & \\ 0 & \longrightarrow & c_0 & \longrightarrow & C_0(\wedge_{\mathcal{D}}) & \longrightarrow & c_0(\mathcal{D}) & \longrightarrow & 0 \\ & & \| & & \uparrow & & \uparrow{\scriptstyle \imath_p} & & \\ 0 & \longrightarrow & c_0 & \longrightarrow & \mathsf{JL}_p & \longrightarrow & \ell_p(\mathcal{D}) & \longrightarrow & 0 \end{array} \tag{2.38}$$

The lower pullback space / sequence

$$\mathsf{JL}_p = \left\{(\xi, z)\colon \xi \in \ell_\infty, z \in \ell_p(\mathcal{D})\colon \xi + c_0 = \imath\,\imath_p(z)\right\}$$

is called the Johnson–Lindenstrauss space / sequence.

2.12.6 *The Johnson–Lindenstrauss sequence is non-trivial.*

Indeed, as in 2.2.10, no injective operator $\ell_p(\mathcal{D}) \longrightarrow \ell_\infty$ exists. In particular, JL_p is not WCG. The spaces JL_p were obtained in [225] and have the following surprising property:

Lemma 2.12.7 *JL_p and ℓ_∞ do not have isomorphic non-separable subspaces. Moreover, $\mathsf{JL}_p^* \simeq \ell_1 \times \ell_p^*(\mathcal{D})$ with duality given by $\langle (y^*, z^*), (y, z)\rangle = y^*(y) + z^*(z)$.*

Proof Let us show first that JL_p cannot have a countable norming set of functionals, which means that it cannot be a subspace of ℓ_∞. To this end, let (y_n^*, z_n^*) be a sequence of norm 1 functionals. Observe that $(1_{M_\alpha}, e_\alpha) \in \mathsf{JL}_p$. Consider the set M_0 formed by all $\gamma \in \mathcal{D}$ appearing in the support of all z_n^*.

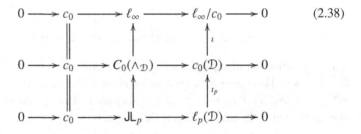

Take a sequence γ such that $\gamma(k) \notin M_0$ for all k and consider, for each $\gamma(k)$, the element $(1_{M_{\gamma(k)}}, e_{\gamma(k)})$. Take $N \in \mathbb{N}$ such that

$$\left\| \left(\frac{1}{\sqrt{k}} \sum_{k=1}^{N} 1_{M_{\gamma(k)}}, \frac{1}{\sqrt{k}} \sum_{k=1}^{N} e_{\gamma(k)} \right) \right\| = 1$$

This yields

$$\left| \left\langle (y_n^*, z_n^*), \left(\frac{1}{\sqrt{k}} \sum_{k=1}^{N} 1_{M_{\gamma(k)}}, \frac{1}{\sqrt{k}} \sum_{k=1}^{N} e_{\gamma(k)} \right) \right\rangle \right| = \left\langle y_n^*, \frac{1}{\sqrt{k}} \sum_{k=1}^{N} 1_{M_{\gamma(k)}} \right\rangle \leq \frac{1}{\sqrt{k}}$$

and thus the functionals cannot norm the space. A few minor changes make the previous argument work for arbitrary non-separable subspaces of JL_p that contain the canonical copy of c_0. If, however, X is non-separable but does not contain that c_0, then form the exact sequence $0 \longrightarrow X \longrightarrow [X+c_0] \longrightarrow c_0 \longrightarrow 0$ and use a straightforward 3-space argument to get that if X is a subspace of ℓ_∞ then $[X + c_0]$ must also be a subspace of ℓ_∞, which has been shown to be impossible. □

Proposition 8.7.18 asserts that the spaces $C_0(\wedge_{\mathcal{M}})$ can be different or equal depending on which cardinal axioms are assumed. It is our belief that different $C_0(\wedge_{\mathcal{M}})$ generate different JL_p spaces, but what happens after taking new pullbacks is unclear. Moving on in a different direction, Yost [461] constructs a twisted sum of c_0 and $\ell_p(\aleph_1)$ with properties completely different to those of the Johnson–Lindenstrauss space(s):

2.12.8 *For each $1 < p < \infty$, there exists an exact sequence*

$$0 \longrightarrow c_0 \longrightarrow \mathscr{Y}_p \longrightarrow \ell_p(\aleph_1) \longrightarrow 0$$

in which $\mathscr{Y}_p$ is a subspace of ℓ_∞.

Pick a continuous surjection $\gamma\colon \mathbb{N}^* \longrightarrow B^*_{\ell_p(\aleph_1)}$ and form the diagram

$$\begin{array}{ccccccccc}
0 & \longrightarrow & c_0 & \longrightarrow & \ell_\infty & \longrightarrow & \ell_\infty/c_0 & \longrightarrow & 0 \\
 & & \| & & \uparrow & & \uparrow{\scriptstyle \gamma^\circ} & & \\
0 & \longrightarrow & c_0 & \longrightarrow & \mathrm{PB} & \longrightarrow & C(B^*_{\ell_p(\aleph_1)}) & \longrightarrow & 0 \\
 & & \| & & \uparrow & & \uparrow{\scriptstyle \delta} & & \\
0 & \longrightarrow & c_0 & \longrightarrow & \mathscr{Y}_p & \longrightarrow & \ell_p(\aleph_1) & \longrightarrow & 0
\end{array}$$

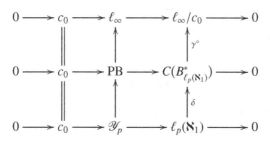

The map γ exists by Parovičenko's theorem since the dual unit ball of $\ell_p(\aleph_1)$ is, with its weak topology, a compact space of weight $\aleph_1$. Thus, γ° as well as $\gamma^\circ\delta$ is an isometry, which is what makes $\mathscr{Y}_p$ isometric to a subspace of ℓ_∞. The lower sequence cannot split since $c_0 \times \ell_p(\aleph_1)$ is not a subspace of ℓ_∞. A more careful analysis of Johnson–Lindenstrauss spaces can be found in [460; 396].

Twisted Sums of c_0 and ℓ_∞

2.12.9 *There exist non-trivial twisted sums of c_0 and ℓ_∞.*

Pick a weakly compact operator $\tau\colon \ell_\infty \longrightarrow \ell_\infty/c_0$ with non-separable range and form the pullback diagram

$$\begin{array}{ccccccccc} 0 & \longrightarrow & c_0 & \longrightarrow & \ell_\infty & \longrightarrow & \ell_\infty/c_0 & \longrightarrow & 0\\ && \| && \uparrow && \uparrow{\scriptstyle\tau} && \\ 0 & \longrightarrow & c_0 & \longrightarrow & \mathsf{CC}(\tau) & \longrightarrow & \ell_\infty & \longrightarrow & 0 \end{array} \tag{2.39}$$

It is clear that the lower pullback sequence does not split since τ cannot be lifted to ℓ_∞ because any weakly compact operator $\ell_\infty \longrightarrow \ell_\infty$ has separable range. There are different ways to get such a τ:

- A composition $\tau = \imath \jmath_2 \pi$, where $\imath\colon c_0(I) \longrightarrow \ell_\infty/c_0$ is an embedding, $\jmath_2\colon \ell_2(I) \longrightarrow c_0(I)$ is the canonical inclusion and $\pi\colon \ell_\infty \longrightarrow \ell_2(I)$ is a quotient map. Such a quotient map exists when $\aleph_1 \leq |I| \leq \mathfrak{c}$: (a) observe that $\ell_1(\mathfrak{c})$ is a subspace of ℓ_∞ – a proof can be seen in [22, Claim 3, p. 138]; (b) any quotient map $Q\colon \ell_1(\mathfrak{c}) \longrightarrow \ell_2(\mathfrak{c})$ must be 2-summing [153, Theorem 3.1]; (c) extend Q to ℓ_∞. Thus, the lower sequence in the pullback diagram

$$\begin{array}{ccccccccc} 0 & \longrightarrow & c_0 & \longrightarrow & \mathsf{JL}_2 & \longrightarrow & \ell_2(I) & \longrightarrow & 0\\ && \| && \uparrow && \uparrow && \\ 0 & \longrightarrow & c_0 & \longrightarrow & \mathsf{CC}(\imath\jmath_2\pi) & \longrightarrow & \ell_\infty & \longrightarrow & 0 \end{array}$$

  is non-trivial. It is likely that different choices of τ generate different spaces $\mathsf{CC}(\tau)$. It is not known if there is a twisted sum of c_0 and ℓ_∞ that is not isomorphic to a $\mathscr{C}$-space.
- Use [300, Section 4].

Twisted Sums of $\mathscr{C}$-Spaces

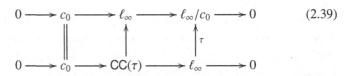

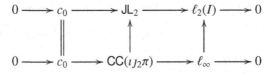

A twisted sum of two $\mathscr{C}$-spaces can fail to have Pełczyński's property (V), as we show in Section 10.5, thus it does not have even to be isomorphic to a $\mathscr{C}$-space. The first example of such a phenomenon appears in [102, 3.5] and is worked out in detail in [22, 2.2.6]. The reason we talk about this here is because it is based on a wonderfully clever construction of Benyamini [39] whose bricks and mortar are, however, simple pullback. Indeed, the counterexample depends on Benyamini's argument that given a multiplication operator $x \mapsto \theta x$, the pullback space PB_θ in the diagram

$$\begin{array}{ccccccccc} 0 & \longrightarrow & c_0 & \longrightarrow & \ell_\infty & \longrightarrow & \ell_\infty/c_0 & \longrightarrow & 0 \\ & & \| & & \uparrow & & \uparrow{\scriptstyle \theta} & & \\ 0 & \longrightarrow & c_0 & \longrightarrow & \mathrm{PB}_\theta & \longrightarrow & \ell_\infty/c_0 & \longrightarrow & 0 \end{array} \tag{2.40}$$

which is clearly a renorming of ℓ_∞, is such that $\|u\|\|u^{-1}\|\|P\| \geq \theta^{-1}$ for every compact K, every embedding $u\colon \mathrm{PB}_\theta \longrightarrow C(K)$ and every projection $P\colon C(K) \longrightarrow u[\mathrm{PB}_\theta]$. The rest is oil steadily flowing down an inclined plane: both $\ell_\infty(\mathbb{N}, \mathrm{PB}_{1/n})$ and $c_0(\mathbb{N}, \mathrm{PB}_{1/n})$ are twisted sums of two $\mathscr{C}$-spaces, for instance

$$0 \longrightarrow c_0(\mathbb{N}, c_0) \longrightarrow c_0(\mathbb{N}, \mathrm{PB}_{1/n}) \longrightarrow c_0(\mathbb{N}, C(\mathbb{N}^*)) \longrightarrow 0$$

thus they cannot be complemented in any $\mathscr{C}$-space. Whether an anodyne space such as $c_0(I)$ can play the role of $C(\mathbb{N}^*)$ is a topic to discuss around the bonfire.

2.13 The Device

As we know, given an embedding $\jmath\colon Y \longrightarrow A$ and an operator $\tau\colon Y \longrightarrow B$, the pushout space can be understood as a superspace of B such that τ admits an extension $A \longrightarrow \mathrm{PO}$. Can we do the same with a family $(\imath_i\colon Y_i \longrightarrow A_i)_{i\in I}$ of embeddings and a family $(\tau_i\colon Y_i \longrightarrow B_i)_{i\in I}$ of operators? Under some reasonable restrictions, we certainly can. First condition: I must be a set, and all spaces involved must be p-normed spaces for some fixed p. Otherwise, the amalgamation of the spaces becomes complicated. Second condition: the operators must be uniformly bounded. Otherwise, their amalgamation becomes a nice linear map, not an operator. Third condition: when we work with embeddings (or quotients), they must be uniformly open. Otherwise, their amalgamation becomes a nice operator, not an embedding (or quotient). Having accepted those conditions, we can paste all embeddings into one single embedding $\prod \jmath_i\colon \ell_p(I, Y_i) \longrightarrow \ell_p(I, A_i)$, form the operator $\oplus\tau_i\colon \ell_p(I, Y_i) \longrightarrow B$ and obtain the pushout in $p\mathbf{B}$:

$$\begin{array}{ccc} \ell_p(I, Y_i) & \xrightarrow{\ \prod \jmath_i\ } & \ell_p(I, A_i) \\ {\scriptstyle \oplus\tau_i}\downarrow & & \downarrow{\scriptstyle \overline{\oplus\tau_i}} \\ B & \xrightarrow{\ \overline{\prod \jmath_i}\ } & \mathrm{PO} \end{array} \tag{2.41}$$

Observe that $\overline{\oplus\tau_i}$ provides an extension $\overline{\oplus\tau_i}|_{A_k}$ of each τ_k since the following diagram is commutative:

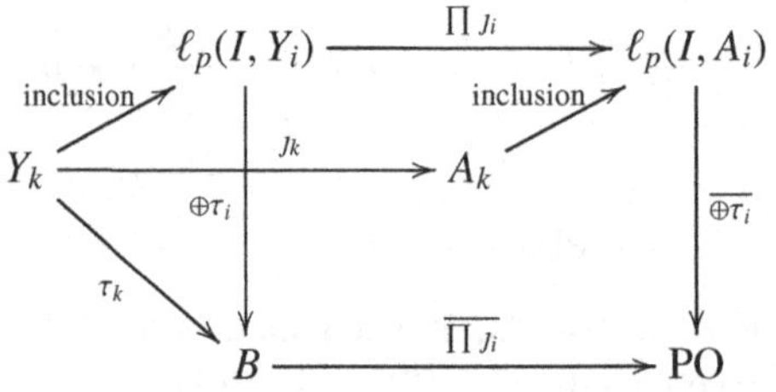

The iteration of this construction yields a rather flexible device to construct spaces with additional properties. Examples? Consequences? *Il catalogo è questo*: the p-Gurariy spaces, the p-Kadets spaces, the Bourgain–Pisier spaces, the Kubiś space, many other spaces of universal disposition, Lindenstrauss and $\mathscr{L}_{\infty,\lambda}$-envelopes etc. All these topics and examples will be treated from various angles in this book. Let us now present a detailed account of how this technique works. Even if the Device looks like a Turing machine, it actually works more like a Thermomix: with adequately chosen ingredients (data), a recipe and continuous attention will produce a wholesome (quasi-) Banach space.

Ingredients and recipe

- Pick a scalar $0 < p \leq 1$ and a p-Banach space X_0. These indicate the p-Banach category in which we will be working and the 'initial' object.
- Fix an ordinal μ. This indicates the length of the iteration and, to some extent, the size of the space to be obtained. Usually, μ is a limit ordinal.
- Construct an inductive system of p-Banach spaces $(X_\alpha)_{0\leq\alpha\leq\mu}$ by transfinite induction on α, starting with X_0. We will do that as follows: first assume that $(X_\alpha)_{0\leq\alpha<\beta}$ has been constructed for all $\alpha < \beta$. If β is a limit ordinal, we set $X_\beta = \lim_{\alpha<\beta} X_\alpha$, namely the completion of $\bigcup_{\alpha<\beta} X_\alpha$. If $\beta = \alpha + 1$, we will perform a pushout as described before, which we explain now in detail. We need to add two new ingredients: a uniformly bounded family $\mathfrak{J}_\alpha$ of embeddings between p-Banach spaces and a uniformly bounded family of operators $\mathfrak{L}_\alpha$ with values in X_α. Both $\mathfrak{J}_\alpha$ and $\mathfrak{L}_\alpha$ have to be *sets*. The space $X_{\alpha+1}$ that we will construct next will allow us to extend any operator $u \in \mathfrak{L}_\alpha$ through any embedding $v \in \mathfrak{J}_\alpha$ whenever this makes sense. Consider the set $I_\alpha = \{(u, v) \in \mathfrak{L}_\alpha \times \mathfrak{J}_\alpha\colon \operatorname{dom}(u) = \operatorname{dom}(v)\}$ and the ℓ_p-sums $\ell_p(I_\alpha, \operatorname{dom}(v))$ and $\ell_p(I_\alpha, \operatorname{cod}(v))$. There is an obvious operator $\prod \mathfrak{J}_\alpha\colon \ell_p(I_\alpha, \operatorname{dom}(v)) \longrightarrow \ell_p(I_\alpha, \operatorname{cod}(v))$ sending $(x_{(u,v)})_{(u,v)\in I_\alpha}$ to $(v(x_{(u,v)}))_{(u,v)\in I_\alpha}$. There is another obvious operator $\bigoplus \mathfrak{L}_\alpha\colon \ell_p(I_\alpha, \operatorname{dom}(u)) \longrightarrow X_\alpha$ sending $(x_{(u,v)})_{(u,v)\in I_\alpha}$ to

$\sum_{(u,v)\in I_\alpha} u(x_{(u,v)})$. The notation is slightly imprecise because those operators depend, not only on $\mathfrak{L}_\alpha$ and $\mathfrak{J}_\alpha$, but also on I_α. Now we take the pushout

$$\begin{array}{ccc} \ell_p(I_\alpha, \mathrm{dom}(u)) = \ell_p(I_\alpha, \mathrm{dom}(v)) & \xrightarrow{\ \prod \mathfrak{J}_\alpha\ } & \ell_p(I_\alpha, \mathrm{cod}(v)) \\ \Big\downarrow {\scriptstyle \bigoplus \mathfrak{L}_\alpha} & & \Big\downarrow \\ X_\alpha & \longrightarrow & \mathrm{PO} \end{array} \tag{2.42}$$

set $\mathrm{PO} = X_{\alpha+1}$ and use the lower arrow to embed X_α into $X_{\alpha+1}$.

- The recipe is finished by iterating the construction until μ. The output will be a p-Banach space X_μ plus an isometry $X_0 \longrightarrow X_\mu$.

Some Device constructions are Fraïssé limits, a topic to which Chapter 6 is entirely devoted. As a product demo of the Device in action, let us construct p-Banach spaces U having the following extension property, which is usually said 'to be of universal disposition for separable spaces' (SUD): given an isometry $v: A \longrightarrow B$ between separable p-Banach spaces and an isometry $u: A \longrightarrow U$, there is an isometry $w: B \longrightarrow U$ such that $u = wv$. To be able to proceed, let us confirm that there is a *set*, $\mathscr{S}_p = \{\ell_p/Y : Y \text{ is a closed subspace of } \ell_p\}$, containing an isometric copy of every separable p-Banach space. In fact, $|\mathscr{S}_p| = \mathfrak{c}$.

2.13.1 Recipe for Spaces of Separable Universal Disposition

- Work in the category $p\mathbf{B}$; pick as X your favourite p-Banach space of dimension up to $\mathfrak{c}$ and pick any ordinal $\mu \leq \mathfrak{c}$ of uncountable cofinality, say ω_1.
- Set $X_0 = X$; once X_α is constructed, fix $\mathfrak{L}_\alpha$ as the set of isometries $u: A \longrightarrow X_\alpha$, with domain in $\mathscr{S}_p$; the set $\mathfrak{J} = \mathfrak{J}_\alpha$ is the same for all α: the set of all isometries with domain and codomain in $\mathscr{S}_p$.

Let us observe the output space X_μ in some detail. It is of SUD: let $v: A \longrightarrow B$ an isometry between separable p-Banach spaces and let $u: A \longrightarrow X_\mu$ be an isometry. We can assume that v belongs to $\mathfrak{J}$ since there are surjective isometries $a: A' \longrightarrow A$ and $b: B' \longrightarrow B$ with A', B' in $\mathscr{S}_p$, and so $v' = b^{-1}va$ belongs to $\mathfrak{J}$. Now, letting $u' = ua$, it is clear that if w' is an isometry such that

$u' = w'v'$ then $w = w'b^{-1}$ is the required extension of u. Don't believe us, just watch:

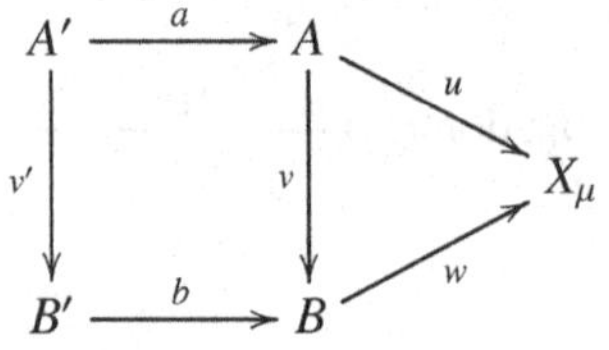

As μ has uncountable cofinality, $X_\mu = \bigcup_{\alpha<\mu} X_\alpha$, since this last space is already complete. It follows that $u[A] \subset X_\alpha$ for some $\alpha < \mu$ and so $u \in \mathfrak{L}_\alpha$, when u is interpreted as an isometry $A \longrightarrow X_\alpha$. Therefore, the pair (u, v) belongs to I_α, and thus there is $w\colon B \longrightarrow X_{\alpha+1}$ such that $wv = \imath_{\alpha,\alpha+1}u$, where $\imath_{\alpha,\alpha+1}\colon X_\alpha \longrightarrow X_{\alpha+1}$ is the inclusion map. So, X_μ is of SUD, and, since $\mathfrak{c}^{\aleph_0} = \mathfrak{c}$, $\dim X_\mu = \mathfrak{c}$ (actually $\dim X_1$ is already $\mathfrak{c}$ by the proposition in Note 6.5.1). The interested chef can find inspiration for their own elaborations and variations on this recipe in either [22, Chapter 3] or Proposition 7.3.2. Here, we prove:

Proposition 2.13.2

(a) *Any p-Banach space of SUD is* 1*-separably injective in p***B**.
(b) *Any p-Banach space of SUD contains isometric copies of all p-Banach spaces of dimension up to* $\aleph_1$.
(c) [CH] *All p-Banach spaces of SUD with dimension* $\aleph_1$ *are isometric.*

Proof (a) Let B be a separable p-Banach space and let $\tau\colon A \longrightarrow U$ be a contractive operator, where A is a subspace of B. If U is of separable universal disposition, there is a commutative diagram

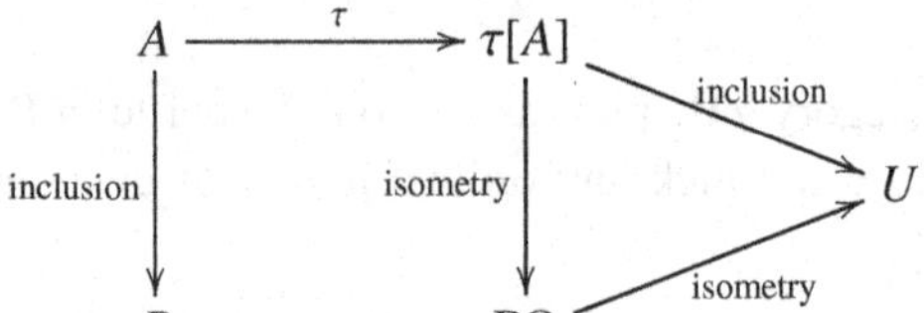

and so τ has a 1-extension to B. (b) Every p-Banach space of density $\aleph_1$ or less can be written as the union of a continuous ω_1-chain of separable subspaces: $X = \bigcup_{\alpha<\omega_1} X_\alpha$. This means that $X_\alpha \subset X_\beta$ for $\alpha < \beta$ and that $X_\beta = \overline{\bigcup_{\alpha<\beta} X_\alpha}$ when β is a limit ordinal. We may assume that $X_0 = 0$. If U is of SUD, we can construct a compatible system ($u_\beta|_{X_\alpha} = u_\alpha$ for $\alpha < \beta < \omega_1$) of isometries $u_\alpha\colon X_\alpha \longrightarrow U$ as follows: u_0 must be 0; then, assuming that u_α has been defined for all $\alpha < \beta$, we define u_β by continuity if β is a limit ordinal and using the SUD of U when $\beta = \alpha + 1$ to extend $u_\alpha\colon X_\alpha \longrightarrow U$ to an isometry $u_{\alpha+1}\colon X_{\alpha+1} \longrightarrow U$. After that, just define $u\colon X \longrightarrow U$ by declaring $u(x) = u_\alpha(x)$ if $x \in X_\alpha$. (c) Assume U and V are of SUD and density $\aleph_1$ and write them as $U = \bigcup_{\alpha<\omega_1} U_\alpha$ and $V = \bigcup_{\beta<\omega_1} V_\beta$, where (U_α) and (V_β) are continuous ω_1-chains of separable subspaces with $U_0 = V_0 = 0$. Set $U_{\omega_1} = U$ and $V_{\omega_1} = V$. Consider the set S of all triples (α, β, f), where $\alpha, \beta \in [0, \omega_1]$ and $f\colon U_\alpha \longrightarrow V_\beta$ is a surjective isometry. Declare $(\alpha, \beta, f) \le (\gamma, \delta, g)$ if $\alpha \le \gamma, \beta \le \delta$ and $g|_{U_\alpha} = f$. The set S is not empty since $(0, 0, 0) \in S$. A maximal element (α, β, f) exists by Zorn's lemma since every chain admits an upper bound. We end the proof by showing that $\alpha = \beta = \omega_1$. Otherwise, we inductively define sequences $\alpha \le \alpha_1 \le \alpha_2 \le \ldots$ and $\beta \le \beta_1 \le \beta_2 \le \ldots$ and isometries $f_n\colon U_{\alpha_n} \longrightarrow V_{\beta_n}$ and $f_n^{-1}\colon V_{\beta_n} \longrightarrow U_{\alpha_{n+1}}$ with $f_{n+1}|_{U_{\alpha_n}} = f_n$ in the obvious way: assuming α_n, β_n, f_n have been obtained, extend f_n^{-1} to an isometry $g_n\colon V_{\beta_n} \longrightarrow g_n[V_{\beta_n}]$ and set α_{n+1} to be the smallest ordinal such that $g_n[V_{\beta_n}] \subset U_{\alpha_{n+1}}$ and then extend f_n to an isometry $f_{n+1}\colon U_{\alpha_{n+1}} \longrightarrow f_{n+1}[U_{\alpha_{n+1}}]$ and set β_{n+1} to be the smallest ordinal such that $f_{n+1}[U_{\alpha_{n+1}}] \subset V_{\beta_{n+1}}$. Set $\alpha' = \sup_n \alpha_n$ and $\beta' = \sup_n \beta_n$ and let the continuity of the chain produce a surjective isometry $f'\colon U_{\alpha'} \longrightarrow V_{\beta'}$ extending f, in flagrant contradiction of the alleged maximality. □

The Bourgain–Pisier Construction

Bourgain and Pisier showed in [52] that, for each $\lambda > 1$, every separable Banach space X can be embedded into some $\mathscr{L}_{\infty,\lambda}$-space $\mathscr{L}_\infty^{\mathsf{BP}}(X)$ in such a way that the corresponding quotient space $\mathscr{L}_\infty^{\mathsf{BP}}(X)/X$ has the Schur property and the RNP. Their construction is a clever iteration of pushouts in which the embeddings are no longer isometries. Assume that $X = \overline{\bigcup_{n\ge1} X_n}$, where (X_n) is a chain of finite-dimensional subspaces. Fix $\lambda > 1$ and then take $\eta \in (\lambda^{-1}, 1)$. Set $a(1) \in \mathbb{N}$ such that there is a subspace $S_1 \subset \ell_\infty^{a(1)}$ and an isomorphism $u_1\colon S_1 \longrightarrow X_1$ with $\|u_1\| \le \eta$ and $\|u_1^{-1}\| \le \lambda$. Form the consecutive pushouts

$$\begin{array}{ccc} S_1 & \xrightarrow{\text{inclusion}} & \ell_\infty^{a(1)} \\ {\scriptstyle u_1}\downarrow & & \downarrow{\scriptstyle \overline{u_1}} \\ X_1 & \longrightarrow & \mathrm{PO}_1 \end{array} \qquad\qquad \begin{array}{ccc} X_1 & \longrightarrow & \mathrm{PO}_1 \\ {\scriptstyle \text{inclusion}}\downarrow & & \downarrow \\ X_2 & \longrightarrow & \mathrm{PO}_1' \end{array}$$

and continue inductively: set $a(n) \in \mathbb{N}$ such that there is $S_n \subset l_\infty^{a(n)}$ and an isomorphism $u_n\colon S_n \longrightarrow \mathrm{PO}_n'$ with $\|u_n\| \le \eta$ and $\|u_n^{-1}\| \le \lambda$ and form new pushouts

$$\begin{array}{ccc} S_n & \xrightarrow{\text{inclusion}} & \ell_\infty^{a(n)} \\ {\scriptstyle u_n}\downarrow & & \downarrow{\scriptstyle \overline{u_n}} \\ \mathrm{PO}'_n & \longrightarrow & \mathrm{PO}_{n+1} \end{array}$$

The space $\mathscr{L}_\infty^{\mathsf{BP}}(X) = \lim \mathrm{PO}_n$ is an $\mathscr{L}_{\infty,\lambda}$-space since it is the inductive limit of the spaces PO_n, which are λ-isomorphic to $\ell_\infty^{a(n)}$. A diagram might illuminate the construction

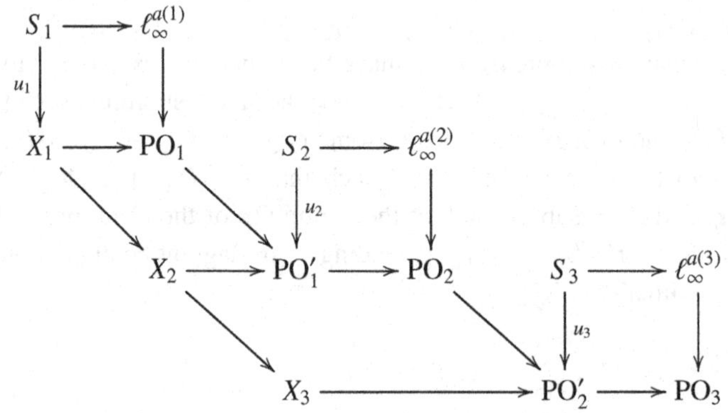

(2.43)

The trickier part of the proof is showing that $\mathscr{L}_\infty^{\mathsf{BP}}(X)/X$ has the Schur and RNP. Both properties follow from an imaginative idea. An isometry $C \longrightarrow D$ is *η-admissible*, where $0 < \eta < 1$, if it can be placed in a pushout diagram

$$\begin{array}{ccc} A & \xrightarrow{b} & B \\ {\scriptstyle c}\downarrow & & \downarrow \\ C & \longrightarrow & D \end{array}$$

in which b is an isometry and $\|c\| \leq \eta$. The following result [52, Theorem 1.6], not proved here, is the key to the argument:

2.13.3 *Let $0 \leq \eta < 1$. The direct limit of a sequence of η-admissible isometries has the Schur and RNP.*

To be able to apply the result to our situation, let us complete Diagram (2.43) with new pushouts into X:

$$\begin{array}{ccc} S_1 & \longrightarrow & \ell_\infty^{a(1)} \\ {\scriptstyle u_1}\downarrow & & \downarrow \\ X_1 & \longrightarrow & \mathrm{PO}_1 \\ \downarrow & & \downarrow \\ X & \longrightarrow & E_n \end{array}$$

Here $X \longrightarrow E_1$ is η-admissible. Now observe that E_1 contains PO_2', thus

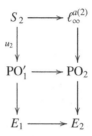

$$\begin{array}{ccc} S_2 & \longrightarrow & \ell_\infty^{a(2)} \\ {\scriptstyle u_2}\downarrow & & \downarrow \\ \mathrm{PO}_1' & \longrightarrow & \mathrm{PO}_2 \\ \downarrow & & \downarrow \\ E_1 & \longrightarrow & E_2 \end{array}$$

yields the new η-admissible map $E_1 \longrightarrow E_2$. Continue this way and consider the direct limit $Y = \lim_n E_n$. This Y is a limit of η-admissible maps, but not between finite-dimensional spaces. However, $Y/X = \lim_n E_n/X$ is a limit of η-admissible maps between finite-dimensional spaces, and thus Y/X is a Schur space with the RNP, as is its subspace $\mathscr{L}_\infty^{\mathsf{BP}}(X)/X$ since, obviously, Y contains $\mathscr{L}_\infty^{\mathsf{BP}}(X)$ and those properties pass to subspaces [155, III. Theorem 2]. The properties of the embedding $X \longrightarrow \mathscr{L}_\infty^{\mathsf{BP}}(X)$ will be studied between Propositions 10.6.9 and 10.6.11. López-Abad extends the Bourgain–Pisier construction to arbitrary Banach spaces in [340].

2.14 Extension and Lifting of Operators

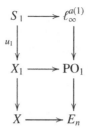

The extension of operators is one of the most classical problems. It consists of determining when, given an operator $\tau\colon Y \longrightarrow E$ and an isomorphic embedding $\jmath\colon Y \longrightarrow X$, there exists an extension $T\colon X \longrightarrow E$; i.e. an operator T such that $T\jmath = \tau$. The lifting problem is entirely dual in its formulation, although maybe not in the results we get: determine when, given a quotient operator $\rho\colon X \longrightarrow Z$ and an operator $\tau\colon E \longrightarrow Z$, there exists a lifting for τ; i.e. an operator $T\colon E \longrightarrow X$ such that $\rho T = \tau$. Homological techniques provide tools to treat these questions (which is what we are doing throughout this book). As a rule, operators between quasi-Banach spaces do not extend. The simplest example to mention is the identity operator: it cannot be extended unless the subspace is complemented. The extension problem admits a natural formulation in homological terms: an operator $\tau\colon Y \longrightarrow E$ and an embedding $\jmath\colon Y \longrightarrow Z$ form the diagram

$$\begin{array}{ccccccccc} 0 & \longrightarrow & Y & \stackrel{\jmath}{\longrightarrow} & Z & \longrightarrow & X & \longrightarrow & 0 \\ & & \downarrow{\scriptstyle \tau} & & \downarrow & & \| & & \\ 0 & \longrightarrow & E & \stackrel{\bar{\jmath}}{\longrightarrow} & \mathrm{PO} & \longrightarrow & X & \longrightarrow & 0 \end{array}$$

The splitting criterion for pushout sequences Lemma 2.6.3 yields that τ can be extended to Z if and only if the lower sequence is trivial. If we consider instead the question of when all operators $Y \longrightarrow E$ can be extended through the embedding $\jmath$ then we see that this happens if and only if the restriction operator $j^\circ : \mathfrak{L}(Z, E) \longrightarrow \mathfrak{L}(Y, E)$ is surjective. A different question, considered throughout this book, is that of when all operators $Y \longrightarrow E$ can be extended through any embedding $Y \longrightarrow Z$ such that $Z/Y = X$. As a rule, operators between quasi-Banach spaces cannot be lifted (a quotient map can be lifted if and only if its kernel is complemented). The lifting problem admits a natural formulation in homological terms: a quotient map $\rho\colon Z \longrightarrow X$ and an operator $\tau\colon E \longrightarrow X$ can be assembled in a pullback diagram

$$\begin{array}{ccccccccc} 0 & \longrightarrow & Y & \longrightarrow & Z & \stackrel{\rho}{\longrightarrow} & X & \longrightarrow & 0 \\ & & \| & & \uparrow & & \uparrow{\scriptstyle \tau} & & \\ 0 & \longrightarrow & Y & \longrightarrow & \mathrm{PB} & \longrightarrow & E & \longrightarrow & 0 \end{array}$$

for which we know (Lemma 2.8.3) that τ can be lifted through ρ if and only if the pullback sequence splits. That all operators $E \longrightarrow X$ can be lifted to Z through ρ can be reformulated as follows: the operator $\rho_\circ : \mathfrak{L}(E, Z) \longrightarrow \mathfrak{L}(E, X)$ is surjective. We will also consider the question whether all operators $E \longrightarrow X$ can be lifted to Z through any quotient $Z \longrightarrow X$ whose kernel is Y.

Extension: $\mathscr{A}$-Trivial Sequences

The following notation unifies different ideas scattered through the literature and, most importantly, is useful. Let $\mathscr{A}$ be a class of quasi-Banach spaces.

Definition 2.14.1 An emdedding $\jmath\colon Y \longrightarrow Z$ is $\mathscr{A}$-trivial if, for every $A \in \mathscr{A}$, every operator $\tau\colon Y \longrightarrow A$ has an extension $T\colon Z \longrightarrow A$. If this can be achieved with $\|T\| \le \lambda\|\tau\|$, we say that $\jmath$ is $(\lambda, \mathscr{A})$-trivial.

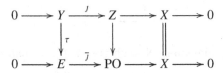

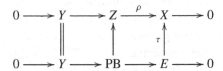

This is really a property of the inclusion of $\jmath[Y]$ in Z, but some flexibility is convenient here. These definitions extend to short exact sequences by declaring sequences to be $\mathscr{A}$-trivial when their embeddings are. If the reader wonders whether $\mathscr{A}$-trivial sequences and homology like to mingle, just observe that given a fixed space A, the (contravariant; see Chapter 4 for details) functor $\mathfrak{L}(\cdot, A)$ takes each exact sequence $0 \longrightarrow Y \xrightarrow{\jmath} Z \xrightarrow{\rho} X \longrightarrow 0$ into the exact sequence

$$0 \longrightarrow \mathfrak{L}(X, A) \xrightarrow{\rho^\circ} \mathfrak{L}(Z, A) \xrightarrow{\jmath^\circ} \mathfrak{L}(Y, A)$$

(open end!); $\jmath^\circ$ is surjective (so the diagram can be completed with a right 0) if and only if the sequence is A-trivial in the obvious sense. Chapter 8 is devoted to $\mathscr{C}$-trivial sequences. The behaviour of $\mathscr{A}$-trivial sequences under pullback/pushout constructions is as follows:

Proposition 2.14.2 *Pullbacks and pushouts preserve $\mathscr{A}$-trivial sequences.*

Proof The case of pullbacks is obvious. Assume we have a pushout diagram

$$\begin{array}{ccccccccc} 0 & \longrightarrow & Y & \xrightarrow{\jmath} & Z & \longrightarrow & X & \longrightarrow & 0 \\ & & \downarrow{\scriptstyle \alpha} & & \downarrow{\scriptstyle \overline{\alpha}} & & \| & & \\ 0 & \longrightarrow & Y' & \xrightarrow{\overline{\jmath}} & \mathrm{PO} & \longrightarrow & X & \longrightarrow & 0 \end{array}$$

in which the upper row is A-trivial. Given an operator $\tau \colon Y' \longrightarrow A$, the composition $\tau\alpha$ extends to an operator $T \colon Z \longrightarrow A$ through $\jmath$. As $T\jmath = \tau\alpha$ the universal property of the pushout yields an operator $\gamma \colon \mathrm{PO} \longrightarrow A$ such that $\gamma\overline{\jmath} = \tau$ and $\gamma\overline{\alpha} = T$. □

Lemma 2.14.3 *In each of the following commutative diagrams, in which any three aligned points represent a short exact sequence of quasi-Banach spaces,*

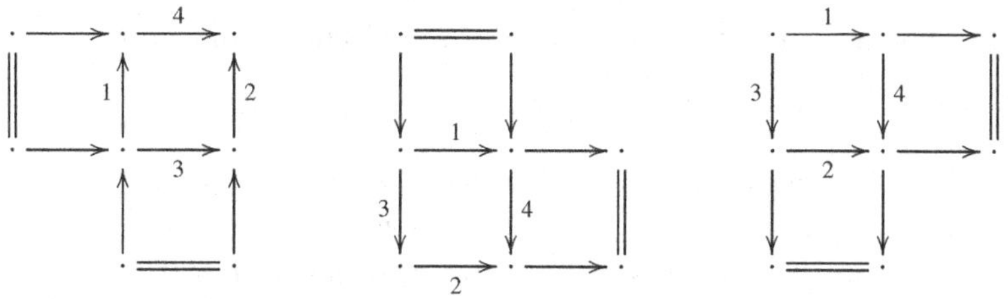

the sequences 1 *and* 3 *are $\mathscr{A}$-trivial if and only if the sequences* 2 *and* 4 *are $\mathscr{A}$-trivial.*

Proof The proof is a mere diagram chase keeping Proposition 2.14.2 in mind. A more homologically flavoured proof can be given. Fix $A \in \mathscr{A}$ and apply the functor $\mathfrak{L}(\cdot, A)$. This transforms the three previous diagrams into

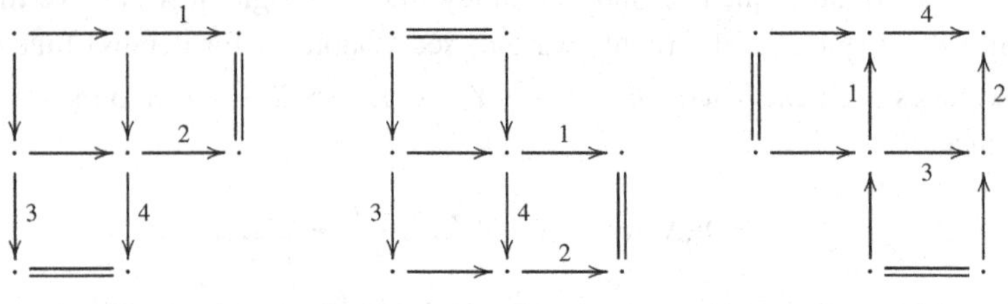

with open right ends now since, we already mentioned, $\mathfrak{L}(\cdot, A)$ is not necessarily exact at the subspace. Which is why we have passed the identification numbers to the 'quotient' maps. It is now easy to check that 1 and 3 are surjective if and only if 2 and 4 are surjective. □

Lifting: M-Ideals

According to the Yellow Book of M-idealism [209, Definition I.1.1], a closed subspace J of a Banach space X is an M-ideal if the annihilator $J^\perp = \{x^* \in X^* : \langle x^*, x\rangle = 0\ \forall x \in J\}$ is an L-summand in X^*, which means that it is the range of an L-projection. Just in case, recall that a projection $P \in \mathfrak{L}(X)$ is called an L-projection if $\|x\| = \|P(x)\| + \|x - P(x)\|$ for all $x \in X$, and it is called an M-projection if $\|x\| = \max(\|P(x)\|, \|x - P(x)\|)$ for all $x \in X$. The simplest examples of M-ideals are the ideals $J = \{f \in C(K) : f|_S = 0\}$ for some closed $S \subset K$ in a $C(K)$-space, as a direct consequence of the Riesz representation of $C(K)^*$. But there are many more:

Lemma 2.14.4 *If $(X_n)_n$ is an increasing sequence of subspaces of X with $X = \overline{\bigcup_n X_n}$ then $c_0(\mathbb{N}, X_n)$ is an M-ideal in $c(\mathbb{N}, X_n)$. If $(A_i)_{i\in I}$ is a family of Banach spaces and $\mathfrak{U}$ is a free ultrafilter on I then $c_0(I, A_i)$ and $c_0^{\mathfrak{U}}(I, A_i)$ are M-ideals in $\ell_\infty(I, A_i)$.*

Proof For the first part, define the desired L-projection in the form

$$\langle P\mu, (x_n)\rangle = \lim_n \langle \mu, (0, \dots, 0, x_n, x_{n+1}, \dots)\rangle$$

for $\mu \in c(\mathbb{N}, X_n)^*$ and $(x_n)_n \in c(\mathbb{N}, X_n)$, which makes sense because μ is applied to a weakly Cauchy sequence. P is a projection onto $c_0(\mathbb{N}, X_n)^\perp$, and if one picks $\varepsilon > 0$ and normalised sequences $(y_n)_n$ and $(z_n)_n$ such that $\langle P\mu, (y_n)\rangle > \|P\mu\| - \varepsilon$ and $\langle \mu - P\mu, (z_n)\rangle > \|\mu - P\mu\| - \varepsilon$, then

$$\|\mu\| \geq \sup_n \langle \mu, (z_1, \dots, z_n, y_{n+1}, y_{n+2}, \dots)\rangle \geq \|P\mu\| + \|\mu - P\mu\| - 2\varepsilon.$$

The second part requires a different approach avoiding duality since the dual of $\ell_\infty(I, A_i)$ is unmanageable. It turns out [209, Theorem I.2.2] that J is an M-ideal in X if and only if, for every finite family of closed balls $B(x^k, r_k)$ in X such that $B(x^k, r_k) \cap J \neq \emptyset$ for all k and every $\varepsilon > 0$, we have

$$\bigcap_k B(x^k, r_k) \neq \emptyset \quad \Longrightarrow \quad \bigcap_k B(x^k, r_k + \varepsilon) \cap J \neq \emptyset.$$

Let us check this condition for $c_0^{\mathcal{U}}(I, A_i)$, the case of $c_0(I, A_i)$ being simpler. Let $B(x^k, r_k)$ be the corresponding balls, and take $x = (x_i)$ in their intersection. Also, for each k, pick $y^k \in B(x^k, r_k) \cap c_0^{\mathcal{U}}(I, A_i)$. Now, given $\varepsilon > 0$, as $\|y_i^k\| \longrightarrow 0$ along $\mathcal{U}$, we may find I_ε in $\mathcal{U}$ such that $\|y_i^k\| \leq \varepsilon$ for all k and all $i \in I_\varepsilon$. If we define $y = (y_i)$, setting $y_i = 0$ for $i \in I_\varepsilon$ and $y_i = x_i$ otherwise, it is clear that $y \in \bigcap_k B(x^k, r_k + \varepsilon) \cap c_0^{\mathcal{U}}(I, A_i)$. □

The following remarkable connection between M-ideals and lifting properties was discovered by Ando [11] and, almost simultaneously and independently, in a slightly weaker form, by Choi and Effros [138].

Theorem 2.14.5 *Let J be an M-ideal in the Banach space Z, and let Y be a separable Banach space with the λ-AP. Every operator $T\colon Y \longrightarrow Z/J$ admits a lifting $L\colon Y \longrightarrow Z$ such that $\|L\| \leq \lambda\|T\|$.*

The result can be seen as a wide generalisation of the Borsuk–Dugundji theorem. A complete proof can be found in [209, Theorem II.2.1]. What we present here is a modulo ε proof that suffices for our qualitative purposes. A few comments on the $\varepsilon = 0$ case can be found at the end of this section. A simple observation to warm up is that if Q is an M-projection on a Banach space X then

$$\|Q(x) + (\mathbf{1}_X - Q)(y)\| \leq \max\{\|x\|, \|y\|\}. \tag{2.44}$$

The crucial step in the proof of Theorem 2.14.5 is the following magical

Lemma 2.14.6 *Let J be an M-ideal in the Banach space Z. Let E be a finite-dimensional space, and let F be a 1-complemented subspace of E. Let $T\colon E \longrightarrow Z/J$ be an operator. Then, for every lifting $L_F\colon F \longrightarrow Z$ of $T|_F$ and every $\varepsilon > 0$, there is a lifting $L_E\colon E \longrightarrow Z$ of T such that $L_E = L_F$ on F and $\|L_E\| < (1 + \varepsilon)\max(\|T\|, \|L_F\|)$.*

Proof The lifting we are looking for is a point of small norm in the space $\mathfrak{L}(E, Z)$. The action takes place in the bidual $\mathfrak{L}(E, Z)^{**}$, identified with $\mathfrak{L}(E, Z^{**})$ thanks to Dean's identity [145]. The simplest linear functionals on $\mathfrak{L}(E, Z)$ have the form $S \longmapsto \langle z^*, S(e)\rangle$, for fixed $z^* \in Z^*, e \in E$, and these generate the dual of $\mathfrak{L}(E, Z)$. Thus, typical neighbourhoods of zero in the weak topology of $\mathfrak{L}(E,Z)$ and the weak* topology of $\mathfrak{L}(E,Z^{**})$ are given, respectively, by $\{S \in \mathfrak{L}(E,Z) \colon |\langle x^*, Se\rangle| \le 1\}$ and $\{S \in \mathfrak{L}(E,Z^{**}) \colon |\langle x^*, Se\rangle| \le 1\}$, where $x^* \in X^*$ and $e \in E$. We consider the following subspaces of $\mathfrak{L}(E,Z)$:

- $W = \{S \in \mathfrak{L}(E,Z) \colon S[E] \subset J\} = \mathfrak{L}(E,J)$.
- $V = \{S \in W \colon S|_F = 0\}$.

Clearly, $W^{\perp\perp} = \{S \in \mathfrak{L}(E,Z^{**}) \colon S[E] \subset J^{\perp\perp}\} = \mathfrak{L}(E,J^{\perp\perp})$ and $V^{\perp\perp} = \{S \in W^{\perp\perp} \colon S|_F = 0\}$. The operator $Q_\circ$ sending T to QT is a weak* continuous M-projection on $\mathfrak{L}(E,Z)^{**}$ of range $W^{\perp\perp}$; in fact, W is an M-ideal in $\mathfrak{L}(E,Z)$, but this fact is not to be used in the ensuing argument. Let $T \in \mathfrak{L}(E,Z/J)$ be an operator and let $L_F \colon F \longrightarrow Z$ be a lifting of $T|_F$. Fix a contractive projection $\pi \colon E \longrightarrow F$. Let $L \in \mathfrak{L}(E,Z)$ be *any* lifting of T such that $L|_F = T_F$ and, using the two projections at hand, write

$$L = (\mathbf{1}_{Z^{**}} - Q)L + QL = (\mathbf{1}_{Z^{**}} - Q)L + QL\pi + QL(\mathbf{1}_E - \pi). \tag{2.45}$$

Obviously, $QL(\mathbf{1}_E - \pi) \in V^{\perp\perp}$. With an eye on (2.44), we want to bound the other two chunks: we have $QL\pi = QL_F\pi$, so $\|QL\pi\| \le \|L_F\|$. As for the first summand, since $\mathbf{1}_{Z^{**}} - Q = P^*$ vanishes on $J^{\perp\perp}$, we have a commutative diagram

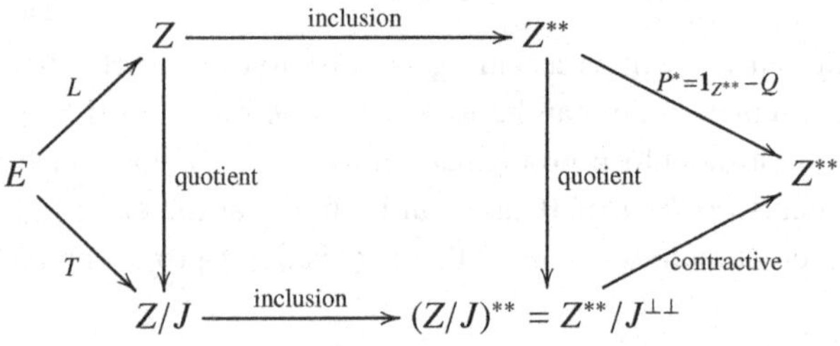

and so $\|(\mathbf{1}_{Z^{**}} - Q)L\| \le \|T\|$. Thus, $\|(\mathbf{1}_{Z^{**}} - Q)L + QL\pi\| \le \max(\|(\mathbf{1}_{Z^{**}} - Q)L\|, \|QL\pi\|) \le \max(\|T\|, \|L_F\|)$. Hence, letting $r = \max(\|T\|, \|L_F\|)$,

$$L \in rB_{\mathfrak{L}(E,Z^{**})} + V^{\perp\perp} = \overline{rB_{\mathfrak{L}(E,Z)}}^{\text{weak*}} + \overline{V}^{\text{weak*}} = \overline{rB_{\mathfrak{L}(E,Z)} + V}^{\text{weak*}},$$

and since $L \in \mathfrak{L}(E,Z)$, the weak* topology of $\mathfrak{L}(E,Z^{**})$ restricted to $\mathfrak{L}(E,Z)$ is its weak topology, and the weak and norm closures of convex sets coincide, $L \in \overline{rB_{\mathfrak{L}(E,Z)} + V}$. Therefore, given $\varepsilon > 0$, there exist $L' \in \mathfrak{L}(E,Z)$ with $\|L'\| \le r$ and $S \in V$ such that $\|L - L' - S\| < \varepsilon$, and so $L_E = L - S$ is a lifting of T that extends L_F, with $\|L_E\| \le r + \varepsilon$, which is enough. □

Proof of Theorem 2.14.5 It suffices to prove the result assuming that Y has a 1-FDD by Lemma 2.2.20. Write $Y = \overline{\bigcup_n Y_n}$ for an increasing chain of finite-dimensional subspaces $(Y_n)_{n\ge 0}$ with $Y_0 = 0$ and Y_n 1-complemented in Y_{n+1}. Fix $\varepsilon > 0$, and pick a sequence $(\varepsilon_n)_n$ such that $\prod_n(1 + \varepsilon_n) < 1 + \varepsilon$. Use the preceding lemma inductively to find a sequence of operators $L_n\colon Y_n \longrightarrow Z$ such that (a) L_n is a lifting of $T|_{Y_n}$; (b) L_{n+1} is an extension of L_n and (c) $\|L_n\| \le \prod_{1\le k\le n}(1+\varepsilon_k)$. These operators define the required lifting of T. □

The hypotheses on Y of Theorem 2.14.5 cannot be removed without asking something else. The separability is necessary: consider the family of Hilbert spaces $(\ell_2^n)_{n\ge 1}$, a free ultrafilter $\mathcal{U}$ on $\mathbb{N}$ and the exact sequence defining the ultraproduct $0 \longrightarrow c_0^{\mathcal{U}}(\mathbb{N},\ell_2^n) \longrightarrow \ell_\infty(\mathbb{N},\ell_2^n) \longrightarrow [\ell_2^n]_{\mathcal{U}} \longrightarrow 0$. Since $[\ell_2^n]_{\mathcal{U}}$ is a Hilbert space of density $\mathfrak{c}$, every subspace Y has 1-AP, but if Y is non-separable then the inclusion $Y \longrightarrow [\ell_2^n]_{\mathcal{U}}$ cannot be lifted to $\ell_\infty(\mathbb{N},\ell_2^n)$ because this space can be separated by a countable family of functionals, something that Y cannot. The BAP is also necessary, as shown by Proposition 2.2.19. But one could simply ask that Z/J has BAP (it is in this form that the result will be used in Section 10.1) or, in general, that the operator $T\colon Y \longrightarrow Z/J$ factorises through a space with BAP. Another variation of the result is possible by asking for J to be a Lindenstrauss space: in that case, $J^{\perp\perp}$ is an 1-injective Banach space, and $QL_F\colon F \longrightarrow J^{\perp\perp}$ has an extension $\Lambda\colon E \longrightarrow J^{\perp\perp}$ with $\|\Lambda\| \le \|L_F\|$. Now use the decomposition $L = ((\mathbf{1}_{Z^{**}} - Q)L + \Lambda) + (QL - \Lambda)$ instead of (2.45) and proceed as in the proof. The final estimate in this case is better: $\|L\| = \|T\|$. The *lifting* Theorem 2.14.5 has the following application to the *extension* of operators:

Corollary 2.14.7 *Let J be an M-ideal in A, and let $\tau\colon Y \longrightarrow J$ be an operator, where Y is a subspace of the Banach space Z such that Z/Y is separable. We denote the inclusion maps by $\jmath\colon Y \longrightarrow Z$ and $\imath\colon J \longrightarrow A$. Assume that*

- *there is an operator $\tau_A\colon Z \longrightarrow A$ such that $\tau_A \jmath = \imath\tau$, with $\|\tau_A\| \le \mu\|\tau\|$,*
- *Z/Y or A/J has the λ-AP.*

Then τ has a $\mu(1+\lambda)$ extension $T\colon Z \longrightarrow J$.

Proof Assume $\|\tau\| \le 1$, and display all the information at hand in the diagram

$$\begin{array}{ccccccccc} 0 & \longrightarrow & Y & \xrightarrow{\ \jmath\ } & Z & \xrightarrow{\ \rho\ } & Z/Y & \longrightarrow & 0 \\ & & \downarrow{\scriptstyle \tau} & & \downarrow{\scriptstyle \tau_A} & & \downarrow{\scriptstyle \tau'} & & \\ 0 & \longrightarrow & J & \xrightarrow{\ \imath\ } & A & \xrightarrow{\ \pi\ } & A/J & \longrightarrow & 0 \end{array}$$

where τ' is induced by the fact that $\pi\tau_A\jmath = \pi\imath\tau = 0$. The operator τ' has a lifting $L\colon Z/Y \longrightarrow A$ with $\|T\| \le \lambda\|\tau'\| \le \lambda\mu$. The required extension is $T = \tau_A - L\rho$: this operator takes values in J since $\pi T = \pi\tau_A - \pi L\rho = \pi\tau_A - \tau'\rho = 0$. Obviously, $T|_Y = \tau$ and, finally, $\|T\| \le \|\tau_A\| + \|L\rho\| \le \mu(1+\lambda)$. □

The pervasive chronicle of the quest for Sobczyk's theorem, as related in this book, departs from Sections 1.7 and 1.8.4, crosses through this section here with the obtention of the following optimal vector-valued version and will meander through 5.2.5 and its consequences, to arrive at 10.1 in a rather satisfactory conclusion.

2.14.8 Vector-valued Sobczyk's theorem *Let Y be a subspace of a separable Banach space Z such that Z/Y has the λ-AP. Let $\tau\colon Y \longrightarrow c_0(\mathbb{N}, E_n)$ be an operator. If each $\pi_n\tau\colon Y \longrightarrow E_n$ admits a μ-extension to Z then τ admits a $\mu(1+\lambda)$-extension to Z.*

Proof Let $\imath\colon c_0(\mathbb{N}, E_n) \longrightarrow \ell_\infty(\mathbb{N}, E_n)$ denote the canonical inclusion. Since each $\pi_n\tau$ admits a λ-extension to Z, the operator $\imath\tau\colon Y \longrightarrow \ell_\infty(\mathbb{N}, E_n)$ admits a μ-extension $T\colon Z \longrightarrow \ell_\infty(\mathbb{N}, E_n)$ as in the diagram:

$$\begin{array}{ccccccccc}
0 & \longrightarrow & Y & \longrightarrow & Z & \longrightarrow & Z/Y & \longrightarrow & 0\\
 & & \downarrow{\scriptstyle \tau} & & \downarrow{\scriptstyle T} & & \downarrow & & \\
0 & \longrightarrow & c_0(\mathbb{N}, E_n) & \xrightarrow{\ \imath\ } & \ell_\infty(\mathbb{N}, E_n) & \longrightarrow & \ell_\infty(\mathbb{N}, E_n)/c_0(\mathbb{N}, E_n) & \longrightarrow & 0
\end{array}$$

Since Z/Y has the λ-AP, τ admits a $\mu(1+\lambda)$-extension $Z \longrightarrow c_0(\mathbb{N}, E_n)$. □

We are not that interested in removing the ε appearing in the proof of Theorem 2.14.5. But a sharp proof in the simplest conceivable case of rank one operators is definitely worthwhile:

2.14.9 M-ideals are proximinal *If J is an M-ideal in Z then for every $z \in Z$, there is $y \in J$ such that $\|z - y\| = \|z + J\|$.*

Proof It is clear that if J is an M-ideal in Z and $J \subset X \subset Z$ then J is an M-ideal in X. Hence, it suffices to prove the result when J is a hyperplane of Z. Write $J = \ker\rho$, where ρ is normalised in Z^*, and observe that what one must show is that ρ attains the norm on B_Z. Write $Z^* = J^\sharp \oplus_1 J^\perp$, where $J^\sharp = \{f \in Z^*\colon \|f\| = \|f|_J\|\}$. By the Bishop–Phelps theorem, not every norm-attaining functional can be in $J^\sharp$. Pick a normalised, norm-attaining $f \in Z^*\backslash J^\sharp$ and then $z \in Z$ such that $\langle f, z\rangle = 1$. Obviously, $z \notin J$; writing $f = g + h$, with $g \in J^\sharp$ and h non-zero in $J^\perp$, we have $1 = \langle f, z\rangle \leq |\langle g, z\rangle| + |\langle g, z\rangle| \leq \|g\| + \|h\| = 1$, hence h attains the norm at z, and so does ρ. □

2.15 Notes and Remarks

2.15.1 Categorical Limits

The simplest diagram is formed by just two points $\bullet \quad \bullet$ and no arrows, except identities that we will not draw. The limit of this diagram in a category is an object $\prod$ of the category and two arrows $\bullet \leftarrow \prod \rightarrow \bullet$ with the universal property: for any other object $\diamond$ yielding a similar diagram $\bullet \leftarrow \diamond \rightarrow \bullet$ there is a unique arrow $\diamond \longrightarrow \prod$ filling a commutative diagram

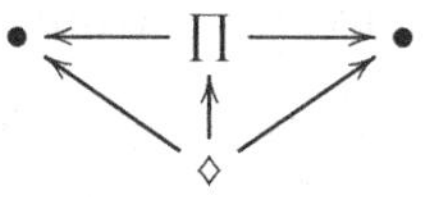

Anyone who does not find it evident that the limit of diagram $\bullet \quad \bullet$ is the product (and its colimit, the coproduct $\coprod$, usually named the direct sum in topological surroundings) in the time a chestnut takes to drop from a stool can skip this section with no permanent harm. Let's nonetheless pretend we already know what a category and a functor are. An abstract diagram **D** made with points and arrows is itself a category, a *small* category for what it is worth, that can be depicted by means of a directed graph. For instance, the diagram on the left in

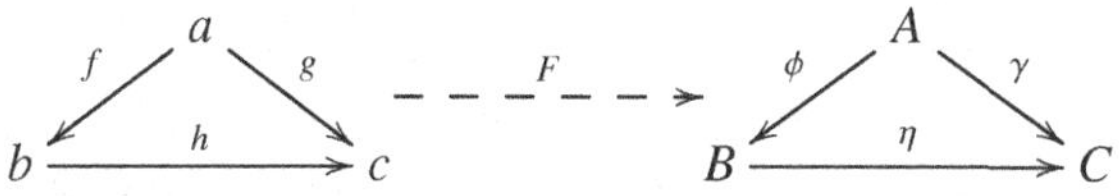

represents the not-so-entertaining category having three objects a, b, c and the following sets of morphisms: $\mathrm{Hom}(a, b) = \{f\}$, $\mathrm{Hom}(a, c) = \{g, hf\}$, $\mathrm{Hom}(b, c) = \{h\}$. The other sets of morphisms are either empty or consist of the corresponding identities. Given another category **C**, a functor $F\colon \mathbf{D} \longrightarrow \mathbf{C}$ means just filling **D** with objects and arrows of **C**, as in the preceding diagram. Think of Banach spaces and operators instead of points and arrows, as represented on the right side of the drawing. The limit $\lim F$ of F is an object of **C** together with a family of arrows $(\alpha_d\colon F(d) \longrightarrow \lim F)_d$ parametrised by the points of **D** satisfying the following conditions:

- Compatibility: if $s\colon d \longrightarrow e$ is an arrow of **D** then $\alpha_d = \alpha_e F(s)$.
- Universality: for any other object $X \in \mathbf{C}$ and a system of arrows $(\alpha_d\colon F(d) \longrightarrow X)_d$ with the same property, there is a unique arrow $\xi\colon \lim F \longrightarrow X$ such that $\xi\alpha_d = \beta_d$ for all d.

The colimit (also called inverse limit) is defined by considering 'arrows from' instead of 'arrows into'. The universal mapping property of these objects guarantees that they are unique, up to isomorphism in the corresponding category. It is obvious that even the simplest infinite diagram cannot have a limit in **B** since the fact that operators have norms prevents it. The groundbreaking result in this regard is from Semadeni and Zidenberg [431]: *Every diagram in* $\mathbf{B}_1$ *admits a limit and a colimit.* All mathematical constructions are (co)limits, or so the Eilenberg–MacLane programme [165] says.

2.15.2 How to Draw More Diagrams

Working with diagrams is simple, extremely rewarding and a little addictive. Powerful, too, sometimes. This is so because the mere drawing of an exact sequence $0 \longrightarrow Y \longrightarrow Z \longrightarrow X \longrightarrow 0$ encodes much more information than its Banach space counterpart 'Y is a subspace of Z in such a way that Z/Y is isomorphic to X'. It also contains the assertions '$Y \longrightarrow Z$ is the kernel of $Z \longrightarrow X$' and '$Z \longrightarrow X$ is the cokernel of $Y \longrightarrow X$'. In turn, the former of these two assertions contains more information than $Y = \{x \in Z \colon \rho x = 0\}$, while the latter contains more information than $X \simeq Z/Y$. And the amount of information grows exponentially with each new arrow we add. Working with diagrams requires us to follow some rules too. The rule that drawings should be *complete*, meaning they should start and end in 0, means that if yours is not then something has been overlooked. And the rule is subtly demoed when there is some (categorical) construction that makes the diagram complete. For instance, an operator $T \colon A \longrightarrow B$ is an incomplete diagram $A \longrightarrow B$, and thus kernels and cokernels are there to complete it as $0 \longrightarrow \ker T \longrightarrow A \longrightarrow B \longrightarrow \operatorname{coker} T \longrightarrow 0$. To complete more complex diagrams, one will need more complex constructions. Let us see some.

The diamond lemma. This is an elementary result in linear algebra: if A and B are linear subspaces of V then the quotients $(A + B)/A$ and $B/(A \cap B)$ are isomorphic. The name comes from the figure

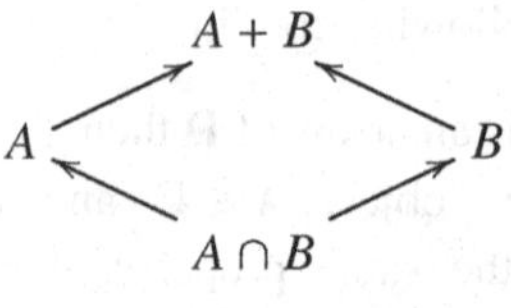

where the arrows mean containment. To transform this linear isomorphism into a linear *homeomorphism* when working in **Q** requires an additional hypothesis to ensure all the spaces are complete: if A, B are complete, then $A + B$ is complete, assuming that either A or B is finite-dimensional or A and B are *totally incomparable* (do not have infinite-dimensional isomorphic subspaces) [410].

The snake lemma. To complete the commutative diagram

$$\begin{array}{ccccccccc}
0 & \longrightarrow & A & \xrightarrow{\jmath} & B & \xrightarrow{\rho} & C & \longrightarrow & 0 \\
 & & \downarrow{\scriptstyle\alpha} & & \downarrow{\scriptstyle\beta} & & \downarrow{\scriptstyle\gamma} & & \\
0 & \longrightarrow & A' & \xrightarrow{\jmath'} & B' & \xrightarrow{\rho'} & B' & \longrightarrow & 0
\end{array}$$

with exact rows, we can start by drawing the sequence of kernels and cokernels. It is easy to check that one gets the following commutative diagram with exact rows and columns:

$$\begin{array}{ccccccccc}
 & & 0 & & 0 & & 0 & & \\
 & & \downarrow & & \downarrow & & \downarrow & & \\
0 & \longrightarrow & \ker\alpha & \longrightarrow & \ker\beta & \longrightarrow & \ker\gamma & & \\
 & & \downarrow & & \downarrow & & \downarrow & & \\
0 & \longrightarrow & A & \longrightarrow & B & \longrightarrow & C & \longrightarrow & 0 \\
 & & \downarrow{\scriptstyle\alpha} & & \downarrow{\scriptstyle\beta} & & \downarrow{\scriptstyle\gamma} & & \\
0 & \longrightarrow & A' & \longrightarrow & B' & \longrightarrow & C' & \longrightarrow & 0 \\
 & & \downarrow & & \downarrow & & \downarrow & & \\
 & & \operatorname{coker}\alpha & \longrightarrow & \operatorname{coker}\beta & \longrightarrow & \operatorname{coker}\gamma & \longrightarrow & 0 \\
 & & \downarrow & & \downarrow & & \downarrow & & \\
 & & 0 & & 0 & & 0 & &
\end{array} \tag{2.46}$$

But, still, this is not complete: have you seen the open end at the topmost right corner? And the open start at the lowest left corner? The snake lemma says that the diagram can be completed with a *connecting morphism* ω: $\ker\gamma \longrightarrow$

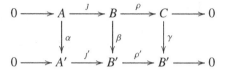

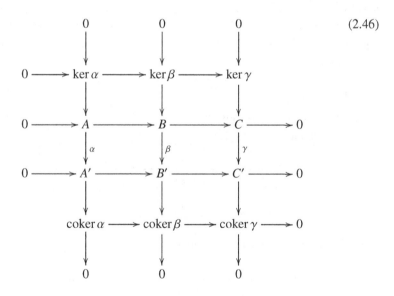

cokerα, whose construction is mere diagram chasing (promise), yielding a 'long exact sequence'

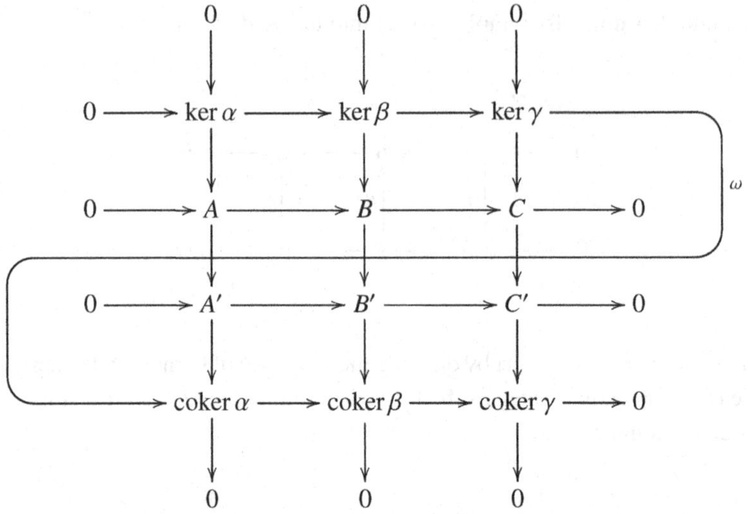

In applications, it shall be expected that α, β and γ shall be either embeddings or quotient maps, so that either ker or coker will be 0. This means that the sequence of quotient spaces (case of embeddings) or kernels (case of quotient maps) in the completed diagram is also exact. In particular, we have verified that the test Diagram (2.1) in the introduction exercise of the chapter was correctly drawn.

More pullback and pushout diagrams. Diagram (2.32) = (2.33) is the diagram we obtain by completing

$$\begin{array}{ccccccccc} 0 & \longrightarrow & Y & \xrightarrow{\jmath} & Z & \xrightarrow{\rho} & X & \longrightarrow & 0 \\ & & & & \uparrow{\scriptstyle \beta} & & & & \\ & & & & Z' & & & & \end{array}$$

via pullback when β is a quotient map to obtain

$$\begin{array}{ccccccccc} 0 & \longrightarrow & Y & \xrightarrow{\jmath} & Z & \xrightarrow{\rho} & X & \longrightarrow & 0 \\ & & \uparrow{\scriptstyle \underline{\beta}} & & \uparrow{\scriptstyle \beta} & & \uparrow{\scriptstyle \overline{\beta}} & & \\ 0 & \longrightarrow & \mathrm{PB} & \longrightarrow & Z' & \longrightarrow & Z'/\mathrm{PB} & \longrightarrow & 0 \end{array}$$

If, however, β is an embedding then no simple outcome exists. Indeed, imagine for simplicity's sake that $\jmath, \beta$ are natural inclusions. Then $\mathrm{PB} = Y \cap Z'$.

The diamond lemma yields that, if $Y + Z'$ is closed then $Z'/\mathrm{PB} = Z'/(Y \cap Z') = (Y + Z')/Y$ is a subspace of X. But if not then $\overline{\beta}$ is just an injective operator. Assuming that $Y + Z'$ is closed, the complete diagram is

$$\begin{array}{ccccccccc} & & 0 & & 0 & & 0 & & \\ & & \uparrow & & \uparrow & & \uparrow & & \\ 0 & \longrightarrow & Y/\mathrm{PB} & \longrightarrow & Z/Z' & \longrightarrow & Q & \longrightarrow & 0 \\ & & \uparrow & & \uparrow & & \uparrow & & \\ 0 & \longrightarrow & Y & \xrightarrow{\jmath} & Z & \xrightarrow{\rho} & X & \longrightarrow & 0 \\ & & \uparrow & & \uparrow{\scriptstyle\beta} & & \uparrow & & \\ 0 & \longrightarrow & \mathrm{PB} & \longrightarrow & Z' & \longrightarrow & Z'/\mathrm{PB} & \longrightarrow & 0 \\ & & \uparrow & & \uparrow & & \uparrow & & \\ & & 0 & & 0 & & 0 & & \end{array} \tag{2.47}$$

Completing the dual diagram

$$\begin{array}{ccccccccc} 0 & \longrightarrow & Y & \xrightarrow{\jmath} & Z & \xrightarrow{\rho} & X & \longrightarrow & 0 \\ & & & & \downarrow{\scriptstyle\beta} & & & & \\ & & & & Z' & & & & \end{array}$$

via a pushout, when β is moreover surjective, yields

$$\begin{array}{ccccccccc} & & 0 & & 0 & & 0 & & \\ & & \downarrow & & \downarrow & & \downarrow & & \\ 0 & \longrightarrow & \ker \underline{\beta} & \xrightarrow{\underline{\jmath}} & \ker \beta & \xrightarrow{\underline{\rho}} & \ker \overline{\beta} & \longrightarrow & 0 \\ & & \downarrow{\scriptstyle\iota_-} & & \downarrow{\scriptstyle\iota} & & \downarrow{\scriptstyle\iota^-} & & \\ 0 & \longrightarrow & Y & \xrightarrow{\jmath} & Z & \xrightarrow{\rho} & X & \longrightarrow & 0 \\ & & \downarrow{\scriptstyle\underline{\beta}} & & \downarrow{\scriptstyle\beta} & & \downarrow{\scriptstyle\overline{\beta}} & & \\ 0 & \longrightarrow & \ker \overline{\rho} & \longrightarrow & Z' & \xrightarrow{\overline{\rho}} & \mathrm{PO} & \longrightarrow & 0 \\ & & \downarrow & & \downarrow & & \downarrow & & \\ & & 0 & & 0 & & 0 & & \end{array}$$

where ι_-, ι, ι^- are the corresponding inclusions. The reader might doubt that $\underline{\beta}$ is surjective. But take $x_1 \in \ker \overline{\rho}$ and find $x \in Z$ such that $\beta x = x_1$. This and the commutativity of the pushout square imply $\overline{\beta}\rho x = \overline{\rho}\beta x = 0$, and thus

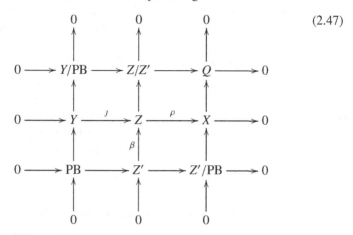

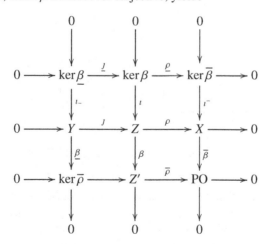

$(0, \rho x) \in \Delta(\beta, \rho)$. Thus, there exists $x' \in X$ such that $(0, \rho x) = (\beta x', \rho x')$, thus giving $x - x' \in Y$ and $\beta(x') = 0$. Therefore, $\beta(x - x') = \beta(x) = x_1$. The equality $\ker \underline{\beta} = \ker \underline{\rho}$ is standard, and the exactness of the upper sequence of kernels is due to the snake lemma. An attentive reader should have realised that we have already encountered this diagram: turn Diagram (2.47) upside down, and there it is! In fact, the top-most left corner space $\ker \underline{\beta}$ is the pullback space of the two operators ι, $\jmath$ that point at Z; to see this, assume that for some space A and arrows $a\colon A \longrightarrow \ker \beta$ and $b\colon A \longrightarrow Y$, we have $\jmath b = \iota a$. Then $\beta \jmath b = \beta \iota a = 0$, and thus $\underline{\beta} b = \beta \jmath b = 0$, which means that b factorises through $\ker \underline{\beta}$ as $b = \underline{\iota} u$ for some operator $u\colon A \longrightarrow \ker \underline{\beta}$. The other equality $\underline{\jmath} u = a$ just follows from $\iota \underline{\jmath} u = \jmath \underline{\iota} u = \jmath b = \iota a$ and the injectivity of ι. When β is an embedding then $\underline{\beta}$ is an embedding, and we get – turn the diagram upside down – Diagram (2.47) again.

2.15.3 Amalgamation of Sequences

The much less frequently used multiple pullback also exists [110; 365]. Other diagonals concealed in complete pullback / pushout diagrams, such as

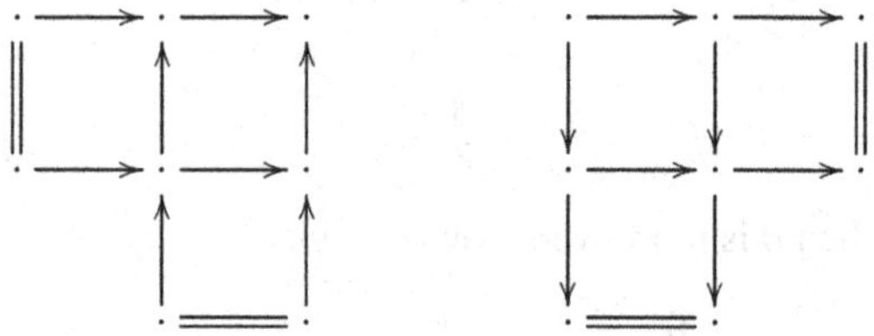

exist too:

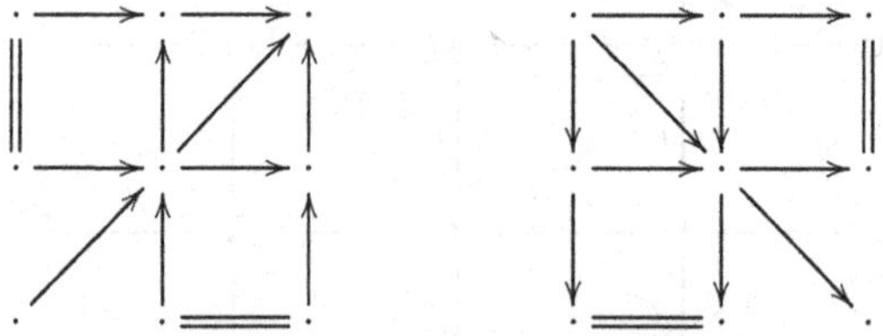

And they exist in complete multiple pushput / pullback diagrams as well. The left-upwards diagonal can be understood as the product of the two (or many) data sequences $0 \longrightarrow Y_i \longrightarrow \diamond_i \longrightarrow X \longrightarrow 0$ in the appropriate category, and it will typically have the form $0 \longrightarrow \ell_\infty(I, Y_i) \longrightarrow \mathrm{PB} \longrightarrow X \longrightarrow 0$; while the right-downwards diagonal can be understood as the coproduct of the two (or many) data sequences $0 \longrightarrow Y \longrightarrow \diamond_i \longrightarrow X_i \longrightarrow 0$ and will typically have the form $0 \longrightarrow Y \longrightarrow \mathrm{PO} \longrightarrow \ell_p(I, X_i) \longrightarrow 0$ in $p\mathbf{B}$.

2.15.4 Categories of Short Exact Sequences

We dealt with categories, we deal with exact sequences and we will deal with commutative diagrams

$$\begin{array}{ccccccccc} 0 & \longrightarrow & \cdot & \longrightarrow & \cdot & \longrightarrow & \cdot & \longrightarrow & 0 \\ & & \downarrow{\scriptstyle\alpha} & & \downarrow{\scriptstyle\beta} & & \downarrow{\scriptstyle\gamma} & & \\ 0 & \longrightarrow & \cdot & \longrightarrow & \cdot & \longrightarrow & \cdot & \longrightarrow & 0 \end{array} \tag{2.48}$$

And, call us Ishmael, we are just one whale away from sinking into the ocean of forming our own category of exact sequences. Indeed, the preceding diagram can be regarded as a 'morphism of sequences', an idea that is implicit in the notion of isomorphic sequences. A natural first attempt is to set exact sequences as objects and triples of arrows (α,β,γ), as in Diagram (2.48), as morphisms. If one does so, *isomorphic* objects are isomorphic exact sequences, and we can keep sailing. But we would also like to consider as objects equivalence classes of exact sequences, in which case triples of arrows are no longer well suited (after all, *where* should one define β?). So, let us declare that morphisms in this category are pairs (α,γ) such that $[\alpha\,\mathsf{z}] = [\mathsf{z}'\gamma]$, as we did in fact, in (2.28), where it was shown that if (α,β,γ) represents a morphism from z to z' then $[\alpha\,\mathsf{z}] = [\mathsf{z}'\gamma]$. With these objects and morphisms, our class of isomorphisms has changed [110, Proposition 3.1]: two objects are *isomorphic in this category* if and only if they have isomorphic multiples (understanding that the multiples of a sequence z are the sequences $E \times \mathsf{z}$ and $\mathsf{z} \times E$). There are other possibilities for forming categories of short exact sequences, such as: following the uses of the theory of complexes in homology and considering *homotopic* triples [213] – a slightly different form of equivalence that is pointless to define here: we will encounter it but one more time, long after all this is over, and the outcome of that meeting will not be satisfactory for either of us – or by fixing the start or the end spaces in exact sequences [110] and determining equality in terms of pullback/pushout.

Sources

The (arguably) more d(iag)ramatic than necessary example leading to Diagram (2.1) appeared in [459], albeit with a far more innocuous purpose. The notion of isomorphic sequences in 2.1.6 appears perhaps for the first time in [65] and [107]. The analysis of the Foiaş–Singer sequences is, broadly speaking, as in [73], even though these ideas can be traced back to Ditor [158; 159]. Proposition 2.2.5 provides (many) continuous surjections $\Delta \longrightarrow [0,1]$ without averaging operators. Typical examples are Cantor's dyadic expansion

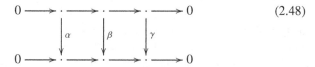

$\varepsilon \in \{0,1\}^{\mathbb{N}} \longmapsto \sum_n \varepsilon_n 2^{-n-1} \in [0,1]$ and Lebesgue's ternary one $\{-1,0,1\}^{\mathbb{N}} \longrightarrow [-1,1]$ given by $\varepsilon \longmapsto \sum 2\varepsilon_n 3^{-n}$; see [430, 8.3.2], [458, III.D.Ex. 4], [5, Proposition 4.4.6]. The crux in Milutin's theorem is to prove that surjections $\Delta \longrightarrow [0,1]$ admitting averaging operators do exist: each of them provides a complemented copy of $C[0,1]$ in $C(\Delta)$. This was shown by Milutin in [364] with a rather involved construction (see also [377, Lemma 5.5]); some simplifications are available: see [5, Lemma 4.4.7] or [458, III.D.18 Proposition]. Argyros and Arvanitakis stablish a clean criterion for a surjection $\Delta \longrightarrow [0,1]$ to admit an averaging operator and conclude that all maps $\varphi_r(\varepsilon) = (1-r)\sum_{n\geq 1} \varepsilon_n r^{n-1}$ admit one when $r \in (\frac{1}{2}, 1)$; see [15, Theorems 2 and 12]. Proposition 2.2.19 (and its proof) is from Lusky [347]. Whether or not it could be attributed to Pełczyński [382] is left to the reader's opinion. Read from an expert about the origin and development of the notion of categorical limit in [350, p.76]. Assertion 2.7.1 is due to Ortyński [372, Theorem 2]. There is a continued long-standing tradition in functional analysis of reinventing the pushout: we can mention, in chronological order, Gurariy [203], Dierolf [149], Kalton [247; 248], Kisliakov [294], Lusky [346], Pisier [389] and Kuchment [313], but admission to this club is still open. A very welcome more recent way to join the club is to reinvent the pushout in a different related category, such as the different *complemented pushouts* of [116] modeled on ideas of Garbulińska and W. Kubiś [184; 308], or to work in the category of Banach spaces and *pairs* (see Chapter 6 and Section 10.7). The Device presented in Section 2.13 is, in a more or less recognisable form, in Kalton [248, Lemma 4.2], Lusky [346, Lemma] (applied to an infinite set of operators) and Pisier [389, Corollary 2.3] (applied to *three* operators and including estimates for the norms of the involved operators). The parallel principles appeared in [121], while the diagonal principles appeared in [109] with the declared purpose of understanding the Lindenstrauss–Rosenthal theorem. Lemma 2.12.2 is from Benyamini and Lindenstrauss [41] and obtained with the purpose of doing what the title says. The 3-space problem for the Dunford–Pettis property has seen a few false positive answers, a few interesting partial solutions and a counterexample [102]. The general result in Proposition 2.12.5 is from [124]. Vogt's duality problem was treated by Díaz, Dierolf, Domański and Fernández in [150], providing a nice counterexample in the context of Fréchet spaces. Reformulated in diagrams, their solution can be read as follows: everything stems from the existence of a Fréchet–Montel space FM admitting ℓ_1 as a quotient (see [303]). Let N be a subspace of ℓ_1 that is not complemented in its bidual (say, the kernel of a quotient map $\ell_1 \longrightarrow L_1$). Form the pullback diagram

$$
\begin{array}{ccccccccc}
 & & & & 0 & & 0 & & \\
 & & & & \uparrow & & \uparrow & & \\
 & & & & \mathsf{FM}/\mathrm{PB} & = & \ell_1/N & & \\
 & & & & \uparrow & & \uparrow & & \\
0 & \longrightarrow & K & \longrightarrow & \mathsf{FM} & \longrightarrow & \ell_1 & \longrightarrow & 0 \\
 & & \| & & \uparrow & & \uparrow & & \\
0 & \longrightarrow & K & \longrightarrow & \mathrm{PB} & \longrightarrow & N & \longrightarrow & 0 \\
 & & & & \uparrow & & \uparrow & & \\
 & & & & 0 & & 0 & &
\end{array}
$$

with diagonal pullback sequence $0 \longrightarrow \mathrm{PB} \longrightarrow \mathsf{FM} \times N \longrightarrow \ell_1 \longrightarrow 0$. Since PB is a closed subspace of FM, it is itself a Fréchet–Montel space, hence reflexive and thus a dual space; as is ℓ_1. On the other hand, N is not complemented in its bidual, so the same happens to $\mathsf{FM} \times N$. It is clear that this approach cannot work for Banach spaces. The Banach space solution is from [65]. The construction in 2.12.9 appeared in [67]; we taught it to Nigel on the blackboard one fine day when he was curious 'about all this funny diagram stuff', and he used it in [276, Proposition 6.3], graciously calling the resulting space CC. The material on M-ideals is taken from [209] with the exception of the gorgeus proof of 2.14.9, which is due to Indumathi and Lalithambigai [215]. The disquisitions on categories of short exact sequences are taken from [365], later developed in [107; 110].

We conclude with a remark: homology is not category theory. It is just a part. Therefore, homological Banach space theory is not the same as categorical Banach space theory, even if Manuel González finds a logical contradiction in this sentence. Probably Categorical Banach Space Theory has still to be created for good; see [387, 6.9.7.1] or [92]. There are sound arguments to maintain that (quasi-) Banach spaces is a very interesting test category to work with even if you care only about category theory. Even if it is not an Abelian category (whatever that means), it is a forgiving place to work, since it is 'almost Abelian' and *exact* in Quillen's sense, which amounts to saying, more or less, that, one way or another, most homological and categorical constructions can be used there.

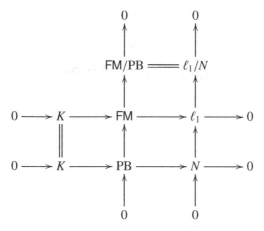

3

Quasilinear Maps

In this chapter, we plunge into the non-linear aspects of the theory of twisted sums. Why should we do so? Because it turns out that exact sequences of quasi-Banach spaces correspond to certain non-linear maps, called quasilinear maps, that offer a convenient, useful and relatively simple tool to construct, describe and study such sequences. Of course a lazy reader (or author) may argue that a considerable stock of exact sequences has already been presented in Chapter 2. And that is indeed the case. However, a sober look at those constructions soon reveals that they all depend on the knowledge of one single operator: the embedding of the quotient map. And, more often than not, knowing an embedding entails no control over the quotient space, and knowing a quotient map does not provide much control on its kernel. Sometimes, but only sometimes, a stroke of luck makes the third space manageable, but this is not to be expected in general. We deplore to say that nothing in Chapter 2 is of great help for constructing a single non-trivial exact sequence $0 \longrightarrow \ell_2 \longrightarrow \cdot \longrightarrow \ell_2 \longrightarrow 0$ or proving that such sequence does not exist. One may also argue that, at the end of the day, all short exact sequences arise as, say, pushouts, and thus one only need consider a quotient map $Q\colon \ell_1 \longrightarrow \ell_2$ and study whether all operators $\ker Q \longrightarrow \ell_2$ extend to ℓ_1. That may be true but is, for various reasons, unfeasible: as a rule, kernels of projective presentations are a complete mystery, and the extension of operators is a problem of its own. Thus, one of our objectives in this chapter is to provide the reader with practical ways to construct non-trivial exact sequences $0 \longrightarrow Y \longrightarrow \cdot \longrightarrow X \longrightarrow 0$ when only the spaces Y and X are known. The central idea here is that such exact sequences correspond to quasilinear maps $\Phi\colon X \longrightarrow Y$.

The chapter has been organised so that the reader can reach at an early stage a number of important applications. It begins with an informal discussion of quasilinear maps in order to immediately work through two classical examples:

- Ribe's solution to the 3-space problem for local convexity, namely a quasilinear functional $\varrho\colon \ell_1 \longrightarrow \mathbb{R}$ yielding a non-trivial exact sequence of the form $0 \longrightarrow \mathbb{R} \longrightarrow \cdot \longrightarrow \ell_1 \longrightarrow 0$. The middle space, obviously, cannot be locally convex.
- The Kalton–Peck spaces: a family of quasilinear maps $\Phi\colon X \longrightarrow X$ that generate non-trivial exact sequences $0 \longrightarrow X \longrightarrow \cdot \longrightarrow X \longrightarrow 0$ for most quasi-Banach spaces X with unconditional basis, including the spaces ℓ_p for $0 < p < \infty$. The reader who welcomes a challenge is invited to reflect on the meaning of the cases $0 < p \leq 1$ and $p = 2$.

Other applications of quasilinear maps, many of them presented in this chapter, but not all, include finding pairs of quasi-Banach spaces X, Y such that all exact sequences $0 \longrightarrow Y \longrightarrow \cdot \longrightarrow X \longrightarrow 0$ split; natural representations (natural equivalences is the right word) for the functor Ext; getting valuable insight into the structure of exact sequences and twisted sum spaces; simplifications for pullback, pushout and other homological constructions; a duality theory for exact sequences; uniform boundedness principles for exact sequences; a local theory for exact sequences ... Enough talk. Let's dive in.

3.1 An Introduction to Quasilinear Maps

Let us try to explain why quasilinear maps are so useful for both the description of twisted sums and the construction of relevant examples. Suppose that for some reason we are given a short exact sequence of quasi-Banach spaces $0 \longrightarrow Y \longrightarrow Z \longrightarrow X \longrightarrow 0$. We ask the reader to imagine that X and Y are the data (that is, they are known) while the space Z is not: one only knows that it fits in the exact sequence above. As vector spaces, all exact sequences split; this quickly follows from the existence of a Hamel basis in X. This enables us to regard Z as the direct product $Y \times X$ equiped with some quasinorm transferred from Z. The embedding is $\imath(y) = (y, 0)$ and the quotient map is $\pi(y, x) = x$. Replacing the original quasinorms of X and Y by suitable equivalent ones, we may assume that

- $\|y\| = \|(y, 0)\|$ for all $y \in Y$.
- $\|x\| = \min_{y \in Y} \|(y, x)\|$ for all $x \in X$.

For each $x \in X$, let us choose $y = \Phi(x)$ such that $\|x\| = \|(\Phi(x), x)\|$. We may assume that the map $\Phi\colon X \longrightarrow Y$ is homogenous since $\|\lambda x\| = \|(\lambda\Phi(x), \lambda x)\|$ for any $\lambda \in \mathbb{K}$. Inspecting Φ more closely reveals that if one compares $\Phi(x + x')$ and $\Phi(x) + \Phi(x')$ for two different $x, x' \in X$, the choice of Φ yields

$$\begin{aligned}\|\Phi(x+x')-\Phi(x)-\Phi(x')\| &= \|(\Phi(x+x')-\Phi(x)-\Phi(x'),0)\|\\ &= \|(\Phi(x+x'),x+x')-(\Phi(x),x)-(\Phi(x'),x')\|\\ &\le (\Delta_Y+\Delta_Y^2)(\|x+x'\|+\|x\|+\|x'\|)\\ &\le (\Delta_Y+\Delta_Y^2)(1+\Delta_X)(\|x\|+\|x'\|).\end{aligned}$$

Recapitulating; a certain map $\Phi\colon X \longrightarrow Y$ arises as soon as one considers an extension of X by Y. Most probably, this map is not linear, or bounded, but it is homogeneous and obeys the estimate

$$\|\Phi(x+x')-\Phi(x)-\Phi(x')\| \le M(\|x\|+\|x'\|). \tag{3.1}$$

Moreover, Φ can be used to describe an equivalent quasinorm on Z: indeed, each point of the product space can be decomposed as $(y,x)=(y-\Phi(x),0)+(\Phi(x),x)$ and thus $\|(y,x)\| \le \Delta(\|(y-\Phi(x),0)\|+\|(\Phi(x),x)\|)=\Delta(\|y-\Phi(x)\|+\|x\|)$. But since $\|(y,x)\|\ge\|x\|$ and $\|y-\Phi x\|=\|(y,x)-(\Phi x,x)\|\le\Delta(\|(y,x)\|+\|(\Phi x,x)\|)=\Delta(\|(y,x)\|+\|x\|)$, we get $\|y-\Phi(x)\|+\|x\|\le(1+\Delta)\|(y,x)\|$, and thus we have

$$\Delta^{-1}\|(y,x)\| \le \|y-\Phi(x)\|+\|x\| \le (1+\Delta)\|(y,x)\|.$$

Since any positively homogeneous functional that is equivalent to a quasinorm is itself a quasinorm, what has been shown is that the functional

$$\|y,x)\|_\Phi = \|y-\Phi(x)\|+\|x\| \tag{3.2}$$

gives an equivalent quasinorm on Z. Hence, for most practical purposes, the space Z is just the product space $Y\times X$ endowed with that quasinorm. If we agree to call a homogeneous map $\Phi\colon X\longrightarrow Y$ satisfying (3.1) *quasilinear*, we hit the nail on the head, since now we can follow the chain of inequalities backwards to get that if $\Phi\colon X\longrightarrow Y$ is a quasilinear map then $\|(y,x)\|_\Phi = \|y-\Phi(x)\|+\|x\|$ is a quasinorm on $Y\times X$: the homogeneity is trivial and

$$\begin{aligned}\|(y+y',x+x')\|_\Phi &= \|(y+y'-\Phi(x+x')\|+\|x+x'\|\\ &\le \|(y-\Phi(x)+y'-\Phi(x')-\Phi(x+x')+\Phi(x)+\Phi(x')\|+\|x+x'\|\\ &\le M(\|y-\Phi x\|+\|y'-\Phi x'\|+\|\Phi(x+x')-\Phi x-\Phi x'\|+\|x\|+\|x'\|)\\ &\le M(\|(y,x)\|_\Phi+\|(y',x')\|_\Phi),\end{aligned}$$

as required. Let us write $Y\oplus_\Phi X$ for the product space $Y\times X$ equipped with the quasinorm $\|(\cdot,\cdot)\|_\Phi$. We have the exact sequence

$$0 \longrightarrow Y \stackrel{\imath}{\longrightarrow} Y\oplus_\Phi X \stackrel{\pi}{\longrightarrow} X \longrightarrow 0,$$

in which $\imath(y)=(y,0)$, and $\pi(y,x)=x$. It is obvious that the kernel of π and the image of $\imath$ agree, that $\imath$ is an isometric embeding and that π maps the unit ball of $Y \oplus_\Phi X$ onto that of X. The middle space is automatically complete, thus a quasi-Banach space, in view of Proposition 2.3.4.

3.2 Quasilinear Maps in Action

The moral of the preceding discussion is clear: if we want to construct an extension of X by Y, we have to define a quasilinear map $\Phi\colon X \longrightarrow Y$. We must warn the reader that this can be a nigh impossible task. we cannot 'explicitly' define a quasilinear map on the whole quasi-Banach space unless it is bounded (in which case the corresponding extension simply splits), for the same reason that we cannot explicitly define a linear map on a quasi-Banach space unless it is bounded: Banach proved in [30, Théorème 4, Chapitre 1] that Borel linear maps between F-spaces are continuous. Fortunately, this blow is not fatal since in practice it suffices to define quasilinear maps on some dense subspace. Let us make it official:

Definition 3.2.1 Let X and Y be quasinormed spaces. A map $\Phi\colon X \longrightarrow Y$ is quasilinear if it is homogeneous and there is a constant Q such that

$$\|\Phi(x+y) - \Phi(x) - \Phi(y)\| \leq Q\,(\|x\| + \|y\|)$$

for every $x, y \in X$. The quasilinearity constant of Φ, denoted by $Q(\Phi)$, is the infimum of the numbers Q above.

The functional $\|(y,x)\|_\Phi = \|y - \Phi(x)\| + \|x\|$ is a quasinorm on $Y \times X$, and if we denote by $Y \oplus_\Phi X$ the corresponding quasinormed space, then $0 \longrightarrow Y \stackrel{\imath}{\longrightarrow} Y \oplus_\Phi X \stackrel{\pi}{\longrightarrow} X \longrightarrow 0$ with $\imath(y) = (y, 0)$ and $\pi(y,x) = x$ is an isometrically exact sequence. Assume now that X and Y are quasi-Banach spaces, that X_0 is a dense subspace of X so that $X_0 \longrightarrow X$ is a completion of X_0 and that $\Phi\colon X_0 \longrightarrow Y$ is quasilinear. Let us form the exact sequence $0 \longrightarrow Y \longrightarrow Y \oplus_\Phi X_0 \longrightarrow X_0 \longrightarrow 0$. Let $\kappa\colon Y \oplus_\Phi X_0 \longrightarrow Z(\Phi)$ denote *a* completion of $Y \oplus_\Phi X_0$ (please peek back at Note 1.8.1 if the word 'a' came as a surprise). We know from 2.3.5 that there is a commutative diagram

$$\begin{array}{ccccccccc} 0 & \longrightarrow & Y & \longrightarrow & Y \oplus_\Phi X_0 & \longrightarrow & X_0 & \longrightarrow & 0 \\ & & \| & & \downarrow{\scriptstyle \kappa} & & \downarrow{\scriptstyle \text{inclusion}} & & \\ 0 & \longrightarrow & Y & \longrightarrow & Z(\Phi) & \longrightarrow & X & \longrightarrow & 0 \end{array} \tag{3.3}$$

where the lower sequence is also exact. There are good reasons to say that this lower sequence is generated by Φ. To be honest, this exact sequence depends also on the completion of $Y \oplus_\Phi X_0$ we choose, but it is clear that all completions provide equivalent exact sequences. We shall see soon (Section 3.3) that the space $Z(\Phi)$ can be obtained as $Y \oplus_{\widetilde{\Phi}} X$, where $\widetilde{\Phi}\colon X \longrightarrow Y$ is a quasilinear map extending Φ. We now pause our study of quasilinear maps to follow Pełzyński's advice: examples first.

Ribe's Map

Ribe's map is a real-valued quasilinear map ϱ defined on the subspace ℓ_1^0 of finitely supported sequences of ℓ_1. It generates a non-trivial exact sequence $0 \longrightarrow \mathbb{R} \longrightarrow Z(\varrho) \longrightarrow \ell_1 \longrightarrow 0$. The construction depends on the properties of the function of a single variable $\omega(t) = t\log|t|$ (assuming that $\omega(0) = 0 \cdot \log 0 = 0$).

Lemma 3.2.2 *For all $s, t \in \mathbb{R}$, we have $|\omega(s+t)-\omega(s)-\omega(t)| \leq (\log 2)(|s|+|t|)$.*

Proof Let us consider first the case in which s and t have the same sign. Assuming for instance that $s, t \geq 0$, we have

$$\begin{aligned}|\omega(s+t) - \omega(s) - \omega(t)| &= |(s+t)\log(s+t) - s\log s - t\log t|\\ &= |s\log(s+t) + t\log(s+t) - s\log s - t\log t|\\ &= \left|s\log\left(\frac{s}{s+t}\right) + t\log\left(\frac{t}{s+t}\right)\right|\\ &\leq (|s|+|t|)\underbrace{\left|\frac{s}{s+t}\log\left(\frac{s}{s+t}\right) + \frac{t}{s+t}\log\left(\frac{t}{s+t}\right)\right|}_{(\star)}.\end{aligned}$$

To maximise $(\star)$, simply write it as

$$|s'\log s' + t'\log t'|, \quad \text{with} \quad s' = \frac{s}{s+t} \quad \text{and} \quad t' = \frac{t}{s+t},$$

and observe that, since $s' + t' = 1$, the maximum value is $\log 2$ (attained at $s' = t' = 1/2$). It follows that $|\omega(s+t) - \omega(s) - \omega(t)| \leq (\log 2)(|s| + |t|)$. Now, if s and t have distinct signs, we may assume that s is positive, t is negative and $s + t > 0$. Taking into account that ω is an odd map, the proof concludes with

$$\begin{aligned}|\omega(s+t) - \omega(s) - \omega(t)| &= |\omega(s) - \omega(-t) - \omega(s+t)|\\ &\leq (\log 2)(|-t| + |s+t|) \leq (\log 2)(|s| + |t|).\end{aligned}$$ □

Proposition 3.2.3 *The map $\varrho\colon \ell_1^0 \longrightarrow \mathbb{R}$ given by*

$$\varrho(x) = \sum_i x(i)\log|x(i)| - \left(\sum_i x(i)\right)\log\left|\sum_i x(i)\right|$$

is quasilinear with quasilinearity constant $2\log 2$ *and induces a nontrival sequence*

$$0 \longrightarrow \mathbb{R} \longrightarrow Z(\varrho) \longrightarrow \ell_1 \longrightarrow 0.$$

Proof The homogeneity of ϱ is obvious. Writing $S(x) = \sum_i x(i)$ for $x \in \ell_1$, we have $\varrho(x) = S(\omega \circ x) - \omega(S(x))$, so

$$\begin{aligned}
&|\varrho(x+y) - \varrho(x) - \varrho(y)| \\
&= |S(\omega \circ (x+y)) - \omega(S(x+y)) - S(\omega \circ x) + \omega(Sx) - S(\omega \circ y) + \omega(Sy)| \\
&\leq |\omega(Sx+Sy)) - \omega(Sx) - \omega(Sy))| + |S(\omega \circ (x+y)) - S(\omega \circ x) - S(\omega \circ y)| \\
&\leq (\log 2)\left(|S(x)| + |S(y)| + \sum_i (|x(i)| + |y(i)|)\right) \\
&\leq (2\log 2)(\|x\| + \|y\|).
\end{aligned}$$

To prove that the sequence does not split, observe that the estimate $\left\| \sum_i x_i \right\| \leq C(\sum_i \|x_i\|)$ in Lemma 1.1.2 is impossible in $\mathbb{R} \oplus_\varrho \ell_1^0$ because if (e_i) is the unit basis of ℓ_1 then $\|(0, e_i)\|_\varrho = 1$ for all $i \in \mathbb{N}$, which makes $\sum_{i=1}^n \|(0, e_i)\|_\varrho = n$, while

$$\left\| \left(0, \sum_{i=1}^n e_i\right) \right\|_\varrho = n \log n. \qquad \square$$

This construction from Ribe [401] shows that local convexity is not a 3-space property in the domain of quasi-Banach spaces. Other counterexamples were presented by Kalton [251] and Roberts [403], independently and more or less simultaneously; Smirnov and Sheikhman [437] gave one more, entirely alien, example. Roberts' example is obtained by a different technique not involving quasilinear maps. It is not coincidence that all four counterexamples are twisted sums of $\mathbb{R}$ and ℓ_1: one of the conclusions of Section 3.4 is that *any* counterexample for the 3-space problem for local convexity leads, by simple algebraic manipulations, to one in which the subspace is 1-dimensional and the quotient space is ℓ_1.

Kalton–Peck Maps

Kalton and Peck found a way to transform Ribe's scalar map into a vector-valued quasilinear map that can be defined on every quasi-Banach space with unconditional basis. The ground field can be either $\mathbb{R}$ or $\mathbb{C}$ from now on.

3.2.4 By a (quasinormed) sequence space, we understand a linear space X of functions $x\colon \mathbb{N} \longrightarrow \mathbb{K}$ equipped with a quasinorm $\|\cdot\|$ such that

- if $|y| \leq |x|$ and $x \in X$, then $y \in Y$ and $\|y\| \leq \|x\|$,
- the unit vectors are normalised and the finitely supported sequences form a dense subspace of X.

Of course, all that means that the unit vectors form a 1-unconditional basis of X. If X is complete, we call it a quasi-Banach sequence space. We shall invariably denote by X^0 the dense subspace of finitely supported sequences in X. The simplest sequence spaces are ℓ_p for $0 < p < \infty$ and c_0. Every quasi-Banach space with a (normalised) 1-unconditional basis can also be seen as a sequence space in the obvious way. The space ℓ_∞ acts on every sequence space by pointwise multiplication in such a way that $\|ax\| \leq \|a\|\|x\|$ for every $a \in \ell_\infty$ and every $x \in X$ and such that $\|ax\| = \|x\|$ when a is unitary, that is, when $|a(k)| = 1$ for all $k \in \mathbb{N}$. Let $\mathrm{Lip}_0(\mathbb{R})$ be space of Lipschitz functions $\varphi\colon \mathbb{R} \longrightarrow \mathbb{K}$ vanishing at zero, and let $\mathrm{Lip}(\varphi)$ denote its Lipschitz constant. Similar conventions apply to $\mathrm{Lip}_0(\mathbb{R}^+)$. Lipschitz functions can be used to produce a variety of Ribe-like functions that share the quasiadditivity property appearing in Lemma 3.2.2:

Lemma 3.2.5 *Let $\varphi\colon \mathbb{R} \longrightarrow \mathbb{K}$ be a Lipschitz function vanishing at zero, and let $\omega_\varphi\colon \mathbb{K} \longrightarrow \mathbb{K}$ be the map $\omega_\varphi(z) = z\varphi(-\log|z|)$. For all scalars z, z', one has*

$$|\omega_\varphi(z+z') - \omega_\varphi(z) - \omega_\varphi(z)| \leq 2\mathrm{Lip}(\varphi)e^{-1}(|z| + |z'|). \tag{3.4}$$

Proof The proof is based on the trivial fact that $|t\log t| \leq e^{-1}$ for $0 \leq t \leq 1$. We first assume that $|z+z'| \geq \max(|z|,|z'|)$. Then

$$\begin{aligned}
&\frac{|\omega_\varphi(z+z') - \omega_\varphi(z) - \omega_\varphi(z')|}{|z| + |z'|}\\
&\leq \frac{|z(\varphi(-\log|z+z'|) - \varphi(-\log|z|) - z'(\varphi(-\log|z+z'|) - \varphi(-\log|z'|)|}{|z+z'|}\\
&\leq \frac{|z|}{|z+z'|}\mathrm{Lip}(\varphi)\log\frac{|z+z'|}{|z|} + \frac{|z'|}{|z+z'|}\mathrm{Lip}(\varphi)\log\frac{|z+z'|}{|z'|}\\
&\leq 2\mathrm{Lip}(\varphi)e^{-1}.
\end{aligned}$$

If, on the contrary, $|z+z'| < \max(|z|,|z'|)$ then we may assume that $|z| \geq \max(|z+z'|,|z'|)$. Replacing z by $z+z'$ and z' by $-z'$ and taking into account that ω_φ is odd, we have

$$\begin{aligned}
|\omega_\varphi(z+z') - \omega_\varphi(z) - \omega_\varphi(z')| &= |\omega_\varphi(z) - \omega_\varphi(-z') - \omega_\varphi(z+z')|\\
&\leq 2\mathrm{Lip}(\varphi)e^{-1}(|z'| + |z+z'|)\\
&\leq 2\mathrm{Lip}(\varphi)e^{-1}(|z| + |z'|).
\end{aligned}$$

□

Given a sequence space X and $\varphi \in \mathrm{Lip}_0(\mathbb{R}^+)$, we define the map $\mathsf{KP}_\varphi\colon X^0 \longrightarrow X$ by

$$\mathsf{KP}_\varphi(x) = x \cdot \varphi\left(\log \frac{\|x\|}{|x|}\right) \tag{3.5}$$

and call it the Kalton–Peck map induced by φ on X. For the avoidance of doubt, the definition means

$$\mathsf{KP}_\varphi(x)(k) = \begin{cases} x(k) \cdot \varphi\left(\log(\|x\|/|x(k)|)\right) & \text{if } x(k) \neq 0 \\ 0 & \text{if } x(k) = 0. \end{cases}$$

These maps commute with the action of the unitary group: if u is unitary, then $\mathsf{KP}_\varphi(ux) = u\mathsf{KP}_\varphi(x)$ for all $x \in X^0$ and thus $\|(uy, ux)\|_{\mathsf{KP}_\varphi} = \|(y, x)\|_{\mathsf{KP}_\varphi}$. This property will be used over and over.

Proposition 3.2.6 *The map $\mathsf{KP}_\varphi\colon X^0 \longrightarrow X$ is quasilinear and $Q(\mathsf{KP}_\varphi)$ depends only on the Lipschitz constant of φ and the modulus of concavity of X.*

Proof We can assume that φ is defined on the whole line by taking $\varphi(t) = 0$ for all $t < 0$; this extension has the same Lipschitz constant as the original φ. Consider the non-homogeneous map $\mathsf{kp}_\varphi\colon X^0 \longrightarrow X$ given by

$$\mathsf{kp}_\varphi(x) = x \cdot \varphi(-\log|x|),$$

with the same meaning as above. Since $\mathsf{kp}_\varphi(x)(k) = \omega_\varphi(x(k))$, the preceding Lemma provides the pointwise estimate

$$|\mathsf{kp}_\varphi(x+y)(k) - \mathsf{kp}_\varphi(x)(k) - \mathsf{kp}_\varphi(y)(k)| \leq 2\,\mathrm{Lip}(\varphi)e^{-1}(|x(k)| + |y(k)|); \tag{3.6}$$

hence,

$$\|\mathsf{kp}_\varphi(x+y) - \mathsf{kp}_\varphi(x) - \mathsf{kp}_\varphi(y)\| \leq \frac{2\,\mathrm{Lip}(\varphi)}{e}\big\|\,|x| + |y|\,\big\| \leq \frac{2\Delta\,\mathrm{Lip}(\varphi)}{e}(\|x\| + \|y\|),$$

where Δ is the modulus of concavity of X. To complete the proof, observe that

$$\begin{aligned} \|\mathsf{KP}_\varphi(x) - \mathsf{kp}_\varphi(x)\| &= \left\| x \cdot \left\{ \varphi\left(\log\frac{\|x\|}{|x|}\right) - \varphi(-\log|x|) \right\} \right\| \\ &\leq \big\|\mathrm{Lip}(\varphi) \cdot \log\|x\| \cdot x\big\|, \end{aligned} \tag{3.7}$$

since the term between braces is pointwise dominated by $\mathrm{Lip}(\varphi)\log\|x\|$. In particular, for $\|x\| \leq 1$, we have

$$\|\mathsf{KP}_\varphi(x) - \mathsf{kp}_\varphi(x)\| \leq e^{-1}\,\mathrm{Lip}(\varphi). \tag{3.8}$$

If $\|x\|, \|y\| \leq (2\Delta)^{-1}$ then $\|x + y\| \leq 1$, so adding and substracting $\mathsf{kp}_\varphi(x+y) - \mathsf{kp}_\varphi(x) - \mathsf{kp}_\varphi(y)$, applying the quasinorm inequality twice and applying (3.8) to the first three chunks, we get

$$\begin{aligned}
&\|\mathsf{KP}_\varphi(x+y) - \mathsf{KP}_\varphi x - \mathsf{KP}_\varphi y\| \\
&\le \|(\mathsf{KP}_\varphi - \mathsf{kp}_\varphi)(x+y) - (\mathsf{KP}_\varphi - \mathsf{kp}_\varphi)x - (\mathsf{KP}_\varphi - \mathsf{kp}_\varphi)y + \mathsf{kp}_\varphi(x+y) - \mathsf{kp}_\varphi x - \mathsf{kp}_\varphi y\| \\
&\le \Delta^2\,(3\,\mathrm{Lip}(\varphi)/e + 2\,\mathrm{Lip}(\varphi)/e)\,,
\end{aligned}$$

and thus $\|\mathsf{KP}_\varphi(x+y) - \mathsf{KP}_\varphi(x) - \mathsf{KP}_\varphi(y)\| \le 5\Delta^2\,\mathrm{Lip}(\varphi)e^{-1}$ whenever $\|x\|, \|y\| \le (2\Delta)^{-1}$. Being KP_φ homogeneous, for any $x, y \in X^0$, we have

$$\begin{aligned}
&\frac{\|\mathsf{KP}_\varphi(x+y) - \mathsf{KP}_\varphi(x) - \mathsf{KP}_\varphi(y)\|}{2\Delta(\|x\| + \|y\|)} \\
&= \left\| \mathsf{KP}_\varphi\left(\frac{x+y}{2\Delta(\|x\|+\|y\|)}\right) - \mathsf{KP}_\varphi\left(\frac{x}{2\Delta(\|x\|+\|y\|)}\right) - \mathsf{KP}_\varphi\left(\frac{y}{2\Delta(\|x\|+\|y\|)}\right)\right\| \\
&\le \frac{5\Delta^2\,\mathrm{Lip}(\varphi)}{e},
\end{aligned}$$

hence KP_φ is quasilinear, with constant at most $10\Delta^3\,\mathrm{Lip}(\varphi)e^{-1}$. □

We now form the exact sequences and twisted sum spaces generated by the Kalton–Peck maps KP_φ. The twisted sum space $Z(\mathsf{KP}_\varphi)$ will be denoted by $X(\varphi)$ to emphasise that it depends on both the Lipschitz function φ and the space X where KP_φ acts. An especially interesting case is when $X = \ell_p$, in which we obtain the exact sequences

$$0 \longrightarrow \ell_p \longrightarrow \ell_p(\varphi) \longrightarrow \ell_p \longrightarrow 0 \tag{3.9}$$

The following result bluntly precludes the possibility of these sequences being trivial for unbounded φ.

Proposition 3.2.7 *If φ is unbounded on $\mathbb{R}^+$ then $\ell_p(\varphi)$ is not isomorphic to ℓ_p for any $p \in (0, \infty)$.*

Proof Set $s_n = \sum_{1\le i\le n} e_i$ such that $\mathsf{KP}_\varphi(s_n) = \varphi(\|s_n\|)s_n = \varphi(\log(n^{1/p})s_n$. The Lipschitz condition on φ implies that $\sup_n |\varphi(\log(n^{1/p})| = \infty$ since the points of the form $\log n^{1/p}$ form a $1/p$-net on $\mathbb{R}^+$. Now, the proof is different for different values of p: we show that $\ell_p(\varphi)$ is not (isomorphic to) a p-Banach space for $p \in (0, 1]$, that it does not have type p for $p \in (1, 2]$ and that it does not have cotype p for $p \in [2, \infty)$.

Case $p \in (0, 1]$. Since $\mathsf{KP}_\varphi(e_i) = 0$, $\|(0, e_i)\|_{\mathsf{KP}_\varphi} = 1$ for every $i \in \mathbb{N}$. However,

$$\left\|\sum_{i=1}^{n}(0, e_i)\right\|_{\mathsf{KP}_\varphi} = \|(0, s_n)\|_{\mathsf{KP}_\varphi} = \left(\left|\varphi(\log(n^{1/p})\right| + 1\right)n^{1/p}.$$

The estimate in Lemma 1.1.2 shows that $\ell_p(\varphi)$ cannot be isomorphic to a p-normed space.

Case $p \in (1,2]$. We use a randomised version of the preceding argument. For $\varepsilon_i = \pm 1$, one has $\mathsf{KP}_\varphi(\sum_{1\le i\le n}\varepsilon_i e_i) = |\varphi(\log(n^{1/p})| \sum_{1\le i\le n}\varepsilon_i e_i$. Hence, if r_i are the Rademacher functions, then

$$\int_0^1 \left\| \sum_{i=1}^n r_i(t)(0,e_i) \right\|_{\mathsf{KP}_\varphi}^p dt = \int_0^1 \left(\left\| \mathsf{KP}_\varphi\left(\sum_{i=1}^n r_i(t)e_i \right) \right\|_p + \left\| \sum_{i=1}^n r_i(t)e_i \right\|_p \right)^p dt$$
$$= \left(\left|\varphi(\log(n^{1/p})\right| + 1 \right)^p n, \tag{3.10}$$

and thus $\ell_p(\varphi)$ does not have type p and cannot be isomorphic to ℓ_p.
Case $p \in [2,\infty)$. We show that $\ell_p(\varphi)$ does not have cotype p by exploiting the symmetries of KP_φ. We have

$$\mathsf{KP}_\varphi(s_n) = \varphi(\log n^{1/p})s_n \qquad \Longrightarrow \qquad \mathsf{KP}_\varphi\left(\frac{s_n}{\varphi(\log n^{1/p})} \right) = s_n.$$

For $i \in \mathbb{N}$, set $z_i = (e_i, \varphi(\log n^{1/p})^{-1} e_i)$ so that $\|z_i\|_{\mathsf{KP}_\varphi} = 1 + \varphi(\log n^{1/p})^{-1} \ge 1$ and $\left(\sum_{1\le i\le n} \|z_i\|^p \right)^{1/p} \ge n^{1/p}$, while

$$\left(\int_0^1 \left\| \sum_{i=1}^n r_i(t)z_i \right\|_{\mathsf{KP}_\varphi}^p dt \right)^{1/p} = \left\| \left(s_n, \frac{s_n}{\varphi(\log n^{1/p})} \right) \right\| = \frac{n^{1/p}}{\varphi(\log n^{1/p})}. \qquad \square$$

Observe that by taking $p = 1$, we obtain new counterexamples to the 3-space problem for local convexity, namely $\ell_1(\varphi)$ is not a Banach space. But, wait! Are $\ell_p(\varphi)$ Banach spaces when $p > 1$? The topic is studied in Section 3.4, and an affirmative and non-trivial answer is given in Proposition 3.4.5: any twisted sum of ℓ_p can be renormed to be a Banach space when $1 < p < \infty$. Thus, in the end, the following notion refers only to Banach spaces:

Definition 3.2.8 A twisted Hilbert space is a twisted sum of two Hilbert spaces.

The choice $p = 2$ in Proposition 3.2.7 produces our first non-trivial (i.e. not isomorphic to a Hilbert space) twisted Hilbert spaces. The spaces $\ell_p(\varphi)$ obtained from the simplest unbounded Lipschitz function $\varphi(t) = t$ play a very special role in the theory and in this book. We have reserved for them their original name: Z_p. The twisted Hilbert Z_2 space stands apart: it is so special that it will close this book by explaining its many hidden talents. Summing up, none of the sequences (3.9) for $0 < p < \infty$ is trivial when φ is unbounded. In general, it is hard to find isomorphic invariants capable of distinguishing a particular twisted sum from the corresponding direct sum. Actually, it may be an impossible task since there are non-trivial exact sequences $0 \longrightarrow A \longrightarrow B \longrightarrow C \longrightarrow 0$ in which $B \simeq A \times C$ (see the comments after Definition 2.1.4), the Foiaş–Singer sequence (2.5) being perhaps the most natural example. In other cases, however, an isomorphism $B \simeq A \times C$ immediately forces the splitting of the sequence, as is the case for sequences $0 \longrightarrow \ell_p \longrightarrow \cdot \longrightarrow \ell_p \longrightarrow 0$ with $p = 2$ or $0 < p \leq 1$, due to the peculiarities of those spaces; see also Proposition 7.5.2 for another result along these lines. It is an open problem to determine whether non-trivial sequences $0 \longrightarrow \ell_p \longrightarrow \ell_p \longrightarrow \ell_p \longrightarrow 0$ exist for $p \in (1, \infty)$ different from 2.

3.3 Quasilinear Maps versus Exact Sequences

We need to develop an operational theory of quasilinear maps. As the reader may expect, the first order of business is to find criteria that detect when a quasilinear map induces a trivial extension and when two quasilinear maps induce the same extension. The following delightful result yields both criteria showing that two quasilinear maps generate equivalent extensions if and only if the extension generated by their difference splits. Moreover, it hints towards an underlying vector space structure.

Equivalence and Triviality

Let X, Y be quasi-Banach spaces and let X_0 be a dense subspace of X. The following assertion is somehow implicit in 2.3.5:

Lemma 3.3.1 *If $\Phi\colon X \longrightarrow Y$ is quasilinear map and X_0 is a dense subspace of X_0 then $Y \oplus_\Phi X_0$ is a dense subspace of $Y \oplus_\Phi X$.*

Proof Fix $(y, x) \in Y \oplus_\Phi X$ and $\varepsilon > 0$. Take $x' \in X_0$ such that $\|x - x'\| < \varepsilon$ and then set $y' = y - \Phi(x - x')$. Then, even if y' might be far from y in Y, we have

$$\|(y, x) - (y', x')\|_\Phi = \|y - y' - \Phi(x - x')\| + \|x - x'\| = \|x - x'\| < \varepsilon. \qquad \square$$

In other words, the canonical inclusion $Y \oplus_\Phi X_0 \longrightarrow Y \oplus_\Phi X$ is a completion of $Y \oplus_\Phi X_0$. Now assume Φ and Ψ are quasilinear maps $X_0 \longrightarrow Y$ (defined only on the dense subspace X_0).

Lemma 3.3.2 *There is a commutative diagram*

$$\begin{array}{ccccccccc} 0 & \longrightarrow & Y & \longrightarrow & Z(\Phi) & \longrightarrow & X & \longrightarrow & 0 \\ & & \| & & \downarrow & & \| & & \\ 0 & \longrightarrow & Y & \longrightarrow & Z(\Psi) & \longrightarrow & X & \longrightarrow & 0 \end{array} \tag{3.11}$$

if and only if $\Phi - \Psi = B + L$, *where* $B\colon X_0 \longrightarrow Y$ *is (homogeneous) bounded and* $L\colon X_0 \longrightarrow Y$ *is linear. Consequently, the extension induced by* Φ *is trivial if and only if* $\Phi = B + L$, *where* B *is homogenous bounded and* L *is linear.*

Proof The proof is a simple recipe: one grain of the universal property of the completion (Diagram (1.4)) and three drops of the behaviour of its associated exact sequence 2.3.5, plus the fact that $Y \oplus_\Phi X_0 \longrightarrow Z(\Phi)$ and $Y \oplus_\Psi X_0 \longrightarrow Z(\Psi)$ are completions, show that if the commutative diagram (3.11) exists then there is also a commutative diagram

$$\begin{array}{ccccccccc}
0 & \longrightarrow & Y & \longrightarrow & Y \oplus_\Phi X_0 & \longrightarrow & X_0 & \longrightarrow & 0 \\
 & & \| & & \downarrow & & \downarrow & & \\
0 & \longrightarrow & Y & \longrightarrow & Z(\Phi) & \longrightarrow & X & \longrightarrow & 0 \\
 & & \| & & \downarrow{\scriptstyle u} & & \| & & \\
0 & \longrightarrow & Y & \longrightarrow & Z(\Psi) & \longrightarrow & X & \longrightarrow & 0 \\
 & & \| & & \uparrow & & \uparrow & & \\
0 & \longrightarrow & Y & \longrightarrow & Y \oplus_\Psi X_0 & \longrightarrow & X_0 & \longrightarrow & 0
\end{array}$$

in which all vertical unnamed arrows are inclusions. The operator u necessarily maps $Y \oplus_\Phi X_0$ onto $Y \oplus_\Psi X_0$ and makes the following diagram commute:

$$\begin{array}{ccccccccc}
0 & \longrightarrow & Y & \longrightarrow & Y \oplus_\Phi X_0 & \longrightarrow & X_0 & \longrightarrow & 0 \\
 & & \| & & \downarrow{\scriptstyle u} & & \| & & \\
0 & \longrightarrow & Y & \longrightarrow & Y \oplus_\Psi X_0 & \longrightarrow & X_0 & \longrightarrow & 0
\end{array}$$

On the other hand, a linear map u makes the preceding diagram commute if and only if it has the form $u(y, x) = (y - L(x), x)$ for some linear $L\colon X_0 \longrightarrow Y$, in which case u is invertible, with inverse $u^{-1}(y, x) = (y + L(x), x)$. We then just need to show that such a u is continuous if and only if $\|\Phi - \Psi - L\| < \infty$. The two implications are easy. If u is continuous, then $\|u(y, x)\|_\Psi \le \|u\|\,\|(y, x)\|_\Phi$, that is, $\|y - L(x) - \Psi(x)\| + \|x\| \le \|u\|(\|y - \Phi(x)\| + \|x\|)$. Taking $y = \Phi(x)$, we get $\|\Phi(x) - L(x) - \Psi(x)\| + \|x\| \le \|u\|\,\|x\|$, and thus $\|\Phi - \Psi - L\| \le \|u\| - 1$. To get the converse,

$$\begin{aligned}
\|u(y, x)\|_\Psi &= \|(y - L(x), x)\|_\Psi \\
&= \|y - L(x) - \Psi(x)\| + \|x\| \\
&= \|y - \Phi(x) + \Phi(x) - L(x) - \Psi(x)\| + \|x\| \\
&\le \Delta_Y(\|y - \Phi(x)\| + \|\Phi(x) - L(x) - \Psi(x)\|) + \|x\|,
\end{aligned}$$

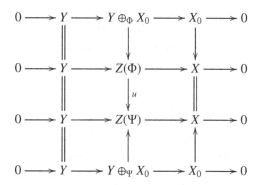

hence $\|u\| \leq 1 + \Delta_Y\|\Phi - \Psi - L\|$. Reversing the roles of Φ and Ψ and using $u^{-1}(y, x) = (y + L(x), x)$, we obtain the bound $\|u^{-1}\| \leq 1 + \Delta_Y\|\Phi - \Psi - L\|$. □

Definition 3.3.3 A quasilinear map $\Phi\colon X \longrightarrow Y$ is said to be trivial if it is the sum of a bounded homogeneous map $B\colon X \longrightarrow Y$ and a linear map $L\colon X \longrightarrow Y$. Two quasilinear maps Φ, Ψ are said to be equivalent, written $\Phi \sim \Psi$, if $\Phi - \Psi$ is trivial.

Thus, a quasilinear map is trivial if and only if it is equivalent to the zero map. Lemma 3.3.2 can be rephrased as saying that a quasilinear map is trivial if and only if the induced sequence is trivial, while two quasilinear maps defined between the same quasinormed spaces are equivalent if and only if they induce equivalent exact sequences. When necessary, we will say that $\Phi\colon X \longrightarrow Y$ is μ-trivial to mean that $\|\Phi - L\| \leq \mu$ for some linear map $L\colon X \longrightarrow Y$.

A timely intermission about twisted sums generated by bounded or linear maps: The second part of Lemma 3.3.2 provides a remarkably simple criterion for the splitting of the sequence induced by a quasilinear map; namely that the quasilinear map be the sum of a bounded and a linear map. It is thus obvious that bounded and linear maps are the simplest quasilinear maps there are. Let us see what occurs when Φ is taken to be one of them.

- If L is linear then $Y \oplus_L X$ is isometric to the direct sum $Y \oplus_1 X$ under the isometry $(y, x) \mapsto (y - L(x), x)$. The induced extension splits, but beware: the projection $Y \oplus_L X \longrightarrow Y$ is not the obvious one $(y, x) \longmapsto y$, which turns out to be discontinuous unless L is bounded, in which case $\|(\cdot,\cdot)\|_L$ is even equivalent to the sum quasinorm. No, the correct projection is $P(y, x) = y - L(x)$, which is even contractive.
- If B is bounded then $\|(\cdot,\cdot)\|_B$ is merely equivalent to $\|(\cdot,\cdot)\|_1$. The sequence splits, and $(y, x) \longmapsto y$ is a bounded projection, although its norm depends on $\|B\|$.

If $\Phi = B + L$ is the sum of both then it takes a few seconds of pondering to realise that the induced exact sequence also splits through the retraction $(y, x) \longrightarrow y - L(x)$ or the section $x \longmapsto (L(x), x)$. More precisely:

Lemma 3.3.4 *Let $\Phi\colon X \longrightarrow Y$ be a quasilinear map and let $L\colon X \longrightarrow Y$ be a linear map. Then $\Phi - L$ is bounded $\iff$ $P(y, x) = y - L(x)$ is bounded from $Y \oplus_\Phi X$ to Y $\iff$ $S(x) = (L(x), x)$ is bounded from X to $Y \oplus_\Phi X$. Moreover, $\|\Phi - L\| \leq \|P\| \leq \max(\Delta_Y, \|\Phi - L\|)$ and $\|S\| = 1 + \|\Phi - L\|$.*

Proof It clearly suffices to check the part involving the bounds. As $\Phi(x) - L(x) = P(\Phi(x), x)$ and $\|(\Phi(x), x)\|_\Phi = \|x\|$, we have $\|\Phi - L\| \leq \|P\|$. The other inequality follows from

$$\|y - L(x)\| \leq \Delta_Y(\|y - \Phi(x)\| + \|\Phi(x) - L(x)\|) \leq \Delta_Y(\|y - \Phi(x)\| + \|\Phi - L\|\,\|x\|).$$

The equality is just that $\|S(x)\| = \|(L(x), x)\|_\Phi = \|L(x) - \Phi(x)\| + \|x\|$. □

Non-triviality of the Kalton–Peck Maps

Here is a serious application of the splitting criterion:

Proposition 3.3.5 *Let X be a sequence space that is either*

(a) *a Banach space different from c_0, or*
(b) *a quasi-Banach space such that no subsequence of (e_n) is equivalent to the unit basis of c_0.*

Then $\mathsf{KP}_\varphi : X^0 \longrightarrow X$ is trivial if and only if φ is bounded on $\mathbb{R}^+$.

Proof The proof is different depending on whether one assumes (a) or (b). We begin with (a). Assume there is a linear map $\ell : X^0 \longrightarrow X$ for which $\mathsf{KP}_\varphi - \ell$ is bounded. The idea is to use the symmetries of X and KP_φ to replace ℓ by a bounded operator L with the same symmetries in order to conclude that KP_φ is bounded, which makes φ bounded. Consider the real unitary group $U = \{\pm 1\}^{\mathbb{N}}$ with the product topology and let m denote the Haar measure on U, which is actually a probability. Since X is a Banach space, every continuous function $f : U \longrightarrow X$ can be averaged using the Bochner integral $\int_U f(u)dm(u)$. The convexity of the norm yields the bound

$$\left\| \int_U f(u)dm(u) \right\| \leq \int_U \|f(u)\|dm(u) \leq \max_{u\in U} \|f(u)\|,$$

while the invariance of the Haar measure yields

$$\int_U f(u)dm(u) = \int_U f(uv)dm(u).$$

We will go ahead and use the hypothesised ℓ to define a new map $L : X^0 \longrightarrow X$,

$$L(x) = \int_U u^{-1}\ell(ux)dm(u).$$

The map L is clearly linear and correctly defined since for each $x \in X^0$ the orbit $\{ux : u \in U\}$ is finite, thus it spans a finite-dimensional space, on which the restriction of any linear map, as well as the map $u \in U \longmapsto u^{-1}\ell(ux) \in X$, must be continuous. Actually, if $x(k) = 0$ for $k > n$ then

$$L(x) = \int_U u^{-1}\ell(ux)dm(u) = \frac{1}{2^n}\sum\nolimits_{u\in U_n} u\ell(ux),$$

where $U_n = \{u\colon u(k) = 1 \text{ for all } k > n\}$. To estimate $\|\mathsf{KP}_\varphi - L\|$, pick $x \in X^0$ and observe that $\mathsf{KP}_\varphi(ux) = u\mathsf{KP}_\varphi(x)$ for every $u \in U$. Hence

$$\begin{aligned}\|\mathsf{KP}_\varphi(x) - L(x)\| &= \left\|\int_U u^{-1}\mathsf{KP}_\varphi(ux)dm(u) - \int_U u^{-1}\ell(ux)dm(u)\right\| \\ &\leq \int_U \|u^{-1}\mathsf{KP}_\varphi(ux) - u^{-1}\ell(ux)\|dm(u) \\ &\leq \int_U \|\mathsf{KP}_\varphi(ux) - \ell(ux)\|dm(u) \\ &\leq \int_U \|\mathsf{KP}_\varphi - \ell\|\|ux\|dm(u) \\ &= \|\mathsf{KP}_\varphi - \ell\|\|x\|,\end{aligned}$$

which yields $\|\mathsf{KP}_\varphi - L\| \leq \|\mathsf{KP}_\varphi - \ell\|$. Moreover, by the invariance of m, for each $v \in U$ we have

$$L(vx) = \int_U u\ell(uvx)dm(u) = v\int_U vu\ell(uvx)dm(u) = vL(x).$$

Writing $e_n = \frac{1}{2}(u + v)$ with $u, v \in U$ (say, $u = 1$ and $v = -1 + 2e_n$), we get

$$L(e_n) = L\left(\frac{u+v}{2}e_n\right) = \frac{u+v}{2}L(e_n) = e_nL(e_n).$$

Hence $L(e_n) = a_ne_n$ for some scalar a_n. Since $\mathsf{KP}_\varphi(e_n) = 0$ for all n, we have $|a_n| \leq \|\mathsf{KP}_\varphi - \ell\|$, which means that the sequence $(a_n)_{n\geq 1}$ is bounded. The map L is therefore bounded and thus KP_φ must also be bounded. Set $s_n = \sum_{1\leq i\leq n} e_i$. Since $\mathsf{KP}_\varphi(s_n) = \varphi(\log\|s_n\|)s_n$, we get that the sequence $\varphi(\log\|s_n\|)$ is bounded. But if $X \neq c_0$ then $\|s_n\| \to \infty$, and since $\|s_n\| \leq \|s_{n+1}\| \leq \|s_n\| + 1$, the Lipschitz condition implies that φ is bounded on $\mathbb{R}^+$, which proves the result under the assumption (a).

We now work under assumption (b). If X fails to be locally convex then we cannot reduce the complexity of the approximating linear map so easily, and it is necessary to fight for longer to arrive at the same point as before. So, before entering into the details, let us indicate how the hypotheses will be used to get the result:

(1) At a certain stage of the proof we will have to consider a subsequence $(e_{n(k)})$ of the basis, the differences $e_{n(1)} - e_{n(2)}, e_{n(3)} - e_{n(4)}, \ldots$ and the quasinorms

$$\|e_{n(1)} - e_{n(2)} + \cdots + e_{n(2k-1)} - e_{n(2k)}\| = \left\|\sum_{i=1}^{2k}(-1)^{i+1}e_{n(i)}\right\| = \left\|\sum_{i=1}^{2k}e_{n(i)}\right\|.$$

The hypothesis (b) guarantees that these go to infinity as k increases (otherwise $(e_{n(k)})$ is equivalent to the unit basis of c_0).
(2) It is plain that any linear map $X^0 \longrightarrow Y$ is entirely defined by the sequence $(y_n) = (Le_n)$. If, moreover, Y is a quasi-Banach space and $\|y_n\| \leq 2^{-n}$ for all n, then L is bounded. Indeed, by the Aoki–Rolewicz theorem, we may assume that Y carries a p-norm and check that L actually extends to a bounded operator from ℓ_∞ to Y, since for $f \in \ell_\infty$ we have

$$\left\|\sum_n f(n)Le_n\right\| \leq \left(\sum_n |f(n)|^p\|y_n\|^p\right)^{1/p} \leq \|f\|_\infty\left(\sum_{n\geq 1} 2^{-pn}\right)^{1/p} = \left(\frac{2^{-p}}{1-2^{-p}}\right)^{1/p}\|f\|.$$

(3) The maps KP_φ do not increase supports: $\operatorname{supp}\mathsf{KP}_\varphi(x) \subset \operatorname{supp} x$; consequently, if $\mathsf{KP}_\varphi(x)$ is close to y then it is closer to $y1_{\operatorname{supp} x}$.

(4) If X is a sequence space, $A \subset \mathbb{N}$, and Y denotes the subspace of those sequences of X with support contained in A, then we can regard Y as a sequence space (which satisfies (b) if X does) and KP_φ maps Y^0 to Y. Moreover, if KP_φ is trivial (as a quasilinear map from X^0 to X) then so is the restriction $\mathsf{KP}_\varphi\colon Y^0 \longrightarrow Y$: indeed, if $\ell\colon X^0 \longrightarrow X$ is a linear map at finite distance from KP_φ, then since the projection $P\colon X \longrightarrow Y$ given by $P(x) = 1_A x$ is contractive, we have

$$\|(\mathsf{KP}_\varphi - P\ell)\colon Y^0 \to Y\| \leq \|(\mathsf{KP}_\varphi - \ell)\colon X^0 \to X\|.$$

Ok, a little less conversation and a little more action, please. Assume there is a linear map $\ell\colon X^0 \longrightarrow X$ such that $\mathsf{KP}_\varphi - \ell$ is bounded. The difficulty to overcome is that ℓ need not preserve disjointness, and thus we need to pass to a certain subsequence of the basis where ℓ behaves better. First of all, observe that $\mathsf{KP}_\varphi(e_n) = 0$ for all n, so

$$\|\ell(e_n)\|_\infty \leq \|\ell(e_n)\| = \|\mathsf{KP}_\varphi(e_n) - \ell(e_n)\| \leq \|\mathsf{KP}_\varphi - \ell\|.$$

Hence $(\ell(e_n))$ has a subsequence that converges coordinatewise to some element of ℓ_∞. By our last remark, we can replace X by the subspace spanned by that subsequence and assume that $\ell(e_n)$ is pointwise convergent to a bounded sequence, which implies that $(\ell(e_{2n-1} - e_{2n}))_{n\geq 1}$ is bounded in X and converges to zero in every coordinate. Next we 'disjointify' the values of ℓ on a subsequence (f_k) of $(e_{2n-1} - e_{2n})$ using a gliding hump argument as follows. We start with $f_1 = e_1 - e_2$ and look at $\ell(f_1)$ to choose $p(1) > 2$ such that

$$\|\ell(f_1) - 1_{[1,p(1))}\ell(f_1)\| \leq 1/2.$$

Since $\ell(e_{2n-1} - e_{2n})$ converges to zero in every coordinate, we can select $2k-1 \geq p(1)$ and $p > 2k$ such that

$$\|\ell(e_{2k-1}-e_{2k}) - 1_{[p(1),p(2))}\ell(f_{k(2)})\| \le 2^{-2}.$$

We then set $f_2 = (e_{2k-1} - e_{2k})$ and continue this way, yielding two sequences of integers $(k(n))_{n\ge1}$ and $(p(n))_{n\ge0}$ such that

- $k(1) = p(0) = 1$ and $\operatorname{supp} f_n \subset [p(n-1), p(n))$ for $n \ge 1$,
- $\|\ell(f_n) - 1_{[p(n-1),p(n))}\ell(f_n)\| \le 2^{-n}$ for $n \ge 1$,

where $f_n = e_{2k(n)-1} - e_{2k(n)}$. Let F denote the linear subspace spanned by $(f_n)_{n\ge1}$ in X^0 and let $L\colon F \longrightarrow X$ be the linear map given by $L(f_n) = 1_{[p(n-1),p(n))}\ell(f_n)$. By the remark made in (2), we know that the difference $\ell - L$ is bounded (from F to X), and so is $\mathsf{KP}_\varphi - L$. Since the elements of the sequence (Lf_n) have mutually disjoint supports, the map $\tilde{\ell}\colon F \longrightarrow X$ given by $\tilde{\ell}(f) = 1_{\operatorname{supp} f}L(f)$ is linear. Note that the action of $\tilde{\ell}$ on the elements of the basis of F is

$$\tilde{\ell}(f_n) = |f_n|\ell(f_n) = \ell(f_n)(2k(n)-1)e_{2k(n)-1} + \ell(f_{k(n)})(2k(n))e_{2k(n)}.$$

By (3), we get that $\|\mathsf{KP}_\varphi - \tilde{\ell}\| \le \|\mathsf{KP}_\varphi - L\|$, so $(\tilde{\ell}(f_n))_{n\ge1}$ is bounded in X, and thus $\tilde{\ell}(f) = af$ for some $a \in \ell_\infty$ and every $f \in F$. Now set $\tilde{s}_n = \sum_{k\le n} f_k$. Then since $\mathsf{KP}_\varphi(\tilde{s}_n) = \varphi(\log\|\tilde{s}_n\|)\tilde{s}_n$ and

$$\|\mathsf{KP}_\varphi(\tilde{s}_n) - \tilde{\ell}(\tilde{s}_n)\| = \|\varphi(\log\|\tilde{s}_n\|)\tilde{s}_n - a\tilde{s}_n\| \le M\|\tilde{s}_n\|,$$

we see that the numerical sequence $\varphi(\log\|\tilde{s}_n\|)$ is bounded and that φ is bounded on $\mathbb{R}^+$, by (1). □

From the observation $\mathsf{KP}_\varphi - \mathsf{KP}_\gamma = \mathsf{KP}_{\varphi-\gamma}$ we immediately have:

Corollary 3.3.6 *Under the same hypotheses on X as Proposition 3.3.5, let $\varphi, \gamma \in \mathrm{Lip}_0(\mathbb{R}^+)$. Then KP_φ and KP_γ are equivalent on X if and only if* $\sup_{t>0} |\varphi(t) - \gamma(t)| < \infty$.

While Proposition 3.3.5 shows that the Kalton–Peck maps produce non-trivial sequences $0 \longrightarrow X \longrightarrow X(\varphi) \longrightarrow X \longrightarrow 0$ for most quasi-Banach sequence spaces X and unbounded φ, the conclusion fails when $X = c_0$ for the trivial reason that all Kalton–Peck maps are bounded on c_0, which follows easily from the fact that $|t\log t|$ is bounded for $0 \le t \le 1$. Sobczyk's theorem gives a deeper explanation of why one cannot expect to see a non-trivial self-extension of c_0. However, the question remains whether there exist exact sequences $0 \longrightarrow c_0 \longrightarrow \cdot \longrightarrow c_0 \longrightarrow 0$ in which the middle space is not locally convex. There do not, but this is a *really* deep result of Kalton and Roberts stated later in 3.4.6. It has a surprisingly large number of connections, ramifications and applications through the Maharam problem on exhaustive submeasures, which go far beyond the scope of this volume. The hungry reader can find nutritious information in [257] and [181].

Existence of Quasilinear Maps

It is implicit in the discussion in Section 3.1 that every short exact sequence of quasi-Banach spaces arises from and gives rise to a quasilinear map. Let us make this statement precise:

Proposition 3.3.7 *Every exact sequence of quasi-Banach spaces is generated, up to equivalence, by a quasilinear map. More precisely, for every exact sequence* $0 \longrightarrow Y \longrightarrow Z \longrightarrow X \longrightarrow 0$*, there is a quasilinear map* $\Phi\colon X \longrightarrow Y$ *and a commutative diagram*

$$\begin{array}{ccccccccc} 0 & \longrightarrow & Y & \xrightarrow{\imath} & Y \oplus_\Phi X & \xrightarrow{\pi} & X & \longrightarrow & 0 \\ & & \| & & \downarrow & & \| & & \\ 0 & \longrightarrow & Y & \xrightarrow{\jmath} & Z & \xrightarrow{\rho} & X & \longrightarrow & 0 \end{array} \tag{3.12}$$

Proof Let $B\colon X \longrightarrow Z$ be a homogeneous bounded section of ρ, and let $L\colon X \longrightarrow Z$ be a linear (possibly discontinuous) section. The difference $B - L$ takes values in $\ker \rho = \jmath[Y]$ since $\rho(B(x) - L(x)) = x - x = 0$, and composing with the inverse of $\jmath$, we get the map $\Phi = \jmath^{-1} \circ (B - L)$ from X to Y. The quasilinear character of Φ is obvious: $B - L$ has the same Cauchy differences as B and $\jmath^{-1}$ is bounded. Anyway, let us estimate $Q(\Phi)$. Pick $x, y \in X$. If $\Delta_X, \Delta_Y, \Delta_Z$ denote the respective concavity constants then

$$\begin{aligned} \|\Phi(x+y) - \Phi(x) - \Phi(y)\| &\le \|\jmath^{-1}\| \, \|(B-L)(x+y) - (B-L)(x) - (B-L)(y)\| \\ &= \|\jmath^{-1}\| \, \|B(x+y) - B(x) - B(y)\| \\ &\le \|\jmath^{-1}\|(\Delta_Z\|B(x+y)\| + \Delta_Z^2(\|B(x)\| + \|B(y)\|)) \\ &\le \|\jmath^{-1}\| \, \|B\| \, (\Delta_Z\Delta_X + \Delta_Z^2) \, (\|x\| + \|y\|). \end{aligned}$$

The linear map $u(y, x) = \jmath(y) + L(x)$ makes Diagram (3.12) commutative, as can be easily seen. We conclude the proof by showing that it is bounded:

$$\begin{aligned} \|u(y,x)\| &= \|\jmath(y) - \jmath(\Phi(x)) + \jmath(\Phi(x)) + L(x)\| \\ &\le \Delta_Z(\|\jmath(y) - \jmath(\Phi(x))\| + \|\jmath\Phi(x) + L(x)\|) \\ &= \Delta_Z(\|\jmath(y - \Phi(x))\| + \|B(x)\|) \\ &\le \Delta_Z(\|\jmath\| \, \|y - \Phi(x)\| + \|B\| \, \|x\|) \\ &\le \Delta_Z \max(\|\jmath\|, \|B\|) \cdot \|(y,x)\|_\Phi. \end{aligned}$$

□

We know from Roelcke's lemma that the map u is an isomorphism. A direct computation shows that its inverse is given by $u^{-1}(z) = (\jmath^{-1}(z - L(\rho(z)), \rho(z))$ and satisfies

$$\begin{aligned}\|u^{-1}(z)\|_\Phi &= \|\jmath^{-1}(z - L(\rho(z)) - \Phi(\rho(z))\| + \|\rho z\| \\ &= \|\jmath^{-1}(z - L(\rho(z)) - \jmath^{-1}(B - L)(\rho(z))\| + \|\rho z\| \\ &= \|\jmath^{-1}(z - B(\rho(z))\| + \|\rho z\| \\ &\le \left(\|\jmath^{-1}\|\Delta_Z(1 + \|B\|\,\|\rho\|) + \|\rho\|\right)\|z\|.\end{aligned}$$

Admittedly, the preceding computations are a bit nitpicky. In their defense, they provide very thin bounds for the Banach–Mazur distance between the space Z and the twisted sum $Y \oplus_\Phi X$:

Corollary 3.3.8 *If $0 \longrightarrow Y \longrightarrow Z \longrightarrow X \longrightarrow 0$ is an isometrically exact sequence of p-Banach spaces, then for every $\varepsilon > 0$, there is a quasilinear map $\Phi\colon X \longrightarrow Y$ with $Q(\Phi) < 2^{2/p-1} + \varepsilon$ and a commutative diagram*

$$\begin{array}{ccccccccc} 0 & \longrightarrow & Y & \longrightarrow & Y \oplus_\Phi X & \longrightarrow & X & \longrightarrow & 0 \\ & & \| & & \downarrow{\scriptstyle u} & & \| & & \\ 0 & \longrightarrow & Y & \longrightarrow & Z & \longrightarrow & X & \longrightarrow & 0 \end{array} \qquad (3.13)$$

where u is an isomorphism such that $\|u\| < 2^{1/p-1} + \varepsilon$ and $\|u^{-1}\| < 2^{1/p} + \varepsilon$. In particular, the Banach–Mazur distance between Z and the twisted sum space $Y \oplus_\Phi Z$ is at most $2^{2/p-1}$.

Proof Just follow the proof of Proposition 3.3.7, taking into account that in the isometric setting we have $\|\jmath\| = \|\jmath^{-1}\| = \|\rho\| = 1$, that for every $\varepsilon > 0$ the bounded selection $B\colon X \longrightarrow Z$ can be chosen such that $\|B\| < 1 + \varepsilon$, and that the concavity of any p-normed space is at most $2^{1/p-1}$. □

In general, almost everything involving quasinormed spaces depends on the modulus of concavity of the quasinorm. The quasilinearity constant is not an exception.

Lemma 3.3.9 *Let $\Phi\colon X \longrightarrow Y$ be a quasilinear map acting between quasinormed spaces. Then $Q(\Phi) - 1 \le \Delta_{Y\oplus_\Phi X} \le \max\left(\Delta_Y^2, Q(\Phi)\Delta_Y + \Delta_X\right)$.*

Proof Indeed, if $\|(y,x) + (y',x')\|_\Phi \le \Delta\left(\|(y,x)\|_\Phi + \|(y',x')\|_\Phi\right)$ for some Δ then, setting $y = \Phi(x), y' = \Phi(x')$, we obtain

$$\|\Phi(x + x') - \Phi(x) - \Phi(x')\| + \|x + x'\| \le (\Delta - 1)(\|x\| + \|x'\|).$$

In particular, $Q(\Phi) \leq \Delta - 1$. Conversely,

$$\begin{aligned}
&\|(y,x) + (y',x')\|_\Phi \\
&= \|y + y' - \Phi(x+x')\| + \|x + x'\| \\
&\leq \Delta_Y \left(\|y + y' - \Phi x - \Phi x'\| + \|\Phi x + \Phi x' - \Phi(x+x')\|\right) + \Delta_X \left(\|x\| + \|x'\|\right) \\
&\leq \Delta_Y^2 \left(\|y_1 - \Phi(x)\| + \|y_2 - \Phi(x')\|\right) + (Q(\Phi)\Delta_Y + \Delta_X)(\|x\| + \|x'\|) \\
&\leq \max\left(\Delta_Y^2, Q(\Phi)\Delta_Y + \Delta_X\right) \cdot \left(\|(y,x)\|_\Phi + \|(y',x')\|_\Phi\right).
\end{aligned}$$

□

Extension of Quasilinear Maps from Dense Subspaces

And so we arrive to the core of the non-linear approach. While the equivalence between exact sequences and quasilinear maps explained in Proposition 3.3.7 is in a sense the final word concerning the theory of quasilinear maps, in practice there is still a loose end to be tied: namely, the question of how to obtain such quasilinear maps, since, as we remarked in Section 3.2, quasilinear maps $X \longrightarrow Y$ usually only come defined on a dense subspace of X. This is why Lemma 3.3.2 took the form it did. The introduction to Section 3.2 also claimed that 'fortunately, this blow is not fatal,' and concluded with an obscure reference to a certain 'quasilinear map $\widetilde{\Phi}$ extending Φ'. Time has come to put all this in order and show exactly how things must be done. Let us assume that X and Y are quasi-Banach spaces and that we are given a quasilinear map $\Phi\colon X_0 \longrightarrow Y$, where X_0 is a dense subspace of X. Recall Diagram (3.3),

$$\begin{array}{ccccccccc}
0 & \longrightarrow & Y & \xrightarrow{\imath} & Y \oplus_\Phi X_0 & \xrightarrow{\pi} & X_0 & \longrightarrow & 0 \\
 & & \| & & \downarrow{\scriptstyle\kappa} & & \downarrow{\scriptstyle\text{inclusion}} & & \\
0 & \longrightarrow & Y & \xrightarrow{\imath} & Z(\Phi) & \xrightarrow{\widetilde{\pi}} & X & \longrightarrow & 0
\end{array}$$

and call Proposition 3.3.7 onstage so that the lower sequence can be represented by a quasilinear map $\widetilde{\Phi}\colon X \longrightarrow Y$. It is already clear that the restriction of $\widetilde{\Phi}$ to X_0 is equivalent to Φ. The point, however, is that $\widetilde{\Phi}$ can be chosen to be a true extension of Φ, and this information is contained in the proof of Proposition 3.3.7: all maps that are equivalent to $\widetilde{\Phi}$ are obtained as the difference of a bounded homogeneous section $B\colon X \longrightarrow Z(\Phi)$ and a linear section $L\colon X \longrightarrow Z(\Phi)$ for $\widetilde{\pi}$. So pick B as a homogeneous bounded true extension of $x_0 \longmapsto \kappa(\Phi(x_0), x_0)$ and take for L a true linear extension of $x_0 \longmapsto \kappa(0, x_0)$. Now, yes, $\imath^{-1} \circ (B - L)$ is our true quasilinear extension $\widetilde{\Phi}$ of Φ. Die-hard sceptics can have great fun checking the commutativity of the diagram

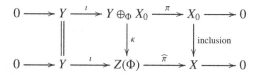

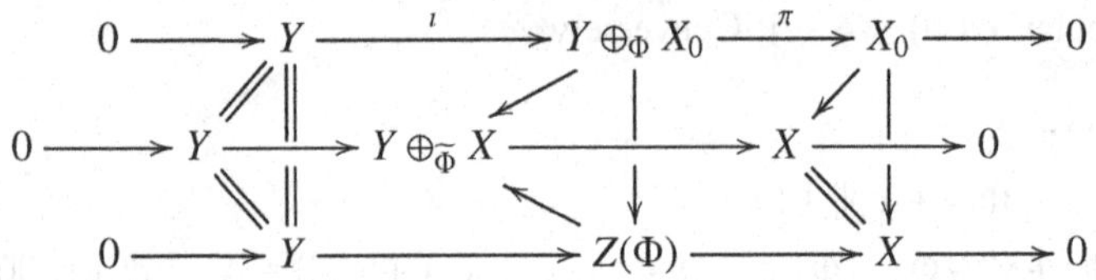

A direct consequence is:

Corollary 3.3.10 *A quasilinear map is trivial if and only if its restriction to some (any) dense subspace is trivial.*

Proof It is clear that the restriction of a trivial quasilinear map to any subspace, dense or not, is again trivial. To prove the converse, let $\Phi\colon X \longrightarrow Y$ be quasilinear and let X_0 be a dense subspace of X. If $\Phi|_{X_0}$ is trivial then the lower quotient map in the diagram

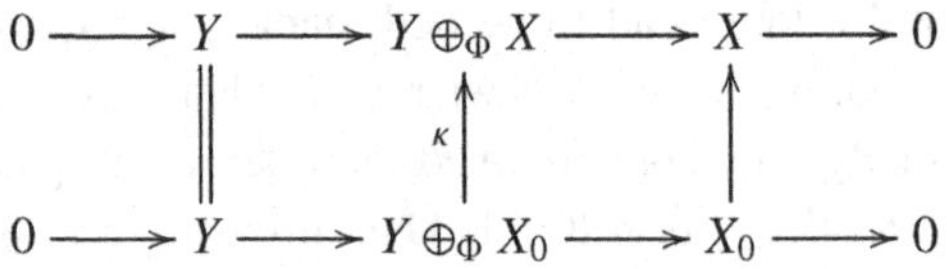

has a linear continuous section $s\colon X_0 \longrightarrow Y \oplus_\Phi X_0$. Any linear continuous extension of κs to X shows that the upper sequence splits. □

A quantisation is always welcome:

Proposition 3.3.11 *For every $\Delta \geq 1$, there is a constant C such that if X and Y are quasi-Banach spaces with moduli of concavity Δ or less and X_0 is a dense subspace of X then every quasilinear map $\Phi\colon X_0 \longrightarrow Y$ has an extension $\widetilde{\Phi}\colon X \longrightarrow Y$ with $Q(\widetilde{\Phi}) \leq CQ(\Phi)$.*

Proof We need a simple amalgamation argument, once we know that individual maps extend from dense subspaces. Let $(X^i)_{i\in I}$ and $(Y^i)_{i\in I}$ be families of quasi-Banach spaces whose moduli of concavity are at most Δ. For each $i \in I$, let X_0^i be a dense subspace of X^i and let $\Phi_i\colon X_0^i \longrightarrow Y^i$ be a quasilinear map with $Q(\Phi_i) \leq 1$. Their c_0-sums $c_0(I, X^i)$ and $c_0(I, Y^i)$ are quasi-Banach spaces with moduli Δ. Let $c_0^0(I, X_0^i)$ be the dense subspace of finitely supported sequences of $c_0(I, X^i)$. The map $\Phi\colon c_0^0(I, X_0^i) \longrightarrow c_0(I, Y^i)$ defined by $\Phi((x_i)_{i\in I}) = (\Phi(x_i))_{i\in I}$ is quasilinear with $Q(\Phi) \leq 1$ and can be extended to a quasilinear map $\widetilde{\Phi}\colon c_0(I, X^i) \longrightarrow c_0(I, Y^i)$. If $\jmath_i\colon X_i \longrightarrow c_0(I, X^i)$ and $\pi_i\colon c_0(I, Y^i) \longrightarrow Y_i$ denote the obvious embedding and projection then $\pi_i \circ \widetilde{\Phi} \circ \jmath_i$ is an extension of Φ_i to X^i with quasilinearity constant at most $Q(\widetilde{\Phi})$. □

3.4 Local Convexity of Twisted Sums and $\mathscr{K}$-Spaces

We can no longer avoid the hard truth that no result proved so far guarantees that any of the twisted sums constructed in this chapter are Banach spaces. Quite the contrary, both the Ribe space and the Kalton–Peck spaces $\ell_1(\varphi)$ clearly show that twisted sums of Banach spaces can be non-locally convex. Thus, the topic that must be considered is the 3-space problem for local convexity; when we undertake this study, we encounter the following notions at the center of it all:

Definition 3.4.1 A minimal extension of X is a short exact sequence of the form $0 \longrightarrow \mathbb{K} \longrightarrow Z \longrightarrow X \longrightarrow 0$. A quasi-Banach space X is said to be a $\mathscr{K}$-space if every minimal extension is trivial, i.e. $\operatorname{Ext}(X, \mathbb{K}) = 0$.

Non-trivial minimal extensions appear whenever the Hahn–Banach theorem fails: if Z has trivial dual then every non-zero point $z \in Z$ gives rise to a minimal extension of $Z/[z]$. A minimal extension $0 \longrightarrow \mathbb{K} \longrightarrow Z \longrightarrow X \longrightarrow 0$ of a Banach space X splits if and only if the twisted sum space Z is locally convex; that is, isomorphic to a Banach space. Indeed, if Z is a Banach space, the Hahn–Banach theorem applied to $\mathbf{1}_{\mathbb{K}}$ yields a projection of Z onto $\mathbb{K}$. And, conversely, if the sequence splits then Z is isomorphic to $\mathbb{K} \times X$, which is locally convex.

Definition 3.4.2 A subspace Y of a quasi-Banach space X is said to have the Hahn–Banach extension property (HBEP) if each linear continuous functional on Y can be extended to a linear continuous functional on X.

Kalton proved in [246] that a quasi-Banach space all of whose subspaces have the HBEP must be locally convex. More modestly one has:

Lemma 3.4.3 *Let $0 \longrightarrow Y \longrightarrow Z \longrightarrow X \longrightarrow 0$ be a short exact sequence in which X and Y are Banach spaces. Then Z is a Banach space if and only if Y has the HBEP in Z.*

Proof The 'only if' part is straightforward from the Hahn–Banach extension theorem. For the converse, observe that a quasi-Banach space A is (isomorphic to) a Banach space if and only if the natural evaluation map $\delta_A : A \longrightarrow A^{**}$ is an isomorphic embedding. Now, if Y has the HBEP in Z then the sequence $0 \longrightarrow X^* \longrightarrow Z^* \longrightarrow Y^* \longrightarrow 0$ is exact, and so is the bidual sequence $0 \longrightarrow Y^{**} \longrightarrow Z^{**} \longrightarrow X^{**} \longrightarrow 0$. Thus, we have a commutative diagram

$$\begin{array}{ccccccccc} 0 & \longrightarrow & Y & \longrightarrow & Z & \longrightarrow & X & \longrightarrow & 0 \\ & & \downarrow{\scriptstyle \delta_Y} & & \downarrow{\scriptstyle \delta_Z} & & \downarrow{\scriptstyle \delta_X} & & \\ 0 & \longrightarrow & Y^{**} & \longrightarrow & Z^{**} & \longrightarrow & X^{**} & \longrightarrow & 0 \end{array}$$

in which the vertical arrows are the natural evaluation maps. Since δ_Y and δ_X are isomorphic embeddings, so is δ_Z by Roelcke's lemma 2.1.8. □

Thus, whenever one has an exact sequence $0 \longrightarrow Y \longrightarrow Z \longrightarrow X \longrightarrow 0$ in which both Y and X are Banach spaces and Z is not, there exists $y^* \in Y^*$ that does not extend to a bounded linear functional on Z; this means that the lower row in the pushout diagram

$$\begin{array}{ccccccccc} 0 & \longrightarrow & Y & \longrightarrow & Z & \longrightarrow & X & \longrightarrow & 0 \\ & & \downarrow{\scriptstyle y^*} & & \downarrow & & \| & & \\ 0 & \longrightarrow & \mathbb{K} & \longrightarrow & \mathrm{PO} & \longrightarrow & X & \longrightarrow & 0 \end{array}$$

does not split. Hence, any counterexample for the 3-space problem for local convexity leads to a non-trivial minimal extension of the quotient space. Such is the context of the following classical result of Dierolf [149]:

3.4.4 Dierolf's Theorem *A Banach space X is a $\mathscr{K}$-space if and only if every extension of X by a Banach space is locally convex.*

Is there some obvious example of $\mathscr{K}$-space in sight? No, because proving that a given Banach or quasi-Banach space is a $\mathscr{K}$-space requires some work ... or, rather, a lot of work. Actually, the only large class of Banach $\mathscr{K}$-spaces we can isolate at this moment is:

Proposition 3.4.5 *Superreflexive Banach spaces are $\mathscr{K}$-spaces.*

Proof Let $\phi\colon X \longrightarrow \mathbb{K}$ be a quasilinear funcional on a Banach space X. For each finite-dimensional subspace $E \in \mathscr{F}(X)$, set $d_E = \operatorname{dist}(\phi|_E, E^*)$. Since these infima are attained, one can pick $L_E \in E^*$ such that $\|\phi|_E - L_E\| = d_E$. Now, if $\sup_E d_E < \infty$, pick $\mathfrak{U}$ an ultrafilter on $\mathscr{F}(X)$ refining the order filter and define $L\colon X \longrightarrow \mathbb{K}$ by $L(x) = \lim_{\mathfrak{U}(E)} L_E(x)$, which makes sense, since for each $x \in X$, the family $(L_E(x))_E$ is bounded because $|L_E(x)| \le |\phi|_E(x)| + d_E$. The map L is clearly linear and $\|\Phi - L\| \le \sup_E d_E$. If, on the contrary, $\sup_E d_E = \infty$, consider the function $f_E\colon E \longrightarrow \mathbb{K}$ given by

$$f_E(x) = \frac{\phi(x) - L_E(x)}{d_E}$$

(set $f_E = 0$ in the innocuous case $d_E = 0$). These maps are all bounded, with $\|f_E\| \le 1$. If $\mathfrak{U}$ is once again an ultrafilter refining the order filter on $\mathscr{F}(X)$ then $\sup_E d_E = \infty$ necessarily implies $\lim_{\mathfrak{U}(E)} Q(f_E) = \lim_{\mathfrak{U}(E)} d_E^{-1} Q(\phi) = 0$. Form the ultraproduct $\mathscr{F}(X)_{\mathfrak{U}}$ and tentatively define a mapping $f\colon \mathscr{F}(X)_{\mathfrak{U}} \longrightarrow \mathbb{K}$ by the formula $f[(x_E)] = \lim_{\mathfrak{U}(E)} f_E(x_E)$.

Claim f is a bounded linear functional on $\mathscr{F}(X)_{\mathfrak{U}}$.

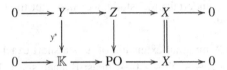

Proof of the claim We must check that f is correctly defined. First assume $[(x_E)] = 0$, that is, $\|x_E\| \longrightarrow 0$ along $\mathfrak{U}$. Then,

$$\Big| \lim_{\mathfrak{U}(E)} f_E(x_E) \Big| = \lim_{\mathfrak{U}(E)} |f_E(x_E)| \leq \lim_{\mathfrak{U}(E)} \|f_E\| \, \|x_E\| = 0.$$

Now, if $[(x_E)] = [(y_E)]$ then

$$\begin{aligned}\lim_{\mathfrak{U}(E)} \big|f_E(x_E) - f_E(y_E)\big| &= \lim_{\mathfrak{U}(E)} \big|f_E(x_E) - f_E(y_E) - f_E(x_E - y_E)\big| \\ &\leq \lim_{\mathfrak{U}(E)} Q(f_E)(\|y_E\| + \|x_E - y_E\|) = 0.\end{aligned}$$

The map f is obviously bounded, and it is linear since it is homogeneous, and given bounded families (x_E) and (y_E), we have

$$\begin{aligned}\big|f([(x_E + y_E)]) - f[(x_E)] - f[(y_E)]\big| &= \lim_{\mathfrak{U}(E)} |f_E(x_E + y_E) - f_E(x_E) - f_E(y_E)| \\ &\leq \lim_{\mathfrak{U}(E)} Q(f_E)(\|x_E\| + \|y_E\|) = 0. \qquad \square\end{aligned}$$

The hypothesis on X enters now: if X is superreflexive, the ultraproduct $\mathscr{F}(X)_{\mathfrak{U}}$ is reflexive and its dual agrees with the ultraproduct of the dual family $(E^*)_{E\in\mathscr{F}(X)}$ with respect to $\mathfrak{U}$. Thus, $f \in (\mathscr{F}(X)_{\mathfrak{U}})^*$ is represented by a bounded family of functionals $g_E \in E^*$ in the form

$$f[(x_E)] = \lim_{\mathfrak{U}(E)} f_E(x_E) = \lim_{\mathfrak{U}(E)} \langle g_E, x_E \rangle$$

for every bounded family (x_E). This clearly implies $\lim_{\mathfrak{U}(E)} \|f_E - g_E\| = 0$ and thus the set $\{E \in \mathscr{F}(X) \colon \|f_E - g_E\| < \frac{1}{2} \text{ and } d_E > 0\}$ belongs to $\mathfrak{U}$ and cannot be empty. Pick some element E in there to get

$$\Big\| \frac{\phi|_E - L_E}{d_E} - g_E \Big\| < \frac{1}{2} \quad \Longrightarrow \quad \big\| \phi|_E - L_E - d_E g_E \big\| < \frac{d_E}{2},$$

in clear contradiction of the definition of d_E. $\square$

The wonderful consequence we get is that all twisted sums $X(\varphi)$ are isomorphic to Banach spaces when X is a superreflexive sequence space, in particular if $X = \ell_p$ for $p \in (1, \infty)$. In Proposition 3.11.3, it will be proved that B-convex spaces are also $\mathscr{K}$-spaces. We can now record the *really deep* result of Kalton and Roberts [285, Theorem 6.3] mentioned previously.

3.4.6 Kalton–Roberts theorem *$\mathscr{L}_\infty$-spaces are $\mathscr{K}$-spaces.*

And what about non-$\mathscr{K}$-spaces? Producing non-$\mathscr{K}$-spaces is much easier when working with quasi-Banach spaces because, as we already said, the quotient of any space with trivial dual by a line fails to be a $\mathscr{K}$-space. In a sense, this is the only way there is to not be a $\mathscr{K}$-space:

Proposition 3.4.7 *Let X be quasi-Banach space. The following are equivalent:*

(i) *X is a $\mathcal{K}$-space.*
(ii) *Whenever $Q\colon Z \longrightarrow X$ is a quotient map,* $\ker Q$ *has the HBEP in Z.*
(iii) *Whenever F is a finite-dimensional subspace of a quasi-Banach space Z, every operator $X \longrightarrow Z/F$ lifts to Z.*

Proof To prove (i) $\implies$ (ii), consider a quotient map $Q\colon Z \longrightarrow X$ and a bounded linear functional f on $\ker Q$ and form the pushout diagram

$$\begin{array}{ccccccccc}
0 & \longrightarrow & \ker Q & \longrightarrow & Z & \xrightarrow{Q} & X & \longrightarrow & 0\\
 & & \downarrow{\scriptstyle f} & & \downarrow & & \| & & \\
0 & \longrightarrow & \mathbb{K} & \longrightarrow & \mathrm{PO} & \longrightarrow & X & \longrightarrow & 0
\end{array}$$

If X is a $\mathcal{K}$-space, the lower sequence splits and f extends to a bounded linear functional on Z; i.e. $\ker Q$ has the HBEP in Z.

(ii) $\implies$ (iii) Let F be a finite-dimensional subspace of Z and $u\colon X \longrightarrow Z/F$ an operator. By (ii), F must have the HBEP in PB. Since finite-dimensional subspaces with the HBEP are complemented (just extend the coordinate functionals of a basis), the lower sequence in the diagram

$$\begin{array}{ccccccccc}
0 & \longrightarrow & F & \longrightarrow & Z & \longrightarrow & Z/F & \longrightarrow & 0\\
 & & \| & & \uparrow & & \uparrow{\scriptstyle u} & & \\
0 & \longrightarrow & F & \longrightarrow & \mathrm{PB} & \longrightarrow & X & \longrightarrow & 0
\end{array}$$

splits, and by the splitting criterion for pullbacks, u has a lifting to Z, which proves (iii). The implication (iii) $\implies$ (i) is trivial: every exact sequence $0 \longrightarrow \mathbb{K} \longrightarrow Z \longrightarrow X \longrightarrow 0$ splits since the hypothesis allows the identity on X to be lifted to an operator $X \longrightarrow Z$. □

Proposition 3.4.8 *Let Z be a quasi-Banach space and let $Q\colon Z \longrightarrow X$ be a quotient map. If Z is a $\mathcal{K}$-space and* $\ker Q$ *has the HBEP in Z then X is a $\mathcal{K}$-space. In particular, quotients of Banach $\mathcal{K}$-spaces are $\mathcal{K}$-spaces.*

Proof Consider a minimal extension of X and form the pullback diagram

$$\begin{array}{ccccccccc}
0 & \longrightarrow & \mathbb{K} & \longrightarrow & \cdot & \xrightarrow{\rho} & X & \longrightarrow & 0\\
 & & \| & & \uparrow & & \uparrow{\scriptstyle Q} & & \\
0 & \longrightarrow & \mathbb{K} & \longrightarrow & \mathrm{PB} & \longrightarrow & Z & \longrightarrow & 0
\end{array}$$

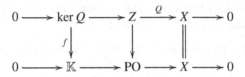

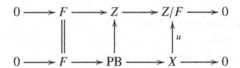

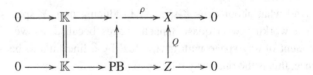

Since Z is a $\mathscr{K}$-space the lower sequence splits, and so Q lifts to a map $\widetilde{Q}\colon Z \longrightarrow \cdot$ such that $Q = \rho\,\widetilde{Q}$. There is therefore a commutative diagram

$$\begin{array}{ccccccccc} 0 & \longrightarrow & \mathbb{K} & \longrightarrow & \cdot & \stackrel{\rho}{\longrightarrow} & X & \longrightarrow & 0 \\ & & \uparrow{\scriptstyle \widetilde{Q}|_{\ker Q}} & & \uparrow{\scriptstyle \widetilde{Q}} & & \| & & \\ 0 & \longrightarrow & \ker Q & \longrightarrow & Z & \stackrel{Q}{\longrightarrow} & X & \longrightarrow & 0 \end{array}$$

Since $\ker Q$ has the HBEP in Z, the functional $\widetilde{Q}|_{\ker Q}$ extends to Z and the upper (pushout) sequence splits. □

We are ready to conclude the analysis of the 3-space problem for local convexity, which we initiated with Dierolf's theorem 3.4.4, by explaining, as promised, why counterexamples for the locally convex 3-space problem were obtained as minimal extensions of ℓ_1. If X is a Banach space and some non-trivial sequence

$$0 \longrightarrow \mathbb{K} \stackrel{\jmath}{\longrightarrow} Z \longrightarrow X \longrightarrow 0 \tag{3.14}$$

exists then Z is non-locally convex. Pick a non-locally convex separable subspace $\widetilde{Z}$ of Z containing $\jmath[\mathbb{K}]$ and form the commutative diagram

$$\begin{array}{ccccccccc} 0 & \longrightarrow & \mathbb{K} & \longrightarrow & Z & \longrightarrow & X & \longrightarrow & 0 \\ & & \| & & \uparrow & & \uparrow & & \\ 0 & \longrightarrow & \mathbb{K} & \longrightarrow & \widetilde{Z} & \longrightarrow & \widetilde{X} & \longrightarrow & 0 \end{array}$$

The lower sequence of the diagram does not split, because $\widetilde{Z}$ is non-locally convex, and $\widetilde{X} = \widetilde{Z}/\jmath[\mathbb{K}]$ is isomorphic to a separable subspace of X. Pick any quotient map $Q\colon \ell_1 \longrightarrow \tilde{X}$ and form the commutative pullback diagram

$$\begin{array}{ccccccccc} 0 & \longrightarrow & \mathbb{K} & \longrightarrow & \widetilde{Z} & \longrightarrow & \widetilde{X} & \longrightarrow & 0 \\ & & \| & & \uparrow & & \uparrow{\scriptstyle Q} & & \\ 0 & \longrightarrow & \mathbb{K} & \longrightarrow & \mathrm{PB} & \longrightarrow & \ell_1 & \longrightarrow & 0 \end{array}$$

The space PB cannot be locally convex since the map $\mathrm{PB} \longrightarrow \widetilde{Z}$ is surjective. Thus, any exact sequence $0 \longrightarrow Y \longrightarrow Z \longrightarrow X \longrightarrow 0$ in which X and Y, but not Z, are Banach spaces leads, after a pushout and pullback, to a minimal extension of ℓ_1.

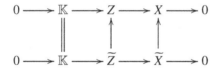

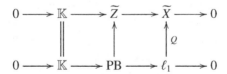

3.5 The Pullback and Pushout in Quasilinear Terms

As is only natural in a book with the word *homological* in the title, we now study the functorial properties of the assignment $(X, Y) \rightsquigarrow \mathsf{Q}(X, Y)$. In practice, this means studying how quasilinear maps and operators compose:

3.5.1 *If $\Phi\colon X \longrightarrow Y$ is quasilinear and $T\colon Y \longrightarrow Y'$ and $S : X' \longrightarrow X$ are operators, then $T \circ \Phi \circ S : X' \longrightarrow Y'$ is quasilinear with $Q(T \circ \Phi \circ S) \leq \|T\| Q(\Phi) \|S\|$.*

Can we identify the exact sequences induced by $T \circ \Phi$ and $\Phi \circ S$? Of course we can: they are, respectively, the pushout and the pullback sequences. To check this, simply contemplate the commutative diagrams

$$\begin{array}{ccccccccc} 0 & \longrightarrow & Y & \longrightarrow & Y \oplus_\Phi X & \longrightarrow & X & \longrightarrow & 0 \\ & & \downarrow{\scriptstyle T} & & \downarrow{\scriptstyle T\times \mathbf{1}_X} & & \| & & \\ 0 & \longrightarrow & Y' & \longrightarrow & Y' \oplus_{T\Phi} X & \longrightarrow & X & \longrightarrow & 0 \end{array} \tag{3.15}$$

where $(T \times \mathbf{1}_X)(y, x) = (Ty, x)$ and

$$\begin{array}{ccccccccc} 0 & \longrightarrow & Y & \longrightarrow & Y \oplus_\Phi X & \longrightarrow & X & \longrightarrow & 0 \\ & & \| & & \uparrow{\scriptstyle \mathbf{1}_Y\times S} & & \uparrow{\scriptstyle S} & & \\ 0 & \longrightarrow & Y & \longrightarrow & Y \oplus_{\Phi S} X & \longrightarrow & X' & \longrightarrow & 0 \end{array}$$

where $(\mathbf{1}_X \times S)(y, x') = (y, Sx')$. The operators in the middle are clearly bounded:

$$\|(Ty, x)\|_{T\Phi} = \|Ty - T\Phi(x)\| + \|x\| \leq \max(\|T\|, 1)\|(y, x)\|_\Phi,$$
$$\|(y, Sx')\|_\Phi = \|y - \Phi S x'\| + \|Sx'\| \leq \max(1, \|S\|)\|(y, x')\|_{\Phi S}.$$

Thus, the quasilinear representation of the pushout and pullback constructions is much simpler than the original: plain left and right composition! And, better yet, the same will happen for the rest of homological constructions studied in Chapter 2 and for others that will be introduced in Chapter 4. Let us give the suspicious reader a taste: the fact that pushout and pullback operations commute required a proof in Chapter 2, but it is completely obvious now – in both cases, we arrive at the exact sequence generated by $T \circ \Phi \circ S$ (if we are being pedantic, in the first case, we arrive at $(T \circ \Phi) \circ S$ and in the second, at $T \circ (\Phi \circ S)$). Another more formal example: remember the diagonal sequence in a pushout diagram? Assertion 2.10.2 reformulates now as:

3.5.2 *The diagonal pushout sequence is induced by $\Phi \circ \overline{p}$.*

The proof consists of observing that there is a commutative diagram

$$\begin{array}{ccccccccc}
0 & \longrightarrow & Y & \xrightarrow{\jmath} & Z(\Phi) & \xrightarrow{\rho} & X & \longrightarrow & 0\\
 & & \| & & \uparrow{\scriptstyle T} & & \uparrow{\scriptstyle \overline{\rho}} & & \\
0 & \longrightarrow & Y & \xrightarrow{(-\tau,\overline{\jmath})} & Y'\oplus Z & \xrightarrow{\mathbf{1}_{Y'}\oplus\tau} & \mathrm{PO} & \longrightarrow & 0
\end{array}$$

simply putting $T(y', z) = z$. Analogously, assertion 2.10.1 becomes

3.5.3 *The diagonal pullback sequence is induced by* $\underline{\jmath}\circ\Phi$.

The proof consists of observing that there is a commutative diagram

$$\begin{array}{ccccccccc}
0 & \longrightarrow & Y & \xrightarrow{\jmath} & Z(\Phi) & \xrightarrow{\rho} & X & \longrightarrow & 0\\
 & & \downarrow{\scriptstyle \underline{\jmath}} & & \downarrow{\scriptstyle T} & & \| & & \\
0 & \longrightarrow & \mathrm{PB} & \xrightarrow{(\underline{\tau},\underline{\rho})} & Z\oplus X' & \xrightarrow{\rho\oplus(-\tau)} & X & \longrightarrow & 0
\end{array}$$

where $T(z) = (z, 0)$.

We conclude quantifying the knot between triviality, extension and lifting: its proof clearly follows from Lemma 3.3.4, taking into account that for $\|T\| \leq 1$, the map $T \times \mathbf{1}_X$ in Diagram (3.15) is contractive; and similarly with S.

Lemma 3.5.4 *Let* $\Phi\colon X \longrightarrow Y$ *be a quasilinear map,* $T\colon Y \longrightarrow Y'$ *and* $S: X' \longrightarrow X$ *contractive operators and* $\mu \geq 0$.

- *If* $T \circ \Phi$ *is* μ*-trivial,* T *has an extension to* $Y \oplus_\Phi X$ *bounded by* $\max(\Delta_{Y'}, \mu)$. *If* T *has a* μ*-extension then* $T \circ \Phi$ *is* μ*-trivial.*
- $\Phi \circ S$ *is* μ*-trivial if and only if* S *has a lifting* $L\colon X' \longrightarrow Y \oplus_\Phi X$ *such that* $\|L\| \leq 1 + \mu$.

3.6 Spaces of Quasilinear Maps

Spaces of quasilinear maps deserve to be studied on their own, and that is what we will do now. Given a pair of quasinormed spaces X, Y, we can consider the following vector spaces:

- the space of quasilinear maps $\mathsf{Q}(X, Y)$,
- the space of bounded homogeneous maps $\mathsf{B}(X, Y)$,
- the space of linear maps $\mathsf{L}(X, Y)$.

When the spaces X, Y are fixed and there is no possibility of confusion, the spaces will be omitted and we will just write Q, L and B. Recalling from Lemma 3.3.2 that two quasilinear maps acting between the same spaces induce equivalent exact sequences if and only if their difference belongs to $\mathsf{B}(X, Y) + \mathsf{L}(X, Y)$, we are especially interested in the quotient space

$$\mathsf{Q}_{\mathsf{LB}}(X, Y) = \frac{\mathsf{Q}(X, Y)}{\mathsf{L}(X, Y) + \mathsf{B}(X, Y)},$$

which appeared, without a name, inside Definition 3.3.3 and Proposition 3.3.7. If we denote the class of Φ in the quotient space $\mathsf{Q}_{\mathsf{LB}}(X, Y)$ by $[\Phi]$ then the quasilinearity constant naturally induces the semi-quasinorm

$$Q[\Phi] = \inf\{Q(\Phi + L + B) : L \in \mathsf{L}, B \in \mathsf{B}\}$$

so that 3.5.1 can be upgraded to:

Proposition 3.6.1 *If X, X', Y, Y' are quasinormed spaces then the mapping $\mathfrak{L}(X', X) \times \mathsf{Q}_{\mathsf{LB}}(X, Y) \times \mathfrak{L}(Y, Y') \longrightarrow \mathsf{Q}_{\mathsf{LB}}(X', Y')$ given by $(u, [\Phi], v) \mapsto [v \circ \Phi \circ u]$ is well defined, trilinear and contractive.*

Proof The map is well defined: if $[\Phi] = [\Psi]$ then $v \circ \Phi \circ u - v \circ \Psi \circ u = v \circ (\Phi - \Psi) \circ u$ is in $\mathsf{B} + \mathsf{L}$. The only other not entirely obvious point is that $[\Phi \circ (u + u')] = [\Phi \circ u + \Phi \circ u']$, which follows from the quasilinearity of Φ since $\Phi \circ (u + u') - \Phi \circ u - \Phi \circ u'$ is bounded from X' to Y. □

Completeness

We now consider in some detail $\mathsf{Q}(X, Y)$ as a topological space under the semi-quasinorm defined by the quasilinearity constant. Readers who have not yet looked at Note 1.8.1 are advised to do so. Observe that when Y is a p-normed space, the quasilinearity constant is p-subadditive: if $\Phi, \Psi \in \mathsf{Q}(X, Y)$ then $Q(\Phi + \Psi)^p \leq Q(\Phi)^p + Q(\Psi)^p$, as clearly follows from the formula

$$Q(\Phi) = \sup_{x,y \in X} \frac{\|\Phi(x + y) - \Phi(x) - \Phi(y)\|}{\|x\| + \|y\|}.$$

It is obvious that $Q(\Phi) = 0$ if and only if Φ is linear. Hence the main properties of $(\mathsf{Q}(X, Y), Q(\cdot))$ depend only on the Hausdorff quotient

$$\mathsf{Q}_{\mathsf{L}}(X, Y) = \frac{\mathsf{Q}(X, Y)}{\ker Q(\cdot)} = \frac{\mathsf{Q}(X, Y)}{\mathsf{L}(X, Y)},$$

where $Q(\cdot)$ becomes a genuine quasinorm since $Q(\Phi) = Q(\Phi + L)$ when L is a linear map. We are going to prove that $\mathsf{Q}(X, Y)$ is complete when Y is. We need the following estimate, whose proof is a straightforward induction argument, taking into account that $(s + t)^p \leq s^p + t^p$ for all $s, t \geq 0$ and $p \in (0, 1]$.

Lemma 3.6.2 *Let X and Y be p-normed spaces. Then, for every $\Phi \in \mathsf{Q}(X,Y)$ and every N one has*

$$\left\| \Phi\left(\sum_{n=1}^{N} x_n\right) - \sum_{n=1}^{N} \Phi x_n \right\|^p \le Q(\Phi)^p \left(\sum_{n=1}^{N} n\|x_n\|^p\right).$$

Theorem 3.6.3 *If X is a quasinormed space and Y is a quasi-Banach space, then $(\mathsf{Q}(X,Y), Q(\cdot))$ is complete.*

Proof We must prove that if $(\Phi_n)_{n\ge1}$ is a Cauchy sequence in Q then there is a quasilinear map $\Phi\colon X \longrightarrow Y$ such that $\lim_n Q(\Phi_n - \Phi) = 0$. There is no loss of generality if we assume that X and Y are p-normed spaces, so that the preceding lemma applies. Let $\mathscr{H}$ be a normalised Hamel basis for X, so that each $x \in X$ can be written as a finite sum $x = \sum_{h\in\mathscr{H}} x_h h$. This allows us to introduce a 'control function' $\varrho\colon X \longrightarrow \mathbb{R}^+$ given by $\varrho(x) = \sum_n n(x^*(n))^p$, where $x^*(n)$ is the decreasing rearrangement of the coefficients of x with respect to the basis $\mathscr{H}$. It follows from Lemma 3.6.2 that for *any* $\Phi \in \mathsf{Q}$, one has the estimate

$$\left\|\Phi(x) - \sum_{h\in\mathscr{H}} x_h\Phi(h)\right\|^p \le Q(\Phi)^p\varrho(x). \tag{3.16}$$

For each n, we consider the linear map $L_n\colon X \longrightarrow Y$ defined by $L_n(h) = \Phi_n(h)$ for all $h \in \mathscr{H}$ so that $\Phi_n - L_n$ vanishes on $\mathscr{H}$. Applying (3.16) to $(\Phi_n - L_n) - (\Phi_k - L_k)$, one gets

$$\|(\Phi_n - L_n)(x) - (\Phi_k - L_k)(x)\|^p \le Q((\Phi_n - L_n) - (\Phi_k - L_k))^p\varrho(x) = Q(\Phi_n - \Phi_k)^p\varrho(x),$$

which means that $((\Phi_n - L_n)(x))_{n\ge1}$ is a Cauchy sequence in Y for every $x \in X$. Set $\Phi(x) = \lim_n(\Phi_n - L_n)(x)$ and let us check that Φ is quasilinear and $Q(\Phi - \Phi_n) \longrightarrow 0$. It is quite straightforward that the constants $Q(\Phi_n)$ are uniformly bounded. Indeed, there is a k such that $Q(\Phi_k - \Phi_n) \le 1$ for every $n \ge k$, so for these n, we have $Q(\Phi_n - L_n) = Q(\Phi_n) \le (1 + Q(\Phi_k)^p)^{1/p}$. Hence Φ is quasilinear since it is obvious that

$$Q(\Phi) \le \liminf_{n\to\infty} Q(\Phi_n - L_n) = \liminf_{n\to\infty} Q(\Phi_n) < \infty.$$

To prove that $Q(\Phi - \Phi_n) = Q(\Phi - (\Phi_n - L_n)) \to 0$, pick $\varepsilon > 0$ and let k be large enough that $Q(\Phi_m - \Phi_n) \le \varepsilon$ for $n, m \ge k$. Suppose $n \ge k$ and take $x, y \in X$ such that

$$\frac{\|(\Phi - \Phi_n)(x+y) - (\Phi - \Phi_n)(x) - (\Phi - \Phi_n)(y)\|}{\|x\| + \|y\|} > \frac{Q(\Phi - \Phi_n)}{2}.$$

Then, for $m \ge k$ large enough, we still have

$$\frac{\|(\Phi_m - \Phi_n)(x+y) - (\Phi_m - \Phi_n)(x) - (\Phi_m - \Phi_n)(y)\|}{\|x\| + \|y\|} > \frac{Q(\Phi - \Phi_n)}{2},$$

whence $\frac{1}{2}Q(\Phi - \Phi_n) < Q(\Phi_m - \Phi_n) \le \varepsilon$. □

Nice, isn't it? Unlike $\mathsf{Q}_{\mathsf{L}}(X, Y)$, which is always Hausdorff, $\mathsf{Q}_{\mathsf{LB}}(X, Y)$ is very often not Hausdorff (find full details in Section 4.5). Since the spaces $\mathsf{Q}_{\mathsf{L}}(X, Y)$ and $\mathsf{Q}_{\mathsf{LB}}(X, Y)$ are quotients of $\mathsf{Q}(X, Y)$ one has:

Corollary 3.6.4 *If X is a quasinormed space and Y is a quasi-Banach space, then $(\mathsf{Q}_{\mathsf{L}}(X, Y), Q(\cdot))$ and $(\mathsf{Q}_{\mathsf{LB}}(X, Y), Q[\cdot])$ are complete.*

This completeness result carries important consequences with it in the form of uniform boundedness principles for quasilinear maps, a harvest of ideas that will be carefully reaped in Chapter 5. The bare facts behind those ideas can be formulated as:

Theorem 3.6.5 *Let X be a quasinormed space and let Y be a quasi-Banach space. Assume that every quasilinear map $X \longrightarrow Y$ is trivial. Then, there is a constant K such that for each quasilinear map $\Phi\colon X \longrightarrow Y$ there exists a linear map $L\colon X \to Y$ satisfying $\|\Phi - L\| \le KQ(\Phi)$.*

Proof The hypothesis implies that the function $D(\Phi) = \operatorname{dist}(\Phi, \mathsf{L})$ is a semi-quasinorm on Q. Note that if L is linear and $\|\Phi - L\| < \infty$ then

$$\begin{aligned}\|\Phi(x+y) - \Phi(x) - \Phi(y)\| &= \|\Phi(x+y) - L(x+y) - \Phi(x) + Lx - \Phi(y) + Ly\|\\ &\le M\|\Phi - L\| \,(\|x\| + \|y\|),\end{aligned}$$

where M is a constant depending only on the moduli of concavity of the quasinorms of X and Y. Therefore $Q(\Phi) \le MD(\Phi)$ and so $D(\cdot)$ has the same kernel as $Q(\cdot)$. Thus $D(\cdot)$ and $Q(\cdot)$ define genuine quasinorms on the quotient space Q_{L}. Since $D(\Phi)$ and $Q(\Phi)$ depend only on the class of Φ in Q_{L}, there is no need to change names. Thus, the proof will immediately follow from the open mapping theorem once we are guaranteed that $(\mathsf{Q}_{\mathsf{L}}, Q(\cdot))$ is complete (as has been already proved) and also that $(\mathsf{Q}_{\mathsf{L}}, D(\cdot))$ is complete, which is true under the additional hypothesis of the theorem: indeed, if every quasilinear map $X \longrightarrow Y$ is trivial then $(\mathsf{Q}_{\mathsf{L}}(X, Y), D(\cdot))$ is complete. To prove this and conclude the proof, just observe that the hypothesis is $\mathsf{Q} = \mathsf{B} + \mathsf{L}$, and since $\mathsf{B} \cap \mathsf{L} = \mathfrak{L}(X, Y)$, the Diamond lemma applied to

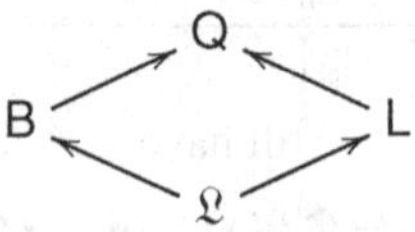

yields $\mathsf{Q}_{\mathsf{L}} = \mathsf{Q}/\mathsf{L} = \mathsf{B}/\mathfrak{L}$. □

The result just proved allows us to attach a parameter to each pair of quasi-Banach spaces X and Y:

$$K[X, Y] = \sup\left\{\frac{D(\Phi)}{Q(\Phi)} : \Phi \in \mathsf{L}(X, Y)\right\}. \tag{3.17}$$

We then have $\operatorname{Ext}(X, Y) = 0$ if and only if $K[X, Y] < \infty$, which quantifies the fact that every extension of X by Y splits.

Exact Sequences of p-Banach Spaces and p-Linear Maps

For each $0 < p \leq 1$ there are exact sequences $0 \longrightarrow Y \longrightarrow Z \longrightarrow X \longrightarrow 0$ in which both X and Y are p-Banach spaces but Z is not, and we have already encountered some of them: the $\ell_p(\varphi)$ spaces in Proposition 3.2.7. Since short exact sequences correspond to quasilinear maps, it is natural to seek some condition on a quasilinear map Φ defined between p-Banach spaces that ensures that the twisted sum $Y \oplus_\Phi Z$ is isomorphic to a p-Banach space. Here it is:

Definition 3.6.6 A homogeneous map $\Phi\colon X \longrightarrow Y$ is said to be p-linear if there is $K > 0$ such that for every $n \in \mathbb{N}$ and every $x_1, \ldots, x_n \in X$, we have

$$\left\| \Phi\left(\sum_{i=1}^{n} x_i\right) - \sum_{i=1}^{n} \Phi(x_i) \right\| \leq K \left(\sum_{i=1}^{n} \|x_i\|^p\right)^{1/p}. \tag{3.18}$$

The least possible constant K above shall be referred to as the p-linearity constant of Φ and denoted by $Q^{(p)}(\Phi)$. It is clear that each p-linear map is quasilinear (isn't it?), and indeed, p-linearity can be seen as a stronger form of quasilinearity involving an arbitrary number of variables instead of two. The choice $p = 1$ in Banach spaces is, by far, the most interesting one, and Section 3.8 explains why. Before going any further, let us denote by $\mathsf{Q}^{(p)}(X, Y)$ the space of p-linear maps from X to Y and observe that if X, Y are p-Banach spaces and $0 < r < p \leq 1$, we have the following containments:

$$\begin{array}{ccccccc} & & \mathsf{B}(X,Y) & & & & \\ & \nearrow & & \searrow & & & \\ \mathfrak{L}(X,Y) = \mathsf{B} \cap \mathsf{L} & & & & \mathsf{Q}^{(p)}(X,Y) \longrightarrow \mathsf{Q}^{(r)}(X,Y) \longrightarrow \mathsf{Q}(X,Y) \\ & \searrow & & \nearrow & & & \\ & & \mathsf{L}(X,Y) & & & & \end{array}$$

And here is the promised characterisation:

Proposition 3.6.7 *Let $\Phi\colon X \longrightarrow Y$ be a quasilinear map acting between p-normed spaces. Then Φ is p-linear if and only if $Y \oplus_\Phi X$ is isomorphic to a p-normed space.*

Proof Recall from Lemma 1.1.2 that a quasinormed space Z is locally p-convex if and only if there is a constant M such that

$$\Big\|\sum_{i=1}^{n} z_i\Big\| \le M\Big(\sum_{i=1}^{n} \|z_i\|^p\Big)^{1/p} \tag{3.19}$$

for every finite set $z_1, \dots, z_n \in Z$. If Φ is p-linear and we momentarily write $K = Q^{(p)}(\Phi)$ to ease notation then, since $s^p + t^p \le 2^{1-p}(s+t)^p$ for $0 < p \le 1$, we have

$$\begin{aligned}
\Big\|\sum_i (y_i, x_i)\Big\|_\Phi &= \Big\|\sum_i y_i - \Phi\Big(\sum_i x_i\Big)\Big\| + \Big\|\sum_i x_i\Big\| \\
&\le \left(\Big\|\sum_i y_i - \sum_i \Phi x_i\Big\|^p + \Big\|\Phi\Big(\sum_i x_i\Big) - \sum_i \Phi x_i\Big\|^p + \Big\|\sum_i x_i\Big\|^p\right)^{1/p} \\
&\le \left(\sum_i (\|y_i - \Phi x_i\|^p + \|x_i\|^p) + K^p \sum_i \|x_i\|^p\right)^{1/p} \\
&\le (1+K^p)^{1/p}\left(\sum_i (\|y_i - \Phi x_i\|^p + \|x_i\|^p)\right)^{1/p} \\
&\le (1+K^p)^{1/p}\left(\sum_i 2^{1-p}(\|y_i - \Phi x_i\| + \|x_i\|)^p\right)^{1/p} \\
&\le (1+K^p)^{1/p} 2^{1/p-1}\left(\sum_i \|(y_i, x_i)\|_\Phi^p\right)^{1/p}.
\end{aligned}$$

Suppose now that $Y \oplus_\Phi X$ is locally p-convex. Then

$$\Big\|\sum_i (\Phi x_i, x_i)\Big\|_\Phi \le M\left(\sum_i \|(\Phi x_i, x_i)\|^p\right)^{1/p}$$

for every finite set $x_1, \dots, x_n \in X$, and we have

$$\Big\|\sum_i \Phi x_i - \Phi\Big(\sum_i x_i\Big)\Big\| \le M\left(\sum_i \|x_i\|^p\right)^{1/p}. \qquad \square$$

The reader might feel some trepidation at yet another new class of maps. We can all, however, sigh in relief: all that has been done for quasilinear maps translates verbatim to the p-world, just replacing everywhere 'quasi' by 'p', including Q by $\mathsf{Q}^{(p)}$. In particular, $Q^{(p)}(\cdot)$ is a semi-p-norm on $\mathsf{Q}^{(p)}(X, Y)$ whose kernel is the subspace of linear maps. Also, if we set

$$\mathsf{Q}^{(p)}_{\mathsf{LB}}(X, Y) = \frac{\mathsf{Q}^{(p)}(X, Y)}{\mathsf{L}(X, Y) + \mathsf{B}(X, Y)}$$

endowed with the semi-p-norm $Q^{(p)}[\Phi] = \inf\{Q^{(p)}(\Phi + L + B)\colon L \in \mathsf{L}, B \in \mathsf{B}\}$, it turns out that the following p-versions of Theorems 3.6.3 and 3.6.5 are true and even have simpler proofs:

Theorem 3.6.8 *Let X be a p-normed space and Y be a p-Banach space.*

(a) *$(\mathsf{Q}^{(p)}(X,Y), Q^{(p)}(\cdot))$ is complete, as are its quotients $(\mathsf{Q}^{(p)}_{\mathsf{L}}(X,Y), Q^{(p)}(\cdot))$ and $(\mathsf{Q}^{(p)}_{\mathsf{LB}}(X,Y), Q^{(p)}(\cdot))$.*
(b) *If every p-linear map $X \longrightarrow Y$ is trivial then there is a constant K such that every p-linear map $\Phi\colon X \longrightarrow Y$ admits a linear map $L\colon X \longrightarrow Y$ such that $\|\Phi - L\| \le KQ^{(p)}(\Phi)$.*

The road to quantifying the splitting is paved: given p-Banach spaces X, Y, set

$$K^{(p)}[X,Y] = \sup\left\{\frac{D(\Phi)}{Q^{(p)}(\Phi)} : \Phi \in \mathsf{L}(X,Y)\right\}. \tag{3.20}$$

Then, if we denote the set of classes of exact sequences of p-Banach spaces $0 \longrightarrow Y \longrightarrow \cdot \longrightarrow X \longrightarrow 0$ modulo equivalence (full details will be given in Section 4.1) by $\mathrm{Ext}_{p\mathbf{B}}(X,Y)$, we have:

Corollary 3.6.9 $\mathrm{Ext}_{p\mathbf{B}}(X,Y) = 0$ *if and only if* $K^{(p)}[X,Y] < \infty$.

3.7 Homological Properties of ℓ_p and L_p When $0 < p \le 1$

As we said in 2.7.1, the spaces $\ell_p(I)$ are the only projective p-Banach spaces. Thus, any other p-Banach space X, in particular L_p, can be placed in a non-trivial sequence of p-Banach spaces $0 \longrightarrow Y \longrightarrow Z \longrightarrow X \longrightarrow 0$; in fact, any projective presentation $0 \longrightarrow \kappa(X) \longrightarrow \ell_p \longrightarrow X \longrightarrow 0$ serves this purpose. However, having the same local structure as ℓ_p, the spaces L_p exhibit a partially projective character that we now describe.

Theorem 3.7.1 *Let $p \in (0,1]$. Assume Y is a p-Banach ultrasummand and that X has a directed set of finite-dimensional subspaces uniformly isomorphic to the ℓ_p space of the corresponding dimension whose union is dense in X. Then every exact sequence of p-Banach spaces $0 \longrightarrow Y \longrightarrow Z \longrightarrow X \longrightarrow 0$ splits.*

Proof It is clear that if Y is a p-normed space then for every p-linear map $\Psi\colon \ell_p^n \longrightarrow Y$, there is a linear map $L\colon \ell_p^n \longrightarrow Y$ such that $\|\Psi - L\| \le Q^{(p)}(\Psi)$, namely, the linear map that agrees with Ψ on the unit basis. It follows that, if E is λ-isomorphic to ℓ_p^n, then for every p-linear map $\Psi\colon E \longrightarrow Y$, there is a linear map $L: E \longrightarrow Y$ such that $\|\Psi - L\| \leq \lambda Q^{(p)}(\Psi)$. Now let $I \subset \mathscr{F}(X)$ be the hypothesised set of subspaces for which we assume

- $\sup\{d(E, \ell_p^{\dim E}) : E \in I\} = \lambda < \infty$
- $X_0 = \bigcup_{E \in I} E$ is dense in X.

According to Proposition 3.6.7, it suffices to check that every p-linear map $\Phi: X_0 \longrightarrow Y$ is trivial. To this end, let $\mathfrak{U}$ be an ultrafilter refining the order filter on I, and, since Y is an ultrasummand, let $P: Y_{\mathfrak{U}} \longrightarrow Y$ be a bounded projection along the diagonal embedding. For each $E \in I$, pick a linear map $L_E: E \longrightarrow Y$ such that $\|\Phi|_E - L_E\| \leq \lambda Q^{(p)}(\Phi)$. Form a mapping $L: X_0 \longrightarrow Y$ as follows: given $x \in X_0$, we consider the bounded family $(L_E(x))_{E \in I}$ (understood to take the value 0 when $x \notin E$) and set $L(x) = P[(L_E(x))_{E \in I}]$. We have to show two things: that L is linear, for which it obviously suffices to observe that $x \in X_0 \longmapsto [(L_E(x))] \in Y_{\mathfrak{U}}$ is linear, and that $\|\Phi - L\| \leq \lambda \|P\| Q^{(p)}(\Phi)$. For the second statement, note that for normalised $x \in X_0$, we have

$$\begin{aligned}\|\Phi x - Lx\| &= \|P([(\Phi|_E(x) - L_E(x))_E])\| \leq \|P\| \lim_{\mathfrak{U}(E)} \|\Phi(x) - L_E(x)\| \\ &\leq \|P\| \sup_{E \in I} \sup_{x \in E} \|\Phi(x) - L_E(x)\| \leq \lambda \|P\| Q^{(p)}(\Phi).\end{aligned}$$ □

The proof actually gives $K^{(p)}[X_0, Y] \leq \lambda\|P\|$ and, as we announced just before Theorem 3.7.1, it establishes a homological property of L_p-spaces in $p\mathbf{B}$. The hypotheses of the theorem are satisfied by L_p taking $I = (E_n)_{n\geq 1}$, where E_n is the subspace spanned by the characteristic functions of the intervals $[(k-1)/2^n, k/2^n]$ for $1 \leq k \leq 2^n$, which is isometric to $\ell_p^{2^n}$. Thus

Corollary 3.7.2 $\operatorname{Ext}_{p\mathbf{B}}(L_p, Y) = 0$ *whenever Y is a p-Banach ultrasummand.*

The case $p = 1$ is the popular

3.7.3 Lindenstrauss' lifting *If X is an $\mathscr{L}_1$-space and Y is an ultrasummand then every exact sequence of Banach spaces $0 \longrightarrow Y \longrightarrow Z \longrightarrow X \longrightarrow 0$ splits.*

A number of homological properties of the spaces ℓ_p and L_p hold not only in $p\mathbf{B}$ but even in $\mathbf{Q}$:

Theorem 3.7.4 *Let I be any index set, $0 < p < q \leq 1$ and let Y be a q-Banach space. Then* $\operatorname{Ext}(\ell_p(I), Y) = 0$. *In particular, ℓ_p is a $\mathscr{K}$-space for all $0 < p < 1$.*

Proof Actually, if $\Omega: \ell_p^0(I) \longrightarrow Y$ is quasilinear, then the linear map $L(\sum_{i \in I} \lambda_i e_i) = \sum_i \lambda_i \Omega(e_i)$ is at finite distance from Ω on $\ell_p^0(I)$, as it follows from the next lemma, taking into account that the series $\sum_{n=1}^{\infty} n^r$ converges for $r < -1$. □

Lemma 3.7.5 *Let Y be a q-normed space and let $\Omega\colon \ell_p^0(I) \longrightarrow Y$ be a quasilinear map. If f_i have disjoint supports then*

$$\left\| \Omega\left(\sum_{i=1}^{n} f_i\right) - \sum_{i=1}^{n} \Omega(f_i) \right\| \leq Q(\Omega)\cdot\left(\sum_{i=1}^{n}\left(\frac{2}{i}\right)^{q/p}\right)^{1/q} \left\|\sum_{i=1}^{n} f_i\right\|_p . \tag{3.21}$$

Proof The proof is by induction on n, the number of summands in (3.21). The initial case is less than a tautology. Assume that (3.21) holds for $n-1$ summands, and let us check it for n summands. Let $f = \sum_{i=1}^n f_i$. Since $\|f\|^p = \sum_{i=1}^n \|f_i\|^p$, there exist k, l such that $\|f_k\|^p + \|f_l\|^p \leq 2\|f\|^p/n$. Now, as desired,

$$\begin{aligned}
&\left\|\Omega(f) - \sum_{i=1}^{n}\Omega(f_i)\right\|^q \\
&= \left\|\Omega(f) - \Omega(f_k+f_l) - \sum_{i\neq k,l}\Omega(f_i) + \Omega(f_k+f_l) - \Omega(f_k) - \Omega(f_l)\right\|^q \\
&\leq \underbrace{\left\|\Omega(f) - \Omega(f_k+f_l) - \sum_{i\neq k,l}\Omega(f_i)\right\|^q}_{n-1 \text{ summands}} + \|\Omega(f_k+f_l) - \Omega(f_k) - \Omega(f_l)\|^q \\
&\leq Q(\Omega)^q\cdot\left(\sum_{i=1}^{n-1}\left(\frac{2}{i}\right)^{q/p}\right)\cdot\|f\|^q + Q(\Omega)^q\,(\|f_k\| + \|f_l\|)^q \\
&\leq Q(\Omega)^q\cdot\left[\left(\sum_{i=1}^{n-1}\left(\frac{2}{i}\right)^{q/p}\right)\cdot\|f\|^q + (\|f_k\|^p + \|f_l\|^p)^{q/p}\right] \\
&\leq Q(\Omega)^q\cdot\left[\left(\sum_{i=1}^{n-1}\left(\frac{2}{i}\right)^{q/p}\right)\cdot\|f\|^q + \left(\frac{2}{n}\right)^{q/p}\|f\|^q\right].
\end{aligned}$$

□

Proposition 3.7.6 *Let X be a p-Banach space and let Y be a q-Banach space. Every twisted sum of Y and X is (isomorphic to) an r-Banach space for all $r < \min(p, q)$. If, moreover, $0 < p < q \leq 1$ then every twisted sum of Y and X is p-convex.*

Proof Assume Z is a twisted sum of Y and X. Let $Q\colon \ell_r(I) \longrightarrow X$ be any quotient map. By Theorem 3.7.4, the lower row in the pullback diagram

$$\begin{array}{ccccccccc}
0 & \longrightarrow & Y & \longrightarrow & Z & \longrightarrow & X & \longrightarrow & 0 \\
 & & \| & & \uparrow & & \uparrow{\scriptstyle Q} & & \\
0 & \longrightarrow & A & \longrightarrow & \mathrm{PB} & \longrightarrow & \ell_r(I) & \longrightarrow & 0
\end{array}$$

splits. Therefore, Z is a quotient of $Y \times \ell_r(I)$, and thus it is r-convex. As for the second part, just take $r = p$ and proceed. □

The raison d'être of the following application is withheld until Chapter 4, but we can enjoy it now:

Corollary 3.7.7 *Let $0 < r < p \leq 1$. There is $K(p,r) > 0$ such that if X, Y are p-normed spaces and $\Phi : X \longrightarrow Y$ is quasilinear then Φ is r-linear and $Q^{(r)}(\Phi) \leq K(p,r)Q(\Phi)$.*

Proof Proposition 3.7.6 gives that every quasilinear map between p-normed spaces is r-linear for $r < p$. If the qualitative conclusion that every quasilinear map is r-linear fails, there must be a sequence of quasilinear maps $\Phi_n : X_n \longrightarrow Y_n$, where X_n and Y_n are p-Banach spaces, such that $Q(\Phi_n) \leq 1$ and $Q^{(r)}(\Phi_n) \longrightarrow \infty$. These maps can be amalgamated into a single quasilinear map $\Phi : c_0^0(\mathbb{N}, X_n) \longrightarrow c_0^0(\mathbb{N}, Y_n)$ given by $\Phi((x_n)_n) = (\Phi_n(x_n))_n$ with $Q(\Phi) \leq 1$ and $Q^{(r)}(\Phi) = \infty$, a contradiction. □

Theorem 3.7.8 $\operatorname{Ext}(L_p, Y) = 0$ *whenever Y is a q-Banach space with $0 < p < q \leq 1$. In particular, L_p is a $\mathscr{K}$-space when $0 < p < 1$.*

Proof Let $0 \longrightarrow Y \longrightarrow Z \stackrel{\rho}{\longrightarrow} L_p \longrightarrow 0$ be an extension, with Y a q-Banach space. The second part of Proposition 3.7.6 shows that Z is p-convex. For ease of notation, for each $n \in \mathbb{N}$ and $1 \leq k \leq 2^n$, let χ_k^n be the characteristic function of the interval $[(k-1)/2^n, k/2^n]$ and let E_n be the subspace of L_p spanned by χ_k^n, $1 \leq k \leq 2^n$. Since E_n is isometric to $\ell_p^{2^n}$, there is a linear map $s_n : E_n \longrightarrow Z$ such that $\rho s_n = \mathbf{1}_{E_n}$ with $\|s_n\| \leq C$ for some constant C independent of n. Let us check that $s(f) = \lim_n s_n(f)$ exists for $f \in \bigcup_n E_n$, in which case it extends to an operator $S : L_p \rightarrow Z$ which is a section of the quotient map $\rho : Z \longrightarrow L_p$. By linearity, it suffices to verify that $(s_n \chi_k^j)_n$ is a Cauchy sequence in Z. Suppose $j \leq m < n$ and $1 \leq k \leq 2^j$. The difference $s_m \chi_k^j - s_n \chi_k^j$ lies in Y and

$$\|s_m\chi_k^j - s_n\chi_k^j\| = \left\|(s_m - s_n)\left(\sum_{i=1}^{2^{m-j}} \chi^m_{2^{m-j}k+i}\right)\right\| = \left\|\sum_{i=1}^{2^{m-j}} (s_m - s_n)\chi^m_{2^{m-j}k+i}\right\|$$

$$\leq C\left(\sum_{i=1}^{2^{m-j}} \|(s_m - s_n)\chi^m_{2^{m-j}k+i}\|^q\right)^{1/q} \leq C\left(\sum_{i=1}^{2^{m-j}} \|\chi^m_{2^{m-j}k+i}\|_p^q\right)^{1/q}$$

$$= C2^{-j/q}2^{m(1/q-1/p)}.$$

□

The section S appearing in the above proof is unique: if $\tilde{S}$ is another right inverse for ρ in $\mathfrak{L}(L_p, Z)$ then $\tilde{S} = S$ as the difference $\tilde{S} - S$ is an operator from L_p to Y. In a sense, S attracts the local sections s_n despite the arbitrariness of their choice. The full force of the proof is unnecessary to establish that L_p is a $\mathscr{K}$-space since this follows from the p-convexity of Z, Proposition 3.7.6 and Theorem 3.7.1, taking into account that $\mathbb{K}$ is, obviously, an ultrasummand.

Our next result is a link in the chain formed by Corollary 4.2.8, Corollary 4.5.3 and Proposition 7.2.16.

Theorem 3.7.9 *Let $0 < p < 1$ and let A, A' be closed subspaces of L_p such that $L_p/A \simeq L_p/A'$. Assume that each of the spaces A, A' is either q-normable for some $0 < p < q \leq 1$ or an ultrasummand. Then there is an automorphism U of L_p such that $U[A] = A'$.*

Proof The proof depends on the fact that if A is either a q-Banach space for some $0 < p < q \leq 1$ or a p-Banach ultrasummand then $\mathfrak{L}(L_p, A) = 0$ by 1.1.5 and the corollary in Section 1.8.3, and also that $\operatorname{Ext}_{p\mathbf{B}}(L_p, Y) = 0$ by the preceding theorem. Let $u\colon L_p/A \longrightarrow L_p/A'$ be an isomorphism and consider the diagram

$$\begin{array}{ccccccccc}
0 & \longrightarrow & A' & \longrightarrow & L_p & \xrightarrow{\pi'} & L_p/A' & \longrightarrow & 0 \\
 & & & & & & \uparrow{\scriptstyle u} & & \\
0 & \longrightarrow & A & \longrightarrow & L_p & \xrightarrow{\pi} & L_p/A & \longrightarrow & 0
\end{array} \tag{3.22}$$

Since the lower sequence in the pullback diagram

$$\begin{array}{ccccccccc}
0 & \longrightarrow & A' & \longrightarrow & L_p & \xrightarrow{\pi'} & L_p/A' & \longrightarrow & 0 \\
 & & \| & & \uparrow & & \uparrow{\scriptstyle u\pi} & & \\
0 & \longrightarrow & A' & \longrightarrow & \mathrm{PB} & \longrightarrow & L_p & \longrightarrow & 0
\end{array}$$

splits because $\operatorname{Ext}(L_p, A') = 0$, the splitting criterion for pullback sequences yields a lifting $U\colon L_p \longrightarrow L_p$ for $u\pi$ through π'. This lifting is unique since the difference of two liftings should map L_p to A' and $\mathfrak{L}(L_p, A') = 0$. Thus, U maps A to A and we have the commutative diagram

$$\begin{array}{ccccccccc}
0 & \longrightarrow & A' & \longrightarrow & L_p & \xrightarrow{\pi'} & L_p/A' & \longrightarrow & 0 \\
 & & \uparrow{\scriptstyle U|_A} & & \uparrow{\scriptstyle U} & & \uparrow{\scriptstyle u} & & \\
0 & \longrightarrow & A & \longrightarrow & L_p & \xrightarrow{\pi} & L_p/A & \longrightarrow & 0
\end{array}$$

The same argument applies to u^{-1}, thus showing that there is exactly one operator $V\colon L_p \longrightarrow L_p$ such that $\pi V = u^{-1}\pi'$ sitting in a commutative diagram

$$\begin{array}{ccccccccc}
0 & \longrightarrow & A & \longrightarrow & L_p & \xrightarrow{\pi} & L_p/A & \longrightarrow & 0 \\
 & & \uparrow{\scriptstyle V|_{A'}} & & \uparrow{\scriptstyle V} & & \uparrow{\scriptstyle u^{-1}} & & \\
0 & \longrightarrow & A' & \longrightarrow & L_p & \xrightarrow{\pi'} & L_p/A' & \longrightarrow & 0 \\
 & & \uparrow{\scriptstyle U|_{A}} & & \uparrow{\scriptstyle U} & & \uparrow{\scriptstyle u} & & \\
0 & \longrightarrow & A & \longrightarrow & L_p & \xrightarrow{\pi} & L_p/A & \longrightarrow & 0
\end{array}$$

Clearly, VU is the identity of L_p since $\mathbf{1}_{L_p} - VU = 0$ because it maps L_p to A and, for the same reason, $UV = \mathbf{1}_{L_p}$. □

Theorem 3.7.9 shows that taking quotients of L_p when $0 < p < 1$ by q-normable or ultrasummand subspaces is extremely sensitive to the choice of subspace. Thus, not only is the quotient of L_p by a line not isomorphic to L_p; but in fact L_p/A and L_p/A' are not isomorphic for any A and A' with finite but different dimensions (the converse is also true; see Proposition 7.2.16), or if there are $p < q < r \leq 1$ such that A is q-normable but not r-normable and A' is r-normable. It is perhaps worth recalling that L_p contains isometric copies of each L_q for $p < q \leq 2$. Beyond that, the lines are blurry: L_p is isomorphic to its quotient by the subspace $L_p[0, \frac{1}{2}]$. A more interesting example springs from the interaction between the Hardy class H_p and $L_p(\mathbb{T})$. Recall that H_p can be seen as a subspace of $L_p(\mathbb{T})$ by means of the boundary values. Put $\overline{H}_p = \{f \in L_p(\mathbb{T}) \colon \overline{f} \in H_p\}$. An important result of Aleksandrov establishes that $L_p(\mathbb{T}) = H_p + \overline{H}_p$ for $0 < p < 1$ (combine Theorems 2.4 and 3.2 of [6] for a much more general result). Set $J_p = H_p \cap \overline{H}_p$ and consider the diagram

$$\begin{array}{ccccccccc}
 & & 0 & & 0 & & & & \\
 & & \downarrow & & \downarrow & & & & \\
0 & \longrightarrow & J_p & \xrightarrow{\text{diag}} & H_p \times \overline{H}_p & \xrightarrow{\text{diff}} & L_p(\mathbb{T}) & \longrightarrow & 0 \\
 & & \downarrow{\scriptstyle \text{inclusion}} & & \downarrow & & \| & & \\
0 & \longrightarrow & L_p(\mathbb{T}) & \longrightarrow & \text{PO} & \longrightarrow & L_p(\mathbb{T}) & \longrightarrow & 0 \\
 & & \downarrow & & \downarrow & & & & \\
 & & L_p/J_p & = & \text{PO}/(H_p \times \overline{H}_p) & & & & \\
 & & \downarrow & & \downarrow & & & & \\
 & & 0 & & 0 & & & &
\end{array}$$

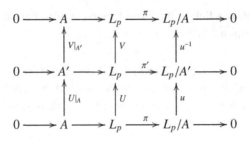

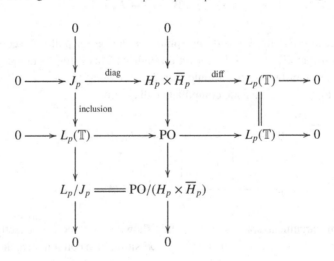

where $\text{diag}(f) = (f, f)$ and $\text{diff}(f, g) = f - g$. The upper exact sequence does not split because $H_p \times \overline{H}_p$ has separating dual, and so every operator $L_p(\mathbb{T}) \longrightarrow H_p \times \overline{H}_p$ is zero. This implies that J_p is not an ultrasummand since $\text{Ext}_{p\mathbf{B}}(L_p, J_p) \neq 0$; in particular, J_p is not isomorphic to the ultrasummand $H_p \times \overline{H}_p$. The middle horizontal sequence splits because the inclusion of J_p into $L_p(\mathbb{T})$ factors through H_p and thus $\text{PO} \simeq L_p(\mathbb{T})^2 \approx L_p$. Therefore, L_p contains two subspaces, one isomorphic to J_p and the other to $H_p \times \overline{H}_p$, whose corresponding quotients are isomorphic. Moreover, the Diamond lemma in combination with Aleksandrov's equality yields

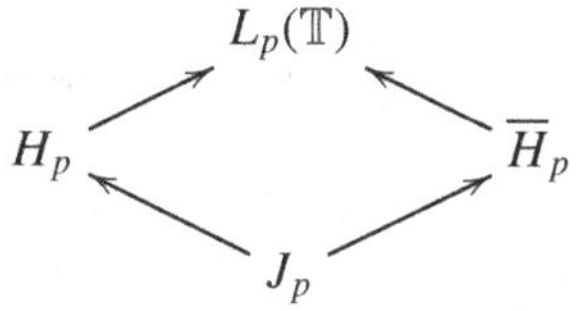

such that $L_p(\mathbb{T})/H_p \simeq \overline{H}_p/J_p$. In fact, the three quotient spaces $L_p(\mathbb{T})/J_p$, $L_p(\mathbb{T})/H_p$, H_p/J_p are isomorphic; see [255, Section 9].

3.8 Exact Sequences of Banach Spaces and Duality

Proposition 3.6.7 says that exact sequences of Banach spaces correspond to 1-linear maps. Since exact sequences of Banach spaces have dual sequences, 1-linear maps should admit dual 1-linear maps. This, by the way, is one of the main differences between p-linear maps for $0 < p < 1$ and 1-linear maps. About the question of how to find that dual 1-linear map, the standard method of taking differences between a bounded homogeneous and a linear selection for the quotient map works as well here as elsewhere. But if the question is how exactly to construct that dual 1-linear map, recall that virtually every non-trivial fact about Banach space duality ultimately depends upon the Hahn–Banach theorem, and this construction, even if it is non-linear, is not an exception.

A Non-Linear Hahn–Banach Theorem

Applying Proposition 3.6.7, the Hahn–Banach theorem and Lemma 3.3.2, in that order, shows that each 1-linear map $\phi\colon X \longrightarrow \mathbb{K}$ admits a linear functional $\ell\colon X \longrightarrow \mathbb{K}$ at finite distance. But ultimately, we should not need to use a sledgehammer to crack a nut. So, let us present a direct proof for that fact, yielding, as a bonus, the optimal distance from a quasilinear map to the space of linear functionals. Given a homogeneous mapping $\Phi\colon X \longrightarrow Y$ acting between Banach spaces, define

$$Q_0^{(1)}(\Phi) = \sup \frac{\|\sum_i \Phi(x_i)\|}{\sum_i \|x_i\|},$$

where the sup is taken over all $x_1, \ldots, x_n \in X$ whose sum is zero. Clearly, $Q_0^{(1)}(\Phi) \leq Q^{(1)}(\Phi) \leq 2Q_0^{(1)}(\Phi)$.

Lemma 3.8.1 *Let $\phi\colon X \longrightarrow \mathbb{K}$ be a homogeneous function. Then there is a linear function $\ell\colon X \longrightarrow \mathbb{K}$ such that $\|\phi - \ell\| = \operatorname{dist}(\phi, \mathsf{L}(X, \mathbb{K})) = Q_0^{(1)}(\phi)$.*

Proof We first observe that if ℓ is any linear functional and $\sum_i x_i = 0$, then

$$\left|\sum_i \phi(x_i)\right| = \left|\sum_i \phi(x_i) - \sum_i \ell(x_i)\right| \leq \|\phi - \ell\| \left(\sum_i \|x_i\|\right),$$

so we certainly have $Q_0^{(1)}(\phi) \leq \operatorname{dist}(\phi, \mathsf{L}(X, \mathbb{K}))$. To complete the proof we must find a linear map whose distance to ϕ is exactly $Q_0^{(1)}(\phi)$. The proof of this part depends on which ground we are working over.

Real case. This proof goes as in the classical proof via Zorn's lemma of the Hahn–Banach theorem. The main difficulty to overcome is that an induction hypothesis such as 'there is a linear functional ℓ defined on a subspace U of X such that $|\phi(x) - \ell(x)| \leq Q_0^{(1)}(\phi)\|x\|$ holds for all $x \in U$' is not strong enough to ensure that ℓ can be extended to a larger subspace, say $W = [w] \oplus U$, in such a way that the previous estimate still holds for $x \in W$. Our strategy, then, is to use as the induction hypothesis that there is a linear mapping $\ell\colon U \longrightarrow \mathbb{R}$ satisfying

$$\left|\sum_i \phi(x_i) - \ell\left(\sum_i x_i\right)\right| \leq Q_0^{(1)}(\phi)\left(\sum_i \|x_i\|\right) \tag{3.23}$$

for every finite set $\{x_i\} \subset X$ such that $\sum x_i \in U$. This plainly implies that $\|\phi - \ell\| \leq Q_0^{(1)}(\phi)$ when $U = X$. Assume that a linear mapping ℓ has been defined on a subspace $U \subset X$ such that (3.23) holds when $x = \sum x_i \in U$. Fixing $w \notin U$, we want to see that it is possible to define $\ell(w) = \lambda \in \mathbb{R}$ in such a way that (3.23) holds for $x = \sum x_i \in [w] \oplus U$. Since U is a linear subspace, one can assume that $x = w - u$, with $u \in U$. In this case, it suffices to prove that there exists a number λ satisfying

$$\left|\lambda - \ell(u) - \sum_i \phi(z_i)\right| \leq Q_0^{(1)}(\phi)\left(\sum_i \|x_i\|\right)$$

when $w - u = \sum x_i$ and $u \in U$. This is equivalent to

$$\ell(u) + \sum_i \phi(x_i) - Q_0^{(1)}(\phi)\left(\sum_i \|x_i\|\right) \leq \lambda \leq \ell(u) + \sum_i \phi(x_i) + Q_0^{(1)}(\phi)\left(\sum_i \|x_i\|\right).$$

So, the question is whether

$$\ell(u) + \sum_i \phi(x_i) - Q_0^{(1)}(\phi)\left(\sum_i \|x_i\|\right) \leq \ell(v) + \sum_j \phi(s_j) + Q_0^{(1)}(\phi)\left(\sum_j \|s_j\|\right)$$

whenever $w - u = \sum x_i$, $w - v = \sum s_j$, $u, v \in U$ and $x_i, s_j \in X$. The preceding inequality can be written as

$$\ell(u) - \ell(v) + \sum_i \phi(x_i) - \sum_j \phi(s_j) \leq Q_0^{(1)}(\phi)\left(\sum_i \|x_i\| + \sum_j \|s_j\|\right)$$

and follows from the induction hypothesis, which yields

$$\left|\ell(u - v) - \left(\sum_i \phi(-x_i) + \sum_j \phi(s_j)\right)\right| \leq Q_0^{(1)}(\phi)\left(\sum_i \| - x_i\| + \sum_j \|s_j\|\right),$$

since $u - v = -(w - u) + (w - u) = \sum_i -x_i + \sum_j s_j$ and $u - v \in U$. Observe that so far we have only used the homogeneity of ϕ. In fact, the induction step holds independently of the value or meaning of $Q_0^{(1)}(\phi)$. The 1-linearity of ϕ only appears as the condition we need to start the induction: when $U = 0$ the inequality of the proof states that whenever (x_i) is a finite collection of points of X with $\sum x_i = 0$,

$$\left|\sum \phi(x_i)\right| \leq Q_0^{(1)}(\phi)\left(\sum \|x_i\|\right).$$

The rest of the proof is a typical application of Zorn's lemma. Of course, there is no need for Zorn when the space X is finite-dimensional.

Complex case. The proof for complex-valued functions depends on the familiar decomposition of a complex function into real and imaginary parts. Given a complex function f on X, we define the real part $f_r\colon X \longrightarrow \mathbb{R}$ by $f_r(x) = \frac{1}{2}(f(x) + \overline{f(x)})$. It is clear that if f is (complex) homogeneous, then f_r is (real) homogeneous and $\|f\| = \|f_r\|$. Also, if $g\colon X \longrightarrow \mathbb{R}$ is (real) homogeneous, then there is a unique complex homogeneous $g_c\colon X \longrightarrow \mathbb{C}$ whose real part is g, namely $g_c(x) = g(x) - ig(ix)$. Needless to say, g_c is complex linear if and only if g is real linear. Now, let $\phi\colon X \longrightarrow \mathbb{C}$ be complex 1-linear. Then the real part of ϕ is (real) 1-linear, with $Q_0^{(1)}(\phi_r) \leq Q_0^{(1)}(\phi)$, so there is a (real) linear map $\ell\colon X \longrightarrow \mathbb{R}$ such that $\|\phi_r - \ell\| = Q_0^{(1)}(\phi_r)$. The map ℓ_c is a complex linear functional on X, and since the assignment $g \longmapsto g_c$ is real linear and $\phi = (\phi_r)_c$ we have

$$\|\phi - \ell_c\| = \|(\phi_r - \ell)_c\| = \|\phi_r - \ell\| = Q_0^{(1)}(\phi_r) \leq Q_0^{(1)}(\phi).$$

This completes the proof and shows that $Q_0^{(1)}(\phi) = Q_0^{(1)}(\phi_r)$. □

Dual Sequence and Dual 1-Linear Map

If $\Phi\colon X \longrightarrow Y$ is a 1-linear map acting between Banach spaces then $0 \longrightarrow Y \longrightarrow Z(\Phi) \longrightarrow X \longrightarrow 0$ is an exact sequence of Banach spaces (after renorming) with a dual sequence $0 \longrightarrow X^* \longrightarrow Z(\Phi)^* \longrightarrow Y^* \longrightarrow 0$ by virtue of the Hahn–Banach theorem. This sequence can be generated by some 1-linear map $\Psi\colon Y^* \longrightarrow X^*$ that could quite judiciously be called the dual of Φ. Its explicit construction is implicit in the proof of Lemma 3.8.1 and will be made still more explicit in this section. The following lemma explains how a map Φ and its dual Ψ yoke together:

Lemma 3.8.2 *Let X, Y be Banach spaces and let X_0, Y_0^* be dense subspaces of X and Y^*, respectively. Let $\Phi\colon X_0 \longrightarrow Y$ and $\Psi\colon Y_0^* \longrightarrow X^*$ be quasilinear maps. The following statements are equivalent:*

(i) *There is a commutative diagram*

$$\begin{array}{ccccccccc} 0 & \longrightarrow & X^* & \longrightarrow & Z(\Psi) & \longrightarrow & Y^* & \longrightarrow & 0 \\ & & \| & & \downarrow{\scriptstyle u} & & \| & & \\ 0 & \longrightarrow & X^* & \longrightarrow & Z(\Phi)^* & \longrightarrow & Y^* & \longrightarrow & 0 \end{array} \tag{3.24}$$

(ii) *There is a bilinear form $\beta\colon Y_0^* \times X_0 \longrightarrow \mathbb{K}$ such that*

$$|\langle \Psi(y^*), x\rangle + \langle y^*, \Phi(x)\rangle - \beta(y^*, x)| \leq C\|y^*\|\,\|x\| \tag{3.25}$$

for some constant C, all $y^* \in Y_0^*$ and all $x \in X_0$.

Proof Before the alert reader panics, let us note that the inequality in (ii) already implies that both Φ and Ψ are 1-linear. Indeed, pick $x_1, \ldots, x_n \in X$ and choose a normalised $y^* \in Y^*$ almost norming $\Phi(\sum_i x_i) - \sum_i \Phi(x_i)$. Since

$$\left|\left\langle \Psi(y^*), \sum_i x_i\right\rangle + \left\langle y^*, \Phi\left(\sum_i x_i\right)\right\rangle - \beta\left(y^*, \sum_i x_i\right)\right| \leq \left\|\sum_i x_i\right\|$$

and $|\langle \Psi(y^*), x_i\rangle + \langle y^*, \Phi(x_i)\rangle - \beta(y^*, x_i)| \leq C\|x_i\|$ for each i, we have

$$\left\|\Phi\left(\sum_i x_i\right) - \sum_i \Phi x_i\right\| \leq (1+\varepsilon)\left|\left\langle y^*, \Phi\left(\sum_i x_i\right) - \sum_i \Phi x_i\right\rangle\right| \leq 2(1+\varepsilon)C\sum_i \|x_i\|,$$

so Φ is 1-linear, with $Q^{(1)}(\Phi) \leq 2C$. Interchanging the roles of $\sum_i x_i$ and y^*, we get that also $Q^{(1)}(\Psi) \leq 2C$. We now prove the implication (ii) $\Longrightarrow$ (i). Since $Y \oplus_\Phi X_0$ is dense in $Z(\Phi)$ by Lemma 3.3.1, both spaces have the same dual. Assuming that (ii) holds, define $u\colon X^* \oplus_\Psi Y_0^* \longrightarrow (Y \oplus_\Phi X_0)^*$ by

$$u(x^*, y^*)(y, x) = \langle x^*, x\rangle + \langle y^*, y\rangle - \beta(y^*, x).$$

This pairing is correctly defined. We check now that it is also bounded and makes Diagram (3.24) commute.

- u is bounded:

$$\begin{aligned}
|u(x^*, y^*)(y, x)| &= |\langle x^*, x\rangle + \langle y^*, y\rangle - \beta(y^*, x)| \\
&\leq |\langle x^*, x\rangle + \langle y^*, y\rangle - \langle \Psi(y^*), x\rangle - \langle (y^*, \Phi(x)\rangle| + C\|y^*\|\,\|x\| \\
&\leq |\langle x^* - \Psi(y^*), x\rangle| + |\langle y^*, y - \Phi(x)\rangle| + C\|y^*\|\,\|x\| \\
&\leq \|x^* - \Psi(y^*)\|\,\|x\| + \|y^*\|\,\|y - \Phi(x)\| + C\|y^*\|\,\|x\| \\
&\leq \max(1, C)\,\|(x^*, y^*)\|_\Psi\,\|(y, x)\|_\Phi.
\end{aligned}$$

- The left-hand square of (3.24) is commutative since $u(x^*, 0)(y, x) = \langle x^*, x\rangle$.
- The right-hand square is commutative since $u(x^*, y^*)(y, 0) = \langle y^*, y\rangle$.

The implication (i) $\Longrightarrow$ (ii) is also easy. Indeed, assume that u is linear and makes (3.24) commute, and let us take a look at the action of $u(x^*, y^*)$ on $Y \oplus_\Phi X_0$. We have $u(x^*, 0)(y, x) = \langle x^*, x\rangle$ and $u(x^*, y^*)(y, 0) = \langle y^*, y\rangle$. The function $\beta\colon Y_0^* \times X_0 \longrightarrow \mathbb{K}$ defined by $\beta(y^*, x) = -u(0, y^*)(0, x)$ is bilinear and it is obvious that

$$u(x^*, y^*)(y, x) = \langle x^*, x\rangle + \langle y^*, y\rangle - \beta(y^*, x).$$

Finally, let us assume u is bounded: since $\|(\Psi y^*, y^*)\|_\Psi = \|y^*\|$ and $\|(\Phi x, x)\|_\Phi = \|x\|$, from

$$u(\Psi y^*, y^*)(\Phi x, x) = \langle \Psi(y^*), x\rangle + \langle y^*, \Phi(x)\rangle - \beta(y^*, x)$$

we get the estimate in (ii) with $C = \|u\|$. □

We now obtain the dual map of a given 1-linear map and the dual pairing. Let X be a normed space, Y a Banach space and $\Phi\colon X \longrightarrow Y$ a 1-linear map. Glancing out the corner of our eye at (3.25), it is clear that we need to assign a linear functional $\Lambda(y^*)$ to each $y^* \in Y^*$ such that $\langle \Lambda(y^*), x\rangle$ 'almost cancels' $\langle y^*, \Phi(x)\rangle$. And to do that, we form the 1-linear composition $y^* \circ \Phi$ for which $Q^{(1)}(y^* \circ \Phi) \leq \|y^*\| Q^{(1)}(\Phi)$ and $Q_0^{(1)}(y^* \circ \Phi) \leq \|y^*\| Q_0^{(1)}(\Phi)$. Lemma 3.8.1 yields a linear map $\Lambda(y^*)\colon X \longrightarrow \mathbb{K}$ such that $\|\Lambda(y^*) + y^* \circ \Phi\| \le \|y^*\| Q_0^{(1)}(\Phi)$. We thus have

$$|\langle \Lambda(y^*), x\rangle + \langle y^*, \Phi(x)\rangle| \le Q_0^{(1)}(\Phi)\|y^*\|\|x\|,$$

which is quite close to, if not exactly, what we wanted. Part of the problem is that the map $\Lambda\colon Y^* \longrightarrow \mathsf{L}(X, \mathbb{K})$ takes values in the wrong space. To push its values down into X^* what we will do is to make it vanish on a Hamel basis of Y^*. Let us assume that $\Lambda(y^*)$ depends homogeneously on y^*. Let $\mathscr{H}$ be a Hamel basis of Y^* and let $L\colon Y^* \longrightarrow \mathsf{L}(X, \mathbb{K})$ be the linear map that agrees with Λ on $\mathscr{H}$. Define $\Phi^*\colon Y^* \longrightarrow \mathsf{L}(X, \mathbb{K})$ by $\Phi^*(y^*) = \Lambda(y^*) - L(y^*)$ to get:

Theorem 3.8.3 *The map $\Phi^*\colon Y^* \longrightarrow X^*$ is correctly defined and 1-linear, with $Q^{(1)}(\Phi^*) \le Q^{(1)}(\Phi)$ and $Q_0^{(1)}(\Phi^*) \le Q_0^{(1)}(\Phi)$. Moreover, there is a commutative diagram*

$$\begin{array}{ccccccccc}
0 & \longrightarrow & X^* & \longrightarrow & X^* \oplus_{\Phi^*} Y^* & \longrightarrow & Y^* & \longrightarrow & 0\\
 & & \| & & \downarrow & & \| & & \\
0 & \longrightarrow & X^* & \longrightarrow & (Y \oplus_\Phi X)^* & \longrightarrow & Y^* & \longrightarrow & 0
\end{array} \tag{3.26}$$

In particular, the extension generated by Φ^* is equivalent to the dual of that generated by Φ.

Proof First of all, we have to check that the linear functional $\Phi^*(y^*)$ is bounded for every $y^* \in Y^*$, which is obvious once we notice that

- $\Phi^*(y^*) = 0$ when $y^* \in \mathscr{H}$.
- The set $\{y^* \in Y^* : \|\Phi^*(y^*)\| < \infty\}$ is a linear subspace of Y^*.

Since Φ^* is homogeneous, in order to check the second point, it suffices to see that if $\Phi^*(y_1^*)$ and $\Phi^*(y_2^*)$ are bounded then $\Phi^*(y_1^* + y_2^*)$ is bounded as well:

$$\begin{aligned}
\|\Phi^*(y_1^* + y_2^*) - \Phi^*(y_1^*) - \Phi^*(y_2^*)\| &= \|\Lambda(y_1^* + y_2^*) - \Lambda(y_1^*) - \Lambda(y_2^*)\|\\
&= \|\Lambda(y_1^* + y_2^*) + (y_1^* + y_2^*) \circ \Phi - y_1^* \circ \Phi - \Lambda(y_1^*) - y_2^* \circ \Phi - \Lambda(y_2^*)\|\\
&\le \|\Lambda(y_1^* + y_2^*) + (y_1^* + y_2^*) \circ \Phi\| + \|y_1^* \circ \Phi + \Lambda(y_1^*)\| + \|y_2^* \circ \Phi + \Lambda(y_2^*)\|\\
&< \infty.
\end{aligned}$$

That Φ^* is 1-linear is also straightforward: pick finitely many $y_i^* \in Y^*$ such that $\sum_i y_i^* = 0$; then

$$\left\|\sum_i \Phi^* y_i^*\right\| = \left\|\sum_i \Lambda(y_i^*)\right\| = \left\|\sum_i \Lambda(y_i^*) + \sum_i y_i^* \circ \Phi\right\| \le Q_0^{(1)}(\Phi)\left(\sum_i \|y_i^*\|\right).$$

It remains to determine the form of the duality between $Y \oplus_\Phi X$ and $X^* \oplus_{\Phi^*} Y^*$. By the preceding lemma, it suffices to identify a bilinear map $\beta\colon Y^* \times X \longrightarrow \mathbb{K}$ satisfying an estimate of the form

$$|\langle \Phi^*(y^*), x\rangle + \langle y^*, \Phi(x)\rangle - \beta(y^*, x)| \le C\|y^*\|\,\|x\|.$$

The linear map $L\colon Y^* \longrightarrow \mathsf{L}(X,\mathbb{K})$ can be also regarded as a bilinear map $\beta\colon Y^* \times X \longrightarrow \mathbb{K}$ simply by letting $\beta(y^*, x) = \langle L(y^*), x\rangle$. We have

$$\begin{aligned}|\langle \Phi^*(y^*), x\rangle + \langle y^*, \Phi(x)\rangle - \beta(y^*, x)| &= |\langle \Phi^*(y^*) - L(y^*), x\rangle + \langle y^*, \Phi(x)\rangle|\\ &\le \|\Lambda(y^*) + y^* \circ \Phi\|\,\|x\|\\ &\le C\|y^*\|\,\|x\|.\end{aligned}$$

This completes the proof and shows, in passing, that the map defined by

$$u(x^*, y^*)(y, x) = \langle x^*, x\rangle + \langle y^*, y\rangle - L(y^*)(x)$$

is a linear homeomorphism making Diagram (3.26) commute. □

'The' 1-linear map $\Phi^*\colon Y^* \longrightarrow X^*$ dual to Φ provided by Theorem 3.8.3 can be written as $\Phi^* = \Lambda - L$, where $\Lambda\colon Y^* \longrightarrow \mathsf{L}(X,\mathbb{K})$ satisfies $\|\Lambda(y^*) + y^* \circ \Phi\| \le M\|y^*\|$ and $L\colon Y^* \longrightarrow \mathsf{L}(X,\mathbb{K})$ is a linear map such that $\Lambda - L$ takes values in X^*. Actually, every version of Φ^* can be written in this way since if $\Psi = \Phi^* + L' + B$ with $L'\colon Y^* \longrightarrow X^*$ linear and $B\colon Y^* \longrightarrow X^*$ bounded then $\Psi = (\Lambda + B) - (L - L')$ is the desired decomposition.

The Dual of a Kalton–Peck Map

In this section, $1 < p, q < \infty$ are always conjugate exponents ($p^{-1} + q^{-1} = 1$), and we identify the dual of ℓ_p with ℓ_q in the usual way: if $g \in \ell_q, f \in \ell_p$ then $\langle g, f\rangle = \sum_n g(n)f(n)$. We proceed to compute the duals of the Kalton–Peck maps when acting on ℓ_p. This allows us to describe the dual of the twisted sum space $\ell_p(\varphi)$ and to establish that it is locally convex without needing to invoke Proposition 3.4.5. We will find that the dual of each Kalton–Peck map KP_φ is just a 'scaled' version of KP_φ acting on the dual space. To achieve an optimal matching of the maps, throughout this section we will consider the map $\mathsf{KP}_{p,\varphi}\colon \ell_p^0 \longrightarrow \ell_p$ defined as

$$\mathsf{KP}_{p,\varphi}(f) = f\varphi\left(p \log \frac{\|f\|_p}{|f|}\right). \tag{3.27}$$

If $\varphi(t) = t$, we omit it. It is clear that the twisted sum generated by $\mathsf{KP}_{p,\varphi}$ is just $\ell_p(\varphi_p)$, where $\varphi_p(t) = \varphi(pt)$. The main estimate that we need is a consequence of $|t \log t| \le e^{-1}$ for $0 < t \le 1$.

Lemma 3.8.4 *For every $s, t \in \mathbb{C}$, we have* $\left|st \log\left(|s|^q/|t|^p\right)\right| \le e^{-1}(p|s|^q + q|t|^p)$.

Proof We may assume that s and t are real and positive. Now, if $s^q \le t^p$, then

$$\left|st\log\frac{s^q}{t^p}\right| = qst\left|\log\frac{s}{t^{p/q}}\right| = q\frac{s}{t^{p/q}}t\left|\log\frac{s}{t^{p/q}}\right|t^{p/q} \le \frac{q}{e}t^{1+p/q} \le \frac{q}{e}t^p.$$

Otherwise $s^q \ge t^p$ and, reversing the roles of s, q and t, p, we get

$$\left|st\log\frac{s^q}{t^p}\right| \le \frac{p}{e}s^q. \qquad \square$$

To illustrate how the inequality we just proved will produce $\mathsf{KP}^*_{p,\varphi}$, let us rewrite it as

$$\left|stp\log\frac{1}{|t|} - stq\log\frac{1}{|s|}\right| \le \frac{p|s|^q + q|t|^p}{e},$$

which tell us that if f, g are finitely supported and $\|f\|_p = \|g\|_q = 1$ then

$$\left|\langle g, \mathsf{KP}_p(f)\rangle - \langle \mathsf{KP}_q(g), f\rangle\right| \le \frac{p+q}{e}.$$

Since the left-hand side of the preceding inequality is positively homogeneous in f and g, we actually have

$$\left|\langle g, \mathsf{KP}_p(f)\rangle - \langle \mathsf{KP}_q(g), f\rangle\right| \le \frac{p+q}{e}\|f\|_p\|g\|_q$$

for finitely supported f, g, and Lemma 3.8.2 shows that $-\mathsf{KP}_q$ can be used as the dual of KP_p. The general case is similar:

Proposition 3.8.5 *Let $p, q > 1$ be such that $p^{-1} + q^{-1} = 1$ and $\varphi \in \mathrm{Lip}_0(\mathbb{R}^+)$. Then*

(a) *there is $C > 0$ such that, for every finitely supported f and g, we have*

$$\left|\langle g, \mathsf{KP}_{p,\varphi}(f)\rangle - \langle \mathsf{KP}_{q,\varphi}(g), f\rangle\right| \le C\|g\|_q\|f\|_p,$$

(b) *there is a commutative diagram*

$$\begin{array}{ccccccccc} 0 & \longrightarrow & \ell_q & \longrightarrow & \ell_q(-\varphi_q) & \longrightarrow & \ell_q & \longrightarrow & 0 \\ & & \| & & \downarrow u & & \| & & \\ 0 & \longrightarrow & \ell_p^* & \longrightarrow & \ell_p(\varphi_p)^* & \longrightarrow & \ell_p^* & \longrightarrow & 0 \end{array}$$

(c) *in particular, $\ell_q(-\varphi_q)$ and $\ell_q(\varphi_q)$ are isomorphic to $\ell_p(\varphi_p)^*$.*

Proof Since $-\mathsf{KP}_{q,\varphi} = \mathsf{KP}_{q,-\varphi}$ and $Z(\Phi)$ and $Z(-\Phi)$ are always isometric via $(y, x) \leftrightarrow (-y, x)$ or $(y, x) \leftrightarrow (y, -x)$, it suffices to check (a) because this already implies (b) and (c). Pick f, g with $\|f\|_p = \|g\|_q = 1$. If $I = \{n\colon f(n)g(n) \neq 0\}$, we have

$$\begin{aligned}
\left|\left\langle g, \mathsf{KP}_{p,\varphi}(f)\right\rangle - \langle \mathsf{KP}_{q,\varphi}(g), f\rangle\right| &= \left|\left\langle g, f\varphi\left(p\log\frac{1}{|f|}\right)\right\rangle - \left\langle g\varphi\left(q\log\frac{1}{|g|}\right), f\right\rangle\right| \\
&\leq \sum_{n\in I} |g(n)f(n)| \left|\varphi\left(p\log\frac{1}{|f(n)|}\right) - \varphi\left(q\log\frac{1}{|g(n)|}\right)\right| \\
&\leq \sum_{n\in I} |g(n)|\cdot|f(n)|\cdot \mathrm{Lip}(\varphi)\cdot\left|\log\frac{|g(n)|^q}{|f(n)|^p}\right| \\
&\leq \frac{1}{e}\sum_n (p|g(n)|^q + q|f(n)|^p) \\
&= \frac{p+q}{e}. \qquad \square
\end{aligned}$$

When $p = 2$, we have the further simplification that $\ell_2^* = \ell_2$, so $\varphi_q = \varphi_p$, and since $\varphi = (\varphi_{1/2})_2$, the manifest consequence is:

Corollary 3.8.6 *For every $\varphi \in \mathrm{Lip}_0(\mathbb{R}^+)$, the space $\ell_2(\varphi_2)^*$ is isomorphic to $\ell_2(-\varphi_2)$ and, therefore, the space $\ell_2(\varphi)$ is isomorphic to its dual via the map $D\colon \ell_2(\varphi_2) \longrightarrow \ell_2(\varphi_2)^*$ given by $D(y, x)(y', x') = \langle y, x'\rangle - \langle x, y'\rangle$.*

The appearance of one minus sign is unavoidable: if we want an isomorphism between $\ell_2(\varphi)$ and $\ell_2(\varphi)^*$ then the form of the duality has to be $\langle y, x'\rangle - \langle x, y'\rangle$ (or $\langle x, y'\rangle - \langle y, x'\rangle$, of course); if we insist on keeping the 'straight' duality $\langle y, x'\rangle + \langle x, y'\rangle$ then $\ell_2(\varphi)^*$ must be represented as $\ell_2(-\varphi)$.

We leave the duality issues with a remark on the intertwining bilinear form β in Lemma 3.8.2. Note that for the Kalton–Peck map we have $\beta(g, f) = 0$ for $g \in \ell_q^0$; even so:

3.8.7 *Let X, Y be Banach spaces and let $\Phi\colon X \longrightarrow Y, \Psi\colon Y^* \longrightarrow X^*$ be quasilinear maps for which there is a bounded bilinear map $\beta\colon Y^* \times X \longrightarrow \mathbb{K}$ such that for some constant C independent on $y^* \in Y^*$ and $x \in X$, we have*

$$|\langle \Psi(y^*), x\rangle + \langle y^*, \Phi(x)\rangle - \beta(y^*, x)| \leq C\|y^*\|\,\|x\|.$$

Then Φ and Ψ are both bounded.

Indeed, if β is continuous then we have $|\langle \Psi(y^*), x\rangle + \langle y^*, \Phi(x)\rangle| \leq C'\|y^*\|\,\|x\|$, which implies that the composition $y^* \circ \Phi$ is bounded for each $y^* \in Y^*$. Hence

$$Y^* = \bigcup_{n\in\mathbb{N}} \{y^* \in Y^*\colon \|y^* \circ \Phi\| \leq n\}.$$

The Baire category theorem and the homogeneity of Φ yield that the closure of $\{y^* \in Y^*\colon \|y^* \circ \Phi\| \leq 1\}$ must contain an open set of Y^* and therefore an open ball centered at the origin. If this ball has radius r then $\|y^* \circ \Phi\| \leq 1$ whenever $\|y^*\| \leq r$, and thus $\|\Phi\| \leq 1/r$.

3.9 Different Versions of a Quasilinear Map

A *version* of a quasilinear map is another equivalent quasilinear map. Quasilinear maps have many different versions. What is not immediate at this stage is that some versions are better than others depending on the purpose we have in mind. Here we present three versions, each with some additional useful property.

Killing a Quasilinear Map on a Hamel Basis

If $\Phi\colon X \longrightarrow Y$ is a quasilinear map and $\mathscr{H}$ is a Hamel basis of X then there is exactly one version of Φ modulo L vanishing on $\mathscr{H}$: let $L\colon X \longrightarrow Y$ be the only linear map such that $La = \Phi a$ for all $a \in \mathscr{H}$ and set $\Phi - L$. We will denote by $\mathsf{Q}_{\mathscr{H}}(X, Y)$ the space of all quasilinear maps vanishing on $\mathscr{H}$. We have that $\mathsf{Q}_{\mathscr{H}}(X, Y)$ is isometric to $\mathsf{Q}_{\mathsf{L}}(X, Y)$, and there is a decomposition $\mathsf{Q}(X, Y) = \mathsf{L}(X, Y) \oplus \mathsf{Q}_{\mathscr{H}}(X, Y)$. These versions have already appeared: during the proof of Theorem 3.6.3, for instance.

Quasilinear Maps on Finite-Dimensional Spaces

It is easy to give examples of quasilinear maps on finite-dimensional spaces having infinite-dimensional range: consider $\mathbb{C}$ as a real space of dimension 2 and define $\Phi\colon \mathbb{C} \to C[0, 1]$ by sending $e^{i\theta}$ to the function x^θ for $0 \leq \theta < \pi$, and by homogeneity on the rest. Sometimes we need to guarantee that the image of a given finite-dimensional subspace spans a finite-dimensional subspace in the target space. We present here a construction of 'finite-dimensional versions' of a quasilinear map suitable for applications.

Lemma 3.9.1 *Let F be a finite-dimensional p-Banach space, let Y be a p-normed space and let $\Phi\colon F \longrightarrow Y$ be a p-linear map. For each $\varepsilon > 0$, there is a p-linear map $\Phi'\colon F \longrightarrow Y$ such that*

- $\|\Phi - \Phi'\| \leq (1 + \varepsilon) Q^{(p)}(\Phi)$,
- $\Phi'[F]$ *spans a finite-dimensional subspace of Y,*
- $Q^{(p)}(\Phi') \leq 3^{1/p}(1 + \varepsilon) Q^{(p)}(\Phi)$.

Proof The key point is to associate with every $f \in F$ a good-natured p-convex decomposition. To this end, let S be the unit sphere of F. As S is compact, for fixed $\delta > 0$, select a δ-net $f_1, \dots, f_n \in S$. For $f \in S$, pick f_i in the net such that $\|f - f_i\| \leq \delta$ (take the smallest i if the minimun is attained on several elements of the net). Let us set $f_0 = 0$ and consider the new point $g = (f - f_i)/\|f - f_i\|$. Now, choose $0 \leq j \leq n$ minimising $\|g - f_j\|$, with the same tie-break rule as before, to get $\big\|f - f_i - (\|f - f_i\|)f_j\big\| < \delta^2$. If $f = f_i$, then we take $j = 0$. In any case, continuing in this way we can select sequences $i(k)$ and λ_k such that

- $0 \leq i(k) \leq n$ and $0 \leq \lambda_k < \delta^k$ for all $k = 0, 1, 2, \dots$,
- for every $m \in \mathbb{N}$, we have $\|f - \sum_{0 \leq k \leq m} \lambda_k f_{i(k)}\| \leq \delta^m$.

Grouping the terms in the obvious way and taking into account that the ℓ_1-norm is dominated by the ℓ_p-quasinorm, we can write $f = \sum_{i=1}^n c_i f_i$ with $\left(\sum_{i=1}^n |c_i|^p\right)^{1/p} \leq \left(\sum_{k=0}^\infty \lambda_k^p\right)^{1/p} \leq (1 - \delta^p)^{-1/p} < 1 + \varepsilon$ for δ sufficiently small. This 'greedy algorithm' specifies a unique decomposition for each f in the unit sphere of F. However, it does not guarantee any kind of homogeneity in these decompositions. To amend that, let S_0 be a subset of the unit sphere of F that is maximal with respect to the property that any two points of S_0 are linearly independent (of course, when the ground field is $\mathbb{R}$, this just means that S_0 does not contain 'antipodal' points). Equivalently, S_0 is a subset of the sphere such that every non-zero $f \in F$ can be written in a unique way as $f = cx$, with $c \in \mathbb{K}$ and $x \in S_0$. Now we define $\Phi' : F \longrightarrow Y$ as follows: if $f \in S_0$, we set

$$\Phi'(f) = \sum_{i=1}^{n} c_i \Phi(f_i),$$

where $f = \sum_{i=1}^n c_i f_i$ is the decomposition provided by the algorithm. We extend the map to the whole of F by homogeneity; that is, for arbitrary $f \in F$, we write $x = cf$, with $c \in \mathbb{K}$ and $f \in S_0$, in the only way that this can be done, and we set $\Phi'(f) = c\Phi'(f)$. It is clear that the resulting map is homogeneous. Let us check that Φ' does the job. Let $Q^{(p)}$ denote the p-linearity constant of the starting map Φ. For $f \in S_0$, we have

$$\|\Phi(f) - \Phi'(f)\| = \left\|\Phi(f) - \sum_{i=1}^{n} c_i \Phi(f_i)\right\| \leq Q^{(p)} \left(\sum_{i=1}^{n} |c_i|^p\right)^{1/p} < (1 + \varepsilon) Q^{(p)},$$

and for arbitrary f, just use the homogeneity of both maps. It is obvious that the range of Φ' is contained in $[\Phi(f_1), \dots, \Phi(f_n)]$, which is a finite-dimensional subspace of Y. Finally, to estimate the new quasilinearity constant $Q^{(p)}(\Phi')$, note that

$$\begin{aligned}
&\left\|\Phi'\left(\sum_{i=1}^{k} x_i\right) - \sum_{i=1}^{k} \Phi'(x_i)\right\|^p \\
&= \left\|\Phi'\left(\sum_{i=1}^{k} x_i\right) - \Phi\left(\sum_{i=1}^{k} x_i\right) + \Phi\left(\sum_{i=1}^{k} x_i\right) - \sum_{i=1}^{k} \Phi x_i + \sum_{i=1}^{k} \Phi x_i - \sum_{i=1}^{k} \Phi' x_i\right\|^p \\
&\le \left((1+\varepsilon)Q^{(p)}\right)^p \left\|\sum_{i=1}^{k} x_i\right\|^p + \left(Q^{(p)}\right)^p \sum_{i=1}^{k} \|x_i\|^p + \left((1+\varepsilon)Q^{(p)}\right)^p \sum_{i=1}^{k} \|x_i\|^p \\
&\le 3(1+\varepsilon)^p \left(Q^{(p)}\right)^p \sum_{i=1}^{k} \|x_i\|^p
\end{aligned}$$

for all $x_1, \ldots, x_k \in F$, hence $Q^{(p)}(\Phi') \le 3^{1/p}(1+\varepsilon)Q^{(p)}$. □

A subtler version is still possible:

Lemma 3.9.2 *Let $\Phi: X \longrightarrow Y$ be a p-linear map, where X and Y are p-normed spaces. Let F be a finite-dimensional subspace of X, and let $x_1, \ldots, x_n$ be points in the unit sphere of F. Then, for each $\varepsilon > 0$, there is a p-linear map $\Phi_F : X \longrightarrow Y$ such that*

- $\|\Phi - \Phi_F\| \le (1+\varepsilon)Q^{(p)}(\Phi)$,
- $\Phi_F[F]$ *spans a finite-dimensional subspace of Y,*
- $\Phi_F(x_i) = \Phi(x_i)$ *for $1 \le i \le n$,*
- $Q^{(p)}(\Phi_F) \le 3^{1/p}(1+\varepsilon)Q^{(p)}(\Phi)$.

Proof Fix $\varepsilon > 0$, and let us consider the map $\Phi|_F$ as a p-linear map from F to Y. Let us apply Lemma 3.9.1 to $\Phi|_F$. We ensure that the δ-net $f_1, \ldots, f_n$ appearing in the third line of the previous proof contains the set $\{x_1, \ldots, x_m\}$ simply by adding these points, if necessary, and then observe that the resulting map $\Phi' : F \longrightarrow Y$ agrees with Φ on every f_i. We define $\Phi_F : X \longrightarrow Y$ by

$$\Phi_F(x) = \begin{cases} \Phi'(x) & \text{if } x \in F \\ \Phi(x) & \text{otherwise.} \end{cases}$$

Observe that $\|\Phi - \Phi_F\| \le (1+\varepsilon)Q^{(p)}(\Phi)$, and then repeat the arguments in the proof of Lemma 3.9.1 to check the remaining properties of Φ_F. □

Corollary 3.9.3 *Let $\Phi: X \longrightarrow Y$ be a quasilinear map acting between quasinormed spaces. There is a constant M such that if F is a finite-dimensional subspace of X then there is a quasilinear map $\Phi_F : F \longrightarrow Y$ such that*

- $\|\Phi|_F - \Phi_F\| \le M$,

- *$Q(\Phi_F) \leq M$,*
- *$\Phi_F[F]$ is contained in some finite-dimensional subspace of Y.*

Proof Consider the twisted sum $Y \oplus_\Phi X$. By the Aoki–Rolewicz theorem, there is $0 < p \leq 1$ such that $Y \oplus_\Phi X$ is p-normable and, according to Proposition 3.6.7, the constant $Q^{(p)}(\Phi)$ is finite for this p. Fix some ε and use Lemma 3.9.1. □

Quasilinear Maps on Separable Spaces

Let $\Phi \colon X \longrightarrow Y$ be a quasilinear map on a separable space X. There is no guarantee that $\Phi[X]$ is separable (unless Y itself is separable). In fact, unless X has dimension 1, each quasilinear map $\Phi \colon X \longrightarrow Y$ into a non-separable space Y has a version with non-separable range (it should be clear why). To discipline this possibly bad behaviour, observe that given a quotient map $\rho \colon Z \longrightarrow X$ between quasi-Banach spaces, there is a closed subspace $Z' \subset Z$ with the same dimension as X such that $\rho \colon Z' \longrightarrow X$ is still a quotient map (it should be even clearer why). Thus, if $0 \longrightarrow Y \xrightarrow{\jmath} Z \xrightarrow{\rho} X \longrightarrow 0$ is an exact sequence of quasi-Banach spaces, taking Z' as before and then forming $Y' = \jmath^{-1}[\ker \rho \cap Z']$, we obtain the pushout diagram

$$\begin{array}{ccccccccc} 0 & \longrightarrow & Y' & \longrightarrow & Z' & \longrightarrow & X & \longrightarrow & 0 \\ & & \downarrow & & \downarrow & & \| & & \\ 0 & \longrightarrow & Y & \xrightarrow{\jmath} & Z & \xrightarrow{\rho} & X & \longrightarrow & 0 \end{array}$$

from which the desired version emerges:

Lemma 3.9.4 *Let $\Phi \colon X \longrightarrow Y$ be a quasilinear map. There is a version of Φ whose range is contained in a subspace of Y having dimension at most* $\dim(X)$.

3.10 Linearisation of Quasilinear Maps

What we present here is the construction of 'non-linear envelopes', which linearise quasilinear functions in a similar way as tensor products linearise bilinear mappings. A passing glance at Note 4.6.1, 'Linearisation constructions', will probably clarify the general schema. In this section, we will work with p-Banach spaces and p-linear maps for some fixed p. Let I be the set of all closed subspaces Y of $\ell_p(\kappa)$, and form the ℓ_p-amalgam $U_\kappa = \ell_p(I, \ell_p(\kappa)/Y)$.

Lemma 3.10.1 *For every cardinal κ, there is a p-Banach space U_κ containing an isometric copy of every p-Banach space whose cardinality is at most κ.*

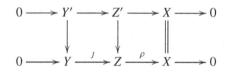

Now, let X be a p-Banach space, for which we fix a Hamel basis $\mathscr{H}$. Let κ be the cardinality (not dimension) of X. Consider the spaces $\mathsf{Q}^{(p)}(X, U_\kappa)$ and let $\mathsf{Q}^{(p)}_{\mathscr{H}}(X, U_\kappa)$. We can take as $\mathrm{co}^{(p)}(X)$ the closed subspace of $\mathfrak{L}\left(\mathsf{Q}^{(p)}_{\mathscr{H}}(X, U_\kappa), U_\kappa\right)$ generated by the evaluation maps $(\delta_x)_{x\in X}$, once we have checked that the maps δ_x are bounded: since Φ vanishes on $\mathscr{H}$ then, writing $x = \sum_{h\in\mathscr{H}} x_h h$, we have

$$\|\delta_x(\Phi)\| = \Big\|\Phi(x) - \sum_h x_h\Phi(h)\Big\| \le Q^{(p)}(\Phi)\Big(\sum_h |x_h|^p\|h\|^p\Big)^{1/p}.$$

The universal property behind this construction is uncovered now:

Theorem 3.10.2 *Let $\mho\colon X \longrightarrow \mathrm{co}^{(p)}(X)$ be the map $\mho(x) = \delta_x$.*

(a) *$\mho$ is p-linear, with $Q^{(p)}(\mho) \le 1$.*
(b) *For every p-Banach space Y and every p-linear map $\Phi\colon X \longrightarrow Y$ vanishing on $\mathscr{H}$, there exists a unique operator $\phi\colon \mathrm{co}^{(p)}(X) \longrightarrow Y$ such that $\Phi = \phi\circ\mho$.*
(c) *$\|\phi\| = Q^{(p)}(\Phi)$.*

Proof A proof is only needed because the definition of $\mathrm{co}^{(p)}(X)$ is kind of a tongue-twister. It is clear that $\mho$ is homogeneous since

$$\mho(\lambda x)(\Phi) = \delta_{\lambda x}\Phi = \Phi(\lambda x) = \lambda\Phi(x) = (\lambda\mho x)(\Phi).$$

The map $\mho$ is p-linear because if one picks finitely many $x_i \in X$ then

$$\begin{aligned}\Big\|\mho\Big(\sum_i x_i\Big) - \sum_i \mho(x_i)\Big\| &= \sup_{Q^{(p)}(\Phi)\le 1}\Big\|\Big(\mho\Big(\sum_i x_i\Big) - \sum_i \mho(x_i)\Big)(\Phi)\Big\| \\ &= \sup_{Q^{(p)}(\Phi)\le 1}\Big\|\Phi\Big(\sum_i x_i\Big) - \sum_i \Phi(x_i)\Big\| \le \Big(\sum_i \|x_i\|^p\Big)^{1/p},\end{aligned}$$

which, in particular, yields $Q^{(p)}(\mho) \le 1$. To prove (b), let us first consider the case $Y = U_\kappa$ and assume $\Phi\colon X \longrightarrow U_\kappa$ is a p-linear map vanishing on $\mathscr{H}$. Since $\mathrm{co}^{(p)}(X) \subset \mathfrak{L}\left(\mathsf{Q}^{(p)}_{\mathscr{H}}(X, U_\kappa), U_\kappa\right)$, it is clear that the required operator $\phi\colon \mathrm{co}^{(p)}(X) \longrightarrow U_\kappa$ is just the restriction to $\mathrm{co}^{(p)}(X)$ of the evaluation map δ_Φ given by $\delta_\Phi(u) = u(\Phi)$. It is obvious that $\|\phi\| \le \|\delta_\Phi\| = \sup_{\|u\|\le 1}\|u(\Phi)\| \le Q^{(p)}(\Phi)$. That $\Phi = \phi\circ\mho$ is easy as well: $\phi(\mho(x)) = \phi(\delta_x) = \delta_\Phi(\delta_x) = \delta_x(\Phi) = \Phi(x)$ for every $x \in X$. This also shows that $\|\phi\| = Q^{(p)}(\Phi)$ since $\Phi = \phi\circ\mho$ implies $Q^{(p)}(\Phi) \le \|\phi\| Q^{(p)}(\mho) \le \|\phi\|$. To complete the proof, we treat the general case. Let $\Phi\colon X \longrightarrow Y$ be a p-linear map that vanishes on $\mathscr{H}$. Let Y' be the closed subspace spanned by $\Phi[X]$ in Y, which necessarily has $|Y'| \le \kappa$. Recall that what we are considering here is *cardinality*, not dimension. Fix an isometry $\imath\colon Y' \longrightarrow U_\kappa$ and form the composition $\imath\circ\Phi\colon X \longrightarrow U_\kappa$ to get an operator $\phi\colon \mathrm{co}^{(p)}(X) \longrightarrow U_\kappa$ such that $\phi\circ\mho = \imath\circ\Phi$. Since $\phi[\mathrm{co}^{(p)}(X)] \subset \imath[Y']$, the desired operator is $\imath^{-1}\phi$. □

The exact sequence

$$0 \longrightarrow \mathrm{co}^{(p)}(X) \xrightarrow{\ \imath\ } \mathrm{co}^{(p)}(X) \oplus_{\mho} X \xrightarrow{\ \pi\ } X \longrightarrow 0 \quad (3.28)$$

behaves, in many respects, as a projective presentation of X (except that we do not know that $\mathrm{co}^{(p)}(X) \oplus_{\mho} X$ is projective):

Corollary 3.10.3 *Every exact sequence* $0 \longrightarrow Y \longrightarrow Z \longrightarrow X \longrightarrow 0$ *of p-Banach spaces is a pushout of* (3.28). *Therefore* (3.28) *is semi-equivalent to any projective presentation of* X *in* **pB**.

3.11 The Type of Twisted Sums

We already know that the type of a twisted sum of two spaces does not necessarily match that of their direct sum; see Proposition 3.2.7. We now perform a systematic study of the type of twisted sums of spaces – mostly Banach spaces, but quasi-Banach spaces as well. We then immediately derive no fewer than three remarkable consequences; further applications are spread throughout later chapters. We start with the observation that since type passes to subspaces and quotients, the type of a twisted sum is at best the worst type between those of the subspace and quotient. To deal with the type of a twisted sum, let us formulate the randomised version of Proposition 3.6.7, whose (randomised) proof is left to the reader.

Lemma 3.11.1 *Let* X *and* Y *be quasi-Banach spaces having type* p *and* $\Phi\colon X \longrightarrow Y$ *a quasilinear map. Then* $Z(\Phi)$ *has type* p *if and only if there is a constant* K *such that for every finite set* $x_1, \ldots, x_n \in X$ *we have*

$$\left(\int_0^1 \Big\|\Phi\Big(\sum_{1\le i\le n} r_i(t)x_i\Big) - \sum_{1\le i\le n} r_i(t)\Phi(x_i)\Big\|^p dt\right)^{1/p} \le K\Big(\sum_{1\le i\le n} \|x_i\|^p\Big)^{1/p}, \quad (3.29)$$

where $(r_i)_{i\ge1}$ *is the sequence of Rademacher functions.*

We will need a probabilistic trick to handle randomised sums. Suppose that $k\colon \{1, \ldots, n\} \longrightarrow \{1, \ldots, m\}$ is any surjection. Note that such a k can be regarded as a partition of $\{1, \ldots, n\}$ into m many sets $(A_j)_{1\le j\le m}$, where $A_j = k^{-1}(j)$. Then, if $x_1, \ldots, x_n$ are points of X, the X^n-valued functions

$$t \in [0,1] \longmapsto (r_1(t)x_1, \ldots, r_i(t)x_i, \ldots, r_n(t)x_n)$$
$$(s,t) \in [0,1]^2 \longmapsto (r_{k(1)}(s)r_1(t)x_1, \ldots, r_{k(i)}(s)r_i(t)x_i, \ldots, r_{k(n)}(s)r_n(t)x_n)$$

have the same distribution. Therefore, for every function $f\colon X^n \longrightarrow \mathbb{C}$,

$$\int_0^1 f(r_1(t)x_1,\dots,r_n(t)x_n)\,dt = \int_0^1\int_0^1 f(r_{k(1)}(s)r_1(t)x_1,\dots,r_{k(n)}(s)r_n(t)x_n)\,dt\,ds.$$

We also need to fix some notation: given a mapping $\Phi\colon X \longrightarrow Y$ and finitely many points $(x_i)_{1\le i\le n}$ of X, we set

$$\nabla\Phi(x_1,\dots,x_n) = \Phi\left(\sum_{i=1}^n x_i\right) - \sum_{i=1}^n \Phi(x_i).$$

Theorem 3.11.2 *Let $0 < p < q \le 2$. Every twisted sum of a space having type p and a space having type q has type p.*

Proof Let X by a space having type p with type constant $T_p(X)$ and let Y by a space having type q with type constant $T_q(Y)$. We set $T = \max(T_p(X), T_q(Y))$. Let K be the Kahane constant of the 'L_q versus L_p' estimate of X, so that for $(x_i)_{1\le i\le n}$ in X, we have

$$\left(\int_0^1 \Big\|\sum_{1\le i\le n} r_i(t)x_i\Big\|^q dt\right)^{1/q} \le K\left(\int_0^1 \Big\|\sum_{1\le i\le n} r_i(t)x_i\Big\|^p dt\right)^{1/p}. \tag{3.30}$$

Finally, let $\Phi\colon X \longrightarrow Y$ be a quasilinear map, and, for each $n \in \mathbb{N}$, let c_n be the least constant for which

$$\left(\int_0^1 \Big\|\nabla\Phi(r_1(t)x_1,\dots,r_n(t)x_n)\Big\|^q dt\right)^{1/q} \le c_n\left(\sum_{1\le i\le n} \|x_i\|^p\right)^{1/p}$$

for all finite sets $x_1,\dots,x_n \in X$. Our immediate goal is to prove that the sequence $(c_n)_{n\ge1}$ is bounded. Suppose $\|x_1\|^p + \cdots + \|x_n\|^p = 1$ and that $\|x_n\|^p \ge N^{-1}$. Put $u(t) = \sum_{1\le i<n} r_i(t)x_i$. Then

$$\nabla\Phi(r_1x_1,\dots,r_nx_n) = \nabla\Phi(u, r_nx_n) + \nabla\Phi(r_1x_1,\dots,r_{n-1}x_{n-1}).$$

Now, since the Y-valued functions

$$t \in [0,1] \mapsto \nabla\Phi(u(t), r_n(t)x_n) = \Phi\left(\sum_{i\le n} r_i(t)x_i\right) - \Phi\left(\sum_{i<n} r_i(t)x_i\right) - r_n(t)\Phi(x_n)$$

$$(s,t) \in [0,1]^2 \mapsto \nabla\Phi(r_1(s)u(t), r_2(s)r_n(t)x_n)$$

$$= \Phi\left(\sum_{i<n} r_1(s)r_i(t)x_i + r_2(s)r_n(t)x_n\right) - \Phi\left(\sum_{i<n} r_1(s)r_i(t)x_i\right) - r_2(s)r_n(t)\Phi(x_n)$$

have the same distribution, it follows that

$$\int_0^1 \|\nabla\Phi(u(t), r_n(t)x_n)\|^q dt = \int_0^1\int_0^1 \|\nabla\Phi(r_1(s)u(t), r_2(s)r_n(t)x_n)\|^q ds\, dt$$
$$\leq c_2^q \int_0^1 (\|u(t)\|^p + \|x_n\|^p)^{q/p}\, dt.$$

Hence, by Minkowski's inequality (recall that $q/p > 1$),

$$\begin{aligned}\left(\int_0^1 \|\nabla\Phi(u, r_n x_n)\|^q dt\right)^{p/q} &\leq c_2^p \left(\int_0^1 (\|u(t)\|^p + \|x_n\|^p)^{q/p}\, dt\right)^{p/q}\\ &\leq c_2^p \left(\left(\int_0^1 \|u(t)\|^q dt\right)^{p/q} + \|x_n\|^p\right)\\ &\leq c_2^p \left(K^p \left(\int_0^1 \|u(t)\|^p dt\right)^p + \|x_n\|^p\right)\\ &\leq c_2^p \left(K^p T_p(X)^p \left(\sum_{1\leq i<n} \|x_i\|^p\right) + \|x_n\|^p\right).\end{aligned}$$

Thus, if $\|x_1\|^p + \cdots + \|x_n\|^p = 1$ and $\max_i \|x_i\|^p \geq N^{-1}$, then

$$\left(\int_0^1 \left\|\nabla\Phi(r_1(t)x_1, \ldots, r_n(t)x_n)\right\|^q dt\right)^{1/q} \leq c_{n-1}\left(1 - \frac{1}{N}\right)^{1/p} + Kc_2c. \tag{3.31}$$

Now assume $\|x_1\|^p + \cdots + \|x_n\|^p = 1$ and $\|x_i\|^p < N^{-1}$ for $1 \leq i \leq n$. Then we can partition $\{1, \ldots, n\}$ into N subsets $(A_j)_{1\leq j\leq N}$ such that $\sum_{i\in A_j} \|x_i\|^p \leq 2/N$ for all j. Then, letting $u_j(t) = \sum_{i\in A_j} r_i(t)x_i$, we have

$$\nabla\Phi(r_1x_1, \ldots, r_nx_n) = \nabla\Phi(u_1, \ldots, u_N) + \sum_{j=1}^N \nabla(\Phi)((r_ix_i)_{i\in A_j}).$$

Using the same argument as before,

$$\left(\int_0^1 \|\nabla\Phi(u_1(t), \ldots, u_N(t))\|^q dt\right)^{p/q} \leq c_N^p \left(\int_0^1 \left(\sum_{1\leq j\leq N} \|u_i(t)\|^p\right)^{q/p} dt\right)^{p/q}$$
$$\leq c_N^p \sum_{1\leq j\leq N} \left(\int_0^1 \|u_i(t)\|^q dt\right)^{p/q} \leq c_N^p \sum_{1\leq j\leq N} K^p T_p(X) \sum_{i\in A_j} \|x_i\|^p \leq c_N^p K^p c^p.$$

We are almost there:

$$\left(\int_0^1 \Big\| \sum_{j=1}^N \nabla\Phi((r_i x_i)_{i\in A_j}) \Big\|^q dt\right)^{1/q} \le c\left(\sum_{j=1}^N \int_0^1 \|\nabla\Phi((r_i x_i)_{i\in A_j})\|^q dt\right)^{1/q}$$

$$\le c c_n \left(\sum_{j=1}^N \left(\sum_{i\in A_j} \|x_i\|^q\right)^{p/q}\right)^{1/q} \le c c_n \left(\sum_{j=1}^N \left(\frac{2}{N}\right)^{q/p-1} \sum_{i\in A_j} \|x_i\|^p\right)^{1/q} = \left(\frac{2}{N}\right)^{1/p-1/q} c c_n.$$

All together now,

$$\left(\int_0^1 \Big\|\delta(\Phi)(r_1(t)x_1,\ldots,r_n(t)x_n)\Big\|^q dt\right)^{1/q} \le cKc_N + \left(\frac{2}{N}\right)^{1/p-1/q} c\, c_n.$$

Thus, for all N,

$$c_n \le \max\left(c_{n-1}\left(1-\frac{1}{N}\right)^{1/p} + Kc_2c,\ cKc_N + \left(\frac{2}{N}\right)^{1/p-1/q} c\, c_n\right).$$

Choosing N so that $c\left(\frac{2}{N}\right)^{1/p-1/q} < 1$ yields a bound for (c_n). □

The many consequences of this result make the considerable effort invested in its proof worth it. The first consequence is the extension of Proposition 3.4.5 to B-convex spaces.

Corollary 3.11.3 *Banach spaces having type $p > 1$ are $\mathscr{K}$-spaces.*

Indeed, since the ground field has type 2, every minimal extension of a space having type $p > 1$ also has type strictly greater than 1, thus it has to be a Banach space (Proposition 1.4.4).

Corollary 3.11.4 *Twisted sums of Banach spaces having type p have type $p - \varepsilon$ for every $\varepsilon > 0$. If X and Y have type $p > 1$ and cotype q then every twisted sum of Y and X has cotype $q + \varepsilon$ for every $\varepsilon > 0$. In particular, twisted Hilbert spaces are near-Hilbert.*

Moving to the border of the Banach zone, we encounter:

Proposition 3.11.5 *Minimal extensions of Banach spaces have type* 1.

This is surprising since, for instance, the non-locally convex Ribe space has type 1, despite estimate (3.10) clearly showing that its vector-valued versions $\ell_1(\varphi)$ do not have type 1 when φ is unbounded. It is even possible to derive Theorem 3.7.4 from Theorem 3.11.2: indeed, if $0 < p < q \le 1$, then every twisted sum of a q-Banach space and $\ell_p(I)$ has type p, hence it is a p-Banach space, and the result follows just by lifting the unit basis of $\ell_p(I)$.

3.12 A Glimpse of Centralizers

We introduce centralizers on function spaces. Centralizers combine two seemingly contradictory ideas: a relaxation of the 'approximate additivity' property of quasilinear maps and a strengthening of their homogeneity. To explain how the first idea is used, observe that what really matters about a quasilinear map are the Cauchy differences $\Phi(x + x') - \Phi(x) - \Phi(x')$ rather than the values of Φ themselves. The reader who is sceptical of this point should skip forward to Section 3.13.2, where the argument is taken to its extreme. So, assume that X, Y are quasinormed spaces and W is a (not necessarily topological) linear space containing Y. A homogeneous mapping $\Phi\colon X \longrightarrow W$ is *quasilinear* (no need to change the name) from X to Y if

(a) for every $x, x' \in X$, the difference $\Phi(x + x') - \Phi(x) - \Phi(x')$ lies in Y,

(b) there is a constant Q such that $\|\Phi(x + x') - \Phi(x) - \Phi(x')\| \leq Q(\|x\| + \|x'\|)$ for all $x, x' \in X$.

This definition generalises the standard one in which $W = Y$. We have encountered this situation already: when $\Lambda\colon Y^* \longrightarrow \mathsf{L}(X, \mathbb{K})$ appeared during the construction of the dual 1-linear map in Section 3.8 or, more implicitly, when amalgams in the proof of Lindenstrauss p-lifting (Theorem 3.7.1) occurred in an ultrapower. In those cases we hastened to push the values down to the right space. We can now take a more relaxed attitude: if $\Phi\colon X \longrightarrow W$ is quasilinear from X to Y, it still generates a twisted sum space $Y \oplus_\Phi X = \{(w, x) \in W \times X\colon w - \Phi(x) \in Y\}$, which is a linear subspace of $W \times X$ thanks to (a) and, if endowed with the functional $\|(w, x)\|_\Phi = \|w - \Phi(x)\| + \|x\|$, which is a quasinorm by (b), we get an isometrically exact sequence

$$0 \longrightarrow Y \xrightarrow{\ \imath\ } Y \oplus_\Phi X \xrightarrow{\ \pi\ } X \longrightarrow 0$$

in which $\imath(y) = (y, 0)$ and $\pi(w, x) = x$ as always. In particular, $Y \oplus_\Phi X$ is complete when X and Y are. The criterion for triviality is almost the same as Lemma 3.3.2: Φ generates a trivial extension if and only if $\Phi = B + L$, where $L\colon X \longrightarrow W$ is linear and $B\colon X \longrightarrow Y$ is homogeneous bounded. The reader should not have any difficulty in adapting the proof of Lemma 3.3.2 to this situation. In fact, this is a consequence of the comparison criterion:

3.12.1 *Let $\Phi, \Psi\colon X \longrightarrow W$ be quasilinear maps from X to Y. Then Φ and Ψ generate equivalent extensions if and only if $\Phi - \Psi = B + L$, where $L\colon X \longrightarrow W$ is linear and $B\colon X \longrightarrow Y$ is homogeneous bounded.*

There is a loose end to tie up: if, by pure bad luck, $\Phi\colon X \longrightarrow W$ and $\Psi\colon X \longrightarrow W'$ are quasilinear from X to Y but taking values in different spaces (even if $W = W'$ in most practical situations), just form the pushout

$$\begin{array}{ccccc} & & W & & \\ & \nearrow & & \searrow & \\ Y & & & & W'' \\ & \searrow & & \nearrow & \\ & & W' & & \end{array}$$

(in the linear category) and consider that both Φ, Ψ take values in W''.

Centralizers on Function Spaces

Centralizers are fussy objects: they require the presence of a Banach algebra and a comfortable ambient space to exist. A measure space (S,μ) provides both: the algebra $L_\infty(\mu)$ and the ambient space $L_0(\mu)$. It is clear that if X is a function space on μ then for every $a \in L_\infty(\mu)$ and every $x \in X$, we have $ax \in X$ and $\|ax\| \leq \|a\|_\infty\|x\|$, that is, X is a module over $L_\infty(\mu)$ under the pointwise operations.

Definition 3.12.2 Let X and Y be function spaces. A homogeneous mapping $\Phi\colon X \longrightarrow L_0(\mu)$ is said to be a centralizer from X to Y if, for every $a \in L_\infty(\mu)$ and every $x \in X$, the difference $\Phi(ax) - a\Phi(x)$ lies in Y and obeys, for some constant C, an estimate of the form $\|\Phi(ax) - a\Phi(x)\| \leq C\|a\|_\infty\|x\|$.

When $Y = X$, we just say that Φ is a centralizer on X.

Lemma 3.12.3 *Every centralizer is quasilinear.*

Proof This is very easy. Let $\Phi\colon X \longrightarrow L_0(\mu)$ be a centralizer from X to Y. Pick $x, y \in X$, and set $z = |x| + |y|$ such that $z \in X$ and $\|z\| \leq \Delta_X(\|x\| + \|y\|)$. As $|x|, |y| \leq z$, there are $a, b \in L_\infty(\mu)$, with $\max\{\|a\|_\infty, \|b\|_\infty, \|a + b\|_\infty\} \leq 1$, such that $x = az$ and $y = bz$ and $x + y = (a + b)z$. We thus have

$$\|\Phi(x + y) - (a + b)\Phi(z)\| \leq C(\Phi)\|a + b\|_\infty\|z\|,$$
$$\|\Phi(x) - a\Phi(z)\| \leq C(\Phi)\|a\|_\infty\|z\|,$$
$$\|\Phi(y) - b\Phi(z)\| \leq C(\Phi)\|b\|_\infty\|z\|$$

(and the differences belong to Y). Therefore,

$$\|\Phi(x + y) - \Phi(x) - \Phi(y)\| \leq \Delta_Y^2 C(\Phi)4\|z\| \leq 4\Delta_Y^2\Delta_X C(\Phi)(\|x\| + \|y\|). \qquad \square$$

The following remark highlights the main feature of twisted sums generated by centralizers:

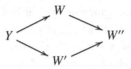

Lemma 3.12.4 *If $\Phi \longrightarrow L_0(\mu)$ is a centralizer from X to Y and $a \in L_\infty(\mu)$ then the map $(y, x) \longmapsto (ay, ax)$ is a bounded endomorphism of $Y \oplus_\Phi X$.*

Proof The pair (ay, ax) belongs to $Y \oplus_\Phi X$ since $y - \Phi(x), a(y - \Phi x)$ and $\Phi(ax) - a\Phi(x)$ are all in $Y \oplus_\Phi X$. Moreover,

$$\begin{aligned}\|(ay, ax)\|_\Phi &= \|ay - \Phi(ax)\| + \|ax\|\\ &\le \Delta_Y(\|ay - a\Phi x\| + \|\Phi x - \Phi(ax)\|) + \|ax\|\\ &\le \max(\Delta_Y C(\Phi), 1)\|a\|_\infty \|(y, x)\|_\Phi.\end{aligned}$$

□

The Kalton–Peck Maps Are Centralizers

The following result provides the continuous version of the Kalton–Peck maps. Note that the Lipschitz function must now be defined on the whole line $\mathbb{R}$.

Proposition 3.12.5 *Let X be a function space and $\varphi \in \mathrm{Lip}_0(\mathbb{R})$. The map $\Phi\colon X \longrightarrow L_0(\mu)$ defined by*

$$\mathsf{KP}_\varphi(x) = x \cdot \varphi\left(\log \frac{\|x\|}{|x|}\right) \tag{3.32}$$

is a centralizer on X, and $C(\Phi)$ depends only on $\mathrm{Lip}(\varphi)$ *and Δ_X.*

Proof We write KP instead of KP_φ. As in the proof of Proposition 3.2.6, we consider the non-homogeneous map $\mathsf{kp}_\varphi\colon X \longrightarrow L_0(\mu)$ defined by $\mathsf{kp}_\varphi(x) = x\varphi(-\log|x|)$. From

$$\mathsf{KP}_\varphi(x) - \mathsf{kp}_\varphi(x) = x\left(\varphi\left(\log\frac{\|x\|}{|x|}\right) - \varphi\left(\log\frac{1}{|x|}\right)\right),$$

$$\mathsf{kp}_\varphi(ax) - a\mathsf{kp}_\varphi(x) = ax\left(\varphi\left(\log\frac{1}{|ax|}\right) - \varphi\left(\log\frac{1}{|x|}\right)\right),$$

we obtain the pointwise estimates

$$\begin{aligned}|\mathsf{KP}_\varphi(x) - \mathsf{kp}_\varphi(x)| &\le \mathrm{Lip}(\varphi)\big|x\log\|x\|\big|,\\ |\mathsf{kp}_\varphi(ax) - a\mathsf{kp}_\varphi(x)| &\le \mathrm{Lip}(\varphi)\big|ax\log|a|\big|\end{aligned}$$

so that $\mathsf{KP}_\varphi(x) - \mathsf{kp}_\varphi(x)$ and $\mathsf{kp}_\varphi(ax) - a\mathsf{kp}_\varphi(x)$ belong to Y and

$$\begin{aligned}\|\mathsf{KP}_\varphi(x) - \mathsf{kp}_\varphi(x)\| &\le \mathrm{Lip}(\varphi)\,\Big|\,\|x\|\log\|x\|\,\Big|,\\ \|\mathsf{kp}_\varphi(ax) - a\mathsf{kp}_\varphi(x)\| &\le \mathrm{Lip}(\varphi)\,\Big|\,\|a\|_\infty\log|a|_\infty\,\Big|\,\|x\|.\end{aligned}$$

To compute the centralizer constant of KP_φ, it suffices to consider the case $\|x\| = 1$, so that $\mathsf{KP}_\varphi x = \mathsf{kp}_\varphi x$ and $\|a\|_\infty \le 1$. We have

$$\begin{aligned}\|\mathsf{KP}_\varphi(ax) - a\mathsf{KP}_\varphi(x)\| &= \|\mathsf{KP}_\varphi(ax) - \mathsf{kp}_\varphi(ax) + \mathsf{kp}_\varphi(ax) - a\mathsf{kp}_\varphi(x)\|\\ &\le \Delta_Y(\|\mathsf{KP}_\varphi(ax) - \mathsf{kp}_\varphi(ax)\| + \|\mathsf{kp}_\varphi(ax) - a\mathsf{kp}_\varphi(x)\|)\\ &\le 2e^{-1}\Delta_X \operatorname{Lip}(\varphi).\end{aligned}$$

Hence, $C(\mathsf{KP}_\varphi) \le 2\Delta_X \operatorname{Lip}(\varphi)e^{-1}$ and $Q(\mathsf{KP}_\varphi) \le 8\Delta_X^4 \operatorname{Lip}(\varphi)e^{-1}$, according to Lemma 3.12.3. □

From now on, we write

$$X(\varphi) = \left\{(y,x) \in L_0(\mu) \times X : y - \mathsf{KP}_\varphi(x) \in X\right\}$$

for the twisted sum defined by the centralizer $\mathsf{KP}_\varphi \colon X \longrightarrow L_0(\mu)$ endowed with the quasinorm $\|(y,x)\|_{\mathsf{KP}_\varphi} = \|y - \mathsf{KP}_\varphi(x)\| + \|x\|$. If X is complete then so is $X(\varphi)$, and no further action is required. Recalling that every sequence space is a function space on $\mathbb{N}$, we see that if X is a sequence space then $X(\varphi)$ is a delightfully concrete completion of $X \oplus_{\mathsf{KP}_\varphi} X^0$, and there is no conflict with the notation of Section 3.2. The analysis of the sequences $0 \longrightarrow X \longrightarrow X(\varphi) \longrightarrow X \longrightarrow 0$ in the continuous case is much more involved and requires techniques specific to centralizers. Anyway, we can exhibit some connections between the discrete and the continuous constructions to prove in passing that:

Proposition 3.12.6 *Let $\varphi \in \operatorname{Lip}_0(\mathbb{R})$ and $0 < p < \infty$.*

(a) *$L_p(\mathbb{R}^+)(\varphi)$ contains an isometric copy of $\ell_p(\varphi)$.*
(b) *KP_φ is trivial on L_p if and only if φ is bounded on $(-\infty, 0]$.*
(c) *KP_φ is trivial on $L_p(\mathbb{R}^+)$ if and only if φ is bounded on $\mathbb{R}$.*

Proof (a) Let $(A_i)_{i\ge1}$ be a sequence of disjoint measurable subsets of measure 1. We define $\alpha \colon \mathbb{R}^{\mathbb{N}} \longrightarrow L_0(\mathbb{R}^+)$ by $\alpha(x) = \sum_{i=1}^\infty x(i)1_{A_i}$. It is clear that α restricts to an isometry from ℓ_p into $L_p(\mathbb{R}^+)$. If $\mathsf{KP}_{p,\varphi} \colon \ell_p \longrightarrow \mathbb{R}^{\mathbb{N}}$ is the Kalton–Peck map on ℓ_p then it is clear that $\alpha(\mathsf{KP}_{p,\varphi}(x)) = \mathsf{KP}_\varphi(\alpha(x))$. This implies that the map $\alpha \times \alpha \colon \mathbb{R}^{\mathbb{N}} \times \mathbb{R}^{\mathbb{N}} \longrightarrow L_0(\mathbb{R}^+) \times L_0(\mathbb{R}^+)$ restricts to an isometry $u \colon \ell_p(\varphi) \longrightarrow L_p(\mathbb{R}^+)(\varphi)$ fitting in the commutative diagram

$$\begin{array}{ccccccccc} 0 & \longrightarrow & \ell_p & \longrightarrow & \ell_p(\varphi) & \longrightarrow & \ell_p & \longrightarrow & 0\\ & & \downarrow{\scriptstyle\alpha} & & \downarrow{\scriptstyle u} & & \downarrow{\scriptstyle\alpha} & & \\ 0 & \longrightarrow & L_p(\mathbb{R}^+) & \longrightarrow & L_p(\mathbb{R}^+)(\varphi) & \longrightarrow & L_p(\mathbb{R}^+) & \longrightarrow & 0\end{array}$$

Indeed, if $(y,x) \in \ell_p(\varphi)$, then $(\alpha y, \alpha x) \in L_p(\mathbb{R}^+)(\varphi)$ since

$$\begin{aligned}\alpha y - \mathsf{KP}_\varphi(\alpha x) &= \alpha y - \alpha \mathsf{KP}_{p,\varphi} x = \alpha(y - \mathsf{KP}_{p,\varphi} x) \in L_p(\mathbb{R}^+),\\ \|(\alpha y, \alpha x)\|_{\mathsf{KP}_\varphi} &= \|\alpha(y - \mathsf{KP}_{p,\varphi} x)\|_{L_p} + \|\alpha x\|_{L_p}\\ &= \|y - \mathsf{KP}_{p,\varphi} x\|_{\ell_p} + \|x\|_{\ell_p}\\ &= \|(y,x)\|_{\ell_p(\varphi)}.\end{aligned}$$

(b) First assume that $|\varphi(s)| \leq M$ for all $s \in (-\infty, 0]$. Let f be normalised in L_p, and consider the sets $A = \{s \in [0,1] : |f(s)| \leq 1\}$ and $A^c = [0,1]\backslash A$. Write $f = g + h$, where $g = 1_A f$ and $h = 1_{A^c}$, and observe that $\|g\|^p + \|h\|^p = 1$. Let us bound $\mathsf{KP}_\varphi(g)$ and $\mathsf{KP}_\varphi(h)$ separately. We have $|\mathsf{KP}_\varphi(h)| = |h\varphi(\log \|h\|/|h|)| \leq M|h|$ as $|h(s)| \geq \|h\|$ for s off A. In particular, $\|\mathsf{KP}_\varphi(h)\| \leq M$. On the other hand, since $\mathsf{KP}_\varphi(1_{[0,1]}) = 0$, we have

$$\|\mathsf{KP}_\varphi(g)\| = \|\mathsf{KP}_\varphi(g) - g\mathsf{KP}_\varphi(1_{[0,1]})\| \leq C(\mathsf{KP}_\varphi)\|g\|_\infty \|1_{[0,1]}\|_p \leq C(\mathsf{KP}_\varphi).$$

Finally, since KP_φ is quasilinear,

$$\|\mathsf{KP}_\varphi(f) - \mathsf{KP}_\varphi(g) - \mathsf{KP}(h)\| \leq Q(\mathsf{KP}_\varphi)(\|g\| + \|h\|) \leq 2Q(\mathsf{KP}_\varphi),$$

and this leads to a bound for $\|\mathsf{KP}_\varphi(f)\|$. To prove the converse we use the same idea as in Proposition 3.2.7 to show that $L_p(\varphi)$ is not isomorphic to L_p if φ is unbounded on the half-line $(-\infty, 0]$. First of all, note that the sequence $(\varphi(\log n^{-1/p}))_{n\geq 1}$ cannot be bounded, that $\mathsf{KP}_\varphi(1_A) = 1_A\varphi(\log |A|^{1/p})$ for every measurable $A \subset [0,1]$ and that KP_φ vanishes at every unitary function. Fix $n \in \mathbb{N}$ and let $(A_i)_{1\leq i\leq n}$ be a partition of $[0,1]$ into sets of equal measure. Clearly, $\|1_{A_i}\| = n^{-1/p}$ for $1 \leq i \leq n$ such that $\sum_{i\leq n} \|1_{A_i}\|^p = 1$. If (r_i) is the sequence of Rademacher functions, then

$$\begin{aligned}\int_0^1 \Big\|\mathsf{KP}_\varphi\Big(\overbrace{\sum_{i\leq n} r_i(t)1_{A_i}}^{\text{unitary}}\Big) - \sum_{i\leq n} r_i(t)\mathsf{KP}_\varphi(1_{A_i})\Big\|^p dt &= \int_0^1 \Big\|\sum_{i\leq n} r_i(t)\mathsf{KP}_\varphi(1_{A_i})\Big\|^p dt\\ &= \int_0^1 \Big\|\sum_{i\leq n} r_i(t)\varphi(\log n^{-1/p})1_{A_i}\Big\|^p dt = |\varphi(\log n^{-1/p})|^p.\end{aligned}$$

This shows that $L_p(\varphi)$ does not have type p and concludes the proof for $p \in (0,2]$. For $p \in [2,\infty)$, we show that $L_p(\varphi)$ does not have cotype p using the vectors $(0, 1_{A_i})$. To see this, note that

$$\|(0, 1_{A_i})\|_{\mathsf{KP}_\varphi} = \|1_{A_i}\varphi(\log n^{-1/p})\| + \|1_{A_i}\| = (\varphi(\log n^{-1/p}) + 1)n^{-1/p}$$

so that

$$\sum_{i\leq n} \|(0, 1_{A_i})\|^p_{\mathsf{KP}_\varphi} = (\varphi(\log n^{-1/p}) + 1)^p,$$

while

$$\int_0^1 \Big\| \sum_{i\le n} r_i(t)(0, 1_{A_i}) \Big\|_{\mathsf{KP}_\varphi}^p dt = \int_0^1 \Big\| \Big(0, \sum_{i\le n} r_i(t) 1_{A_i}\Big) \Big\|_{\mathsf{KP}_\varphi}^p dt$$

$$= \int_0^1 \left(\Big\| \mathsf{KP}_\varphi \Big(\sum_{i\le n} r_i(t) 1_{A_i} \Big) \Big\| + \Big\| \sum_{i\le n} r_i(t) 1_{A_i} \Big\| \right)^p dt = 1.$$

(c) If φ is bounded on $\mathbb{R}$, then KP_φ is homogenenous bounded. For the converse, if φ is unbounded on $\mathbb{R}^+$ then $L_p(\mathbb{R}^+)(\varphi)$ cannot be isomorphic to $L_p(\mathbb{R}^+)$ in view of what was proved in Proposition 3.2.7 and (a). Otherwise, φ cannot be bounded on $(-\infty, 0]$, and the result follows from (b) since the restriction of KP_φ to L_p produces a complemented copy of $L_p(\varphi)$ in $L_p(\mathbb{R}^+)(\varphi)$. □

3.13 Notes and Remarks

3.13.1 Domański's Work on Quasilinear Maps

This note reports on Domański's [161] work about quasilinear maps on general topological vector spaces (TVS), which we find most interesting. A mapping $\Phi\colon X \longrightarrow Y$, acting between TVS, is *now* called quasilinear if $\Phi(0) = 0$ and satisfies the following properties:

- Quasiadditivity: $\Phi(x + x') - \Phi(x) - \Phi(x') \to 0$ in Y as $(x, x') \to (0, 0)$ in $X \times X$.
- Quasihomogeneity: $\Phi(cx) - c\Phi(x) \to 0$ in Y as $(c, x) \to (0, 0)$ in $\mathbb{K} \times X$.

Such a map can be used to construct a TVS denoted $Y \oplus_\Phi X$ by endowing $Y \times X$ with the linear topology for which the sets $W(V, U) = \{(y, x) \in Y \times X \colon x \in U, y - \Phi(x) \in V\}$ with $U \in \mathcal{O}_X$, $V \in \mathcal{O}_Y$ form a neighbourhood base at zero. The resulting sequence $0 \longrightarrow Y \longrightarrow Y \oplus_\Phi X \longrightarrow X \longrightarrow 0$ is topologically exact and splits if and only if Φ is approximable, i.e. there is a linear map $L\colon X \longrightarrow Y$ such that $\Phi(x) - L(x) \to 0$ in Y as $x \to 0$ in X. The question of which topologically exact sequences can be generated by quasilinear maps has a simple answer:

Lemma *A topologically exact sequence is equivalent to a sequence generated by a quasilinear map* $\iff$ *its quotient map has a section that is continuous at zero.*

Proof One implication is clear: if $\Phi\colon X \longrightarrow Y$ is quasilinear, the map $X \longrightarrow Y \oplus_\Phi X$ given by $x \longmapsto (\Phi(x), x)$ is the required section. For the converse, let

$$0 \longrightarrow Y \longrightarrow Z \xrightarrow{\ \rho\ } X \longrightarrow 0$$

be a topologically exact sequence in which we may assume that $Y = \ker \rho$. Let $\Gamma\colon X \longrightarrow Z$ be a section of the quotient map that is continuous at 0. Without loss of generality, we may assume that $\Gamma(0) = 0$; otherwise, we can replace Γ by $\Gamma - \Gamma(0)$. Let $\Lambda\colon X \longrightarrow Z$ be any linear section of ρ. The difference $\Gamma - \Lambda$ takes values in Y and is quasilinear, so we can form the space $Y \oplus_\Phi X$, where $\Phi = \Gamma - \Lambda$. The map $u\colon Z \longrightarrow Y \oplus_\Phi X$ given by $u(z) = (z - \Lambda(z), \rho(z))$ is clearly continuous and makes the diagram

$$\begin{array}{ccccccccc} 0 & \longrightarrow & Y & \longrightarrow & Z & \xrightarrow{\rho} & X & \longrightarrow & 0 \\ & & \| & & \downarrow{\scriptstyle u} & & \| & & \\ 0 & \longrightarrow & Y & \longrightarrow & Y \oplus_\Phi X & \xrightarrow{\rho} & X & \longrightarrow & 0 \end{array}$$

commute. In fact, u is a linear homeomorphism, by Roelcke's lemma. □

In [161], Domański exhibits a locally convex space Z with a quotient map $\rho\colon Z \longrightarrow \mathbb{K}^{\mathbb{N}}$ that has no section that is continuous at zero (and another example having continuous sections but no homogeneous section that is continuous at zero). Thus, not all extensions come from quasilinear maps. The situation is more favourable when one works with F-spaces (metrisable and complete TVS) [161, Lemma 2.2 (a)].

Proposition *All short exact sequences of F-spaces are generated by quasilinear maps.*

Proof Exact sequences of F-spaces are topologically exact, by the open mapping theorem. It suffices to show that if Z is an F-space and $\rho\colon Z \longrightarrow X$ is a quotient map then there is a section $\Gamma\colon X \longrightarrow Z$ that is continuous at zero and has $\Gamma(0) = 0$. Let $(U_n)_{n\geq 0}$ be a (decreasing) base of neighbourhoods of the origin in Z, with $U_0 = Z$. If $V_n = \rho[U_n]$, then $(V_n)_{n\geq 0}$ is a (decreasing) base of neighbourhoods of the origin in X, with $V_0 = X$. For each n, let $\Gamma_n\colon V_n \longrightarrow U_n$ be *any* mapping such that $\rho(\Gamma_n(x)) = x$ for all $x \in V_n$. Finally, we define $\Gamma\colon X \longrightarrow Z$ by $\Gamma(0) = 0$ and $\Gamma(x) = \Gamma_n(x)$, where $n = \sup\{k\colon x \in V_k\}$. □

These ideas can be used to winkle out some homological properties of L_0:

Corollary *If Y is a quasi-Banach space then every topologically exact sequence $0 \longrightarrow Y \longrightarrow Z \longrightarrow L_0 \longrightarrow 0$ splits. In particular, L_0 is a $\mathscr{K}$-space.*

Proof It suffices to see that every quasilinear map $\Phi\colon L_0 \longrightarrow Y$ is approximable. We treat the real case. The crucial property of L_0 is that if U is a neighbourhood of zero, then there exist subspaces $X_1, \dots, X_k \subset U$ such that $L_0 = X_1 \oplus \cdots \oplus X_k$. This is so because $|f|_0 \leq |\operatorname{supp} f|$, no matter which values

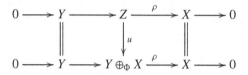

f assumes. For instance, if $U = \{f\colon |f|_0 \le \delta\}$, we may take a partition of $[0,1]$ into $k = \lceil \delta^{-1} \rceil$ subintervals $I_1, \dots, I_k$ of measure at most δ and then write $L_0 = L_0(I_1) \oplus \cdots \oplus L_0(I_k)$. If $\Phi\colon L_0 \longrightarrow Y$ is quasilinear then there is $\delta > 0$ such that $\|\Phi(f+g) - \Phi(f) - \Phi(g)\| \le 1$ for $|f|_0, |g|_0 \le \delta$. Taking $I_1, \dots, I_k$ as before and letting $\Phi_i = \Phi|_{L_0(I_i)}$ for $1 \le i \le k$, we now have $\|\Phi_i(f+g) - \Phi_i(f) - \Phi_i(g)\| \le 1$ for *all* $f, g \in L_0(I_i)$. We shall show that each Φ_i is approximable, from which follows the same for Φ. Let us state and prove this fact separately:

★ Let $\Phi\colon X \longrightarrow Y$ be a quasilinear map, where X is a TVS and Y is a quasi-Banach space. Assume that there is $U \in \mathcal{O}_X$ such that $\|\Phi(x + x') - \Phi(x) - \Phi(x')\| \le \varepsilon$ for some (possibly large) $\varepsilon > 0$ and all $x, x' \in U$. Then Φ is approximable.

Where does the approximating linear map come from? The following argument, due to Hyers, is a celebrity in certain circles: assuming that Y is a p-Banach space, given $x \in X$, we consider the sequence $(\Phi(2^n x)/2^n)_{n \ge 1}$. A straightforward induction argument yields $\|\Phi(2^n x) - 2^n \Phi(x)\|^p \le (2^{pn} - 1)\varepsilon^p$ for all n. Thus, for $n, m \in N$, we have $\|\Phi(2^{n+m} x) - 2^m \Phi(2^n x)\|^p \le 2^{pm} \varepsilon^p$. Dividing by 2^{n+m}, we obtain the estimate $\|\Phi(2^{n+m} x)/2^{n+m} - \Phi(2^n x)/2^n\| \le \varepsilon/2^m$ so that $(\Phi(2^n x)/2^n)_{n \ge 1}$ is a Cauchy sequence and $\|\Phi(2^n x)/2^n - \Phi(x)\| \le \varepsilon$. Put $L(x) = \lim_n \Phi(2^n x)/2^n$. Let us check that L is additive. Pick $x, y \in X$:

$$\|L(x+y) - L(x) - L(y)\| = \lim_n \left\| \frac{\Phi(2^n(x+y))}{2^n} - \frac{\Phi(2^n x)}{2^n} - \frac{\Phi(2^n y)}{2^n} \right\| \le \lim_n \frac{\varepsilon}{2^n} = 0.$$

Wow! To complete the proof, we must see that L is linear (and not merely additive) and that $\Phi - L$ is continuous at zero. The second assertion is contained in:

★★ Let $\Phi\colon X \longrightarrow Y$ be a quasiadditive map, where X is a TVS and Y is a quasi-Banach space. If there is M such that $\|\Phi(x)\| \le M$ for all $x \in X$ then Φ is continuous at zero.

Fix $\varepsilon > 0$ and take $U \in \mathcal{O}_X$ such that $\|\Phi(x + x') - \Phi(x) - \Phi(x')\| < \varepsilon$ for $x, x' \in U$. Choose n such that $M/2^n < \varepsilon$, and let $V \in \mathcal{O}_X$ such that $V + \cdots + V \subset U$ (2^n times). Induction on $k = 1, \dots, n$ yields $\|\Phi(2^k x) - 2^k \Phi(x)\| \le 2^k \varepsilon$ for all $x \in V$. In particular, $\|\Phi(2^n x)/2^n - \Phi(x)\| \le \varepsilon$, and thus $\|\Phi(x)\| \le 2^{1-1/p}(\|\Phi(x) - \Phi(2^n x)/2^n\| + \|\Phi(2^n x)/2^n\|) \le 2^{2-1/p}\varepsilon$ provided $x \in V$. This already implies that L is linear since $\Phi - L$ is continuous at zero, L is quasihomogeneous and, in particular, for every $x \in X$, we have $L(tx) \to 0$ as $t \to 0$ in $\mathbb{R}$. Thus, the additive map $t \in \mathbb{R} \longmapsto L(tx) \in Y$ is continuous, and therefore it is real-linear [161, Lemma 3.1]. □

A still open problem, to which Domański was very attached in the 1980s, is whether Dierolf's theorem extends to the locally convex setting. Stated precisely, if X is a locally convex $\mathscr{K}$-space, must every topological extension of X by a locally convex space be locally convex? This problem is discussed in depth in the papers [160; 161], which contain a number of partial results.

3.13.2 A Cohomological Approach to Quasilinearity

The guiding idea for this chapter has been to use quasilinear maps $\Phi: X \longrightarrow Y$ to twist the topology of the product space $Y \times X$ while retaining the underlying linear structure. There is a classical procedure in group theory that proceeds the other way around. A homogeneous bounded mapping $\phi: X \times X \longrightarrow Y$ is a cocycle if, for every $x, x', x'' \in X$ and $\lambda \in \mathbb{K}$, one has $\phi(\lambda x, x) = 0$, $\phi(x, x') = \phi(x', x)$ and $\phi(x, x') - \phi(x, x' + x'') = \phi(x', x'') - \phi(x + x', x'')$. Keeping the sum quasinorm $\|(y, x)\| = \|y\| + \|x\|$ and the multiplication by scalars on $Y \times X$, we can define a new sum by the formula

$$(y, x) +_\phi (y', x') = (y + y' + \phi(x, x'), x + x').$$

Using the cocycle properties of ϕ, we easily verify that this is a true sum (associative, commutative...) and satisfies the weak triangle estimate $\|(y, x) +_\phi (y', x')\| \le M(\|(y, x)\| + \|(y', x')\|)$. In particular, the resulting quasinormed space $Y \times_\phi X$ is actually quasi-Banach since we have an isometrically exact sequence

$$0 \longrightarrow Y \xrightarrow{\ I\ } Y \times_\phi X \xrightarrow{\ Q\ } X \longrightarrow 0$$

where $I(y) = (y, 0)$ and $Q(y, x) = x$. The only non-trivial point here is to realise that Q is additive with respect to the new sum. Thus, each cocycle induces an extension. All extensions arise in this way: if $0 \longrightarrow Y \longrightarrow Z \longrightarrow X \longrightarrow 0$ is an extension and $B: X \longrightarrow Z$ is a bounded homogeneous section of the quotient map, then $\phi(x, x') = B(x) + B(x') - B(x + x')$ takes values in Y and is a cocycle and the sequence $0 \longrightarrow Y \longrightarrow Y \times_\phi X \longrightarrow X \longrightarrow 0$ is equivalent to the starting extension. Note that a homogeneous mapping $\Phi: X \longrightarrow Y$ is quasilinear if and only if $\phi(x, x') = \Phi(x) + \Phi(x') - \Phi(x + x')$ is a cocycle.

Unlike quasilinear maps, cocycles are very sensitive to the 'quality' of the bounded section $B: X \longrightarrow Z$ that generates them: for instance, it is clear that if B is continuous, uniformly continuous or Lipschitz, then so is ϕ. It follows from classical results of Michael that if Z is an F-space and $Y \subset Z$ is a locally convex, closed subspace, then the natural quotient map $Z \longrightarrow Z/Y$ admits a continuous section which can moreover be taken to be homogeneous when Z is a quasi-Banach space. At no point in this chapter have we taken advantage of this fact. We do not know if every quotient map between quasi-Banach spaces has a continuous section.

3.13.3 Table of Correspondences between Diagrams and Quasilinear Maps

Exact sequence $0 \to Y \to Z \to X \to 0$	Quasilinear map $\Phi\colon X \to Y$
Trivial (equivalent to direct sum sequence)	Trivial (bounded plus linear)
Equivalent sequences $\begin{array}{ccccccccc} 0 & \longrightarrow & Y & \longrightarrow & Z_1 & \longrightarrow & X & \longrightarrow & 0 \\ & & \Vert & & \downarrow & & \Vert & & \\ 0 & \longrightarrow & Y & \longrightarrow & Z_2 & \longrightarrow & X & \longrightarrow & 0 \end{array}$	$\Phi_2 - \Phi_1$ is trivial
Let $S : X' \to X$ and $T : Y \to Y'$ be operators	
Pushout $\begin{array}{ccccccccc} 0 & \longrightarrow & Y & \longrightarrow & Z & \xrightarrow{\rho} & X & \longrightarrow & 0 \\ & & {\scriptstyle T}\downarrow & & \downarrow{\scriptstyle \overline{T}} & & \Vert & & \\ 0 & \longrightarrow & Y' & \longrightarrow & \mathrm{PO} & \xrightarrow{\overline{\rho}} & X & \longrightarrow & 0 \end{array}$	Left composition $T \circ \Phi$ $\begin{array}{ccc} Y & \xleftarrow{\Phi} & X \\ {\scriptstyle T}\downarrow & & \\ Y' & & \end{array}$
Pullback $\begin{array}{ccccccccc} 0 & \longrightarrow & Y & \xrightarrow{\jmath} & Z & \longrightarrow & X & \longrightarrow & 0 \\ & & \Vert & & {\scriptstyle S'}\uparrow & & {\scriptstyle S}\uparrow & & \\ 0 & \longrightarrow & Y & \xrightarrow{\underline{\jmath}} & \mathrm{PB} & \longrightarrow & X' & \longrightarrow & 0 \end{array}$	Right composition $\Phi \circ S$ $\begin{array}{ccc} Y & \xleftarrow{\Phi} & X \\ & & \uparrow{\scriptstyle S} \\ & & X' \end{array}$
Commutativity of pullback and pushout	Associativity of composition $(S \circ \Phi) \circ T = S \circ (\Phi \circ T)$
Baer's sum	Pointwise sum
Diagonal pushout $0 \longrightarrow Y \longrightarrow Y' \oplus Z \longrightarrow \mathrm{PO} \longrightarrow 0$	Composition $\Phi \circ \overline{\rho}$ $Y \leftarrow X \leftarrow \mathrm{PO}$
Diagonal pullback $0 \longrightarrow \mathrm{PB} \longrightarrow X' \oplus Z \longrightarrow X \longrightarrow 0$	Composition $\underline{\jmath} \circ \Phi$ $\mathrm{PB} \leftarrow Y \leftarrow X$
Exact sequences of p-Banach spaces	p-linear maps
Dual exact sequence of Banach spaces	Dual 1-linear map
$\begin{array}{ccccccccc} 0 & \longrightarrow & \cdot & \longrightarrow & \cdot & \longrightarrow & \cdot & \longrightarrow & 0 \\ & & {\scriptstyle \alpha}\downarrow & & \downarrow & & \downarrow{\scriptstyle \gamma} & & \\ 0 & \longrightarrow & \cdot & \longrightarrow & \cdot & \longrightarrow & \cdot & \longrightarrow & 0 \end{array}$	$\alpha\Omega \sim \Phi\gamma$

Sources

Quasilinear techniques burst into Banach space theory in Enflo, Lindenstrauss and Pisier's paper [167], where a quasilinear map is used for the first time to construct a twisted Hilbert space, thus solving a problem they attributed to Palais. Incidentally, and according to Pietsch [387], Palais was unaware that a 3-space problem had been associated with his name. It is clear from [388] that the idea of using a non-linear map to construct an extension of Banach spaces is due to Lindenstrauss.

Enough mathematical gossip. The connection between quasilinearity and the twisted sums in [167] is provided by stipulating that the unit ball of the norm of $Y \times X$ has to be the convex hull of the set $\{(y, 0) \colon y \in B_Y\} \bigcup \{(\Phi(x), x) \colon x \in B_X\}$, which yields the formula

$$|(y, x)|_\Phi = \inf\left\{\sum_i \|y_i\| + \|x_i\| : x = \sum_i x_i, y = \sum_i \Phi(x_i) + \sum_i x_i\right\}.$$

Kalton [251] adapted this construction for p-Banach spaces. Ribe's paper [401], from where the construction in Section 3.2 is taken, was an important advance in the area. In fact, the clean formula $\|y - \Omega x\| + \|x\|$ for the quasinorm appeared there for the first time and was quickly adopted by Kalton and Peck [280] and has been widely used ever since. The Kalton–Peck construction can be exploited with different levels of depth and generality. A complete account of Kalton's findings, first with Peck and subsequently solo, requires much more time and space than these comments to uncover the wonderful connections between centralizers and complex interpolation theory. In this chapter, we dealt with the simplest of those levels. Most of Sections 3.2 and 3.3 are from [280]; Kalton's map [251] that solves the 3-space problem for local convexity also admits a 'centralizer version' on each ℓ_p that provides a non-trivial twisted Hilbert space; but it is only in the theory of centralizers that it finds its home. The first part of Proposition 3.3.5 appeared in [78]. Although Corollary 3.4.4 makes the study of locally convex $\mathscr{K}$-spaces especially rewarding, the notion of a $\mathscr{K}$-space was motivated by non-locally convex considerations. Actually, $\mathscr{K}$-spaces were introduced by Kalton and Peck in [281] to show that L_p is not isomorphic to its quotient by a line when $0 \leq p < 1$. As far as we know, this was the first time that a homological invariant was used to distinguish between two quasi-Banach spaces. In truth, the paper does not contain the full proof that L_p is a $\mathscr{K}$-space when $0 < p < 1$, but instead the fact that every minimal extension $0 \longrightarrow \mathbb{K} \longrightarrow Z \longrightarrow L_p \longrightarrow 0$ in which Z is a p-Banach space splits. Needless to say, the quotient of L_p by a line (or by any subspace Y with non-trivial dual) cannot have that property, as the sequence

$0 \longrightarrow Y \longrightarrow L_p \longrightarrow L_p/Y \longrightarrow 0$ shows. The material of Section 3.6, and in particular the 'uniform boundedness principle for quasilinear maps', is due to Kalton [251] and enhanced with Ribe's formula. Section 3.9 is an adaptation of [75], from where the construction of the spaces $\mathrm{co}^{(p)}(\cdot)$ was taken too. The first two parts of Section 3.4 are basically as in Kalton–Peck [281]. The third part is from [251]. The treatment of Section 3.8 follows [65]; however, Lemma 3.8.2 appeared in [112], and the computations leading to the identification of the duals of the Kalton–Peck spaces are taken from the paper of their legitimate owners. Theorem 3.11.2 is taken from [252], where Kalton performs a rather complete study of the type of twisted sums: he considers both the case where the subspace has better type than the quotient space, which corresponds to Theorem 3.11.2, and also the reverse situation, with similar conclusions (twisted sums retain the type of the 'worst' summand, although its proof is different). He also considers the subtler case in which the summands have the same type. The corollary in Note 3.13.1 is due to Kalton and Peck [281, Theorem 3.6], although the proof we present is taken from [60]. The observations on cocycles are a straightforward adaptation of [436], which is in turn based on a classical construction in group theory that can be found in [53, Chapter IV]. The paper [312] describes the cocycles acting between Banach spaces that produce locally convex extensions.

4

The Functor Ext and the Homology Sequences

Many of the results treated in previous chapters can be lighted from a categorical perspective. Contrary to its notorious reputation, category theory helps in understanding concrete constructions, leads to the right questions and, oftentimes, suggests answers. And this is because, paradoxical though it may seem, the understanding of concrete facts is frequently obscured by the facts themselves. So as not to discourage us (readers *and* authors), categories will be used in an elementary (some might say naive) way, but without sacrificing rigour. Despite this easy-going approach, the results of this chapter deserve to be in the spotlight. They will gradually illuminate, right in front of our eyes, the full power of the following categorical and homological ideas, which have so far remained in the shadows:

- The assignment $(X, Y) \rightsquigarrow \operatorname{Ext}(X, Y)$ is a functor.
- Ext acts on operators by pullback and pushout.
- The spaces $\operatorname{Ext}(X, Y)$ carry a natural linear structure.
- The structure of $\operatorname{Ext}(X, Y)$ is enriched with a natural, locally bounded topology which, more often than not, refuses to be Hausdorff.
- The correspondence between extensions and quasilinear maps studied in Chapter 3 is a natural equivalence of functors.
- The categorical meaning of 'natural' in all previous sentences.
- The way all the pieces fit together in *longer* exact sequences that appear once an exact sequence $0 \longrightarrow Y \longrightarrow Z \longrightarrow X \longrightarrow 0$ of (quasi-) Banach spaces and a (quasi-) Banach space E have been fixed:

$$0 \longrightarrow \mathfrak{L}(E, Y) \longrightarrow \mathfrak{L}(E, Z) \longrightarrow \mathfrak{L}(E, X) \longrightarrow \operatorname{Ext}(E, Y) \longrightarrow \operatorname{Ext}(E, Z) \longrightarrow \operatorname{Ext}(X, E)$$

$$0 \longrightarrow \mathfrak{L}(X, E) \longrightarrow \mathfrak{L}(Z, E) \longrightarrow \mathfrak{L}(Y, E) \longrightarrow \operatorname{Ext}(X, E) \longrightarrow \operatorname{Ext}(Z, E) \longrightarrow \operatorname{Ext}(Y, E)$$

Time to turn off the dark:

4.1 The Functor Ext

Definition 4.1.1 Given two quasi-Banach spaces X, Y, we denote the set of exact sequences $0 \longrightarrow Y \longrightarrow \cdot \longrightarrow X \longrightarrow 0$ modulo equivalence by $\operatorname{Ext}(X, Y)$.

We want to prove now that the assignment $(X, Y) \rightsquigarrow \operatorname{Ext}(X, Y)$ is a functor and understand the meaning of such a statement. After subscribing to every word said at the beginning of Section 2.4, let us pass to what a functor is.

Functors

Definition 4.1.2 A covariant (resp. contravariant) functor $F\colon \mathbf{C} \longrightarrow \mathbf{D}$ between two categories is a correspondence assigning objects to objects and arrows to arrows in such a way that

- if $f\colon A \longrightarrow B$ is an arrow of $\mathbf{C}$ then $F(f)\colon F(A) \longrightarrow F(B)$ (resp., $F(f)\colon F(B) \longrightarrow F(A)$) is an arrow of $\mathbf{D}$;
- if C is an object of $\mathbf{C}$ then $F(\mathbf{1}_C) = \mathbf{1}_{F(C)}$;
- if $f\colon A \longrightarrow B$ and $g\colon B \longrightarrow C$ are arrows of $\mathbf{C}$, then $F(g \circ f) = F(g) \circ F(f)$ (resp., $F(g \circ f) = F(f) \circ F(g)$).

The opposite category trick Assume we have a contravariant functor $F\colon \mathbf{C} \longrightarrow \mathbf{D}$ and, for some reason, it is important that F becomes covariant. Replacing the category $\mathbf{C}$ by its opposite category $\mathbf{C}^{\mathrm{op}}$, namely the category which has the same objects as $\mathbf{C}$ with the arrows going in the opposite direction, it is clear that $F\colon \mathbf{C}^{\mathrm{op}} \longrightarrow \mathbf{D}$ is now covariant! From now on, all functors are assumed to be covariant unless otherwise stated.

Product category If $\mathbf{C}, \mathbf{D}$ are categories, then the product $\mathbf{C} \times \mathbf{D}$ is the category whose objects are ordered pairs (C, D), where C and D are objects of $\mathbf{C}$ and $\mathbf{D}$, respectively, and the arrows are pairs of arrows of $\mathbf{C}$ and $\mathbf{D}$.

Bifunctor A bifunctor (we give it that name mostly for psychological reasons) is a functor acting on a product category. We already encountered would-be bifunctors such as $\mathfrak{L}(\cdot, \cdot)$, which sends (X, Y) to $\mathfrak{L}(X, Y)$ and assigns to an arrow (u, v), namely a pair of operators $u\colon X' \longrightarrow X$ and $v\colon Y \longrightarrow Y'$, the arrow $\mathfrak{L}(u, v)\colon \mathfrak{L}(X, Y) \longrightarrow \mathfrak{L}(X', Y')$ that sends $\tau \in \mathfrak{L}(X, Y)$ to the composition $v\tau u$. If we are being completely honest, this action is contravariant in the first argument and covariant in the second, so it is only a bifunctor in the form $\mathfrak{L}\colon \mathbf{Q}^{\mathrm{op}} \times \mathbf{Q} \longrightarrow \mathbf{Q}$. Its functorial character is due to the equality $\mathfrak{L}(u\,u', v'v) = \mathfrak{L}(u', v') \circ \mathfrak{L}(u, v)$.

Similar considerations apply to $\mathsf{Q}_{\mathsf{LB}}\colon \mathbf{Q}^{\mathrm{op}} \times \mathbf{Q} \longrightarrow \mathbf{sQ}$, if we agree that the action on a pair of operators $u\colon X' \longrightarrow X$ and $v\colon Y \longrightarrow Y'$ is given by $[\Phi] \longmapsto [v \circ \Phi \circ u]$. We suggest that the reader check that $\mathfrak{L}$ and Q_{LB} are bifunctors, mainly to become aware of the crucial role played by the associativity rules $v(\tau u) = (v\tau)u$ and $v \circ (\Phi \circ u) = (v \circ \Phi) \circ u$. In principle, it is not guaranteed that a correspondence $F: \mathbf{C} \times \mathbf{D} \longrightarrow \mathbf{E}$ that is a functor on each variable separately must also be a functor. To amend this, we have:

4.1.3 Bifunctor lemma *$F: \mathbf{C} \times \mathbf{D} \longrightarrow \mathbf{E}$ is a (covariant) bifunctor if and only if, for every object $D \in \mathbf{D}$ and every object $C \in \mathbf{C}$,*

- *the assignment $F(\cdot, D): \mathbf{C} \longrightarrow \mathbf{E}$ is a covariant functor;*
- *the assignment $F(C, \cdot): \mathbf{D} \longrightarrow \mathbf{E}$ is a covariant functor;*
- *if $f: C \longrightarrow C'$ and $g: D \longrightarrow D'$ are arrows in $\mathbf{C}$ and $\mathbf{D}$, respectively, then $F(C', g) \circ F(f, D) = F(f, D') \circ F(C, g)$; that is, the following square is commutative:*

$$\begin{array}{ccc} F(C,D) & \xrightarrow{F(C,g)} & F(C,D') \\ {\scriptstyle F(f,D)}\downarrow & & \downarrow{\scriptstyle F(f,D')} \\ F(C',D) & \xrightarrow{F(C',g)} & F(C',B') \end{array}$$

Proof Simple, after realising that each arrow $(f, g): (C, D) \longrightarrow (C', D')$ in $\mathbf{C} \times \mathbf{D}$ factors as $(f, g) = (\mathbf{1}_{C'}, g) \circ (f, \mathbf{1}_D) = (f, \mathbf{1}_{D'}) \circ (\mathbf{1}_C, g)$. □

We are thus ready to blurt out:

Ext Is a Functor

Let us consider Ext as a (would-be) bifunctor taking values in the category of sets. Even if the *class* of all extensions of X by Y is too large to be a set, every such extension is equivalent to one in which the middle space is the product space $Y \times X$ equipped with a suitable quasinorm, and the quasinorms on a fixed linear space clearly form a set. Thus, Ext is well defined on the objects of $\mathbf{Q} \times \mathbf{Q}$. Next we need to define Ext on arrows of $\mathbf{Q}^{\mathrm{op}} \times \mathbf{Q}$, formed by a pair of operators $u: X' \longrightarrow X$ (the contravariant component) and $v: Y \longrightarrow Y'$ (the covariant component). Let $[\mathrm{z}]$ be the equivalence class of an exact sequence z. We tentatively define $\operatorname{Ext}(u, v)[\mathrm{z}]$ to be the (class of the) lower row in the first-pullback-and-then-pushout diagram

$$
\begin{array}{ccccccccc}
0 & \longrightarrow & Y & \longrightarrow & Z & \longrightarrow & X & \longrightarrow & 0 \qquad [\mathrm{z}] \\
 & & \| & & \uparrow & & \uparrow{\scriptstyle u} & & \\
0 & \longrightarrow & Y & \longrightarrow & \mathrm{PB} & \longrightarrow & X' & \longrightarrow & 0 \\
 & & {\scriptstyle v}\downarrow & & \downarrow & & \| & & \\
0 & \longrightarrow & Y' & \longrightarrow & \mathrm{PO(PB)} & \longrightarrow & X' & \longrightarrow & 0
\end{array}
$$

We have $[\mathrm{z}'] = \mathrm{Ext}(u, v)[\mathrm{z}]$ if and only if there is a commutative diagram

$$
\begin{array}{ccccccccc}
0 & \longrightarrow & Y & \longrightarrow & Z & \longrightarrow & X & \longrightarrow & 0 \qquad [\mathrm{z}] \\
 & & \| & & \uparrow & & \uparrow{\scriptstyle u} & & \\
0 & \longrightarrow & Y & \longrightarrow & \cdot & \longrightarrow & X' & \longrightarrow & 0 \\
 & & {\scriptstyle v}\downarrow & & \downarrow & & \| & & \\
0 & \longrightarrow & Y' & \longrightarrow & Z' & \longrightarrow & X' & \longrightarrow & 0 \qquad [\mathrm{z}']
\end{array}
$$

The second thing we learnt in Section 2.10 was that pushout and pullback commute, so we have

$$
\begin{array}{ccccccccc}
0 & \longrightarrow & Y & \longrightarrow & Z & \longrightarrow & X & \longrightarrow & 0 \qquad [\mathrm{z}] \\
 & & {\scriptstyle v}\downarrow & & \downarrow & & \| & & \\
0 & \longrightarrow & Y' & \longrightarrow & \cdot & \longrightarrow & X & \longrightarrow & 0 \\
 & & \| & & \uparrow & & \uparrow{\scriptstyle u} & & \\
0 & \longrightarrow & Y' & \longrightarrow & Z' & \longrightarrow & X' & \longrightarrow & 0 \qquad [\mathrm{z}']
\end{array}
$$

The plan is to use the bifunctor lemma to check that this definition works and that we have a genuine functor. Fixing Y, the arrow $\mathrm{Ext}(u, Y) = \mathrm{Ext}(u, \mathbf{1}_Y)$ sends $[\mathrm{z}]$ into any lower row in a diagram

$$
\begin{array}{ccccccccc}
0 & \longrightarrow & Y & \longrightarrow & Z & \longrightarrow & X & \longrightarrow & 0 \\
 & & \| & & \uparrow & & \uparrow{\scriptstyle u} & & \\
0 & \longrightarrow & Y & \longrightarrow & K & \longrightarrow & X' & \longrightarrow & 0
\end{array}
$$

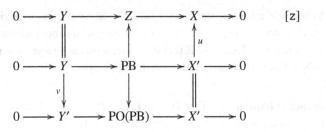

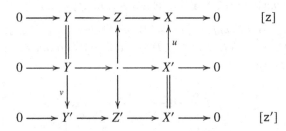

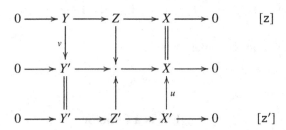

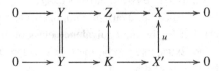

This is clearly a contravariant functor since $\mathrm{Ext}(u\,u', Y) = \mathrm{Ext}(u', Y) \circ \mathrm{Ext}(u, Y)$ for any other operator $u'\colon X'' \longrightarrow X'$, as the diagram

$$\begin{array}{ccccccccc}
0 & \longrightarrow & Y & \longrightarrow & Z & \longrightarrow & X & \longrightarrow & 0 \\
 & & \| & & \uparrow & & \uparrow{\scriptstyle u} & & \\
0 & \longrightarrow & Y & \longrightarrow & \mathrm{PB} & \longrightarrow & X' & \longrightarrow & 0 \\
 & & \| & & \uparrow & & \uparrow{\scriptstyle u'} & & \\
0 & \longrightarrow & Y & \longrightarrow & \mathrm{PB}' & \longrightarrow & X'' & \longrightarrow & 0
\end{array}$$

shows. In a similar vein, fixing X, we have $[\mathrm{k}] = \mathrm{Ext}(X, v)[\mathrm{z}]$ if and only if there is a commutative diagram

$$\begin{array}{cccccccccc}
0 & \longrightarrow & Y & \longrightarrow & Z & \longrightarrow & X & \longrightarrow & 0 & \qquad [\mathrm{z}] \\
 & & {\scriptstyle v}\downarrow & & \downarrow & & \| & & & \\
0 & \longrightarrow & Y' & \longrightarrow & K & \longrightarrow & X & \longrightarrow & 0 & \qquad [\mathrm{k}]
\end{array}$$

It should be clear that $\mathrm{Ext}(X, v'v) = \mathrm{Ext}(X, v') \circ \mathrm{Ext}(X, v)$ for any other operator $v'\colon Y' \longrightarrow Y''$. Hence, $\mathrm{Ext}(X, \cdot)\colon \mathbf{Q} \longrightarrow \mathbf{S}$ is a covariant functor. It only remains to check that $\mathrm{Ext}(\cdot, Y)$ and $\mathrm{Ext}(X, \cdot)$ have the compatibility property $\mathrm{Ext}(X', v) \circ \mathrm{Ext}(u, Y)[\mathrm{z}] = \mathrm{Ext}(u, Y') \circ \mathrm{Ext}(X, v)[\mathrm{z}]$ appearing in the bifunctor lemma. But we already know this by the commutativity of the pullback and pushout operations.

All this can be summarised:

Theorem 4.1.4 $\mathrm{Ext}\colon \mathbf{Q}^{\mathrm{op}} \times \mathbf{Q} \longrightarrow \mathbf{S}$ *is a functor.*

A Natural Equivalence

Natural transformations are the categorical way to make precise the idea that two functors vary in the same way.

Definition 4.1.5 Given two functors F, G from $\mathbf{C}$ to $\mathbf{D}$, a natural transformation $\eta\colon F \longrightarrow G$ is a correspondence that assigns to each object A of $\mathbf{C}$ an arrow $\eta_A\colon F(A) \longrightarrow G(A)$ in $\mathbf{D}$ such that for every arrow $f\colon A \longrightarrow B$, there is a commutative diagram:

$$\begin{array}{ccc}
F(A) & \xrightarrow{\ \eta_A\ } & G(A) \\
{\scriptstyle F(f)}\downarrow & & \downarrow{\scriptstyle G(f)} \\
F(B) & \xrightarrow{\ \eta_B\ } & G(B)
\end{array}$$

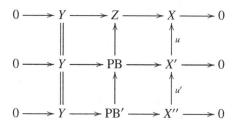

A natural transformation is said to be a natural equivalence if the arrows η_A are isomorphisms in **D**. Two functors are called naturally equivalent when there exists a natural equivalence between them.

Let us test the definition by having a look at the correspondence between quasilinear maps and exact sequences developed in Chapter 3. Proposition 3.3.7 contained no explicit statement '$\mathsf{Q}_{\mathsf{LB}}\colon \mathbf{Q}^{\mathrm{op}} \times \mathbf{Q} \longrightarrow \mathbf{S}$ is a functor', for obvious reasons. But it is also clear that that is the content of Section 3.5: given operators $u\colon X' \longrightarrow X$ and $v\colon Y \longrightarrow Y'$, the map $\mathsf{Q}_{\mathsf{LB}}(u,v)\colon \mathsf{Q}_{\mathsf{LB}}(X,Y) \longrightarrow \mathsf{Q}_{\mathsf{LB}}(X',Y')$ is defined by sending $[\Phi]$ to $[v \circ \Phi \circ u]$. So without further ado, let us consider the functors Ext, $\mathsf{Q}_{\mathsf{LB}}\colon \mathbf{Q}^{\mathrm{op}} \times \mathbf{Q} \longrightarrow \mathbf{S}$.

Proposition 4.1.6 Ext *and* Q_{LB} *are naturally equivalent functors.*

Proof For each pair of quasi-Banach spaces (X, Y), we define $\eta_{(X,Y)}\colon \mathsf{Q}_{\mathsf{LB}} \longrightarrow \operatorname{Ext}(X, Y)$, sending (the class of) a quasilinear map $\Phi \in \mathsf{Q}(X, Y)$ into the (class of the) induced extension $0 \longrightarrow Y \longrightarrow Y \oplus_\Phi X \longrightarrow X \longrightarrow 0$. We know from Proposition 3.3.7 that each $\eta_{(X,Y)}$ is a bijection. To be finally convinced that η is a natural transformation, we just need to check that all diagrams

$$\begin{array}{ccc} \mathsf{Q}_{\mathsf{LB}}(X,Y) & \xrightarrow{\eta_{(X,Y)}} & \operatorname{Ext}(X,Y) \\ {\scriptstyle \mathsf{Q}_{\mathsf{LB}}(u,v)}\downarrow & & \downarrow{\scriptstyle \operatorname{Ext}(u,v)} \\ \mathsf{Q}_{\mathsf{LB}}(X',Y') & \xrightarrow{\eta_{(X',Y')}} & \operatorname{Ext}(X',Y') \end{array}$$

are commutative, and this was proved in Section 3.5. □

Linear Structure on Ext

It is not very surprising that Ext is a functor since it is very difficult to find an honest mathematical construction that is not one. The sets $\operatorname{Ext}(X, Y)$ can be enriched with a linear structure which turns Ext into a functor with values in the category of vector spaces. This still shouldn't raise any eyebrows in view of the natural equivalence between Ext and Q_{LB} and the obvious fact that $\mathsf{Q}_{\mathsf{LB}}(X, Y)$ is a vector space with pointwise operations. So, a linear structure can be transplanted from $\mathsf{Q}_{\mathsf{LB}}(X, Y)$ onto $\operatorname{Ext}(X, Y)$. The fact is, however, that the operations of $\operatorname{Ext}(X, Y)$ can be intrinsically defined through pullbacks and pushouts without any reference to the natural equivalence above. To do that, pick exact sequences of quasi-Banach spaces

$$0 \longrightarrow Y_i \xrightarrow{\jmath_i} Z_i \xrightarrow{\rho_i} X_i \longrightarrow 0 \qquad [\mathsf{z}_i]$$

for $i = 1, 2$ and form their direct product $[\mathsf{z}_1 \times \mathsf{z}_2]$,

$$0 \longrightarrow Y_1 \times Y_2 \xrightarrow{\jmath_1 \times \jmath_2} Z_1 \times Z_2 \xrightarrow{\rho_1 \times \rho_2} X_1 \times X_2 \longrightarrow 0$$

If, in addition, $X_1 = X_2 = X$ and $Y_1 = Y_2 = Y$ then we can consider the sum operator $Y \times Y \longrightarrow Y$ sending (y_1, y_2) to $y_1 + y_2$ and the diagonal embedding $\mathrm{diag}\colon X \longrightarrow X \times X$ sending x to (x, x) and form the pullback/pushout diagram

$$\begin{array}{ccccccccc}
0 & \longrightarrow & Y \times Y & \longrightarrow & Z_1 \times Z_2 & \longrightarrow & X \times X & \longrightarrow & 0 \\
 & & \| & & \uparrow & & \uparrow{\scriptstyle \mathrm{diag}} & & \\
0 & \longrightarrow & Y \times Y & \longrightarrow & \mathrm{PB} & \longrightarrow & X & \longrightarrow & 0 \\
 & & {\scriptstyle \mathrm{sum}}\downarrow & & \downarrow & & \| & & \\
0 & \longrightarrow & Y & \longrightarrow & \mathrm{PO(PB)} & \longrightarrow & X & \longrightarrow & 0
\end{array} \qquad (4.1)$$

It goes without saying that the outcome of the preceding diagram is well defined, up to equivalence. The lower sequence in Diagram (4.1) is the so-called Baer sum of the two sequences, namely

$$[\mathsf{z}_1] + [\mathsf{z}_2] = \mathrm{Ext}(\mathrm{diag}, \mathrm{sum})[\mathsf{z}_1 \times \mathsf{z}_2].$$

The product of $\lambda \in \mathbb{K}$ by $[\mathsf{z}]$ is the (equivalence class of the) lower sequence in the pullback diagram

$$\begin{array}{ccccccccc}
0 & \longrightarrow & Y & \longrightarrow & Z & \longrightarrow & X & \longrightarrow & 0 \\
 & & \| & & \uparrow & & \uparrow{\scriptstyle \lambda} & & \\
0 & \longrightarrow & Y & \longrightarrow & \mathrm{PB} & \longrightarrow & X & \longrightarrow & 0
\end{array} \qquad [\mathsf{z}]$$

that is, $\lambda[\mathsf{z}] = \mathrm{Ext}(\lambda, \mathbf{1}_Y)[\mathsf{z}]$. What else is needed to proclaim that the Baer operations yield a vector space structure on $\mathrm{Ext}(X, Y)$? The (typically eight) axioms of a vector space that have to be checked one by one. This could be (painfully) done just using the universal properties of pullbacks and pushouts. Instead, we switch to quasilinear mode to avoid some irritating difficulties that appear while checking the vector space conditions one by one and dispatch the matter with two easily checkable facts:

- The direct product $[\mathsf{z}_1 \times \mathsf{z}_2]$ of two sequences induced by the quasilinear maps $\Phi_i\colon X_i \longrightarrow Y_i$, $i = 1, 2$, is (equivalent to) the sequence induced by $\Phi_1 \times \Phi_2\colon X_1 \times X_2 \longrightarrow Y_1 \times Y_2$.
- The Baer operations on $\mathrm{Ext}(X, Y)$ correspond to the pointwise operations on quasilinear maps.

Namely, $\Phi_1 + \Phi_2 = \mathrm{sum} \circ (\Phi_1 \times \Phi_2) \circ \mathrm{diag}$ and $\lambda\Phi = \Phi \circ \lambda$. The rest,

$$\eta_{(X,Y)}[c_1\Phi_1 + c_2\Phi_2] = c_1\eta_{(X,Y)}[\Phi_1] + c_2\eta_{(X,Y)}[\Phi_2],$$

is a mere tautology. We can thus upgrade Proposition 4.1.6 to

Proposition 4.1.7 $\mathrm{Ext}, \mathsf{Q}_{\mathsf{LB}}\colon \mathbf{Q}^{\mathrm{op}} \times \mathbf{Q} \longrightarrow \mathbf{V}$ *are naturally equivalent.*

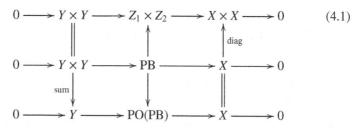

The enrichment concludes with the transfer of the topology from Q_{LB} to Ext setting in these spaces the semi-quasinorm $\|[\mathsf{z}]\|_{\mathrm{Ext}} = \inf\{Q(\Omega)\colon \Omega$ generates a sequence equivalent to $\mathsf{z}\}$. This yields the final version of Propositions 4.1.6 and 4.1.7:

Theorem 4.1.8 Ext, $\mathsf{Q}_{\mathsf{LB}}\colon \mathbf{Q}^{\mathrm{op}} \times \mathbf{Q} \longrightarrow \mathbf{sQ}$ *are naturally equivalent.*

Deeper aspects of the topological vector space structure of $\mathrm{Ext}(X, Y)$ are deferred until Section 4.5. One more result here leaves us on the threshold of homology sequences:

4.1.9 *Let X, X', Y, Y' be quasi-Banach spaces. The map $\mathfrak{L}(X', X)\times\mathrm{Ext}(X, Y)\times \mathfrak{L}(Y, Y') \longrightarrow \mathrm{Ext}(X', Y')$ defined by $(u, [\mathsf{z}], v) \longmapsto \mathrm{Ext}(u, v)[\mathsf{z}]$ is trilinear and bounded.*

Proof If $\Phi\colon X \longrightarrow Y$ is a quasilinear map such that $\eta_{(X,Y)}[\Phi] = [\mathsf{z}]$ then $\mathrm{Ext}(u, v)[\mathsf{z}] = \eta_{(X',Y')}[v \circ \Phi \circ u]$ with $Q(v \circ \Phi \circ u) \leq \|v\|Q(\Phi)\|u\|$. □

Theorem 4.1.8 adapts immediately to $p\mathbf{B}$ by cobbling together the following facts:

- Every short exact sequence of p-Banach spaces is also a short exact sequence of quasi-Banach spaces.
- The notion of equivalence for exact sequences of p-Banach spaces is the same as for exact sequences of quasi-Banach spaces.
- Products, pushouts and pullbacks exist in $p\mathbf{B}$.

Given two p-Banach spaces X, Y, we denote by $\mathrm{Ext}_{p\mathbf{B}}(X, Y)$ the set of classes of exact sequences $0 \longrightarrow Y \longrightarrow \cdot \longrightarrow X \longrightarrow 0$ of p-Banach spaces modulo equivalence. It is clear that $\mathrm{Ext}_{p\mathbf{B}}(X, Y)$ is a subset of $\mathrm{Ext}(X, Y)$ and that by restricting the operations, we obtain a bifunctor $p\mathbf{B}^{\mathrm{op}} \times p\mathbf{B} \longrightarrow \mathbf{V}$ that, thanks to Theorem 3.6.7, turns out to be naturally equivalent to $\mathsf{Q}^{(p)}_{\mathsf{LB}}$. Besides, if we define a semi-p-norm on $\mathrm{Ext}_{p\mathbf{B}}(X, Y)$ by letting $\|[\mathsf{z}]\|_{\mathrm{Ext}_{p\mathbf{B}}} = \inf\{Q^{(p)}(\Omega)\colon \Omega$ generates a sequence equivalent to $\mathsf{z}\}$ then

Theorem 4.1.10 $\mathrm{Ext}_{p\mathbf{B}}, \mathsf{Q}^{(p)}_{\mathsf{LB}}\colon p\mathbf{B}^{\mathrm{op}} \times p\mathbf{B} \longrightarrow \mathbf{s}(p\mathbf{B})$ *are naturally equivalent.*

4.2 The Homology Sequences

Homology, sometimes regarded as the differential calculus of the twentieth century, considers the functor Ext as the measure of the failure of the functor $\mathfrak{L}$

to be exact. A functor is said to be *exact* if it transforms short exact sequences into short exact sequences; while some functors are exact, many others are not. The idea is that what one now 'differentiates' is a functor instead of a function so that in the same way differentiation yields a new function that somehow measures the variation of the former, derivation of a functor yields a new functor that somehow measures its deviation from exactness. The reader is invited to look at Note 4.6.2 to learn more about derivation of functors. In the particular case of $\mathfrak{L}$, neither $\mathfrak{L}(\cdot, E)$ nor $\mathfrak{L}(E, \cdot)$ is exact. The former transforms an exact sequence, say $0 \longrightarrow Y \stackrel{\jmath}{\longrightarrow} Z \stackrel{\rho}{\longrightarrow} X \longrightarrow 0$, into the exact sequence

$$0 \longrightarrow \mathfrak{L}(X, E) \xrightarrow{\rho^{\circ}} \mathfrak{L}(Z, E) \xrightarrow{\jmath^{\circ}} \mathfrak{L}(Y, E) \tag{4.2}$$

which, unless $\jmath^{\circ}$ is surjective, is not short exact. The latter transforms it into

$$0 \longrightarrow \mathfrak{L}(E, Y) \xrightarrow{\jmath_{\circ}} \mathfrak{L}(E, Z) \xrightarrow{\rho_{\circ}} \mathfrak{L}(E, X) \tag{4.3}$$

which is exact but, again, not short exact, unless $\rho_{\circ}$ is surjective. The information the derived functors provide is encoded in the so-called homology sequence of derived functors: an exact sequence involving all the iterated derived functors of a given functor, each one measuring how much the previous one moves away from exactness. The rest of this chapter deals with the construction and use of homology sequences.

Don't panic.

The Standard Approach

Let us make a tactical move: to try to figure out how to extend the sequence (4.3) while maintaining exactness. This means finding a vector space V and a linear map $L\colon \mathfrak{L}(E, X) \longrightarrow V$ whose kernel agrees with the set of elements of $\mathfrak{L}(E, X)$ that lift to Z. The first solution that comes to our minds is that L should send $u \in \mathfrak{L}(E, X)$ to the lower exact sequence in the pullback diagram

$$\begin{array}{ccccccccc}
0 & \longrightarrow & Y & \longrightarrow & Z & \longrightarrow & X & \longrightarrow & 0 \\
 & & \| & & \uparrow & & \uparrow{\scriptstyle u} & & \\
0 & \longrightarrow & Y & \longrightarrow & \mathrm{PB} & \longrightarrow & E & \longrightarrow & 0
\end{array}$$

because $L(u) = 0$ (the lower sequence splits) precisely when u can be lifted to an operator $\mathfrak{L}(E, Z)$. Assertion 4.1.9 even stamps an official name on this map: $\mathrm{Ext}(-, \mathbf{1}_Y)[\mathtt{z}]$. The arachnid sense is now tingling: the map that extends (4.2) keeping exactness should be $\mathrm{Ext}(\mathbf{1}_X, -)[\mathtt{z}]$, which acts via pushouts. If this discussion has been fruitful, the following statement should not sound too amazing. Nor too obvious.

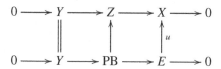

Theorem 4.2.1 *Let E be a quasi-Banach space, and let*

$$0 \longrightarrow Y \xrightarrow{\ \jmath\ } Z \xrightarrow{\ \rho\ } X \longrightarrow 0 \qquad [z]$$

be a short exact sequence of quasi-Banach spaces. Then,

$$\begin{array}{l} 0 \longrightarrow \mathfrak{L}(E,Y) \xrightarrow{\ \jmath\circ\ } \mathfrak{L}(E,Z) \xrightarrow{\ \rho\circ\ } \mathfrak{L}(E,X) \\ \xrightarrow{\operatorname{Ext}(-,\mathbf{1}_E)[z]} \operatorname{Ext}(E,Y) \xrightarrow{\operatorname{Ext}(\jmath,\mathbf{1}_E)} \operatorname{Ext}(E,Z) \xrightarrow{\operatorname{Ext}(\rho,\mathbf{1}_E)} \operatorname{Ext}(E,X) \end{array} \qquad (4.4)$$

is an exact sequence in **sQ**, *and so is*

$$\begin{array}{l} 0 \longrightarrow \mathfrak{L}(X,E) \xrightarrow{\ \rho^\circ\ } \mathfrak{L}(Z,E) \xrightarrow{\ \jmath^\circ\ } \mathfrak{L}(Y,E) \\ \xrightarrow{\operatorname{Ext}(\mathbf{1}_E,-)[z]} \operatorname{Ext}(X,E) \xrightarrow{\operatorname{Ext}(\mathbf{1}_E,\rho)} \operatorname{Ext}(Z,E) \xrightarrow{\operatorname{Ext}(\mathbf{1}_E,\jmath)} \operatorname{Ext}(Y,E) \end{array} \qquad (4.5)$$

Proof Let us proceed with the covariant case (4.4). Exactness at $\mathfrak{L}(E,Y)$ and $\mathfrak{L}(E,Z)$ is obvious, and exactness at $\mathfrak{L}(E,X)$ is just a rewording of the splitting criterion for pullback sequences in Lemma 2.8.3, which was our inspiration a few moments ago: $\operatorname{Ext}(u,\mathbf{1}_E)[z]=0$ if and only if u lifts to Z.

Exactness at Ext(*E*, *Y*). The map $\operatorname{Ext}(\mathbf{1}_E,\jmath)$ acts by forming the pushout with the inclusion $\jmath\colon Y \longrightarrow Z$. To check that the composition $\operatorname{Ext}(\mathbf{1}_E,\jmath)\operatorname{Ext}(-,\mathbf{1}_Y)[z]$ is zero, pick $u \in \mathfrak{L}(E,X)$ and observe that the lower sequence in the diagram

$$\begin{array}{ccccccccc} 0 & \longrightarrow & Y & \longrightarrow & Z & \longrightarrow & X & \longrightarrow & 0 \\ & & \| & & \uparrow{\scriptstyle \underline{u}} & & \uparrow{\scriptstyle u} & & \\ 0 & \longrightarrow & Y & \longrightarrow & \mathrm{PB} & \longrightarrow & E & \longrightarrow & 0 \\ & & {\scriptstyle \jmath}\downarrow & & \downarrow & & \| & & \\ 0 & \longrightarrow & Z & \longrightarrow & \mathrm{PO} & \longrightarrow & E & \longrightarrow & 0 \end{array}$$

splits since $\underline{u}$ is an extension of $\jmath$ to PB. Conversely, if the lower row in a pushout diagram

$$\begin{array}{ccccccccc} 0 & \longrightarrow & Y & \longrightarrow & A & \longrightarrow & E & \longrightarrow & 0 \\ & & {\scriptstyle \jmath}\downarrow & & \downarrow & & \| & & \\ 0 & \longrightarrow & Z & \longrightarrow & \mathrm{PO} & \longrightarrow & E & \longrightarrow & 0 \end{array} \qquad [a]$$

splits, and $J\colon A \longrightarrow Z$ is an extension of $\jmath$ witnessing it, then [a] fits in a commutative diagram

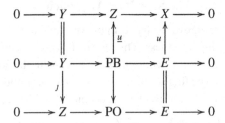

$$\begin{array}{ccccccccc} 0 & \longrightarrow & Y & \longrightarrow & Z & \longrightarrow & X & \longrightarrow & 0 \qquad [\mathrm{z}] \\ & & \| & & \big\uparrow{\scriptstyle \jmath} & & \big\uparrow{\scriptstyle u} & & \\ 0 & \longrightarrow & Y & \longrightarrow & A & \longrightarrow & E & \longrightarrow & 0 \qquad [\mathrm{a}] \end{array}$$

and therefore $[\mathrm{a}] = \operatorname{Ext}(u, \mathbf{1}_Y)[\mathrm{z}]$.

Exactness at Ext(*E*, *Z*). It is clear that $\operatorname{Ext}(\mathbf{1}_E, \rho)\ \operatorname{Ext}(\mathbf{1}_E, \jmath) = \operatorname{Ext}(\mathbf{1}_E, \rho\,\jmath) = 0$ since $\rho\,\jmath = 0$. Conversely, if the lower extension in the pushout diagram

$$\begin{array}{ccccccccc} 0 & \longrightarrow & Z & \longrightarrow & B & \longrightarrow & E & \longrightarrow & 0 \qquad [\mathrm{b}] \\ & & \big\downarrow{\scriptstyle \rho} & & \big\downarrow & & \| & & \\ 0 & \longrightarrow & X & \longrightarrow & \mathrm{PO} & \longrightarrow & E & \longrightarrow & 0 \end{array}$$

splits and $R\colon B \longrightarrow X$ is an extension of ρ then the complete diagram

$$\begin{array}{ccccccccc} & & 0 & & 0 & & & & \\ & & \big\downarrow & & \big\downarrow & & & & \\ 0 & \longrightarrow & Y & \longrightarrow & \ker R & \longrightarrow & E & \longrightarrow & 0 \qquad [\mathrm{k}] \\ & & \big\downarrow{\scriptstyle \jmath} & & \big\downarrow & & \| & & \\ 0 & \longrightarrow & Z & \longrightarrow & B & \longrightarrow & E & \longrightarrow & 0 \qquad [\mathrm{b}] \\ & & \big\downarrow{\scriptstyle \rho} & & \big\downarrow{\scriptstyle R} & & & & \\ & & X & = & X & & & & \\ & & \big\downarrow & & \big\downarrow & & & & \\ & & 0 & & 0 & & & & \end{array}$$

shows that $[\mathrm{b}] = \operatorname{Ext}(\mathbf{1}_E, \jmath)[\mathbf{k}]$. A similar kind of proof can be found in any homology book, so we will not suffer by omitting the contravariant case (which anyway follows by categorical duality). Moreover, a complete quasilinear proof will be presented in Theorem 4.3.2. □

At this point, we only have to formulate the p-Banach version.

Theorem 4.2.2 *Let $0 < p \leq 1$ and fix a p-Banach space E. Let*

$$0 \longrightarrow Y \xrightarrow{\ \jmath\ } Z \xrightarrow{\ \rho\ } X \longrightarrow 0 \qquad [\mathrm{z}]$$

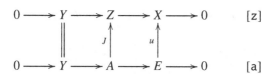

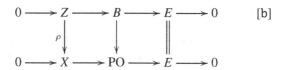

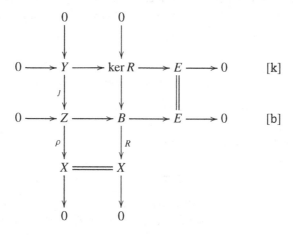

be a short exact sequence of p-Banach spaces. Then,

$$0 \longrightarrow \mathfrak{L}(E,Y) \xrightarrow{\jmath\circ} \mathfrak{L}(E,Z) \xrightarrow{\rho\circ} \mathfrak{L}(E,X)$$

$$\xrightarrow{\mathrm{Ext}(\cdot,\mathbf{1}_Y)[z]} \mathrm{Ext}_{p\mathbf{B}}(E,Y) \xrightarrow{\mathrm{Ext}(\mathbf{1}_E,\jmath)} \mathrm{Ext}_{p\mathbf{B}}(E,Z) \xrightarrow{\mathrm{Ext}(\mathbf{1}_E,\rho)} \mathrm{Ext}_{p\mathbf{B}}(E,X)$$

is an exact sequence in **s**(*p***B**)*, and so is*

$$0 \longrightarrow \mathfrak{L}(X,E) \xrightarrow{\rho^\circ} \mathfrak{L}(Z,E) \xrightarrow{\jmath^\circ} \mathfrak{L}(Y,E)$$

$$\xrightarrow{\mathrm{Ext}(\mathbf{1}_X,\cdot)[z]} \mathrm{Ext}_{p\mathbf{B}}(X,E) \xrightarrow{\mathrm{Ext}(\rho,\mathbf{1}_E)} \mathrm{Ext}_{p\mathbf{B}}(Z,E) \xrightarrow{\mathrm{Ext}(\jmath,\mathbf{1}_E)} \mathrm{Ext}_{p\mathbf{B}}(Y,E)$$

When $p = 1$, we simply write $\mathrm{Ext}_{\mathbf{B}}$. This seems to be a right place to mention that higher-order derived functors $\mathrm{Ext}^n_{p\mathbf{B}}$ exist and that all of them can be strung together in a long exact sequence. The study of the nature of those spaces and their applications to (quasi-) Banach space theory goes far beyond the scope of this book if only because, in this area, there are only a few known knowns but many known unknowns and unknown unknowns. Note 4.6.3 contains a few unknown knowns.

Uses of the Homology Sequences

The homology sequences convey a considerable amount of information in a quite compact form. They can be used in many different ways; perhaps the simplest is to realise that when one has an exact sequence

$$A \xrightarrow{f} B \xrightarrow{g} C$$

in which $f = 0$ and $g = 0$ (in particular, if $A, C = 0$), then $B = 0$. If one merely knows that $g = 0$ (in particular, when $C = 0$), then f is onto. Let us apply these simple ideas to the contravariant sequence. Given an extension $0 \longrightarrow Y \longrightarrow Z \longrightarrow X \longrightarrow 0$ and a target space E, we obtain a six-term exact sequence

$$0 \longrightarrow \mathfrak{L}(X,E) \longrightarrow \mathfrak{L}(Z,E) \longrightarrow \mathfrak{L}(Y,E) \longrightarrow \mathrm{Ext}(X,E) \longrightarrow \mathrm{Ext}(Z,E) \longrightarrow \mathrm{Ext}(Y,E),$$

which immediately yields:

4.2.3 *(Vanishing of spaces):*

(1) $\mathfrak{L}(X,E) = 0$ *and* $\mathfrak{L}(Y,E) = 0$ *imply* $\mathfrak{L}(Z,E) = 0$.
(2) $\mathfrak{L}(Z,E) = 0$ *and* $\mathrm{Ext}(X,E) = 0$ *imply* $\mathfrak{L}(Y,E) = 0$.

(3) $\mathfrak{L}(Y, E) = 0$ *and* $\mathrm{Ext}(Z, E) = 0$ *imply* $\mathrm{Ext}(X, E) = 0$.
(4) $\mathrm{Ext}(X, E) = 0$ *and* $\mathrm{Ext}(Y, E) = 0$ *imply* $\mathrm{Ext}(Z, E) = 0$.

That results of type (1) hold is probably too simple and well known to deserve to be called an application of homology. The following complements to Proposition 3.4.8 are of type (2) and type (3), respectively:

- Quotients of $\mathscr{K}$-spaces by subspaces with trivial dual are $\mathscr{K}$-spaces.
- If Y is a subspace of a quasi-Banach space X with trivial dual and X/Y is a $\mathscr{K}$-space then Y has trivial dual.

All in all, the most interesting results are those of type (4), among which we will soon encounter 3-space results for properties of a space X having the form $\mathrm{Ext}(X, \diamond) = 0$ for $\diamond$ belonging to a certain class of spaces. For instance,

- X is a $\mathscr{K}$-space if $\mathrm{Ext}(X, \mathbb{K}) = 0$;
- a Banach space X is $\aleph$-projective if $\mathrm{Ext}_\mathbf{B}(X, E) = 0$ for every Banach space E with $\dim E < \aleph$ – it is unknown whether $\aleph_1$-projective implies projective;
- a Banach space X has the Kalton–Pełczyński property (KPP, from Kalton and Pełczyński [284]) if $\mathrm{Ext}_\mathbf{B}(X, \ell_2) = 0$.

We have:

Proposition 4.2.4

(a) *If E is a quasi-Banach space, $\mathrm{Ext}(\cdot, E) = 0$ is a 3-space property on* **Q**. *If E is a p-Banach space then $\mathrm{Ext}_{p\mathbf{B}}(\cdot, E) = 0$ is a 3-space property on p**B**.*
(b) *Being a $\mathscr{K}$-space is a 3-space property.*
(c) *Being $\aleph$-projective is a 3-space property.*
(d) *The KPP is a 3-space property.*

The simplest examples of Banach spaces with the KPP are the $\mathscr{L}_1$-spaces, by Lindenstrauss' lifting, but there are more:

Lemma 4.2.5 *Any quotient of two $\mathscr{L}_1$-spaces has the KPP.*

Proof Consider an exact sequence $0 \longrightarrow Y \overset{\jmath}{\longrightarrow} Z \longrightarrow X \longrightarrow 0$. Fix the second variable as ℓ_2 so that the homology sequence becomes

$$0 \longrightarrow \mathfrak{L}(X, \ell_2) \longrightarrow \mathfrak{L}(Z, \ell_2) \longrightarrow \mathfrak{L}(Y, \ell_2) \overset{\jmath^\circ}{\longrightarrow} \mathrm{Ext}_\mathbf{B}(X, \ell_2) \longrightarrow 0$$

because X is an $\mathscr{L}_1$-space and thus $\mathrm{Ext}_\mathbf{B}(Z, \ell_2) = 0$, by Lindenstrauss' lifting. The surjectivity of $\jmath^\circ$ means that every exact sequence of $\mathrm{Ext}_\mathbf{B}(X, \ell_2)$ can be placed in a pushout diagram

$$\begin{array}{ccccccccc}
0 & \longrightarrow & Y & \longrightarrow & Z & \longrightarrow & X & \longrightarrow & 0 \\
 & & \downarrow & & \downarrow & & \| & & \\
0 & \longrightarrow & \ell_2 & \longrightarrow & \cdot & \longrightarrow & X & \longrightarrow & 0
\end{array}$$

Now, all operators $Y \longrightarrow \ell_2$ are 2-summing when Y is an $\mathscr{L}_1$-space, by Grothendieck's theorem, and 2-summing operators extend anywhere (in particular, to Z). Thus the lower sequence splits. □

Bourgain's sequence (2.10) provides the only known example of a non-$\mathscr{L}_1$-space $\mathcal{B}^*$ with the KPP. Another application of Proposition 4.2.4 is that $\operatorname{Ext}(X, E) = 0$ for every twisted sum $X = L_p \oplus_\Omega L_p$ when E is a q-Banach space $0 < p < q \leq 1$, since that is what occurs with L_p, by Theorem 3.7.8; of course, one also has $\mathfrak{L}(X, E) = 0$, by 1.1.5. A kind of amusing exercise is to derive Proposition 3.4.7 from the homology sequence using that the vanishing of an arrow in an exact sequence forces the preceding arrow to be onto. More sophisticated ways of extracting information from the homology sequence appear after observing that if

$$A \xrightarrow{\ 0\ } B \xrightarrow{\ g\ } C \xrightarrow{\ 0\ } D$$

is exact (in particular, if $A, D = 0$), then g has to be a linear bijection. Looking through that lens at the sequence

$$0 \longrightarrow \mathfrak{L}(X, E) \longrightarrow \mathfrak{L}(Z, E) \longrightarrow \mathfrak{L}(Y, E) \longrightarrow \operatorname{Ext}(X, E) \longrightarrow \operatorname{Ext}(Z, E) \longrightarrow \operatorname{Ext}(Y, E)$$

generated from $0 \longrightarrow Y \longrightarrow Z \longrightarrow X \longrightarrow 0$, we obtain three types of results:

4.2.6 *(Equality of spaces)*

(5) $\mathfrak{L}(X, E) = 0$ *and* $\operatorname{Ext}(X, E) = 0$ *imply* $\mathfrak{L}(Z, E) = \mathfrak{L}(Y, E)$.
(6) $\mathfrak{L}(Z, E) = 0$ *and* $\operatorname{Ext}(Z, E) = 0$ *imply* $\mathfrak{L}(Y, E) = \operatorname{Ext}(X, E)$.
(7) $\mathfrak{L}(Y, E) = 0$ *and* $\operatorname{Ext}(Y, E) = 0$ *imply* $\operatorname{Ext}(X, E) = \operatorname{Ext}(Z, E)$.

For $E = \mathbb{K}$, the hypotheses of results of type (5) perfectly match those of a $\mathscr{K}$-space with trivial dual: *if Z/Y is a $\mathscr{K}$-space with trivial dual then $Z^* = Y^*$.* Results of type (6) are more captivating:

Proposition 4.2.7 *If Z is a $\mathscr{K}$-space with trivial dual and Y is a subspace of Z then* $\operatorname{Ext}(Z/Y, \mathbb{K}) = Y^*$.

Upon hearing '$\mathscr{K}$-space with trivial dual', we naturally think of L_p with $0 < p < 1$. Specialising Proposition 4.2.7 in this way can reprove that L_p/A and L_p/A' cannot be isomorphic if $A, A' \subset L_p$ have finite, different, dimensions, thus forging the second link in a chain that started with Theorem 3.7.9:

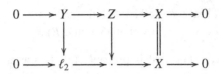

Corollary 4.2.8 *If Y is a finite-dimensional subspace of L_p for $0 < p < 1$, then* $\mathrm{Ext}(L_p/Y, \mathbb{K})$ *has the same dimension as* Y.

In addition, this provides our first examples of quasi-Banach spaces A, B for which $\mathrm{Ext}(A, B)$ is finite-dimensional. It is an open question whether something similar can occur for Banach spaces (see Note 4.6.4). The forthcoming links in that chain, Corollaries 4.5.2 and 4.5.3 and Proposition 7.2.16, still hold many wonders in stock. Moving back to results of type (6); it is easy to believe that the exact sequence obtained in (6), namely

$$0 \longrightarrow \mathfrak{L}(X, F) \longrightarrow 0 \longrightarrow \mathfrak{L}(Y, F) \longrightarrow \mathrm{Ext}(X, F) \longrightarrow 0$$

says that the connecting map is a topological isomorphism; after all, it is a linear continuous injective surjection! However that is not so, because it is not granted that $\mathrm{Ext}(X, F)$ is Hausdorff. Section 4.5 is entirely devoted to this topic.

Finally, we list three types of results for operators in an exact sequence:

$$\cdot \xrightarrow{a_{n-2}} \cdot \xrightarrow{a_{n-1}} \cdot \xrightarrow{a_n} \cdot \xrightarrow{a_{n+1}} \cdot \xrightarrow{a_{n+2}}$$

4.2.9 *(Results for operators):*

(8) $a_n = 0$ *and* $a_{n+2} = 0 \Longrightarrow a_{n+1}$ *is an isomorphism.*
(9) a_n *surjective* $\Longrightarrow a_{n+1} = 0 \Longrightarrow a_{n+2}$ *injective.*
(10) a_n *injective* $\Longrightarrow a_{n-1} = 0 \Longrightarrow a_{n-2}$ *injective.*

The covariant sequence (4.4) is the mirror image of the contravariant sequence (4.5) by categorical duality. It is a fact that contravariant functors tend to be more interesting than covariant ones: observe that while functors $\mathfrak{L}(\cdot, E)$ yield deep insights even for very simple choices of E (look what happens when $E = \mathbb{K}$), functors $\mathfrak{L}(E, \cdot)$ rarely go beyond hovering over the surface of Banach space theory ($\mathfrak{L}(\mathbb{K}, \cdot)$ ranks second to zero among boring functors). Nevertheless, deep applications for the covariant sequence (4.4) are available. To start with, the (meaningful) covariant version of Proposition 4.2.4 deserves to be mentioned: a Banach space X is $\aleph$-injective if $\mathrm{Ext}_{\mathbf{B}}(E, X) = 0$ for every Banach space E with $\dim E < \aleph$. We have:

Proposition 4.2.10

(a) $\mathrm{Ext}(E, \cdot) = 0$ *is a 3-space property on* $\mathbf{Q}$. *If E is a p-Banach space then* $\mathrm{Ext}_{p\mathbf{B}}(E, \cdot) = 0$ *is a 3-space property on* $p\mathbf{B}$.
(b) $\aleph$*-injectivity is a 3-space property.*

There are more results of this type that are far from obvious: Since $\mathrm{Ext}_{\mathbf{B}}(L_1, c_0) = 0$ (by Sobczyk's theorem) and $\mathrm{Ext}_{\mathbf{B}}(L_1, \ell_p(I)) = 0$ (by Lindenstrauss' lifting), $\mathrm{Ext}_{\mathbf{B}}(L_1, \mathsf{JL}_p) = 0$ for the Johnson–Lindenstrauss spaces.

Or else, from Proposition 2.12.3 and Lindenstrauss' lifting, it follows that $\mathrm{Ext}_{\mathbf{B}}(L_1, X) = 0$ for every Banach space X such that X^{**}/X is an ultrasummand with RNP.

So, being the mirror image of the contravariant sequence, the covariant homology sequence (4.4) can be used to prove the same elementary results about vanishing 4.2.3 and equality 4.2.6 of spaces and about operators 4.2.9. These are all now summarised as:

4.2.11

(1) $\mathfrak{L}(E,X) = 0$ *and* $\mathfrak{L}(E,Y) = 0$ *imply* $\mathfrak{L}(E,Z) = 0$.
(2) $\mathfrak{L}(E,Z) = 0$ *and* $\mathrm{Ext}(E,X) = 0$ *imply* $\mathfrak{L}(E,Y) = 0$.
(3) $\mathfrak{L}(E,Y) = 0$ *and* $\mathrm{Ext}(E,Z) = 0$ *imply* $\mathrm{Ext}(E,X) = 0$.
(4) $\mathrm{Ext}(E,X) = 0$ *and* $\mathrm{Ext}(E,Y) = 0$ *imply* $\mathrm{Ext}(E,Z) = 0$.
(5) $\mathfrak{L}(E,X) = 0$ *and* $\mathrm{Ext}(E,X) = 0$ *imply* $\mathfrak{L}(E,Z) = \mathfrak{L}(E,Y)$.
(6) $\mathfrak{L}(E,Z) = 0$ *and* $\mathrm{Ext}(E,Z) = 0$ *imply* $\mathfrak{L}(E,Y) = \mathrm{Ext}(E,X)$.
(7) $\mathfrak{L}(E,Y) = 0$ *and* $\mathrm{Ext}(E,Y) = 0$ *imply* $\mathrm{Ext}(E,X) = \mathrm{Ext}(E,Z)$.
(8) $a_n = 0$, *and* $a_{n+2} = 0 \Longrightarrow a_{n+1}$ *is an isomorphism.*
(9) a_n *surjective* $\Longrightarrow a_{n+1} = 0 \Longrightarrow a_{n+2}$ *injective.*
(10) a_n *injective* $\Longrightarrow a_{n-1} = 0 \Longrightarrow a_{n-2}$ *injective.*

4.3 Homology in Quasilinear Terms

Since Theorem 4.1.8 established that the functors Ext, $\mathsf{Q}_{\mathsf{LB}}\colon \mathbf{Q}^{\mathrm{op}} \times \mathbf{Q} \longrightarrow \mathbf{V}$ are naturally equivalent, it is now clear that if a homology sequence exists for one of them then it also exists for the other. This does not mean that the sequence and the connecting morphism adopt equally simple forms, and actually, the quasilinear construction will be revealed to be more natural, particularly when topologies are introduced. Theorem 4.2.1 reformulates in quasilinear terms as:

Theorem 4.3.1 *Let* $0 \longrightarrow Y \xrightarrow{\jmath} Z \xrightarrow{\rho} X \longrightarrow 0$ *be an exact sequence of quasi-Banach spaces with associated quasilinear map* $\Omega\colon X \longrightarrow Y$, *and let* E *be a quasi-Banach space. The following sequences are exact in* $\mathbf{sQ}$:

$$0 \longrightarrow \mathfrak{L}(E,Y) \xrightarrow{\jmath_\circ} \mathfrak{L}(E,Z) \xrightarrow{\rho_\circ} \mathfrak{L}(E,X) \tag{4.6}$$

$$\xrightarrow{\Omega_\circ} \mathsf{Q}_{\mathsf{LB}}(E,Y) \xrightarrow{\jmath_\circ} \mathsf{Q}_{\mathsf{LB}}(E,Z) \xrightarrow{\rho_\circ} \mathsf{Q}_{\mathsf{LB}}(E,X)$$

and

$$0 \longrightarrow \mathfrak{L}(X,E) \xrightarrow{\rho^\circ} \mathfrak{L}(Z,E) \xrightarrow{\jmath^\circ} \mathfrak{L}(Y,E) \tag{4.7}$$

$$\xrightarrow{\Omega^\circ} \mathsf{Q}_{\mathsf{LB}}(X,E) \xrightarrow{\rho^\circ} \mathsf{Q}_{\mathsf{LB}}(Z,E) \xrightarrow{\jmath^\circ} \mathsf{Q}_{\mathsf{LB}}(Y,E)$$

Proof As promised, the arrows appearing in these sequences are much simpler than before: just plain left (resp. right) composition with the corresponding map. Thus, the maps appearing in (4.6) are

$$\jmath_\circ(u) = \jmath u,\quad \rho_\circ(u) = \rho u,\quad \Omega_\circ(u) = [\Omega \circ u],\quad \jmath_\circ[\Phi] = [\jmath \circ \Phi],\quad \rho_\circ[\Phi] = [\rho \circ \Phi],$$

while those in (4.7) are

$$\rho^\circ(u) = u\rho,\quad \jmath^\circ(u) = u\jmath,\quad \Omega^\circ(u) = [u \circ \Omega],\quad \rho^\circ[\Phi] = [\Phi \circ \rho],\quad \jmath^\circ[\Phi] = [\Phi \circ \jmath].$$

It is pretty clear that all these arrows are well-defined linear maps and that they are all bounded, with $\|\jmath_\circ\|, \|\jmath^\circ\| \leq \|\jmath\|$, $\|\rho_\circ\|, \|\rho^\circ\| \leq \|\rho\|$ and $\|\Omega_\circ\|, \|\Omega^\circ\| \leq Q(\Omega)$. Thus, in no time, we have proved almost everything but the exactness. To do that, take a deep breath and realise that we can assume without loss of generality that the starting sequence is $0 \longrightarrow Y \longrightarrow Y \oplus_\Omega X \longrightarrow X \longrightarrow 0$ so that $Z = Y \oplus_\Omega X$, $\jmath(y) = (y, 0)$ and $\rho(y, x) = x$, which simplifies things considerably. For the rest of the proof, bear in mind also that every operator $Y \oplus_\Omega Z \longrightarrow F$ has the form $(u \oplus L)(y, x) = u(y) + L(x)$ for some linear maps $u\colon Y \longrightarrow E$ and $L\colon X \longrightarrow F$ and that every operator $E \longrightarrow Y \oplus_\Omega X$ has the form $(u, L)(e) = (u(e), L(e))$ for some linear maps $E \longrightarrow Y$ and $E \longrightarrow X$. We proceed first with the contravariant sequence (4.7).

Exactness at $\mathfrak{L}(X,E)$ and $\mathfrak{L}(Z,E)$ is as obvious now as it has always been.

Exactness at $\mathfrak{L}(Y,E)$ means that an operator $u \in \mathfrak{L}(Y,E)$ extends to a bounded operator $u \oplus L\colon Y \oplus_\Omega X \longrightarrow E$ if and only if $u \circ \Omega \sim 0$. But $u \oplus L$ is bounded if and only if $\|u(y) + L(x)\| \leq C(\|y - \Omega(x)\| + \|x\|)$ for some $C \geq 0$ and all y, x. Picking $y = \Omega(x)$, we get $\|u \circ \Omega + L\| \leq C$.

Exactness at $\mathsf{Q}_{\mathsf{LB}}(X,E)$ means that if $\Psi\colon X \longrightarrow E$ is a quasilinear map then $\Psi \circ \rho$ is trivial if and only if there is $u \in \mathfrak{L}(Y,F)$ such that $u \circ \Omega \sim \Psi$. But $\Psi \circ \rho \sim 0$ means that $\|\Psi \circ \rho - u \oplus L\| \leq C$ for some C and some choice of linear maps u and L, namely $\|\Psi(x) - u(y) - L(x)\| \leq C(\|y - \Omega x\| + \|x\|)$, which yields that u is bounded (set $x = 0$) and $\Psi \sim u \circ \Omega$ (set $y = \Omega x$).

Exactness at $\mathsf{Q}_{\mathsf{LB}}(Z,E)$ means that if $\Lambda\colon Y \oplus_\Omega X \longrightarrow F$ is quasilinear then $\Lambda \circ \jmath$ is trivial if and only if there exists a quasilinear map $\Psi\colon X \longrightarrow F$ such that $\Lambda \sim \Psi \circ \rho$. The 'if' part is obviously obvious since $\rho\,\jmath = 0$. To prove the converse, there is no loss of generality in assuming that $\Lambda \circ \jmath = 0$ since both bounded and linear maps can be extended anywhere. By the quasilinearity of Λ, we have

$$\|\Lambda(y,x)-\Lambda(y-\Omega(x),0)-\Lambda(\Omega(x),x)\| \leq Q(\Lambda)(\|y-\Omega x\|+\|x\|) = Q(\Lambda)\|(y,x)\|_\Omega,$$

and, since $\Lambda(y-\Omega(x),0)=0$, if we define $\Psi(x)=\Lambda(\Omega(x),x)$ then $\|\Lambda-\Psi\circ\rho\| \leq Q(\Lambda)$. We just need to verify that $\Psi\colon X \longrightarrow E$ is quasilinear. We have

$$\begin{aligned}
&\|\Psi(x+x')-\Psi(x)-\Psi(x')\| \\
&\quad = \|\Lambda(\Omega(x+x'),x+x')-\Lambda(\Omega x,x)-\Lambda(\Omega x',x')\| \\
&\quad \leq \Delta_F\big(\|\Lambda(\Omega(x+x'),x+x')-\Lambda(\Omega x+\Omega x',x+x')\| \\
&\quad + \|\Lambda(\Omega x+\Omega x',x+x')-\Lambda(\Omega(x),x)-\Lambda(\Omega(x'),x')\|\big).
\end{aligned}$$

The last summand in the right-hand side is bounded by $Q(\Lambda)(\|x\|+\|x'\|)$, by the quasilinearity of Λ. To estimate the first summand, we rewrite it as

$$\begin{aligned}
&\|\Lambda(\Omega(x+x'),x+x')-\Lambda(\Omega x+\Omega x'-\Omega(x+x'),0)-\Lambda(\Omega x+\Omega x',x+x')\| \\
&\leq Q(\Lambda)\big(\|(\Omega(x+x'),x+x')\|_\Omega + Q(\Omega)(\|x\|+\|x'\|)\big) \\
&\leq (Q(\Lambda)Q(\Omega)\Delta_X + Q(\Omega))(\|x\|+\|x'\|).
\end{aligned}$$

And now, for the covariant sequence (4.6):

Exactness at $\mathfrak{L}(E,Y)$ and $\mathfrak{L}(E,Z)$ is clear.

Exactness at $\mathfrak{L}(E,X)$. The pullback splitting criterion yields that an operator $\tau\in\mathfrak{L}(E,X)$ admits a lifting $(L,\tau)\colon E\longrightarrow Y\oplus_\Omega X$ if and only if $\Omega\circ\tau$ is trivial. Since $\|\tau\|\leq\|(L,\tau)\|\leq\|L-\Omega\circ\tau\|+\|\tau\|$ then (L,τ) can be chosen to be continuous if and only if $\Omega\circ\tau$ is trivial.

Exactness at $\mathbf{Q_{LB}}(E,Y)$ means that if $\Psi\colon E\longrightarrow Y$ is a quasilinear map then $\jmath\circ\Psi\sim 0$ if and only if there is $\tau\in\mathfrak{L}(E,X)$ such that $\Psi\sim\tau\circ\Omega$, i.e., $\|\jmath\circ\Psi-(\sigma,\tau)\|\leq C$ for some $C>0$ and some linear maps $\sigma\colon E\longrightarrow Y$ and $\tau\colon E\longrightarrow X$, which happens if and only if $\|\Psi(e)-\sigma(e)+\Omega\tau(e)\|+\|\tau(e)\|\leq C\|e\|$ for every $e\in E$. This implies that both τ and $\Psi-\Omega\circ\tau-\sigma$ are bounded and thus that $\Psi\sim\Omega\circ\tau$.

Exactness at $\mathbf{Q_{LB}}(E,Z)$ means that if $\Lambda\colon E\longrightarrow Y\oplus_\Omega Z$ is a quasilinear map then $\rho\circ\Lambda$ is trivial if and only if $\Lambda\sim\jmath\circ\Psi$ for some quasilinear $\Psi: E\longrightarrow Y$. The 'if' part is obvious. For the converse, write $\Lambda=(\Lambda_1,\Lambda_2)$ and recall that the hypothesis means that $\Lambda_2=B+L$, with $B\colon E\longrightarrow X$ bounded and $L\colon E\longrightarrow X$ linear. Hence, $\Lambda=(\Lambda_1,\Lambda_2)=(\Lambda_1,B+L)\sim(\Lambda_1,B)\sim(\Lambda_1,B)-(\Omega\circ B,B)$, and thus $\Psi=\Lambda_1-\Omega\circ B$ is the quasilinear map we were looking for. □

The p-Banach space counterpart is a carbon copy of Theorem 4.3.1.

Theorem 4.3.2 *Let* $0 \longrightarrow Y \xrightarrow{\jmath} Z \xrightarrow{\rho} X \longrightarrow 0$ *be an exact sequence of p-Banach spaces with associated p-linear map* $\Omega\colon X \longrightarrow Y$, *and let E be a quasi-Banach space. Then, the following sequences of semi-p-Banach spaces are exact:*

$$0 \longrightarrow \mathfrak{L}(E,Y) \xrightarrow{\jmath_\circ} \mathfrak{L}(E,Z) \xrightarrow{\rho_\circ} \mathfrak{L}(E,X) \tag{4.8}$$

$$\xrightarrow{\Omega_\circ} \mathsf{Q}^{(p)}_{\mathsf{LB}}(E,Y) \xrightarrow{\jmath_\circ} \mathsf{Q}^{(p)}_{\mathsf{LB}}(E,Z) \xrightarrow{\rho_\circ} \mathsf{Q}^{(p)}_{\mathsf{LB}}(E,X)$$

and

$$0 \longrightarrow \mathfrak{L}(X,E) \xrightarrow{\rho^\circ} \mathfrak{L}(Z,E) \xrightarrow{\jmath^\circ} \mathfrak{L}(Y,E) \tag{4.9}$$

$$\xrightarrow{\Omega^\circ} \mathsf{Q}^{(p)}_{\mathsf{LB}}(X,E) \xrightarrow{\rho^\circ} \mathsf{Q}^{(p)}_{\mathsf{LB}}(Z,E) \xrightarrow{\jmath^\circ} \mathsf{Q}^{(p)}_{\mathsf{LB}}(Y,E)$$

The next result comes entirely out of the blue; it will be important later.

Lemma 4.3.3 *Let* $\Omega\colon X \longrightarrow Y$ *be a quasilinear map acting between quasi-Banach spaces. Assume that X has a separating dual. If* $\kappa\colon X \longrightarrow X$ *is a compact operator such that* $\Omega \sim \Omega \circ \kappa$, *then* Ω *is trivial.*

Proof If $\Omega\circ\kappa \sim \Omega$ then $\Omega - \Omega\circ\kappa \sim \Omega\circ(\mathbf{1}_X - \kappa) \sim 0$. If 1 is not an eigenvalue of κ then $\mathbf{1}_X - \kappa$ is an automorphism of X, so $\Omega\circ(\mathbf{1}_X - \kappa) \sim 0$ implies $\Omega \sim 0$. If 1 is an eigenvalue of κ then $K_1 = \ker(\mathbf{1}_X - \kappa)$ is finite-dimensional. Since $\Omega|_{K_1} \sim 0$, if we use the notation $X_1 = X/K_1$, there is a quasilinear map $\Omega_1 : X_1 \longrightarrow Y$ such that $\Omega \sim \Omega_1 \circ \rho_1$. Moreover, since X has separating dual the exact sequence

$$0 \longrightarrow K_1 \longrightarrow X \xrightarrow{\rho_1} X_1 \longrightarrow 0$$

splits, and thus one can pick a linear continuous section $s_1 : X_1 \longrightarrow X$ for ρ_1. Now, $\kappa_1 = \rho_1 k s_1$ is a compact operator $X/X_1 \longrightarrow X/X_1$, and also

$$\Omega_1 \circ \rho_1 k s_1 \sim \Omega \circ k s_1 \sim \Omega \circ s_1 \sim \Omega_1 \circ \rho_1 s_1 \sim \Omega_1.$$

In other words, we are back at square one. Repeating the process with κ_1 and Ω_1 in X_1, we encounter the finite-dimensional eigenspace $K_2 = \ker(\mathbf{1}_{X_1} - \kappa_1)$, the quotient $\rho_2\colon X_1 \longrightarrow X_2 = X_1/K_2$, a quasilinear map $\Omega_2\colon X_2 \longrightarrow Y$ such that $\Omega_2 \circ \rho_2 \sim \Omega_1$ and a linear continuous section $s_2\colon X_2 \longrightarrow X_1$ for ρ_2 such that the compact operator $\kappa_2 = \rho_2 \kappa_1 s_2$ satisfies $\Omega_2 \circ \kappa_2 \sim \Omega_2$. And, continuing this way, $K_{n+1}, X_{n+1} = X_n/K_{n+1}, \rho_{n+1}, \Omega_{n+1}, s_{n+1}, \kappa_{n+1} = \rho_{n+1}\kappa_n s_{n+1}$ such that $\Omega_{n+1} \circ \rho_{n+1} \sim \Omega_n$ and $\Omega_{n+1} \circ \kappa_{n+1} \sim \Omega_{n+1}$. So we are always back at square one. Back at square one? Not at all, because it is easy to see (though perhaps not easy to notice) that

- $X_n = X/\ker(\rho_n \cdots \rho_1)$,
- $\ker(\rho_n \cdots \rho_1) = \ker(\mathbf{1}_X - \kappa)^n$.

The finite ascent property of compact operators [305, 8.4-3], which also works in quasi-Banach spaces since it only depends on the Riesz lemma, implies the existence of $N \in \mathbb{N}$ such that $\ker(\mathbf{1}_X - \kappa)^n = \ker(\mathbf{1}_X - \kappa)^N$ for all $n \geq N$. Therefore, $X_n = X_N$ for all $n \geq N$, and the process terminates. □

The additional hypothesis on the separating dual, which covers the Banach case, is likely superfluous: if we go to the other extreme, the proof also works for spaces with trivial dual because in that case, compact operators have no eigenvalues at all [456].

4.4 Alternative Constructions of Ext

Most texts in homological algebra introduce Ext using projective or injective presentations. Let us quickly review those constructions of Ext where they make sense in our context: the construction using projectives in p-Banach spaces and the construction using injectives in Banach spaces.

Ext Using Projectives

As explained already in Chapter 2, the basic fact underpinning the construction of Ext in the category of p-Banach spaces is that $p\mathbf{B}$ has enough projectives: every p-Banach space X admits a projective presentation in $p\mathbf{B}$:

$$0 \longrightarrow \kappa_p(X) \xrightarrow{\ \jmath\ } \ell_p(I) \xrightarrow{\ \rho\ } X \longrightarrow 0$$

Proceeding as in Section 2.7, we find out that any exact sequence of p-Banach spaces $0 \longrightarrow Y \longrightarrow Z \longrightarrow X \longrightarrow 0$ finds its place in a pushout diagram

$$\begin{array}{ccccccccc}
0 & \longrightarrow & \kappa_p(X) & \xrightarrow{\ \jmath\ } & \ell_p(I) & \xrightarrow{\ \rho\ } & X & \longrightarrow & 0 \\
 & & \downarrow & & \downarrow & & \| & & \\
0 & \longrightarrow & Y & \longrightarrow & Z & \longrightarrow & X & \longrightarrow & 0
\end{array}$$

And this is exactly the outcome of applying $\mathfrak{L}(\cdot, Y)$ to the projective presentation and forming the homology sequence

$$0 \longrightarrow \mathfrak{L}(X, Y) \xrightarrow{\ \rho^\circ\ } \mathfrak{L}(\ell_p(I), Y) \xrightarrow{\ \jmath^\circ\ } \mathfrak{L}(\kappa_p(X), Y) \xrightarrow{\text{pushout}} \operatorname{Ext}_{p\mathbf{B}}(X, Y) \longrightarrow 0$$

since $\operatorname{Ext}_{p\mathbf{B}}(\ell_p(I), Y) = 0$. Therefore, we define a 'new' space Ext using projectives by

$$\operatorname{Ext}^{\text{proj}}_{p\mathbf{B}}(X, Y) = \frac{\mathfrak{L}(\kappa_p(X), Y)}{\jmath^\circ[\mathfrak{L}(\ell_p(I), Y)]},$$

where the superscript is there to remind us that the definition depends on the choice of the projective presentation (and yes, Proposition 2.7.3 is there to remind us that it does not). Let us closely examine how functorial the assignment $(X, Y) \rightsquigarrow \operatorname{Ext}^{\text{proj}}_{p\mathbf{B}}(X, Y)$ is. On the 'yes, why not?' side:

- $\operatorname{Ext}^{\text{proj}}_{p\mathbf{B}}(X, Y)$ is a linear space.
- As a (possibly non-Hausdorff) quotient of $\mathfrak{L}(\kappa_p(X), Y)$, it carries a natural topology associated with the semi-p-norm

$$\left\|\tau + \jmath^\circ[\mathfrak{L}(\ell_p(I), Y)]\right\| = \inf\left\{\left\|\tau - T|_{\kappa_p(X)}\right\| : T \in \mathfrak{L}(\ell_p(I), Y)\right\}.$$

- $\operatorname{Ext}^{\text{proj}}_{p\mathbf{B}}(X, \cdot)\colon p\mathbf{B} \longrightarrow \mathbf{s}(p\mathbf{B})$ is a covariant functor: if $u\colon Y \longrightarrow Y'$ is an operator, then $\operatorname{Ext}^{\text{proj}}_{p\mathbf{B}}(X, u)\colon \operatorname{Ext}^{\text{proj}}_{p\mathbf{B}}(X, Y) \longrightarrow \operatorname{Ext}^{\text{proj}}_{p\mathbf{B}}(X, Y')$ maps $[\tau]$ to $[u\tau]$.

Some frowning appears here since the functorial character of $\operatorname{Ext}^{\text{proj}}_{p\mathbf{B}}$ with respect to the first variable is not as clear as it should be. Let us consider an operator $u\colon X' \longrightarrow X$ and see what happens when one picks projective presentations

$$0 \longrightarrow \kappa_p(X) \longrightarrow \ell_p(I) \xrightarrow{\pi} X \longrightarrow 0$$

$$0 \longrightarrow \kappa_p(X') \longrightarrow \ell_p(J) \xrightarrow{\pi'} X' \longrightarrow 0$$

with which to define $\operatorname{Ext}^{\text{proj}}_{p\mathbf{B}}(X, Y)$ and $\operatorname{Ext}^{\text{proj}}_{p\mathbf{B}}(X', Y)$. Form a commutative diagram

$$\begin{array}{ccccccccc} 0 & \longrightarrow & \kappa_p(X) & \longrightarrow & \ell_p(I) & \xrightarrow{\pi} & X & \longrightarrow & 0 \\ & & \uparrow{\scriptstyle w} & & \uparrow{\scriptstyle v} & & \uparrow{\scriptstyle u} & & \\ 0 & \longrightarrow & \kappa_p(X') & \longrightarrow & \ell_p(J) & \xrightarrow{\pi'} & X' & \longrightarrow & 0 \end{array}$$

by finding a lifting v of $u\pi'$ and then setting $w = v|_{\kappa(X')}$. Now, for each fixed Y, define

$$\operatorname{Ext}^{\text{proj}}_{p\mathbf{B}}(u, Y)\colon \operatorname{Ext}^{\text{proj}}_{p\mathbf{B}}(X, Y) \longrightarrow \operatorname{Ext}^{\text{proj}}_{p\mathbf{B}}(X', Y)$$

by $[\tau] \longmapsto [\tau w]$. The definition makes sense because if τ extends to $\ell_p(I)$, then τw extends to $\ell_p(J)$. What is more important, although τw may depend on the lifting of $u\pi'$, the class $[\tau w]$ does not (since the difference between any two liftings of $u\pi'$ maps $\ell_p(J)$ to $\kappa_p(X)$). In conclusion,

- $\mathrm{Ext}^{\mathrm{proj}}_{p\mathbf{B}}(\cdot, Y)\colon p\mathbf{B} \longrightarrow \mathbf{s}(p\mathbf{B})$ is a contravariant functor.

All tidy? Not yet: we wanted to make $\mathrm{Ext}^{\mathrm{proj}}_{p\mathbf{B}}$ a bifunctor and not merely a functor in each variable. So, pick operators $u\colon X' \longrightarrow X$ and $v\colon Y \longrightarrow Y'$ and define $\mathrm{Ext}^{\mathrm{proj}}_{p\mathbf{B}}(u, v)\colon \mathrm{Ext}^{\mathrm{proj}}_{p\mathbf{B}}(X, Y) \longrightarrow \mathrm{Ext}^{\mathrm{proj}}_{p\mathbf{B}}(X', Y')$ as the composition $\mathrm{Ext}^{\mathrm{proj}}_{p\mathbf{B}}(X', v)\,\mathrm{Ext}^{\mathrm{proj}}_{p\mathbf{B}}(u, Y)$. This definition makes

$$\mathrm{Ext}^{\mathrm{proj}}_{p\mathbf{B}} : (p\mathbf{B})^{\mathrm{op}} \times (p\mathbf{B}) \longrightarrow \mathbf{s}(p\mathbf{B})$$

a functor because all the squares

$$\begin{array}{ccc}
\mathrm{Ext}^{\mathrm{proj}}_{p\mathbf{B}}(X, Y) & \xrightarrow{\mathrm{Ext}^{\mathrm{proj}}_{p\mathbf{B}}(u,Y)} & \mathrm{Ext}^{\mathrm{proj}}_{p\mathbf{B}}(X', Y) \\
{\scriptstyle \mathrm{Ext}^{\mathrm{proj}}_{p\mathbf{B}}(X,v)}\downarrow & & \downarrow{\scriptstyle \mathrm{Ext}^{\mathrm{proj}}_{p\mathbf{B}}(X',v)} \\
\mathrm{Ext}^{\mathrm{proj}}_{p\mathbf{B}}(X, Y') & \xrightarrow{\mathrm{Ext}^{\mathrm{proj}}_{p\mathbf{B}}(u,Y')} & \mathrm{Ext}^{\mathrm{proj}}_{p\mathbf{B}}(X', Y')
\end{array}$$

are commutative. Hence, regardless of the projective presentation chosen:

Proposition 4.4.1 $\mathrm{Ext}^{\mathrm{proj}}_{p\mathbf{B}} : (p\mathbf{B})^{\mathrm{op}} \times p\mathbf{B} \longrightarrow \mathbf{s}(p\mathbf{B})$ *is a functor.*

$\mathrm{Ext}_{\mathbf{B}}$ Using Injectives

A remarkable difference between the categories **B** and p**B** when $p < 1$ is that while the Hahn–Banach theorem guarantees the existence of enough injectives in **B**, by 2.9.1, injective spaces do not exist at all in p**B**. This section is therefore about Banach spaces, for which every space Y admits an injective presentation $0 \longrightarrow Y \xrightarrow{\imath} \ell_\infty(I) \longrightarrow c\kappa(Y) \longrightarrow 0$ for some index set I. Here, as in Section 2.9, $c\kappa(Y) = \ell_\infty(I)/\imath[Y]$ is the cokernel of $\imath$ and, as in Diagram (2.22), every exact sequence with Y as the subspace fits in a pullback diagram

$$\begin{array}{ccccccccc}
0 & \longrightarrow & Y & \xrightarrow{\imath} & \ell_\infty(I) & \longrightarrow & c\kappa(Y) & \longrightarrow & 0 \\
 & & \| & & \uparrow & & \uparrow & & \\
0 & \longrightarrow & Y & \longrightarrow & Z & \longrightarrow & X & \longrightarrow & 0
\end{array}$$

That exactly corresponds to the outcome of applying $\mathfrak{L}(X, \cdot)$ to the injective presentation and forming the homology sequence

$$0 \longrightarrow \mathfrak{L}(X, Y) \xrightarrow{\imath_\circ} \mathfrak{L}(X, \ell_\infty(I)) \xrightarrow{\pi_\circ} \mathfrak{L}(X, c\kappa(Y)) \xrightarrow{\text{pullback}} \mathrm{Ext}_{\mathbf{B}}(X, Y) \longrightarrow 0$$

since $\operatorname{Ext}_{\mathbf{B}}(X, \ell_\infty(I)) = 0$. Therefore, we define a 'new' space Ext using injectives by

$$\operatorname{Ext}_{\mathbf{B}}^{\text{inj}}(X, Y) = \frac{\mathfrak{L}(X, c\kappa(Y))}{\pi_\circ[\mathfrak{L}(X, \ell_\infty(I)]}.$$

The superscript is there to remind us that the definition depends on the choice of the injective presentation (though actually it does not; Theorem 2.11.7). The assignment $(X, Y) \rightsquigarrow \operatorname{Ext}_{\mathbf{B}}^{\text{inj}}(X, Y)$ is functorial and, as above, the arguments in favour are:

- $\operatorname{Ext}_{\mathbf{B}}^{\text{inj}}(X, Y)$ is a linear space;
- as a (possibly non-Hausdorff) quotient of $\mathfrak{L}(X, c\kappa(X))$, it carries a natural topology, associated with the semi-p-norm

$$\left\|\tau + \pi_\circ[\mathfrak{L}(X, \ell_\infty(I))]\right\| = \inf\left\{\left\|\tau - \pi T\right\| : T \in \mathfrak{L}(X, \ell_\infty(I))\right\};$$

- $\operatorname{Ext}_{\mathbf{B}}^{\text{inj}}(\cdot, Y)\colon \mathbf{B} \longrightarrow \mathbf{sB}$ is a contravariant functor, where it is understood that given $u\colon X' \longrightarrow X$, we define $\operatorname{Ext}_{\mathbf{B}}^{\text{inj}}(u, Y)\colon \operatorname{Ext}_{\mathbf{B}}^{\text{inj}}(X, Y) \longrightarrow \operatorname{Ext}_{\mathbf{B}}^{\text{inj}}(X', Y)$ as $[\tau] \longmapsto [\tau u]$. Note that $\|\tau u\|_{\operatorname{Ext}_{\mathbf{B}}^{\text{inj}}(X',Y)} \leq \|u\| \, \|\tau\|_{\operatorname{Ext}_{\mathbf{B}}^{\text{inj}}(X,Y)}$.

On the other hand, the functorial character of $\operatorname{Ext}_{\mathbf{B}}^{\text{inj}}$ with respect to the second variable is in doubt. To put things straight, let $v\colon Y \longrightarrow Y'$ be an operator, and form the commutative diagram

$$\begin{array}{ccccccccc} 0 & \longrightarrow & Y & \stackrel{\imath}{\longrightarrow} & \ell_\infty(I) & \stackrel{\pi}{\longrightarrow} & \varrho(Y) & \longrightarrow & 0 \\ & & \downarrow{\scriptstyle v} & & \downarrow{\scriptstyle V} & & \downarrow{\scriptstyle w} & & \\ 0 & \longrightarrow & Y' & \stackrel{\imath'}{\longrightarrow} & \ell_\infty(J) & \stackrel{\pi'}{\longrightarrow} & \varrho(Y') & \longrightarrow & 0 \end{array} \tag{4.10}$$

where $V\imath = \imath' v$. Then define $\operatorname{Ext}_{\mathbf{B}}^{\text{inj}}(X, v)\colon \operatorname{Ext}_{\mathbf{B}}^{\text{inj}}(X, Y) \longrightarrow \operatorname{Ext}_{\mathbf{B}}^{\text{inj}}(X, Y')$ by $[\tau] \longmapsto [w\tau]$. The definition makes sense by an argument similar to the one used in the projective case. Now, if $v'\colon Y' \longrightarrow Y''$ is another operator, then

$$\operatorname{Ext}_{\mathbf{B}}^{\text{inj}}(X, v'v) = \operatorname{Ext}_{\mathbf{B}}^{\text{inj}}(X, v') \operatorname{Ext}_{\mathbf{B}}^{\text{inj}}(X, v).$$

The boundedness is also clear since $\|w\tau\|_{\operatorname{Ext}_{\mathbf{B}}^{\text{inj}}(X,Y')} \leq \|w\| \, \|\tau\|_{\operatorname{Ext}_{\mathbf{B}}^{\text{inj}}(X,Y)}$. To conclude that $\operatorname{Ext}_{\mathbf{B}}^{\text{inj}}$ is a genuine functor, set

$$\operatorname{Ext}_{\mathbf{B}}^{\text{inj}}(u, v) = \operatorname{Ext}_{\mathbf{B}}^{\text{inj}}(u, Y') \operatorname{Ext}_{\mathbf{B}}^{\text{inj}}(X, v) = \operatorname{Ext}_{\mathbf{B}}^{\text{inj}}(X', v) \operatorname{Ext}_{\mathbf{B}}^{\text{inj}}(u, Y)$$

for any given pair of operators $u\colon X' \longrightarrow X, v\colon Y \longrightarrow Y'$. In conclusion, independently of the injective presentations chosen:

Proposition 4.4.2 $\operatorname{Ext}_{\mathbf{B}}^{\text{inj}}\colon \mathbf{B}^{\text{op}} \times \mathbf{B} \longrightarrow \mathbf{sB}$ *is a functor.*

Natural Equivalences of the Functor Ext (Take p)

When homology people define $\mathrm{Ext}^{\mathrm{inj}}$ or $\mathrm{Ext}^{\mathrm{proj}}$ functors, they must guarantee that the outcome is the same no matter which injective or projective presentation has been used. We shrugged this off for two reasons: the Diagonal principles 2.11.7 and 2.11.6 already ensure that; and, more importantly, we have already secured an Ext to compare with, which is the business of this section. While dealing with p-Banach spaces there are three functors to compare:

$$\mathrm{Ext}_{p\mathbf{B}}, \qquad \mathsf{Q}^{(p)}_{\mathsf{LB}}, \qquad \mathrm{Ext}^{\mathrm{proj}}_{p\mathbf{B}}\,.$$

As the reader can anticipate, we have:

Proposition 4.4.3 *For each fixed* $p \in (0,1]$, $\mathrm{Ext}_{p\mathbf{B}}$, $\mathrm{Ext}^{\mathrm{proj}}_{p\mathbf{B}}$ *and* $\mathsf{Q}^{(p)}_{\mathsf{LB}}$ *are naturally equivalent functors* $(p\mathbf{B})^{\mathrm{op}} \times p\mathbf{B} \longrightarrow \mathbf{s}(p\mathbf{B})$.

Proof The natural equivalence between $\mathrm{Ext}_{p\mathbf{B}}$ and $\mathsf{Q}^{(p)}_{\mathsf{LB}}$ was established in Theorem 4.1.10. Hence it suffices to see that $\mathsf{Q}^{(p)}_{\mathsf{LB}}$ and $\mathrm{Ext}^{\mathrm{proj}}_{p\mathbf{B}}$ are naturally equivalent. Fix p-Banach spaces X and Y, and let

$$0 \longrightarrow \kappa_p(X) \longrightarrow \ell_p(I) \stackrel{\pi}{\longrightarrow} X \longrightarrow 0 \tag{4.11}$$

be the projective presentation of X that defines $\mathrm{Ext}^{\mathrm{proj}}_{p\mathbf{B}}$. Let $\Upsilon\colon X \longrightarrow \kappa_p(X)$ be a quasilinear map associated to (4.11); it is actually p-linear and makes $\ell_p(I) \simeq \kappa_p(X) \oplus_\Upsilon X$. This identification will simplify the exposition. We define $\eta_{(X,Y)}\colon \mathrm{Ext}^{\mathrm{proj}}_{p\mathbf{B}}(X,Y) \longrightarrow \mathsf{Q}^{(p)}_{\mathsf{LB}}(X,Y)$ in the form $\eta_{(X,Y)}[\tau] = [\tau \circ \Upsilon]$. By examining the diagram

$$\begin{array}{ccccccccc} 0 & \longrightarrow & \kappa_p(X) & \longrightarrow & \kappa_p(X)\oplus_\Upsilon X & \longrightarrow & X & \longrightarrow & 0 \\ & & \downarrow{\scriptstyle\tau} & & \downarrow & & \| & & \\ 0 & \longrightarrow & Y & \longrightarrow & Y \oplus_{\tau\circ\Upsilon} X & \longrightarrow & X & \longrightarrow & 0 \end{array} \tag{4.12}$$

we can immediately prove that η enjoys the following properties:

- $\eta_{(X,Y)}$ is well defined and injective: by the splitting criterion for pushout sequences, τ extends to $\ell_p(I) \simeq \kappa_p(X)\oplus_\Upsilon X$ if and only if the lower sequence splits, which happens if and only if the quasilinear map $\tau \circ \Upsilon$ is trivial.
- $\eta_{(X,Y)}$ is linear and bounded since $Q^{(p)}(\tau \circ \Upsilon) \leq \|\tau\| Q^{(p)}(\Upsilon)$.
- If $v\colon Y \longrightarrow Y'$ is an operator, then $\eta_{(X,Y)}[v\tau] = [(v\tau)\circ\Upsilon]$.

Thus, all squares of the form

$$\begin{array}{ccc} \operatorname{Ext}^{\mathrm{proj}}_{p\mathbf{B}}(X,Y) & \xrightarrow{\eta_{(X,Y)}} & \mathsf{Q}^{(p)}_{\mathsf{LB}}(X,Y) \\ \Big\downarrow{\scriptstyle \operatorname{Ext}^{\mathrm{proj}}_{p\mathbf{B}}(\mathbf{1}_X,v)} & & \Big\downarrow{\scriptstyle \mathsf{Q}^{(p)}_{\mathsf{LB}}(\mathbf{1}_X,v)} \\ \operatorname{Ext}^{\mathrm{proj}}_{p\mathbf{B}}(X,Y') & \xrightarrow{\eta_{(X,Y')}} & \mathsf{Q}^{(p)}_{\mathsf{LB}}(X,Y') \end{array}$$

are commutative. We must also prove that the squares

$$\begin{array}{ccc} \operatorname{Ext}^{\mathrm{proj}}_{p\mathbf{B}}(X,Y) & \xrightarrow{\eta_{(X,Y)}} & \mathsf{Q}^{(p)}_{\mathsf{LB}}(X,Y) \\ \Big\downarrow{\scriptstyle \operatorname{Ext}^{\mathrm{proj}}_{p\mathbf{B}}(u,\mathbf{1}_Y)} & & \Big\downarrow{\scriptstyle \mathsf{Q}^{(p)}_{\mathsf{LB}}(u,\mathbf{1}_Y)} \\ \operatorname{Ext}^{\mathrm{proj}}_{p\mathbf{B}}(X',Y) & \xrightarrow{\eta_{(X',Y)}} & \mathsf{Q}^{(p)}_{\mathsf{LB}}(X',Y) \end{array}$$

are commutative for any operator $u\colon X' \longrightarrow X$. To do so, recover the commutative diagram

$$\begin{array}{ccccccccc} 0 & \longrightarrow & \kappa_p(X) & \longrightarrow & \ell_p(I) & \xrightarrow{\pi} & X & \longrightarrow & 0 \\ & & \Big\uparrow{\scriptstyle w} & & \Big\uparrow{\scriptstyle v} & & \Big\uparrow{\scriptstyle u} & & \\ 0 & \longrightarrow & \kappa_p(X') & \longrightarrow & \ell_p(J) & \xrightarrow{\pi'} & X' & \longrightarrow & 0 \end{array} \tag{4.13}$$

from the proof of Proposition 4.4.1, whose lower row is the prescribed presentation of X', and recall that $[\tau w] = \operatorname{Ext}^{\mathrm{proj}}_{p\mathbf{B}}(u,\mathbf{1}_Y)[\tau]$. Thus, if $\Upsilon'\colon X' \longrightarrow \kappa_p(X')$ is a quasilinear map associated with the lower projective presentation of X' then

$$\mathsf{Q}^{(p)}_{\mathsf{LB}}(u,\mathbf{1}_Y)\,\eta_{(X,Y)}[\tau] = [(\tau\circ\Upsilon)\circ u] = [\tau\circ(\Upsilon\circ u)],$$
$$\eta_{(X',Y)}\operatorname{Ext}^{\mathrm{proj}}_{p\mathbf{B}}(u,\mathbf{1}_Y)[\tau] = \eta_{(X',Y)}[\tau w] = [(\tau w)\circ\Upsilon'] = [\tau\circ(w\circ\Upsilon')].$$

This shows that η is a natural transformation since Diagram (4.13) can be read as $\Upsilon\circ u \sim w\circ\Upsilon'$. It remains to prove that $\eta_{(X,Y)}$ is a linear homeomorphism, which is an unfettered headache: we already have a proof that $\eta_{(X,Y)}$ is a linear bijection (just apply the second part of Theorem 4.3.2 to the sequence (4.11) and Y), but there is no open mapping theorem for semi-quasi-Banach spaces, as was remarked in Note 1.8.1. So, checking that $\eta_{(X,Y)}$ has a continuous inverse must be done by hand. To simplify the notation, we shall omit the subscripts whenever possible. What we have to do is attach to each p-linear map $\Phi\colon X \longrightarrow Y$ an operator $\tau\colon \kappa_p(X) \longrightarrow Y$ in such a way that $\Phi \sim \tau\circ\Upsilon$ *and* maintain some control on $\|\tau\|$ through the p-linearity constant of Φ. This dependence will be examined later. Pick a p-linear $\Phi\colon X \longrightarrow Y$ and form the twisted sum $Y\oplus_\Phi X$, which is isomorphic to a p-Banach space. So the quotient

$\pi\colon \ell_p(I) \longrightarrow X$ admits a lifting $T\colon \ell_p(I) \longrightarrow Y \oplus_\Phi X$, and it is obvious that $\tau = T|_{\kappa_p(X)}\colon \kappa_p(X) \longrightarrow Y$ is going to be the operator we are looking for: the commutative diagram

$$\begin{array}{ccccccccc} 0 & \longrightarrow & \kappa_p(X) & \longrightarrow & \ell_p(I) & \xrightarrow{\pi} & X & \longrightarrow & 0 \\ & & \downarrow{\scriptstyle\tau} & & \downarrow{\scriptstyle T} & & \| & & \\ 0 & \longrightarrow & Y & \longrightarrow & Y \oplus_\Phi X & \longrightarrow & X & \longrightarrow & 0 \end{array} \tag{4.14}$$

shows that $\Phi \sim \tau \circ \Upsilon$. This reasoning shows that η is surjective, and since we already know that it is one to one and linear, η^{-1} exists. To prove that η^{-1} is bounded, we define a wannabe lifting by the formula

$$T\left(\sum_{i\in I} \lambda_i e_i\right) = \sum_{i\in I} \lambda_i(\Phi(\pi e_i), \pi e_i). \tag{4.15}$$

The right-hand series converges because $Y \oplus_\Phi X$ is p-normable and complete and because

$$\sum_i \|\lambda_i(\Phi(\pi e_i), \pi e_i)\|_\Phi^p \leq \|\pi\|^p \sum_i |\lambda_i|^p < \infty.$$

A meticulous reader might argue that it is not yet justified that T is an operator at all; in fact, (4.15) only defines a linear map on $\ell_p(I)$ that is bounded on the subspace of finitely supported families. We agree. To amend this, observe that Theorem 3.6.7 in tandem with Lemma 1.1.2 implies that there is a p-norm on $Y \oplus_\Phi X$, say, $|\cdot|$, such that

$$|(y,x)| \leq \|(y,x)\|_\Phi \leq 2^{1/p-1}(1 + Q^{(p)}(\Phi))^{1/p}|(y,x)|.$$

Now, if $f = \sum_{i\in I} \lambda_i e_i$, then

$$\begin{aligned} \|T(f)\|_\Phi &\leq 2^{1/p-1}(1 + Q^{(p)}(\Phi))^{1/p}\Big|\sum_{i\in I} \lambda_i(\Phi(\pi e_i), \pi e_i)\Big| \\ &\leq 2^{1/p-1}(1 + Q^{(p)}(\Phi))^{1/p}\Big(\sum_{i\in I} \lambda_i\big\|(\Phi(\pi e_i), \pi e_i)\big\|^p\Big)^{1/p} \\ &\leq 2^{1/p-1}(1 + Q^{(p)}(\Phi))^{1/p}\|\pi\|\,\|f\|. \end{aligned}$$

There is therefore a constant C, depending only on the projective presentation of X, such that if $\Phi\colon X \longrightarrow Y$ is p-linear with $Q^p(\Phi) \leq 1$, then $\|T\| \leq C$. □

A warning seems to be in order here: the obvious projection $Y \oplus_\Phi X \longrightarrow Y$, namely $(y,x) \longmapsto y$, is not continuous unless Φ is bounded. And this means that even if $x_n \to 0$ (in X), the sequence $(0, x_n)$ is not necessarily convergent to $(0,0)$ in $Y \oplus_\Phi X$. Therefore, even if Φ and T are linked by (4.15), the restriction of T to $\kappa_p(X)$ is not $\sum_i \lambda_i(\Phi\pi e_i, 0)$. This mischievous fact is in part responsible for making a straight proof that η^{-1} is linear difficult: actually, the operator τ does not linearly depend on Φ, although the dependence becomes linear when one passes to the quotient structures.

Natural Equivalences of the Functor Ext (Take 1)

If we turn our attention to $\mathrm{Ext}_{\mathbf{B}}\colon \mathbf{B}^{\mathrm{op}} \times \mathbf{B} \longrightarrow \mathbf{sB}$, there is one more sibling,

$$\mathrm{Ext}_{\mathbf{B}}, \qquad \mathsf{Q}^{(1)}_{\mathsf{LB}}, \qquad \mathrm{Ext}^{\mathrm{proj}}_{\mathbf{B}}, \qquad \mathrm{Ext}^{\mathrm{inj}}_{\mathbf{B}} \tag{4.16}$$

in the family photo. However, we still have:

Proposition 4.4.4 *All functors appearing in* (4.16) *are naturally equivalent.*

Sketch of the proof The first three are naturally equivalent by Proposition 4.4.3. It then suffices to find a natural equivalence between $\mathrm{Ext}^{\mathrm{inj}}_{\mathbf{B}}$ and any other in the list. And we have to choose: let us pick $\mathsf{Q}^{(1)}_{\mathsf{LB}}$ and then follow the track of the proof of Proposition 4.4.3 plus some eventual dualisation. It is, however, much more entertaining to display the equivalence between the injective and the projective constructions of Ext. So, let X and Y be Banach spaces, and let

$$0 \longrightarrow \kappa(X) \longrightarrow \mathcal{P} \xrightarrow{\ \pi\ } X \longrightarrow 0$$

$$0 \longrightarrow Y \xrightarrow{\ \varepsilon\ } \mathcal{I} \xrightarrow{\ \varpi\ } c\kappa(Y) \longrightarrow 0$$

be the projective and injective presentations of X and Y involved in the definition of $\mathrm{Ext}^{\mathrm{proj}}_{\mathbf{B}}(X,\cdot)$ and $\mathrm{Ext}^{\mathrm{inj}}_{\mathbf{B}}(\cdot, Y)$, respectively. We have:

$$\mathrm{Ext}^{\mathrm{inj}}_{\mathbf{B}}(X,Y) = \frac{\mathfrak{L}(X, c\kappa(Y))}{\varpi_\circ[\mathfrak{L}(X,\mathcal{I})]} \quad \text{and} \quad \mathrm{Ext}^{\mathrm{proj}}_{\mathbf{B}}(X,Y) = \frac{\mathfrak{L}(\kappa(X), Y)}{\jmath^\circ[\mathfrak{L}(\mathcal{P}, Y)]}.$$

Now, given operators $v \in \mathfrak{L}(\kappa(X), Y)$ and $u \in \mathfrak{L}(X, c\kappa(Y))$, declare $[u] \leftrightarrow [v]$ if and only if there is $w\colon \mathcal{P} \longrightarrow \mathcal{I}$, making the following diagram commute:

$$\begin{array}{ccccccccc} 0 & \longrightarrow & \kappa(X) & \longrightarrow & \mathcal{P} & \xrightarrow{\ \pi\ } & X & \longrightarrow & 0 \\ & & \big\downarrow{\scriptstyle v} & & \big\downarrow{\scriptstyle w} & & \big\downarrow{\scriptstyle u} & & \\ 0 & \longrightarrow & Y & \xrightarrow{\ \varepsilon\ } & \mathcal{I} & \xrightarrow{\ \varpi\ } & c\kappa(Y) & \longrightarrow & 0 \end{array}$$

Take your time, revisit (2.30), reread Note 2.15.4 and flamboyantly conclude that $[u] \leftrightarrow [v]$ is the required natural equivalence. □

We close this section with a duality formula for extensions of Banach spaces that can be considered as the 'first-derived' version of the identity $\mathfrak{L}(A, B^*) = \mathfrak{L}(B, A^*)$, which, for obvious reasons, does not have a counterpart for any

$p < 1$. We first observe that, if A, B are Banach spaces, the spaces $\mathfrak{L}(A, B^*)$ and $\mathfrak{L}(B, A^*)$ are naturally isometric since both can be identified with the space of bounded bilinear forms $A \times B \longrightarrow \mathbb{K}$. In this way, $u\colon A \longrightarrow B^*$ corresponds to $v\colon B \longrightarrow A^*$ if and only if $\langle u(a), b\rangle = \langle v(b), a\rangle$ for all $a \in A, b \in B$, which happens if and only if $v = u^*|_B$ or, equivalently, $u = v^*|_A$. The crucial point here is that if E is another Banach space and $\jmath\colon A \longrightarrow E$ is an embedding, so that $\jmath^*\colon E^* \longrightarrow A^*$ is a quotient map, then $U\colon E \longrightarrow B^*$ is an extension of $u\colon A \longrightarrow B^*$ through $\jmath$ if and only if the corresponding operator $V\colon B \longrightarrow E^*$ is a lifting of v through $\jmath^*$. Indeed, since $\langle U(e), b\rangle = \langle V(b), e\rangle$ for all $e \in E, b \in B$, we have $u = U\jmath \iff v = \jmath^* V$.

Corollary 4.4.5 *For all Banach spaces X, Y, we have a natural isomorphism between* $\operatorname{Ext}_{\mathbf{B}}(X, Y^*)$ *and* $\operatorname{Ext}_{\mathbf{B}}(Y, X^*)$.

Proof Let $0 \longrightarrow \kappa(X) \stackrel{\imath}{\longrightarrow} \ell_1(I) \longrightarrow X \longrightarrow 0$ be a projective presentation of X in $\mathbf{B}$. The dual sequence is an injective presentation of X^*, and since the operators in $\mathfrak{L}(\kappa(X), Y^*)$ that extend to $\ell_1(I)$ correspond to those operators in $\mathfrak{L}(Y, \kappa(X)^*)$ that lift to $\ell_\infty(I)$, by identifying $\operatorname{Ext}_{\mathbf{B}}(X, Y^*)$ with $\operatorname{Ext}^{\text{proj}}_{\mathbf{B}}(X, Y^*)$ and $\operatorname{Ext}_{\mathbf{B}}(Y, X^*)$ with $\operatorname{Ext}^{\text{inj}}_{\mathbf{B}}(X, Y^*)$, we have

$$\operatorname{Ext}_{\mathbf{B}}(X, Y^*) = \frac{\mathfrak{L}(\kappa(X), Y^*)}{\imath^\circ[\mathfrak{L}(\ell_1(I), Y^*)]} = \frac{\mathfrak{L}(Y, \kappa(X)^*)}{\imath^*_\circ[\mathfrak{L}(Y, \ell_\infty(I))]} = \operatorname{Ext}_{\mathbf{B}}(Y, X^*). \qquad \square$$

An alternative, intrinsic description of the isomorphism between $\operatorname{Ext}_{\mathbf{B}}(X, Y^*)$ and $\operatorname{Ext}_{\mathbf{B}}(Y, X^*)$ is as follows: given an extension $0 \longrightarrow Y^* \longrightarrow Z \longrightarrow X \longrightarrow 0$, its mate in $\operatorname{Ext}_{\mathbf{B}}(Y, X^*)$ is obtained taking the dual sequence and forming the pullback with the inclusion of Y into Y^{**}.

4.5 Topological Aspects of Ext

There is no way out: $\operatorname{Ext}(X, Y)$ and $\operatorname{Ext}_{p\mathbf{B}}(X, Y)$ are topological vector spaces. We longed for a (p-) Banach space structure but had to be content with a dismaying 'semi'. And to make things worse, we have to deal with the mob

$$\operatorname{Ext}(X, Y),\quad \mathsf{Q}_{\mathsf{LB}}(X, Y),\quad \operatorname{Ext}_{p\mathbf{B}}(X, Y),\quad \mathsf{Q}^{(p)}_{\mathsf{LB}}(X, Y),\quad \operatorname{Ext}_{\mathbf{B}}(X, Y),\quad \mathsf{Q}^{(1)}_{\mathsf{LB}}(X, Y).$$

The good news is that when two of those spaces coincide, the topology smuggles from one to the other. In this regard, it was already proved in Chapter 3 that

- $\operatorname{Ext}(X, Y) = \operatorname{Ext}_{r\mathbf{B}}(X, Y)$ with equivalent semi-quasinorms when X is p-Banach, Y is q-Banach and $r < \min(p, q)$; see Corollary 3.7.7.

- $\operatorname{Ext}(X, Y) = \operatorname{Ext}_{\mathbf{B}}(X, Y)$ with equivalent norms if X, Y are Banach spaces and X is a $\mathscr{K}$-space, by Dierolf's Theorem 3.4.4.

Topology in the Homology Sequence

The categorical point of view is that things can only be done one way or must remain undone. Accepting this, the continuity of the maps appearing in the homology sequences is the confirmation that the topologies considered are the right ones, at least once it is accepted that the operator (quasi)-norm is the right topology on $\mathfrak{L}(X, Y)$. We turn our attention to application 4.2.6 (6). The discussion after Corollary 4.2.8 suggests we should treat the natural belief that the continuous bijection Ω° is open with caution, because Ext spaces could be non-Hausdorff. The discussion concluded with 'we will return to this issue in Section 4.5', i.e. now. So let us study when the connecting map in the homology sequence is not only bijective but a topological isomorphism.

Theorem 4.5.1 *Let* $0 \longrightarrow Y \longrightarrow Z \longrightarrow X \longrightarrow 0$ *be an exact sequence of quasi-Banach spaces, and let E be another quasi-Banach space. If* $\mathfrak{L}(Z, E) = 0$ *and* $\operatorname{Ext}(Z, E) = 0$ *then* $\operatorname{Ext}(X, E)$ *is linearly homeomorphic to* $\mathfrak{L}(Y, E)$.

Proof What we actually prove is that $\mathsf{Q}_{\mathsf{LB}}(X, E)$ is linearly homeomorphic to $\mathfrak{L}(Y, E)$. As we have so often done when working with quasilinear maps, we will assume that the starting sequence is

$$0 \longrightarrow Y \xrightarrow{\;\jmath\;} Y \oplus_\Omega X \xrightarrow{\;\rho\;} X \longrightarrow 0$$

In particular, $Z = Y \oplus_\Omega X$. Set $B(x) = (\Omega(x), x)$ and $L(x) = (0, x)$. Then B is bounded, L is linear, both are sections of ρ and $\jmath \circ \Omega = B + L$. Taking the hypotheses into account, elementary use (6) of the homology sequence implies that the next sequence is exact:

$$0 \longrightarrow \mathfrak{L}(Y, E) \xrightarrow{\;\Omega^\circ\;} \mathsf{Q}_{\mathsf{LB}}(X, E) \longrightarrow 0$$

We have then to identify the inverse of Ω° and prove that it is continuous, for which it suffices to show that the map is relatively open; i.e. there is a constant C such that if $\Phi\colon X \longrightarrow E$ is quasilinear then there is an operator $\tau\colon Y \longrightarrow E$ such that $\tau \circ \Omega \sim \Phi$ and $\|\tau\| \leq CQ(\Phi)$. Pick $\Phi\colon X \longrightarrow E$ a quasilinear map, and form the induced sequence

$$0 \longrightarrow E \xrightarrow{\;\imath\;} E \oplus_\Phi X \xrightarrow{\;\pi\;} X \longrightarrow 0$$

As before, we define $B', L'\colon X \longrightarrow E \oplus_\Phi X$ by letting $B'(x) = (\Phi(x), x)$ and $L'(x) = (0, x)$, with a similar conclusion: $\imath \circ \Phi = B' + L'$. The hypothesis $\operatorname{Ext}(Z, E) = 0$ implies that

$$\Phi \circ \rho = B_\Phi + L_\Phi, \tag{4.17}$$

with $B_\Phi\colon Z \longrightarrow E$ homogeneous bounded and $L_\Phi\colon Z \longrightarrow E$ linear. Since $\mathfrak{L}(Z,E) = 0$, the decomposition above is unique: if $\Phi \circ \rho = \tilde{B}_\Phi + \tilde{L}_\Phi$ is another decomposition then $\tilde{B}_\Phi - B_\Phi = L_\Phi - \tilde{L}_\Phi$ is simultaneously bounded and linear, so it is 0. Moreover, $\|B_\Phi\| \le KQ(\Phi \circ \rho) \le KQ(\Phi)$ for some constant K independent of Φ; this follows from Theorem 3.6.5 (since *all* quasilinear maps $Z \longrightarrow E$ are trivial). Let us rewrite (4.17) as $(B' + L') \circ \rho = \imath \circ (B_\Phi + L_\Phi)$ and consider the map $T\colon Z \longrightarrow E \oplus_\Phi X$ given by

$$T = \overbrace{L'\rho - \imath L_\Phi}^{\text{linear}} = \overbrace{\imath \circ B_\Phi - B' \circ \rho}^{\text{bounded}},$$

which is actually an operator whose norm (if it mattered) is bounded by

$$\|B' \circ \rho - \imath \circ B_\Phi\| \le \Delta(E \oplus_\Phi X)(\|B' \circ \rho\| + \|\imath \circ B_\Phi\|) \le K'Q(\Phi)(1 + KQ(\Phi))$$

according to (the proof of) Theorem 3.6.7. But what really matters is the norm of $T\jmath$, namely $\|T\jmath\| \le K'Q(\Phi)KQ(\Phi)$. It remains to check that $[\Omega^\circ(T\jmath)] = [\Phi]$ or, equivalently, $T\jmath \circ \Omega \sim \Phi$. Since

$$T\jmath \circ \Omega - \Phi = T \circ (B + L) - (B' + L') = \overbrace{T \circ B - B'}^{\text{bounded}} + \overbrace{(TL - L')}^{\text{linear}},$$

we just have to show that $T \circ B - B'$ and $TL - L'$ take values in E (well, in $\imath[E] = \ker \pi$), which is easy since

$$\pi(T \circ B - B') = \pi(B'\rho - \imath B_\Phi)B - \pi B' = \pi B'\rho B - \pi\imath B_\Phi B - \pi B' = \mathbf{1}_X^2 - \mathbf{1}_X = 0$$
$$\pi(TL - L') = \pi(L'\rho + \imath L_\Phi)L - \pi L' = \mathbf{1}_X - \mathbf{1}_X = 0. \qquad \square$$

In this way, we can upgrade Proposition 4.2.7 and, rather unexpectedly, obtain that every quasi-Banach space arises as a space of extensions:

Corollary 4.5.2 *If E is a q-Banach space, then $E \simeq$* Ext$(L_p/\mathbb{K}, E)$ *for $p < q$. If Y is a subspace of a $\mathscr{K}$-space Z with trivial dual then $Y^* \simeq$* Ext$(Z/Y, \mathbb{K})$.

Here is the third link in the chain that began with Theorem 3.7.9:

Corollary 4.5.3 *Let $0 < p < 1$, and let A, B subspaces of L_p such that $L_p/A \simeq L_p/B$. Then $\mathfrak{L}(A,E) \simeq \mathfrak{L}(B,E)$ for every quasi-Banach space E satisfying $\mathfrak{L}(L_p, E) = 0$ and* Ext$(L_p, E) = 0$, *in particular when E is r-Banach for some $r > p$.*

When Is Ext Hausdorff?

It is enticing to believe that Ext(X, Y) is Hausdorff. Moreover, the results in the preceding section may give the impression that, give or take a couple of counterexamples, that is how things are. However, nothing could be further from the truth. To understand how non-zero elements of size zero can arise, let us consider a quasilinear map $\Phi\colon X \longrightarrow Y$; since

$$Q[\Phi] = \inf\{Q(\Phi + B + L) : B \in \mathsf{B}(X, Y), L \in \mathsf{L}(X, Y)\} = \inf\{Q(\Phi + B) : B \in \mathsf{B}\},$$

$Q[\Phi] = 0$ means that for every $\varepsilon > 0$, there is a bounded map $B\colon X \longrightarrow Y$ such that $Q(\Phi - B) \le \varepsilon$, while $[\Phi] = 0$ means that the preceding condition can be achieved even for $\varepsilon = 0$. When working in $p\mathbf{B}$, it is possible to translate this into the language of operators: given $\tau\colon \kappa_p(X) \longrightarrow Y$, the class $[\tau]$ has quasinorm zero in $\mathrm{Ext}^{\mathrm{proj}}_{p\mathbf{B}}(X, Y)$ if and only if τ is the uniform limit of operators that admit extension to $\ell_p(I)$, while $[\tau] = 0$ if and only if τ is the restriction of an operator $\ell_p(I) \longrightarrow Y$. The catch, of course, is that the quasinorms of the extensions tend to infinity; otherwise, the uniform limit of a sequence of operators admitting uniformly bounded extensions must also admit an extension. A rewording of Proposition 3.3.11 marries theory and practice.

Lemma 4.5.4 *Let X, Y be quasi-Banach spaces, and let X_0 be a dense subspace of X. Then the restriction map $\mathsf{Q}(X, Y) \longrightarrow \mathsf{Q}(X_0, Y)$ induces a topological isomorphism between $\mathsf{Q}_{\mathsf{LB}}(X, Y)$ and $\mathsf{Q}_{\mathsf{LB}}(X_0, Y)$. If X and Y are p-Banach spaces, the same is true after replacing Q_{LB} by $\mathsf{Q}^{(p)}_{\mathsf{LB}}$.*

In other words, if $\Phi_0\colon X_0 \longrightarrow Y$ has $Q[\Phi_0] = 0$ and $\Phi\colon X \longrightarrow Y$ is a quasilinear extension of Φ_0, then also $Q[\Phi] = 0$. At this point, awful suspicions creep into our minds: do there exist non-zero Φ with $Q[\Phi] = 0$? Do there exist non-zero Φ with $Q[\Phi] \neq 0$? The truth is out there: both are possible, and with the friendliest of our quasilinear friends, the Kalton–Peck maps.

Proposition 4.5.5 *Let $\varphi \in \mathrm{Lip}_0(\mathbb{R}_+)$, let X be a quasi-Banach sequence space and let $\mathsf{KP}_\varphi\colon X^0 \longrightarrow X$ be the corresponding Kalton–Peck map. If $\mathrm{Lip}(\varphi|_{[t,\infty)}) \to 0$ as $t \to \infty$ then $Q[\mathsf{KP}_\varphi] = 0$.*

Proof Fix $t > 0$ and write $\varphi = a[t] + b[t]$, where

$$a[t](s) = \begin{cases} \varphi(s), & 0 \le s \le t, \\ \varphi(t), & s \ge t, \end{cases} \qquad \text{and} \qquad b[t](s) = \begin{cases} 0, & 0 \le s \le t, \\ \varphi(s) - \varphi(t), & s \ge t. \end{cases}$$

Note that $a[t]$ is bounded and $b[t]$ Lipschitz, with $\mathrm{Lip}(b[t]) = \mathrm{Lip}(\varphi|_{[t,\infty)})$. Thus, $\mathsf{KP}_\varphi = \mathsf{KP}_{a[t]} + \mathsf{KP}_{b[t]}$ with $\mathsf{KP}_{b[t]}$ bounded and, by Proposition 3.2.6,

$$\lim_{t\to\infty} Q(\mathsf{KP}_{b[t]}) \le C_X \lim_{t\to\infty} \mathrm{Lip}(b[t]) = 0. \qquad \square$$

A particularly noteworthy instance of φ satisfying the hypothesis is when $\varphi'(t) \to 0$ as $t \to \infty$. Thus, if either $\varphi(t) = t^\alpha$ for $0 < \alpha < 1$ or $\varphi(t) = \log^+(t)$ then $Q[\mathsf{KP}_\varphi] = 0$, despite that KP_φ is non-trivial when no subsequence of the basis of X spans an isomorphic copy of c_0. In the following result, KP_p is the Kalton–Peck map associated with $\varphi(t) = t$ acting on ℓ_p.

Proposition 4.5.6 *For* $1 \le p < \infty$, $Q[\mathsf{KP}_p] \ge (\log 2)/(p2^{1/p})$.

Proof We need to prove that the quasilinearity constant of the Kalton–Peck map cannot be substantially improved by adding a bounded homogeneous map – any bounded homogeneous map. To do so, we will average bounded maps, a technique that requires local convexity and some compactness assumption, which explains the restrictions in the range of p. For each integer $k \ge 1$, consider the following subgroup of the group of all permutations of $\mathbb{N}$: $S_k = \{\sigma : \sigma(n) = n \text{ for all } n > k\}$, and set $S = \bigcup_{k\ge1} S_k$. We also consider the following subgroups of the unitary group of ℓ_∞:

$$U_k = \{u \in \ell_\infty : u(n) = \pm 1 \text{ for all } n \text{ and } u(n) = 1 \text{ for } n > k\},$$

and set $U = \bigcup_{k\ge1} U_k$. Note that S and U act on every sequence space by the rules $x \longmapsto x \circ \sigma$ and $x \longmapsto ux$, respectively, and that both actions preserve finitely supported sequences. We will say that a map $H : \mathbb{K}^{(\mathbb{N})} \longrightarrow \mathbb{K}^{\mathbb{N}}$ *commutes with the action of U (resp. of S)* if $H(ux) = uH(x)$ for every $u \in U$ (resp. $H(x \circ \sigma) = H(x) \circ \sigma$ for every $\sigma \in S$) and every finitely supported x. We have:

Lemma 4.5.7 *If H commutes with the actions of S and U then*

- *H preserves supports; i.e.* $\operatorname{supp} H(x) \subset \operatorname{supp} x$ *for every* $x \in \mathbb{K}^{(\mathbb{N})}$,
- *for every n, there is a constant c_n such that if $I \subset \mathbb{N}$ contains exactly n integers, then* $H\big(\sum_{i\in I} e_i\big) = c_n \sum_{i\in I} e_i$.

Proof The first part is clear: if there is $k \in \operatorname{supp} H(x) \setminus \operatorname{supp} x$, we can consider the unitary $u \in U$ that takes the value -1 at k, and 1 elsewhere. Then $ux = x$, while $H(x) = H(ux) \neq uH(x)$. For the second part, let I be a finite subset of $\mathbb{N}$. We know from the first part that $H\big(\sum_{i\in I} e_i\big) = \sum_{i\in I} \lambda_i e_i$ for some λ_i depending on I. If $\sigma \in S$ is the identity off I, then σ leaves $\sum_{i\in I} e_i$ invariant, so $\sum_{i\in I} \lambda_i e_i$ is also invariant under the action of σ. This is possible only if all the λ_i agree. Therefore, for every finite I, there is a constant $c(I)$ such that $H\big(\sum_{i\in I} e_i\big) = c(I) \sum_{i\in I} e_i$. If $|J| = |I|$, then there is $\sigma \in S$ such that $\sigma[I] = J$, hence $\sum_{i\in I} e_i = \big(\sum_{j\in J} e_j\big) \circ \sigma$, which gives $c(I) = c(J)$. □

The next step is to develop a symmetrisation procedure for bounded mappings acting on sequence spaces. Since it is intended that the resulting functions commute with the actions of U and S, it will be convenient to 'merge' U and S into a unique group acting on sequences. Given $\sigma \in S$ and $u \in U$, we consider (u, σ) acting on $\mathbb{K}^{\mathbb{N}}$ by the rule $(u, \sigma)x = u(\sigma^{\circ} x) = u(x \circ \sigma)$. Note that

$$(v, \tau)(u, \sigma)x = (v, \tau)(u(x \circ \sigma)) = v(u \circ \tau)(x \circ \sigma \circ \tau)$$

such that $(v, \tau)(u, \sigma) = (v(\tau^{\circ}(u)), \sigma \circ \tau)$. Following the notation of group theory, the resulting group will be denoted by $U \rtimes S$. The official name for this construction is a semidirect product. Note that the inverse of (u, σ) in $U \rtimes S$ is $(u^{-1} \circ \sigma^{-1}, \sigma^{-1})$, not (u^{-1}, σ^{-1}). Also note that the superscript -1 has at least two different meanings here. If we denote

$$(U \rtimes S)_k = \{(u, \sigma) : u(n) = 1, \sigma(n) = n \text{ for } n > k\}$$

then it is clear that $(U \rtimes S)_k$ is a subgroup of order $k!2^k$ which 'agrees' with $U_k \rtimes S_k$ in the obvious sense. The following trick will be quite useful: assume Γ is a finite group acting on a linear space V and let $f : V \longrightarrow V$ be an arbitrary function. Then the function $h : V \longrightarrow V$ defined by

$$h(x) = \frac{1}{|\Gamma|} \sum_{\gamma \in \Gamma} \gamma^{-1} f(\gamma x)$$

commutes with the action of Γ in the sense that $h(\gamma x) = \gamma h(x)$ for every $\gamma \in \Gamma$ and every $x \in V$. Indeed, pick $\beta \in \Gamma$. Then, letting $\alpha = \gamma\beta$ such that $\gamma^{-1} = \beta\alpha^{-1}$, and rearranging the sum, we have

$$h(\beta x) = \frac{1}{|\Gamma|} \sum_{\gamma \in \Gamma} \gamma^{-1} f(\gamma\beta x) = \frac{1}{|\Gamma|} \sum_{\alpha \in \Gamma} \beta\alpha^{-1} f(\alpha x) = \beta(h(x)). \tag{4.18}$$

This argument cannot be used for the infinite group $U \rtimes S$ but can be easily adapted if the target space carries a linear topology having enough compact convex sets. A Banach sequence space X will be called symmetric if $\|x\| = \|x \circ \sigma\|$ for every permutation σ of $\mathbb{N}$. Let $\mathfrak{U}$ be a free ultrafilter on $\mathbb{N}$. If a map $B : X^0 \longrightarrow X$ is bounded on bounded sets and the unit ball of X is closed in ℓ_∞ with respect to the topology of pointwise convergence, then we can define a mapping $H : X^0 \longrightarrow X$ by the pointwise limit

$$H(x) = \lim_{\mathfrak{U}(k)} \frac{1}{k!2^k} \sum_{(u,\sigma) \in U_k \rtimes S_k} (u, \sigma)^{-1}(B((u, \sigma)x)). \tag{4.19}$$

In these conditions:

Lemma 4.5.8 *The map H has the following properties:*

- *for every $r > 0$, one has* $\sup_{\|x\| \le r} \|H(x)\| \le \sup_{\|x\| \le r} \|B(x)\|$;

- *in particular, if B is bounded homogeneous, then so is H and $\|H\| \leq \|B\|$;*
- *H commutes with S and U.*

Proof The hypotheses on X are equivalent to X being the dual of a symmetric sequence space. The first and second points are due to the convexity and lower semicontinuity of the norm and the fact that $U \rtimes S$ acts by isometries on X. The third is clear once we realise that the action of $U \rtimes S$ is continuous for the topology of pointwise convergence. Indeed, each $(v, \zeta) \in U \rtimes S$ belongs to some $(U \rtimes S)_n$ for some $n \in \mathbb{N}$, and for $k \geq n$, we can apply (4.18) to get

$$(v,\zeta)\left(\sum_{(u,\sigma)\in(U\rtimes S)_k}(u,\sigma)^{-1}B((u,\sigma)x)\right) = \sum_{(u,\sigma)\in(U\rtimes S)_k}(u,\sigma)^{-1}B((u,\sigma)(v,\zeta)x).$$

Thus,

$$\begin{aligned}(v,\zeta)H(x) &= (v,\zeta)\left(\lim_{\mathfrak{U}(k)}\frac{1}{k!2^k}\sum_{(u,\sigma)\in U_k\rtimes S_k}(u,\sigma)^{-1}B((u,\sigma)x)\right)\\ &= \lim_{\mathscr{U}(k)}\frac{1}{k!2^k}\sum_{(u,\sigma)\in U_k\rtimes S_k}(u,\sigma)^{-1}B((u,\sigma)(v,\zeta)x)\\ &= H((v,\zeta)x).\end{aligned}$$

□

We are ready to conclude the proof of Proposition 4.5.6 and show that $\inf Q(\mathsf{KP}_p + B) > 0$, where the infimum is taken over all bounded homogeneous functions $B\colon \ell_p^0 \longrightarrow \ell_p$. The first step is to reduce the class of functions under consideration. We claim that:

- If X satisfies the hypotheses of Lemma 4.5.8, in particular if $X = \ell_p$ for $1 \leq p < \infty$,
- $\Phi\colon X^0 \longrightarrow X$ is a quasilinear map commuting with U and S,
- $B\colon X^0 \longrightarrow X$ is bounded homogeneous

and H is given by (4.19) then $Q(\Phi + H) \leq Q(\Phi + B)$. Indeed, let $x, y \in X$. Then, since

$$\Phi(x) = \lim_{\mathscr{U}(k)}\frac{1}{k!2^k}\sum_{\gamma\in U_k\rtimes S_k}\gamma^{-1}(\Phi(\gamma x)),$$

and similarly with y and $x + y$, using the lower semicontinuity of the norm with respect to the topology of pointwise convergence and taking into account that $U \rtimes S$ acts by isometries on X, we have

$$
\begin{aligned}
&\left\|(\Phi + H)(x+y) - (\Phi + H)(x) - (\Phi + H)(y)\right\| \\
&= \left\|\Phi(x+y) - \Phi x - \Phi y + \lim_{\mathcal{U}(k)} \frac{1}{k!2^k} \sum_{\gamma \in U_k \rtimes S_k} \gamma^{-1}\Big(B(\gamma(x+y)) - B\gamma x - B\gamma y\Big)\right\| \\
&= \left\| \lim_{\mathcal{U}(k)} \frac{1}{k!2^k} \sum_{\gamma \in U_k \rtimes S_k} \gamma^{-1}\Big((\Phi + B)(\gamma(x+y)) - (\Phi + B)(\gamma x) - (\Phi + B)(\gamma y)\Big)\right\| \\
&\le \lim_{\mathcal{U}(k)} \frac{1}{k!2^k} \sum_{\gamma \in U_k \rtimes S_k} \left\|(\Phi + B)(\gamma(x+y)) - (\Phi + B)(\gamma x) - (\Phi + B)(\gamma y)\right\| \\
&\le \lim_{\mathcal{U}(k)} \frac{1}{k!2^k} \sum_{\gamma \in U_k \rtimes S_k} Q(\Phi + B)\Big(\|\gamma x\| + \|\gamma y\|\Big) \\
&= \lim_{\mathcal{U}(k)} \frac{1}{k!2^k} \sum_{\gamma \in U_k \rtimes S_k} Q(\Phi + B)(\|x\| + \|y\|) \\
&= Q(\Phi + B)(\|x\| + \|y\|).
\end{aligned}
$$

Thus, it suffices to estimate $Q(\mathsf{KP}_p + H)$ when H is homogeneous bounded and commutes with U and S, in which case there is a sequence $(c_n)_{n\ge1}$ such that $H\big(\sum_{i\in I} e_i\big) = c_n \sum_{i\in I} e_i$ provided $|I| = n$. Note also that since the norm of $\sum_{i\in I} e_i$ in ℓ_p equals $|I|^{1/p}$, if we assume that $Q(\mathsf{KP}_p + H) \le d$, then letting $x = \sum_{1\le i\le n} e_i$ and $y = \sum_{n+1\le i\le 2n} e_i$ in the inequality

$$\left\|(\mathsf{KP}_p + H)(x+y) - (\mathsf{KP}_p + H)(x) - (\mathsf{KP}_p + H)(y)\right\| \le d(\|x\| + \|y\|)$$

yields

$$\left\|\left(\frac{\log(2n)}{p} + c_{2n}\right)(x+y) - \left(\frac{\log(n)}{p} + c_n\right)(x+y)\right\| \le d(\|x\| + \|y\|) \le d2^{\frac{1}{p}}\|x+y\|,$$

hence $\left|p^{-1}\log 2 + c_{2n} - c_n\right| \le d2^{1/p}$, which is definitely incompatible with the boundedness of the subsequence $(c_{2^k})_{k\ge0}$ if $p^{-1}\log 2 < d2^{1/p}$. □

The situation for Ribe's maps is the same:

Corollary 4.5.9 *Let $\varrho_\varphi\colon \ell_1^0 \longrightarrow \mathbb{K}$ be the scalar-valued map*

$$\varrho_\varphi(x) = \sum_{n\ge1} x(n)\varphi\left(\log\frac{\|x\|}{|x(n)|}\right),$$

where $\varphi\colon [0,\infty) \longrightarrow \mathbb{K}$ is a Lipschitz function vanishing at zero.

(a) *If φ is differentiable and* $\lim_{t\to\infty}\varphi'(t) = 0$, *then* $Q[\varrho_\varphi] = 0$.

(b) *If $\varphi(t) = t$ then* $Q[\varrho_\varphi] > 0$.

Proof The first part follows from Proposition 4.5.5 since ϱ_φ is quite clearly the Kalton–Peck map $\mathsf{KP}_\varphi\colon \ell_1^0 \longrightarrow \ell_1$ followed by the sum functional on ℓ_1. The second part follows after the lines of the proof of Proposition 4.5.6. □

Ok, we are now relieved knowing that the topology of Ext can be non-trivial ($Q[\Phi] \neq 0$ for *some* Φ). Could it be Hausdorff ($Q[\Phi] \neq 0$ for *all* non-zero Φ)? Despite what Theorem 4.5.1 says, Ext spaces will rarely be Hausdorff. For instance, Propositions 3.3.5 and 4.5.5 together imply that $\operatorname{Ext}(X, X)$ fails to be Hausdorff if X is a sequence space without copies of c_0, while Corollary 4.5.9 implies that $\operatorname{Ext}(\ell_1, Y)$ fails to be Hausdorff if $Y^* \neq 0$. There is a simple criterion to detect whether Ext spaces are Hausdorff in terms of the behaviour of bounded homogeneous maps. Let us introduce the bounded versions of the parameters $K[X, Y]$ in Equation (3.17) and $K^{(p)}[X, Y]$ in Equation (3.20): given two quasi-Banach spaces X, Y, we set

$$K_0[X, Y] = \sup\left\{\frac{D(\Phi)}{Q(\Phi)} : \Phi \in \mathsf{B}(X, Y)\right\}, \tag{4.20}$$

and, when X, Y are p-Banach spaces,

$$K_0^{(p)}[X, Y] = \sup\left\{\frac{D(\Phi)}{Q^{(p)}(\Phi)} : \Phi \in \mathsf{B}(X, Y)\right\}. \tag{4.21}$$

These parameters do not vary when B is replaced by $\mathsf{B}+\mathsf{L}$. If K and $K^{(p)}$ detect when the corresponding Ext spaces vanish, their 'bounded variations' K_0 and $K_0^{(p)}$ detect when they are Hausdorff.

Lemma 4.5.10 $\operatorname{Ext}(X, Y)$ *is Hausdorff if and only if* $K_0[X, Y] < \infty$.

Proof Throughout this proof, we will omit the spaces X, Y to lighten the notation. Observe that B is also a quasi-Banach space endowed with the standard sup quasinorm $\|\cdot\|$. The obvious operator $J\colon (\mathsf{B}, \|\cdot\|) \longrightarrow \mathsf{Q}_{\mathsf{L}}$ sending B to $[B + \mathsf{L}]$ is bounded, with $\|J\|$ depending only on the moduli of concavity of X and Y. Since $\mathsf{Q}_{\mathsf{LB}} = \mathsf{Q}_{\mathsf{L}}/J[\mathsf{B}]$, we have (a quick look at Note 2.15.2 could help)

$$\mathsf{Q}_{\mathsf{LB}} \text{ is Hausdorff} \iff J \text{ has closed range} \iff J \text{ is relatively open.}$$

But the last condition is a rewording of $K_0[X, Y] < \infty$: indeed, J relatively open means that there is a constant C such that whenever $\Phi \in \mathsf{Q}$, there is $B \in \mathsf{B}$ with $\|B\| \leq C$ such that $\Phi = B + L$ for some $L \in \mathsf{L}$. So, C is $K_0[X, Y]$ in disguise. □

Corollary 4.5.11 $\operatorname{Ext}_{p\mathbf{B}}(X, Y)$ *is Hausdorff if and only if* $K_0^{(p)}[X, Y] < \infty$.

Now we need to do the first of several time warps, invoking Theorem 5.3.13 to obtain:

Corollary 4.5.12 *Let X be a quasi-Banach space and Y an ultrasummand. Assume that X either has the BAP or is a $\mathscr{K}$-space. Then* $\operatorname{Ext}(X, Y)$ *is Hausdorff if and only if it is* 0.

This result will appear in Chapter 5 as $K_0[X, Y] < \infty \Rightarrow K[X, Y] < \infty$ and will be understood then as a positive result that allows approximation of non-trivial extensions by means of trivial ones. In our present circumstances, it appears as a negative result, if only because one would prefer to have Hausdorff topologies. Oh well, *c'est la vie*. We bid farewell to problems about the topology of Ext spaces by providing some pairs of Banach spaces for which $\operatorname{Ext}(X, Y)$ and $\operatorname{Ext}_{\mathbf{B}}(X, Y)$ is non-zero *and* Hausdorff. Note that the examples of this phenomenon that can be derived from Theorem 4.5.1 and Corollaries 4.5.2 and 4.5.3 are based on pairs of quasi-Banach spaces satisfying $\mathfrak{L}(X, Y) = 0$, something that cannot occur in Banach spaces.

Proposition 4.5.13 *If X is a non-separable WCG Banach space,* $\operatorname{Ext}_{\mathbf{B}}(X, c_0)$ *is non-zero and Hausdorff.*

Proof We need to do another time warp here: future Lemma 8.7.1 will show that $\operatorname{Ext}_{\mathbf{B}}(X, c_0) \neq 0$, even if past Sections 2.2.10 and 2.12.6 obtained that same result for some non-separable WCG spaces X such as $c_0(\Gamma)$ and $\ell_p(\Gamma)$. To prove that $\operatorname{Ext}_{\mathbf{B}}(X, c_0)$ is Hausdorff, we will show that $K_0^{(1)}(X, c_0) \leq 4$. Let $B\colon X \longrightarrow c_0$ be a bounded homogeneous map, with $Q^{(1)}(B) \leq 1$, so that $c_0 \oplus_B X \simeq c_0 \times X$ is WCG. We know from Proposition 1.7.7 that c_0 is 2-complemented in any WCG Banach space, but we require a way to control the norm of the projection since the quasinorm $\|\cdot\|_B$ is not a norm, only equivalent to one. This is the point where the hypothesis on the 1-linearity constant of B comes into play. Indeed, set $p = 1$ in Theorem 3.6.7 and Lemma 1.1.2 to obtain a true norm $|\cdot|_{\mathrm{co}}$ on $c_0 \oplus_B X$ that is 2-equivalent to $\|(\cdot,\cdot)\|_B$. Therefore, the canonical copy of c_0 inside $c_0 \oplus_B X$ yields a 2-copy of c_0 inside $(c_0 \oplus_B X, |\cdot|_{\mathrm{co}})$, which is 2-complemented, so that the canonical copy is 4-complemented in $(c_0 \oplus_B X, |\cdot|_{\mathrm{co}})$ and thus also 4-complemented in $c_0 \oplus_B X$ since $|\cdot|_{\mathrm{co}} \leq \|\cdot\|_B$. This projection must have the form $P(y, x) = y - L(x)$, thus we conclude that $\|B - L\| \leq 4$. □

If, moreover, X is a $\mathscr{K}$-space then $\operatorname{Ext}(X, c_0)$ is non-zero and Hausdorff. The $\mathscr{K}$-space hypothesis cannot be dropped: $\operatorname{Ext}(\ell_2(I) \oplus \ell_1, c_0)$ contains $\operatorname{Ext}(\ell_1, \mathbb{K})$ complemented, and thus, according to Corollary 4.5.9, it cannot be Hausdorff.

4.6 Notes and Remarks

4.6.1 Adjoint Functors

The notion of adjoint functors, discovered by Kan in the 1950s, is one of the landmarks of category theory. It captures a phenomenon that pervades modern mathematics yet is virtually imperceptible outside category theory.

Definition Let $F\colon \mathbf{C} \longrightarrow \mathbf{D}$ and $G\colon \mathbf{D} \longrightarrow \mathbf{C}$ be functors. Then F is said to be left-adjoint to G (and G right-adjoint to F), written $F \dashv G$, if, given objects $C \in \mathbf{C}$ and $D \in \mathbf{D}$, there is an isomorphism between $\mathrm{Hom}_{\mathbf{D}}(F(C), D)$ and $\mathrm{Hom}_{\mathbf{C}}(C, G(D))$ that is natural in both C and D, in the sense that $(C, D) \rightsquigarrow \mathrm{Hom}_{\mathbf{D}}(F(C), D)$ and $(C, D) \rightsquigarrow \mathrm{Hom}_{\mathbf{C}}(C, G(D))$ are naturally equivalent set-valued functors on $\mathbf{C}^{\mathrm{op}} \times \mathbf{D}$.

Here you will find relevant examples en regalia, even if a few details are left to the reader. Please be sure that arrows stick with the pointy end.

(a) **The duality functor versus itself.** Let $*\colon \mathbf{B} \longrightarrow \mathbf{B}^{\mathrm{op}}$ be the duality functor taking each Banach space X to its dual X^* and each operator $u\colon X \longrightarrow Y$ to its adjoint $u^*\colon Y^* \longrightarrow X^*$. If we abuse notation by also writing $*\colon \mathbf{B}^{\mathrm{op}} \longrightarrow \mathbf{B}$ then the natural isomorphism between $\mathfrak{L}(Y, X^*)$ and $\mathfrak{L}(X, Y^*) = \mathfrak{L}^{\mathrm{op}}(Y^*, X)$ provided by $u \longmapsto u^*|_X$ with inverse $v \longmapsto v^*|_Y$ shows that $* \dashv *$.

(b) **The p-convex envelope versus the forgetful functor.** The construction of the p-convex envelope $X_{(p)}$ of a quasi-Banach space X as described in Section 1.1 is quite clearly a functor $(\cdot)_{(p)}\colon \mathbf{Q} \longrightarrow p\mathbf{B}$. If $\Box\colon p\mathbf{B} \longrightarrow \mathbf{Q}$ is the forgetful functor (formal inclusion) then $\mathfrak{L}(X_{(p)}, Y) = \mathfrak{L}(X, Y)$ means $(\cdot)_{(p)} \dashv \Box$.

(c) **Boolean algebras versus Stone compacta.** The duality between Boolean algebras and Stone compacta acts as follows. The contravariant functor $\mathrm{ult}\colon \mathbf{A} \longrightarrow \mathbf{K_0}$ assigns to each Boolean algebra $\mathfrak{B}$ its Stone compact $\mathrm{ult}(\mathfrak{B}) = \mathrm{Hom}_{\mathbf{A}}(\mathfrak{B}, \mathbf{2})$ and to each Boolean morphism $\alpha\colon \mathfrak{A} \longrightarrow \mathfrak{B}$ the continuous function $\alpha^\circ\colon \mathrm{ult}(\mathfrak{B}) \longrightarrow \mathrm{ult}(\mathfrak{A})$ of simple right composition $f \longmapsto fa$. The contravariant functor $\mathfrak{cl}\colon \mathbf{K_0} \longrightarrow \mathbf{A}$ assigns to a (0-dimensional) compactum K the Boolean algebra $\mathfrak{cl}(K)$ of its clopen sets and to a continuous function $f\colon K \longrightarrow S$ the Boolean morphism $\mathfrak{cl}(S) \longrightarrow \mathfrak{cl}(K)$ that sends C to $f^{-1}[C]$. Considering $\mathrm{ult}\colon \mathbf{A} \longrightarrow \mathbf{K}_0^{\mathrm{op}}$ and $\mathfrak{cl}\colon \mathbf{K}_0^{\mathrm{op}} \longrightarrow \mathbf{A}$ as covariant functors, we have $\mathfrak{cl} \dashv \mathrm{ult}$: a function $f\colon K \longrightarrow \mathrm{ult}(\mathfrak{B}) \subset 2^{\mathfrak{B}}$ can obviously be understood as a map $K \times \mathfrak{B} \longrightarrow \mathbf{2}$, and thus the correspondence $\phi \longleftrightarrow f \iff f(\cdot, \phi(C)) = 1$ on every clopen $C \subset K$ is clear, and therefore $\mathrm{Hom}_{\mathbf{A}}(\mathfrak{cl}(K), \mathfrak{B}) = C(K, \mathrm{ult}(\mathfrak{B}))$.

(d) **Continuous functions versus the dual ball functor.** Consider the (contravariant) functor $C\colon \mathbf{K} \longrightarrow \mathbf{B}_1$ that associates to each compact space

K the Banach space $C(K)$ and to each continuous mapping $\varphi\colon K \longrightarrow S$ the contraction $\varphi^\circ\colon C(T) \longrightarrow C(K)$ given by $\varphi^\circ(f) = f \circ \varphi$. This functor has a right-adjoint, namely the dual ball functor $B^*_\bullet\colon \mathbf{B_1} \longrightarrow \mathbf{K}$ that assigns to each Banach space X the compact space B^*_X and to each contractive operator $\tau\colon X \longrightarrow Y$ the restriction $\tau^*\colon B^*_Y \longrightarrow B^*_X$. To check that $C \dashv B^*_\bullet$, we just need to identify $C(K, B^*_Y)$ with the contractive operators in $\mathfrak{L}(Y, C(K))$, which occurs by the formula $\tau(y)(k) = \langle\varphi(k), y\rangle$, where $\tau \in \mathfrak{L}(Y, C(K))$ is contractive, $\varphi\colon K \longrightarrow B^*_Y$ is continuous and $y \in Y, k \in K$. The composition $X \rightsquigarrow B^*_X \rightsquigarrow C(B^*_X)$ is called the Banach–Mazur functor [430, p. 180]. It is a covariant 'endofunctor' in the contractive category of Banach spaces. Lemma 2.12.1 is desperately trying to say that the correspondence $X \rightsquigarrow \delta_X$ defines a natural transformation between the 'identity functor' on $\mathbf{B_1}$ and the Banach–Mazur functor $C(B^*_\bullet)$, which is now obvious in view of the commutativity of the diagram

$$\begin{array}{ccc} X & \xrightarrow{\ \delta_X\ } & C(B^*_X) \\ \big\downarrow{\scriptstyle \tau} & & \big\downarrow{\scriptstyle \left(\tau^*|_{B^*_Y}\right)^\circ} \\ Y & \xrightarrow{\ \delta_Y\ } & C(B^*_Y) \end{array}$$

Now please pay Lemma 2.12.1 a second visit. After that, do the time warp again to peruse Section 8.1 and return back to the present. Everything is now clear, right? And you can always go back to Semadeni [429; 430] for further developments.

(e) **Linearisation of Lipschitz maps.** Let us consider the category $\mathbf{M_0}$ of metric spaces with a distinguished point (pointed spaces) and Lipschitz maps acting between them that preserve the distinguished points. Every Banach space can be seen as a pointed space by taking the origin as the distinguished point. For every pointed metric space (M, m), there exists a Banach space $\mathcal{F}(M, m)$ and a contractive Lipschitz mapping $\delta^M\colon M \longrightarrow \mathcal{F}(M, m)$ vanishing at m with the following universal property: for every Banach space Y and every Lipschitz function $L\colon M \longrightarrow Y$ vanishing at m, there exists a unique operator $\phi_L\colon \mathcal{F}(M, m) \longrightarrow Y$ such that $L = \phi_L \circ \delta^M$; pictorially,

$$\begin{array}{ccc} M & \xrightarrow{\ \delta^M\ } & \mathcal{F}(M, m) \\ & {\scriptstyle L}\searrow \quad \swarrow{\scriptstyle \phi_L} & \\ & Y & \end{array}$$

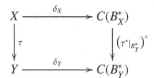

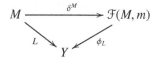

Before going any further, we ask the reader to check that, no matter how the Banach spaces $\mathcal{F}(M, m)$ are to be determined, the construction defines a functor $\mathcal{F}\colon \mathbf{M_0} \longrightarrow \mathbf{B}$ assigning to each 'pointed' Lipschitz map $f\colon (M,m) \longrightarrow (N,n)$ the operator that forms a commutative square

$$\begin{array}{ccc} M & \xrightarrow{\ f\ } & N \\ {\scriptstyle \delta^M}\downarrow & & \downarrow{\scriptstyle \delta^N} \\ \mathcal{F}(M,m) & \xrightarrow{\ \phi\ } & \mathcal{F}(N,n) \end{array}$$

The fact that $\mathrm{Lip}_0((M,m),Y) = \mathfrak{L}(\mathcal{F}(M,m),Y)$ for every pointed M and every Banach space Y means that $\mathcal{F}\dashv\Box$ for the obvious forgetful functor $\Box\colon \mathbf{B} \longrightarrow \mathbf{M_0}$. The following description of $\mathcal{F}(M,m)$, called the Arens–Eells space of M or the Lipschitz-free space of M, depending on the context, is basically the second version given by Weaver [455, 2.2]. Fix (M,m), and consider the space $\mathrm{Lip}_0(M,m)$ of Lipschitz functions $f\colon M \longrightarrow \mathbb{K}$ vanishing at m; it is a Banach space normed by the Lipschitz constant. The mapping $\delta^M\colon M \longrightarrow \mathrm{Lip}_0(M,m)^*$ is the obvious evaluation map $\langle \delta^M(x), f\rangle = f(x)$, which linearises scalar-valued functions since it provides a commutative diagram

$$\begin{array}{ccc} M & \xrightarrow{\quad \delta^M \quad} & \mathrm{Lip}_0(M,m)^* \\ & {\scriptstyle f}\searrow \quad \swarrow{\scriptstyle \delta_f} & \\ & \mathbb{K} & \end{array}$$

Let $\mathcal{F}(M,m) = \overline{\delta^M[M]} \subset \mathrm{Lip}_0(M,m)^*$ such that, if $F\colon M \longrightarrow Y$ is a Lipschitz map with $F(m) = 0$, there is an operator ϕ acting by $\phi(\mu)(y^*) = \delta_{y^*F}(\mu)$ for $\mu \in \mathrm{Lip}_0(M,m)^*$, which makes the following diagram commute:

$$\begin{array}{ccccc} M & \xrightarrow{\ \delta^M\ } & \mathcal{F}(M,m) & \xrightarrow{\text{inclusion}} & \mathrm{Lip}_0(M,m)^* \\ & {\scriptstyle F}\searrow & \downarrow & & \downarrow{\scriptstyle \phi} \\ & & Y & \xrightarrow{\text{inclusion}} & Y^{**} \end{array}$$

Curiously enough, the restriction of ϕ to $\mathcal{F}(M,m)$ takes values in Y: indeed, $\phi(\delta^M x) = F(x)$ since $\phi(\delta^M x)(y^*) = \delta_{y^*F}(\delta^M x) = y^*F(x) = \langle y^*, F(x)\rangle$ and $\delta^M[M]$ spans a dense subspace of $\mathcal{F}(M,m)$. In [189] Godefroy and Kalton consider the case in which the metric space is a Banach space, say X: applying the universal property of $\mathcal{F}(X)$ to the Lipschitz map $\mathbf{1}_X$, we obtain a quotient operator $\beta\colon \mathcal{F}(X) \longrightarrow X$, called the baryentric map. Since $\beta \circ \delta^X = \mathbf{1}_X$, the map δ^X is a Lipschitz section for the exact sequence

$$0 \longrightarrow \ker\beta \longrightarrow \mathcal{F}(X) \xrightarrow{\ \beta\ } X \longrightarrow 0$$

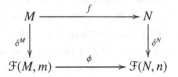

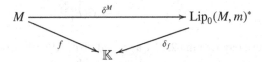

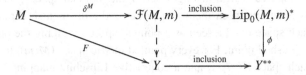

Their deep result [189, Theorem 3.1] establishes that when X is separable, *every* exact sequence of Banach spaces $0 \longrightarrow Y \longrightarrow Z \longrightarrow X \longrightarrow 0$ that admits a Lipschitz section splits.

4.6.2 Derived Functors

Functors can be derived; rather, *certain* functors on *certain* categories can be derived. And no one can describe the process more clearly than Mac Lane [351, p. 389]: 'A standard method is: Take a resolution, apply a covariant functor, take homology of the resulting complex. This gives a connected sequence of functors, called the derived functors'. Entering into specifics, to perform right derivation, we need additive categories (those admitting 0, finite products and coproducts and such that the spaces $\mathrm{Hom}(A, B)$ are Abelian groups and composition is additive in each variable), a covariant additive functor F (this means that F must preserve finite products and coproducts and that the induced map $\mathrm{Hom}(A, B) \longrightarrow \mathrm{Hom}(FA, FB)$ is a group homomorphism) and enough injective objects in the category. Given an object X, we can form an injective resolution, i.e., an exact sequence

$$0 \longrightarrow X \longrightarrow \mathfrak{I}_1 \xrightarrow{\partial_1} \mathfrak{I}_2 \xrightarrow{\partial_2} \mathfrak{I}_3 \xrightarrow{\partial_3} \cdots$$

in which each $\mathfrak{I}_n$ is injective. Applying the functor F, we cease to have an exact sequence and just get, when F is exact on the left, a *complex* which we will call $\mathbf{FX}$:

$$0 \longrightarrow FX \longrightarrow F\mathfrak{I}_1 \xrightarrow{F\partial_1} F\mathfrak{I}_2 \xrightarrow{F\partial_2} F\mathfrak{I}_3 \xrightarrow{F\partial_3} \cdots$$

A complex $\mathbf{C}$ is a diagram $0 \longrightarrow C_1 \xrightarrow{\partial_1} C_2 \xrightarrow{\partial_2} C_3 \xrightarrow{\partial_3} \cdots$ in which $\partial_{n+1}\partial_n = 0$ (but exactness is not required). Having a complex in an additive category allows us to form the *cohomology groups* $H^n(\mathbf{C}) = \ker \partial_{n+1}/\operatorname{coker} \partial_n$. The groups $R_nF(X) = H^n(\mathbf{FX})$ of $\mathbf{FX}$ are the values of the (right-) derived functors of F at X. Thus, a suitable functor $F\colon \mathbf{A} \longrightarrow \mathbf{B}$ generates a sequence of functors $R_nF\colon \mathbf{A} \longrightarrow \mathbf{B}$. A similar procedure with projective presentations and right exact functors produces the left derived functors L_nF. To work with contravariant functors, use the opposite trick. The 'connected sequence' of derived functors MacLane mentions means that given an exact sequence $0 \longrightarrow Y \longrightarrow Z \longrightarrow X \longrightarrow 0$, we obtain a long exact sequence

$$
\begin{array}{l}
0 \longrightarrow FY \longrightarrow FZ \longrightarrow FX \longrightarrow \\
\quad\longrightarrow R_1FY \longrightarrow R_1FZ \longrightarrow R_1FX \longrightarrow \\
\quad\longrightarrow R_2FY \longrightarrow R_2FZ \longrightarrow R_2FX \longrightarrow \dots
\end{array}
$$

Its construction requires two things to be explained: (a) that R_nF is a functor so that given an arrow $A \longrightarrow B$, we have a morphism $R_nF(A) \longrightarrow R_nF(B)$, namely $H^n(\mathbf{FA}) \longrightarrow H^n(\mathbf{FB})$; and (b) the construction of the connecting morphism $R_nF(X) \longrightarrow R_{n+1}F(Y)$, namely $H_n(\mathbf{FX}) \longrightarrow H_{n+1}(\mathbf{FY})$. To obtain (a), the point is to understand that H^n is a functor from the category of complexes to the category of chains (like complexes, but without the condition $\partial_{n+1}\partial_n = 0$), in which a morphism $\varphi\colon \mathbf{C} \longrightarrow \mathbf{D}$ between two complexes is a family φ of maps $\varphi_n\colon C_n \longrightarrow D_n$ such that $\varphi_n\partial_n = \partial_n\varphi_n$. The connecting morphism in (b) appears by taking an exact sequence $0 \longrightarrow Y \longrightarrow Z \longrightarrow X \longrightarrow 0$ and constructing projective resolutions of Y, Z, X connected as in Lemma 2.9.4,

$$
\begin{array}{ccccccccc}
 & & 0 & & 0 & & 0 & & \\
 & & \downarrow & & \downarrow & & \downarrow & & \\
0 & \longrightarrow & Y & \longrightarrow & Z & \longrightarrow & X & \longrightarrow & 0 \\
 & & \downarrow & & \downarrow & & \downarrow & & \\
0 & \longrightarrow & \mathcal{J}_1^Y & \longrightarrow & \mathcal{J}_1^Z & \longrightarrow & \mathcal{J}_1^X & \longrightarrow & 0 \\
 & & \downarrow & & \downarrow & & \downarrow & & \\
0 & \longrightarrow & \mathcal{J}_2^Y & \longrightarrow & \mathcal{J}_2^Z & \longrightarrow & \mathcal{J}_2^X & \longrightarrow & 0 \\
 & & \downarrow & & \downarrow & & \downarrow & & \\
 & & \dots & & \dots & & \dots & &
\end{array}
$$

to obtain an exact sequence of complexes $0 \longrightarrow \mathbf{Y} \longrightarrow \mathbf{Z} \longrightarrow \mathbf{X} \longrightarrow 0$. Now, each map ∂_n in a complex $\mathbf{C}$ induces a map $\overline{\partial}_n : \operatorname{coker} \partial_n \longrightarrow \ker \partial_{n+1}$ such that $\ker \overline{\partial}_n = H^n(\mathbf{C})$ and $\operatorname{coker} \overline{\partial}_n = H^{n-1}(\mathbf{C})$, as is easy to see from [213, IV Lemma 2.2.]. Being already well aware that in a diagram like (2.48),

$$
\begin{array}{ccccccccc}
0 & \longrightarrow & \cdot & \longrightarrow & \cdot & \longrightarrow & \cdot & \longrightarrow & 0 \\
 & & \downarrow & & \downarrow & & \downarrow & & \\
0 & \longrightarrow & \cdot & \longrightarrow & \cdot & \longrightarrow & \cdot & \longrightarrow & 0
\end{array}
$$

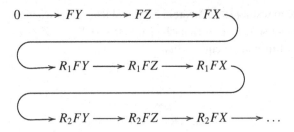

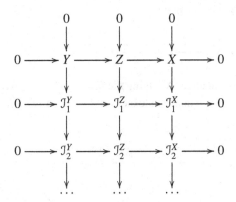

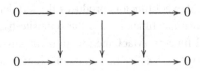

the kernels and cokernels of the vertical arrows form exact sequences,

$$
\begin{array}{ccccccccc}
0 & \longrightarrow & \cdot & \longrightarrow & \cdot & \longrightarrow & \cdot & & \\
 & & \downarrow & & \downarrow & & \downarrow & & \\
0 & \longrightarrow & \cdot & \longrightarrow & \cdot & \longrightarrow & \cdot & \longrightarrow & 0 \\
 & & \downarrow & & \downarrow & & \downarrow & & \\
0 & \longrightarrow & \cdot & \longrightarrow & \cdot & \longrightarrow & \cdot & \longrightarrow & 0 \\
 & & \downarrow & & \downarrow & & \downarrow & & \\
 & & \cdot & \longrightarrow & \cdot & \longrightarrow & \cdot & \longrightarrow & 0
\end{array}
$$

we thus get a commutative diagram

$$
\begin{array}{ccccccccc}
 & & H^{n-1}(\mathbf{Y}) & \longrightarrow & H^{n-1}(\mathbf{Z}) & \longrightarrow & H^{n-1}(\mathbf{X}) & & \\
 & & \downarrow & & \downarrow & & \downarrow & & \\
 & & \operatorname{coker} \partial_n & \longrightarrow & \operatorname{coker} \partial_n & \longrightarrow & \operatorname{coker} \partial_n & \longrightarrow & 0 \\
 & & \downarrow{\scriptstyle \overline{\partial}_n} & & \downarrow{\scriptstyle \overline{\partial}_n} & & \downarrow{\scriptstyle \overline{\partial}_n} & & \\
0 & \longrightarrow & \ker \partial_{n+1} & \longrightarrow & \ker \partial_{n+1} & \longrightarrow & \ker \partial_{n+1} & & \\
 & & \downarrow & & \downarrow & & \downarrow & & \\
 & & H^n(\mathbf{Y}) & \longrightarrow & H^n(\mathbf{Z}) & \longrightarrow & H^n(\mathbf{X}) & &
\end{array}
$$

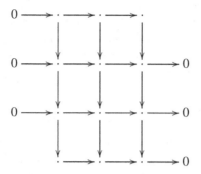

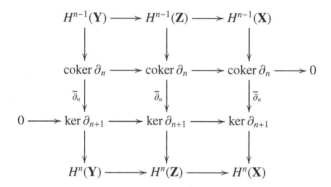

Here the snake lemma provides the connecting morphism $H^n(\mathbf{X}) \longrightarrow H^{n-1}(\mathbf{Y})$. The whole process is perfectly described in [351, XII, 9], done in [213, Chapter IV] and depicted in [423, Chapter 6]. The injective presentations we choose have no real effect because taking cohomology groups preserves homotopy equivalence (the reader was warned in Note 2.15.4 that this second encounter will not be satisfactory for either of us) between complexes. Left derivation is analogously obtained, only that using projective resolutions. When both projective and injective presentations are available, the two derivations are naturally equivalent, and the derived functor is unique (even under some more general conditions; see [186, III.7]). The point of all this is that the derivation of functors in $\mathbf{B}$ (or even in $p\mathbf{B}$) can be considered kosher: the left-derived functor of $\mathfrak{L}$ is $\operatorname{Ext}_{p\mathbf{B}}$, higher-order derived functors are $\operatorname{Ext}^n_{p\mathbf{B}}$ and all of them form a long homology sequence. But that is not the case in $\mathbf{Q}$. The case of $\mathbf{Q}$ is startling: it is a category that exists in nature, where not many things work properly but still it does everything it is expected to do. In particular, one could perform Yoneda-like constructions of Ext and Ext^n that somehow work fine. From a more sceptical point of view, the *long* homology sequence we constructed in the text had only six terms, which, we must admit, is not that long after all. What comes after the first six terms? What about those phantomatic Ext^n (see the next section)? At the categorical core of these considerations, we encounter the question; is Ext a (the) derived functor of $\mathfrak{L}$ in $\mathbf{Q}$ for some reasonable sense of the word?

4.6.3 Unknown Knowns about Ext^2

This note is not about the things we know about Ext^2 or the many things we do not know about Ext^2: it is about a few things that we do not know we know about Ext^2. Let us focus on Banach spaces, which is secure land against the quicksands of $\mathbf{Q}$. What is the second derived functor of $\mathfrak{L}$? Fix two Banach spaces X, Y. Then $\mathrm{Ext}^2(X, Y)$ is the set (vector space) of exact sequences (thus, complexes) E,

$$0 \longrightarrow Y \longrightarrow E_1 \longrightarrow E_2 \longrightarrow X \longrightarrow 0$$

modulo a non-immediate relation of equivalence: the smallest making E equivalent to F if there is a morphism of complexes $E \longrightarrow F$ that is the identity on the end spaces X, Y. A four-term exact sequence E is *trivial* if it is equivalent to the zero-sequence

$$0 \longrightarrow Y = Y \xrightarrow{0} X = X \longrightarrow 0$$

As before, $\mathrm{Ext}^2(X, Y) = 0$ means that all elements in $\mathrm{Ext}^2(X, Y)$ are trivial. We were not aware so far that those notions held any interest for us. One reason to be interested is to know whether $\mathrm{Ext}^2(X, Y) = 0$ (or not) provides information about important spaces associated with X or Y that are too complicated for a direct inspection. Another is that in the same way that deep questions about $\mathfrak{L}$ can be formulated using Ext (e.g. Zippin's Problem 8.6.1, which is seminal for the whole theory developed in this book, can be reformulated as asking for which separable spaces X we have $\mathrm{Ext}(X, C[0, 1]) = 0$), deep questions about Ext can be formulated using Ext^2, and the responsible is the fantabulous representations of $\mathrm{Ext}(X, Y)$ as $\mathfrak{L}(\kappa(X), Y)$ (modulo 'extendable' operators) or $\mathfrak{L}(X, c\kappa(Y))$ (modulo 'liftable' operators). We now have

$$\mathrm{Ext}^2(X, Y) = \mathrm{Ext}(\kappa(X), Y) = \frac{\mathfrak{L}(\kappa^2(X), Y)}{\text{extendable}} = \frac{\mathfrak{L}(X, c\kappa^2(Y))}{\text{liftable}} = \mathrm{Ext}(X, c\kappa(Y)).$$

After Proposition 8.6.4, we will know that $\mathrm{Ext}(X, C[0, 1]) = 0$ implies that X is Schur, for a separable Banach space X. The obvious question of whether $\mathrm{Ext}(X, C[0, 1]) = 0$ for every Schur space, or at least for every subspace of ℓ_1, appears. Kalton worked on this question, yelling 'not yet' when asked whether he already had a counterexample. The question can be reformulated as for which separable spaces X one has $\mathrm{Ext}^2(X, C[0,1]) = 0$. This is one of the things we did not know we knew. It is hard to resist mentioning what we *do* know at the moment of writing these lines: $\mathrm{Ext}^2(X, Y) = 0$ *in each of the following cases:* (a) *Y is injective or X is projective;* (b) *Sobczyk's theorem: $Y = c_0$ and X is separable;* (c) *Lindenstrauss lifting: X is an $\mathscr{L}_1$-space and Y is an ultrasummand;* (d) *Johnson–Zippin theorem: X is an $\mathscr{L}_\infty$-space and Y is a subspace of c_0;* (e) $\mathrm{Ext}^2(X, X) \neq 0$ *for every B-convex Banach space; in particular* $\mathrm{Ext}^2(\ell_p, \ell_p) \neq 0$ *for* $1 < p < \infty$. Assertions (a)–(d) are in [71], and (e) is in [70].

4.6.4 Open Problems about the Topology of Ext

Questions about the topological nature of Ext can be somehow shepherded back to quasi-Banach space theory. Anyway, there is no more suitable way to say farewell to the topology of Ext than with a note on open problems. We know that (a) non-zero Φ with $Q[\Phi] = 0$ exist; (b) non-zero Φ with $Q[\Phi] \neq 0$ exist; (c) non-zero Ext spaces where $Q[\Phi] \neq 0$ for every non-zero Φ exist. It would be interesting to know if (d) non-zero $\mathrm{Ext}(X, Y)$ spaces where $Q[\Phi] = 0$ for every Φ also exist. In particular, we do not know if that happens with $X = \ell_p(\ell_1^n)$ and $Y = \mathbb{K}$ for $1 < p < \infty$. We do not know either if there are separable Banach spaces for which Ext (or $\mathrm{Ext}_{\mathbf{B}}$) is non-zero and Hausdorff or if $\mathrm{Ext}(X, Y)$ or $\mathrm{Ext}_{\mathbf{B}}(X, Y)$ can be a 'classical' space, in particular finite-dimensional or 1-dimensional. Simple algebraic manipulations show that if either X or Y is isomorphic to its square then $\mathrm{Ext}(X, Y)$ is either zero or infinite-dimensional. Indeed, $\mathrm{Ext}(A \oplus B, Y) = \mathrm{Ext}(A, Y) \times \mathrm{Ext}(B, Y)$, hence if $X \simeq X^2$, then $\mathrm{Ext}(X, Y)$ is isomorphic to its square for all Y. The case where $Y \simeq Y^2$ is similar, using the natural isomorphism $\mathrm{Ext}(X, A \times B) = \mathrm{Ext}(X, A) \times \mathrm{Ext}(X, B)$ instead. This was observed by Kuchment long ago [312] with a different, much nicer proof; see [313, Lemma 4.6]. The ideas of Kuchment [313, Section 4] can be easily adapted to show the following:

Proposition *Let X be a quasi-Banach space and let Y be an ultrasummand such that* $\mathrm{Ext}(X, Y)$ *is finite-dimensional. Then* $\mathrm{Ext}(X, Y) = 0$ *in the following cases:*

(a) *X has the BAP.*
(b) *X is a $\mathscr{K}$-space and Y has the BAP.*

However, the vril power of coming Theorem 5.3.13 reduces the proof to show that $\mathrm{Ext}(X, Y)$ is Hausdorff as long as it is finite-dimensional. This fact, which we prove next, relies on the following obvious consequence of the open mapping theorem that Kuchment attributes to Krein: *if* $\tau: A \longrightarrow B$ *is an operator between quasi-Banach spaces and* $B/\tau[A]$ *is finite-dimensional then* τ *has closed range.* With this in hand, assume X, Y are q-Banach spaces for some $q \in (0,1]$, pick any $p \in (0,q)$ and use the identification $\operatorname{Ext}(X,Y) = \operatorname{Ext}_{p\mathbf{B}}^{\mathrm{proj}}(X,Y)$ from Section 4.4. Consider the projective presentation of X implicit in the definition of $\operatorname{Ext}_{p\mathbf{B}}^{\mathrm{proj}}(X,Y)$ and observe that

$$\operatorname{Ext}_{p\mathbf{B}}^{\mathrm{proj}}(X,Y) = \frac{\mathfrak{L}(\kappa_p(X),Y)}{\jmath^\circ[\mathfrak{L}(\ell_p(I),Y)]},$$

as semi-quasi-Banach spaces, where $\jmath: \kappa_p(X) \longrightarrow \ell_p(I)$ is the inclusion map. Now, if $\operatorname{Ext}(X,Y)$ is finite-dimensional then $\jmath^\circ$ has closed range, and so $\operatorname{Ext}(X,Y)$ is Hausdorff.

The fact that $\operatorname{Ext}(X,\mathbb{K})$ has dimension 1 if X is the quotient of L_p by a line and $0 < p < 1$ (Corollary 4.2.8) shows that the hypotheses in (a) and (b) cannot be removed. If X and Y are Banach spaces, we can replace Ext by $\operatorname{Ext}_{\mathbf{B}}$ in the proposition and omit the requirement that X be a $\mathscr{K}$-space in (b); see Corollary 5.3.14.

Sources

There are several excellent books on category theory. Three of them, with quite different approaches, are [350; 27; 402]. Two more with different goals in which the basic definitions used in this chapter can be found are [213, Chapter II] and [430, Chapter III]. Some texts in homological algebra *first* introduce Ext via projectives (or injectives), *then* prove that the definition does not depend on the particular presentation and *finally* establish the (natural) equivalence with exact sequences, which is often called *Yoneda-Ext*. We have taken the opposite route since our primary interest is the study of short exact sequences and because the category of quasi-Banach spaces does not have (infinite-dimensional) projectives or injectives (at all). Our treatment of (equivalence classes of) quasilinear maps as a functor follows [68]. The description of the homology sequences, also known as Hom-Ext sequences, using pushouts and pullbacks is a commonplace in homological algebra; see [213, Chapter III, Theorems 5.2 and 5.3] or the very advisable notes by Frerik and Sieg [180, Chapter 6, Theorems 6.17 and 6.18]. The bulk of Section 4.2 is taken from [68]. The classical part of Section 4.4 can be considered homological folklore, with the sole exception of Proposition 4.4.3, which elaborates on [69]. Propositions 4.5.5 and 4.5.6 and their garnish appear here for the first time. The basic criterion in Lemma 4.5.10 was proved in an unnecessarily tortuous way in [69], from which the examples in Proposition 4.5.13 are also taken.

5

Local Methods in the Theory of Twisted Sums

Just as there is a local theory of (quasi-) Banach spaces, nowadays an essential part of (quasi-) Banach space theory, we also have a local theory of extensions of (quasi-) Banach spaces. In this chapter we explain what it means and how it can be used. Following the usage of Banach space theory (and in contrast with its usage in most other areas in mathematics), 'local' refers to finite-dimensional objects. Accordingly, let us think about those exact sequences that split at the finite-dimensional level (i.e., *locally*). Such as? Such as the sequence $0 \longrightarrow c_0 \longrightarrow \ell_\infty \longrightarrow \ell_\infty/c_0 \longrightarrow 0$, which may not split, but no matter which finite-dimensional subspace of ℓ_∞/c_0 one chooses, the resulting pullback sequence 2-splits thanks to Sobczyk's theorem (but not thanks to Sobczyk's theorem, really). And the same occurs no matter which finite-dimensional subspace of ℓ_∞ one adds to c_0 (this time due to Sobczyk's theorem, for real). The paramount examples of locally split sequences are those that involve $\mathscr{L}_\infty$ or $\mathscr{L}_1$ spaces, to the point that those classes can be characterised by the facts that all exact sequences of Banach spaces $0 \longrightarrow \mathscr{L}_\infty \longrightarrow \cdot \longrightarrow \cdot \longrightarrow 0$ and $0 \longrightarrow \cdot \longrightarrow \cdot \longrightarrow \mathscr{L}_1 \longrightarrow 0$ split locally.

The material of the chapter is divided into three sections: The first contains the definition and characterisations of locally split sequences (including the just-mentioned characterisations of $\mathscr{L}_\infty$ and $\mathscr{L}_1$ spaces) and their connections with the extension and lifting of operators. The second presents the uniform boundedness theorem for exact sequences. The general form of a uniform boundedness principle is that 'if something happens then it happens uniformly'. In this case, *what* happens is

$$\operatorname{Ext}(X, Y) = 0,$$

and the meaning of *uniformly* depends on the way one interprets $\operatorname{Ext}(X, Y)$, leading to different formulations of the principle. In its plain Banach space form it says that if all extensions of X by Y split then there is a constant λ such that Y is λ-complemented in every enlargement Z such that $Z/Y \approx X$. In its not-so-intuitive quasilinear form it says that if all quasilinear maps $X \longrightarrow Y$ are trivial then their distances to linear maps can be controlled by the quasilinearity constant. Oddly, the plain Banach space form is the more difficult to prove, while obtaining the quasilinear form is really simple. There are other two forms: the projective and injective forms. All of them will be explained and relentlessly compared. From the applications side, the different forms of the principle will be used to show that $\operatorname{Ext}(X, Y) \neq 0$ implies that also $\operatorname{Ext}(X', Y') \neq 0$ when X' has the same local structure as X and Y' has the same local structure as Y, in a sense to be specified. For instance, X' could be an ultrapower of X or its bidual. From here one can easily obtain that $\operatorname{Ext}(A, B) \neq 0$ (or not) for many pairs X, Y of spaces, both classical and exotic, or that $\operatorname{Ext}(X, Y) \neq 0$ if X, Y are B-convex Banach spaces, or that whenever X is an $\mathscr{L}_\infty$-space and Y is an $\mathscr{L}_1$-space, $\operatorname{Ext}_{\mathbf{B}}(X, Y) \neq 0$.

The third section of the chapter studies concrete applications of the local approach to Banach spaces. In this study, the Maleficent role, that of a powerful uninvited guest, is played by the bounded approximation property, which, with its ability of decomposing infinite-dimensional objects into finite-dimensional pieces in a controlled way, allows obtaining global information when only local information is given to us. In this line, we present general forms of Lusky's result (kernels of projective presentations of BAP spaces have the BAP), its dual version of Figiel, Johnson and Pełczyński, solutions to the duality or ultrapower splitting problem and applications to the extension or lifting of operators, all of which casts new light on the often overlooked homological nature of the BAP.

5.1 Local Splitting

We want to consider short exact sequences with trivial behaviour at the finite-dimensional level. There are at least two ways to interpret 'finite-dimensional level' in this sentence: by looking at the middle twisted sum or, alternatively, by looking at the quotient space. The following result says that it makes no difference.

Lemma 5.1.1 *Let* $0 \longrightarrow Y \xrightarrow{\jmath} Z \xrightarrow{\rho} X \longrightarrow 0$ *be a short exact sequence of quasi-Banach spaces. The following are equivalent:*

(i) *There is a constant* λ *such that if* F *is a finite-dimensional subspace of* Z*, then there is* $\tau_F \in \mathfrak{L}(F, Y)$ *with* $\|\tau_F\| \leq \lambda$ *such that* $\jmath\, \tau_F(f) = f$ *for every* $f \in F \cap \jmath[Y]$.

(ii) *There is a constant* μ *such that if* G *is a finite-dimensional subspace of* X *then there is an operator* $s_G: G \longrightarrow Z$ *with* $\|s_G\| \leq \mu$ *such that* $\rho s_G = \mathbf{1}_G$.

If the sequence is isometrically exact and Z *is a p-Banach space then* (ii) *implies* (i) *with* $\lambda = (1 + \mu^p)^{1/p}$, *while* (i) *implies* (ii) *for any* $\mu > (1 + \lambda^p)^{1/p}$.

Proof It suffices to prove the last statement. There is no loss of generality in assuming that Y is a subspace of Z and $X = Z/Y$, in which case the condition on τ_F becomes $\tau_F f = f$ for $f \in F \cap Y$. We prove (ii) $\Longrightarrow$ (i): given a finite-dimensional subspace $F \subset Z$, set $G = \rho[F]$ and let $s_G: G \longrightarrow Z$ be an operator with $\|s_G\| \leq \mu$ such that $\rho s_G = \mathbf{1}_G$. The operator $\tau_F: F \longrightarrow Y$ given by $\tau_F(f) = f - s_G\rho(f)$ has norm at most $(1 + \mu^p)^{1/p}$ and is the identity on $F \cap Y$. To prove (i) $\Longrightarrow$ (ii), suppose G is a finite-dimensional subspace of X. Fixing $\varepsilon > 0$, an elementary compactness argument shows that there is a finite-dimensional $F \subset \rho^{-1}[G] \subset Z$ such that for every $g \in G$, there is $f \in F$ with $\rho(f) = g$ and $\|f\| \leq (1 + \varepsilon)\|g\|$. The restriction of ρ to F allows us to identify G and $F/(F \cap Y)$ as sets and thus $\|g\|_G \leq \|g\|_{F/(F\cap Y)} \leq (1 + \varepsilon)\|g\|$ for every $g \in G$. If $\tau_F: F \longrightarrow Y$ is the operator provided by (i), the map $f \mapsto f - \tau_F(f)$ is an operator $F \longrightarrow Y$ with norm at most $(1 + \lambda^p)^{1/p}$ that vanishes on $F \cap Y$; so, it induces an operator $s_G: F/(F \cap Y) \longrightarrow Z$ such that $\rho s_G = \mathbf{1}_G$. Clearly, $\|s_G: F/(F\cap Y) \longrightarrow Z\| \leq (1+\lambda^p)^{1/p}$, and so $\|s_G: G \longrightarrow Z\| \leq (1+\varepsilon)(1+\lambda^p)^{1/p}$, as required. □

Definition 5.1.2 An exact sequence splits locally if it satisfies the equivalent conditions of the preceding lemma. A subspace $Y \subset Z$ is locally complemented when the sequence $0 \longrightarrow Y \longrightarrow Z \longrightarrow Z/Y \longrightarrow 0$ splits locally.

The operators τ_F appearing in Lemma 5.1.1 are called local projections; the operators s_G will be called local sections. The constant λ appearing in Lemma 5.1.1 (i) shall be referred to as the local splitting constant of the sequence or the local complementation constant of the subspace, depending on the context, and it is clear what should be understood by a locally λ-split sequence and a locally λ-complemented subspace. The constant μ has no specific name. Local splitting can also be described neatly in quasilinear terms.

Lemma 5.1.3 *An exact sequence* $0 \longrightarrow Y \longrightarrow Y \oplus_\Phi X \longrightarrow X \longrightarrow 0$ *splits locally if and only if there is a constant* M *such that for every finite-dimensional* $G \subset X$ *there is a linear map* $L: G \longrightarrow Y$ *such that* $\|\Phi|_G - L\| \leq M$.

The reader should not have any difficulty in writing down a complete proof just following a simple pattern of quiet and curiosity. It is clear that complemented subspaces are locally complemented and that 'to be a locally complemented subspace' is a transitive relation. Natural situations in which local complementation appears include:

Proposition 5.1.4

(a) *Every Banach space is locally* 1^+*-complemented in its bidual.*
(b) *Every p-Banach space is locally* 1^+*-complemented in its ultrapowers.*
(c) *Given an ultrafilter* $\mathcal{U}$ *on a set* I, *the space* $c_0^{\mathcal{U}}(I, X_i)$ *is locally complemented in* $\ell_\infty(I, X_i)$.
(d) *If* Y *is a locally complemented subspace of a p-Banach space* Z *then* Z/Y *is isomorphic to a locally complemented subspace of some ultrapower of* Z.
(e) *The Pełczyński–Lusky sequence* $0 \longrightarrow c_0(\mathbb{N}, X_n) \longrightarrow c(\mathbb{N}, X_n) \longrightarrow X \longrightarrow 0$ *splits locally when* (X_n) *an increasing sequence of finite-dimensional subspaces of the p-Banach space* $X = \overline{\bigcup_n X_n}$.

Proof (a) is contained in the principle of local reflexivity. To check (b), let X be a p-Banach space , let $\delta\colon X \longrightarrow X_{\mathcal{U}}$ be the canonical inclusion and let F be a finite-dimensional subspace of $X_{\mathcal{U}}$. Choose $x_1, \ldots, x_m \in X$ and $f^1, \ldots, f^n \in F$ such that $\delta(x_1), \ldots, \delta(x_m)$ is a basis of $F \cap \delta[X]$ and $\delta(x_1), \ldots, \delta(x_m), f^1, \ldots, f^n$ is a basis of F. For each $1 \le k \le n$, we fix a bounded family $(f_i^k)_i$ in X such that $[(f_i^k)] = f^k$. Now, for each $i \in I$, we define an operator $T_i\colon F \longrightarrow X$ taking $T_i(\delta(x_j)) = x_j$ for $1 \le j \le m$ and $T_i(f^k) = f_i^k$ for $1 \le k \le n$. Clearly, each T_i is a local projection in the sense that $\delta T_i = \delta$ on $\delta[X] \cap F$ and $\|T_i\| \to 1$ along $\mathcal{U}$. Therefore, for each fixed $\varepsilon > 0$, there are (many) $i \in I$ such that $\|T_i\| \le 1 + \varepsilon$. To prove (c), we will find local sections for the quotient map of the sequence

$$0 \longrightarrow c_0^{\mathcal{U}}(I, X_i) \longrightarrow \ell_\infty(I, X_i) \xrightarrow{[\cdot]} [X_i]_{\mathcal{U}} \longrightarrow 0$$

Let G be a finite-dimensional subspace of $[X_i]_{\mathcal{U}}$ and let $g^1, \ldots, g^n$ be a basis of G. For each $1 \le k \le n$, let (g_i^k) be a representative of g^k. For each $i \in I$, we define an operator $u_i\colon G \longrightarrow X_i$ by setting $u_i(g) = g_i^k$ and extending linearly on the rest of G. The definition of the ultrapower p-norm guarantees that for each $\varepsilon > 0$, the set $I_\varepsilon = \{i \in I\colon u_i \text{ is an } \varepsilon\text{-isometry from } G \text{ into } X_i\}$ belongs to $\mathcal{U}$. If we fix $\varepsilon \in (0, 1)$ and define $s_G\colon G \longrightarrow \ell_\infty(I, X_i)$ as $s_G(f)(i) = u_i(f)1_{I_\varepsilon}(i)$ then $[\cdot] \circ s_G = \mathbf{1}_G$ and $\|s_G\| \le 1 + \varepsilon$. To obtain (d), put $X = Z/Y$ and let $\pi\colon Z \longrightarrow X$ be the natural quotient map. For each $G \in \mathscr{F}(X)$, let $s_G\colon G \longrightarrow Z$ be a local section with $\|s_G\| \le \mu$. Let $\mathcal{U}$ be an ultrafilter refining the order filter on $\mathscr{F}(X)$ and define $S : X \longrightarrow Z_{\mathcal{U}}$ by $S(x) = [s_G(x 1_G(x))]$; here 1_G is the characteristic function of G so that the Gth entry of the family that defines $S(x)$ is $s_G(x)$ if $x \in G$, and zero otherwise. This S is well defined, is linear and satisfies $\|S\| \le \mu$; moreover, the composition $\pi_{\mathcal{U}} S : X \longrightarrow Z_{\mathcal{U}} \longrightarrow X_{\mathcal{U}}$ agrees with the diagonal embedding, which implies that S is an embedding (actually $\|S(x)\| \ge \|x\|$).

Finally, $S[X]$ is locally complemented in $Z_{\mathfrak{U}}$ by (b). The proof of (e), which somehow is contained in that of (c), is simple: given a finite-rank operator $\tau\colon Y \longrightarrow X$, there is no loss of generality in assuming that $\tau[Y] \subset X_N$ for some N. The operator $T\colon Y \longrightarrow c(\mathbb{N}, X_n)$ given by $T(y) = (0, \dots, 0, \tau(y), \tau(y), \dots)$ provides an equal norm lifting of τ. □

When X is a Banach space, the dual sequence $0 \longrightarrow X^* \longrightarrow c(\mathbb{N}, X_n)^* \longrightarrow \ell_1(\mathbb{N}, X_n^*) \longrightarrow 0$ splits since the operator $s((x_n^*))(x_n) = \sum x_n^*(x_n)$ is a section of the quotient map. Thus, if only we could be now aware of Corollary 5.1.8, we would get an alternative proof for $p = 1$. The following simple result yields quantitative estimates connecting local and global splitting of exact sequences when either the subspace or the quotient space is finite-dimensional.

Lemma 5.1.5 *Let Z be a p-Banach space, Y a subspace of Z and $\varepsilon > 0$.*

(a) *If Y is locally λ-complemented in Z and Z/Y is finite-dimensional then Y is $((2 + \lambda^p)^{1/p} + \varepsilon)$-complemented in Z.*
(b) *If Y is finite-dimensional and the quotient map $Z \longrightarrow Z/Y$ has local sections bounded by μ then it has a section with norm at most $(2+\mu^p)^{1/p}+\varepsilon$.*

Proof We just prove (a) since (b) is analogous. By Lemma 5.1.1 the sequence $0 \longrightarrow Y \longrightarrow Z \longrightarrow Z/Y \longrightarrow 0$ has local (hence global) sections of norm $(1 + \lambda^p + \varepsilon^p)^{1/p}$ for every $\varepsilon > 0$. This yields a (global) projection of Z onto Y of norm at most $(2 + \lambda^p + \varepsilon^p)^{1/p}$, and the result follows. □

The stability properties of locally split sequences are simple: local splitting is preserved under isomorphisms of sequences, pullback and pushout. The first assertion is obvious and the other two are easy to check. Consider, for instance, a pushout diagram

$$\begin{array}{ccccccccc}
0 & \longrightarrow & Y & \longrightarrow & Z & \stackrel{\rho}{\longrightarrow} & X & \longrightarrow & 0 \\
 & & \downarrow{\scriptstyle \alpha} & & \downarrow{\scriptstyle \overline{\alpha}} & & \| & & \\
0 & \longrightarrow & A & \longrightarrow & \mathrm{PO} & \stackrel{\overline{\rho}}{\longrightarrow} & X & \longrightarrow & 0
\end{array}$$

If G is a finite-dimensional subspace of X and $s_G\colon G \longrightarrow Z$ is a local section for ρ then $\overline{\alpha} s_G\colon F \longrightarrow \mathrm{PO}$ is a local section for $\overline{\rho}$. As for pullback diagrams,

$$\begin{array}{ccccccccc}
0 & \longrightarrow & Y & \longrightarrow & Z & \longrightarrow & X & \longrightarrow & 0 \\
 & & \| & & \uparrow{\scriptstyle \underline{\beta}} & & \uparrow{\scriptstyle \beta} & & \\
0 & \longrightarrow & Y & \longrightarrow & \mathrm{PB} & \longrightarrow & B & \longrightarrow & 0
\end{array}$$

if F is a finite-dimensional subspace of PB then the composition of $\overline{\beta}$ with a local projection $\tau\colon \underline{\beta}[F] \longrightarrow Y$ yields a local projection $\tau\underline{\beta}\colon F \longrightarrow Y$.

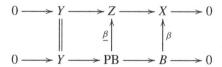

Under what conditions does a locally split sequence split? There is a rather satisfactory answer to this question.

Proposition 5.1.6 *Every locally split sequence in which the subspace is an ultrasummand splits.*

Proof What we will actually prove is that every operator from a locally complemented subspace $Y \subset Z$ to an ultrasummand extends to Z. Let λ be the local complementation constant, let A be an ultrasummand and let $u\colon Y \longrightarrow A$ be an operator. For each finite-dimensional $F \subset Z$, let $\tau_F\colon F \longrightarrow Y$ be a local projection with $\|\tau_F\| \leq \lambda$. We define a mapping $\tilde{u}\colon Z \longrightarrow \ell_\infty(\mathscr{F}(Z), A)$ by $\tilde{u}(x)_F = u(\tau_F(x1_F(x)))$ that, most likely, is not linear. Pick an ultrafilter $\mathcal{U}$ refining the Fréchet filter on $\mathscr{F}(Z)$. Define a mapping $U\colon Z \longrightarrow A_{\mathcal{U}}$ as the composition $[\cdot] \circ \tilde{u}\colon Z \longrightarrow \ell_\infty(\mathscr{F}(Z), A) \longrightarrow A_{\mathcal{U}}$. This is indeed an operator: it is obviously homogeneous and bounded by λ. Besides, if $x_1, x_2 \in Z$, we have, as long as $F \in \mathscr{F}(Z)$ contains both x_1 and x_2,

$$\tilde{u}(x_1 + x_2)_F = u(\tau_F(x_1 + x_2)) = u\tau_F(x_1) + u\tau_F(x_2) = \tilde{u}(x_1)_F + \tilde{u}(x_2)_F,$$

hence U is additive. Moreover, the restriction of U to Y is just the composition of u with the diagonal embedding of A into $A_{\mathcal{U}}$. Thus, if $P\colon A_{\mathcal{U}} \longrightarrow A$ is a projection, then the composition PU is an extension of u to Z. □

In Banach spaces, the local splitting notion interacts particularly well with duality:

Lemma 5.1.7 *Let Y be a subspace of a Banach space Z. The following are equivalent:*

(i) *Y is locally λ^+-complemented in Z.*
(ii) *There is a linear extension operator $E\colon Y^* \longrightarrow Z^*$ such that $\|E\| \leq \lambda$.*
(iii) *The inclusion $Y \longrightarrow Y^{**}$ has a λ-extension to Z.*
(iv) *Y^{**} is λ-complemented in Z^{**}.*

Proof (i) $\Longrightarrow$ (ii) Assume Y is a locally λ^+-complemented subspace of Z. We introduce an order on $\mathscr{F}(Z) \times (0, \infty)$ by declaring $(G, \varepsilon) \leq (F, \delta)$ if $G \subset F$ and $\varepsilon \geq \delta$ and let $\mathcal{U}$ be an ultrafilter containing the order filter. Now, given $F \in \mathscr{F}(Z)$ and $\varepsilon > 0$, let us choose a local projection $P_{(F,\varepsilon)}\colon F \longrightarrow Y$ with $\|P_{(F,\varepsilon)}\| \leq \lambda + \varepsilon$ and define $E\colon Y^* \longrightarrow Z^*$ by $E(y^*)(x) = \lim_{\mathcal{U}(F,\varepsilon)} \langle y^*, P_{(F,\varepsilon)}(x)\rangle$. Such E is obviously linear, $E(y^*)$ extends y^* and $\|E\| \leq \lambda$. (ii) $\Longrightarrow$ (iii) The λ-extension is $E^*|_Z$. (iii) $\Longrightarrow$ (ii) If $T\colon Z \longrightarrow Y^{**}$ extends the inclusion $Y \longrightarrow Y^{**}$ then $T^*|_{Y^{**}}$ is an extension operator. (ii) $\Longrightarrow$ (iv) $E^*\colon Z^{**} \longrightarrow Y^{**}$ is a projection. (iv) $\Longrightarrow$ (i) is obvious: Y is locally 1^+-complemented in Y^{**}, which in turn is λ-complemented in Z^{**}. Hence Y is λ^+-complemented in Z^{**} and so in Z. □

The immediate consequence is:

Corollary 5.1.8 *A short exact sequence of Banach spaces splits locally if and only if the dual sequence splits if and only if the bidual sequence splits.*

A striking application is that in any pullback diagram of Banach spaces

$$\begin{array}{ccccccccc} 0 & \longrightarrow & Y & \longrightarrow & X & \longrightarrow & \mathscr{L}_\infty & \longrightarrow & 0 \\ & & \| & & \uparrow & & \uparrow{\scriptstyle \tau} & & \\ 0 & \longrightarrow & Y & \longrightarrow & E & \longrightarrow & \ell_2(I) & \longrightarrow & 0 \end{array} \tag{5.1}$$

the lower sequence splits locally because the lower sequence in the dual diagram does:

$$\begin{array}{ccccccccc} 0 & \longrightarrow & \mathscr{L}_1 & \longrightarrow & X^* & \longrightarrow & Y^* & \longrightarrow & 0 \\ & & {\scriptstyle \tau^*}\downarrow & & \downarrow & & \| & & \\ 0 & \longrightarrow & \ell_2(I) & \longrightarrow & E^* & \longrightarrow & Y^* & \longrightarrow & 0 \end{array}$$

Indeed, the operator τ^* is 2-summing by Grothendieck's theorem, and thus it extends anywhere. The next result reveals the operator ideal fabric out of which the local splitting notion is made.

Proposition 5.1.9 *Let Y be a subspace of a Banach space Z. The following are equivalent:*

(i) *Y is locally complemented in Z.*
(ii) *There is a constant $\lambda > 0$ such that every finite-rank operator $\tau\colon Y \longrightarrow E$ with values in a Banach space has an λ-extension to Z.*
(iii) *For every Banach space E, every approximable, compact or weakly compact operator $\tau\colon Y \longrightarrow E$ can be extended to an operator $T\colon Z \longrightarrow E$ of the same type.*

Proof The equivalence (i) $\Longleftrightarrow$ (ii) follows from the duality formula $\mathfrak{L}(A, B^*) = \mathfrak{L}(B, A^*)$. Thus, each finite-rank operator $v\colon E \longrightarrow Y^*$ corresponds to a finite-rank operator $u\colon Y \longrightarrow E^*$ in such a way that $\langle v(e), y\rangle = \langle u(y), e\rangle$ for every $e \in E$ and every $y \in Y$. Moreover, $V \in \mathfrak{L}(E, Z^*)$ is a lifting of v if and only if its mate $U \in \mathfrak{L}(Z, E^*)$ is an extension of u, as illustrated in the diagrams

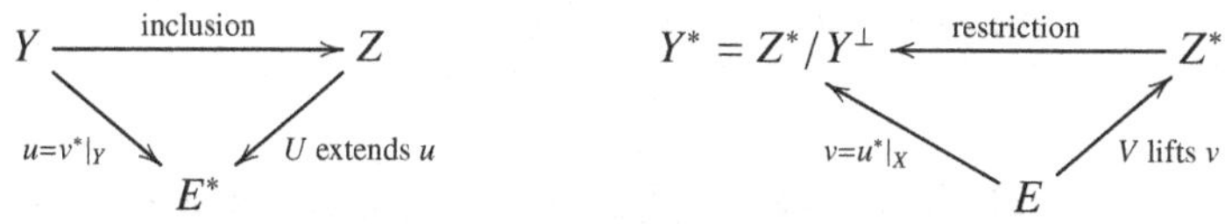

If Y is locally complemented in Z then the dual sequence splits, and thus operators v lift ‘uniformly’ to Z^*, which means that operators u extend uniformly to

Z, which is (ii). Conversely, if (ii) holds then the dual sequence splits locally, hence it splits by Proposition 5.1.6, and Y is locally complemented in Z by the preceding corollary. (ii) implies the approximable assertion of (iii) by a limit process. We prove that (i) implies (iii) for compact operators: given a compact $\tau: Y \longrightarrow E$, for each finite-dimensional $F \subset Z$, we consider a 'local projection' $P_F: F \longrightarrow Y$ with $\|P_F\| \leq \lambda$ and the net of operators $\tau P_F: F \longrightarrow Y \longrightarrow E$. Pick an ultrafilter $\mathfrak{U}$ refining the order filter on $\mathscr{F}(Z)$ and form the operator $T(z) = \lim_{\mathfrak{U}(F)} \tau(P_F(z))$, taking advantage of the relative compactness of $\tau[\lambda B_Y]$. The proof for weakly compact τ is the same, using the weak topology instead of the norm topology. The reason no uniform bound appears in (iii) is that the operator ideals $\mathfrak{G}, \mathfrak{K}, \mathfrak{W}$ are closed in the operator norm; the fact that all operators in the ideal can be extended automatically entails the existence of a constant controlling the norm of the extension ... depending a priori on the target space E. To obtain a 'universal constant', consider, for instance, the case of compact operators. Let $\tau_i: Y \longrightarrow E_i$ be any family of compact operators indexed by I. Form the Banach space $E = \ell_1(I, E_i)$ and let λ be the associated constant so that each compact operator $\tau: Y \longrightarrow E$ has a compact λ-extension to Z. Since each τ_i can be seen as a compact operator in $\mathfrak{L}(Y, E)$, each τ_i has a compact λ-extension $T_i: Z \longrightarrow E$. Composition with the obvious projection $\pi_i: E \longrightarrow E_i$ does the trick. It is now clear that any version of (iii) implies (ii). The ideal $\mathfrak{F}$ is, however, not closed, and thus only the existence of a uniform bound for the extension guarantees local splitting. □

Ultrapowers of Exact Sequences

The connections between an exact sequence and its ultrapowers are like a spider's web: almost invisible, stronger than it seems and difficult to get rid of. One might shrewdly conjecture that ultrapowers of locally split sequences that split locally must split. But this is not a riddle in which only the sound of words matters: it is true that a sequence splits locally if and only if its ultrapowers do, but there are locally split sequences without trivial ultrapowers. Examples include any exact sequence $0 \longrightarrow \mathsf{G} \longrightarrow C(K) \longrightarrow \cdot \longrightarrow 0$, since no ultrapower of the Gurariy space G is complemented in a $\mathscr{C}$-space [22, Section 3.3.4] and because of the Foiaş–Singer sequence which remains non-trivial after applying ultrapowers (Lemma 10.5.5).

Locally Injective and Locally Projective Spaces

Local splitting suggests that injectivity and projectivity can also be localised.

Definition 5.1.10 A p-Banach space Y is said to be locally injective if every exact sequence of p-Banach spaces $0 \longrightarrow Y \longrightarrow \cdot \longrightarrow \cdot \longrightarrow 0$ splits locally.

It should be almost clear that every locally injective space Y is locally λ-injective for some λ, with the meaning that every isometrically exact sequence $0 \longrightarrow Y \longrightarrow \cdot \longrightarrow \cdot \longrightarrow 0$ of p-Banach spaces λ-splits locally: the 'right-downwards' diagonal sequence mentioned in Section 2.15.3 allows us to obtain from sequences $0 \longrightarrow Y \longrightarrow Z_n \longrightarrow X_n \longrightarrow 0$ that do not n-split locally an exact sequence $0 \longrightarrow Y \longrightarrow \mathrm{PO} \longrightarrow \ell_p(\mathbb{N}, X_n) \longrightarrow 0$ that cannot, obviously, split locally, in contradiction with the local injectivity of Y. Locally injective spaces are easy to characterise but difficult to find. Easy to characterise:

Lemma 5.1.11 *A p-Banach space Y is locally λ-injective if and only if every operator $\tau\colon F \longrightarrow Y$ defined on a subspace of a finite-dimensional p-Banach space E has a λ-extension $T\colon E \longrightarrow Y$.*

Proof Suppose not. If E is finite-dimensional, the diagram

$$\begin{array}{ccccccccc} 0 & \longrightarrow & F & \longrightarrow & E & \longrightarrow & E/F & \longrightarrow & 0 \\ & & \downarrow{\scriptstyle \tau} & \swarrow & \downarrow & & \| & & \\ 0 & \longrightarrow & Y & \longrightarrow & \mathrm{PO} & \longrightarrow & E/Y & \longrightarrow & 0 \end{array}$$

conveys the idea that local λ-splitting of the lower sequence yields a λ-extension for τ. Conversely, if Y has the property in the statement and $E \subset Z$ is finite-dimensional, a λ-extension of the inclusion $Y \cap E \longrightarrow Y$ to E provides the required local projection. □

And difficult to find, at least for $p < 1$: we just obtained one in 2.13.1. The Banach space tree of locally injective spaces is much leafier thanks to the following two facts: (a) the existence of enough injectives and (b) that locally complemented subspaces of $\mathscr{L}_\infty$-spaces are $\mathscr{L}_\infty$-spaces, whose proof can be found in [331] (a streamlined version for (b) is in [22, Appendix 2]). One has:

Proposition 5.1.12 *A Banach space is an $\mathscr{L}_\infty$-space if and only if it is locally injective. In particular, an $\mathscr{L}_{\infty,\lambda}$-space X is $(2+\lambda)^+$-complemented in every superspace Z such that Z/X is finite-dimensional.*

Proof The 'if' part is an obvious consequence of the Hahn–Banach theorem. The other implication is clear since X has to be locally λ-complemented in some $\ell_\infty(I)$. The coda is a consequence of the Hahn–Banach argument above: $\mathscr{L}_{\infty,\lambda}$-spaces are locally λ-complemented in every superspace and therefore Lemma 5.1.5 applies. □

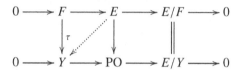

The projective case is much more controversial. Observe why:

Definition 5.1.13 A p-Banach space X is locally projective if every exact sequence of p-Banach spaces $0 \longrightarrow \cdot \longrightarrow \cdot \longrightarrow X \longrightarrow 0$ splits locally. A p-Banach space X is finitely λ-projective if for every finite-dimensional space F and every subspace $E \subset F$, every operator $\tau : X \longrightarrow F/E$ has a lifting $T \colon X \longrightarrow F$ with $\|T\| \leq \lambda\|\tau\|$. We say that a p-Banach space is finitely projective if it is finitely λ-projective for some λ.

It is plain that locally projective implies finitely projective. What catches us unprepared is that any space with trivial dual is finitely projective. So it seems that the right notion to consider is local projectivity, which will lead us to Kalton's definition of $\mathscr{L}_p$-spaces for $0 < p \leq 1$ in the next section. We have already seen that $\mathscr{L}_\infty$-spaces are the locally injective Banach spaces, and thus it could not be strange that $\mathscr{L}_1$-spaces are the locally projective Banach spaces. But it is important to remark that this depends on a non-trivial fact from [331]: a Banach space is an $\mathscr{L}_\infty$-space if and only if its dual is an $\mathscr{L}_1$-space. With this in hand, the situation for Banach spaces becomes neat:

Proposition 5.1.14 *In the category of Banach spaces, the following are equivalent:*

(i) *X is locally projective.*
(ii) *X is finitely projective.*
(iii) *X is an $\mathscr{L}_1$-space.*

Proof Keep in mind the proof of Proposition 5.1.9: how the identity $\mathfrak{L}(A, B^*) = \mathfrak{L}(B, A^*)$ works, how an extension $U \colon F \longrightarrow X^*$ of an operator $u \colon E \longrightarrow X^*$ from a subspace $E \subset F$ of a finite-dimensional space corresponds to a lifting $V \in \mathfrak{L}(X, E^*)$ of the corresponding operator $v = u^*|_X \in \mathfrak{L}(X, F^*)$ and the diagrams

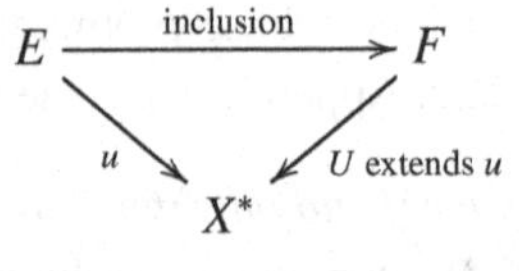

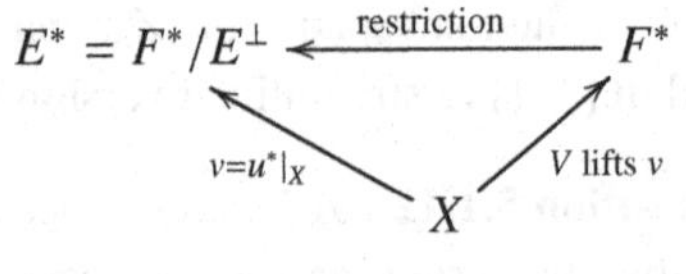

This and Lemma 5.1.11 render almost obvious that X is finitely projective if and only if X^* is locally injective since $\langle U(f), z\rangle = \langle u(f), z\rangle$ for every $f \in E$ and every $z \in X$ if and only if $\langle V(z), f\rangle = \langle v(z), f\rangle$ for every $f \in E$ and every $z \in X$. Additionally, $\|u\| = \|v\|$ and $\|U\| = \|V\|$. □

We can now round off Lindenstrauss' lifting:

5.1.15 Lindenstrauss' lifting and its converse *A Banach space X is an $\mathscr{L}_1$-space if and only if* $\mathrm{Ext}_{\mathbf{B}}(X, U) = 0$ *for every ultrasummand U.*

Proof The 'only if' part is Lindenstrauss' lifting. The converse is a combination of $\mathrm{Ext}_{\mathbf{B}}(X, A^*) = 0$ for every Banach space A; Corollary 4.4.5, which yields $\mathrm{Ext}_{\mathbf{B}}(A, X^*) = \mathrm{Ext}_{\mathbf{B}}(X, A^*) = 0$ for every Banach space A; and the fact that X^* is therefore injective. Thus, X^* is an $\mathscr{L}_\infty$-space and X must be an $\mathscr{L}_1$-space. □

And obtain the following companion to Proposition 5.1.12:

Proposition 5.1.16 *A Banach space X is an $\mathscr{L}_1$-space if and only if there is a constant μ such that every isometrically exact sequence of Banach spaces $0 \longrightarrow F \longrightarrow \cdot \longrightarrow X \longrightarrow 0$ in which F is finite-dimensional admits a linear section bounded by μ.*

Proof Every short exact sequence of Banach spaces whose quotient is an $\mathscr{L}_{1,\mu}$-space has local sections of norm μ^+. Thus, the only if part is clear from Lemma 5.1.5 (b). To prove the other implication, we shall establish that X^* is an $\mathscr{L}_\infty$-space. If $0 \longrightarrow X^* \longrightarrow \cdot \longrightarrow G \longrightarrow 0$ is an isometrically exact sequence of Banach spaces with G finite-dimensional, let $0 \longrightarrow G^* \longrightarrow \cdot \longrightarrow X \longrightarrow 0$ be the exact sequence arising from the formula $\mathrm{Ext}_{\mathbf{B}}(G, X^*) = \mathrm{Ext}_{\mathbf{B}}(X, G^*)$. Since the latter sequence admits a μ-section, the former admits a μ-projection. Hence, X^* is an $\mathscr{L}_\infty$-space and X is an $\mathscr{L}_1$-space. □

A final quantitative observation: if X is locally λ-complemented in, say, $\ell_\infty(I)$, its bidual X^{**} is λ-complemented in $\ell_\infty(I)^{**}$. But we only know that a λ-complemented subspace of a $\mathscr{C}$-space is an $\mathscr{L}_{\infty,\mu}$-space for some μ (not necessarily λ) (see [22, Lemma A.12]), except when $\lambda = 1^+$, in which case we know that a 1^+-complemented subspace of a 1-injective space is an $\mathscr{L}_{\infty,1^+}$-space; see [466]. The principle of local reflexivity then yields in all cases that if X^{**} is an $\mathscr{L}_{\infty,\mu}$-space then X is an $\mathscr{L}_{\infty,\mu^+}$-space. Thus, Lindenstrauss spaces, which are the $\mathscr{L}_{\infty,1^+}$-spaces, coincide with the locally 1^+-injective spaces.

For dessert, a few enjoyable consequences for 3-space properties:

Corollary 5.1.17 *Being an $\mathscr{L}_\infty$-space or an $\mathscr{L}_1$-space are 3-space properties of Banach spaces. Moreover, given an exact sequence $0 \longrightarrow Y \longrightarrow Z \longrightarrow X \longrightarrow 0$ of Banach spaces, if Y and Z are $\mathscr{L}_\infty$-spaces then so is X, and if Z and X are $\mathscr{L}_1$-spaces then so is Y.*

The two remaining cases? Both false: Bourgain's examples (2.9) and (2.10) show that Z, X can be both $\mathscr{L}_\infty$-spaces, but Y not, and that Y, Z can be both $\mathscr{L}_1$-spaces, and X not.

The $\mathscr{L}_p$-Spaces for $0 < p < 1$

We now study the locally projective p-Banach spaces. We know that the spaces $\ell_p(I)$ are locally projective; the next most likely candidates are the $L_p(\mu)$ spaces:

Lemma 5.1.18 *Let X be a p-Banach space. Assume that there is $\lambda \geq 1$ and a system of finite-dimensional subspaces $(X_i)_{i\in I}$, directed by inclusion, with $d(X_i, \ell_p^n) \leq \lambda$ for $n = \dim(X_i)$ and such that $X = \overline{\bigcup_{i\in I} X_i}$. Then X is locally λ^+-projective among p-Banach spaces.*

Proof We must check that every quotient map $\rho\colon Z \longrightarrow X$ from a p-Banach space Z has uniformly bounded local sections. By replacing the p-norm of Z by an equivalent one, if necessary, we can assume that ρ maps the open ball of Z onto that of X. Let $G \subset X$ be a finite-dimensional subspace. If $G \subset X_i$ for some $i \in I$ then there is a local section $s\colon G \longrightarrow Z$ with $\|s\| \leq \lambda$, as follows from the lifting property of ℓ_p^n and the hypothesis on X_i. Otherwise, a perturbation argument is needed. Let $g_1, \ldots, g_k$ be a normalised basis of G and take $C > 0$ such that $\|(c_n)\|_p \leq C\|\sum_{1\leq n\leq k} c_n g_n\|$ for all $c_k \in \mathbb{K}$. Fix $\varepsilon > 0$ and take X_i large enough so that for each $1 \leq n \leq k$, there is $f_n \in X_i$ such that $\|f_n - g_n\| < \varepsilon/C$. Let $F = [f_1, \ldots, f_k]$, and let $\ell \in \mathfrak{L}(F, Z)$ be a local section, with $\|\ell\| \leq \lambda$. For each $1 \leq n \leq k$, take $z_n \in Z$ such that $\rho(z_n) = g_n - f_n$ with $\|z_n\| < \varepsilon/C$ and define a linear map $s\colon G \longrightarrow Z$ by $sg_n = \ell f_n + z_n$. It is clear that $\rho s = \mathbf{1}_G$. It only remains to obtain a (uniform) bound for $\|s\|$. Take $g = \sum_{n\leq k} c_n g_n$, with $\|g\| \leq 1$. Then, if $f = \sum_{n\leq k} c_n f_n$, we have $\|g - f\|^p \leq \sum_n |c_n|^p \|f_n - g_n\|^p \leq \varepsilon^p$. In particular, $\|f\|^p \leq 1 + \varepsilon^p$. Hence

$$\|sg\|^p \leq \|Lf\|^p + \sum_n |c_n|^p \|z_n\|^p \leq \|L\|^p(1 + \varepsilon^p) + \varepsilon^p. \qquad \square$$

The spaces $L_p(\mu)$ satisfy the hypothesis of the lemma (think of the simple functions), and therefore they are locally projective in $p\mathbf{B}$. The same is true of their locally complemented subspaces:

Lemma 5.1.19 *Every locally complemented subspace of a locally projective space is locally projective.*

Proof Let Y be a locally complemented subspace of a locally projective p-Banach space Z. Let $\pi\colon \ell_p(I) \longrightarrow Z$ and $\varpi\colon \ell_p(J) \longrightarrow Y$ be quotient maps. The following diagram, whose lower rows are apparently unrelated,

$$\begin{array}{ccccccccc}
0 & \longrightarrow & \kappa_p(Z) & \longrightarrow & \ell_p(I) & \xrightarrow{\pi} & Z & \longrightarrow & 0\\
 & & \| & & \uparrow & & \uparrow & & \\
0 & \longrightarrow & \kappa_p(Z) & \longrightarrow & \pi^{-1}[Y] & \xrightarrow{\pi} & Y & \longrightarrow & 0\\
 & & & & & & \| & & \\
0 & \longrightarrow & \kappa_p(Y) & \longrightarrow & \ell_p(J) & \xrightarrow{\varpi} & Y & \longrightarrow & 0
\end{array}$$

should be looked at very suspiciously, as though it were something sticky we stepped in on board the Nostromo. Our plan is then to show that the lowest row is a pushout of the middle one, which only requires that we lift π through ϖ. For each finite-dimensional $G \subset Z$, we pick a local section $s_G \in \mathfrak{L}(G, Y)$ with $\|s_G\| \leq M$ for some M independent of G. Let $\mathcal{U}$ be an ultrafilter refining the order filter on $\mathscr{F}(Z)$. We form a diagram from the 'diagonal' embedding $\delta\colon Z \longrightarrow \mathscr{F}(Z)_{\mathcal{U}}$ given by $\delta(z) = [z1_G(z)]$, the ultraproduct operators $[s_G]_{\mathcal{U}}\colon \mathscr{F}(Z)_{\mathcal{U}} \longrightarrow Y_{\mathcal{U}}$ and $\varpi_{\mathcal{U}}\colon (\ell_p(J))_{\mathcal{U}} \longrightarrow Y_{\mathcal{U}}$, a lifting $L\colon \ell_p(I) \longrightarrow Y_{\mathcal{U}}$ of $[s_G]_{\mathcal{U}}\delta$ through $\varpi_{\mathcal{U}}$ that exists because $\ell_p(I)$ is projective plus the two unlabelled diagonal embeddings:

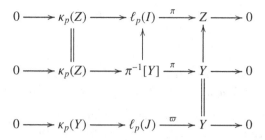

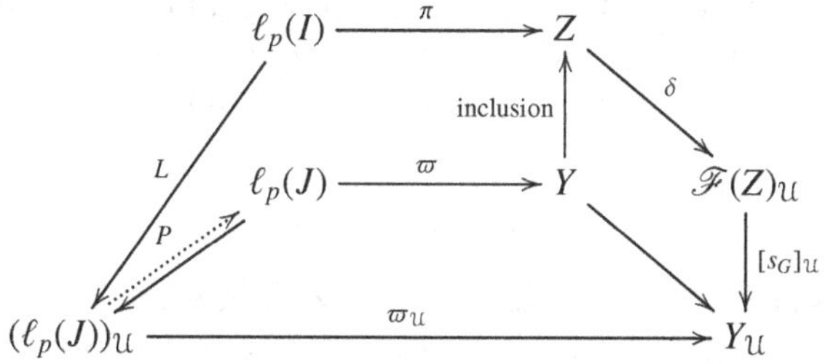

Since $\ell_p(J)$ is an ultrasummand, there is a projection P along the diagonal embedding, and thus the restriction of PL to $\pi^{-1}[Y]$ yields the much sought map. □

And that is all:

Lemma 5.1.20 *Every locally projective p-Banach space is isomorphic to a locally complemented subpace of some $L_p(\mu)$.*

Proof This is an easy consequence of Proposition 5.1.4(d) once we are told that every ultrapower of $L_p(\mu)$ is isometric to some $L_p(\nu)$ [427, Proposition 3.3]. Thus, if X is locally projective, then it is locally complemented in an ultrapower of some $\ell_p(I)$ just considering a quotient map $\pi\colon \ell_p(I) \longrightarrow X$. □

We have thus obtained the following characterisation of local projectivity:

Theorem 5.1.21 *In the category of p-Banach spaces, the following are equivalent:*

(i) *X is locally projective.*
(ii) *Some (all) projective presentations of X split locally.*
(iii) *X is isomorphic to a locally complemented subspace of some* $L_p(\mu)$.

Proof The implication (i) $\Longrightarrow$ (ii) is trivial; the converse, used implicitly in the proof of Lemma 5.1.19, follows from the fact that all extensions of X are pushouts of any projective presentation and that pushouts preserve the local triviality of sequences. (ii) $\Longrightarrow$ (iii) is Lemma 5.1.20. (iii) $\Longrightarrow$ (ii) is Lemma 5.1.18 plus Lemma 5.1.19. □

Statement (iii) corresponds to Kalton's definition of $\mathscr{L}_p$-spaces when $0 < p < 1$ in [255]. The definition also makes sense for $1 \leq p \leq \infty$ and would provide an alternative to the more popular definition of Lindenstrauss and Rosenthal, which, in turn, *could* be 'extended' to $0 < p < 1$. Both definitions are equivalent for $p = 1, 2, \infty$. However, when $p \in (1, \infty), p \neq 2$, a 'Kalton $\mathscr{L}_p$-space', would simply be a complemented subspace of some $L_p(\mu)$, while an $\mathscr{L}_p$-space is a complemented subspace of some $L_p(\mu)$ that is not isomorphic to an infinite-dimensional Hilbert space. The grafting of Lindenstrauss–Rosenthal's definition to Kalton's zone is more problematic since it seems to be unknown if even L_p satisfy it. Theorem 5.1.21 and Proposition 5.1.6 yield:

5.1.22 Lindenstrauss' *p*-lifting *Let* $p \in (0, 1]$. *For every p-Banach ultrasummand U and every* $\mathscr{L}_p$*-space X, we have* $\operatorname{Ext}_{p\mathbf{B}}(X, U) = 0$.

We do not know if this characterises the $\mathscr{L}_p$-spaces for $0 < p < 1$. It can be proved that a p-Banach space having the BAP is a (necessarily discrete, see below) $\mathscr{L}_p$-space if $\operatorname{Ext}_{p\mathbf{B}}(X, U) = 0$ for every p-Banach ultrasummand U.

$\mathscr{L}_p$-Subspaces of ℓ_p for $0 < p \leq 1$

In classifying $\mathscr{L}_1$ and $\mathscr{L}_\infty$ spaces, one of the basic questions one can ask is how many isomorphism types there are. For $\mathscr{L}_\infty$-spaces, the first answer that occurs would be 'a lot', since there are uncountably many non-isomorphic separable $\mathscr{C}$-spaces based on ordinals, just for starters. A more thoughtful answer would then be 'a hell of a lot' since more and more exotic types of $\mathscr{L}_\infty$-spaces keep coming into life. The question for $\mathscr{L}_1$-spaces was posed by Lindenstrauss, who partially solved it in [327], showing that there are infinitely many. Johnson and Lindenstrauss showed later [226] that there is actually a continuum of non-isomorphic $\mathscr{L}_1$-subspaces of ℓ_1. The existence of infinitely many isomorphism types can be achieved through the following partial converse of the Lindenstrauss–Rosenthal theorem, due to Lindenstrauss:

Lemma 5.1.23 *$\kappa(\mathscr{L}_1) \simeq \kappa(\mathscr{L}_1')$ if and only if $\mathscr{L}_1 \simeq \mathscr{L}_1'$.*

Proof If $\kappa(\mathscr{L}_1)$ and $\kappa(\mathscr{L}_1')$ are isomorphic, the exact sequences

$$\begin{array}{ccccccccc} 0 & \longrightarrow & \kappa(\mathscr{L}_1) & \longrightarrow & \ell_1 & \longrightarrow & \mathscr{L}_1 & \longrightarrow & 0 \\ & & \| & & & & & & \\ 0 & \longrightarrow & \kappa(\mathscr{L}_1') & \longrightarrow & \ell_1 & \longrightarrow & \mathscr{L}_1' & \longrightarrow & 0 \end{array}$$

are semi-equivalent since $\operatorname{Ext}_{\mathbf{B}}(\mathscr{L}_1, \ell_1) = \operatorname{Ext}_{\mathbf{B}}(\mathscr{L}_1', \ell_1) = 0$. The diagonal principles therefore yield $\mathscr{L}_1 \times \ell_1 \simeq \mathscr{L}_1' \times \ell_1$. Since every $\mathscr{L}_1$-space contains ℓ_1 complemented, $\mathscr{L}_1 \simeq \mathscr{L}_1 \times \ell_1 \simeq \mathscr{L}_1' \times \ell_1 \simeq \mathscr{L}_1'$. □

Proposition 5.1.24 *There are infinitely many separable non-isomorphic $\mathscr{L}_1$-spaces.*

Proof An infinite sequence of non-isomorphic separable $\mathscr{L}_1$-spaces is given by $\kappa^n(L_1)$, where we inductively define $\kappa^{n+1}(X) = \kappa(\kappa^n(X))$ for $n \geq 1$. If $\kappa^m(L_1) \simeq \kappa^n(L_1)$ for $m > n$ then $\kappa^{m-n}(L_1) \simeq L_1$, which is impossible since L_1 is not a subspace of ℓ_1. □

The same idea works for $0 < p < 1$ and produces the family of (Kalton) $\mathscr{L}_p$-spaces mentioned after Corollary 5.3.5. Call an $\mathscr{L}_p$-space *discrete* when it has the BAP. The $\ell_p(I)$ spaces are prototypes, and since the kernels of projective presentations of $\mathscr{L}_p$-spaces are locally complemented and the BAP passes to locally complemented subspaces (obvious and explicitly stated in Proposition 5.3.2 (b)), one has:

Lemma 5.1.25 *Let $0 < p < 1$. If X is an $\mathscr{L}_p$-space and $\pi\colon \ell_p(I) \longrightarrow X$ is a quotient map then* $\ker \pi$ *is a discrete $\mathscr{L}_p$-space.*

Now, given a separable p-Banach space X, the isomorphism type of $\ell_p \times \kappa_p(X)$ is well defined by Corollary 2.7.4. So we can pretend to be working with the sequence $(\kappa_p^n(X))_{n\geq 1}$ of subspaces of ℓ_p defined by $\kappa_p^1(X) = \kappa_p(X)$ and $\kappa_p^{n+1}(X) = \kappa_p(\kappa_p^n(X))$. We have:

Proposition 5.1.26 *The spaces $\kappa_p^n(L_p)$ are pairwise non-isomorphic discrete $\mathscr{L}_p$-spaces.*

Indeed, observe that $\mathfrak{L}(L_p, \kappa_p^n(L_p)) = 0$ for all n and use the same argument as before, taking into account that $\operatorname{Ext}_{p\mathbf{B}}(\mathscr{L}_p, \ell_p) = 0$. In [255, Section 7], Kalton produces a continuum of mutually non-isomorphic $\mathscr{L}_p$-spaces with trivial dual and observes that they lead, as one might guess, to a continuum of non-isomorphic discrete $\mathscr{L}_p$-subspaces of ℓ_p.

5.2 Uniform Boundedness Principles for Exact Sequences

We thus arrive at the second milestone of the chapter: the statement of uniform boundedness principles for exact sequences of (quasi-) Banach spaces. A number of results of the type 'if something happens then it happens uniformly' have already been encountered. In the case we are considering now, *what* happens is $\operatorname{Ext}(X, Y) = 0$. The first two possible interpretations of that fact directly follow from the open mapping theorem.

Projective form. If Ext is interpreted via projective presentations as $\operatorname{Ext}^{\mathrm{proj}}_{p\mathbf{B}}$ then $\operatorname{Ext}(X, Y) = 0$ means that once a projective presentation $0 \longrightarrow \kappa_p(X) \longrightarrow \mathcal{P} \longrightarrow X \longrightarrow 0$ has been chosen, all operators $\kappa_p(X) \longrightarrow Y$ extend to $\mathcal{P}$, and 'uniformly' in this context refers to the ratio between the norm of the extension and the norm of the operator.

Injective form. If Ext is interpreted via injective presentations as $\operatorname{Ext}^{\mathrm{inj}}_{\mathbf{B}}$ then $\operatorname{Ext}(X, Y) = 0$ means that once an injective presentation $0 \longrightarrow Y \longrightarrow \mathcal{J} \longrightarrow c\kappa(Y) \longrightarrow 0$ has been chosen, all operators $X \longrightarrow c\kappa(Y)$ lift to $\mathcal{J}$, and 'uniformly' refers to the ratio between the norm of the lifting and the norm of the operator.

Although clean and simple, these forms are not very manageable because, in general, one does not have explicit projective or injective presentations. Also, as explained in Section 4.4, while $\operatorname{Ext}^{\mathrm{proj}}$ behaves very well with pushouts, it does not with pullbacks; $\operatorname{Ext}^{\mathrm{inj}}$ exhibits the opposite behaviour. Understanding Ext in its basic form as the space of exact sequences makes $\operatorname{Ext}(X, Y) = 0$ mean that all sequences $0 \longrightarrow Y \longrightarrow \cdot \longrightarrow X \longrightarrow 0$ split, and the principle should then say that all of them split 'uniformly'. And here the problematic point arises: it is necessary to find a way to measure the 'degree of exactness' and the 'degree of splitting' of the sequence, all of which will be done very soon in Lemmata 5.2.3 and 5.2.4. We thus pass to the:

Quasilinear form. Interpreting Ext in the naturally equivalent form Q_{LB}, we know by now that $\mathsf{Q}_{\mathsf{LB}}(X, Y) = 0$ means that each quasilinear map $\Phi\colon X \longrightarrow Y$ admits a linear map $L\colon X \longrightarrow Y$ at finite distance. Now 'uniformly' refers to the ratio between the distance $D(\Phi)$ to linear maps and the quasilinearity constant $Q(\Phi)$ of the map.

5.2.1 Uniform boundedness principle for quasilinear maps *Let X be a quasinormed space and Y be a quasi-Banach space. If every quasilinear map from X to Y is trivial then there is a constant K such that, whenever $\Phi\colon X \longrightarrow Y$ is a quasilinear map, there is a linear map $L\colon X \longrightarrow Y$ such that $\|\Phi - L\| \leq KQ(\Phi)$.*

The proof, Theorem 3.6.5, derives from the fact that the semiquasinorms $D(\cdot)$ and $Q(\cdot)$ on $\mathsf{Q}(X, Y)$ are comparable, become norms on $\mathsf{Q}_{\mathsf{L}}(X, Y)$ and make this space complete when all the elements of $Q(X, Y)$ are trivial. The p-normed version was Theorem 3.6.8.

5.2.2 Uniform boundedness principle for p-linear maps *Let X be a p-normed space and Y a p-Banach space. If every p-linear map from X to Y is trivial then there is a constant K such that, whenever $\Phi\colon X \longrightarrow Y$ is a p-linear map, there is a linear map $L\colon X \longrightarrow Y$ such that $\|\Phi - L\| \leq KQ^{(p)}(\Phi)$.*

Observe that the case $p = 1$ of 5.2.2 is definitely not 5.2.1. The bonuses accrued from working with quasilinear maps are the usual ones: both pullbacks and pushouts, as well as the vector space operations, are easy to handle. So let us move forward by focusing on applications, specialisations and so on of the quasilinear form of the uniform boundedness principle. To begin with, observe that Theorems 3.6.5 and 3.6.8 were followed by their quantifications through the parameters $K[\cdot,\cdot]$ and $K^{(p)}[\cdot,\cdot]$. To properly do the same with the uniform principles, we need to keep track of the concavity constants involved:

Lemma 5.2.3 *Let X, Y be quasi-Banach spaces such that* $\operatorname{Ext}(X, Y) = 0$ *and let $0 \longrightarrow Y \longrightarrow Z \longrightarrow X \longrightarrow 0$ be an isometrically exact sequence. If Δ is the modulus of concavity of Z then for every $\varepsilon > 0$, there is a linear section $S : X \longrightarrow Z$ for the quotient map such that $\|S\| \leq \Delta(1 + 2\Delta^2 K[X, Y]) + \varepsilon$ and a projection $P\colon Z \longrightarrow Y$ such that $\|P\| \leq (1 + 2\Delta)\max(\Delta, 2\Delta^2 K[X, Y]) + \varepsilon$.*

Proof The proof is just a combination of two known results. Fix $\varepsilon > 0$ and observe that $\Delta_Y, \Delta_X \leq \Delta$. Use Proposition 3.3.7 to obtain a quasilinear $\Phi\colon X \longrightarrow Y$ with $Q(\Phi) \leq 2(1 + \varepsilon)\Delta^2$, yielding a commutative diagram

$$\begin{array}{ccccccccc} 0 & \longrightarrow & Y & \xrightarrow{\imath} & Y \oplus_\Phi X & \xrightarrow{\pi} & X & \longrightarrow & 0 \\ & & \| & & \downarrow{\scriptstyle u} & & \| & & \\ 0 & \longrightarrow & Y & \xrightarrow{\jmath} & Z & \xrightarrow{\rho} & X & \longrightarrow & 0 \end{array}$$

with $\|u\| \leq (1 + \varepsilon)\Delta$ and $\|u^{-1}\| \leq (2 + \varepsilon)\Delta + 1$. Use now Lemma 3.3.4 to get that any linear map $L\colon X \longrightarrow Y$ at finite distance from Φ produces a section s of π with $\|s\colon X \longrightarrow Y \oplus_\Phi X\| \leq 1 + \|\Phi - L\|$ as well as a projection $p\colon Y \oplus_\Phi X \longrightarrow Y$ with $\|p\| \leq \max(\Delta, \|\Phi - L\|)$. Clearly, $S = us$ is a section of ρ, and $P = pu^{-1}\colon Z \longrightarrow Y$ is a projection along $\jmath$. Choosing L such that $\|\Phi - L\| \leq K[X, Y]Q(\Phi)$, one obtains:

$$\|S\| \leq \|u\|\,\|s\| \leq (1 + \varepsilon)\Delta(1 + 2(1 + \varepsilon)\Delta^2 K[X, Y]);$$
$$\|P\| \leq \|u^{-1}\|\,\|p\| \leq (1 + (2 + \varepsilon)\Delta)\max(\Delta, 2(1 + \varepsilon)\Delta^2 K[X, Y]). \qquad \square$$

We now state and prove a version of the preceding result for p-Banach spaces where the concavity of the middle space is controlled through the

Banach–Mazur distance: p-Banach spaces have modulus of concavity at most $2^{1/p-1}$, and thus a space at distance λ from a p-Banach space has modulus of concavity not larger than $\lambda 2^{1/p-1}$.

Lemma 5.2.4 *Let X, Y be p-Banach spaces. If every exact sequence $0 \longrightarrow Y \longrightarrow Z \longrightarrow X \longrightarrow 0$ in which Z is isomorphic to a p-Banach space splits, then there are increasing functions $\mu, \nu\colon [1,\infty) \longrightarrow \mathbb{R}$ such that every isometrically exact sequence $0 \longrightarrow Y \longrightarrow Z \longrightarrow X \longrightarrow 0$ of quasi-Banach spaces in which Z is λ-isomorphic to a p-Banach space admits a linear section S of the quotient map such that $\|S\| \leq \mu(\lambda)$ and a projection $P\colon Z \longrightarrow Y$ such that $\|P\| \leq \nu(\lambda)$.*

Proof The hypothesis implies that $K^{(p)}[X, Y]$ is finite. The proof now goes as before: if $0 \longrightarrow Y \longrightarrow Z \longrightarrow X \longrightarrow 0$ is isometrically exact and Φ is a quasilinear map obtained as the difference of a homogeneous section $B\colon X \longrightarrow Z$ with $\|B\| \leq 1+\varepsilon$ and a linear section of the quotient map, then $Q^{(p)}(\Phi) = Q^{(p)}(B) \leq 2^{1/p}(1+\varepsilon)\lambda$, provided Z is λ-isomorphic to a p-normed space, since in that case, for every finite set of points $z_1, \ldots, z_n \in Z$, we have $\left\|\sum_{i\leq n} z_i\right\|^p \leq \lambda^p \sum_{i\leq n} \|z_i\|^p$. Now, proceed as before, replacing $K[X, Y]$ by $K^{(p)}[X, Y]$ and $Q(\Phi)$ by $Q^{(p)}(\Phi)$, taking into account that $\Delta_Z \leq \lambda 2^{1/p-1}$ when necessary. □

The assumption on the Banach–Mazur distance is necessary by the following argument. On one hand, $\mathrm{Ext}_{\mathbf{B}}(\ell_1, \mathbb{K}) = 0$. On the other hand, using the n-dimensional versions $\varrho_n\colon \ell_1^n \longrightarrow \mathbb{K}$ of Ribe's map (Proposition 3.2.3), we obtain isometrically exact sequences $0 \longrightarrow \mathbb{K} \longrightarrow Z(\varrho_n) \longrightarrow \ell_1^n \longrightarrow 0$ in which $\Delta(Z(\varrho_n)) \leq 2$ and, since the sequences are necessarily trivial, $Z(\varrho_n) \simeq \mathbb{R} \oplus \ell_1^n$. However, $\|s\| \geq \frac{1}{2}\log n$ for any linear section. In particular, the Banach–Mazur distance between $Z(\rho_n)$ and the Banach spaces is larger than $\frac{1}{2}\log n$.

The Ext Form of the Vector-Valued Sobczyk Theorem

Now that we have the language, we can state (compare with 2.14.8):

5.2.5 Vector-valued Sobczyk theorem *Let X be a separable Banach space and let (E_n) be a sequence of Banach spaces. If* $\mathrm{Ext}_{\mathbf{B}}(X, E_n) = 0$ *uniformly on n then* $\mathrm{Ext}_{\mathbf{B}}(X, c_0(\mathbb{N}, E_n)) = 0$ *in either of the following situations:*

(a) *X has the BAP.*
(b) *The sequence (E_n) has the joint-UAP.*

The proof of (a) is just the proof of 2.14.8. And the same is true for (b) since the hypothesis forces $\ell_\infty(\mathbb{N}, E_n)/c_0(\mathbb{N}, E_n)$ to have the BAP: since the space

$\ell_\infty(\mathbb{N}, E_n)$ has the λ-UAP and the sequence $0 \longrightarrow c_0(\mathbb{N}, E_n) \longrightarrow \ell_\infty(\mathbb{N}, E_n) \longrightarrow Q \longrightarrow 0$ splits locally, $Q^{**} \simeq \ell_\infty(\mathbb{N}, E_n)^{**}/c_0(\mathbb{N}, E_n)^{**}$ has the λ-UAP, as well as Q. Observe that everything in this plot works because of the UAP and falls flat for the mere BAP; the Pełczyński–Lusky sequence explains why. A remarkable instance of 5.2.5 follows; the next section has another.

Corollary 5.2.6 *Let X be a separable Banach space. If (E_n) is a sequence of $\mathscr{L}_{\infty,\lambda}$-spaces and* $\mathrm{Ext}_{\mathbf{B}}(X, E_n) = 0$ *uniformly on n then* $\mathrm{Ext}_{\mathbf{B}}(X, c_0(\mathbb{N}, E_n)) = 0$. *In particular, if each E_n is λ-separably injective then $c_0(\mathbb{N}, E_n)$ is separably injective.*

Why doesn't the BAP occur in this formulation? Well, the long version of the explanation is in Section 10.1. The short version is: because it is! Admittedly, it is disguised by Propositions 5.3.1 and 4.2.10: every separable Banach space is a twisted sum of two spaces with the BAP and, therefore, a space X is separably injective if and only if $\mathrm{Ext}_{\mathbf{B}}(S, X) = 0$ for all separable spaces S with the BAP. Now apply the proof 2.14.8 to those BAP spaces. Quantitative estimates are not so easy to obtain because of the required BAP decomposition. The result that $c_0(\mathbb{N}, E_n)$ is $f(\lambda)$-separably injective when all the spaces E_n are λ-separably injective has been independently obtained by Rosenthal [416] using operator techniques, by Johnson and Oikhberg [230], using M-ideals, by Cabello Sánchez [61] using a topological approach and by Castillo and Moreno [114] with non-linear techniques. Each of them comes with its own estimate for $f(\lambda)$: Rosenthal obtains $f(\lambda) = \lambda(1 + \lambda)^+$, Johnson and Oikhberg get $f(\lambda) = 2\lambda^2$ and Cabello Sánchez obtains $f(\lambda) = 3\lambda^2$, while Castillo and Moreno get $f(\lambda) = 6\lambda^+$.

$\mathscr{L}_\infty$ and $\mathscr{L}_p$ Spaces, $0 < p \leq 1$ Revisited

The second instance of 5.2.5 is as follows:

Corollary 5.2.7 *If X is an $\mathscr{L}_1$-space and all E_n are μ-complemented in their biduals then* $\mathrm{Ext}_{\mathbf{B}}(X, c_0(\mathbb{N}, E_n)) = 0$.

There is also a version for extension of operators:

Corollary 5.2.8 *Let Y be a subspace of the Banach space Z such that Z/Y is an $\mathscr{L}_{1,\lambda}$-space and let E_n be Banach spaces μ-complemented in their biduals. Any operator $\tau\colon Y \longrightarrow c_0(\mathbb{N}, E_n)$ admits a $\lambda\mu(1 + \lambda)$-extension to Z.*

Proof The proof of Theorem 3.7.1 in combination with Lemma 3.5.4 says that the components $Y \longrightarrow E_n$ have $\lambda\mu$-extensions to Z. Corollary 2.14.7 yields that τ admits a $\lambda\mu(1 + \lambda)$-extension. □

Combining this with the decomposition provided by Proposition 5.3.1 (every subspace of c_0 is a twisted sum of two spaces $c_0(\mathbb{N}, F_n)$ with all F_n finite-dimensional) plus the fact that, for fixed X, properties of the form $\mathrm{Ext}(X, \cdot) = 0$ are 3-space properties, yields part (a) in the following:

Proposition 5.2.9

(a) *A separable Banach space X is an $\mathcal{L}_1$-space if and only if* $\mathrm{Ext}_{\mathbf{B}}(X, H) = 0$ *for every subspace H of c_0.*

(b) *A Banach space X is an $\mathcal{L}_\infty$-space if and only if* $\mathrm{Ext}_{\mathbf{B}}(H^*, X) = 0$ *for every subspace $H \subset c_0$.*

Proof Part (b) follows by duality: the dual of a subspace of c_0 is a twisted sum of two spaces having the form $\ell_1(\mathbb{N}, F_n)$ with F_n finite-dimensional and, consequently, that it is enough to prove that $\mathrm{Ext}_{\mathbf{B}}(\ell_1(\mathbb{N}, F_n), X) = 0$. This amounts to saying that all exact sequences $0 \longrightarrow X \longrightarrow \cdot \longrightarrow F \longrightarrow 0$ in which F is finite-dimensional split uniformly, which is precisely the characterisation of $\mathcal{L}_\infty$-spaces given in Proposition 5.1.12. □

One might wonder about minimal classes $\mathscr{V}$ of Banach spaces that can replace the class of subspaces of c_0 in the characterisation (a) of $\mathcal{L}_1$-spaces above: X is an $\mathcal{L}_1$-space if and only if $\mathrm{Ext}_{\mathbf{B}}(X, V) = 0$ for all $V \in \mathscr{V}$. Let us show that the class of reflexive spaces is a valid choice:

Proposition 5.2.10 *A Banach space X is an $\mathcal{L}_1$-space if and only if* $\mathrm{Ext}_{\mathbf{B}}(X, C_r^{(1)}) = 0$ *for some (all) $1 \le r \le \infty$ or $r = 0$.*

Proof Lindenstrauss' lifting 5.1.15 yields one implication since the spaces $C_r^{(1)}$ are ultrasummands for all $1 \le r < \infty$. The case $r = \infty$ merely rewrites Proposition 5.1.16 (b). As for the other implication, since $C_r^{(1)}$ contains isometric 1-complemented copies of all the F_n and (F_n) is dense in $\mathscr{F}^{(1)}$, we have

$$\sup_{F \in \mathscr{F}^{(1)}} K^{(1)}[X, F] = \sup_n K^{(1)}[X, F_n] \le K^{(1)}[X, C_2^{(1)}] < \infty.$$

To conclude, invoke Proposition 5.1.16 (b) again. All this proves the case $r \neq 0$. For $r = 0$, use the same argument combined with Corollary 5.2.7. □

Even if $\mathrm{Ext}_{\mathbf{B}}(X, C_2^{(1)}) = 0$ implies that X is an $\mathcal{L}_1$-space, Hilbert spaces alone do not suffice since $\mathcal{B}^*$ has the KPP. Whether the class of super-reflexive spaces suffices is open and difficult. The p-versions of those results are simple, up to a point: $\mathrm{Ext}_{p\mathbf{B}}(L_p, C_q^{(r)}) = 0$ for all $p \le q, r \le \infty$ because ℓ_r is an r-Banach space ultrasummand when $r \le 1$, as well as $C_q^{(r)}$ (same proof). Readers who dare to go off-limits should inspect Section 10.1.

Extensions of Spaces with the Same Local Structure

We now present a technique to show that there exist non-trivial twisted sums of two spaces X, Y provided the existence of non-trivial twisted sums of other spaces X', Y' such that X and X' (resp. Y and Y') have the same local structure, in a sense to be determined next. Let $\mathscr{E}$ be a family of quasi-Banach spaces.

Definition 5.2.11 A quasi-Banach space X is said to be λ-locally $\mathscr{E}$ if every finite-dimensional subspace of X is contained in another finite-dimensional subspace $F \subset X$ such that $d(F, E) \leq \lambda$ for some $E \in \mathscr{E}$. We say that X is locally $\mathscr{E}$ if it is λ-locally $\mathscr{E}$ for some $\lambda \geq 1$. The space X is said to contain the class $\mathscr{E}$ uniformly (complemented) if there is λ such that every element of $\mathscr{E}$ is λ-isomorphic to some (λ-complemented) subspace of X.

We have already encountered examples of these notions: for instance, the $\mathscr{L}_p$-spaces are the locally ℓ_p^n spaces and the B-convex spaces contain ℓ_2^n uniformly complemented. Much more sophisticated is Bourgain's example [49]: the space $\ell_\infty(L_1)$ is locally $\ell_\infty^n(\ell_1^m)$. As for more general examples, we have:

Lemma 5.2.12 *Let X be a Banach space.*

(a) *If X^{**} is λ-locally $\mathscr{E}$ then X is λ^+-locally $\mathscr{E}$.*
(b) *Suppose $\mathscr{X}$ is a net of finite-dimensional subspaces of X whose union is dense in X. Then X is 1^+-locally $\mathscr{X}$.*

Proof (a) is obvious from the principle of local reflexivity. (b) seems obvious, but it is not: the complete proof can be found in Lacey [316, Theorem 6, p. 168] and uses local convexity in an essential way. □

Time to launch the idea of *uniform* splitting for families.

Definition 5.2.13 Given two families of quasi-Banach spaces $\mathscr{X}$ and $\mathscr{Y}$, we write $\operatorname{Ext}(\mathscr{X}, \mathscr{Y}) = 0$ to mean $\operatorname{Ext}(X, Y) = 0$ for every $X \in \mathscr{X}, Y \in \mathscr{Y}$. We shall say that $\operatorname{Ext}(\mathscr{X}, \mathscr{Y}) = 0$ uniformly if

$$K[\mathscr{X}, \mathscr{Y}] = \sup\{K[X, Y] \colon X \in \mathscr{X}, Y \in \mathscr{Y}\} < \infty.$$

If $\mathscr{X}$ and $\mathscr{Y}$ consist of p-Banach spaces only, we write $\operatorname{Ext}_{p\mathbf{B}}(\mathscr{X}, \mathscr{Y}) = 0$ to mean $\operatorname{Ext}_{p\mathbf{B}}(X, Y) = 0$ for every $X \in \mathscr{X}, Y \in \mathscr{Y}$ and say $\operatorname{Ext}_{p\mathbf{B}}(\mathscr{X}, \mathscr{Y}) = 0$ uniformly if

$$K^{(p)}[\mathscr{X}, \mathscr{Y}] = \sup\{K^{(p)}[X, Y] : X \in \mathscr{X}, Y \in \mathscr{Y}\} < \infty.$$

The uniform splitting is much stronger than $\operatorname{Ext}(\mathscr{X}, \mathscr{Y}) = 0$. For instance, $\operatorname{Ext}(\mathscr{F}, \mathscr{F}) = 0$ rather obviously, but the splitting is not uniform by far. It is clear that to get $\operatorname{Ext}(\mathscr{X}, Y) = 0$ uniformly, it is sufficient that $\operatorname{Ext}(X, Y) = 0$

for some space X containing $\mathscr{X}$ be uniformly complemented, and the same for the other variable. Now, Lemma 5.2.3 forces uniform splitting for families to behave well; that is, assertion (a) below holds:

Proposition 5.2.14 *Let $\mathscr{X}, \mathscr{Y}$ be families of quasi-Banach spaces and let $0 \longrightarrow Y \longrightarrow Z \longrightarrow X \longrightarrow 0$ be an isometrically exact sequence with $X \in \mathscr{X}, Y \in \mathscr{Y}$ in which Z has modulus of concavity Δ_Z, and let $\varepsilon > 0$.*

(a) *If* $\operatorname{Ext}(\mathscr{X}, \mathscr{Y}) = 0$ *uniformly then there is a linear section $S : X \longrightarrow Z$ of the quotient map such that $\|S\| \leq \Delta(1+2\Delta^2 K[\mathscr{X}, \mathscr{Y}])+\varepsilon$ and a projection $P \colon Z \longrightarrow Y$ such that $\|P\| \leq (1 + 2\Delta) \max(\Delta, 2\Delta^2 K[\mathscr{X}, \mathscr{Y}]) + \varepsilon$.*
(b) *If the families $\mathscr{X}, \mathscr{Y}$ have uniformly bounded moduli of concavity and there is a function f such that for every exact sequence as above there is a linear section $S : X \longrightarrow Z$ such that $\|S\| \leq f(\Delta_Z)$ (or a linear projection $P \colon Z \longrightarrow Y$ such that $\|P\| \leq f(\Delta_Z)$) then* $\operatorname{Ext}(\mathscr{X}, \mathscr{Y}) = 0$ *uniformly.*
(c) *If $\mathscr{X}, \mathscr{Y}$ are formed by p-Banach spaces then* $\operatorname{Ext}_{p\mathbf{B}}(X, Y) = 0$ *uniformly if and only if there is a function f such that whenever $0 \longrightarrow Y \longrightarrow Z \longrightarrow X \longrightarrow 0$ is an isometrically exact sequence in which Z is λ-isomorphic to a p-Banach space, there is a projection $P \colon Z \longrightarrow Y$ (resp. a section $S : X \longrightarrow Z$) bounded by $f(\lambda)$.*

Proof We prove (b). Take $\Delta \geq \Delta_A$ for all A in $\mathscr{X}, \mathscr{Y}$, pick $X \in \mathscr{X}$ and $Y \in \mathscr{Y}$ and let $\Phi \colon X \longrightarrow Y$ be a quasilinear map with $Q(\Phi) \leq 1$. Lemma 3.3.9 yields $\Delta_{Y \oplus_\Phi X} \leq 2\Delta^2$. By hypothesis, the sequence $0 \longrightarrow Y \longrightarrow Y \oplus_\Phi X \longrightarrow X \longrightarrow 0$ admits a linear section $S : X \longrightarrow Y \oplus_\Phi X$ with $\|S\| \leq f(\Delta)$ which necessarily has the form $S(x) = (L(x), x)$ for some linear map $L \colon X \longrightarrow Y$. Then, $\|S\| = \|\Phi - L\| + 1$, and thus $\|\Phi - L\| \leq f(\Delta)$, which yields $K[\mathscr{X}, \mathscr{Y}] \leq f(\Delta) < \infty$. □

One more estimate is necessary before practical results can be harvested:

Lemma 5.2.15 *Let X and Y be quasi-Banach spaces. If X' is λ-isomorphic to a λ'-complemented subspace of X and Y' is μ-isomorphic to a μ'-complemented subspace of Y, then $K[X', Y'] \leq \lambda\lambda'\mu\mu' K[X, Y]$. If all the spaces are p-Banach then one can replace $K[\cdot, \cdot]$ by $K^{(p)}[\cdot, \cdot]$.*

We thus discover an alleyway passing from an individual result to a uniformity result:

Proposition 5.2.16 *Assume that X contains the class $\mathscr{X}$ uniformly complemented and Y contains the class $\mathscr{Y}$ uniformly complemented. If* $\operatorname{Ext}(X, Y) = 0$ *then* $\operatorname{Ext}(\mathscr{X}, \mathscr{Y}) = 0$ *uniformly. If X, Y are p-Banach spaces then one can replace* Ext *by* $\operatorname{Ext}_{p\mathbf{B}}$.

In particular, since B-convex spaces contain ℓ_2^n uniformly complemented and $\sup_n K^{(1)}[\ell_2^n, \ell_2^n] = \infty$, we have:

Corollary 5.2.17 *If X, Y are infinite-dimensional B-convex Banach spaces then* $\mathrm{Ext}_{\mathbf{B}}(X, Y) \neq 0$.

Moving up the hill backwards, we now want to find a passage from $\mathrm{Ext}(\mathscr{X}, \mathscr{Y}) = 0$ uniformly to $\mathrm{Ext}(X, Y) = 0$ for particular spaces X and Y:

Proposition 5.2.18 *Let X, Y be quasi-Banach spaces that are locally $\mathscr{X}$ and $\mathscr{Y}$, respectively. If* $\mathrm{Ext}(\mathscr{X}, \mathscr{Y}) = 0$ *uniformly and Y is an ultrasummand then* $\mathrm{Ext}(X, Y) = 0$. *If $\mathscr{X}$ and $\mathscr{Y}$ consist of p-Banach spaces then one can replace* Ext *by* $\mathrm{Ext}_{p\mathbf{B}}$.

Proof It is easy to guess that the idea behind the proof is to show that every extension of X by Y splits locally and then use the hypothesis on the target space to guarantee global splitting. Assume without loss of generality that X and Y are p-normed spaces and let $\Phi\colon X \longrightarrow Y$ be a quasilinear map with $Q(\Phi) \leq 1$. Use Lemma 3.9.3 to get a family of quasilinear maps $\Phi_F\colon F \longrightarrow Y$, indexed by $F \in \mathscr{F}(X)$, such that $\Phi_F[F]$ spans a finite-dimensional subspace of Y and

$$\sup_{F\in\mathscr{F}(X)} \left(Q(\Phi_F), \|\Phi|_F - \Phi_F\|\right) < \infty.$$

Assume Y is λ-locally $\mathscr{Y}$. For each $F \in \mathscr{F}(X)$, the set $\Phi_F[F]$ lies inside a finite-dimensional subspace of Y that is λ-isomorphic to a certain $G \in \mathscr{Y}$. Since $K[\mathscr{F}(X), \mathscr{Y}] < \infty$, one can select, for each F, a linear map $L_F\colon F \longrightarrow Y$ such that $M = \sup_F \|\Phi_F - L_F\| < \infty$; i.e.,

$$\|\Phi_F(x) - L_F(x)\| \leq M\|x\| \tag{5.2}$$

for all $x \in F$. The remainder of the proof is the dullest thing we can do with those ingredients. Pick $\mathfrak{U}$ an ultrafilter refining the order filter of $\mathscr{F}(X)$ and $P\colon Y_{\mathfrak{U}} \longrightarrow Y$ a projection through the canonical embedding $\delta\colon Y \longrightarrow Y_{\mathfrak{U}}$. Define the mapping $L\colon X \longrightarrow Y_{\mathfrak{U}}$ given by $L(x) = [L_F(x1_F(x))]$ and then check that L is linear to conclude from (5.2) that $\|\delta \circ \Phi - L\| \leq M$. Finally, the composition PL yields $\|\Phi - PL\| \leq M\|P\|$, and this makes Φ trivial. The proof of the second assertion is analogous using p-linear maps. □

Finally, we build a bridge from one particular pair of spaces to another:

Proposition 5.2.19 *Let $\mathscr{X}$ and $\mathscr{Y}$ be families of quasi-Banach spaces and let X, Y, X', Y' be quasi-Banach spaces such that*

- *X contains $\mathscr{X}$ uniformly complemented and X' is locally $\mathscr{X}$.*

- *Y contains $\mathscr{Y}$ uniformly complemented and Y' is an ultrasummand that is locally $\mathscr{Y}$.*

Then $\operatorname{Ext}(X, Y) = 0$ *implies* $\operatorname{Ext}(X', Y) = 0$. *If all the spaces in consideration are p-Banach then we can replace* Ext *with* $\operatorname{Ext}_{p\mathbf{B}}$.

Proof Just go step by step: from $\operatorname{Ext}(X, Y) = 0$ to $\operatorname{Ext}(\mathscr{X}, \mathscr{Y}) = 0$ uniformly via Proposition 5.2.16, and from there to $\operatorname{Ext}(X', Y') = 0$ crossing through Proposition 5.2.18. □

Twisting $\mathscr{L}_p$-Spaces, $1 \leq p \leq \infty$

We now study the existence of non-trivial twisted sums of Banach spaces of type $\mathscr{L}_p$ for equal or different values of p. For obvious reasons, all spaces in this section will be considered infinite-dimensional without further notice. Since all $\mathscr{L}_p$-spaces have the same local structure (for a fixed p), it is a good opportunity to check how far this local approach can go. Observe that the local theory has nothing else to say about $\operatorname{Ext}_{\mathbf{B}}(\cdot, \mathscr{L}_\infty)$ or $\operatorname{Ext}_{\mathbf{B}}(\mathscr{L}_1, \cdot)$, thus those cases must be treated on an individual basis. A perhaps surprising assertion we will have opportunity to assess is that most of the results presented here are, at the end of the day, formal consequences of the fact $\operatorname{Ext}_{\mathbf{B}}(\ell_2, \ell_2) \neq 0$. To avoid annoying repetitions, we will adopt the following (illogical, but not absurd) convention: if $\mathscr{X}$ and $\mathscr{Y}$ are families of Banach spaces, then $\operatorname{Ext}_{\mathbf{B}}(\mathscr{X}, \mathscr{Y}) \neq 0$ means that $\operatorname{Ext}_{\mathbf{B}}(X, Y) \neq 0$ for every $X \in \mathscr{X}$ and every $Y \in \mathscr{Y}$.

Proposition 5.2.20 $\operatorname{Ext}_{\mathbf{B}}(\mathscr{L}_p, \mathscr{L}_q) \neq 0$ *for* $1 \leq p, q \leq \infty$ *unless* $p = 1$ *or* $q = \infty$.

Proof That $\operatorname{Ext}_{\mathbf{B}}(\mathscr{L}_p, \mathscr{L}_q) \neq 0$ for $1 < p, q < \infty$ is contained in Corollary 5.2.17. We pass to $\operatorname{Ext}_{\mathbf{B}}(\mathscr{L}_p, \mathscr{L}_1) \neq 0$ for $1 < p < \infty$, which, as we observed first and will show now, is somehow a consequence of the existence of twisted Hilbert spaces. Fix an isomorphic embedding $u \colon \ell_2 \longrightarrow L_1$ (think of the Rademacher functions, if you prefer to be more specific), pick the Kalton–Peck Z_2 space and form the pushout diagram

$$\begin{array}{ccccccccc} 0 & \longrightarrow & \ell_2 & \longrightarrow & Z_2 & \longrightarrow & \ell_2 & \longrightarrow & 0 \\ & & {\scriptstyle u}\downarrow & & \downarrow & & \| & & \\ 0 & \longrightarrow & L_1 & \longrightarrow & \mathrm{PO} & \longrightarrow & \ell_2 & \longrightarrow & 0 \end{array} \tag{5.3}$$

The pushout sequence cannot split because every subspace of $L_1 \times \ell_2$ has cotype 2, while Z_2 does not. Thus, $\operatorname{Ext}_{\mathbf{B}}(\ell_2, L_1) \neq 0$. Set

$$\begin{array}{lll} \mathscr{X} = (\ell_2^n)_n, & X = \text{any } \mathscr{L}_p\text{-space}, & X' = \ell_2, \\ \mathscr{Y} = (\ell_1^n)_n, & Y = \text{any } \mathscr{L}_1\text{-space}, & Y' = L_1 \end{array}$$

and apply Proposition 5.2.19 to get $\mathrm{Ext}_{\mathbf{B}}(\mathscr{L}_p, \mathscr{L}_1) \neq 0$ for $1 < p < \infty$. Since L_1 is an ultrasummand, the dual of a non-trivial sequence $0 \longrightarrow L_1 \longrightarrow \cdot \longrightarrow \ell_2 \longrightarrow 0$ cannot split, and one also has $\mathrm{Ext}_{\mathbf{B}}(L_\infty, \ell_2) \neq 0$ and thus $\mathrm{Ext}_{\mathbf{B}}(\mathscr{L}_\infty, \ell_2) \neq 0$. To settle the remaining case, namely $\mathrm{Ext}_{\mathbf{B}}(\mathscr{L}_\infty, \mathscr{L}_1) \neq 0$, let us first show how to construct a non-trivial extension of L_∞ by L_1. As one might guess, we start once more with the Kalton–Peck Z_2 sequence and an isomorphic embedding $u\colon \ell_2 \longrightarrow L_1$ to which we add now a quotient map $Q\colon L_\infty \longrightarrow \ell_2$ to form the pushout / pullback diagram

$$\begin{array}{ccccccccc} 0 & \longrightarrow & \ell_2 & \longrightarrow & Z_2 & \longrightarrow & \ell_2 & \longrightarrow & 0 \\ & & {\scriptstyle u}\downarrow & & \downarrow & & \| & & \\ 0 & \longrightarrow & L_1 & \longrightarrow & \mathrm{PO} & \xrightarrow{\overline{\rho}} & \ell_2 & \longrightarrow & 0 \\ & & \| & & \uparrow & & \uparrow{\scriptstyle Q} & & \\ 0 & \longrightarrow & L_1 & \longrightarrow & \mathrm{PB} & \longrightarrow & L_\infty & \longrightarrow & 0 \end{array}$$

The key point is to show that the lower sequence does not split, or, equivalently, that Q cannot be lifted to PO. To this end, let us first point out a special feature of this construction: the quotient map $\overline{\rho}$ is strictly singular because given any infinite-dimensional subspace H of ℓ_2, the lower sequence of the commutative diagram

$$\begin{array}{ccccccccc} 0 & \longrightarrow & \ell_2 & \longrightarrow & \rho^{-1}[H] & \xrightarrow{\rho} & H & \longrightarrow & 0 \\ & & {\scriptstyle u}\downarrow & & \downarrow & & \downarrow{\scriptstyle \text{inclusion}} & & \\ 0 & \longrightarrow & L_1 & \longrightarrow & \overline{\rho}^{-1}[H] & \xrightarrow{\overline{\rho}} & \ell_2 & \longrightarrow & 0 \end{array} \tag{5.4}$$

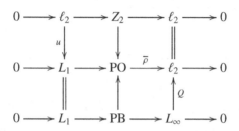

cannot split since $\rho^{-1}[H]$ contains an isomorphic copy of Z_2, as explained in Section 10.9, in the paragraph labelled 'The space Z_2 is "self similar"', and thus it cannot be a subspace of $L_1 \times H$, as would be the case were the lower sequence trivial. Returning to the proof, assume that some linear continuous lifting $L\colon L_\infty \longrightarrow \mathrm{PO}$ for Q exists. Let (x_n) be a bounded sequence in L_∞ such that $Q(x_n) = e_n$. Since PO has finite cotype, we infer from [123, Theorem 2.3] that there is $f \in \mathrm{PO}$ and a subsequence $(L(x_m))_m$ such that $(L(x_m) - f)_m$ is weakly-2-summable, or, equivalently, the continuous image of $(e_m)_m$ [153, Proposition 2.2]. This would imply that the quotient map $\mathrm{PO} \longrightarrow \ell_2$ is invertible on the subspace spanned by the sequence $(e_m)_m$, which is impossible because $\overline{\rho}$ is strictly singular. Thus, $\mathrm{Ext}_{\mathbf{B}}(L_\infty, L_1) \neq 0$ and, setting

$$\begin{array}{lll} \mathscr{X} = (\ell_\infty^n)_n & X = \text{any } \mathscr{L}_\infty\text{-space}, & X' = L_\infty, \\ \mathscr{Y} = (\ell_1^n)_n & Y = \text{any } \mathscr{L}_1\text{-space}, & Y' = L_1 \end{array}$$

in Proposition 5.2.19, we get $\mathrm{Ext}_{\mathbf{B}}(\mathscr{L}_\infty, \mathscr{L}_1) \neq 0$. □

Claim The role of Z_2 can be played by any non-trivial twisted Hilbert space.

Proof of the claim Let $0 \longrightarrow \ell_2 \longrightarrow \diamond \longrightarrow \ell_2 \longrightarrow 0$ be a non-trivial extension. Consider any isomorphic embedding $\jmath\colon \ell_2 \longrightarrow L_1$ and any quotient map $\rho\colon C[0,1] \longrightarrow \ell_2$ and form the commutative diagram

$$\begin{array}{ccccccccc} 0 & \longrightarrow & \ell_2 & \longrightarrow & \diamond & \longrightarrow & \ell_2 & \longrightarrow & 0 \\ & & \downarrow{\scriptstyle \jmath} & & \downarrow & & \| & & \\ 0 & \longrightarrow & L_1 & \xrightarrow{\jmath_2} & \mathrm{PO} & \xrightarrow{\rho_2} & \ell_2 & \longrightarrow & 0 \\ & & \| & & \uparrow & & \uparrow{\scriptstyle \rho} & & \\ 0 & \longrightarrow & L_1 & \longrightarrow & \mathrm{PB} & \xrightarrow{\rho_3} & C[0,1] & \longrightarrow & 0 \end{array}$$

No twisted Hilbert space can have cotype 2 (Section 10.9) and $\diamond$ is no exception, so it cannot be a subspace of the cotype 2 space $L_1 \times \ell_2$. This prevents the middle sequence from splitting. Our goal is to show that the bottom sequence does not split. Suppose it does. Then there exists an operator $s_3\colon C[0,1] \longrightarrow \mathrm{PB}$ such that $\rho_3 s_3 = \mathbf{1}_{C[0,1]}$. Since L_1 has cotype 2, the space PO has cotype q for all $q > 2$, as follows from Corollary 3.11.4. Thus, the operator $\underline{\rho} s_3\colon C[0,1] \longrightarrow \mathrm{PO}$ must factor as $\underline{\rho} s_3 = \beta\alpha$ through some L_r-space with $r > \overline{2}$ [153, Theorem 11.14 (b)]. Form the pullback diagram

$$\begin{array}{ccc} L_1 & \xrightarrow{\jmath_2} & \mathrm{PO} \\ \uparrow & & \uparrow{\scriptstyle \beta} \\ \mathrm{PB}(\jmath_2, \beta) & \longrightarrow & L_r \end{array}$$

to discover that since $\rho_2\beta\alpha = \rho$, the map $\rho_2\beta$ is surjective and thus $\mathrm{PB}(\jmath_2, \beta) = \ker \rho_2\beta$, which yields the commutative diagram

$$\begin{array}{ccccccccc} 0 & \longrightarrow & L_1 & \xrightarrow{\jmath_2} & \mathrm{PO} & \xrightarrow{\rho_2} & \ell_2 & \longrightarrow & 0 \\ & & \uparrow{\scriptstyle u} & & \uparrow{\scriptstyle \beta} & & \| & & \\ 0 & \longrightarrow & \mathrm{PB}(\jmath_2, \beta) & \longrightarrow & L_r & \xrightarrow{\rho_2\beta} & \ell_2 & \longrightarrow & 0 \end{array}$$

Now, since L_r has type 2 and L_1 has cotype 2, Maurey's extension theorem (see the comments after 1.4.10) yields that every operator from a subspace of

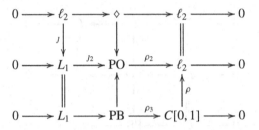

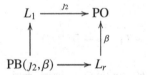

L_r to L_1 extends to L_r. Apply this to u to conclude that the upper sequence must split. □

The case $p = \infty, q = 1$ has already been treated accidentally:

Corollary 5.2.21 *If X, Y are Banach spaces, X contains ℓ_∞^n uniformly and Y contains ℓ_1^n uniformly complemented,* $\mathrm{Ext}_{\mathbf{B}}(X, Y) \neq 0$ *and* $\mathrm{Ext}_{\mathbf{B}}(Y^*, X^*) \neq 0$.

The argument showing that $\mathrm{Ext}_{\mathbf{B}}(\ell_2, L_1) \neq 0$ and $\mathrm{Ext}_{\mathbf{B}}(L_\infty, \ell_2) \neq 0$ that appears in the middle of the proof of Proposition 5.2.20 can be localised in different ways, but only one of them requires a proof:

Proposition 5.2.22 *If X contains ℓ_∞^n uniformly then* $\mathrm{Ext}_{\mathbf{B}}(X, \ell_2) \neq 0$. *If X contains ℓ_1^n uniformly complemented then* $\mathrm{Ext}_{\mathbf{B}}(\ell_2, X) \neq 0$. *If X is an infinite-dimensional Banach space of cotype 2, then* $\mathrm{Ext}_{\mathbf{B}}(\ell_2, X) \neq 0$.

Proof We know that every Banach space contains ℓ_2^n almost isometrically thanks to the Dvoretzky–Rogers theorem [153, Theorem 19.2]. For each $n \in \mathbb{N}$, let $r_n\colon \ell_2^n \longrightarrow X$ be a $\frac{1}{n}$-isometry, let $R_n\colon \ell_2 \longrightarrow X$ be the composition with the projection $\pi_n\colon \ell_2 \longrightarrow \ell_2^n$ onto the first n coordinates and let $R\colon \ell_2 \longrightarrow \ell_\infty(X)$ be the embedding $R(x) = (R_n(x))$. If $\mathcal{U}$ is a free ultrafilter on $\mathbb{N}$ and $[\cdot]\colon \ell_\infty(X) \longrightarrow X_{\mathcal{U}}$ is the natural quotient map, it is clear that the composition $[\cdot]R$ is still an embedding. Now, let $0 \longrightarrow \ell_2 \longrightarrow \diamond \longrightarrow \ell_2 \longrightarrow 0$ be a non-trivial twisted Hilbert space and form the successive pushout diagrams

$$\begin{array}{ccccccccc}
0 & \longrightarrow & \ell_2 & \longrightarrow & \diamond & \longrightarrow & \ell_2 & \longrightarrow & 0\\
 & & \downarrow{\scriptstyle R} & & \downarrow & & \| & & \\
0 & \longrightarrow & \ell_\infty(X) & \longrightarrow & \mathrm{PO}_1 & \longrightarrow & \ell_2 & \longrightarrow & 0\\
 & & \downarrow{\scriptstyle [\cdot]} & & \downarrow & & \| & & \\
0 & \longrightarrow & X_{\mathcal{U}} & \longrightarrow & \mathrm{PO}_2 & \longrightarrow & \ell_2 & \longrightarrow & 0
\end{array}$$

If $\mathrm{Ext}_{\mathbf{B}}(\ell_2, X) = 0$ then $\mathrm{Ext}_{\mathbf{B}}(\ell_2, \ell_\infty(X)) = 0$, and thus the lower sequence splits. This makes $\diamond$ a subspace of $X_{\mathcal{U}} \times \ell_2$, which has cotype 2 since $X_{\mathcal{U}}$ has the same cotype as X: a contradiction. □

The dual result, *If X is a Banach space whose dual has cotype 2, then* $\mathrm{Ext}_{\mathbf{B}}(X, \ell_2) \neq 0$, holds by the duality formula: $\mathrm{Ext}_{\mathbf{B}}(X, \ell_2) = \mathrm{Ext}_{\mathbf{B}}(\ell_2, X^*)$.

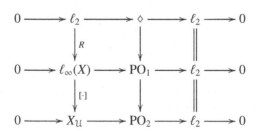

5.3 The Mysterious Role of the BAP

We arrive at the third milestone of the chapter: approximation properties. The BAP is no doubt useful in Banach space theory because it allows us to split large objects into smaller ones using finite-rank operators. However, the only whiffs we have had so far of any homological sniff about the BAP are the Pełczyński–Lusky sequence (2.7) and the scent Proposition 2.2.19 left (the sequence splits if and only if X has the BAP) that it somehow detects the BAP. Is there any other homological connection in sight? Yes: the structural theorem of Johnson, Rosenthal and Zippin we have mentioned so often and which is usually seen in its negative form (the BAP is not a 3-space property):

Proposition 5.3.1 *Every separable Banach space X admits a representation $0 \longrightarrow A \longrightarrow X \longrightarrow B \longrightarrow 0$ in which both A, B have a FDD. Moreover, if X^* is separable then A and B may be chosen having a shrinking FDD. In the particular case in which X is a subspace of c_0, X admits a representation $0 \longrightarrow c_0(\mathbb{N}, A_n) \longrightarrow X \longrightarrow c_0(\mathbb{N}, B_n) \longrightarrow 0$, where A_n and B_n are finite-dimensional spaces.*

A neat proof for the first part can be found in [334, Theorem 1.g.2], while the second part can be deduced from [334, Theorem 2.d.1]; see the proof in [334, Theorem 2.f.6]. This failure of the 3-space property for the BAP has a bright side: a Banach space X is separably injective if and only if $\operatorname{Ext}_{\mathbf{B}}(S, X) = 0$ for all separable spaces S with the BAP. And, more good news, the negative 3-space result is not that negative, since restricted forms of the 3-space property are still available:

Proposition 5.3.2 *Let $0 \longrightarrow Y \longrightarrow Z \xrightarrow{\rho} X \longrightarrow 0$ be a locally split sequence.*

(a) *If Y and X have the BAP then Z has the BAP.*
(b) *If Z has the BAP then Y must have the BAP.*

Proof Assertion (b) is trivial. To prove (a), let $M < \infty$ be such that the sequence M-splits locally and both Y and X have the M-AP. Given a finite-dimensional subspace F of Z,

- there is a finite-rank operator $\tau \in \mathfrak{L}(X)$ of norm at most M fixing $\rho[F]$;
- there is a local section $s\colon \tau[X] \longrightarrow Z$ of norm at most M;
- if $E = (\mathbf{1}_F - s\rho)[F] \subset Y$, there is a finite-rank operator $\omega \in \mathfrak{L}(Y)$ fixing E having norm at most M;
- there is a finite-rank extension $\varpi\colon Z \longrightarrow Y$ of ω with norm at most M since the sequence splits locally.

Thus, the operator $T = \varpi(\mathbf{1}_X - s\tau\rho) + s\tau\rho$ has finite rank, has controlled norm and fixes F since for $f \in F$, we have $Tf = \varpi(f - s\rho f) + s\rho f = f - s\rho f + s\rho f = f$. □

There are counterexamples for the two remaining cases. To get examples (c) in which both Z, X have the BAP but Y does not, we use Szankowski's remark [448] that the classical Enflo–Davies counterexample for the AP provides a subspace H of c_0 without the AP yielding an exact sequence:

$$0 \longrightarrow H \longrightarrow c_0 \longrightarrow c_0(\mathbb{N}, \ell_2^n) \longrightarrow 0 \tag{5.5}$$

(by the way, this sequence cannot split locally). An example (d) in which Y and Z have the BAP but X does not is the Pełczyński–Lusky sequence for a separable space X without the BAP: the space $c(\mathbb{N}, X_n)$ has the BAP, and since the sequence splits locally, so does $c_0(\mathbb{N}, X_n)$. Another example of this type can be obtained by recalling Lindenstrauss' 'outgrowth' [328] of James [218] according to which every separable space X can be written as $X = Y^{**}/Y$ where Y^{**} (hence Y) has the BAP. The sequence $0 \longrightarrow Y \longrightarrow Y^{**} \longrightarrow X \longrightarrow 0$ splits locally, but X may fail the BAP. The next two sections present two rather surprising results in this context: an example like (c) cannot exist when Z is an $\mathscr{L}_1$-space and an example like (d) cannot exist when Z is an $\mathscr{L}_\infty$-space. If the reader is still sceptical of the homological content of the BAP, we find their lack of faith disturbing: the BAP force is just beginning to manifest.

Projective Presentations and the BAP

Projective presentations are the archetype of exact sequences in which the middle space is an $\mathscr{L}_1$-space. Thus, the paradigmatic result in this context is Lusky's theorem [348] that whenever X has the BAP, the kernel of any of its projective presentations $0 \longrightarrow \kappa(X) \longrightarrow \ell_1(I) \longrightarrow X \longrightarrow 0$ must have the BAP as well. Lusky's proof is technically demanding; we will (somehow) vault such difficulties using the $\mathrm{co}^{(p)}$ spaces appearing in Section 3.10.

Lemma 5.3.3 *If a p-Banach space X has the λ-AP, $\mathrm{co}^{(p)}(X)$ has the 3λ-AP.*

Proof Fix a Hamel basis $\mathscr{H}$ for X and let $\mho: X \longrightarrow \mathrm{co}^{(p)}(X)$ be the (version of) the universal p-linear map of Theorem 3.10.2 vanishing on $\mathscr{H}$. Let F be a finite-dimensional subspace of $\mathrm{co}^{(p)}(X)$. We can assume without loss of generality that $F = [\mho(x_1), \ldots, \mho(x_m)]$, where $x_j \in X$ for $1 \leq j \leq m$. Let $\mathscr{H}_0$ be a finite subset of $\mathscr{H}$ whose linear span $[\mathscr{H}_0]$ contains $[x_1, \ldots, x_m]$. Now, let $T \in \mathfrak{L}(X)$ be a finite-rank operator fixing $[\mathscr{H}_0]$ and with $\|T\| \leq \lambda^+$. Since $Q^{(p)}(\mho) = 1$, Lemma 3.9.1 allows us to obtain a small perturbation $\mho': X \longrightarrow \mathrm{co}^{(p)}(X)$ satisfying:

- $\mho'(x_j) = \mho(x_j)$ for $1 \leq j \leq m$,
- $\mho'(b) = \mho(b) = 0$ for $b \in \mathscr{H}$,
- $\|\mho' - \mho\| \leq 1 + \varepsilon$,
- $Q^{(p)}(\mho') \leq 3 + \varepsilon$,
- $\mho\,(T[X])$ spans a finite-dimensional subspace of $\mathrm{co}^{(p)}(X)$.

Let $(\mho' \circ T)_{\mathscr{H}} : X \longrightarrow \mathrm{co}^{(p)}(X)$ be the version of $\mho' \circ T$ that vanishes on $\mathscr{H}$. The universal property of $\mho$ yields an operator $\phi\colon \mathrm{co}^{(p)}(X) \longrightarrow \mathrm{co}^{(p)}(X)$ such that $\phi \circ \mho = (\mho' \circ T)_{\mathscr{H}}$. Let us check that ϕ has the required properties:

- $\|\phi\| = Q^{(p)}((\mho' \circ T)_{\mathscr{H}}) = Q^{(p)}(\mho' \circ T) \leq Q^{(p)}(\mho')\|T\| \leq (3 + \varepsilon)\lambda^{+}$.
- ϕ has finite rank: the image of $\mho' \circ T$, and therefore that of $(\mho' \circ T)_{\mathscr{H}}$, spans a finite-dimensional subspace of $\mathrm{co}^{(p)}(X)$. Hence $\{\phi(\mho(x)) : x \in X\} \subset [(\mho' \circ T)_{\mathscr{H}}]$, and since $\mathrm{co}^{(p)}(X)$ is the closure of the space spanned by the points of the form $\mho(x)$ and ϕ is continuous, we also get that $\phi[\mathrm{co}^{(p)}(X)] \subset [(\mho' \circ T)_{\mathscr{H}}]$.
- ϕ fixes F. Indeed, since $x_j = \sum_{b \in \mathscr{H}_0} \lambda_b b$ for $1 \leq j \leq m$, one has

$$\begin{aligned}\phi(\mho(x_j)) &= (\mho' \circ T)_{\mathscr{H}}(x_j)\\ &= \mho'(Tx_j) - \sum_{b \in \mathscr{H}_0} \lambda_b \mho'(Tb)\\ &= \mho'(x_j) - \sum_{b \in \mathscr{H}_0} \lambda_b \mho'(b)\\ &= \mho(x_j).\end{aligned}$$ □

The stage is set for the proof of the main result. The rest is just throwing balls to the homological wall:

Proposition 5.3.4 *Let $0 < p \leq 1$. If X is a p-Banach space with the BAP then $\kappa_p(X)$ has the BAP.*

Proof The universal property of $\mathrm{co}^{(p)}(X)$ yields a commutative diagram

$$\begin{array}{ccccccccc} 0 & \longrightarrow & \mathrm{co}^{(p)}(X) & \longrightarrow & \diamond & \longrightarrow & X & \longrightarrow & 0\\ & & \downarrow & & \downarrow & & \| & & \\ 0 & \longrightarrow & \kappa_p(X) & \longrightarrow & \ell_p(I) & \longrightarrow & X & \longrightarrow & 0 \end{array}$$

whose diagonal pushout sequence $0 \longrightarrow \mathrm{co}^{(p)}(X) \longrightarrow \diamond \times \kappa_p(X) \longrightarrow \ell_p(I) \longrightarrow 0$ splits. Thus $\mathrm{co}^{(p)}(X) \times \ell_p(I) \simeq \diamond \times \kappa_p(X)$ and, by Lemma 5.3.3, $\kappa_p(X)$ must have the BAP. □

Corollary 5.3.5 *Let $0 \longrightarrow Y \longrightarrow Z \longrightarrow X \longrightarrow 0$ be an exact sequence in which Z is a discrete $\mathscr{L}_p$-space, $0 < p \leq 1$. If X has BAP then also Y has BAP.*

Proof Proceed as before. The diagonal sequence $0 \longrightarrow \mathrm{co}^{(p)}(X) \longrightarrow \diamond\times Y \longrightarrow Z \longrightarrow 0$ splits locally. Thus, Proposition 5.3.2 (a) yields that $\diamond\times Y$ has BAP. □

Focusing on $p = 1$, given a Banach space X, we get that either all kernels of all quotient maps $\mathscr{L}_1 \longrightarrow X$ enjoy the BAP or none of them does (the result has a straightforward version for discrete $\mathscr{L}_p$-spaces which we just skip):

Proposition 5.3.6 *Given exact sequences*

$$\begin{array}{ccccccccc}
0 & \longrightarrow & Y & \longrightarrow & \mathscr{L}_1 & \longrightarrow & X & \longrightarrow & 0 \\
 & & & & & & \| & & \\
0 & \longrightarrow & Y' & \longrightarrow & \mathscr{L}_1' & \longrightarrow & X & \longrightarrow & 0
\end{array}$$

the space Y has BAP if and only if Y' has BAP.

Proof The two sequences passing through PB in the commutative diagram

$$\begin{array}{ccccccccc}
 & & & & 0 & & 0 & & \\
 & & & & \uparrow & & \uparrow & & \\
0 & \longrightarrow & Y & \longrightarrow & \mathscr{L}_1 & \longrightarrow & X & \longrightarrow & 0 \\
 & & \| & & \uparrow & & \uparrow & & \\
0 & \longrightarrow & Y & \longrightarrow & \mathrm{PB} & \longrightarrow & \mathscr{L}_1' & \longrightarrow & 0 \\
 & & & & \uparrow & & \uparrow & & \\
 & & & & Y' & = & Y' & & \\
 & & & & \uparrow & & \uparrow & & \\
 & & & & 0 & & 0 & &
\end{array}$$

split locally. If Y has the BAP then so does PB since $\mathscr{L}_1$-spaces have the BAP. Therefore, the same is true for its locally complemented subspace Y'. □

The sequence $0 \longrightarrow \ell_1(\mathbb{N}, \ell_2^n) \longrightarrow \ell_1 \longrightarrow H^* \longrightarrow 0$, dual of (5.5), shows that ℓ_1 contains a subspace E isomorphic to $\ell_1(\mathbb{N}, \ell_2^n)$ with the BAP such that ℓ_1/E does not have the BAP. Moreover, since every infinite-dimensional $\mathscr{L}_1$-space contains a complemented copy of ℓ_1, taking $\mathscr{L}_1 = \ell_1 \times A$, the sequence $0 \longrightarrow \ell_1(\mathbb{N}, \ell_2^n) \times A \longrightarrow \mathscr{L}_1 \longrightarrow H^* \longrightarrow 0$ shows that every infinite-dimensional $\mathscr{L}_1$-space contains a subspace E' with the BAP such that $\mathscr{L}_1/E'$ does not have the BAP.

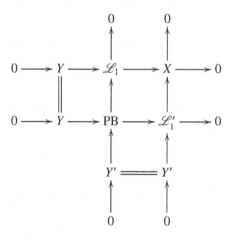

Injective Presentations and the BAP

We now present the corresponding dual results. The key point is the following technical lemma, for which we provide two (two?) proofs.

Lemma 5.3.7 *Let* $0 \longrightarrow Y \longrightarrow X \xrightarrow{\rho} F \longrightarrow 0$ *be an isometrically exact sequence of Banach spaces in which* F *is finite-dimensional and* Y *has the* μ*-AP. Given a finite-dimensional subspace* G *of* X*, there is a finite-rank operator* $T_G \in \mathfrak{L}(X)$ *with norm at most* $3\mu^+$ *fixing* G *and such that* $T_G[Y] \subset Y$.

Proof Fix $\varepsilon > 0$. Enlarging G if necessary, we may assume it almost norms $Y^\perp = F^*$: for every $x^* \in Y^\perp$, we have $\|x^*\| \leq (1+\varepsilon)\sup\{|x^*(g)| : g \in G, \|g\| \leq 1\}$. This implies that for each $x \in X$ there is $g \in G$ with $\|g\| \leq (1+\varepsilon)\|x\|$ such that $\rho(x) = \rho(g)$. Let us consider the norm one sum map $\oplus\colon Y \oplus G \longrightarrow X$. Since for each normalised $x \in X$ we can find $g \in G$ such that $\rho(g) = \rho(x)$ and $\|g\| \leq 1^+$, it follows that $(x - g, g) \in Y \oplus G$ has norm at most 3^+ and $\oplus(x - g, g) = x$. Hence $\oplus$ induces an isomorphism $\sigma\colon (Y \oplus G)/\ker\oplus \longrightarrow X$ satisfying $\|\sigma\| \leq 1$ and $\|\sigma^{-1}\| \leq 3^+$. Let $T \in \mathfrak{L}(Y)$ be a finite-rank operator with norm at most μ^+ that fixes $G \cap Y$. The operator $T \times \mathbf{1}_G\colon Y \oplus G \longrightarrow Y \oplus G$ has finite rank and the same norm as T. Moreover, the restriction of $T \times \mathbf{1}_G$ to $\ker\oplus$ is the identity: if $y + g = 0$ then $y \in Y \cap G$, so $T(y) = y$ and $(T \times \mathbf{1}_G)(y, g) = (y, g)$. Thus, $T \times \mathbf{1}_G$ induces the finite-rank operator $T'\colon (Y \oplus G)/\ker\oplus \longrightarrow (Y \oplus G)/\ker\oplus$ given by $T'((y, g) + \ker\oplus) = (Ty, g) + \ker\oplus$ and with the same norm. Hence $T_G = \sigma T' \sigma^{-1}$ is a finite-rank operator on X with norm at most $3\mu^+$, and T_G fixes G: if $g \in G$, we have

$$T_G(g) = \sigma(T'(\sigma^{-1}(g)) = \sigma(T'((0, g) + \ker\oplus)) = \sigma((0, T(g)) + \ker\oplus)) = g. \qquad \square$$

What has happened here? Well, the argument in the preceding proof should elbow us aware of the pushout construction: indeed, draw Lemma 3.9.1 as follows.

5.3.8 *If* $0 \longrightarrow Y \longrightarrow X \xrightarrow{\rho} F \longrightarrow 0$ *is an isometrically exact sequence of Banach spaces with* F *finite-dimensional then for every* $\varepsilon > 0$ *there is a finite-dimensional subspace* $G \subset X$ *and a commutative diagram*

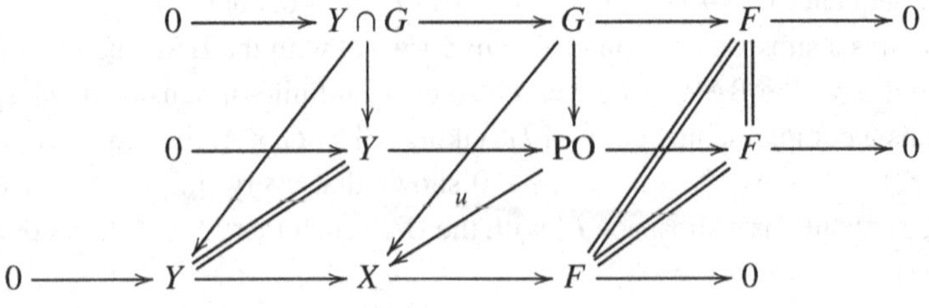

in which u is an isomorphism with $\|u\| \leq 1$ *and* $\|u^{-1}\| \leq 3+\varepsilon$ *and all unlabelled arrows are canonical inclusions.*

To prove it, and keeping the same notation as before, observe that if $T \in \mathfrak{L}(Y)$ is a finite-rank operator fixing $Y \cap G$ with $\|T\| \leq \mu^+$ then the universal property of the pushout yields a commutative the diagram

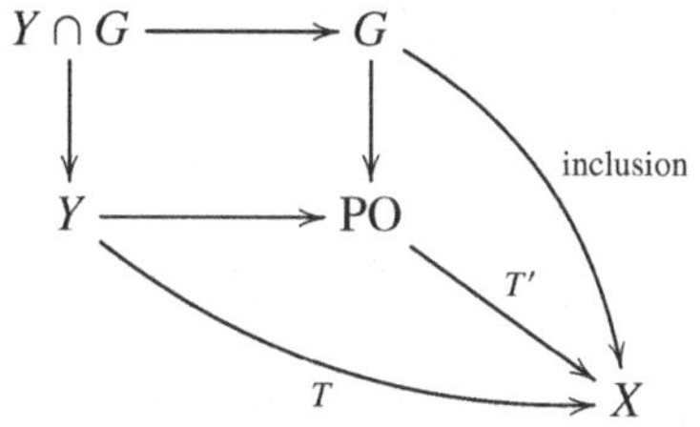

in which $T' : \text{PO} \longrightarrow X$ is a finite-rank operator with $\|T'\| \leq \mu^+$. Set $T_G = T'u^{-1}$ to get the desired operator and estimate.

We are ready to deliver the promised result:

Proposition 5.3.9 *Let Y be a Banach space with the μ-AP and let $Y \longrightarrow \mathscr{L}_{\infty,\lambda}$ be an isometric embedding. Then $\mathscr{L}_{\infty,\lambda}/Y$ has the $3\mu\lambda$-AP.*

Proof What we will show is that, given any finite-dimensional subspace $F \subset \mathscr{L}_{\infty,\lambda}$, there is a finite-rank operator $T \in \mathfrak{L}(\mathscr{L}_{\infty,\lambda})$ fixing F such that $T[Y] \subset Y$ and $\|T\| \leq 3\mu\lambda^+$. This already means that $\mathscr{L}_{\infty,\lambda}/Y$ has the $3\mu\lambda$-AP. To that end, fix F and apply Lemma 5.3.7 to the sequence

$$0 \longrightarrow Y \longrightarrow Y + F \longrightarrow (Y+F)/Y \longrightarrow 0$$

to get a finite-rank operator $\tau \in \mathfrak{L}(Y + F)$ fixing F, leaving Y invariant and having norm at most $3\mu^+$. The finite-dimensional subspace $\tau[Y + F]$ must be contained in a subspace λ^+-isomorphic to some ℓ_∞^m, and therefore there is an extension $T \in \mathfrak{L}(\mathscr{L}_{\infty,\lambda})$ of rank at most m and norm at most $3\lambda\mu^+$ that fixes F and leaves Y invariant. $\qquad\square$

Proposition 5.3.9 can be completed with:

Lemma 5.3.10 *Given an exact sequence $0 \longrightarrow \mathscr{L}_\infty \longrightarrow Z \longrightarrow X \longrightarrow 0$ of Banach spaces in which Z has the BAP, also X has the BAP.*

Proof Form a commutative diagram

$$\begin{array}{ccccccccc} 0 & \longrightarrow & \mathscr{L}_\infty & \longrightarrow & \ell_\infty(I) & \longrightarrow & \ell_\infty(I)/\mathscr{L}_\infty & \longrightarrow & 0 \\ & & \| & & \uparrow & & \uparrow & & \\ 0 & \longrightarrow & \mathscr{L}_\infty & \longrightarrow & Z & \longrightarrow & X & \longrightarrow & 0 \end{array}$$

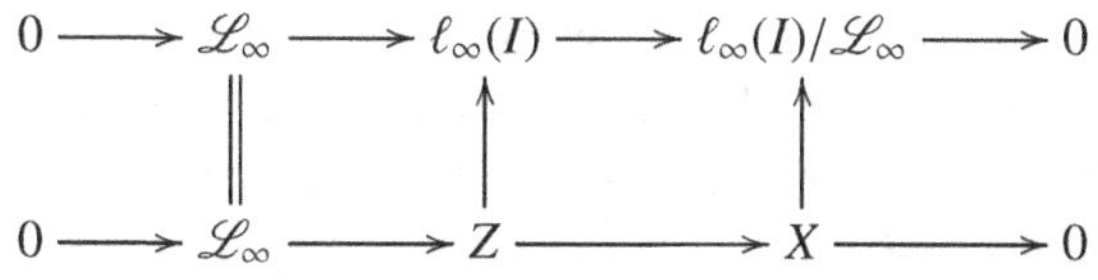

The diagonal pullback sequence $0 \longrightarrow Z \longrightarrow \ell_\infty(I) \times X \longrightarrow \ell_\infty(I)/\mathscr{L}_\infty \longrightarrow 0$ splits locally since it is a pushout of the upper row, which splits locally. And the space $\ell_\infty(I)/\mathscr{L}_\infty$ is an $\mathscr{L}_\infty$-space, as is any quotient of two $\mathscr{L}_\infty$-spaces, hence it has the BAP. Thus, if Z has the BAP then the middle space $\ell_\infty(I) \times X$ has the BAP, and this implies that also X has the BAP. □

Independently of whether Y has the BAP, all the quotients $\mathscr{L}_\infty/Y$ have or fail to have the BAP simultaneously:

Proposition 5.3.11 *Given exact sequences*

$$\begin{array}{ccccccccc} 0 & \longrightarrow & Y & \longrightarrow & \mathscr{L}_\infty & \longrightarrow & X & \longrightarrow & 0 \\ & & \| & & & & & & \\ 0 & \longrightarrow & Y & \longrightarrow & \mathscr{L}'_\infty & \longrightarrow & X' & \longrightarrow & 0 \end{array}$$

the space X has the BAP if and only if X' has the BAP.

Proof Consider the pushout diagram

$$\begin{array}{ccccccccc} & & 0 & & 0 & & & & \\ & & \downarrow & & \downarrow & & & & \\ 0 & \longrightarrow & Y & \longrightarrow & \mathscr{L}_\infty & \longrightarrow & X & \longrightarrow & 0 \\ & & \downarrow & & \downarrow & & \| & & \\ 0 & \longrightarrow & \mathscr{L}'_\infty & \longrightarrow & \mathrm{PO} & \longrightarrow & X & \longrightarrow & 0 \\ & & \downarrow & & \downarrow & & & & \\ & & X' & = & X' & & & & \\ & & \downarrow & & \downarrow & & & & \\ & & 0 & & 0 & & & & \end{array}$$

The two sequences passing through PO split locally. If X' has the BAP then PO must have the BAP, and thus Lemma 5.3.10 applies to the horizontal sequence, allowing us to conclude that also X has the BAP. □

The space ℓ_p contains a subspace without the BAP for every $p \in [1, \infty)$ different from 2 (see [334, Theorem 2.d.6] and [335, Theorem 1.g.4]) and therefore ℓ_p has a quotient E without the BAP. Since ℓ_p is a quotient of $C[0, 1]$ for $p \in [2, \infty)$, the space E is a quotient too. Thus, $C[0, 1]$ contains a subspace Y such that $C[0, 1]/Y = E$ lacks the BAP. This Y cannot have the BAP and no quotient $\mathscr{L}_\infty/Y$ can have the BAP either. Propositions 5.3.4 and 5.3.9 can be forced to cover the UAP case:

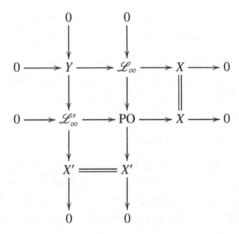

Corollary 5.3.12 *If X is a Banach space with the UAP, the kernel of any surjection $\mathscr{L}_1 \longrightarrow X$ and the cokernel of any embedding $X \longrightarrow \mathscr{L}_\infty$ have the UAP.*

Proof Assume X has the UAP so that, for every ultrafilter $\mathfrak{U}$, the ultrapower $X_\mathfrak{U}$ has the BAP. Let $0 \longrightarrow K \longrightarrow \mathscr{L}_1 \longrightarrow X \longrightarrow 0$ be any exact sequence. The ultrapower sequence $0 \longrightarrow K_\mathfrak{U} \longrightarrow (\mathscr{L}_1)_\mathfrak{U} \longrightarrow X_\mathfrak{U} \longrightarrow 0$ is again exact and since ultrapowers of $\mathscr{L}_1$-spaces are $\mathscr{L}_1$-spaces we can apply Proposition 5.3.4 to conclude that $K_\mathfrak{U}$ has the BAP and, therefore, K has the UAP. The dual version is analogous. □

Trivial Twisting and the BAP

The BAP has been lurking behind the vanishing of Ext(X, Y) spaces at least since Corollary 4.5.12. It is time to reveal its role. To start with, surprising as it may seem, in the presence of the BAP, the existence of non-trivial elements in Ext(X, Y) can always be detected by a careful observation of the trivial ones.

Theorem 5.3.13 *Let X be a quasi-Banach space and Y be a μ-ultrasummand such that $K_0[X, Y] < \infty$.*

(a) *If X has the λ-AP then* Ext$(X, Y) = 0$.
(b) *If X is a $\mathscr{K}$-space and Y has the λ-AP then* Ext$(X, Y) = 0$.

In both cases, $K[X, Y] \leq \lambda^+ \mu K_0[X, Y]$.

Proof By the Aoki–Rolewicz theorem, we may assume that Y is a p-Banach space. To prove (a), let $\mathfrak{U}$ be any ultrafilter refining the order filter on $\mathscr{F}(X)$, and let us consider the corresponding ultrapower $Y_\mathfrak{U}$ and a bounded projection $P\colon Y_\mathfrak{U} \longrightarrow Y$ along the diagonal embedding $\delta\colon Y \longrightarrow Y_\mathfrak{U}$. Since X has the λ-AP, for each $E \in \mathscr{F}(X)$, there is $T_E \in \mathfrak{F}(X)$ fixing E with $\|T_E\| \leq \lambda^+$. Now, let $\Phi\colon X \longrightarrow Y$ be quasilinear, with $Q(\Phi) \leq 1$. We define the map $\phi\colon X \longrightarrow \ell_\infty(\mathscr{F}(X), Y)$ given by $\phi(x) = (\Phi(T_E x))_E$. Observe that $\phi(x)(E) = \Phi(x)$ when $x \in E$, whence it follows that $[\cdot] \circ \phi = \delta \circ \Phi$. In other drawings, there is a commutative diagram

$$\begin{array}{ccc} X & \xrightarrow{\ \Phi\ } & Y \\ {\scriptstyle\phi}\downarrow & & \downarrow{\scriptstyle\delta} \\ \ell_\infty(\mathscr{F}(X), Y) & \xrightarrow{\ [\cdot]\ } & Y_\mathfrak{U} \end{array}$$

and consequently, $\Phi = P[\cdot] \circ \phi$. Since quasilinear maps are bounded on finite-dimensional spaces, $\Phi \circ T_E\colon X \longrightarrow Y$ is bounded. Thus, for each $E \in \mathscr{F}(X)$, there is $\ell_E \in \mathfrak{L}(X, Y)$ such that

$$\|\ell_E - \Phi \circ T_E\| \leq K_0[X,Y]\, Q(\Phi \circ T_E) \leq K_0[X,Y]\, Q(\Phi)\, \|T_E\| \leq K_0[X,Y]\, \lambda^+.$$

This allows us to define a map $\phi' \colon X \longrightarrow \ell_\infty(\mathscr{F}(X), Y)$ by $\phi'(x)(E) = \ell_E(x 1_E(x))$. Of course, we have

$$\|\phi(x) - \phi'(x)\|_\infty = \sup_{x \in E} \|\Phi(T_E(x)) - \ell_E(x)\|_Y \leq K_0[X,Y]\lambda^+ \|x\|.$$

Hence, if one sets $L = [\cdot] \circ \phi'$ then $\|[\cdot] \circ \phi - L\| \leq K_0[X,Y]\lambda^+$. The point is that L is actually linear: it is obviously homogeneous, and moreover, given $x, y \in X$, the set $\{E \in \mathscr{F}(X) \colon x, y \in E\}$ belongs to $\mathfrak{U}$, and thus, if $x, y \in E$, we have

$$\phi'(x+y)(E) = \ell_E(x+y) = \ell_E(x) + \ell_E(y) = \phi'(x)(E) + \phi'(y)(E),$$

which means that $L(x+y) = L(x) + L(y)$. The linear map $PL \colon X \longrightarrow Y$ satisfies

$$\|\Phi - PL\| = \|P[\cdot] \circ \phi - PL\| \leq \|P\|\, K_0[X,Y]\, \lambda^+.$$

Thus, $K[X,Y] \leq \lambda^+ \|P\| K_0[X,Y]$, and every quasilinear map $X \longrightarrow Y$ is trivial.

The proof for (b) is analogous. This time, the index set is $\mathscr{F}(Y)$. For each $E \in \mathscr{F}(Y)$, we pick $T_E \in \mathfrak{F}(Y)$ such that $T_E(y) = y$ for $y \in E$, with $\|T_E\| \leq \lambda^+$. Now, let $\Phi \colon X \longrightarrow Y$ be a quasilinear map with $Q(\Phi) \leq 1$. For each $E \in \mathscr{F}(Y)$, consider the composition $T_E \circ \Phi \colon X \longrightarrow T_E[Y]$. Since X is a $\mathscr{K}$-space, there is a linear map $\ell_E \colon X \longrightarrow T_E[Y]$ at finite distance from $T_E \circ \Phi$. The problem is that we have no bound for that distance. To overcome this difficulty, just consider $T_E \circ \Phi - \ell_E$ as a bounded homogeneous map $X \longrightarrow Y$. Since $Q(T_E \circ \Phi - \ell_E) = Q(T_E \circ \Phi) \leq \lambda^+$, there is a linear map $\ell'_E \colon X \longrightarrow Y$ such that

$$\|(T_E \circ \Phi - \ell_E) - \ell'_E\| = \|T_E \circ \Phi - (\ell_E + \ell'_E)\| \leq \lambda^+ K_0[X,Y].$$

We define $\phi' \colon X \longrightarrow \ell_\infty(\mathscr{F}(Y), Y)$ by $\phi'(x)(E) = \ell_E(x) + \ell'_E(x)$. Observe that in the worst case, i.e. when $\Phi(x) \in E$, we have $T_E(\Phi(x)) = \Phi(x)$, and therefore $\|\phi'(x)(E) - \Phi(x)\| \leq K_0[X,Y]\lambda^+\|x\|$; hence

$$\sup_{E \in \mathscr{F}(Y)} \|\phi'(x)(E)\| \leq \Delta_Y\left(\|\Phi(x)\| + K_0[X,Y]\lambda^+\|x\|\right).$$

The rest goes as before. Let $\mathfrak{U}$ be an ultrafilter refining the Fréchet filter on $\mathscr{F}(Y)$, and form the composition $L = [\cdot] \circ \phi' \colon X \longrightarrow \ell_\infty(\mathscr{F}(Y), Y) \longrightarrow Y_\mathfrak{U}$, which is linear: it is obviously homogeneous and if $x, y \in X$, then as long as E contains $\Phi(x)$, $\Phi(y)$ and $\Phi(x+y)$, we have $\phi'(x+y)(E) = \phi'(x)(E) + \phi'(y)(E)$, which yields $[\phi'(x+y)(E)] = [\phi'(x)(E)] + [\phi'(y)(E)]$. Pick a bounded projection $P \colon Y_\mathfrak{U} \longrightarrow Y$ along the diagonal embedding δ to obtain a linear map $PL \colon X \longrightarrow Y$, which is at finite distance from Φ since $\|\delta \circ \Phi - [\cdot] \circ \phi'\| \leq K_0[X,Y]\lambda^+$, and thus $\|\Phi - PL\| \leq \|P\|\, K_0[X,Y]\, \lambda^+$. □

A Banach space version of Theorem 5.3.13 for 1-linear maps is also true, although the requirement of being a $\mathscr{K}$-space can be omitted since 1-linear maps taking values on finite-dimensional spaces are automatically trivial:

Corollary 5.3.14 *If X and Y are Banach spaces, Y an ultrasummand and either X or Y have the BAP and $K_0^{(1)}[X, Y] < \infty$ then $K^{(1)}[X, Y] < \infty$.*

Now let us consider the question of whether $\mathrm{Ext}(X, Y) = 0$ implies the vanishing of any of the spaces $\mathrm{Ext}(X^{**}, Y), \mathrm{Ext}(X, Y^{**}), \mathrm{Ext}(X_{\mathfrak{U}}, Y)$ or $\mathrm{Ext}(X, Y_{\mathfrak{U}})$. The issue was slightly touched in Section 5.2 since X, X^{**} and $X_{\mathfrak{U}}$ are perhaps the most natural examples of spaces with the same local structure. The novelty here is the use of the BAP to factorise quasilinear maps through finite-dimensional spaces. Thus, it is about time for the BAP to pounce.

Twisted Sums and Biduals

Does $\mathrm{Ext}(X, Y) = 0$ imply $\mathrm{Ext}(X^{**}, Y) = 0$ or $\mathrm{Ext}(X, Y^{**}) = 0$?

Theorem 5.3.15 *If X is a Banach space whose bidual has the BAP and Y is a quasi-Banach ultrasummand such that* $\mathrm{Ext}(X, Y) = 0$, *then* $\mathrm{Ext}(X^{**}, Y) = 0$. *If Y is a Banach space, we can replace* Ext *by* $\mathrm{Ext}_{\mathbf{B}}$.

Proof Suppose on the contrary that there is a non-trivial quasilinear map $\Phi\colon X^{**} \longrightarrow Y$ with $Q(\Phi) \leq 1$. The idea is that, even if the restriction of Φ itself to X can be trivial – it can be zero, in fact – one can use a finite-rank operator to 'push' Φ down to get a non-trivial quasilinear map from X to Y. To this end, assume that X^{**} has the λ-AP, pick $M > 0$ and choose a finite-dimensional subspace $E \subset X^{**}$ such that $\mathrm{dist}(\Phi|_E, \mathfrak{L}(E, Y)) > \lambda M$. Pick $\varepsilon > 0$ and select a finite-rank operator $\tau\colon X^{**} \longrightarrow X^{**}$ such that $\|\tau\| \leq \lambda$ and $\|\tau(x^{**}) - x^{**}\| \leq \varepsilon\|x^{**}\|$ for all $x^{**} \in E$. We will see that $\Phi \circ \tau$ is a 'bad' quasilinear map for sufficiently small ε that will depend on $n = \dim E$ and Y. Before going further, let us indicate how the hypothesis that the bidual of X has the BAP is to be used: since $\mathfrak{F}(X^{**}) = X^{***} \otimes X^{**}$, a finite-rank operator on X^{**} of given norm that ε-fixes a finite-dimensional subspace $E \subset X^{**}$ can be chosen in $X^{*} \otimes X^{**}$ by an obvious application of the Goldstine theorem. So, the preceding τ can be chosen to be an operator $\tau\colon X \longrightarrow X^{**}$ such that $\|\tau\| \leq \lambda$ and

$$\|\tau^{**}(x^{**}) - x^{**}\| \leq \varepsilon\|x^{**}\| \tag{5.6}$$

for all $x^{**} \in E$. Now set $F = \tau^{**}[X^{**}] = \tau[X]$ and apply the following lemma:

Lemma 5.3.16 *Let $\tau\colon X \longrightarrow F$ be a linear operator, where X and F*

are Banach spaces, with F finite-dimensional. Let E be a finite-dimensional subspace of X^{**} and $\varepsilon > 0$. Then there is a subspace $E_0 \subset X$ and a surjective ε-isometry $u\colon E_0 \longrightarrow E$ such that $\tau^{**}(u(x)) = \tau(x)$ for every $x \in E$.

Proof The result follows from the principle of local reflexivity: given E, X as in the statement, G a finite-dimensional subspace of X^* and $\varepsilon > 0$, there is an ε-isometry $v\colon E \longrightarrow X$ such that $x^{**}(g) = g(v(x^{**}))$ for every $x^{**} \in E$ and every $g \in G$. Moreover, v can be chosen such that $v(x) = x$ for every $x \in E \cap X$, but we will not use this fact. Assume that $\tau = \sum_{i=1}^{n} g_i \otimes f_i$, for $g_i \in X^*$ and $f_i \in F$. Fix as G the subspace spanned by $g_1, \ldots, g_n$, and let $\varepsilon > 0$. By the principle of local reflexivity, we obtain an ε-isometry $v\colon E \longrightarrow X$ such that $\tau^{**}(x^{**}) = \tau(v(x^{**}))$ for $x^{**} \in E$. Set $E_0 = v[E]$ and $u = v^{-1}$ to conclude. □

Back to the proof of the theorem, we have obtained a subspace $E_0 \subset X$ together with an ε-isometry $u\colon E_0 \longrightarrow E$ such that $\tau^{**}(u(x)) = \tau(x)$ for $x \in E_0$. Letting $x^{**} = u(x)$ in (5.6), we obtain $\|\tau(x) - u(x)\| \leq \varepsilon\|u(x)\| \leq \varepsilon(1+\varepsilon)\|x\| \leq 2\varepsilon\|x\|$ for all $x \in E_0$. In particular, $\|\tau|_{E_0}\| \leq 2\varepsilon + \|u\| \leq 1 + 3\varepsilon$. We now need to pause to observe that if E, Z are p-Banach spaces and $\dim E = n$ then $K[E, Z]$ can be bounded by a constant $\varkappa(n, p)$ depending only on n and p. We won't spoil the reader's fun here. We also make a detour to obtain a slightly mystifying lemma in which the role of the constant $\varkappa(n, p)$ is finally unmasked.

Lemma 5.3.17 *Let $\Phi\colon X \longrightarrow Y$ be a quasilinear map acting between p-normed spaces, with $Q(\Phi) \leq 1$, and let F be an n-dimensional p-normed space. Given two linear operators $u, v\colon F \longrightarrow X$, we have*

$$\operatorname{dist}(\Phi \circ u, \mathsf{L}(F, Y)) \leq 3^{1/p-1}\left(\operatorname{dist}(\Phi \circ v, \mathsf{L}(F, Y)) + \|v\| + (1 + \varkappa(n, p))\|v - u\|\right).$$

Proof There is no need to freak out about the factor $3^{1/p-1}$: it only appears because we have to sum three chunks to complete the proof. Pick linear maps $L_1, L_2\colon F \longrightarrow Y$ such that

- $D_1 = \|L_1 - \phi \circ v\| \leq \operatorname{dist}(\phi \circ v, \mathsf{L}(F, Y)) + \varepsilon$,
- $D_2 = \|L_2 - \phi \circ (u - v)\| \leq \operatorname{dist}(\phi \circ (u - v), \mathsf{L}(F, Y)) + \varepsilon \leq \varkappa(n, p))\|v - u\| + \varepsilon$

for small $\varepsilon > 0$. Let us estimate $\|\phi \circ u - (L_1 + L_2)\|$. Pick a normalised $f \in F$:

$$\begin{aligned}
&\|\Phi u f - L_1 f - L_2 f\|^p \\
&= \|\Phi u f - \Phi(u - v)f - \Phi v f + \Phi v f - L_1 f + \Phi(u - v)f - L_2 f\|^p \\
&\leq \|\Phi u f - \Phi(u - v)f - \Phi v f\|^p + \|\Phi v f - L_1 f\|^p + \|\Phi(u - v)f - L_2 f\|^p \\
&\leq (\|u - v\| + \|v\|)^p + D_1^p + D_2^p,
\end{aligned}$$

whence, as required,

$$\begin{aligned}&\|\Phi \circ u - (L_1 + L_2)\| \\ &\leq 3^{1/p-1}(\|u - v\| + \|v\| + D_1 + D_2) \\ &\leq 3^{1/p-1}(\|u - v\| + \|v\| + \operatorname{dist}(\Phi \circ v, \mathsf{L}(F, Y)) + \varkappa(n, p)\|v - u\| + 2\varepsilon). \quad \square\end{aligned}$$

We are ready to complete the proof. On account of Lemma 5.3.17, one has

$$\begin{aligned}&\operatorname{dist}(\Phi \circ u, \mathsf{L}(E_0, Y)) \\ &\leq 3^{1/p-1}\left(\operatorname{dist}(\Phi \circ \tau|_{E_0}, \mathsf{L}(E_0, Y)) + \|\tau|_{E_0}\| + (1 + \varkappa(n, p))\|u - \tau|_{E_0}\|\right) \\ &\leq 3^{1/p-1}\left(\operatorname{dist}(\Phi \circ \tau|_{E_0}, \mathsf{L}(E_0, Y)) + 1 + 3\varepsilon + 2\varepsilon(1 + \varkappa(n, p))\right) \\ &\leq 3^{1/p-1}\left(\operatorname{dist}(\Phi \circ \tau|_{E_0}, \mathsf{L}(E_0, Y)) + 6\right),\end{aligned}$$

provided $\varepsilon \leq 1/(1 + \varkappa(n, p))$ – this is therefore the precise value of ε we need to start the proof! On the other hand,

$$\lambda M \leq \operatorname{dist}(\Phi|_E, \mathsf{L}(E, Y)) = \operatorname{dist}(\Phi \circ uu^{-1}, \mathsf{L}(E, Y)) \leq \|u^{-1}\| \operatorname{dist}(\Phi \circ u, \mathsf{L}(E_0, Y)).$$

But $\|u^{-1}\| \leq (1 + \varepsilon) \leq 2$, and so $\operatorname{dist}(\Phi \circ u, \mathsf{L}(E_0, Y)) \geq \lambda M/2$, and therefore $\lambda M/2 \leq 3^{1/p-1}(\operatorname{dist}(\Phi \circ \tau|_{E_0}, \mathsf{L}(E_0, Y)) + 6)$, whence one gets $\operatorname{dist}(\Phi \circ \tau, \mathsf{L}(X, Y)) \geq \operatorname{dist}(\Phi \circ \tau|_{E_0}, \mathsf{L}(E_0, Y)) \geq \lambda M(2 \cdot 3^{1/p-1})^{-1} - 6$, while $Q(\Phi \circ t) \leq Q(\Phi) \cdot \|\tau\| \leq \lambda$, and thus $K_0[X, Y]$ cannot be finite. $\square$

Twisted Sums and Ultrapowers

We tackle the next two cases: when does $\operatorname{Ext}(X, Y) = 0$ imply $\operatorname{Ext}(X_{\mathfrak{U}}, Y) = 0$ or $\operatorname{Ext}(X, Y_{\mathfrak{U}}) = 0$? Here the UAP, which is the approximation property most suited to work with ultraproducts, can go berserk. Let us begin with a companion for Theorem 5.3.15:

Theorem 5.3.18 *Let X be a Banach space, and let Y be a quasi-Banach ultrasummand such that* $\operatorname{Ext}(X, Y) = 0$. *If X has the UAP then* $\operatorname{Ext}(X_{\mathfrak{U}}, Y) = 0$ *for all ultrapowers of X for every countably incomplete ultrafilter $\mathfrak{U}$.*

Proof The proof follows that of Theorem 5.3.15, but it is simpler. Suppose there is a countably incomplete ultrafilter $\mathfrak{U}$, based on I, such that $\operatorname{Ext}(X_{\mathfrak{U}}, Y) \neq 0$, and let $\Phi\colon X_{\mathfrak{U}} \longrightarrow Y$ be a quasilinear map with $Q(\Phi) \leq 1$. Fix $M > 0$ and pick a finite-dimensional subspace F of $X_{\mathfrak{U}}$ such that $\operatorname{dist}(\Phi|_F, L(F, Y)) > \lambda M$. Let $f^1, \ldots, f^n$ be a (normalised) basis of F. Write $f^k = [(f_i^k)]$, and for each $i \in I$, put $F_i = [f_i^1, \ldots, f_i^n]$. For each i, take $\tau_i \in \mathfrak{L}(X)$ such that $\tau_i|_{F_i} = \mathbf{1}_{F_i}$, with $\|\tau_i\| \leq \lambda$ and $\dim(\tau_i[X]) \leq r(n)$, and set $F_i = \tau_i[X]$ and $G = [G_i]_{\mathfrak{U}}$. Obviously, G contains F. Take $f^{n+1}, \ldots, f^m \in F$ such that the enlarged system $f^1, \ldots, f^n, f^{n+1}, \ldots, f^m$ is a basis of G. For $n + 1 \leq k \leq m$, we can write

$f^k = [(f_i^k)]$, where $f_i^k \in G_i$. Given $i \in I$, we define an operator $u_i \colon G \longrightarrow G_i$, taking $u_i(f^k) = f_i^k$ for $1 \le k \le m$. Now, for every $\varepsilon > 0$, the set $\{i \in I : u_i \text{ is an } \varepsilon\text{-isometry}\}$ belongs to $\mathfrak{U}$. Fix $s \in I$ such that u_s is a 1-isometry. In particular, $\|u_s\| \le 2$ and $\|u_s^{-1}\| \le 2$. We will prove that the composition $\Phi \circ u_s^{-1}\tau_s$ is a 'bad' quasilinear map. The following commutative diagram, where unlabelled arrows are plain inclusion maps, can help the reader to visualise the relevant information:

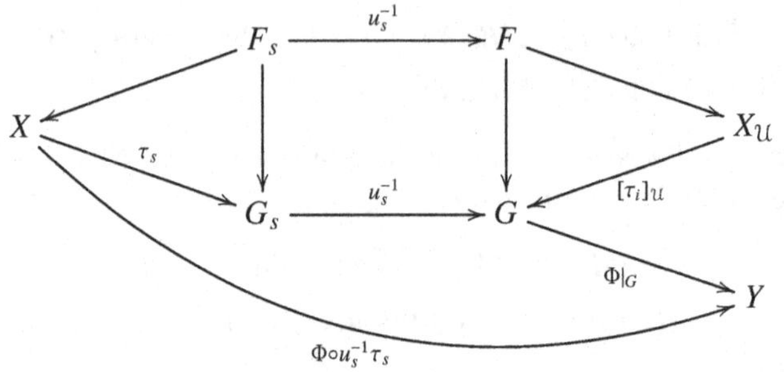

Observe that $Q(\Phi \circ u_s^{-1}\tau_s) \le Q(\Phi)\|u_s^{-1}\|\|\tau_s\| \le 2\lambda$, while

$$\begin{aligned}\operatorname{dist}(\Phi \circ u_s^{-1}\tau_s, \mathsf{L}(X,Y)) &\ge \operatorname{dist}(\Phi \circ u_s^{-1}\tau_s|_{F_s}, \mathsf{L}(F_s, Y))\\ &= \operatorname{dist}(\Phi \circ u_s^{-1}|_{F_s}, \mathsf{L}(F_s, Y))\\ &\ge \frac{1}{\|u_s\|}\operatorname{dist}(\Phi \circ u_s^{-1}u_s, \mathsf{L}(E,Y)) \ge \frac{\lambda M}{2}.\end{aligned}$$

Since M is arbitrary, we get $K_0[X, Y] = \infty$. □

And so we arrive at:

Theorem 5.3.19 *Let X be a separable Banach space and let Y be a Banach space such that* $\operatorname{Ext}_{\mathbf{B}}(X,Y) = 0$. *If X has the BAP or Y has the UAP then* $\operatorname{Ext}_{\mathbf{B}}(X, Y_{\mathfrak{U}}) = 0$ *for all countably incomplete $\mathfrak{U}$.*

Proof Let $0 \longrightarrow \kappa(X) \longrightarrow \ell_1 \longrightarrow X \longrightarrow 0$ be a projective presentation of X and let $\tau : \kappa(X) \longrightarrow Y_{\mathfrak{U}}$ be an operator. We must show that τ extends to ℓ_1. To that end, suppose Y has the UAP so that $Y_{\mathfrak{U}}$ enjoys the BAP. Then the range of τ is contained in a separable subspace of $Y_{\mathfrak{U}}$ with the BAP and, by Theorem 2.14.5, τ lifts to an operator $t \colon \kappa(X) \longrightarrow \ell_\infty(I, Y)$ that can be written as $t = (t_i)$, with $t_i \in \mathfrak{L}(\kappa(X), Y)$. Since $\operatorname{Ext}_{\mathbf{B}}(X, Y) = 0$, each t_i can be extended to an operator $T_i \colon \ell_1 \longrightarrow Y$ with $\|T_i\| \le C\|t_i\|$. Then $T = (T_i)$ is an operator $T \colon \ell_1 \longrightarrow \ell_\infty(I, Y)$, and the composition

$$\ell_1 \xrightarrow{\;T\;} \ell_\infty(I, Y) \xrightarrow{\;[\cdot]\;} Y_{\mathfrak{U}}$$

is an extension of τ. If instead we use the hypothesis that X has the BAP then we use Proposition 5.3.4 to get that $\kappa(X)$ has the BAP, and Theorem 2.14.5 once more yields the required lifting of τ. □

Finally, we treat the remaining case: does $\operatorname{Ext}(X, Y) = 0 \Rightarrow \operatorname{Ext}(X, Y^{**}) = 0$? Since all even-order duals of Y are complemented in suitable ultrapowers of Y, and since $\operatorname{Ext}_{\mathbf{B}}(X, Y^{**}) = \operatorname{Ext}_{\mathbf{B}}(Y^*, X^*)$, Theorem 5.3.15 yields:

Corollary 5.3.20 *Let X, Y be Banach spaces, with X separable. Assume that either X has the BAP or Y has the UAP and that* $\operatorname{Ext}_{\mathbf{B}}(X, Y) = 0$. *Then* $\operatorname{Ext}_{\mathbf{B}}(X, Y^{**}) = \operatorname{Ext}_{\mathbf{B}}(Y^*, X^*) = 0$.

This provides a rather unexpected partial answer for what we might call the duality problem: does $\operatorname{Ext}(X, Y) = 0$ imply $\operatorname{Ext}(Y^*, X^*) = 0$? It would be interesting to know if the approximation properties are truly necessary here. The main difficulty for a direct attack is that there are elements in $\operatorname{Ext}_{\mathbf{B}}(Y^*, X^*)$ that are not duals of elements of $\operatorname{Ext}_{\mathbf{B}}(X, Y)$, as it has been shown in Proposition 2.12.3 and will again be proved in Theorem 10.5.12. Separability cannot be removed in Theorem 5.3.19 because infinite-dimensional ultraproducts via countably incomplete ultrafilters are never injective [22, Theorem 4.6]. Thus, there is some Banach space X for which $\operatorname{Ext}_{\mathbf{B}}(X, (\ell_\infty)_{\mathcal{U}}) \neq 0$, despite having $\operatorname{Ext}_{\mathbf{B}}(X, \ell_\infty) = 0$.

5.4 Notes and Remarks

5.4.1 Which Banach Spaces Are $\mathscr{K}$-Spaces?

From Theorem 5.2.1 we immediately get: *X is a $\mathscr{K}$-space if and only if* $K[X, \mathbb{K}] < \infty$. The following variation there makes sense: X is a $\mathscr{K}_0$-space if $K_0[X, \mathbb{K}] < \infty$; namely, there is a constant C such that for every bounded quasilinear functional $\phi\colon X \longrightarrow \mathbb{K}$ there is $x^* \in X^*$ such that $\|\phi - x^*\| \leq CQ(\phi)$; equivalently, $\operatorname{Ext}(X, \mathbb{K})$ is Hausdorff. Perhaps the most interesting problems on *Banach* $\mathscr{K}$-spaces are deciding whether every $\mathscr{K}_0$-space is a $\mathscr{K}$-space (the converse is obvious) and characterising Banach $\mathscr{K}$- and $\mathscr{K}_0$-spaces. Kalton repeatedly conjectured that 'not containing ℓ_1^n uniformly complemented' is the right characterisation of $\mathscr{K}$-spaces; cf. [285, p. 815], [257, p. 11], [279, Remark on p. 44], [269, Problem 4.2]. In any case, anyone daydreaming about proving this conjecture should take into account that it implies that ultrapowers, thus all even duals, of Banach $\mathscr{K}$-spaces are $\mathscr{K}$-spaces too and also that all Banach $\mathscr{K}_0$-spaces are $\mathscr{K}$-spaces. A first step in this direction follows from Theorems 5.3.15 and 5.3.18:

Corollary *Let X be a Banach $\mathscr{K}$-space. If X^{**} has the BAP then it is a $\mathscr{K}$-space, and if X has the UAP then all ultrapowers of X are $\mathscr{K}$-spaces.*

Not much is known about the nature of Banach $\mathscr{K}$-spaces, and the gap between Kalton's conjecture and the current list of members of the club is indeed oceanic. In particular, we do not know whether the following are or are not $\mathscr{K}$-spaces: Pisier's spaces, i.e. spaces P such that $P \otimes_\pi P = P \otimes_\varepsilon P$ [389]; James' quasireflexive space [216; 217]; the spaces $\mathfrak{R}(\ell_2)$ and $\mathfrak{L}(\ell_2)$; non-commutative L_p spaces built over a von Neumann algebra with no minimal projection and $0 < p < 1$; the p-Gurariy spaces, $0 < p < 1$ in Chapter 6; the Hardy classes H_p for $0 < p < 1$ (see [251, Problem 6]); the spaces of vector-valued functions $\ell_p(E), L_p(E), c_0(E), C(K, E)$ when $p \neq 1$ and E is a $\mathscr{K}$-space, as is the case when $E = \ell_2$ and $p > 1$. Of course, we know no example of a Banach $\mathscr{K}_0$-space whose ultrapowers fail to be $\mathscr{K}_0$-spaces, and the same for $\mathscr{K}$-spaces or for quasi-Banach spaces. On the other hand, if $X_{\mathcal{U}}$ is either a $\mathscr{K}_0$-space or a $\mathscr{K}$-space then so is the base space X.

5.4.2 Twisting a Few Exotic Banach Spaces

There are three methods available for twisting exotic Banach spaces: the local methods developed in this chapter, forming pullbacks / pushouts from other examples and, in the presence of unconditional basis, the quasilinear Kalton–Peck technique as well. In this section, written in a hakuna matata style, we will make all approaches cavort together.

Corollary 5.2.21 implies that if a Banach space X contains ℓ_1^n uniformly complemented then $\operatorname{Ext}_{\mathbf{B}}(X^*, X) \neq 0$. If X contains both ℓ_∞^n and ℓ_1^n uniformly complemented then so does X^*, and thus $\operatorname{Ext}_{\mathbf{B}}(X, X) \neq 0$ and $\operatorname{Ext}_{\mathbf{B}}(X^*, X^*) \neq 0$. Most reflexive spaces X contain uniformly complemented copies of ℓ_p^n for some p, which, reasoning similarly, implies that none of the spaces $\operatorname{Ext}_{\mathbf{B}}(X^*, X)$, $\operatorname{Ext}_{\mathbf{B}}(X, X)$, $\operatorname{Ext}_{\mathbf{B}}(X^*, X^*)$ or $\operatorname{Ext}_{\mathbf{B}}(X, X^*)$ is 0. That B-convex spaces can always be twisted with themselves (Corollary 5.2.17) is perhaps the most staggering result in this line.

Basic information on Schrerier, Baernstein and Tsirelson spaces can be found in [88]. The Schreier space S is likely the *fons et origo* of non-classical spaces and the one that opened the door to Tsirelson-like spaces and these to H.I. spaces. A general construction of Schreier-like spaces can be simply done by fixing a compact family $\mathcal{A} \subset \{0, 1\}^{\mathbb{N}}$ of finite subsets of $\mathbb{N}$ containing the singletons and such that $G \subset F \in \mathcal{A} \implies G \in \mathcal{A}$ and defining the space $\mathrm{S}_{\mathcal{A}}$ to be the completion of the space of finitely supported sequences with respect to the norm $\|x\|_{\mathcal{A}} = \sup_{F \in \mathcal{A}} \|1_F x\|_1$. The space $\mathrm{S}_{\mathcal{A}}$ has a shrinking unconditional basis formed by the unit vectors. It is a subspace of $C(\mathcal{A})$ which, since $\mathcal{A}$ is countable, is c_0-saturated as well as, necessarily, $\mathsf{S}_{\mathcal{A}}$. In particular, $\mathsf{S}_{\mathcal{A}}$ always contains ℓ_∞^n uniformly complemented. If one chooses the family of finite subsets of size at most n for $\mathcal{A}$ then $\mathsf{S}_{\mathcal{A}}$ is just a renorming of c_0. Thus, to obtain something interesting, one needs to assume that $\mathcal{A}$ contains arbitrarily large sets, and that forces $\mathsf{S}_{\mathcal{A}}$ to contain ℓ_1^n uniformly complemented. Since both $\mathsf{S}_{\mathcal{A}}$ and $\mathsf{S}_{\mathcal{A}}^*$ have unconditional basis, the Kalton–Peck map provides non-trivial elements of $\operatorname{Ext}(\mathsf{S}_{\mathcal{A}}, \mathsf{S}_{\mathcal{A}})$ and $\operatorname{Ext}(\mathsf{S}_{\mathcal{A}}^*, \mathsf{S}_{\mathcal{A}}^*)$ that must be non-locally convex by the just mentioned presence of complemented copies of ℓ_1^n. However, the local argument displayed at the beginning of this section yields $\operatorname{Ext}_{\mathbf{B}}(\mathsf{S}_{\mathcal{A}}, \mathsf{S}_{\mathcal{A}}) \neq 0$ as well as $\operatorname{Ext}_{\mathbf{B}}(\mathsf{S}_{\mathcal{A}}^*, \mathsf{S}_{\mathcal{A}}^*)$, $\operatorname{Ext}_{\mathbf{B}}(\mathsf{S}_{\mathcal{A}}^*, \mathsf{S}_{\mathcal{A}})$ and $\operatorname{Ext}_{\mathbf{B}}(\mathsf{S}_{\mathcal{A}}, \mathsf{S}_{\mathcal{A}}^*)$. The first example of a non-trivial family was introduced by Schreier [428], the *admissible* sets – those such that $|A| \leq \min A$. There are many more interesting families in sight (e.g. [7; 104; 105]) generating Schreier-like spaces with their own twisted properties. Baernstein spaces come next as reflexive versions of the Schreier space, and they, too, can be twisted. The next cairn in this road is Tsirelson's space T: a reflexive space with unconditional basis without copies of ℓ_p but containing ℓ_1^n uniformly complemented. Therefore, $\operatorname{Ext}_{\mathbf{B}}(\mathsf{T}, \mathsf{T}^*) \neq 0$ and $\operatorname{Ext}(\mathsf{T}, \mathsf{T}) \neq 0$. We do not, however, know whether $\operatorname{Ext}_{\mathbf{B}}(\mathsf{T}, \mathsf{T}) \neq 0$. Other examples could be given, such as asymptotically ℓ_1-spaces which, containing ℓ_1^n uniformly complemented, are twistable. Or the James Tree space JT, which became famous in the 1960s because the quotient $\mathsf{JT}^{**}/\mathsf{JT}$ is a Hilbert space whose dimension is the continuum. Hence JT and its predual JT_* contain ℓ_2^n uniformly complemented. Since JT_*, moreover, contains ℓ_1^n uniformly complemented, and so JT contains ℓ_∞^n uniformly, these spaces turn out to be very twistable.

Sources

Most of the material opening Section 5.1 is taken from [255], where Kalton introduced $\mathscr{L}_p$-spaces for $0 < p < 1$ and classified them as discrete, continuous and hybrid. Definition 5.1.2 is modelled on the notion of locally complemented subspace of Fakhouri [168]. The trick of Proposition 5.1.6 is often called the 'Lindenstrauss compactness argument' and appears in [323, Proof of Theorem 2.1] and [324]. Lindenstrauss works with Banach spaces complemented in the bidual and uses the weak* topology to paste the pieces together. The quasi-Banach version of Proposition 5.1.6 is reminiscent of the classical proof that biduals are complemented subspaces of suitable ultraproducts. That adaptation already appeared in [255] and has been used several times throughout the chapter (cf. Propositions 5.1.6 and 5.2.18, Theorem 5.3.13). The uniform boundedness principle for quasilinear maps is a gem that Kalton obtained in his first paper on twisted sums [251]. The interpretation Kalton gives in that paper, keen to follow the ideas of Enflo, Lindenstrauss and Pisier [167], is not as clean as it seems nowadays since there was not a clear connection between quasilinear maps and extensions by then. The ideas of Section 5.2 were known for a long time by all those who knew them. The exposition follows [67] from where the examples in Proposition 5.2.20 were taken. The general construction in Proposition 5.2.20 is, however, from [73], while the dual version of Proposition 5.2.22 is from the Kalton–Pełczyński paper [284]. The literature contains several glimmering (or less) variations of Proposition 5.3.4. Indeed, Lusky proved in [348; 349] that if X is a separable Banach space with a basis, then the kernel of any quotient map $\rho\colon \ell_1 \longrightarrow X$ has basis. Now, when X has the BAP, $X \times C_0$ has basis [348]. So, let $\rho\colon \ell_1 \longrightarrow X$ and $\rho'\colon \ell_1 \longrightarrow C_0$ be quotient maps. The operator $\rho \times \rho'\colon \ell_1 \times \ell_1 \longrightarrow X \times C_0$ is a quotient map, and thus $\ker(\rho \times \rho') = \ker\rho \times \ker\rho'$ has a basis. Therefore $\ker\rho$ has the BAP. Another forerunner of Proposition 5.3.4 appears in [173]: Figiel, Johnson and Pełczyński proved that if X^* has the BAP, then the space Y in any exact sequence $0 \longrightarrow Y \longrightarrow \mathscr{L}_1 \longrightarrow X \longrightarrow 0$ must have the BAP. In fact, they establish that Y^* has the BAP through Proposition 5.3.9, which corresponds to their [173, Theorem 2.1.b]. The results about BAP in kernels of projective presentations in Section 5.3 come from [115] and those of Section 5.3 from [173]. Section 5.3 is taken from [115]. Theorems 5.3.15 and 5.3.18 appear here for the first time. Theorem 5.3.19 and its corollary appear in [22] (see also [22, Sections 4.4 and 4.5]). The analogue of Proposition 5.3.4 is due to Figiel, Johnson and Pełczyński [173]. The subsequent paper [172] contains versions of Proposition 5.3.9 and Lemma 5.3.10 for the bare approximation property, namely, that if a Banach space with the AP embeds into an $\mathscr{L}_\infty$-space then the quotient has the AP (Corollary 2 in [172]) and also, that when an $\mathscr{L}_\infty$-space embeds into a Banach space with the AP, the quotient space has the AP too (Corollary 1 in [172]). The twisting of Schreier, Tsirelson and James Tree spaces was first performed in [220] using what the authors called ‘co-local structures’.

6

Fraïssé Limits by the Pound

Fraïssé sequences and their limits are universal constructions whose impact on functional analysis and Banach space theory is not yet well appreciated. There are very good expositions in which one can find the many subtleties and applications of Fraïssé constructions: an introduction to the basic algebraic theory dealing exclusively with countable structures is in Hodges' treatise [214, Chapter 7], but even Pestov [385, Section 6.5] can serve that purpose; Kubiś paper [308] develops a wide variety of examples in various areas, including universal algebra, continuum theory and general topology; Lupini's paper [342] has a more functional analysis orientation. Our rather pedestrian approach is aimed to the construction and study of two concrete examples: the p-Gurariy space G_p, a separable p-Banach space of almost universal disposition, and the p-Kadec space K_p, a separable p-Banach space of almost universal complemented disposition with a 1-FDD. In a sense, they are the same object in different categories: G_p is the Fraïssé limit in the category of finite-dimensional p-Banach spaces and isometric embeddings, and K_p is the Fraïssé limit in a related category whose morphisms are pairs of maps (a contractive embedding and a projection) between finite-dimensional p-Banach spaces whose 'separable' objects (those arising as inductive limits of sequences of finite-dimensional ones) are spaces with 1-FDD. Let us present a comparison table of their similarities and different structural properties, even if we are well aware that some entries might be unintelligible at this moment:

G_p	K_p
AUD	AUCD
Isometrically unique	Isometrically unique
Fraïssé limit of isometries	Fraïssé limit of contractive pairs
Trivial dual if $p < 1$	Separating dual for all p
Locally injective in $p\mathbf{B}$	No
$\mathscr{L}_\infty$-space when $p = 1$	Never
BAP only if $p = 1$	1-FDD for all p
Almost isotropic for all p	Only if $p = 1$
Universal for separable p-Banach spaces	Complementably universal for separable p-Banach with BAP

6.1 Fraïssé Classes and Fraïssé Sequences

A category $\mathbf{C}$ has the amalgamation property if each diagram of the form

$$\begin{array}{ccc} & & B \\ & \nearrow & \\ A & & \\ & \searrow & \\ & & C \end{array} \qquad \text{fits into a commutative diagram} \qquad \begin{array}{ccccc} & & B & & \\ & \nearrow & & \searrow & \\ A & & & & D \\ & \searrow & & \nearrow & \\ & & C & & \end{array}$$

and has the joint embedding property if, given two objects A, B, there is $C \in \mathbf{C}$ such that both A and B have morphisms into C:

$$\begin{array}{ccc} A & & \\ & \searrow & \\ & & C \\ & \nearrow & \\ B & & \end{array}$$

An object of $\mathbf{C}$ is initial if there is a unique morphism from it to any other object in $\mathbf{C}$. Any category with an initial object I and the amalgamation property has the joint embedding property; just amalgamate the morphisms $I \longrightarrow A$ and $I \longrightarrow B$. It is clear the categories $p\mathbf{B}$ and $\mathbf{Q}$ have the joint embedding and amalgamation properties since direct sums and pushouts can be used to construct the required diagrams. Much more relevant for the purposes of this chapter is that the same is true, for each $0 < p \leq 1$, for the 'isometric' subcategory of $p\mathbf{B}$ in which arrows are isometries and for the contractive subcategory $p\mathbf{B}_1$, as Lemma 2.5.2 says. The space 0 is initial in all these categories.

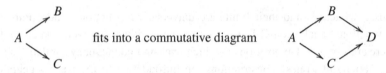

Proposition 6.1.1 *Let* **C** *be a countable category (countable objects, countable arrows) having the amalgamation and joint embedding properties. Then there is a sequence of morphisms* $u_n \colon C_n \longrightarrow C_{n+1}$ *such that*

(a) *if* A *is an object of* **C** *then there is* n *such that* $\mathrm{Hom}(A, C_n) \neq \emptyset$*;*

(b) *if* $v \colon C_n \longrightarrow A$ *is a morphism of* **C** *then there is* $m > n$ *and a morphism* $w \colon A \longrightarrow U_m$ *such that* $w \circ v$ *is the bonding morphism* $U_n \longrightarrow U_m$*.*

Proof Since there are only countable many morphisms in **C**, we can take a sequence (f_n, k_n) passing through all the pairs of the form (f, k), where f is a morphism of **C** and $k \in \mathbb{N}$ is a 'control number', in such a way that each (f, k) appears infinitely many times. The sequence (u_n) is constructed by induction, starting with any morphism. If **C** has an initial object, choose any morphism whose domain is the initial object to start. Having defined $u_{n-1} \colon U_{n-1} \longrightarrow U_n$, we take a look at (f_n, k_n), with $f_n \colon A \longrightarrow B$ and control number k_n. If either $k_n \geq n$ or the 'domain' of f_n (the object A) is not U_{k_n} just wait: set $U_{n+1} = U_n$ and take u_n as the identity of U_n. Otherwise, $k_n < n$ and the domain f_n is U_{k_n}. Thus we have two morphisms with domain $A = U_{k_n}$, namely the 'bonding morphism' $\imath_{(k_n,n)} \colon U_{k_n} \longrightarrow U_{k_n+1} \longrightarrow \cdots \longrightarrow U_n$ and f_n itself. Since **C** has the amalgamation property, these fit into a commutative diagram

$$\begin{array}{ccc} U_{k_n} = A & \xrightarrow{\;f_n\;} & B \\ {\scriptstyle \text{bonding morphism}}\big\downarrow & & \big\downarrow{\scriptstyle \text{whatever}} \\ U_n & \xrightarrow{\;f_n'\;} & C \end{array}$$

Then, setting $U_{n+1} = C$ and $u_n = f_n'$ completes the induction step. Let us check that the resulting sequence $(u_n)_{n \geq 1}$ has the required properties. It is clear that (a) follows from (b) and the joint embedding property, so let us prove (b). Let $f \colon U_n \longrightarrow A$ be a morphism. Take $m > n$ such that $(f_m, k_m) = (f, n)$. Then the $(m-1)$th morphism of the sequence $(u_n)_{n \geq 1}$ arose from the amalgamation diagram

$$\begin{array}{ccc} U_{k_m} = U_n & \xrightarrow{\;f\;} & A \\ {\scriptstyle \imath_{(n,m-1)}}\big\downarrow & & \big\downarrow{\scriptstyle \imath_{(n,m-1)'}} \\ U_{m-1} & \xrightarrow{\;\imath_{m-1} = f'\;} & U_m \end{array}$$

It follows that $\imath'_{(n,m-1)} \circ f = \imath_{(n,m-1)} \circ \imath_{m-1} = \imath_{(n,m)}$ is the bonding morphism $U_n \longrightarrow U_m$. □

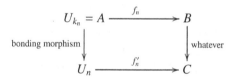

A sequence of morphisms satisfying the conditions of the proposition is called a Fraïssé sequence. The diagram

$$\begin{array}{ccccccccccc} C_1 & \longrightarrow & C_2 & \longrightarrow & \cdots & \longrightarrow & C_n & \longrightarrow \cdots \longrightarrow & C_m & \longrightarrow & \cdots \\ & & & & & & & \searrow \quad \nearrow & & & \\ & & & & & & & A & & & \end{array}$$

illustrates the relevant property of Fraïssé sequences.

6.2 Almost Universal Disposition

Fix $p \in (0, 1]$ once and for all. All reasoning that follows is independent of the actual value of p, but it is required that p be the same everywhere. A p-Banach space X is said to be of *almost universal disposition* (AUD) if, given finite-dimensional p-normed spaces E, F and isometries $u: E \longrightarrow X, v: E \longrightarrow F$, for each $\varepsilon > 0$, there is an ε-isometry $w: F \longrightarrow X$ such that $u = wv$. Diagramatically,

$$\begin{array}{ccc} E & \xrightarrow{\text{isometry}} & F \\ & {}_{\text{isometry}}\searrow \quad \swarrow_{\varepsilon\text{-isometry}} & \\ & X & \end{array} \tag{6.1}$$

To be precise, one should speak of spaces of almost universal disposition for finite-dimensional p-Banach spaces, but let it stand. It is clear that assuming either that E is a subspace of X (and u is plain inclusion) or that E is a subspace of F (and v the inclusion) leads to equivalent formulations, a fact that will be used without further mention. The property of AUD was first considered for Banach spaces by Gurariy, who constructed the separable Banach space G that bears his name in 1966. Its general p-version is:

Theorem 6.2.1 *For each $p \in (0, 1]$ there exists a unique, up to isometries, separable p-Banach space of almost universal disposition.*

This space will be constructed, according to the general plan of the chapter, as the limit of a Fraïssé sequence of isometries between finite-dimensional spaces. It can also be constructed using the Device to obtain a 'countable and finite-dimensional' version of 2.13.1. Spaces of (almost) universal disposition will be encountered again in Section 7.3 and Note 7.5.4.

From Rational p-Norms to Allowable Isometries

We define now a countable category admitting amalgamations and whose morphisms are a family of isometries that is 'dense' among all isometries.

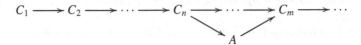

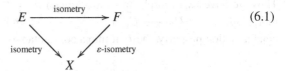

A point $x \in \mathbb{K}^n$ is said to be rational if all its coordinates are rational. When $\mathbb{K} = \mathbb{C}$, this means that both the real and imaginary parts are rational numbers. A linear map $f\colon \mathbb{K}^n \longrightarrow \mathbb{K}^m$ is said to be rational if it carries rational points into rational points. A rational p-norm on $\mathbb{K}^n$ is one whose unit ball is the p-convex hull of a finite set of rational points. Thus, a rational p-norm is given by the formula

$$|x| = \inf\left\{\left(\sum_i |\lambda_i|^p\right)^{1/p} : x = \sum_i \lambda_i x_i\right\}$$

for some finite set $x_1, \ldots, x_n$ of rational points. For each $n \in \mathbb{N}$, let $\mathcal{N}_n$ be the set of all p-norms on $\mathbb{K}^n$, where $\mathbb{K}^0$ is understood as 0, and set $\mathcal{N} = \bigcup_{n\geq 0} \mathcal{N}_n$. We recursively define a class of p-norms which, in the absence of an awe-inspiring name, we call 'allowed p-norms' (formally, a subset of $\mathcal{N}$), as follows:

(a) Each rational p-norm is allowed.
(b) If $f\colon \mathbb{K}^n \longrightarrow \mathbb{K}^m$ is rational and injective and $|\cdot|$ is an allowed p-norm on $\mathbb{K}^m$ then $\|x\| = |f(x)|$ is an allowed p-norm on $\mathbb{K}^n$.
(c) If $f\colon \mathbb{K}^n \longrightarrow \mathbb{K}^m$ is rational and surjective and $|\cdot|$ is an allowed p-norm on $\mathbb{K}^n$ then $\|y\| = \inf |x| \colon y = f(x)$ is allowed on $\mathbb{K}^m$.
(d) If $|\cdot|_1$ and $|\cdot|_2$ are allowed p-norms on $\mathbb{K}^n$ and $\mathbb{K}^m$, respectively, then the p-sum $\|(x,y)\| = \left(|x|_1^p + |y|_2^p\right)^{1/p}$ is allowed on $\mathbb{K}^{n+m}$.
(e) If $|\cdot|_1$ and $|\cdot|_2$ are allowed p-norms on $\mathbb{K}^n$ and $\mathbb{K}^m$, respectively, then the direct product $\|(x,y)\| = \max(|x|_1, |y|_2)$ is an allowed p-norm on $\mathbb{K}^{n+m}$.
(f) If $|\cdot|_1$ and $|\cdot|_2$ are allowed p-norms on $\mathbb{K}^n$ and $\mathbb{K}^m$, respectively, and $f\colon \mathbb{K}^n \longrightarrow \mathbb{K}^m$ is a rational map then the following p-norm is allowed on $\mathbb{K}^m$ for every rational number $\varepsilon > 0$:

$$\|y\| = \inf\left\{\left(|x|_1^p + (1+\varepsilon)^p |z|_2^p\right)^{1/p} : y = f(x) + z, x \in \mathbb{K}^n, g \in \mathbb{K}^m\right\}.$$

An allowed space is just the direct product of finitely many copies of the ground field furnished with an allowed p-norm. Finally, we declare an isometry $u\colon E \longrightarrow F$ allowable if E and F are allowed p-normed spaces and u is rational. Conditions (a) to (f) enable us to perform the basic categorical constructions within the allowable category, as we will see along this chapter.

Lemma 6.2.2 *There is a Fraïssé sequence of allowable isometries.*

Proof We need only check that the countable category of allowable isometries with initial object 0 admits amalgamations. The proof offers a good opportunity to review the pushout construction. Let $f\colon E \longrightarrow F$ and $g\colon E \longrightarrow G$ be allowable isometries. This means that E, F, G are $\mathbb{K}^k, \mathbb{K}^n, \mathbb{K}^m$ equipped with allowed p-norms and with both f and g rational. Condition (d) implies that $F \oplus_p G$ is an allowed space and the map $(f, -g)\colon E \longrightarrow F \oplus_p G$ is rational and injective. Let $(e_i)_{1\le i\le k}$ be the unit basis of E and let

$$(f_1, \dots, f_k, f_{k+1}, \dots, f_n) \quad \text{and} \quad (g_1, \dots, g_k, g_{k+1}, \dots, g_m)$$

be rational bases of F and G with $f_i = f(e_i)$ and $g_i = g(e_i)$ for $1 \le i \le k$. Clearly,

$$(f_1 - g_1, \dots, f_k - g_k, f_1 + g_1, \dots, f_k + g_k f_{k+1}, \dots, f_n, , g_{k+1}, \dots, g_m)$$

is a rational basis of $F \oplus_p G = \mathbb{K}^{n+m}$ which we relabel as $(v_1, \dots, v_k, \dots, v_{n+m})$. We define a rational map $h\colon F \oplus_p G \longrightarrow \mathbb{K}^{n+m-k}$ by $h(\sum_{1\le i\le n+m} c_i v_i) = (c_{k+1}, \dots, c_{n+m})$. Let H be $\mathbb{K}^{n+m-k}$ equipped with the p-norm

$$\|x\| = \inf\left\{\|y\|\colon x = h(y), y \in F \oplus_p G\right\},$$

which is allowed by (c). One has the commutative diagram

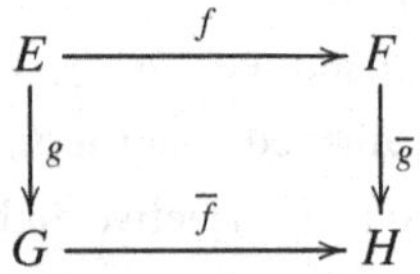

with $\overline{f}$ and $\overline{g}$ allowable since they are the inclusions of G and F into $F \oplus_p G$ followed by h. □

The isometric pushout diagram just constructed has the following additional property: for every pair of rational maps $g'\colon F \longrightarrow \mathbb{K}^r$ and $f'\colon G \longrightarrow \mathbb{K}^r$ such that $g'f = f'g$, there is a unique rational map $h\colon H \longrightarrow \mathbb{K}^r$ such that $h\overline{g} = h\overline{f}$:

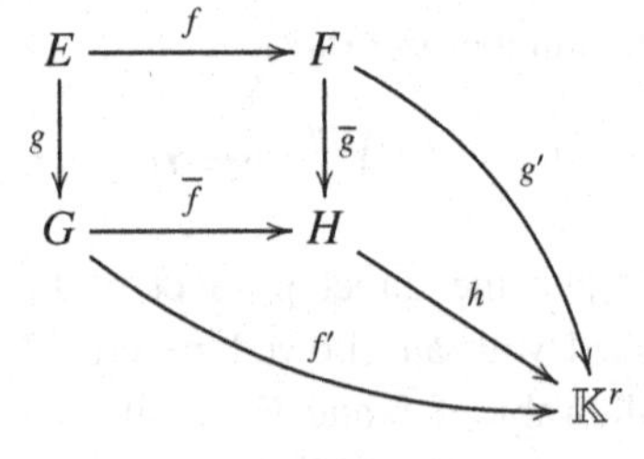

Thus, the allowable category has both amalgamations and pushouts.

Proof of Theorem 6.2.1: Existence

Let us fix a Fraïssé sequence of allowable isometries

$$U_1 \longrightarrow U_2 \longrightarrow \cdots \longrightarrow U_n \longrightarrow U_{n+1} \longrightarrow \cdots \tag{6.2}$$

and prove that the direct limit U of that sequence in $p\mathbf{B}$ is a space of AUD. We can identify each U_k with its image in U so that one can assume $U = \overline{\bigcup_{k\ge1} U_k}$.

To understand why the Fraïssé character of the sequence (6.2) entails the AUD of its limit, pick an isometry $E \longrightarrow F$ between finite-dimensional spaces and an isometry $E \longrightarrow U$. Assume first that we have been so lucky that $v\colon E \longrightarrow F$ is allowable and $E \longrightarrow U$ is the composition of an allowable isometry $u\colon E \longrightarrow U_n$ and the inclusion $U_n \longrightarrow U$. Since (6.2) is Fraïssé, amalgamating u and v within the allowable category yields a commutative diagram

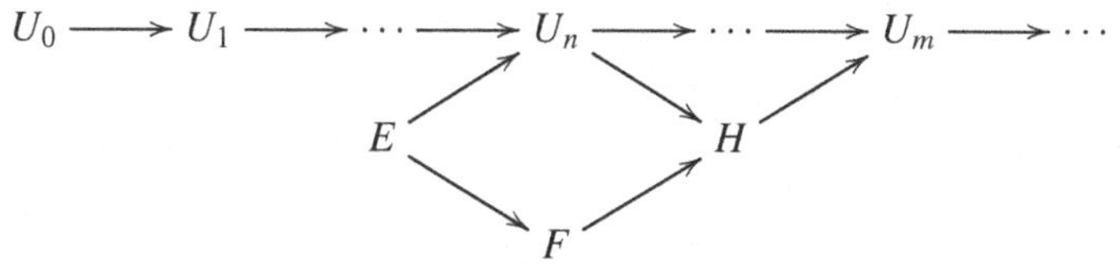

so that the required extension of u is even an isometry in this case. Before passing to the general case, let us perform a couple of mathematical asanas to gain some flexibility. The first one is just to relax the commutativity of Diagram 6.1:

Lemma 6.2.3 *Let E be a finite-dimensional subspace of a p-Banach space X, and let F be a finite-dimensional p-Banach space. Assume that for every $\varepsilon > 0$ and every isometry $v\colon E \longrightarrow F$, there is an ε-isometry $w\colon F \longrightarrow X$ such that $\|w(v(x)) - x\| \leq \varepsilon\|x\|$ for all $x \in X$. Then X is of almost universal disposition.*

Proof This obviously follows from the fact that if $\mathscr{H}$ is a basis of E then for every $\varepsilon > 0$ there is δ (depending on ε and $\mathscr{H}$) such that if $u\colon E \longrightarrow X$ is a linear map with $\|u(b)\| \leq \delta$ for every $b \in \mathscr{H}$ then $\|u\| \leq \varepsilon$. □

The second one is to open the no-brainer chakra: allowable isometries are 'dense' among all isometries between finite-dimensional spaces.

Lemma 6.2.4 *Let $u\colon E \longrightarrow F$ be an isometry where E is an allowed space and F is a finite-dimensional p-normed space. For each $\varepsilon > 0$, there is an allowable isometry $u_0\colon E \longrightarrow F_0$ and a surjective ε-isometry $g\colon F \longrightarrow F_0$ such that $u_0 = g\,u$.*

Proof We may assume that ε is rational. Let $(e_i)_{1\leq i\leq n}$ be the unit basis of $E = \mathbb{K}^n$ and pick $(f_j)_{1\leq j\leq m}$ such that $\{u(e_1), \ldots, u(e_n), f_1, \ldots, f_m\}$ is a basis of F. Let $g\colon F \longrightarrow \mathbb{K}^{n+m}$ be the isomorphism associated to that basis and take a rational p-norm $|\cdot|_0$ on $\mathbb{K}^{n+m}$ making g an ε-isometry such that $(1+\varepsilon)^{-1}\|y\| \leq |g(y)|_0 \leq (1+\varepsilon)\|y\|$. Then $u_0 = g\,u$ is a rational ε-isometry: in fact, $u_0(x) = (x, 0)$. We define a new p-norm on $\mathbb{K}^{n+m}$ by the formula

$$|y| = \inf\left\{\left(\|x\|^p + (1+\varepsilon)^p|z|_0^p\right)^{1/p} : y = u_0(x) + z, x \in \mathbb{K}^n, z \in \mathbb{K}^{n+m}\right\}.$$

Note that the unit ball of $|\cdot|$ is just the p-convex hull of the union of the unit ball of $\|\cdot\|$ and the ball of radius $(1+\varepsilon)^{-1}$ of $\|\cdot\|_0$. This p-norm satisfies the estimate $(1+\varepsilon)^{-1}|y|_0 \le |y| \le (1+\varepsilon)|y|_0$ for $y \in \mathbb{K}^{n+m}$, has to be allowed on $\mathbb{K}^{n+m}$ (by the last allowance rule) and makes u_0 into an isometry, which is therefore allowable. Hence, if F_0 is $\mathbb{K}^{n+m}$ equipped with $|\cdot|$, we have $(1+\varepsilon)^{-2}\|y\| \le |g(y)| \le (1+\varepsilon)^2\|y\|$ for $y \in F$. $\qquad\square$

We are now ready to handle the general case and show that U satisfies the hypothesis of Lemma 6.2.3. Let F be a finite-dimensional p-Banach space, $v: E \longrightarrow F$ an isometry and E a subspace of U. Fix $\varepsilon > 0$. Since $\bigcup_k U_k$ is dense in U, there is a contractive ε-isometry $u_\varepsilon: E \longrightarrow U_n$, for n sufficiently large, such that $\|u_\varepsilon(x) - x\| \le \varepsilon\|x\|$ for all $x \in E$. Form the pushout in $p\mathbf{B}$,

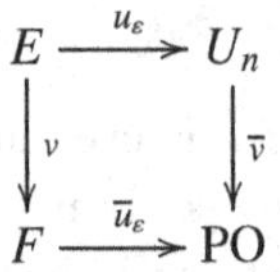

$$
\begin{array}{ccc}
E & \xrightarrow{u_\varepsilon} & U_n \\
\downarrow{v} & & \downarrow{\bar{v}} \\
F & \xrightarrow{\bar{u}_\varepsilon} & \mathrm{PO}
\end{array}
$$

so that $\bar{u}_\varepsilon$ is again a contractive ε-isometry, while $\bar{v}$ is an isometry to which Lemma 6.2.4 can be applied to find an allowable isometry $v_0: U_n \longrightarrow F_0$ together with a surjective ε-isometry $g: \mathrm{PO} \longrightarrow F_0$ such that $v_0 = g\bar{v}$. Finally, the Fraïssé character of (6.2) guarantees that for some $m > n$, there is an allowable $w_0: F_0 \longrightarrow U_m$ such that $w_0 v_0$ is the inclusion of U_n into U_m. The full picture appears in the commutative diagram

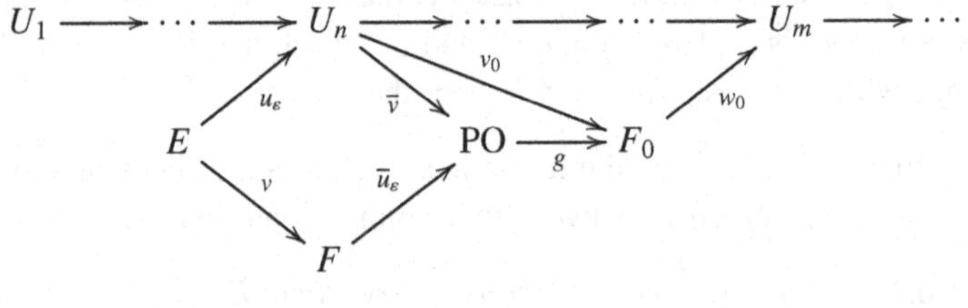

By letting $w = w_0 g \bar{u}_\varepsilon$, we obtain a contractive 3ε-isometry extending u_ε and so $\|w(x) - x\| \le \varepsilon\|x\|$ for $x \in E$.

Proof of Theorem 6.2.1: Uniqueness

Is it not almost obvious that any two separable p-Banach spaces of almost universal disposition are almost isometric? That is, that for each $\varepsilon > 0$, there is a surjective ε-isometry between them. Proposition 6.2.10 provides an explicit proof, just in case it is not clear. Much more surprising is that they are actually isometric, which we are going to prove now. Our approach to isometric properties of spaces of AUD depends one way or another on the following pair of lemmas:

Lemma 6.2.5 *Fix $\varepsilon \in (0,1)$. Let X and Y be p-normed spaces and let $\imath\colon X \longrightarrow X \oplus Y$ and $\jmath\colon Y \longrightarrow X \oplus Y$ be the canonical inclusions. If $f\colon X \longrightarrow Y$ is an ε-isometry then there is a p-norm on $X \oplus Y$ for which $\imath$ and $\imath$ are isometries such that $\|\jmath f - \imath\| \le \varepsilon$.*

Proof The p-norm that does the trick is

$$\|(x,y)\| = \inf\left\{\left(\|x_0\|_X^p + \|y_1\|_Y^p + \varepsilon^p\|x_2\|_X^p\right)^{1/p} : (x,y) = (x_0 + x_2, y_1 - f(x_2))\right\}.$$

We must check that $\|(x,0)\| = \|x\|_X$ for all $x \in X$. The inequality $\|(x,0)\| \le \|x\|_X$ is obvious. For the converse, suppose $x = x_0 + x_2$ and $y_1 = f(x_2)$. Then

$$\begin{aligned}
\|x_0\|_X^p + \|y_1\|_Y^p + \varepsilon^p\|x_2\|_X^p &= \|x_0\|_X^p + \|f(x_2)\|_Y^p + \varepsilon^p\|x_2\|_X^p \\
&\ge \|x_0\|_X^p + (1-\varepsilon)^p\|x_2\|_X^p + \varepsilon^p\|x_2\|_X^p \\
&= \|x_0\|_X^p + \|(1-\varepsilon)x_2\|_X^p + \|\varepsilon x_2\|_X^p \\
&\ge \|x\|_X^p.
\end{aligned}$$

Next we prove that $\|(0,y)\| = \|y\|_Y$ for every $y \in Y$. That $\|(0,y)\| \le \|y\|_Y$ is again obvious. To prove the reversed inequality, assume $x_0 + x_2 = 0$ and $y = y_1 - f(x_2)$. As $t \to t^p$ is subadditive on $\mathbb{R}_+$ for $p \in (0,1]$, we have

$$\begin{aligned}
\|x_0\|_X^p + \|y_1\|_Y^p + \varepsilon^p\|x_2\|_X^p &= \|x_2\|_X^p + \|y_1\|_Y^p + \varepsilon^p\|x_2\|_X^p \\
&= \|y_1\|_Y^p + (1+\varepsilon^p)\|x_2\|_X^p \\
&\ge \|y_1\|_Y^p + (1+\varepsilon)^p\|x_2\|_X^p \\
&\ge \|y_1\|_Y^p + \|f(x_2)\|_Y^p \\
&\ge \|y\|_Y^p.
\end{aligned}$$

Finally, $\|\jmath f - \imath\| = \sup_{\|x\|\le 1} \|\jmath(f(x)) - \imath(x)\| = \sup_{\|x\|\le 1} \|(-x, f(x))\| \le \varepsilon$. □

The indulgent reader will forgive us if, for the remainder of the chapter, we use the notation $X \boxplus_f^\varepsilon Y$ for the space $X \oplus Y$ endowed with the quasinorm defined in the preceding proof. This quasinorm depends on f and ε and also on p, but this should cause no confusion. A linear operator $f\colon X \longrightarrow Y$ will be called a strict ε-isometry if $(1+\varepsilon)^{-1}\|x\|_X < \|f(x)\|_Y < (1+\varepsilon)\|x\|_X$ for $0 < \varepsilon < 1$ and every non-zero $x \in X$. When X is finite-dimensional, every strict ε-isometry is an η-isometry for some $\eta < \varepsilon$.

Lemma 6.2.6 *Let U be a p-Banach space of almost universal disposition. Let Y be a finite-dimensional p-Banach space, X a subspace of U and $\varepsilon \in (0,1)$. If $f\colon X \longrightarrow Y$ is a strict ε-isometry then for each $\delta > 0$, there exists a δ-isometry $g\colon Y \longrightarrow U$ such that $\|g(f(x)) - x\| < \varepsilon\|x\|$ for every non-zero $x \in X$.*

Proof Choose $0 < \eta < \varepsilon$ for which f is an η-isometry. Reducing δ if necessary, we may assume that $\delta^p + (1+\delta)^p\eta^p < \varepsilon^p$. Form the space $X \boxplus_f^\eta Y$ and let $\imath\colon X \longrightarrow X \boxplus_f^\eta Y$ and $\jmath\colon Y \longrightarrow X \boxplus_f^\eta Y$ denote the canonical inclusions so that $\|\jmath f - \imath\| \leq \eta$. If $h\colon X \boxplus_f^\eta Y \longrightarrow U$ is a δ-isometry such that $\|h(\imath(x)) - x\| \leq \delta\|x\|$ for $x \in X$ then $g = h\jmath$ is a δ-isometry from Y into U and

$$\begin{aligned}\|x - g(f(x))\|^p &\leq \|x - h(\imath(x))\|^p + \|h(\imath(x)) - h(\jmath(f(x))\|^p\\ &\leq \delta^p\|x\|^p + (1+\delta)^p\|\imath(x) - \jmath(f(x))\|^p\\ &\leq (\delta^p + (1+\delta)^p\eta^p)\|x\|^p < \varepsilon^p\|x\|^p.\end{aligned}$$ □

We need a technique to 'paste' operators defined on a chain of subspaces. Let A and B be p-Banach spaces and (A_n) a chain of subspaces whose union is dense in A. Let $a_n\colon A_n \longrightarrow B$ be a sequence of operators such that $\|a_{n+1}|_{A_n} - a_n\| \leq \varepsilon_n$, where $\sum_n \varepsilon_n^p < \infty$, with $\sup_n \|a_n\| < \infty$. For each $x \in A_k$, the Cauchy sequence $(a_n(x))_{n\geq k}$ converges in B so there is a unique operator $a\colon A \longrightarrow B$ such that $a(x) = \lim_{n\geq k} a_n(x)$ whenever x is in some A_k. This operator shall be referred to as the *pointwise limit* of the sequence (a_n). The following remarkable result completes the proof of Theorem 6.2.1.

Proposition 6.2.7 *Fix $\varepsilon \in (0,1)$. Let U, V be separable p-Banach spaces of almost universal disposition, and let X be a finite-dimensional subspace of U. If $f\colon X \longrightarrow V$ is a strict ε-isometry then there exists a bijective isometry $h\colon U \longrightarrow V$ such that $\|h(x) - f(x)\| \leq \varepsilon\|x\|$ for every $x \in X$. In particular, U and V are isometric.*

Proof Fix $0 < \varepsilon_0 < \varepsilon$ such that f is an ε_0-isometry. Let $(\varepsilon_n)_{n\geq 1}$ be any decreasing sequence of positive numbers with $\varepsilon_1 < \varepsilon_0$. We inductively define sequences of linear operators $(f_n), (g_n)$ and finite-dimensional subspaces (X_n), (Y_n) of U and V, respectively, such that the following conditions are satisfied for every $n \geq 0$:

(0) $X_0 = X$, $Y_0 = f[X]$, and $f_0 = f$;
(1) $f_n\colon X_n \longrightarrow Y_n$ is an ε_n-isometry;
(2) $g_n\colon Y_n \longrightarrow X_{n+1}$ is an ε_{n+1}-isometry;
(3) $\|g_n f_n(x) - x\| < \varepsilon_n\|x\|$ for every non-zero $x \in X_n$;
(4) $\|f_{n+1}g_n(y) - y\| < \varepsilon_{n+1}\|y\|$ for every non-zero $y \in Y_n$;
(5) $X_n \subset X_{n+1}$, $Y_n \subset Y_{n+1}$, $\bigcup_n X_n$ and $\bigcup_n Y_n$ are dense in U and V, respectively.

We use (0) to start the inductive construction. Suppose that f_i, X_i, Y_i, for $i \leq n$, and g_i for $i < n$, have been constructed. Applying Lemma 6.2.6 twice, we find g_n, X_{n+1}, f_{n+1} and Y_{n+1}. To guarantee that (5) holds, we may start by choosing sequences (x_n) and (y_n) dense in U and V, respectively, and then require first that X_{n+1} contain both x_n and $g_n[Y_n]$ and then that Y_{n+1} contain both y_n and $f_{n+1}[X_{n+1}]$. After that, fix $n \geq 0$ and $x \in X_n$ with $\|x\| = 1$. Using (4) and (1), we get $\|f_{n+1}g_nf_n(x) - f_n(x)\| < \varepsilon_{n+1}\|f_n(x)\| \leq \varepsilon_{n+1}(1+\varepsilon_n)$, while (3) and (2) yield $\|f_{n+1}g_nf_n(x) - f_{n+1}(x)\| \leq \|f_{n+1}\|\,\|g_nf_n(x) - x\| < (1+\varepsilon_{n+1})\varepsilon_n$ Combining,

$$\begin{aligned}\|f_n(x) - f_{n+1}(x)\|^p &\leq \|f_{n+1}g_nf_n(x) - f_n(x)\|^p + \|f_{n+1}g_nf_n(x) - f_{n+1}(x)\|^p\\ &\leq \varepsilon_{n+1}^p(1+\varepsilon_n)^p + (1+\varepsilon_{n+1})^p\varepsilon_n^p. \end{aligned} \tag{6.3}$$

If we agree that $(\varepsilon_n)_{n\geq 1}$ was chosen so that

$$\sum_{n\geq 0}\left(\varepsilon_{n+1}^p(1+\varepsilon_n)^p + (1+\varepsilon_{n+1})^p\varepsilon_n^p\right) < \varepsilon^p, \tag{6.4}$$

then $(f_m(x))_{m\geq n}$ is a Cauchy sequence. We define $h(x) = \lim_{m\geq n} f_m(x)$ for $x \in \bigcup_n X_n$. This h is an isometry since it is an ε_n-isometry for every n. Consequently, it extends to an isometry $U \longrightarrow V$, which we do not relabel. Furthermore, (6.3) and (6.4) imply $\|f(x) - h(x)\|^p \leq \sum_{n=0}^{\infty} \|f_n(x) - f_{n+1}(x)\|^p \leq \varepsilon^p\|x\|^p$ for $x \in X$. It remains to see that h is a bijection. To this end, we check as before that $(g_n(y))_{n\geq m}$ is a Cauchy sequence for every $y \in Y_m$. Once this is done, we obtain an isometry $g\colon V \longrightarrow U$. Conditions (3) and (4) inform us that $gh = \mathbf{1}_U$ and $hg = \mathbf{1}_V$. □

Let us denote (the isometric type of) this unique space G_p and call it the p-Gurariy space; when $p = 1$, we obtain the original Gurariy space, denoted G. Proposition 6.2.7 establishes that the spaces G_p are *almost isotropic*, in the sense that given $x, y \in \mathsf{G}_p$ with $\|x\| = \|y\| = 1$ and $\varepsilon > 0$, there is a bijective isometry f of G_p such that $\|y - f(x)\| \leq \varepsilon$. The next section uncovers some additional properties that G_p shares with all spaces of AUD.

Extension of Operators and Automorphisms

The second lesson we will learn in the forthcoming Section 7.1 is that extending operators to operators does not mean extending isomorphisms to isomorphisms. Even so, the first lesson is that extending isometries means extending operators. Thus, the AUD notion, which is more demanding than local injectivity or the forthcoming UFO (Definition 7.1.3), imposes severe restrictions on the extension of both operators and automorphisms.

Proposition 6.2.8 *Every p-Banach space U of AUD:*

(a) *is locally 1^+-injective; for $p = 1$, this means that it is a Lindenstrauss space,*
(b) *contains an isometric copy of each separable p-Banach space.*
(c) *Moreover, if $p < q \leq 1$ then $\mathfrak{L}(U, Y) = 0$ for all q-Banach spaces Y; in particular, U has trivial dual.*

Proof   Part (a) is a dirty pushout trick. Assume $\tau\colon E \longrightarrow U$ is contractive and that U is of almost universal disposition. Look at the diagram

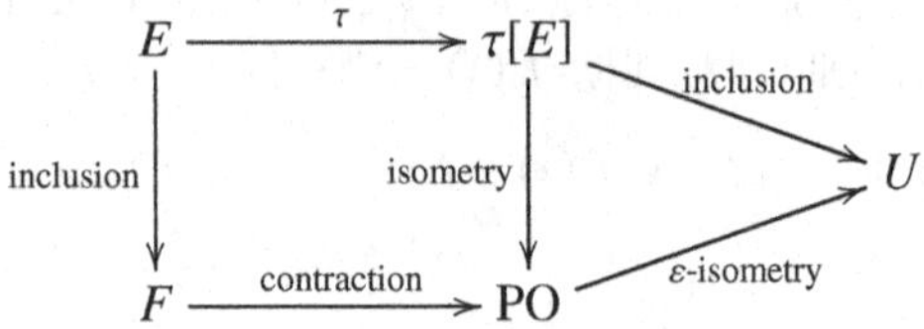

and draw your own conclusions. Part (b) can be derived by iteratively applying Proposition 6.2.6: let X be a separable p-Banach space, and let $(X_n)_{n\geq 1}$ be a chain of finite-dimensional subspaces whose union is dense in X. Then there is a sequence $f_n\colon X_n \longrightarrow U$ in which f_n is a strict 2^{-n}-isometry such that $\|f_{n+1}|_{X_n} - f_n\| < 2^{-n}$. The pointwise limit of these operators is an isometry of X into U. To prove (c), we first prove that, given a normalised $x \in \mathsf{G}_p$ and $\varepsilon > 0$, there are $x'_\varepsilon, x''_\varepsilon \in \mathsf{G}_p$ such that $x = x'_\varepsilon + x''_\varepsilon$ with $\|x'_\varepsilon\|, \|x''_\varepsilon\| \leq (1 + \varepsilon)2^{-1/p}$. Indeed, consider the isometry $u\colon [x] \longrightarrow \mathsf{G}_p$ given by plain inclusion and the isometry $v\colon [x] \longrightarrow \ell_p^2$ given by $v(x) = 2^{-1/p}(1, 1)$. Let $w\colon \ell_p^2 \longrightarrow \mathsf{G}_p$ be any ε-isometry extending u, and set $x'_\varepsilon = 2^{-1/p}w(1, 0)$ and $x''_\varepsilon = 2^{-1/p}w(0, 1)$. That done, the proof goes as in the L_p case in 1.1.5: if $u\colon \mathsf{G}_p \longrightarrow Y$ is an operator and $\|x\| = 1$, then taking $\varepsilon > 0$ and $x'_\varepsilon, x''_\varepsilon \in \mathsf{G}_p$ as before, we have

$$\|ux\| \leq (\|ux'_\varepsilon\|^q + \|ux''_\varepsilon\|^q)^{1/q} \leq (1 + \varepsilon)2^{1-q/p}\|u\|.$$

Since x and ε are arbitrary, $\|u\| \leq 2^{1-q/p}\|u\|$, which is only possible if $u = 0$.   □

Lemma 6.2.9   *Let A be a finite-dimensional subspace of a space U of AUD and let B be finite-dimensional. If $g\colon A \longrightarrow B$ is an embedding then for each $\varepsilon > 0$, there is an embedding $f\colon B \longrightarrow U$ such that $f(g(a)) = a$ for every $a \in A$ with $\|f\| \leq (1 + \varepsilon)\|g^{-1}\|$ and $\|f^{-1}\| \leq (1 + \varepsilon)\|g\|.$*

Proof   We use an even dirtier trick than before. In less than no time, the reader will realise that one can assume $\|g^{-1}\| = 1$. To ease notation, we will write $h = g^{-1}$. Let us take the pushout with the inclusion $g[A] \longrightarrow B$ as follows:

$$
\begin{array}{ccc}
g[A] & \overset{\imath}{\longrightarrow} & B \\
{\scriptstyle h}\downarrow & & \downarrow{\scriptstyle \overline{h}} \\
A & \underset{\overline{\imath}}{\longrightarrow} & \mathrm{PO}
\end{array}
$$

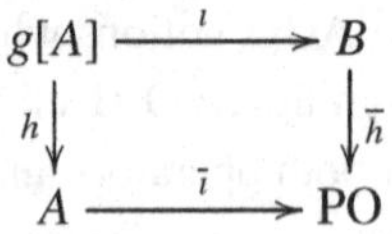

It is clear that $\imath$ is an isometry and $\|h^{-1}\| = \|g\| \geq 1$. By Lemma 2.5.2, $\overline{\imath}$ is an isometry and $\overline{h}$ is an embedding with $\|\overline{h}\| \leq 1$ and $\|(\overline{h})^{-1}\| \leq \|h^{-1}\| = \|g\|$. Now let $w\colon \mathrm{PO} \longrightarrow U$ be an ε-isometry such that $w\overline{\imath}(a) = a$ for all $a \in A$. Then $f = w\overline{h}$ is an embedding which obviously satisfies $fg(a) = a$, for all $a \in A$, and $\|f\| \leq (1 + \varepsilon)$. Moreover, $\|f^{-1}\| \leq \|(\overline{h})^{-1}\|\|w^{-1}\| \leq \|g\|(1 + \varepsilon)$.   □

Proposition 6.2.10 *Let U and V be separable spaces of AUD. Let $A \subset U$ and $B \subset V$ be finite-dimensional subspaces. If $\varphi_0 \colon A \longrightarrow B$ is an isomorphism then, for each $\varepsilon > 0$, there is an isomorphism $\varphi \colon U \longrightarrow V$ extending φ_0 and such that $\|\varphi\| \le (1+\varepsilon)\|\varphi_0\|$ and $\|\varphi^{-1}\| \le (1+\varepsilon)\|\varphi_0^{-1}\|$.*

Proof The result follows from Lemma 6.2.9 and a simple back-and-forth argument. Let $(\varepsilon_n)_{n\ge 0}$ be a sequence of positive numbers such that $\prod_n(1+\varepsilon_n) \le 1+\varepsilon$, and write $U = \overline{\bigcup_n U_n}$, where (U_n) is an increasing sequence of finite-dimensional subspaces of U beginning with $U_0 = A$. Moreover, let (V_n) be an increasing sequence of finite-dimensional subspaces of V such that $V = \overline{\bigcup_n V_n}$, with $V_0 = B$. Let $\varphi_0 \colon A \longrightarrow B$ be an isomorphism. By Lemma 6.2.9, let $\psi_1 \colon V_1 \longrightarrow U$ be an extension of $\varphi_0^{-1} \colon \varphi_0[U_0] \longrightarrow U$, with $\|\psi_1\| \le (1+\varepsilon_1)\|\varphi_0^{-1}\|$ and $\|\psi_1^{-1}\| \le (1+\varepsilon_1)\|\varphi_0\|$. Then let $\varphi_2 \colon \psi_1[V_1] + U_2 \longrightarrow V$ be an extension of $\psi_1^{-1} \colon \psi_1[V_1] \longrightarrow V$ such that $\|\varphi_2\| \le (1+\varepsilon_2)\|\psi_1^{-1}\|$ and $\|\varphi_2^{-1}\| \le (1+\varepsilon_2)\|\psi_1\|$ provided by Lemma 6.2.9. Continuing in this way, one obtains a pair of operators φ, ψ such that $\psi\varphi = \mathbf{1}_U, \varphi\psi = \mathbf{1}_V$, with $\|\varphi\| \le (1+\varepsilon)\|\varphi_0\|$ and $\|\psi\| \le (1+\varepsilon)\|\varphi_0^{-1}\|$ and $\varphi|_A = \varphi_0$. □

6.3 Almost Universal Complemented Disposition

The following notion is a kind of almost universal disposition focused only on 1-complemented subspaces; another possibility, considered in [116], is to additionally require that the projections be, in some sense, 'compatible'.

[Ͽ] If F is a finite-dimensional p-normed space, E is a 1-complemented subspace of F and $u \colon E \longrightarrow X$ is an isometry with 1-complemented range, then for every $\varepsilon > 0$, there is an ε-isometry $F \longrightarrow X$ with $(1+\varepsilon)$-complemented range extending u.

To properly frame it, we will consider the structure of embedding and projection as a whole.

Categories of Pairs

We will find it convenient to use the notation $u \colon E \rightleftarrows F$ for pairs $u = \langle u^\flat, u^\sharp \rangle$ consisting of operators $u^\flat \colon E \longrightarrow F$ and $u^\sharp \colon F \longrightarrow E$ such that $u^\sharp u^\flat = \mathbf{1}_E$. Thus, $u^\flat$ is an embedding of E into F and $u^\sharp$ is a projection along $u^\flat$. It is to be understood that the 'solid' arrow represents the embedding part $u^\flat$ and the 'dotted' arrow is the projection part $u^\sharp$, so that the space E is the 'domain' of u and F is the 'codomain'. Our explanation for this musical notation is that the reader should think of flat and sharp keys on a piano as modulations of the same note (in this case, the arrow). The composition of $u\colon E \rightleftarrows F$ and $v\colon F \rightleftarrows G$ is, as one would expect, $v \circ u = \langle v^\flat u^\flat, u^\sharp v^\sharp \rangle$. We measure the 'size' of a pair by taking $\|u\| = \max\left(\|u^\flat\|, \|u^\sharp\|\right)$. Note that $\|u\| \geq 1$ (unless $E = 0$) and that $\|u\| \leq 1 + \varepsilon$ implies that $u^\flat$ is an ε-isometry. If $\|u\| = 1$ (or $u = 0$), we say that u is contractive. Finally, we declare a contractive pair $u\colon E \rightleftarrows F$ to be allowable if E and F are allowed p-normed spaces and both $u^\flat$ and $u^\sharp$ are rational maps. Clearly, the allowable pairs form a countable category.

Definition 6.3.1 A p-normed space X is said to be of almost universal complemented disposition (AUCD) if, for all contractive pairs $u\colon E \rightleftarrows X$ and $v\colon E \rightleftarrows F$ with F a finite-dimensional p-normed space, and every $\varepsilon > 0$, there exists a pair $w\colon F \rightleftarrows X$ such that $u = w \circ v$ and $\|w\| \leq 1 + \varepsilon$.

The situation is illustrated by the following diagram in which both the solid arrows (embeddings) and the dotted arrows (projections) commute:

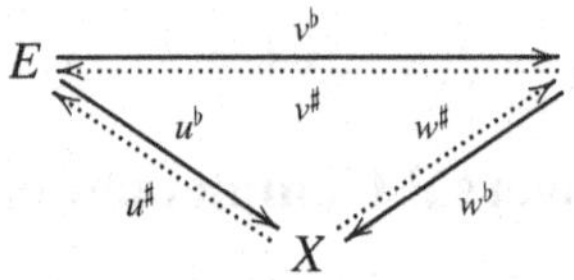

(6.5)

Hence, the AUCD property is formally stronger than [Ↄ]. Note that, according to our definitions, the 'null pair' $0 \rightleftarrows F$ is contractive. Thus, spaces with trivial dual are excluded from Definition 6.3.1 and do not have property [Ↄ].

Amalgamating Pairs

We now establish that pairs have the amalgamation property.

Lemma 6.3.2 *Given pairs $u\colon E \rightleftarrows F$ and $v\colon E \rightleftarrows G$ there are pairs $\overline{u} = \langle \overline{u}^\flat, \overline{u}^\sharp \rangle$ and $\overline{v} = \langle \overline{v}^\flat, \overline{v}^\sharp \rangle$ such that the following diagram commutes:*

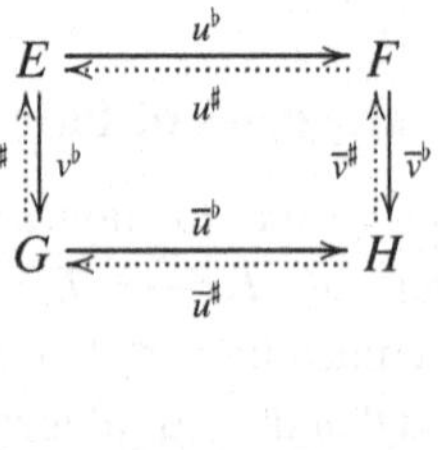

Moreover,

- *if u and v are contractive then so are $\overline{u}$ and $\overline{v}$;*
- *if u is contractive and $\|v^\flat\| \leq 1$ then $\overline{u}$ is contractive and $\|\overline{v}^\sharp\| \leq \|v^\sharp\|$;*

- *if u and v are allowable pairs then $\overline{u}$ and $\overline{v}$ can be taken to be allowable.*

Proof The proof is based on the isometric properties of the pushout construction presented in Section 2.5. We start with $u^\flat$ and $v^\flat$ so that $H = \mathrm{PO}$ is their pushout space and obtain the commutative diagram

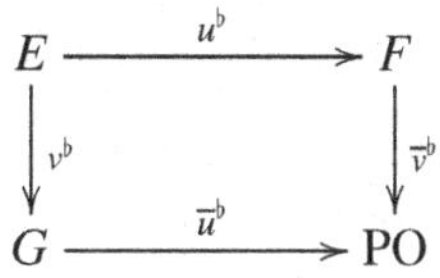

(6.6)

The projection $\overline{u}^\sharp$ is provided by the universal property of the pushout applied to the operators $\mathbf{1}_G, v^\flat u^\sharp$:

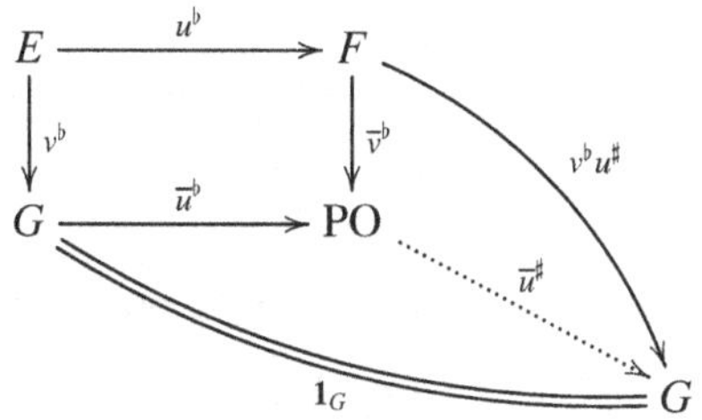

while $\overline{v}^\sharp$ is obtained from $\mathbf{1}_F$ and $u^\flat v^\sharp$. We have (see Lemma 2.5.2) that

- $\|\overline{u}^\flat\|, \|\overline{v}^\flat\| \leq 1$,
- $\|\overline{u}^\sharp\| \leq \|v^\sharp\|\,\|u^\flat\|$,
- $\|\overline{v}^\sharp\| \leq \|u^\sharp\|\,\|v^\flat\|$,
- $\overline{u}^\sharp\overline{u}^\flat = \mathbf{1}_G$, that is, $\overline{u} = \langle\overline{u}^\flat, \overline{u}^\sharp\rangle$ is a pair,
- $\overline{v}^\sharp\overline{u}^\flat = u^\flat v^\sharp$,
- $\overline{v}^\sharp\overline{v}^\flat = \mathbf{1}_F$, that is, $\overline{v} = \langle\overline{v}^\flat, \overline{v}^\sharp\rangle$ is a pair,
- $\overline{u}^\sharp\overline{v}^\flat = v^\flat u^\sharp$.

It only remains to check that the projections commute, that is, $u^\sharp\overline{v}^\sharp = v^\sharp\overline{u}^\sharp$. This follows from the uniqueness part of the universal property of the pushout construction: since $u^\sharp u^\flat = v^\sharp v^\flat$ (they are the identity on E), there must be a unique operator $\gamma\colon \mathrm{PO} \longrightarrow E$ making the following diagram commute:

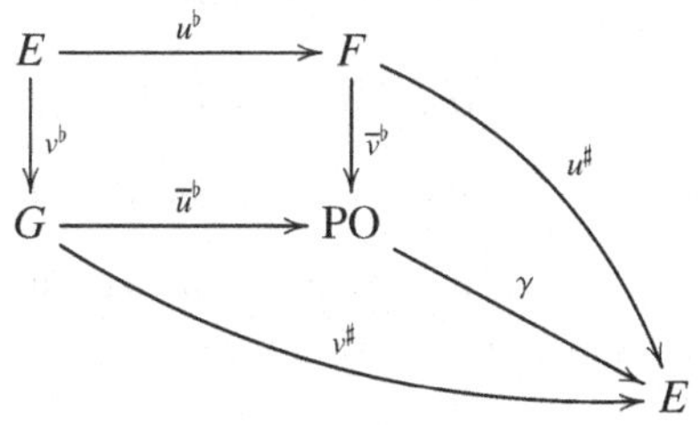

Since both $u^\# \overline{v}^\#$ and $v^\# \overline{u}^\#$ can be chosen for γ, they agree. This also proves the first two 'moreover' statements. To prove the third one, just use the allowable version of the pushout that appears in Lemma 6.2.2. □

Correction and Approximation

Before putting Fraïssé to work, let us state and prove three useful correction and approximation techniques that greatly simplify the manipulation of pairs. Before even that, we make the simple observation that every isomorphism $f\colon X \longrightarrow Y$ can be understood as part of a pair $\langle f, f^{-1}\rangle\colon\ X \rightleftarrows Y$.

Lemma 6.3.3 *Let E be a finite-dimensional subspace that is complemented by a projection P in a p-Banach space X, and let $e_1, \dots, e_k$ be a normalised basis of E. For every $\varepsilon > 0$, there is $\delta > 0$, depending on ε, $\|P\|$ and the chosen basis, such that if $x_i \in X$ satisfy $\|e_i - x_i\| < \delta$ for $1 \le i \le k$ then the linear map $f\colon X \longrightarrow X$ given by*

$$f(x) = \begin{cases} x_i & \text{if } x = e_i \text{ for } 1 \le i \le k \\ x & \text{if } x \in \ker P \end{cases}$$

satisfies $\|f - \mathbf{1}_X\| < \varepsilon$.

Proof Take K so large that $(\sum_i |\lambda_i|^p)^{1/p} \le K \|\sum_i \lambda_i e_i\|$. Pick $x \in X$ and write $x = y + z$ with $y = Px$ and then $y = \sum_i \lambda_i e_i$. Then, since $z \in \ker P$, one has

$$\|fx - x\| = \|fy - y\| = \left\| \sum_i \lambda_i (x_i - e_i) \right\| \le \delta \left(\sum_i |\lambda_i|^p \right)^{1/p} \le \delta K \|y\| \le \delta K \|P\| \|x\|.$$

Hence $\delta = \varepsilon/(K\|P\|)$ suffices. □

In particular, f is an automorphism. The hypothesis that E is complemented is necessary: in a rigid space (where the only endomorphisms are the scalar multiples of the identity), such an f cannot exist.

Lemma 6.3.4 *If $u\colon\ E \rightleftarrows F$ is a pair with $\|u\| \le 1 + \varepsilon$ then there is a p-norm $|\cdot|$ on F such that, for all $f \in F$, one has*

$$(1 + \varepsilon)^{-1} \|f\| \le |f| \le (1 + \varepsilon)\|f\| \tag{6.7}$$

and u becomes contractive when the original p-norm of F is replaced by $|\cdot|$.

Proof The hypotheses imply that $u^\flat$ is an ε-isometry. The unit ball of the new p-norm of F has to be the p-convex hull of the set

$$u^\flat[B_E] \cup (1 + \varepsilon)^{-1} B_F.$$

We thus define $|f| = \inf\{(\|x\|^p + (1+\varepsilon)^p\|g\|^p)^{1/p} : f = u^\flat(x) + g, x \in E, g \in F\}$ and check that everything works with this p-norm. First, taking $x = 0$ and $g = f$, we have $|f| \le (1+\varepsilon)\|f\|$. The other inequality of (6.7) is as follows: if $f = u^\flat(x) + g$, then $|f| \ge (1+\varepsilon)^{-1}\|f\|$ since

$$\|x\|^p + (1+\varepsilon)^p\|g\|^p = \|x\|^p + (1+\varepsilon)^p\|f - u^\flat x\|^p \ge \frac{\|u^\flat x\|^p + \|f - u^\flat x\|^p}{(1+\varepsilon)^p} \ge \frac{\|f\|^p}{(1+\varepsilon)^p}.$$

Let us compute the 'new' quasinorms of $u^\flat$ and $u^\sharp$. Given $x \in E$, one has $|u^\flat x|^p \le \|x\|^p$, so the quasinorm of $u^\flat$ is at most 1. Actually, it is clear that $|u^\flat x| = \|x\|$ for all $x \in E$. Indeed, we have

$$\begin{aligned} |u^\flat x|^p &= \inf\left\{\|y\|^p + (1+\varepsilon)^p\|g\|^p : u^\flat x = u^\flat(y) + g, y \in E, g \in F\right\} \\ &= \inf\left\{\|y\|^p + (1+\varepsilon)^p\|u^\flat(x-y)\|^p : y \in E\right\} \\ &\ge \inf\{\|y\|^p + \|x-y\|^p : y \in E\} \\ &= \|x\|^p. \end{aligned}$$

Finally, we check that $|u^\sharp| = \sup_{|f|\le 1}\|u^\sharp f\| = \sup_{|f|<1}\|u^\sharp f\| \le 1$. If $|f| < 1$, we can write $f = u^\flat(x) + g$, with $\|x\|^p + (1+\varepsilon)^p\|g\|^p < 1$. Hence

$$\begin{aligned} \|u^\sharp f\| &= \|u^\sharp(u^\flat x + g)\| = \|x + u^\sharp g\| \\ &\le \left(\|x\|^p + \|u^\sharp g\|^p\right)^{1/p} \le (\|x\|^p + (1+\varepsilon)^p\|g\|^p)^{1/p} < 1. \end{aligned}$$ □

The following is a version of Lemma 6.2.4 for pairs.

Lemma 6.3.5 *Given a contractive pair $u\colon E \rightleftarrows F$, with allowed domain E, and $\varepsilon > 0$, there is an allowable pair $u_0\colon E \rightleftarrows F_0$ and an ε-isometry $g\colon F \longrightarrow F_0$ making the following diagram commute:*

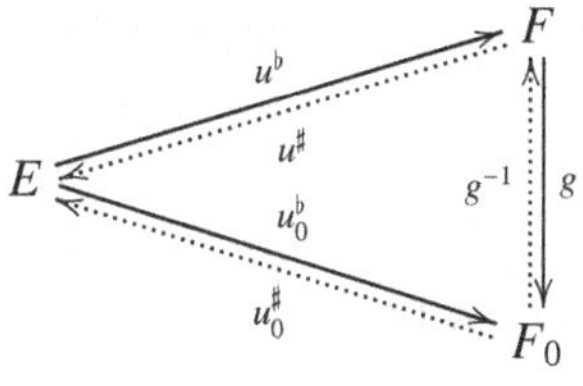

Proof Assume that ε is rational. Let $(e_i)_{1\le i\le n}$ be the unit basis of $E = \mathbb{K}^n$, and let $(f_j)_{1\le j\le m}$ be a basis of $\ker u^\sharp$. Then $\{u^\flat(e_1), \dots, u^\flat(e_n), f_1, \dots, f_m\}$ is a basis of F. Let $g\colon F \longrightarrow \mathbb{K}^{n+m}$ be the induced isomorphism. Take a rational p-norm $|\cdot|_0$ on $\mathbb{K}^{n+m}$ such that $(1+\varepsilon)^{-1}\|y\| \le |g(y)|_0 \le (1+\varepsilon)\|y\|$ for every $y \in F$. Now consider the pair $u_0 = \langle g, g^{-1}\rangle \circ u$. Then u_0 is rational (we have $u_0^\flat(x) = (x, 0)$ and $u_0^\sharp(x, y) = x$) and $\left\| u_0\colon E \rightleftarrows (\mathbb{K}^{n+m}, |\cdot|_0) \right\| \le 1 + \varepsilon$. Finally, we define a new p-norm on $\mathbb{K}^{n+m}$ by the formula

$$|y| = \inf\left\{\left(\|x\|^p + (1+\varepsilon)^p|z|_0^p\right)^{1/p} : y = u_0^\flat(x) + z, x \in \mathbb{K}^n, z \in \mathbb{K}^{n+m}\right\}.$$

This p-norm has to be allowed on $\mathbb{K}^{n+m}$ (by the last condition of the list), satisfies the estimate $(1+\varepsilon)^{-1}|\cdot|_0 \leq |\cdot| \leq (1+\varepsilon)|\cdot|_0$ and makes u_0 into a contractive pair (see the proof of Lemma 6.3.4) which is therefore allowable. Hence, if F_0 is $\mathbb{K}^{n+m}$ equipped with $|\cdot|$ then for every $y \in F$, we have

$$(1+\varepsilon)^{-2}\|y\| \leq |g(y)| \leq (1+\varepsilon)^2\|y\|. \qquad \square$$

A Space of Almost Universal Complemented Disposition

The allowable pairs form a countable category that admits amalgamations (Lemma 6.3.2) and has an initial object. It follows from Proposition 6.1.1 that there exists a Fraïssé sequence

$$U_1 \rightleftarrows U_2 \rightleftarrows \cdots \rightleftarrows U_n \rightleftarrows \cdots \rightleftarrows U_m \rightleftarrows \cdots$$

Define the p-Kadec space K_p to be the direct limit of the inductive system formed by the $(u_n^\flat)$:

$$U_1 \xrightarrow{u_1^\flat} U_2 \longrightarrow \cdots \longrightarrow U_n \xrightarrow{u_n^\flat} U_{n+1} \longrightarrow \cdots$$

Theorem 6.3.6 *K_p is a space of almost universal complemented disposition.*

Proof We identify each U_n with its image in K_p. Let $u\colon E \rightleftarrows \mathsf{K}_p$ and $v\colon E \rightleftarrows F$ be contractive pairs, where F is a finite-dimensional p-normed space, and let $0 < \varepsilon < 1$. We recommend that the reader work out the case in which both u and v are allowable pairs, using amalgamation and the properties of Fraïssé sequences. In the general case, we first push u into some U_n even if this spoils the isometric character of the embedding and the projection is no longer contractive. To this end, note that since the union of the subspaces U_n is dense in K_p, a straighforward application of Lemma 6.3.3 provides an integer n and an automorphism f of K_p such that $f[u^\flat[E]] \subset U_n$ with $\|f - \mathbf{1}_{\mathsf{K}_p}\| < \varepsilon$ and $\max\left(\|f\|, \|f^{-1}\|\right) < 1+\varepsilon$. After dividing f by $\|f\|$ and multiplying f^{-1} by $\|f\|$, we may assume that $\|f\| = 1$ and $\|f^{-1}\| < (1+\varepsilon)^2$. Then $\langle f, f^{-1}\rangle \circ u$ is a pair from E to K_p that 'factors' through the natural pair $\imath_n\colon U_n \rightleftarrows \mathsf{K}_p$ in the sense that $\langle f, f^{-1}\rangle \circ u = \imath_n \circ u_\varepsilon$, where $u_\varepsilon\colon E \rightleftarrows U_n$ is defined as

$$u_\varepsilon^\flat = \imath_n^\sharp f u^\flat, \qquad u_\varepsilon^\sharp = u^\sharp f^{-1} \imath_n^\flat.$$

Indeed $u_\varepsilon^\sharp$ is a projection along $u_\varepsilon^\flat$ since $u_\varepsilon^\sharp u_\varepsilon^\flat = u^\sharp f^{-1} \imath_n^\flat \imath_n^\sharp f u^\flat = \mathbf{1}_E$. Now we work with this u_ε and return to u at the end of the proof. Let us amalgamate u_ε and v in the pushout diagram

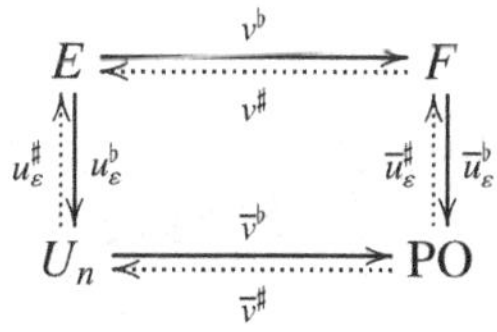

Note that since $\|u_\varepsilon^\flat\| \leq 1$, the lower pair $\overline{v} = \langle \overline{v}^\flat, \overline{v}^\sharp \rangle$ is contractive. Then we apply Lemma 6.3.5 to $\overline{v}$ to obtain an allowed space F_0 together with an ε-isometry $g\colon \mathrm{PO} \longrightarrow F_0$ such that $\overline{v}_0 = \langle g, g^{-1} \rangle \circ \overline{v}$ is an allowable pair. Finally, as the sequence $(u_n)_{n\geq 1}$ is Fraïssé, there is $m > n$ and an allowable pair $w_0\colon\ F_0 \rightleftarrows U_m$ such that $w_0 \circ \overline{v}_0$ is the bonding pair $U_n \rightleftarrows U_m$, so we have the following commutative diagram of pairs:

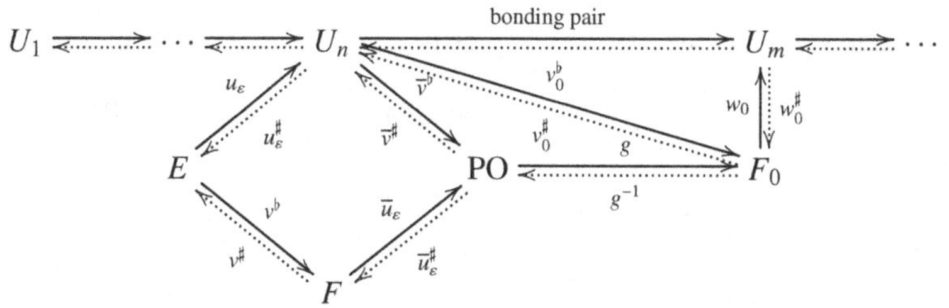

In particular, we have $w_0 \circ \langle g, g^{-1} \rangle \circ \overline{u} \circ v = u_\varepsilon = \langle f, f^{-1} \rangle \circ u$, and letting $w = \langle f^{-1}, f \rangle \circ w_0 \circ \langle g, g^{-1} \rangle \circ \overline{u}_\varepsilon$, we are done, since $w \circ v = u$ and

$$\|w\| \leq \|\langle f^{-1}, f \rangle\| \, \|w_0\| \, \|\langle g, g^{-1} \rangle\| \, \|\overline{u}_*\| \leq (1+\varepsilon)^3 < 1 + 7\varepsilon. \qquad \square$$

Further Properties of K_p

Next we study isometric properties of K_p: universality and uniqueness. There is a key fact that allows us to recover 'approximate pairs' (pairs of operators $f^\dagger\colon X \longrightarrow Y$ and $f^\ddagger\colon Y \longrightarrow X$ whose composition is close to the identity of X) as a composition of the arrows of two pairs with a common, ad hoc codomain.

Lemma 6.3.7 *Let $f^\dagger\colon X \longrightarrow U$ and $f^\ddagger\colon U \longrightarrow X$ be contractive operators such that $\|f^\ddagger f^\dagger - \mathbf{1}_X\| \leq \varepsilon$. There are contractive pairs $\alpha\colon\ X \rightleftarrows X \boxplus_{f^\dagger}^{\varepsilon} U$ and $\beta\colon\ U \rightleftarrows X \boxplus_{f^\dagger}^{\varepsilon} U$ such that $f^\dagger = \beta^\sharp \alpha^\flat$, $f^\ddagger = \alpha^\sharp \beta^\flat$ and $\|\alpha^\flat - \beta^\flat f^\dagger\| \leq \varepsilon$.*

The relevant diagram is

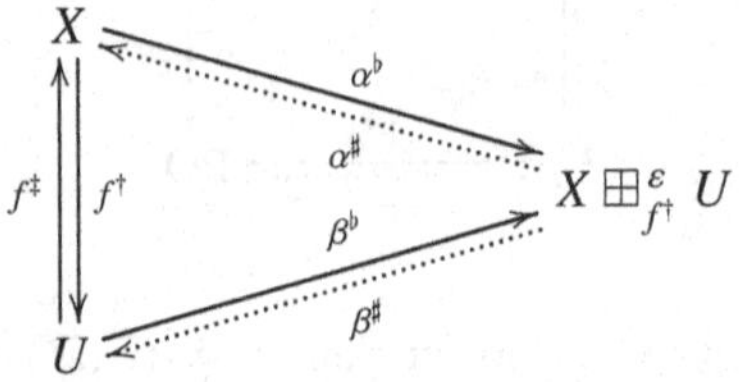

Proof We know from the proof of Lemma 6.2.5 that $\|(x,0)\| = \|x\|$ and $\|(0,y)\| = \|y\|$ for every $x \in X$ and every $y \in U$. Thus, letting $\alpha^\flat(x) = (x,0)$ and $\beta^\flat(y) = (0,y)$, we see that $\|\alpha^\flat - \beta^\flat f^\dagger\| \le \varepsilon$. As for the projections, we are forced to define $\alpha^\sharp(x,y) = x + f^\ddagger(y)$ and $\beta^\sharp(x,y) = y + f^\dagger(x)$. It is then clear that

$$\alpha^\sharp\alpha^\flat = \mathbf{1}_X, \quad \beta^\sharp\beta^\flat = \mathbf{1}_U, \quad f^\dagger = \beta^\sharp\alpha^\flat, \quad f^\ddagger = \alpha^\sharp\beta^\flat.$$

To see that $\alpha^\sharp$ and $\beta^\sharp$ are contractive, pick $(x,y) \in X \boxplus^{\varepsilon}_{f^\dagger} U$ and assume

$$(x,y) = (x_0 + x_2, y_1 - f^\dagger x_2) = (x_0, 0) + (0, y_1) + (x_2, -f^\dagger(x_2)).$$

We then have

$$\|\alpha^\sharp(x,y)\| = \|x_0 + x_2 + f^\ddagger(y_1) - f^\ddagger f^\dagger(x_2)\| \le (\|x_0\|^p + \|y_1\|^p + \varepsilon^p\|x_2\|^p)^{1/p},$$
$$\|\beta^\sharp(x,y)\| = \|f^\dagger(x_0) + y_1\| \le (\|x_0\|^p + \|y_1\|^p)^{1/p} \le (\|x_0\|^p + \|y_1\|^p + \varepsilon^p\|x_2\|^p)^{1/p}$$

and, since $\|(x,y)\|$ is the infimum of the numbers that might appear in the right-hand side, we have $\|\alpha^\sharp\|, \|\beta^\sharp\| \le 1$. □

Universality

A *skeleton* in a quasi-Banach space X is an increasing chain $(X_n)_{n\ge1}$ of finite-dimensional subspaces of X whose union is dense in X and such that each X_n is 1-complemented in X_{n+1}. Those inclusions and projections can be arranged into a sequence of contractive pairs $X_1 \rightleftarrows X_2 \rightleftarrows \cdots$ A quasi-Banach space has a skeleton if and only if it is the direct limit of a sequence of contractive pairs, and 'skeleton' is just a transcription of 1-FDD: if $(Y_n)_{n\ge1}$ is a 1-FDD of X then defining $X_n = Y_1 + \cdots + Y_n$, we obtain a skeleton; conversely, if $(X_n)_{n\ge1}$ is a skeleton then fixing contractive projections $\pi_n : X_{n+1} \longrightarrow X_n$ and letting $Y_1 = X_1$ and $Y_{n+1} = \ker \pi_n$, we get a 1-FDD.

Proposition 6.3.8 *Every p-Banach space with a skeleton is isometric to a 1-complemented subspace of* K_p.

Proof Let $(X_n)_{n\ge1}$ be a skeleton of X. For each integer $n \ge 1$, we denote the 'bonding' pair $X_n \rightleftarrows X_{n+1}$ by ξ_n, that is, $\xi_n^\flat$ is the inclusion of X_n into

X_{n+1} and $\xi_n^\sharp \colon X_{n+1} \longrightarrow X_n$ is a fixed contractive projection. Considering the spaces U_n as subspaces of K_p, we shall construct an increasing sequence of integers $(k(n))_{n\geq 0}$ and a system of contractive operators $f_n^\dagger \colon X_n \longrightarrow U_{k(n)}$ and $f_n^\ddagger \colon U_{k(n)} \longrightarrow X_n$ such that

(1) $\|f_n^\ddagger f_n^\dagger - \mathbf{1}_{X_n}\| < 2^{-n}$,
(2) $\|f_{n+1}^\dagger|_{X_n} - f_n^\dagger\| < 2^{-n}$,
(3) $\|f_{n+1}^\ddagger|_{U_{k(n)}} - f_n^\ddagger\| < 2^{-n}$.

Since $\sum_n 2^{-np} < \infty$, the pointwise limits of the sequences $(f_n^\dagger)$ and $(f_n^\ddagger)$ provide a contractive pair $X \rightleftarrows \mathsf{K}_p$, which will complete the proof. The required sequence is constructed by induction. We can assume $X_1 = 0$ so that $f_1^\dagger = 0$ and $f_1^\ddagger = 0$. Now suppose that $f_n^\dagger \colon X_n \longrightarrow U_{k(n)}$ and $f_n^\ddagger \colon U_{k(n)} \longrightarrow X_n$ have already been constructed and let us see how to get $k(n+1)$ and the maps $f_{n+1}^\dagger \colon X_{n+1} \longrightarrow U_{k(n+1)}$ and $f_{n+1}^\ddagger \colon U_{k(n+1)} \longrightarrow X_{n+1}$. We suggest that the reader fetch a pencil and some paper for a bit of scribbling.

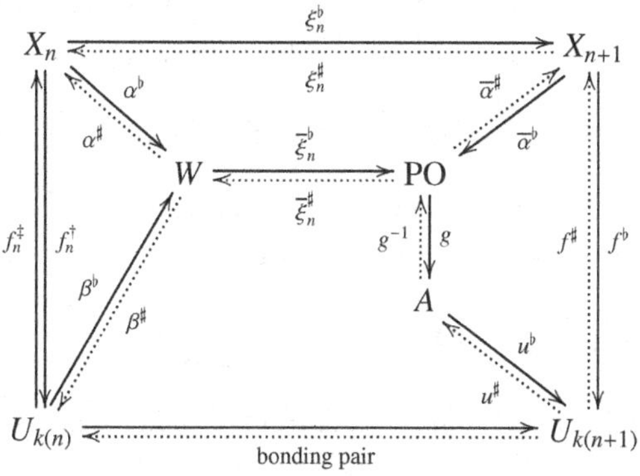

Set $\varepsilon = \|f_n^\ddagger f_n^\dagger - \mathbf{1}_{X_n}\| < 2^{-n}$ and reserve a small $\delta > 0$ of room. The precise value of δ will be specified later.

• First, we apply Lemma 6.3.7 to $f_n^\dagger$ and $f_n^\ddagger$. By doing so, we obtain the space W and the left triangle in the preceding diagram. Note that $\alpha = \langle \alpha^\flat, \alpha^\sharp \rangle$ and $\beta = \langle \beta^\flat, \beta^\sharp \rangle$ are contractive pairs such that

$$\|\beta^\flat f_n^\dagger - \alpha^\flat\| \leq \varepsilon, \qquad f_n^\dagger = \beta^\sharp \alpha^\flat, \qquad f_n^\ddagger = \alpha^\sharp \beta^\flat.$$

• Then, we amalgamate ξ_n and α using Lemma 6.3.2, which yields the upper commutative trapezoid.

• Next, we apply Lemma 6.3.5 to the composition $\overline{\xi}_n \circ \beta$ (which is a contractive pair), thus obtaining a surjective δ-isometry $g \colon \mathrm{PO} \longrightarrow A$ in such a way that the composition $\langle g, g^{-1} \rangle \circ \overline{\xi}_n \circ \beta$ turns out to be an allowable pair.

• Since the sequence of pairs (u_n) is Fraïssé, there must be some $k(n+1) > k(n)$ and an allowable pair $u\colon A \rightleftarrows U_{k(n+1)}$ such that $u \circ \langle g, g^{-1}\rangle \circ \overline{\xi}_n \circ \beta$ is the bonding pair $U_{k(n)} \rightleftarrows U_{k(n+1)}$.

• Now, look at the pair $f = u \circ \langle g, g^{-1}\rangle \circ \overline{\alpha}$. Note that f need not be contractive, as we only have the bound $\|f\| \le \|\langle g, g^{-1}\rangle\| \le 1 + \delta$.

One has:

(4) $f^\sharp f^\flat = \mathbf{1}_{X_{n+1}}$,
(5) $\|f^\flat|_{X_n} - f_n^\dagger\| \le (1+\delta)\varepsilon$,
(6) $f^\sharp|_{U_{k(n)}} = \xi_n^\flat f_n^\ddagger$.

The first identity is trivial. As for (5), note that $f^\flat|_{X_n} = u^\flat\, g\, \overline{\xi}_n^\flat\, \alpha^\flat$, hence

$$\|f^\flat|_{X_n} - f_n^\dagger\| = \|u^\flat\, g\, \overline{\xi}_n^\flat\, \alpha^\flat - \underbrace{u^\flat\, g\, \overline{\xi}_n^\flat \beta^\flat}_{\text{inclusion}} f_n^\dagger\| \le \|g\|\|\beta^\flat f_n^\dagger - \alpha^\flat\| \le (1+\delta)\varepsilon.$$

To check (6), observe that the inclusion of $U_{k(n)}$ into $U_{k(n+1)}$ can be written as $u^\flat\, g\, \overline{\xi}_n^\flat \beta^\flat$. Besides, $f^\sharp = \overline{\alpha}^\sharp\, g^{-1} u^\sharp$, so, recalling that $\overline{\alpha}^\sharp\, \overline{\xi}_n^\flat = \xi^\flat\, \alpha^\sharp$, we have

$$f^\sharp|_{U_{k(n)}} = \overline{\alpha}^\sharp\, g^{-1} u^\sharp\, u^\flat\, g\, \overline{\xi}_n^\flat \beta^\flat = \overline{\alpha}^\sharp\, \overline{\xi}_n^\flat \beta^\flat = \xi_n^\flat \alpha^\sharp \beta^\flat = \xi_n^\flat f_n^\ddagger.$$

As a final touch to render the maps contractive, set $f_{n+1}^\dagger = \frac{f^\flat}{1+\delta}$ and $f_{n+1}^\ddagger = \frac{f^\sharp}{1+\delta}$. Then $f_{n+1}^\ddagger f_{n+1}^\dagger = (1+\delta)^{-2}\mathbf{1}_{X_{n+1}}$. Hence, using (4) for small δ, we get

$$\|f_{n+1}^\ddagger f_{n+1}^\dagger - \mathbf{1}_{X_{n+1}}\| \le 1 - \frac{1}{(1+\delta)^2} < \frac{1}{2^{n+1}}.$$

And also for δ sufficiently small, we get from (5) and (6) that

$$\|f_{n+1}^\dagger|_{X_n} - f_n^\dagger\|^p \le \|f_{n+1}^\dagger - f^\flat\|^p + \|f^\flat|_{X_n} - f_n^\dagger\|^p \le \delta^p + (1+\delta)^p\varepsilon^p < 2^{-pn},$$
$$\|f_{n+1}^\ddagger - \xi_n^\flat f_n^\ddagger\| = \|f_{n+1}^\ddagger - f^\sharp\| \le \delta < 2^{-n}.$$

□

To deduce now that K_p is complementably universal for the spaces with the BAP, we need only firmly grab the trolley of Proposition 6.3.8 and push it resolutely towards Lemma 2.2.20's Platform 9 & 3/4: that we can freely assume that the space Y has a 1-FDD, and actually a skeleton, instead of a mere BAP. Do it without hesitation:

Corollary 6.3.9 *Every separable p-Banach space with the BAP is isomorphic to a complemented subspace of* K_p.

It is difficult to imagine a space peskier than K_p. Indeed, the following spaces are all isomorphic to K_p:

- Products $\mathsf{K}_p \times X$, when X is a separable p-Banach space with the BAP.
- Spaces of K_p-valued sequences $X(\mathsf{K}_p)$, when X is a p-Banach sequence space – in particular, this includes $\ell_q(\mathsf{K}_p)$ for $p \leq q < \infty$ and $c_0(\mathsf{K}_p)$.
- The p-convex envelope of K_q for $0 < q < p$ (see Corollary 6.3.12) and the space $C(\Delta, \mathsf{K}_p)$.

In contrast, if $0 < p < 1$, the space $\mathsf{K}_p \oplus_p L_p$ is of almost universal complemented disposition and not isomorphic to K_p. This assertion will later on be complemented by Propositions 6.3.13 and 10.7.2.

Uniqueness

We now address the uniqueness of K_p. The peak result here is Proposition 6.3.11, the 1-complemented companion of Proposition 6.2.7. We are pleased to make the reader aware that the skeleton assumption is quite natural: it corresponds, in the category of contractive pairs, to standard separability in $p\mathbf{B}$. The route to the proof is now based on a stability result that is interesting in its own right:

Proposition 6.3.10 *Let E be a finite-dimensional p-Banach and $\varepsilon > 0$. Let X be a p-Banach space with a skeleton and that satisfies [Ↄ]. If $f^\dagger : E \longrightarrow X$ and $f^\ddagger : X \longrightarrow E$ are contractive operators such that $\|f^\ddagger f^\dagger - \mathbf{1}_E\| < \varepsilon$ then there is an isometry $f^\flat : E \longrightarrow X$ with 1-complemented range and such that $\|f^\dagger - f^\flat\| < \varepsilon$.*

Proof We fix a skeleton (X_n) of X, and we denote the corresponding pairs of operators by $\xi_n : X_n \rightleftarrows X$ and $\xi_{(n,k)} : X_n \rightleftarrows X_k$. We also fix a sequence $(\varepsilon_n)_{n\geq 0}$ of positive numbers with $\varepsilon_1 < \varepsilon$ such that $\|f^\ddagger f^\dagger - \mathbf{1}_E\| < \varepsilon_1$ and $\sum_{n\geq 0} \varepsilon_n^p < \varepsilon^p$. Note that we must first choose ε_1 and then the other ε_n. Using a small perturbation of the identity of X, we can obtain $n(0)$ and contractive operators $f_0^\dagger : E \longrightarrow X_{n(0)}$ and $f_0^\ddagger : X_{n(0)} \longrightarrow E$ such that

$$\|f^\dagger - f_0^\dagger\| < \varepsilon_0 \qquad \text{and} \qquad \|f_0^\ddagger f_0^\dagger - \mathbf{1}_E\| \leq \varepsilon_1 .$$

Applying Lemma 6.3.7 to $f_0^\dagger, f_0^\ddagger$ and ε_1, we obtain the diagram

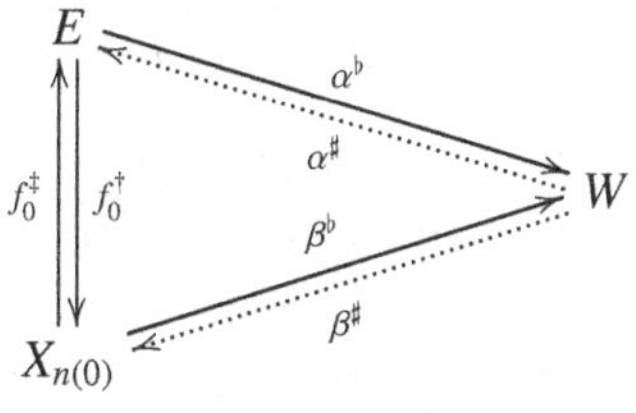

in which α and β are contractive pairs and

$$f_0^\dagger = \beta^\sharp \alpha^\flat, \quad f_0^\ddagger = \alpha^\sharp \beta^\flat, \quad \|\alpha^\flat - \beta^\flat f_0^\ddagger\| \leq \varepsilon_1.$$

Since X has property [∂], after normalising a suitable almost-isometry $W \longrightarrow X$ and the corresponding projection, we obtain $n(1) > n(0)$ and contractive operators $\gamma^\dagger : W \longrightarrow X_{n(1)}$ and $\gamma^\ddagger : X_{n(1)} \longrightarrow W$ satisfying

$$\|\gamma^\ddagger \gamma^\dagger - \mathbf{1}_W\| < \varepsilon_2 \quad \text{and} \quad \|\gamma^\dagger \beta^\flat - \xi^\flat_{(n(0),n(1))}\| < \varepsilon_2.$$

Letting $f_1^\dagger = \gamma^\dagger \alpha^\flat$ and $f_1^\ddagger = \alpha^\sharp \gamma^\ddagger$, we have $\|f_1^\ddagger f_1^\dagger - \mathbf{1}_E\| < \varepsilon_2$ and

$$\begin{aligned}\|f_1^\dagger - f_0^\dagger\|^p &= \|\gamma^\dagger \alpha^\flat - \gamma^\dagger \beta^\flat f_0^\ddagger + \gamma^\dagger \beta^\flat f_0^\dagger - f_0^\dagger\|^p \\ &\leq \|\alpha^\flat - \beta^\flat f_0^\dagger\|^p + \|\gamma^\dagger \beta - \xi^\flat_{(n(0),n(1))}\|^p \\ &< \varepsilon_1^p + \varepsilon_2^p.\end{aligned}$$

Continuing in this way, we obtain an increasing sequence $(n(k))_{k\geq 0}$ and contractive operators $f_k^\dagger : E \longrightarrow X_{n(k)}$ and $f_k^\ddagger : X_{n(k)} \longrightarrow E$ satisfying

- $\|f_k^\ddagger f_k^\dagger - \mathbf{1}_E\| \leq \varepsilon_{k+1}$,
- $\|f_{k+1}^\dagger - f_k^\dagger\| < \left(\varepsilon_{k+1}^p + \varepsilon_{k+2}^p\right)^{1/p}$.

The second estimate implies that $(f_k^\dagger)_k$ is a Cauchy sequence in $\mathfrak{L}(E, X)$ since

$$\|f_{k+m}^\dagger - f_k^\dagger\| \leq \left(\sum_{i=0}^{m-1} \|f_{k+i+1}^\dagger - f_{k+i}^\dagger\|^p\right)^{1/p} \leq \left(\sum_{i=0}^{m-1} \varepsilon_{k+i+2}^p + \varepsilon_{k+i+1}^p\right)^{1/p}.$$

The first estimate then implies that the double sequence $(f_k^\ddagger f_n^\dagger)_{k,n}$ converges to the identity of E in the sense that for every $\delta > 0$ there is m such that $\|f_k^\ddagger f_n^\dagger - \mathbf{1}_E\| < \delta$ whenever $k, n \geq m$. Define $f^\flat : E \longrightarrow X$ as the pointwise limit of the sequence $(f_k^\dagger)_k$. To obtain a suitable projection along $f^\flat$, we can use the local compactness of E: pick a non-trivial ultrafilter $\mathcal{U}$ on $\mathbb{N}$ and set $f^\sharp(x) = \lim_{\mathcal{U}(n)} f_n^\ddagger(x)$ for $x \in \bigcup_k X_k$, and extend by continuity to all of X. It is clear that $f^\flat$ and $f^\sharp$ are contractive. Finally, given $y \in E$, we have

$$f^\sharp f^\flat y = \lim_{\mathcal{U}(n)} f_n^\ddagger \left(f^\flat y\right) = \lim_{\mathcal{U}(n)} f_n^\ddagger \left(\lim_k f_k^\dagger y\right) = \lim_{\mathcal{U}(n)} \left(\lim_k f_n^\ddagger f_k^\dagger y\right) = \lim_{k,n} f_n^\ddagger f_k^\dagger y = y.$$

This shows that $f^\flat$ is an isometry whose range is 1-complemented in X. □

Proposition 6.3.11 *Let X, Y be p-Banach spaces with skeletons and that satisfy* [∂], *let A be a 1-complemented subspace of X and let B be a 1-complemented subspace of Y. If $f_0 : A \longrightarrow B$ is a surjective isometry then for every $\varepsilon > 0$, there is a surjective isometry $f : X \longrightarrow Y$ such that $\|f|_A - f_0\| < \varepsilon$.*

Proof The proof is a typical back-and-forth argument, oiled by Proposition 6.3.10. Fix a sequence of positive real numbers $(\varepsilon_n)_{n\geq 0}$ such that $\sum_{n\geq 0}\varepsilon_n^p < \varepsilon^p$. Let (X_n) and (Y_n) be chains of finite-dimensional 1-complemented subspaces of X and Y, respectively, with dense union. Set $A_1 = A + X_1$. Then f_0^{-1} embeds isometrically B into A_1, as a 1-complemented subspace. Since Y has property [Ә], for each $\delta > 0$, there is a δ-isometry $f_{1/2}\colon A_1 \longrightarrow Y$ whose range is $(1+\delta)$-complemented, extending the inclusion of B. Apply Proposition 6.3.10 to $f_{1/2}$ with δ small enough to obtain an isometry $f_1 : A_1 \longrightarrow Y$ with 1-complemented range such that $\|f_1(f_0^{-1}(y)) - y\| < \varepsilon_1\|y\|$ for all non-zero $y \in B$. Set $B_1 = f_1[A_1] + B + Y_1$ and apply the same argument to obtain an isometry $g_1 : B_1 \longrightarrow X$ with 1-complemented range with $\|g_1(f_1(x)) - x\| < \varepsilon_1\|x\|$ for all non-zero $x \in A_1$. Now set $A_2 = g_1[B_1] + A_1 + X_2$ and let $f_2\colon A_2 \longrightarrow Y$ be an isometry with 1-complemented range such that $\|f_2(g_1(y)) - y\| < \varepsilon_2\|y\|$ for all non-zero $y \in B_1$, and so on. Continuing in this way, we obtain increasing sequences $(A_n)_{n\geq 0}$ and $(B_n)_{n\geq 0}$ of finite-dimensional 1-complemented subspaces of X and Y, respectively, with dense union, where $A_0 = A$ and $B_0 = B$ together with isometries $f_n\colon A_n \longrightarrow B_n$ and $g_n\colon B_n \longrightarrow A_{n+1}$ satisfying

(1) $\|g_n(f_n(x)) - x\| < \varepsilon_n\|x\|$ for all non-zero $x \in A_n$,
(2) $\|f_{n+1}(g_n(y)) - y\| < \varepsilon_n\|y\|$ for all non-zero $y \in B_n$,

where $g_0 = f_0^{-1}$. The situation is illustrated in the following ('almost commutative') diagram

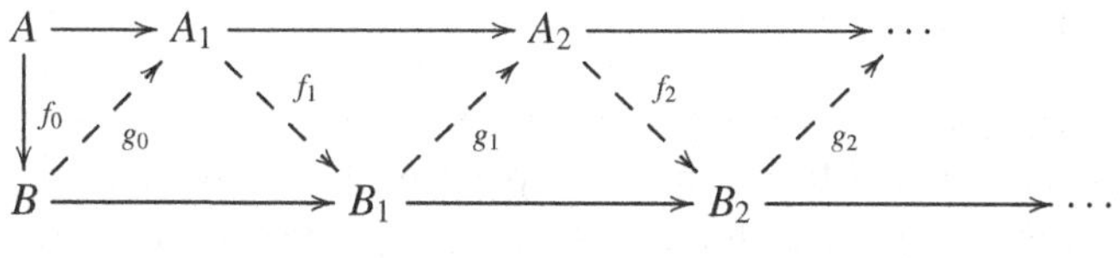

We define an operator $f\colon X \longrightarrow Y$ as follows: if $x \in A_k$, set $f(x) = \lim_{n\geq k} f_n(x)$. The definition makes sense because $(f_n(x))_{n\geq k}$ is a Cauchy sequence. Indeed, for $x \in A_n$, we have

$$\begin{aligned}\|f_{n+1}(x) - f_n(x)\|^p &\leq \|f_{n+1}(x) - f_{n+1}(g_n(f_n(x)))\|^p + \|f_{n+1}(g_n(f_n(x))) - f_n(x)\|^p\\ &\leq \|f_{n+1}\|^p\|x - g_n(f_n(x))\|^p + \varepsilon_n^p\|f_n(x)\|^p\\ &\leq 2\varepsilon_n^p\|x\|^p.\end{aligned}$$

Since $\sum_{n\geq 0}\varepsilon_n^p$ is finite, we see that f is well defined on $\bigcup_n A_n$, and so it extends to a contractive operator on X which we also call f. Besides, for x normalised in $A = A_0$, we have

$$\|f(x) - f_0(x)\| \leq \left(\sum_{n\geq 0}\|f_{n+1}(x) - f_n(x)\|^p\right)^{1/p} \leq \left(\sum_{n\geq 0} 2\varepsilon_n^p\right)^{1/p} \leq 2^{1/p}\varepsilon.$$

Proceeding analogously with the sequence (g_n), we obtain a contractive operator $g\colon Y \longrightarrow X$ given by $g(y) = \lim_{n\geq k} g_n(y)$ for $y \in B_k$. It follows from (1) and (2) that $gf = \mathbf{1}_X$ and $fg = \mathbf{1}_Y$. □

Now that Proposition 6.3.11 is complete, let us stop and smell the roses it has brought to blossom. One result with a fine scent is that any p-Banach space with a skeleton and property [Ↄ] is isometric to K_p and, therefore, of almost universal complemented disposition. (Intrigued by the role of the skeleton? Please move to the next section.) Another fragrant one is:

Corollary 6.3.12 *If $0 < p < q \leq 1$ then the q-Banach envelope of K_p is K_q.*

Proof We only sketch the proof. Fix $0 < p < q \leq 1$. The key point is that a contractive pair $u\colon E \rightleftarrows F$ between finite-dimensional p-normed spaces is also a contractive pair $u\colon E_{(q)} \rightleftarrows F_{(q)}$. Thus, taking the pairs between finite-dimensional q-normed spaces that arise from the q-Banach envelopes of the allowable pairs of p-Banach spaces, the class of which we will momentarily call $\mathfrak{A}$, we obtain a Fraïssé class $\mathfrak{A}_{(q)}$ since the amalgamation property is inherited from $\mathfrak{A}$. These pairs are 'dense' among the contractive pairs of finite-dimensional q-normed spaces because each q-norm is also a p-norm. Moreover, if $U^1 \rightleftarrows U^2 \rightleftarrows \ldots$ is the Fraïssé sequence used to define K_p, it is clear that the q-Banach envelope of K_p arises from the sequence $U^1_{(q)} \rightleftarrows U^2_{(q)} \rightleftarrows \ldots$, which is easily seen to be a Fraïssé sequence for $\mathfrak{A}_{(q)}$. Therefore, its limit is isometric to K_q. □

The Banach space K_1 is almost isotropic too. This follows from Proposition 6.3.11 and the fact that all lines are 1-complemented. It is not isotropic (almost isotropic with $\varepsilon = 0$) since the unit sphere of K_1 contains points where the norm is smooth and points where it is not (think of an isometric copy of, say ℓ_∞^2), while a surjective isometry should preserve each class. In sharp contrast, there is no equivalent p-norm rendering K_p almost isotropic when $p < 1$: if X is almost isotropic and isomorphic to K_p then the functional $|x| = \|x\| + \sup_{\|x^*\|\leq 1} |x^*(x)|$ is another p-norm which must be preserved by every isometry of the original p-norm of X. It quickly follows (see the complete argument in [57, Theorem 3.3]) that $|\cdot| = 2\|\cdot\|$. Thus, $\|x\| = \sup_{\|x^*\|\leq 1} |x^*(x)|$ and X would be locally convex, which is not the case.

Other Spaces of Kadec Type

Let us examine now what occurs when dropping the skeleton assumption:

Proposition 6.3.13 *Every separable p-Banach space is isometric to a* 1*-complemented subspace of a separable p-Banach space with property* [Ə].

The proof is based on a slight weakening of our notion of pair introduced in Section 6.3.

Definition 6.3.14 A λ-pair $u = \langle u^\flat, u^\sharp \rangle$ consists of two contractive operators $u^\flat : E \longrightarrow F$ and $u^\sharp : F \longrightarrow E$ such that $u^\sharp u^\flat = \lambda \mathbf{1}_E$, where $\lambda > 0$. A $\bullet$-pair is a λ-pair for some unspecified λ.

Thus, 1-pairs are the former contractive pairs. Note that if $u = \langle u^\flat, u^\sharp \rangle$ is a λ-pair then $\langle u^\flat, \lambda^{-1} u^\sharp \rangle$ is a pair which is not contractive in general and $u^\flat$ is a contractive ε-isometry, where $\varepsilon = \lambda^{-1} - 1$. Also, if $u = \langle u^\flat, u^\sharp \rangle$ is a pair then the normalisation $\langle u^\flat / \|u^\flat\|, u^\sharp / \|u^\sharp\| \rangle$ is a λ-pair, where $\lambda = (\|u^\flat\| \|u^\sharp\|)^{-1}$. We extend the use of the notation $u \colon E \rightleftarrows F$ to $\bullet$-pairs as well as most of the conventions of Section 6.3. If $u \colon E \rightleftarrows F$ is a λ-pair and $v \colon F \rightleftarrows G$ is a μ-pair, then $v \circ u = \langle v^\flat u^\flat, u^\sharp v^\sharp \rangle$ is a $\lambda\mu$-pair. The distance between two $\bullet$-pairs u, v between the same spaces is defined as $\|u - v\| = \max(\|u^\flat - v^\flat\|, \|u^\sharp - v^\sharp\|)$.

Lemma 6.3.15 *Let X be a p-Banach space and I an index set. For each $i \in I$, let $u_i \colon E_i \rightleftarrows F_i$ be a* 1*-pair and $v_i \colon E_i \rightleftarrows X$ a λ_i-pair. Then there is a p-Banach space X', a* 1*-pair $\xi \colon X \rightleftarrows Y$ and, for each $i \in I$, a λ_i-pair $\overline{v}_i \colon F_i \rightleftarrows X'$ such that $\xi \circ v_i = \overline{v}_i \circ u_i$; i.e. the diagram*

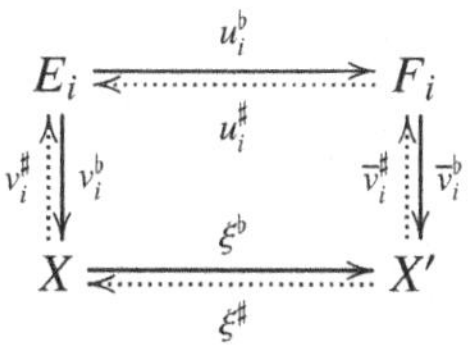

is commutative. Moreover, if I is finite and each F_i is finite-dimensional then $X'/\xi^\flat[X]$ is finite-dimensional.

Proof This is a combination of the Device technique and Lemma 6.3.2. Consider the 1-pair $\Pi \colon \ell_p(I, E_i) \rightleftarrows \ell_p(I, F_i)$ given by $\Pi^\flat = \prod_i u_i^\flat, \Pi^\sharp = \prod_i u_i^\sharp$ and the operator $\Sigma = \bigoplus_i v_i^\flat \colon \ell_p(I, E_i) \longrightarrow X$ and form the pushout

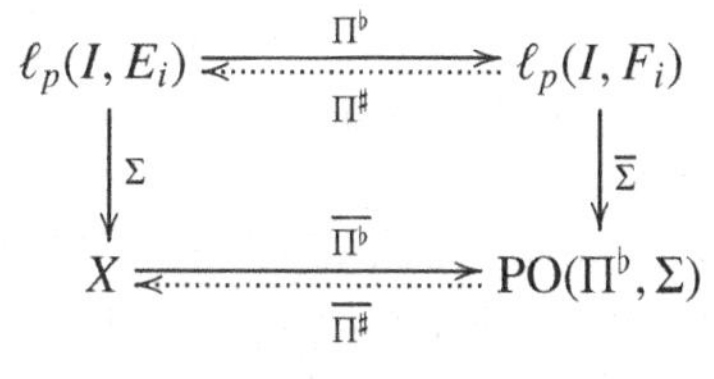

where $\overline{\Pi^\sharp}$ arises from the universal property of PO applied to the pair of operators $(\mathbf{1}_X, \Sigma\Pi^\sharp)$. This provides the 1-pair ξ. As for the $\bullet$-pairs $\overline{v}_j$ we first define $\overline{v}_j^\flat$ as the restriction of $\overline{\Sigma}$ to F_j. To get $\overline{v}_j^\sharp$, just consider the pair of operators $\lambda_j\pi_j\colon \ell_p(I, F_i) \longrightarrow F_j$ and $u_j^\flat v_j^\sharp\colon X \longrightarrow F_j$, where π_j sends $(x_i)_{i\in I}$ to x_j. Note that $u_j^\flat v_j^\sharp \Sigma = \lambda_j\pi_j\Pi^\flat$, as both send $(y_i)_{i\in I}$ to $\lambda_j y_j$. □

Proof of Proposition 6.3.13 Let $\mathfrak{A}$ be the set of allowable pairs of p-normed spaces, and let $(a_n)_{n\geq 1}$ be an enumeration of $\mathfrak{A}$, where $a_1 = \mathbf{1}_{\mathbb{K}}$. We are going to construct a chain of contractive pairs

$$X_1 \underset{\xi_1^\sharp}{\overset{\xi_1^\flat}{\rightleftarrows}} X_2 \underset{\xi_2^\sharp}{\overset{\xi_2^\flat}{\rightleftarrows}} X_3 \rightleftarrows \cdots \tag{6.8}$$

together with a sequence of sets of $\bullet$-pairs $(D^n)_{n\geq 1}$ and enumerations $(u_k^n)_{k\geq 1}$ in such a way that

(1) $X_1 = X$ and $X_{n+1}/\xi_n^\flat[X_n]$ is finite-dimensional;
(2) the elements of D^n are $\bullet$-pairs $u\colon E \rightleftarrows X_n$ with allowed domain;
(3) D^n has the following density property: for every $\bullet$-pair $v : E \rightleftarrows X_n$ with allowed domain and each $\varepsilon > 0$ there is $u \in D^n$ such that $\|v^\flat - u^\flat\| < \varepsilon$ and $\|v^\sharp v^\flat - u^\sharp u^\flat\| < \varepsilon$ (note that we don't care about $\|v^\sharp - u^\sharp\|$);
(4) $D^n \subset D^{n+1}$ in the sense that if $u \in D^n$ then $\xi_n \circ u \in D^{n+1}$;
(5) if $a \in \mathfrak{A}_{\leq n}$ and $u \in \bigcup_{k\leq n} D^k_{\leq n}$ is a λ-pair with the same domain then there is a commutative diagram

$$\begin{array}{ccc} E & \underset{a^\sharp}{\overset{a^\flat}{\rightleftarrows}} & F \\ u^\sharp \uparrow\downarrow u^\flat & & \overline{u}^\sharp \uparrow\downarrow \overline{u}^\flat \\ X_n & \underset{\xi_n^\sharp}{\overset{\xi_n^\flat}{\rightleftarrows}} & X_{n+1} \end{array}$$

where $\overline{u}$ is a λ-pair, $\mathfrak{A}_{\leq n} = \{a_1, \ldots, a_n\}$ and similarly with $D^k_{\leq n}$.

For the initial step, we set $X_1 = X$ and choose D^1 as in (3). This can be done because each $\mathfrak{L}(E, X)$ is separable and there are countably many allowed spaces. Condition (5) is automatic because of our choice of a_1. For the inductive step, assume that one has constructed $X_1, \ldots, X_n$ together with $D^1, \ldots, D^n$ and the corresponding enumerations that satisfy (1)–(5). Then we apply the preceding lemma to the 1-pairs of $\mathfrak{A}_{\leq n}$ and the $\bullet$-pairs in $\bigcup_{k\leq n} D^k_{\leq n}$ that have the same domain, and we set $X_{n+1} = X_n'$ and $\xi_n = \xi$. Note that $X_{n+1}/\xi_n^\flat[X_n]$ is finite-dimensional. Finally, we choose a countable set of $\bullet$-pairs D^{n+1} 'containing' every $\bullet$-pair of the form $\xi_n \circ v$ for $v \in D^n$ and satisfying (3), and we enumerate it. Let $\eth(X)$ be the direct limit of the system (6.8).

Let us verify that $\eth(X)$ has property $[\eth]$. Assume E is 1-complemented in a finite-dimensional space F and that $u^\flat\colon E \longrightarrow \eth(X)$ is an isometry with 1-complemented range. We consider 1-pairs $v\colon E \rightleftarrows F$ and $u\colon E \rightleftarrows X$ in which $v^\flat$ is the inclusion of E into F, $v^\sharp$ is a contractive projection onto E and $u^\sharp$ is a contractive projection along $u^\flat$. Fix $\varepsilon > 0$. Furthermore, $\varepsilon_1, \varepsilon_2, \varepsilon_3$ will appear in the course of the proof, and the only thing that we care about is that $\varepsilon_{n+1} \to 0$ as $\varepsilon_n \to 0$. Pick $\delta > 0$. First, we use Lemma 6.3.3 to obtain a small automorphism f of $\eth(X)$ and n such that $fu^\flat[E] \subset X_n$. Let u_1 be the normalisation of $\langle fu^\flat, u^\sharp f^{-1}\rangle$. This can be done in such a way that $\|f - \mathbf{1}_{\eth(X)}\| < \varepsilon_1$; $\|f\|, \|f^{-1}\| < 1 + \varepsilon_1$; and $\|u_1 - u\| < \varepsilon_1$. We can assume that $E = \mathbb{K}^k$ with some p-norm $\|\cdot\|$. Let $\|\cdot\|_0$ be a small allowed perturbation of the p-norm of E, and let $E_0 = (E, \|\cdot\|_0)$. Let u_2 be the normalisation of the formal identity $\langle \mathbf{1}, \mathbf{1}\rangle\colon E \rightleftarrows E_0$, and let

$$\begin{array}{ccc} E & \overset{v^\flat}{\underset{v^\sharp}{\rightleftarrows}} & F \\ u_2^\sharp \uparrow\downarrow u_2^\flat & & \overline{u_2}^\sharp \uparrow\downarrow \overline{u}_2^\flat \\ E_0 & \overset{\overline{v}^\flat}{\underset{\overline{v}^\sharp}{\rightleftarrows}} & H \end{array}$$

be provided by Lemma 6.3.2. As $\overline{v}$ is a 1-pair with allowed domain, we can immediately activate Lemma 6.2.4 to get an almost isometry $g\colon H \longrightarrow F_0$ such that $v_0 = \langle g, g^{-1}\rangle \circ \overline{v}\colon\ E_0 \rightleftarrows F_0$ is allowable. This can be done in such a way that (a) $\max\left(\|u_2^\sharp u_2^\flat - \mathbf{1}_E\|, \|\overline{u}_2^\sharp \overline{u}_2^\flat - \mathbf{1}_E\|\right) < \varepsilon_2$; (b) $\|g\|, \|g^{-1}\| < 1 + \varepsilon_2$ and (c) $\|x\|_0 \le \|x\| < (1 + \varepsilon_2)\|x\|_0$ for all non-zero $x \in E$. Let $u_3\colon E_0 \rightleftarrows \eth(X)$ be the normalisation of u_1 with respect to the p-norm of E_0. This is clearly a λ-pair, with $1 - \lambda < \varepsilon_3$, provided ε_1 and ε_2 are sufficiently small. By (3), we can find a μ-pair $u_4 \in D^n$ such that $\|u_4^\flat - u_3^\flat\| < \varepsilon_3$, still with $1 - \mu < \varepsilon_3$. By (5), there is $m > n$ and a μ-pair $\overline{u}_4$, making the following diagram commute:

$$\begin{array}{ccc} E_0 & \overset{v_0^\flat}{\underset{v_0^\sharp}{\rightleftarrows}} & F_0 \\ u_4^\sharp \uparrow\downarrow u_4^\flat & & \overline{u}_4^\sharp \uparrow\downarrow \overline{u}_4^\flat \\ X_n & \overset{\text{bonding pair}}{\rightleftarrows} & X_m \end{array}$$

Let us consider the operator $U = \overline{u}_4^\flat g \overline{u}_2^\flat\colon F \longrightarrow X_m$. It should be obvious that if $\varepsilon_1, \varepsilon_2, \varepsilon_3$ are sufficiently small,

- $\|U|_E - u\colon E \longrightarrow \eth(X)\| < \delta$;
- $\|U\| < 1 + \delta$;
- if $P = \overline{u}_2^\sharp g^{-1} \overline{u}_4^\sharp$, then $PU = \eta \mathbf{1}_F$, with $|\eta - 1| < \delta$ and $\|P\| < 1 + \delta$.

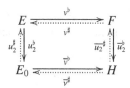

We now write $F = E \oplus \ker v^{\#}$ in order to then set $\tilde{U}(x, y) = u^{\flat}(x) + U(y)$, a true extension of $u^{\flat}$. If δ is sufficiently small, then $\|U - \tilde{U}\| < \varepsilon$ and $\|P\tilde{U} - \mathbf{1}_F\| < \varepsilon$. To obtain a small norm projection of $\mathfrak{D}(X)$ along $\tilde{U}$, we use:

6.3.16 Correcting a defective pair *Let F and Y be p-Banach spaces. Let $f^{\dagger}: F \longrightarrow Y$ and $f^{\ddagger}: Y \longrightarrow F$ be operators satisfying $\|f^{\ddagger} f^{\dagger} - \mathbf{1}_F\| \leq \varepsilon$, where $\varepsilon < 1$. There is an automorphism a of F such that*

- $\|a - \mathbf{1}_F\| \leq \varepsilon(1 - \varepsilon^p)^{-1/p}$,
- $\|a\| \leq (1 - \varepsilon^p)^{-1/p}$,
- $\|a^{-1}\| \leq (1 + \varepsilon^p)^{1/p}$,
- $af^{\ddagger} f^{\dagger} = \mathbf{1}_F$.

The proof is straightforward: set $a = \sum_{n \geq 0}(\mathbf{1}_F - f^{\ddagger} f^{\dagger})^n$ and check the required properties. □

The inexorable conclusion is that when X doesn't have the BAP, the space $\mathfrak{D}(X)$ cannot be isomorphic to K_p since it cannot have the BAP either. Therefore:

6.3.17 *For every $p \in (0, 1]$, there exist non-isomorphic separable p-Banach spaces with property* $[\mathfrak{D}]$.

It is clear that if X has a skeleton then so does $\mathfrak{D}(X)$, and Proposition 6.3.13 provides an alternative construction of K_p *and* a new proof of Proposition 6.3.8. On the other hand, it is clear that the only reason $\mathfrak{D}(X)$ could fail the BAP (or any other approximation property) is because X already lacks it: X is 1-complemented in $\mathfrak{D}(X)$ and $\mathfrak{D}(X)/X$ has a 1-FDD.

6.4 A Universal Operator on G_p

Finding operators on a given quasi-Banach space can be a difficult task. Or an impossible one, since rigid spaces exist. The space G_p is not rigid: Propositions 6.2.7 and 6.2.10 say that it has plenty of automorphisms. Our aim is to construct a contraction $u \in \mathfrak{L}(\mathsf{G}_p)$ with $\ker u \approx \mathsf{G}_p$ and satisfying the following condition:

(♡) For every separable p-Banach space X and every contractive operator $s: X \longrightarrow \mathsf{G}_p$, there exists an isometry $e: X \longrightarrow \mathsf{G}_p$ such that $s = ue$.

This will show that G_p has non-trivial projections since, taking as s the identity of G_p, one obtains an isometric embedding $e: \mathsf{G}_p \longrightarrow \mathsf{G}_p$ and eu is a projection on G_p with kernel and range isometric to G_p. To get the announced construction, we will fix a separable p-Banach space $\mathbb{H}$ and develop some abstract nonsense. Piece by piece, everything will make sense. We start by defining a special category **H** whose objects are contractive operators from finite-dimensional p-Banach spaces into $\mathbb{H}$; a morphism from $e: E \longrightarrow \mathbb{H}$ to $f: F \longrightarrow \mathbb{H}$ is an isometry $\imath: E \longrightarrow F$ such that $e = f\imath$:

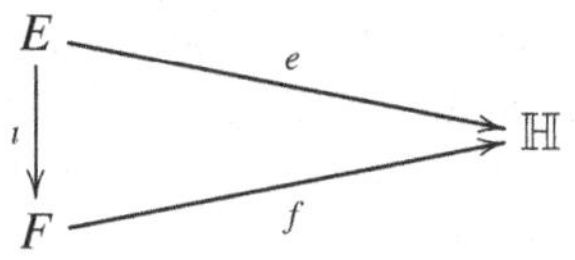

Although this category is conceptually more complex than those used in the preceding sections, our treatment, based on purely formal properties, is similar. Our nonsense training begins with:

Lemma 6.4.1 *The category* **H** *admits amalgamations.*

What does this mean? It means that when one has three objects $e: E \longrightarrow \mathbb{H}, f: F \longrightarrow \mathbb{H}, g: G \longrightarrow \mathbb{H}$ in **H** and morphisms $\imath: e \longrightarrow f, \jmath: e \longrightarrow g$, there is another object $h: H \longrightarrow \mathbb{H}$ and morphisms $\imath': g \longrightarrow h, \jmath': f \longrightarrow h$ such that $\jmath' \circ \imath = \imath' \circ \jmath$. The point is, we *do* know that the lemma is true and *how* to prove it: just stare at the commutative diagram

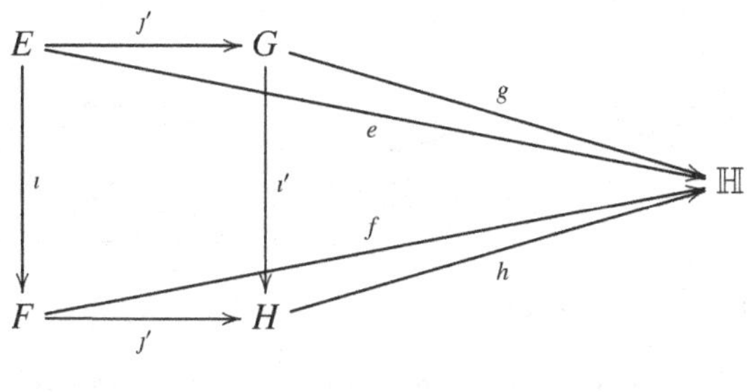

(6.9)

and set $H = \mathrm{PO}$, the pushout of $\imath$ and $\jmath$. Great. We also need an '**H**-version' of Lemma 6.2.5:

Lemma 6.4.2 *Let $f: X \longrightarrow Y$ be an ε-isometry between finite-dimensional p-Banach spaces, and let $r: X \longrightarrow \mathbb{H}$ and $s: Y \longrightarrow \mathbb{H}$ be contractive operators such that $\|sf - r\| \leq \varepsilon$. Let $\imath$ and $\jmath$ be the inclusions of X and Y, respectively, into $X \boxplus_f^\varepsilon Y$. The operator $r \oplus s: X \boxplus_f^\varepsilon Y \longrightarrow \mathbb{H}$ is contractive, and $(r \oplus s)\imath = r$, $(r \oplus s)\jmath = s$. In particular, $\imath: r \longrightarrow (r \oplus s)$ and $\jmath: s \longrightarrow (r \oplus s)$ are morphisms in* **H**.

Proof Fix $(x, y) \in X \oplus Y$ and assume $x = x_0 + x_2$, $y = y_1 - f(x_2)$. Then

$$\|(r \oplus s)(x, y)\|^p = \|r(x_0) + r(x_2) + s(y_1) - s(f(x_2))\|^p \le \|x_0\|_X^p + \|y_1\|_Y^p + \varepsilon^p \|x_2\|_X^p.$$

As $\|(x, y)\|^p$ is the infimum of all expressions that can arise as the right-hand side of the preceding inequality, we see that $\|(r \oplus s)(x, y)\|^p \le \|(x, y)\|^p$. □

To return to Fraïssé's world, we need a countable 'dense' subcategory of **H** having amalgamations. Let D be a dense, countable, linearly independent subset of $\mathbb{H}$ and let $\mathbb{H}_0$ denote the dense subspace of all finite linear combinations of elements of D with rational coefficients. We define a subcategory $\mathbf{H}_0$ of **H** as follows:

- The objects of $\mathbf{H}_0$ are contractive operators $e\colon E \longrightarrow \mathbb{H}$ whose domain is allowed and that send the rational vectors of E into $\mathbb{H}_0$.
- Given objects $e\colon E \longrightarrow \mathbb{H}$ and $f\colon F \longrightarrow \mathbb{H}$ in $\mathbf{H}_0$, an $\mathbf{H}_0$-morphism $\imath\colon e \longrightarrow f$ is an **H**-morphism whose underlying isometry is allowable.

Lemma 6.4.3 $\mathbf{H}_0$ *has amalgamations.*

Proof (Proof of) Lemma 6.2.2 + Diagram 6.9. □

Proposition 6.1.1 says that $\mathbf{H}_0$ has a Fraïssé sequence

$$u_1 \xrightarrow{\ \imath_1\ } u_2 \xrightarrow{\ \imath_2\ } \cdots \tag{6.10}$$

Since each $u_n\colon U_n \longrightarrow \mathbb{H}$ is an object of $\mathbf{H}_0$ and the arrows $\imath_n$ are morphisms in $\mathbf{H}_0$, what one actually has is a commutative diagram

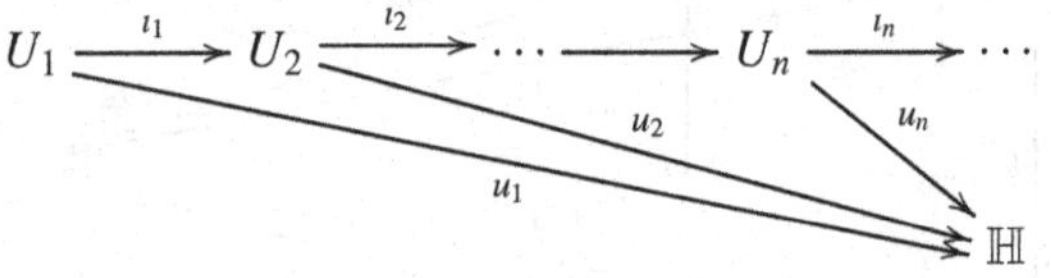

(6.11)

having the following property:

(†) Given a finite-dimensional p-Banach space V, an isometry $e\colon U_n \longrightarrow V$ and a contractive operator $v\colon V \longrightarrow \mathbb{H}$ such that $ve = u_n$, for each $\varepsilon > 0$, there exist $m > n$ and an ε-isometry $e'\colon V \longrightarrow U_m$ such that $\|e'e - \imath_{(n,m)}\| < \varepsilon$ and $\|u_m e' - v\| < \varepsilon$, where $\imath_{(n,m)} = \imath_{m-1} \cdots \imath_n$.

Of course, all this comes preloaded in the definition of a Fraïssé sequence when v and e are in $\mathbf{H}_0$, even with $\varepsilon = 0$. In the general case, we first apply Lemma 6.2.4 to e, thus obtaining an allowable $e_0\colon U_n \longrightarrow V_0$ and a surjective ε-isometry $g\colon V \longrightarrow V_0$ such that $e_0 = ge$. Although $u_n = vg^{-1}e_0$, which is Ok, we cannot apply the preceding case to vg^{-1} because we do not know that vg^{-1}

is contractive or that it takes rational vectors to $\mathbb{H}_0$. It is clear, however, that there is $v_0\colon V_0 \longrightarrow \mathbb{H}_0$ in $\mathbf{H}_0$ such that $\|v_0 - vg^{-1}\| < \varepsilon$:

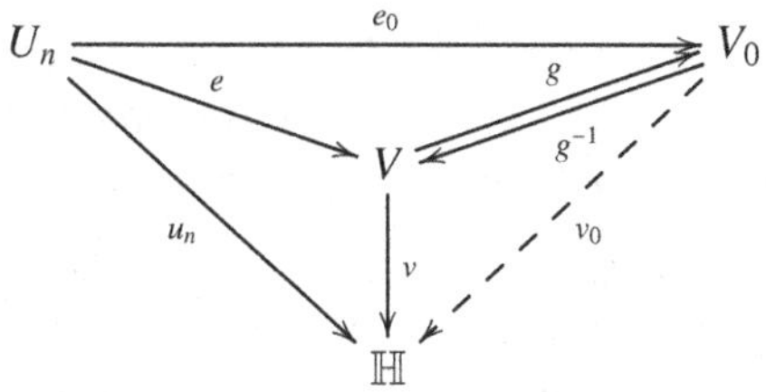

The dashed part of the diagram is there to remind us that it is merely *almost* commutative. Construct the space $U_n \boxplus^{\varepsilon}_{e_0} V_0$, equipped with the direct sum operator $u_n \oplus v_0$, and activate Lemmas 6.2.5 and 6.4.2: if we denote the inclusions of U_n and V_0 into $U_n \boxplus^{\varepsilon}_{e_0} V_0$ by $\imath$ and $\jmath$, we have $\|\jmath e_0 - \imath\| \le \varepsilon$ by Lemma 6.2.5, and since $\|v_0 e_0 - u_n\| \le \varepsilon$, we conclude that $\imath\colon u_n \longrightarrow u_n \oplus v_0$ and $\jmath\colon v_0 \longrightarrow u_n \oplus v_0$ are $\mathbf{H}$-morphisms. But $u_n \oplus v_0$ maps the rational vectors of $U_n \boxplus^{\varepsilon}_{e_0} V_0$ to $\mathbb{H}_0$, and $\imath\colon U_n \longrightarrow U_n \boxplus^{\varepsilon}_{e_0} V_0$ is allowable, by (f). It follows that $\imath\colon u_n \longrightarrow u_n \oplus v_0$ is actually in $\mathbf{H}_0$. Since (6.10) is Fraïssé for $\mathbf{H}_0$, there is $m > n$ and $k\colon u_n \oplus v_0 \longrightarrow u_m$ such that $k \circ \imath = \imath_{(n,m)}$, the bonding morphism $u_n \longrightarrow u_m$. Let us ignore V for a moment and depict the situation in the diagram

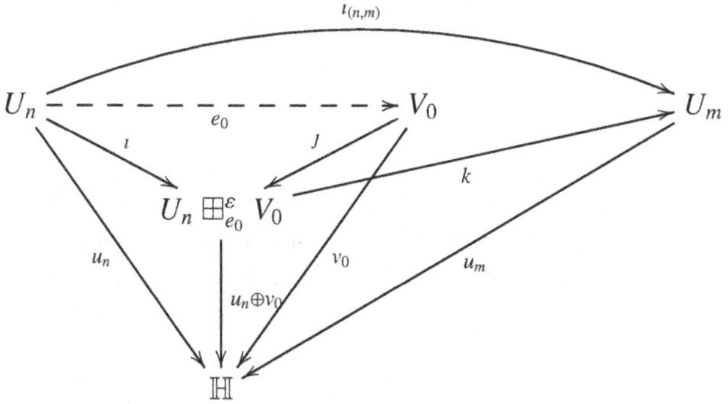

It is now clear that the required map is $e' = k\jmath g$:

- e' is an ε-isometry since g is and $\jmath, k$ are isometries.
- $\|e'e - \imath_{(n,m)}\| = \|k\jmath ge - \imath_{(n,m)}\| = \|k\jmath e_0 - \imath_{(n,m)}\| = \|k\jmath e_0 - k\imath + k\imath - \imath_{(n,m)}\| \le \varepsilon$.
- As $\|v_0 - vg^{-1}\| < \varepsilon$, we have $\|v - v_0 g\| \le \varepsilon\|g\| \le \varepsilon(1+\varepsilon)$. But $u_m e' = u_m k\jmath g = v_0 g$, so $\|u_m e' - v\| \le 2\varepsilon$.

Consider the directed system of p-Banach spaces $U_0 \xrightarrow{\imath_1} U_1 \xrightarrow{\imath_2} Y_2 \longrightarrow \cdots$ underlying the sequence (6.11). Set $U = \varinjlim U_n$ and let $u\colon U \longrightarrow \mathbb{H}$ be the direct limit of the operators u_n. The main properties of these objects can be summarised as follows: we know now that U contains an isometric copy of X as long as X is separable and $\mathfrak{L}(X, \mathbb{H}) \neq 0$, and will know soon (Theorem 6.4.5) that U is isometric to G_p. In the meantime:

Proposition 6.4.4 *For every separable p-Banach space X and every contractive operator* $s\colon X \longrightarrow \mathbb{H}$*, there exists an isometry* $e\colon X \longrightarrow U$ *such that* $s = ue$*. In particular, u is surjective and right-invertible.*

Proof We identify each U_n with its image in U so that $u_n = u|_{U_n}$ and all the bonding maps are plain inclusions. Fix an operator $s\colon X \longrightarrow \mathbb{H}$ with $\|s\| \leq 1$. Let $(X_n)_{n\geq 1}$ be an increasing sequence of finite-dimensional subspaces whose union is dense in X, with $X_1 = 0$. Set $s_n = s|_{X_n}$ and $\varepsilon_n = 2^{-n/p}$. We shall inductively construct an increasing sequence $k\colon \mathbb{N} \longrightarrow \mathbb{N}$ and contractive ε_n-isometries $e_n\colon X_n \longrightarrow U_{k(n)}$ satisfying $\|u_{k(n)}e_n - s_n\| \leq \varepsilon_n$ and also $\|e_{n+1}|_{X_n} - e_n\| \leq (\varepsilon_n^p + \varepsilon_{n+1}^p)^{1/p}$. This clearly implies that the sequence (e_n) converges pointwise to an isometry $e\colon X \longrightarrow U$ such that $s = ue$. We set $k(1) = 1$ and $e_1 = 0$. Having defined $e_n\colon X_n \longrightarrow U_{k(n)}$ with $\|s_n - u_{k(n)}e_n\| \leq \varepsilon_n$, we may apply Lemma 6.4.2 with $f = e_n$ to get the commutative diagram

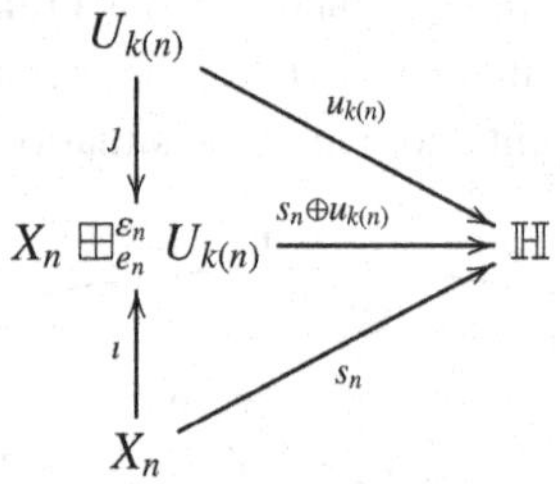

which shows that $\imath$ is a **H**-morphism from s_n to $s_n \oplus u_{k_n}$. On the other hand, the inclusion of X_n into X_{n+1}, which we momentarily denote by ξ, is clearly an **H**-morphism from s_n to s_{n+1}, and amalgamating $\imath$ and ξ, we arrive at the diagram

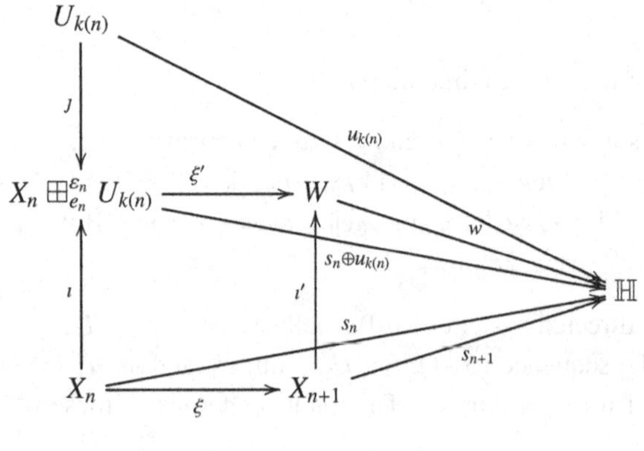

Here, W is a finite-dimensional p-normed space and $\|w\| \leq 1$. Applying (†) to the isometry $\xi' \jmath$, we find $k(n+1) > k(n)$ and obtain a contractive ε_{n+1}-isometry $e' \colon W \longrightarrow U_{k_{n+1}}$ such that

$$\|u_{k(n+1)}e' - w\| \leq \varepsilon_{n+1} \quad \text{and} \quad \|e'\xi' \jmath - \imath_{(k(n),k(n+1))}\| \leq \varepsilon_{n+1}. \tag{6.12}$$

Setting $e_{n+1} = e'\imath' \colon X_{n+1} \longrightarrow W \longrightarrow U_{k(n+1)}$, we complete the induction step. Indeed, e_{n+1} is a contractive ε_{n+1}-isometry. Moreover,

$$\|u_{k(n+1)}e_{n+1} - s_{n+1}\| = \|u_{k(n+1)}e'\imath' - w\imath'\| \leq \|u_{k(n+1)}e' - w\| \leq \varepsilon_{n+1},$$

since $s_{n+1} = w\imath'$, while

$$\|e_n - e_{n+1}|_{X_n}\|^p = \|e_n - e'\imath'\xi\|^p = \|e_n - e'\xi'\imath\|^p \leq \underbrace{\|e_n - e'\xi'\jmath e_n\|^p}_{(\star)} + \underbrace{\|e'\xi'(\imath - \jmath e_n)\|^p}_{(\star\star)}.$$

We have $(\star) = \|\imath_{(k(n),k(n+1))}e_n - e'\xi'\jmath e_n\| \leq \|e'\xi'\jmath - \imath_{(k(n),k(n+1))}\| \leq \varepsilon_{n+1}$ by (6.12) and $(\star\star) \leq \|\imath - \jmath e_n\| \leq \varepsilon_n$ by Lemma 6.2.5. This completes the induction step and the proof. □

Taking $s = \mathbf{1}_{\mathbb{H}}$, we see that u is surjective and right-invertible. Thus, we have a split exact sequence $0 \longrightarrow \ker u \longrightarrow U \stackrel{u}{\longrightarrow} \mathbb{H} \longrightarrow 0$.

Theorem 6.4.5 *Whatever the space $\mathbb{H}$ could be,* $\ker u$ *is isometric to* G_p *and so U is isomorphic to $\mathsf{G}_p \times \mathbb{H}$. If, additionally, $\mathbb{H}$ is a locally 1^+-injective p-Banach space then also U is isometric to G_p and, therefore, G_p is isomorphic to $\mathsf{G}_p \times \mathbb{H}$.*

Proof To prove that $\ker u$ is isometric to G_p we first check that the operator $u \colon U \longrightarrow \mathbb{H}$ has the following additional property:

(‡) If E is a subspace of a finite-dimensional p-Banach space F, $g \colon F \longrightarrow \mathbb{H}$ is contractive and $e \colon E \longrightarrow U$ is an isometry such that $ue = g|_E$ then for each $\delta > 0$, there is a δ-isometry $f \colon F \longrightarrow U$ satisfying $\|f|_E - e\| < \delta$ and $\|uf - g\| < \delta$.

Indeed, after taking a small perturbation, we may assume that $e \colon E \longrightarrow U_n$ is an ε-isometry with $\|ue - g|_E\| < \varepsilon$. Apply Lemma 6.4.2 to $e \colon E \longrightarrow U_n$, $g \colon E \longrightarrow \mathbb{H}$ and $u_n \colon U_n \longrightarrow \mathbb{H}$ to obtain a commutative diagram

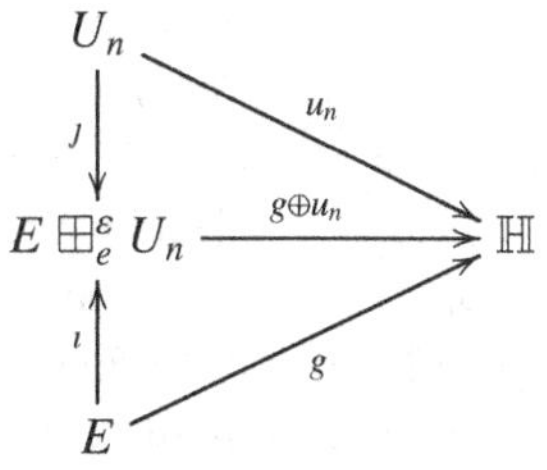

with $\|\jmath e - \imath\| \leq \varepsilon$. Now, amalgamating $\imath\colon E \longrightarrow E \boxplus_e^\varepsilon U_n$, which is a morphism from $g\colon E \longrightarrow \mathbb{H}$ to $g \oplus u_n\colon E \boxplus_e^\varepsilon U_n \longrightarrow \mathbb{H}$, with the inclusion $\xi\colon E \longrightarrow F$ regarded as a morphism from $g\colon E \longrightarrow \mathbb{H}$ to $g\colon F \longrightarrow \mathbb{H}$, we obtain a finite-dimensional p-normed space W and a commutative diagram

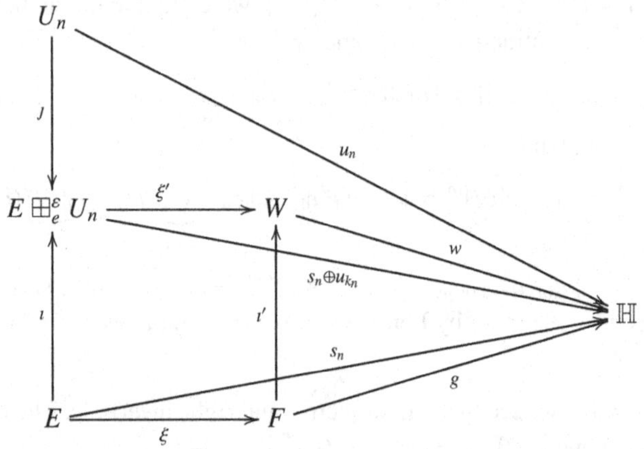

with $\|w\| \leq 1$. Now applying (†) to w and the embedding $\xi'\jmath$, we obtain $m > n$ and an almost isometry $w'\colon W \longrightarrow U_m$ such that $u_m w'$ is close to w and $w'\xi'\jmath$ is close to $\imath_{(n,m)}$. Finally, the composition

$$f : F \xrightarrow{\ \imath'\ } W \xrightarrow{\ w'\ } U_m \longrightarrow U$$

is the desired δ-isometry. Returning to $\ker u$, let F be a finite-dimensional p-normed space; $e\colon E \longrightarrow \ker u$ an isometry, where E is a subspace of F; and $\varepsilon > 0$. We shall construct an ε-isometry $f\colon F \longrightarrow \ker u$ such that $\|f(x) - e(x)\| \leq \varepsilon\|x\|$ for every $x \in E$. This will show that $\ker u$ is of AUD, thus completing the proof. To do so, fix some small δ and apply (‡), taking g as the zero operator from F to $\mathbb{H}$ to get a δ-isometry $f'\colon F \longrightarrow U$ such that $\|f'|_E - e\| < \delta$ and $\|uf'\| < \delta$. Of course, we cannot guarantee that f' takes values in $\ker u$. To amend this, let $r\colon \mathbb{H} \longrightarrow U$ be a right-inverse for u, with $\|r\| \leq 1$, and set $f = (\mathbf{1}_U - ru)f'$, that is, $f(x) = f'(x) - r(u(f'(x)))$. Then f takes values in $\ker u$ since $uf = 0$ and, moreover, $\|f - f'\| = \|ruf'\| \leq \delta$. Thus, for δ sufficiently small, $f\colon F \longrightarrow \ker u$ is an ε-isometry with $\|f|_E - e\| < \varepsilon$ and we are done.

We now assume that $\mathbb{H}$ is locally 1^+-injective among p-Banach spaces and prove that U is isometric to G_p. It suffices to check it is of AUD by showing that it satisfies the hypothesis of Lemma 6.2.3. Let $v\colon E \longrightarrow F$ be an isometry, where E is a subspace of U and F a finite-dimensional p-normed space. Fix

$\delta > 0$, and pick a contractive δ-isometry $u\colon E \longrightarrow U_n$ such that $\|u(x) - x\| \leq \delta\|x\|$ for $x \in E$. Let us form the pushout square

$$\begin{array}{ccc} E & \xrightarrow{\ v\ } & F \\ {\scriptstyle u}\downarrow & & \downarrow{\scriptstyle \overline{u}} \\ U_n & \xrightarrow{\ \overline{v}\ } & \mathrm{PO} \end{array}$$

Here $\overline{v}$ is an isometry and $\overline{u}$ is a contractive δ-isometry as in Lemma 2.5.2. Since $\mathbb{H}$ is locally 1^+-injective and PO is finite-dimensional, there is an operator $\tilde{u}_n\colon \mathrm{PO} \longrightarrow \mathbb{H}$ such that $u_n = \tilde{u}_n\overline{v}$, with $\|\tilde{u}_n\| \leq 1 + \delta$. Next we touch up the p-norm of PO to render $\tilde{u}_n$ contractive: for instance, we may take $|x| = \max(\|x\|_{\mathrm{PO}}, \|\widetilde{u}_n(x)\|_{\mathbb{H}})$. If V denotes the space PO so p-normed then $\overline{v}\colon U_n \longrightarrow V$ is still isometric and $\|\tilde{u}_n\colon V \longrightarrow \mathbb{H}\| \leq 1$ and we may use (†) to get $m > n$ and a δ-isometry $v'\colon V \longrightarrow U_m$ such that $\|v'\overline{v} - \imath_{(n,m)}\| \leq \delta$. Finally, if $\delta > 0$ is sufficiently small, the composition

$$w : F \xrightarrow{\ \overline{u}\ } \mathrm{PO} \xrightarrow{\text{identity}} V \xrightarrow{\ v'\ } U_m \xrightarrow{\text{inclusion}} U$$

is an ε-isometry such that $\|w(v(x)) - x\| \leq \varepsilon\|x\|$ for every $x \in E$. This shows that U is isometric to G_p. □

Time for applications.

Corollary 6.4.6 $\mathsf{G}_p \simeq \mathsf{G}_p \times \mathsf{G}_p \simeq c_0(\mathbb{N}, \mathsf{G}_p) \simeq C(\Delta, \mathsf{G}_p)$.

Proof The theorem just proved yields that if $\mathbb{H}$ is a separable locally 1^+-injective p-Banach space then $\mathsf{G}_p \times \mathbb{H}$ is isomorphic to G_p. Pick $\mathbb{H} = \mathsf{G}_p$, which is locally 1^+-injective according to Proposition 6.2.8(a), to obtain that G_p is isomorphic to $\mathsf{G}_p \times \mathsf{G}_p$ and thus to any finite product $\mathsf{G}_p \times \cdots \times \mathsf{G}_p$. The spaces $c_0(\mathsf{G}_p)$ and $C(\Delta, \mathsf{G}_p)$ can be written as the limit of a chain of subspaces isometric to $\mathsf{G}_p \oplus_\infty \cdots \oplus_\infty \mathsf{G}_p$, and so they are locally 1^+-injective; since $\mathsf{G}_p \simeq \mathsf{G}_p \times c_0(\mathbb{N}, \mathsf{G}_p)$ and $\mathsf{G}_p \simeq \mathsf{G}_p \times C(\Delta, \mathsf{G}_p)$, the Pełczyński decomposition method applies. □

The applications of Theorem 6.4.5 are seriously limited by the scarcity of examples of locally injective p-Banach spaces for $p < 1$, which basically are reduced to … G_p! When $p = 1$, all Lindenstrauss spaces are locally 1^+-injective Banach spaces, and Corollary 6.4.6 can be strengthened to:

Corollary 6.4.7 *Every separable Lindenstrauss space is isometric to a subspace of* G *that is complemented by a contractive projection whose kernel is isometric to* G.

Thus, if X is a separable Lindenstrauss space, then $\mathsf{G} \simeq X \times \mathsf{G} \simeq X \check{\otimes}_\varepsilon \mathsf{G}$. This does not mean that every copy of X is complemented in G. Also, Theorem 6.4.5 shows that some hyperplanes of the almost isotropic space G are isometric to the whole space, since when $p = 1$, the base field is 1-injective, and we can fix $\mathbb{H} = \mathbb{K}$. To the best of our knowledge, Hilbert spaces were the only previously known spaces combining both properties.

6.5 Notes and Remarks

6.5.1 What If $\varepsilon = 0$?

Upon moving ε from *here* to *there* in the definitions of Section 6.3 (and there are various heres and theres to choose), we obtain more or less equivalent variants of the definitions appearing in the text. Actually, the version of property [Ↄ] and the definition of AUCD we used do not match those of [183] or [116]. While 6.3.16 clearly shows that [Ↄ] is equivalent to Garbulińska's property (E) of [183], we cannot ensure that Definition 6.3.1 is equivalent to Definition 2.1 in [116]. And yet, as the following shows, an ε of room is necessary to stay in the separable world.

Proposition *Let X be a p-Banach space containing a 2-dimensional Euclidean subspace E and having the following property: for every 3-dimensional p-normed space F and every isometry $v\colon E \longrightarrow F$ with 1-complemented range, there is an isometry $w\colon F \longrightarrow X$ such that wv is the inclusion of E into X. Then the dimension of X is at least the continuum.*

Proof The proof uses an idea of Haydon, taken from [80; 75]. Let us follow it in the real case. Let E be the Euclidean plane and S the unit sphere of E. For each $u \in S$, we consider the p-norm

$$|(x,t)|_u = \max\left(\|x\|_2, \|(\langle x,u\rangle, t)\|_p\right)$$

on $E \times \mathbb{R}$ and let F_u denote the resulting 3-dimensional space. (The unit ball of F_u is the intersection of a 'vertical' right cylinder and a 'horizontal' right prism whose basis is the 2-dimensional ℓ_p-ball with 'peaks' at $(0,1)$ and $(u,0)$.) Note that $\|(x,0)\|_u = \|x\|_2$ (so E is isometric to a subspace of F_u) and that $|(x,t)|_u \geq \|x\|_2$ for each $(x,t) \in F_u$ (so the obvious projection is contractive). Now we consider E as a subspace of X and assume that for every $u \in S$, we can find an isometry $f_u\colon F_u \longrightarrow X$ such that $f_u(x,0) = x$. Clearly, f_u must have the form $f_u(x,t) = x + te_u$ for some fixed e_u in the unit sphere of X. Now let

S_+ be the 'positive part' of S so that $0 < \langle u, v\rangle < 1$ for different $u, v \in S_+$. We claim that $\|e_u - e_v\| = 1$ for $u, v \in S_+$ unless $u = v$. Pick $\lambda > 0$; we have

$$\|e_u - e_v\|^p \geq \|e_u + \lambda u\|^p - \|e_v + \lambda u\|^p = |(\lambda u, 1)|_u^p - |(\lambda u, 1)|_v^p.$$

But $|(\lambda u, 1)|_u^p = 1 + \lambda^p$, while for large λ, $|(\lambda u, 1)|_v^p = \max(\lambda^p, 1 + \lambda^p\langle u, v\rangle^p) = \lambda^p$. Hence the dimension of X is, at least, the cardinality of S_+. □

Thus any space of 'universal (complemented) disposition for spaces of dimension up to 3' has dimension at least $\mathfrak{c}$.

6.5.2 Before G_p Spaces Fade Out

Shortly hereafter, Fraïssé constructions will fade away in the remainder of this volume, although spaces of (almost) universal (complemented) disposition will not. But first, a few remarks that, once you are told, become very noticeable. The headline is that very few things are known about operators on G_p when $p < 1$. Indeed, the behaviour of operators on G_p is puzzling. On one hand, $\mathfrak{L}(\mathsf{G}_p)$ contains a large number of automorphisms and isometries as well as some projections. It follows from Proposition 6.2.7 that if F is a finite-dimensional subspace of G_p then G_p/F depends only on the dimension of F, up to isomorphisms. Let us denote the isomorphism type of the quotient of G_p by an n-dimensional subspace by $\mathsf{G}_p/(n)$. Since G_p is isomorphic to its square, $\mathsf{G}_p/(n + m) \simeq \mathsf{G}_p/(n) \times \mathsf{G}_p/(m)$ and also $\mathsf{G}_p/(n) \simeq (\mathsf{G}_p/(1))^n$. The sequence $0 \longrightarrow \mathbb{K} \longrightarrow \mathsf{G}_p(1) \longrightarrow 0$ is not trivial because G_p has trivial dual and therefore $\mathsf{G}_p/(1)$ is not a $\mathscr{K}$-space. So, the prickly issue is whether G_p is a $\mathscr{K}$-space. If the answer were yes then G_p could not be isomorphic to $\mathsf{G}_p/(1)$ (something we do not know either). When $p = 1$, both questions have an affirmative answer: G is isomorphic to its hyperplanes and, as for any $\mathscr{L}_\infty$-space, it is a $\mathscr{K}$-space by 3.4.6. However, we do not know whether G_p is prime or primary when $0 < p < 1$ or how to find an uncomplemented copy of G_p in the whole space, which is quite irritating. And since we cannot discard the existence of non-zero separable and separably injective p-Banach spaces, G_p could actually be such a space. In any case, all such spaces must be complemented subspaces of G_p. Ironically, it is the abundance of operators with values in G_p that makes it very difficult to define operators on G_p:

Proposition *If X is a separable p-Banach space and Y is a topological vector space such that $\mathfrak{L}(X, Y) = 0$, then $\mathfrak{L}(\mathsf{G}_p, Y) = 0$.*

Indeed, assume $u\colon \mathsf{G}_p \longrightarrow Y$ is non-zero and take $g \in \mathsf{G}_p$ such that $u(g) \neq 0$. Let $v\colon X \longrightarrow \mathsf{G}_p$ be an embedding, pick $x \in X$ and let $w \in \mathfrak{L}(\mathsf{G}_p)$ be such that $w(v(x)) = g$. Then uwv is a non-zero operator in $\mathfrak{L}(X, Y)$, a contradiction.

Thus, for instance, $\mathfrak{L}(\mathsf{G}_p, Y) = 0$ if Y is either an ultrasummand, by the Corollary in Note 1.8.3, or $Y = L_0$ since $\mathfrak{L}(L_p/H_p, L_0) = 0$ for exactly the same reasons that $\mathfrak{L}(L_p/H_p, L_p) = 0$ for $0 < p < 1$ (see Kalton [250, Theorem 7.2] or Aleksandrov [6, Corollary 4.4 on p. 49]). The same reasoning shows that every non-zero operator defined on G_p must be an isomorphism on some copy of ℓ_2 because this is what happens in L_p when $0 < p < 1$; see [283, Theorem 7.20] for perhaps the simplest proof.

6.5.3 Fraïssé Classes of Banach Spaces

Knowing that a given structure is a Fraïssé limit opens a door to a deeper appreciation of its properties. It is therefore not unproductive to ask which classes of finite-dimensional Banach spaces have the amalgamation property. Two obvious answers are 'all finite-dimensional spaces' (whose Fraïssé limit is G) and 'the Euclidean ones' (whose Fraïssé limit is the separable Hilbert space). To be true, what people knowledgeable about (continuous) Fraïssé structures work with are *separable* classes with *stable* versions of the amalgamation property. What is required is that, given δ-isometries $f: E \longrightarrow F$ and $g: E \longrightarrow G$, there exist some space H in the class and *isometries* $\overline{g}: F \longrightarrow H$, $\overline{f}: G \longrightarrow H$ such that $\|\overline{g}f - \overline{f}g\| < \varepsilon$, with ε depending on δ, and perhaps on $\dim E$. Our naive approach relies on Lemma 6.2.5 to guarantee stability. Very recently [170], the Banach spaces L_p for $p \neq 4, 6, 8\ldots$ have gained access to the elite club of Fraïssé spaces, which means that the class of finite-dimensional subspaces of L_p has a certain (stable) amalgamation property for those values of p. Those amalgamations are not plain pushouts, though. What prevents L_p from being Fraïssé when $p = 4, 6\ldots$ depends on the fact, proved by B. Randrianantoanina, that those L_p contain isometric copies of the same finite-dimensional spaces with very different projection constants [398], which is in turn connected with [345, Theorem 3] where, elaborating earlier work of Plotkin/Rudin, Lusky had shown that if $0 < p < \infty$ is not $4, 6, 8, \ldots$ then, given an isometry $\varphi_0: E \longrightarrow L_p$ from a finite-dimensional subspace E of L_p and $\varepsilon > 0$, there is an automorphism $\varphi \in \mathfrak{L}(L_p)$ extending φ_0 such that $\|\varphi\|, \|\varphi^{-1}\| < 1+\varepsilon$ (the reader can check this with Proposition 6.2.10).

The Kechris–Pestov–Todorcevic (KPT) correspondence [293] provides an unexpected connection between Fraïssé structures and topological dynamics. A topological group is extremely amenable if every continuous action on a compact set has a fixed point. The KPT correspondence states that, given a Fraïssé class **C**, the group of automorphisms of its Fraïssé limit is extremely amenable in the strong operator topology if and only if **C** has the approximate Ramsey property, something that has to do with continuous colorings; see

[385, Section 6.6] for a very readable introduction. Neither implication in the KPT correspondence is trivial. One can count among the applications that the otherwise mysterious isometry group of G is extremely amenable, because the class of all finite-dimensional normed spaces has the approximate Ramsey property; see [32]. Moving in the opposite direction, it implies that finite-dimensional Euclidean spaces have the approximate Ramsey property since the isometry group of a separable Hilbert space is (a Lévy group and thus) extremely amenable; see [385, Section 2.2]. More information on extreme amenability and approximate Ramsey properties can be found in [385] and more examples of Fraïssé classes in functional analysis in [342].

Sources

This chapter's blueprints were drawn in [80; 75; 116], which, in turn, are based on ideas of [310; 183]. The spaces underpinning the chapter are very classical objects in Banach space theory. In [203] Gurariy constructed the space that bears his name, coined the term AUD and proved Proposition 6.2.10 and, in particular, that any two separable AUD Banach spaces are *almost* isometric. The prefix was eliminated by Lusky in [343], a fine paper (which goes without saying when talking about Lusky's papers) which contains the additional result that the isometry group of G acts transitively on the set of smooth points of the unit sphere. More information about G and related constructions can be found in [22, Section 3.4]. A new proof of the uniqueness of Gurariy space was given by Kubiś and Solecki in [310]: the proof basically consists in showing that any separable AUD Banach space is the Fraïssé limit of the class of finite-dimensional spaces and isometries. Given the potential target of their paper, a tactical move was not to pronounce the word 'Fraïssé'. The paper contains the Banach ancestor of the key Lemma 6.2.5 and has (perhaps shared with [308]) the unquestionable merit of introducing Fraïssé structures into the Banach space business. The construction of G_p in Section 6.2 just transplants Kubiś and Solecki's ideas to the soil of p-Banach spaces; the presentation in [80] is more akin to [22, Chapter 3] and uses the Device. A forerunner of G_p appears in [248, Theorem 4.3]. The construction of K_p, taken from the 'related issues' of [75], is an adaptation for quasi-Banach spaces of Garbulińska-Węgrzyn's [183], where the idea of regarding spaces of Kadec type as Fraïssé structures appears for the first time and property $[\Game]$ is introduced as property (E). Categories of embedding-projection pairs had been defined and exploited in [308, Section 6]. Proposition 6.3.13 is the $[\Game]$ version of [116, Theorem 4.1]; there, it is shown that if X is a Banach space with separable dual then the output $\Game(X)$ is a Banach space with the additional property that given contractive pairs

$u\colon E \rightleftarrows X$ and $v\colon E \rightleftarrows F$, where F is a finite-dimensional normed space, for every $\varepsilon > 0$, there exists a pair $w\colon F \rightleftarrows X$ such that $\|u - w \circ v\| < \varepsilon$ and $\|w\| < 1 + \varepsilon$. This property was called 'almost universal complemented disposition' in [116]. We are not as convinced today that it deserves that name, mainly because, as mentioned before, we cannot ensure it is equivalent to AUCD. The topic of complementably universal spaces for a class $\mathscr{A}$ (spaces in the class containing complemented copies of every space in $\mathscr{A}$) emerges in 1969 when Pełczyński [381] exhibits his celebrated 'universal basis' space: a complementably universal space for the class of Banach spaces with basis. In 1971, Kadec [240] obtains the first complementably universal member of the class of separable Banach spaces with the BAP. Back to back with it, the next article in the same issue of *Studia* is from Pełczyński and Wojtaszczyk [384] and shows the existence of a complementably universal space for FDD, necessarily isomorphic to Kadec's. Still in the same volume, Pełczyński proved [382] that every Banach space with the BAP is complemented in a space with a basis, thus making it clear that his own universal space was complementably universal for the BAP and thus isomorphic to Kadec space. Kalton (who else?) performs in [247] a study of universal and complementably universal F-spaces and mentions the existence a complementably universal p-Banach space for the BAP for fixed $0 < p < 1$. He just adds that 'it is easy to duplicate the results for Banach spaces'. It is clear from [247, Theorem 4.1 (b) and Corollary 7.2] that Kalton is alluding to Pełczyński's universal space. He concludes by remarking that 'there are a number of other existence and non-existence results known for other classes of separable spaces'. From the Pełczyński decomposition method, it follows that two separable complementably universal p-Banach spaces for the BAP are isomorphic, and thus it turns out that K_p is isomorphic to Kalton's space, while K_1 is just a renorming of the spaces of Pełczyński, Kadec and Pełczyński and Wojtaszczyk. Several questions can be posed about those spaces, but two especially burning ones are: Does Kadec's space have property [⅁] in its own norm? Are the isometry groups of the spaces K_p extremely (or otherwise) amenable in the SOT? It cannot go unmentioned that no separable complementably universal space exists for the class of separable Banach spaces [233]. Universal operators date back to Rota's celebrated 'model operator' on Hilbert space. The material of Section 6.4 is taken from [80]. The category **H** is a typical *slice* category; see [27, Section 1.6, Example 4]. Theorem 6.4.5 subsumes several results scattered in the literature.

7

Extension of Operators, Isomorphisms and Isometries

Let us begin with a soulful rendition of one of the basic themes of this book: the connection between homological constructions and the extension / lifting of operators on display in Lemmata 2.6.3 and 2.8.3. An operator $\tau\colon Y \longrightarrow E$ can be extended to a superspace Z if and only if the lower sequence in the pushout diagram

$$\begin{array}{ccccccccc} 0 & \longrightarrow & Y & \longrightarrow & Z & \longrightarrow & Z/Y & \longrightarrow & 0 \\ & & \downarrow{\scriptstyle\tau} & & \downarrow & & \| & & \\ 0 & \longrightarrow & E & \longrightarrow & \mathrm{PO} & \longrightarrow & Z/Y & \longrightarrow & 0 \end{array}$$

splits. The dual result is that an operator $\tau\colon E \longrightarrow Z/Y$ can be lifted to Z if and only if the lower sequence in the pullback diagram

$$\begin{array}{ccccccccc} 0 & \longrightarrow & Y & \longrightarrow & Z & \longrightarrow & Z/Y & \longrightarrow & 0 \\ & & \| & & \uparrow & & \uparrow{\scriptstyle\tau} & & \\ 0 & \longrightarrow & Y & \longrightarrow & \mathrm{PB} & \longrightarrow & E & \longrightarrow & 0 \end{array}$$

splits. These results work on an individual basis: τ, Y, Z and E are fixed. A second set of results can then be formulated adding a 'for all' quantifier: if X, Y, E are quasi-Banach spaces, then

- $\mathrm{Ext}(X, E) = 0$ if and only if every operator $\tau\colon Y \longrightarrow E$ can be extended through any embedding $\jmath\colon Y \longrightarrow Z$ such that $Z/\jmath[Y] \simeq X$.
- $\mathrm{Ext}(E, Y) = 0$ if and only if every operator $\tau\colon E \longrightarrow X$ can be lifted through any quotient map $\rho\colon Z \longrightarrow X$ whose kernel is isomorphic to Y.

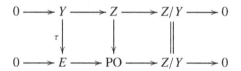

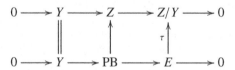

Pairs of Banach spaces A, B for which $\mathrm{Ext}(A, B) = 0$ have already been studied in Chapters 2, 3, 4 and 5. A more expedient action is required when we attempt to extend (or lift) operators of a certain kind (say, belonging to a given operator ideal or semigroup) to operators of the same kind, as in Proposition 5.1.9. We will not tread this path too often. Rather, in this chapter we will focus on a different aspect of the problem: the possibility of extending isomorphisms or isometries to maps of the same type. Let us explain why this can be difficult. All Banach spaces enjoy the property that every isomorphism between two finite-dimensional subspaces can be extended to an automorphism of the space. To prove this, observe that all closed hyperplanes of a given space X are isomorphic (think about their intersection) and, consequently, all closed n-codimensional subspaces are isomorphic. Now, if $\alpha\colon F \longrightarrow G$ is an isomorphism between two finite-dimensional subspaces, their complements H_F, H_G must be isomorphic through some isomorphism γ and thus, since $X = G \oplus H_G = F \oplus H_F$, the map $\alpha \times \gamma\colon F \oplus H_F \longrightarrow G \oplus H_G$ is the desired automorphism extending α. What is galling, though, is that almost any modification one makes to the previous argument leads to disaster. It fails for quasi-Banach spaces, it fails for infinite-dimensional subspaces of Banach spaces, it fails when isometries replace isomorphisms, it fails ... almost always. But it works sometimes: for spaces of AUD (Lemma 6.2.10) and, notably, for L_p. The whole story of this property for the L_p spaces $(0 < p < 1)$ began with Theorem 3.7.9, continued in Corollaries 4.2.8 and 4.5.3 and will arrive at its conclusion in Proposition 7.2.16. The reader will find a scenic view for a snapshot at 7.2.2.

Let us consider the following question: given a space E, a certain type of subspace $A \subset E$ (usually finite-dimensional) and a certain type of operator $A \longrightarrow E$, does there exist an extension $E \longrightarrow E$ of the same type? We will separately consider three cases: arbitrary operators, isomorphisms and isometries.

Extensible. A Banach space E is said to be *extensible* if, for every, subspace $Y \subset E$, every operator $\tau\colon Y \longrightarrow E$ admits an extension $T\colon E \longrightarrow E$. If every operator admits a λ-extension then we will say that E is λ-extensible. The space E is said to be *uniformly extensible* if it is λ-extensible for some λ.

Automorphic. A Banach space is said to be *automorphic* if every isomorphism between two subspaces whose corresponding quotients have the same dimension can be extended to an automorphism of the whole space.

Universal disposition. A Banach space X is said to be of universal disposition if, given two finite-dimensional spaces $F \subset G$, any isometry $F \longrightarrow X$ can be extended to an isometry $G \longrightarrow X$. This can be considered a stronger version (obtained allowing $\varepsilon = 0$) of the notion of almost universal disposition given in Section 6.2.

Knowledge about the first two notions is limited. Hilbert spaces are automorphic but not extensible, and ℓ_∞ is extensible but not automorphic. And none of them is of universal disposition. Ultrapowers of the Gurariy space are of universal disposition but not extensible or automorphic. All this shows that extending operators to operators is not the same as extending isomorphisms to isomorphisms or isometries to isometries. Beyond that, it is not known whether an extensible space must even be uniformly extensible, although there is some partial evidence in Proposition 7.1.2. Lindenstrauss and Rosenthal [332] called a Banach space X *subspace homogeneous* (a notion that coincides with automorphic for separable spaces) if, for every pair of isomorphic subspaces Y and Z of X, both of infinite codimension, there is an automorphism of X which carries Y onto Z; they showed that c_0 is automorphic and conjectured that c_0 and ℓ_2 are the only subspace homogeneous spaces (in particular, the only separable automorphic spaces). We do not even know whether there exist separable extensible spaces different from c_0 and ℓ_2. All this motivates the study of partial / finite-dimensional / approximate versions of the previous properties (say, *partially* automorphic spaces, uniformly *finitely* extensible spaces, spaces of (almost) universal disposition, etc.) which turns out to be fertile ground. Still more rewarding is the fact that the techniques for dealing with *partially* extensible or *partially* automorphic spaces are entirely different.

7.1 Operators: Extensible and UFO Spaces

The first observation to keep in mind is that being able to extending isomorphisms or isometries suffices for extending all operators. At least, with the right quantifiers:

Lemma 7.1.1

(a) *Assume that for every subspace A of E, the inclusion $A \longrightarrow E$ can be extended to any superspace. Then E is injective.*

(b) *Given a subspace A of E, if all embeddings $A \longrightarrow E$ can be extended to E then all operators $A \longrightarrow E$ can be extended to E.*

Proof To prove (a), let $\tau\colon Y \longrightarrow E$ be an operator, where Y is a subspace of X. Assume without loss of generality that $\|\tau\| \leq 1$ and form the pushout

$$\begin{array}{ccc} Y & \xrightarrow{\text{inclusion}} & X \\ \downarrow{\scriptstyle \tau} & & \downarrow{\scriptstyle \overline{\tau}} \\ \tau[Y] & \xrightarrow{\ \overline{\imath}\ } & \mathrm{PO} \end{array}$$

The operator $\bar{\imath}$ is an isometry by Lemma 2.5.2 and thus the inclusion $\tau[Y] \longrightarrow E$ admits an extension $T\colon \mathrm{PO} \longrightarrow E$. Then choose $T\bar{\tau}$ as an extension of τ. To prove (b), let $\imath\colon A \longrightarrow E$ be a subspace of E, and let $\tau\colon A \longrightarrow E$ be an operator. For ε sufficiently small, the operator $\imath - \varepsilon\tau\colon A \longrightarrow E$ is an embedding, which therefore must admit an extension $T\colon E \longrightarrow E$. The operator $\varepsilon^{-1}(\mathbf{1}_E - T)\colon E \longrightarrow E$ is an extension of τ. □

Consequently, automorphic spaces are extensible. Moreover:

Proposition 7.1.2 *An extensible Banach space that is isomorphic to its square is uniformly extensible.*

Proof All X_n, Y_n in this proof are isomorphic copies of X. Write $X = Y_0 \oplus X_1$ and then $X_n = Y_n \oplus X_{n+1}$. Now write $X = Y_0 \oplus Y_1 \oplus \cdots \oplus Y_n \oplus X_{n+1}$ so that $\overline{[Y_i : i \geq n+1]} \subset X_{n+1}$ and let $P_n\colon X \longrightarrow Y_0 + \cdots + Y_n$ be a projection such that $\ker P_n = X_{n+1}$. If X is not uniformly extensible, there are subspaces E_n of Y_n and operators $\tau_n\colon E_n \longrightarrow Y_n$ such that $\|\tau_n\| = 1$ for each n, and the norm of every extension of τ_n to an operator $Y_n \longrightarrow Y_n$ is greater than $2^n(\|P_n\| + \|P_{n-1}\|)n$. Define the operator $\tau\colon \overline{[E_n : n \geq 1]} \longrightarrow X$ by

$$\tau\left(\sum_{x_n \in E_n} x_n\right) = \sum 2^{-n}\tau_n x_n.$$

By construction, $\|\tau\| \leq 1$. If there exists an extension $T\colon X \longrightarrow X$ of τ then $S_n = 2^n(P_n - P_{n-1})T|_{Y_n}$ is an extension of τ_n to an operator $Y_n \longrightarrow Y_n$ with norm $\|S_n\| \leq 2^n(\|P_n\| + \|P_{n-1}\|)\|T\|$, which is impossible for large n. □

All of which leaves open the following question: must separable extensible spaces isomorphic to their squares be automorphic?

UFO Spaces

Operators defined on finite-dimensional spaces can be extended (without any specific control on the norm of the extension beyond a $\sqrt{\dim}$ estimate) to any larger Banach superspace. Imposing a uniform ratio for the norms of the extensions sets a rigid constraint on the structure of the space. This leads to the local version of uniform extendability:

Definition 7.1.3 A Banach space X is said to be λ-uniformly finitely extensible (a λ-UFO, for short) if every operator $\tau\colon E \longrightarrow X$ defined on a finite-dimensional subspace $E \subset X$ admits a λ-extension to X. A Banach space is said to be a UFO if it is a λ-UFO for some λ.

It is clear that $\mathscr{L}_{\infty,\lambda}$ and $\mathscr{L}_{2,\lambda}$ spaces are λ-UFO, though for different reasons. And that is probably the complete list of facts one could call obvious, which also means that none of the following facts are obvious:

Proposition 7.1.4

(a) *Every extensible space is a UFO.*
(b) *Every λ-UFO space that is μ-complemented in its bidual is $\lambda\mu$-uniformly extensible. In particular, if X is a UFO then X^{**} is uniformly extensible.*
(c) *A locally complemented subspace of a UFO is a UFO.*
(d) *Every ultrapower of a λ-UFO is a λ-UFO.*
(e) *Every λ-UFO has the λ-UAP.*
(f) *A space X cannot be a UFO if it contains sequences of finite-dimensional subspaces E_n and F_n such that $\sup_n d(E_n, F_n) < \infty$ and*

$$\lim_{n\to\infty} \frac{\lambda(E_n, X)}{\lambda(F_n, X)} = \infty.$$

Proof To prove (a), assume that X is not a UFO. This implies that its finite-codimensional subspaces cannot be UFO. Let F_1 be a finite-dimensional subspace of X for which there exists a norm 1 operator $T_1 : F_1 \longrightarrow X$ having no extension to the whole X with norm lesser than or equal to 2^2. Let G_1 be a complement for F_1 in X. Since G_1 is of finite codimension in X, it is not UFO, and there must exist a finite-dimensional subspace $F_2 \subset G_1$ and a norm 1 operator $T_2 : F_2 \longrightarrow X$ such that every extension to X has norm at least 2^4. Proceeding inductively, we find a sequence (F_n) of finite-dimensional subspaces of X and norm 1 operators $T_n : F_n \longrightarrow X$ that cannot be extended to X with norm lesser than or equal to 2^{2n}. This sequence necessarily has $\lim \lambda(F_n, X) = \infty$, and, as we show next in Lemma 7.1.5 (u), there is no loss of generality in assuming that it forms an FDD for $\overline{[F_n : n \in \mathbb{N}]}$:

Lemma 7.1.5 *Let X, Y be Banach spaces.*

(u) *Assume that X contains a sequence of finite-dimensional subspaces E_n such that $\lim_n \lambda(E_n, X) = \infty$. Then there are subspaces $E_n' \subset E_n$ such that $(E_n')_n$ is a FDD for $\overline{[E_n' : n \in \mathbb{N}]}$ and $\lim_n \lambda(E_n', X) = \infty$.*
(v) *Assume that X and Y contain finite-dimensional subspaces $E_n \subset X$ and $F_n \subset Y$ such that $\sup_n d(E_n, F_n) < \infty$ and*

$$\lim_{n\to\infty} \frac{\lambda(E_n, X)}{\lambda(F_n, Y)} = \infty .$$

Then there are finite-dimensional subspaces $E_k' \subset X$ and $F_k' \subset Y$ such that (E_k') is a FDD for $\overline{[E_k' : k \in \mathbb{N}]}$, with $\sup_k d(E_k', F_k') < \infty$ and

$$\lim_{k\to\infty} \frac{\lambda(E'_k, X)}{\lambda(F'_k, Y)} = \infty.$$

Proof To prove (u), let $\varepsilon > 0$, and for every n, let V_n be a finite subset of the unit sphere of X^* that $(1-\varepsilon)$-norms $E_1 + \cdots + E_n$. We will inductively choose the subspaces E'_k. First, we set $E'_1 = E_1$. To choose E'_2, we begin by setting $E^\circ_n = E_n \cap V_1^\top$, for $n > 1$. It is clear that $\dim(E_n/E^\circ_n) \le |V_1|$. We now use the following corollary of the Kadec–Snobar theorem [242]: *if E is a finite-dimensional subspace of a Banach space X and E° is a subspace of E then*

$$\frac{\lambda(E, X)}{\sqrt{\dim(E/E^\circ) + 1}} \le \lambda(E^\circ, X) \le \lambda(E, X)\sqrt{\dim(E/E^\circ) + 1}.$$

Applying this to E_n and E°_n, we get $\lambda(E_n, X)(|V_1| + 1)^{-1} \le \lambda(E^\circ_n, X)$. We will choose n_2 such that $\lambda(E_{n_2}, X) \ge 2^2(|V_1| + 1)$ and set $E'_2 = E^\circ_{n_2}$. Observe that $E'_1 \cap E'_2 = 0$ since V_1 is $(1-\varepsilon)$-norming for E_1 while $E'_2 = E_{n_2} \cap V_1^\top$ is orthogonal to V_1. Working now with V_2, we obtain n_3 such that $\lambda(E_{n_3}, X) \ge 2^3(|V_2|+1)$ and set $E'_3 = E^\circ_{n_3}$. As before, points in $E'_1 + E'_2 + E'_3$ have a unique representation as $p = p_1 + p_2 + p_3$ with $p_i \in E'_i$. We continue inductively. The sequence (E'_k) is a FDD for its closed linear span $\overline{[E'_k : k \ge 1]}$ with relative projection constants $\lambda(E'_n, \overline{[E'_k : k \ge 1]}) \le (1-\varepsilon)^{-1}$. The proof of (v) is quite similar: given $\varepsilon > 0$ and V_n as before, let $T_n : E_n \longrightarrow F_n$ be isomorphisms such that $\|T_n\|\,\|T_n^{-1}\| = d(E_n, F_n)$, with $\|T_n\| = 1$. Set $E'_1 = E_1$ and $F'_1 = F_1$. To choose E'_2, F'_2, set $E^\circ_n = E_n \cap V_1^\top$ and $F^\circ_n = T_n[E^\circ_n]$, for $n > 1$. Since $\dim(E_n/E^\circ_n) \le |V_1|$ and $\dim(F_n/F^\circ_n) \le |V_1|$, using the Kadec–Snobar theorem again, applied first to E_n and E°_n and then to F_n and F°_n, we get

$$\frac{\lambda(E_n, X)}{\lambda(F_n, Y)} \frac{1}{|V_1| + 1} \le \frac{\lambda(E^\circ_n, X)}{\lambda(F^\circ_n, Y)}.$$

We choose n_2 such that

$$\frac{\lambda(E_{n_2}, X)}{\lambda(F_{n_2}, Y)} \ge 2^2(|V_1| + 1)$$

and set $E'_2 = E^\circ_{n_2}$ and $F'_2 = F^\circ_{n_2}$. Working now with V_2, we obtain n_3 such that

$$\frac{\lambda(E_{n_3}, X)}{\lambda(F_{n_3}, Y)} \ge 2^3(|V_2| + 1)$$

and set $E'_3 = E^\circ_{n_3}$ and $F'_3 = F^\circ_{n_3}$. Continue inductively. □

We conclude the proof of (a) by defining an operator $\tau : [F_n : n \in \mathbb{N}] \longrightarrow X$:

$$\tau\Big(\sum_n x_n\Big) = \sum_n 2^{-n} T_n x_n\,,$$

where $x_n \in F_n$. The operator τ is compact, and no extension $Y \longrightarrow X$ is possible.

To prove (b), let $\tau\colon E \longrightarrow X$ be an operator, where E is a subspace of X. For each finite-dimensional $F \subset X$, consider an extension $T_F \in \mathfrak{L}(X)$ of $\tau|_{F\cap E}$ such that $\|T_F\| \le \lambda\|\tau|_{F\cap E}\|$. Pick an ultrafilter $\mathcal{U}$ on $\mathscr{F}(X)$ that refines the order filter and define the operator $T\colon X \longrightarrow X^{**}$ by $T(y) = \lim_{\mathcal{U}(F)} T_F(y)$ in the weak*-topology. If $P\colon X^{**} \longrightarrow X$ is a bounded projection then PT is an extension of τ such that $\|PT\| \le \|P\|\lambda$.

To prove (c), let E be a finite-dimensional subspace of a locally λ-complemented subspace Y of a UFO X. Let $\tau\colon E \longrightarrow Y$ be an operator and let $T \in \mathfrak{L}(X)$ be an extension of it. By the definition of a locally complemented subspace, fix $F = \tau[E]$ and find an operator $P\colon F \longrightarrow Y$ such that $P|_{F\cap Y} = \mathbf{1}_{F\cap Y}$ with $\|P\| \le \lambda$. Then $PT|_Y$ is an extension of τ, since $PT(e) = P\tau e = \tau e$ when $e \in E$, with $\|PT\| \le \lambda\|T\|$.

(d) Let $\mathcal{U}$ be an ultrafilter on some index set I, let E be a finite-dimensional subspace of $X_{\mathcal{U}}$ and let $\tau\colon E \longrightarrow X_{\mathcal{U}}$ be an operator. Pick $x^1, \dots, x^n \in X_{\mathcal{U}}$ and $y^1, \dots, y^m \in X_{\mathcal{U}}$ to be bases of E and $\tau[E]$, respectively, and write $x^k = [(x_i^k)]$ and $y^j = [(y_i^j)]$ with $x_i^k, y_i^j \in X$. For each $i \in I$, set $E_i = [x_i^1, \dots, x_i^n]$ and $F_i = [y_i^1, \dots, y_i^m]$ and define linear maps $u_i\colon E \longrightarrow E_i$ sending x^k to x_i^k and $v_i\colon \tau[E] \longrightarrow F_i$ sending y^k to y_i^k. We can assume that all these operators are, say 1-isometries (see the proof of Theorem 5.3.18): in fact, $\|u_i\|\,\|u_i^{-1}\| \to 1$ and $\|v_i\|\,\|v_i^{-1}\| \to 1$ along $\mathcal{U}$. Given $i \in I$, we consider the operator $\tau_i\colon E_i \longrightarrow F_i \subset X$ defined by $\tau_i\colon v_i^{-1}\tau u_i$. Note that $\|\tau_i\| \le 4\|\tau\|$ for all $i \in I$ and that $\|\tau_i\| \to \|\tau\|$ along $\mathcal{U}$. If λ is the constant with which X is a UFO, then for every $i \in I$, we can find $T_i \in \mathfrak{L}(X)$ extending τ_i, with $\|T_i\| \le \lambda\|\tau_i\|$. The ultraproduct operator $[T_i]_{\mathcal{U}}$ is an extension of τ with $\|[T_i]_{\mathcal{U}}\| \le \lambda\|\tau\|$.

Having established (d), to prove (e), it is enough to show:

Lemma 7.1.6 *A λ-UFO has the λ-AP.*

Proof Let X be a λ-UFO and let F be a finite-dimensional subspace of X. Let (x_n) be a finite ε-net of the unit sphere of F, pick normalised functionals (f_n) such that $f_n(x_n) = \|x_n\|$ and form the finite-codimensional subspace $H = \bigcap_n \ker f_n$ of X. It is clear that $F \cap H = 0$: if all f_n vanish on some normalised $x \in F$ then we may take x_k with $\|x - x_k\| \le \varepsilon$ and thus $\|x_k\| = f_k(x_k) = f_k(x_k - x) + f_k(x) \le \varepsilon$, so $\|x\| \le 2\varepsilon$, which is impossible. Since F is finite-dimensional, $F + H$ is closed, so $F + H = F \oplus H$. Given normalised $y \in F$ and $h \in H$, by picking x_k such that $\|y - x_k\| \le \varepsilon$, one has $\|y + h\| \ge f_k(y + h) = f_k(y) = f_k(y - x_k) + f_k(x_k) \ge 1 - \varepsilon$, which means that the natural projection $P\colon F \oplus H \longrightarrow F$ actually has norm at most $(1 - \varepsilon)^{-1}$, which we simplify to $1 + \varepsilon$. Given an operator $\tau\colon F \longrightarrow X$, the operator $\tau P\colon F + H \longrightarrow X$ has norm at most $\|P\|\,\|\tau\|$. Since X is a λ-UFO, X^{**} is uniformly extensible, so τP admits an extension $T \in \mathfrak{L}(X^{**})$ with norm at most $C\,\|P\|\,\|\tau\|$. The restriction of T to X has finite rank since $T[H] = \tau P[H] = 0$ and H has finite codimension in X. So $T[X]$ is a finite-dimensional subspace of X^{**}. Since X is locally complemented in X^{**}, there is an operator $Q\colon T[X] \longrightarrow X$ with norm $1+\varepsilon$ or less such that $Q(y) = y$ for all $y \in T[X] \cap X$. The operator $QT\colon X \longrightarrow X$ has finite rank and $QT(f) = Q\tau P(f) = Q\tau(f) = \tau(f)$ when $f \in F$. Moreover, $\|QT\| \leq (1+\varepsilon)C\|P\|$. □

(f) Let $\tau_n\colon E_n \longrightarrow F_n$ be isomorphisms with $\|\tau_n\| = 1$ and $\|\tau_n^{-1}\| = d(E_n, F_n)$. Let $T_n\colon X \longrightarrow X$ be any extension of τ_n and let $P_n\colon X \longrightarrow F_n$ be projections such that $\|P_n\| = \lambda(F_n, X)$. The operator $Q_n = \tau_n^{-1}P_nT_n$ is a projection onto E_n with $\|\tau_n^{-1}PT_n\| \leq d(E_n, F_n)\lambda(F_n, X)\|T_n\|$ and therefore $\lambda_n(E_n, X) \leq d(E_n, F_n)\lambda(F_n, X)\|T_n\|$, which definitely prohibits X from being UFO. □

7.2 Isomorphisms: the Automorphic Space Problem

For obvious reasons, Hilbert spaces enjoy the property that every isomorphism between two subspaces can be extended to an automorphism of the space – as long as the sizes of the quotients allow it. The 'obvious reasons' are that all closed subspaces are complemented, that all complemented subspaces are Hilbert spaces and that two Hilbert spaces are isomorphic if and only if they have the same dimension. The restriction on the size of the quotient is necessary (Exercise: try sending the whole space into an hyperplane through an automorphism. Hint: you cannot) because if A is a subspace of a quasi-Banach space X and T is an automorphism of X then T induces an isomorphism between X/A and $X/T[A]$. Recall that the dimension of a quasi-Banach space is the least cardinal of a subset spanning a dense subspace.

Definition 7.2.1 A Banach space is said to be automorphic if every isomorphism between two subspaces whose corresponding quotients have the same dimension can be extended to an automorphism of the whole space.

And thus, the question arises, which spaces are automorphic?

Variations on a Theme of Lindenstrauss and Rosenthal

In [332], Lindenstrauss and Rosenthal obtained the first non-trivial examples (cases (a,b,c) below) in which isomorphisms between subspaces / quotients can be extended / lifted to automorphisms of the space.

7.2.2 The Lindenstrauss–Rosenthal Theorem and Its Variations

(a) **(case ℓ_1)** *Let ρ and π be quotient maps from ℓ_1 onto a Banach space X not isomorphic to ℓ_1. There is an automorphism τ of ℓ_1 such that $\rho\tau = \pi$.*

(b) **(case c_0)** *Let $\imath$ and $\jmath$ embeddings of a Banach space H into c_0 such that both $c_0/\imath[H]$ and $c_0/\jmath[H]$ are infinite-dimensional. Then there is an automorphism τ of c_0 such that $\tau\imath = \jmath$.*

(c) **(case ℓ_∞)** *Let $\imath$ and $\jmath$ be embeddings of a Banach space Y into ℓ_∞ such that both $\ell_\infty/\imath[Y]$ and $\ell_\infty/\jmath[Y]$ are not reflexive. Then there is an automorphism τ of ℓ_∞ such that $\tau\imath = \jmath$. If the quotients $\ell_\infty/\imath[Y]$ and $\ell_\infty/\jmath[Y]$ are both reflexive then the automorphism τ exists if and only if the Fredholm index of any extension of $\jmath$ to ℓ_∞ through $\imath$ is 0. If one of the quotients is reflexive but the other is not, no such automorphism exists.*

(d) **(case L_1)** *Let A and B be reflexive subspaces of $L_1(\mu)$ such that $L_1(\mu)/A \simeq L_1(\mu)/B$. Then each of the subspaces A, B is isomorphic to the product of the other one with a finite-dimensional space.*

(e) **(case ℓ_p, $p < 1$)** *Let $0 < p < 1$ and let ρ and π be quotient maps from ℓ_p onto a quasi-Banach space X not isomorphic to ℓ_p. If $\ker\rho$ and $\ker\pi$ contain copies of ℓ_p complemented in ℓ_p then there exists an automorphism τ of ℓ_p such that $\rho\tau = \pi$.*

(f) **(case $L_p, p < 1$, quotients)** *Let $0 < p < 1$ and let ρ and π quotient maps from L_p onto a quasi-Banach space X such that each of the spaces $\ker\rho$ and $\ker\pi$ is either a q-Banach space for some $q > p$ or an ultrasummand. Then there exists an automorphism τ of L_p such that $\rho\tau = \pi$.*

(g) **(case $L_p, p < 1$, subspaces)** *Let $0 < p < 1$. Every isomorphism between two finite-dimensional subspaces of L_p can be extended to an automorphism of L_p.*

We can say, with a bit of sycophancy, that these results all share a common homological root (a feeble one in the case of (g)): they can be derived from the diagonal principles. The reason is that diagonal principles connect the 'extension / lifting of operators to operators' with the 'extension / lifting of isomorphisms to isomorphisms', even if acting between different spaces. An example will clarify this point: the Diagonal principle 2.11.7 can be reformulated saying that given an isomorphism $\alpha\colon Y \longrightarrow Y_1$ between two subspaces $\imath\colon Y \longrightarrow Z$ and $\jmath\colon Y_1 \longrightarrow Z$ of Z such that both α and α^{-1} can be extended to endomorphisms of Z, there is an automorphism $\beta\colon Z\times Z \longrightarrow Z\times Z$

such that $\beta(x,0) = (\alpha(x),0)$ and $\beta^{-1}(x,0) = (\alpha^{-1}(x),0)$ and thus the exact sequences

$$\begin{array}{ccccccccc}
0 & \longrightarrow & Y & \xrightarrow{(\imath,0)} & Z\times Z & \longrightarrow & Z/\imath[Y]\times Z & \longrightarrow & 0\\
 & & \| & & \downarrow{\scriptstyle\beta} & & & & \\
0 & \longrightarrow & Y & \xrightarrow{(\jmath\alpha,0)} & Z\times Z & \longrightarrow & Z/\jmath\alpha[Y]\times Z & \longrightarrow & 0
\end{array}$$

are isomorphic. One will be ready to conclude that α extends to an automorphism of Z provided some additional property of Z could be used to obtain that the sequences

$$\begin{array}{ccccccccc}
0 & \longrightarrow & Y & \xrightarrow{\imath} & Z & \longrightarrow & Z/\imath[Y] & \longrightarrow & 0\\
 & & \| & & & & & & \\
0 & \longrightarrow & Y & \xrightarrow{(\imath,0)} & Z\times Z & \longrightarrow & Z/\imath[Y]\times Z & \longrightarrow & 0
\end{array}$$

are isomorphic. This, and the lifting counterpart on the right side that follows from the Diagonal principle in 2.11.6, is the approach we will use. After such a dithyrambic intro, let the diagonal principles lope, leap, fly and stampede to prove assertions (a) to (g).

An ad hoc approach to (a) was rigged up between Proposition 2.7.3 and Corollary 2.7.4. On the other hand, (a) easily follows from Theorem 2.11.6 using the strategy above since any two exact sequences $0 \longrightarrow A \longrightarrow \ell_1 \longrightarrow X \longrightarrow 0$ and $0 \longrightarrow A\times\ell_1 \longrightarrow \ell_1\times\ell_1 \longrightarrow X \longrightarrow 0$ are isomorphic because every closed infinite-dimensional subspace of ℓ_1 contains a copy of ℓ_1 complemented in ℓ_1.

(e) is (a) relocated in the context of p-Banach spaces, paying simple attention to ℓ_p's anomalies, which is why the additional hypothesis on the kernels is necessary.

(b) is a consequence of 2.11.7 and Sobczyk's theorem: given exact sequences

$$\begin{array}{ccccccccc}
0 & \longrightarrow & H & \xrightarrow{\imath} & c_0 & \longrightarrow & c_0/\imath[H] & \longrightarrow & 0\\
 & & \| & & & & & & \\
0 & \longrightarrow & H & \xrightarrow{\jmath} & c_0 & \longrightarrow & c_0/\jmath[H] & \longrightarrow & 0
\end{array}$$

the sequences

$$\begin{array}{ccccccccc}
0 & \longrightarrow & H & \xrightarrow{(\imath,0)} & c_0\times c_0 & \longrightarrow & c_0/\imath[H]\times c_0 & \longrightarrow & 0\\
 & & \| & & & & & & \\
0 & \longrightarrow & H & \xrightarrow{(\jmath,0)} & c_0\times c_0 & \longrightarrow & c_0/\jmath[H]\times c_0 & \longrightarrow & 0
\end{array}$$

are isomorphic, and Pełczyński's property (V) of c_0 yields that the sequences

$$\begin{array}{ccccccccc} 0 & \longrightarrow & H & \xrightarrow{\imath} & c_0 & \longrightarrow & c_0/\imath[H] & \longrightarrow & 0 \\ & & \| & & & & & & \\ 0 & \longrightarrow & H & \xrightarrow{(\imath,0)} & c_0 \times c_0 & \longrightarrow & c_0/\imath[H] \times c_0 & \longrightarrow & 0 \end{array}$$

are also isomorphic. All together, this yields that the sequences we began with are isomorphic.

Case (c) has to be handled with some care since ℓ_∞ admits strictly singular quotient maps. So, given injective presentations $0 \longrightarrow Y \longrightarrow \ell_\infty \longrightarrow X_1 \longrightarrow 0$ and $0 \longrightarrow Y \longrightarrow \ell_\infty \longrightarrow X_2 \longrightarrow 0$ of a space Y, one has to distinguish three possibilities:

1) *Both quotient operators are not strictly singular.* In this case Rosenthal's property (V) of ℓ_∞ yields that the quotient maps are isomorphisms on some, necessarily complemented, copy of ℓ_∞, from which we conclude that for each $i = 1, 2$, the sequences

$$\begin{array}{ccccccccc} 0 & \longrightarrow & Y & \longrightarrow & \ell_\infty & \longrightarrow & X_i & \longrightarrow & 0 \\ & & \| & & & & & & \\ 0 & \longrightarrow & Y & \longrightarrow & \ell_\infty \times \ell_\infty & \longrightarrow & X_i \times \ell_\infty & \longrightarrow & 0 \end{array}$$

are isomorphic. Since the Diagonal principle 2.11.7 yields that also

$$\begin{array}{ccccccccc} 0 & \longrightarrow & Y & \longrightarrow & \ell_\infty \times \ell_\infty & \longrightarrow & X_1 \times \ell_\infty & \longrightarrow & 0 \\ & & \| & & & & & & \\ 0 & \longrightarrow & Y & \longrightarrow & \ell_\infty \times \ell_\infty & \longrightarrow & \ell_\infty \times X_2 & \longrightarrow & 0 \end{array}$$

are isomorphic, the conclusion is now clear.

2) *One quotient map is strictly singular, but the other is not.* In this case, the two sequences cannot be isomorphic.

3) *Both quotient maps are strictly singular.* Then Pełczyński's property (V) of ℓ_∞ implies that both X_1 and X_2 are reflexive and thus all operators $\ell_\infty \longrightarrow X_i$ are strictly singular for $i = 1, 2$. Since $\ell_\infty \times X_2$ is isomorphic to $\ell_\infty \times X_1$, X_2 must be isomorphic to a complemented subspace of $\ell_\infty \times X_1$. By the Edelstein–Wojtaszczyk decomposition (see 10.6.2), $X_2 \simeq F_2 \times X_1'$, where F_2 is a complemented subspace of ℓ_∞ and X_1' a complemented subspace of X_1. Necessarily, then, F_2 must be finite-dimensional. The same reasoning with X_1 yields $X_1 \simeq F_1 \times X_2'$, and F_1 must be finite-dimensional. The conclusion of the Diagonal principle 2.11.7 can be improved on this time to obtain that the sequences

$$\begin{array}{ccccccccc} 0 & \longrightarrow & Y & \longrightarrow & \ell_\infty \times F_2 & \longrightarrow & X_1 \times F_2 & \longrightarrow & 0 \\ & & \| & & & & & & \\ 0 & \longrightarrow & Y & \longrightarrow & \ell_\infty \times F_1 & \longrightarrow & X_2 \times F_1 & \longrightarrow & 0 \end{array}$$

are also isomorphic. It is now easy to get that the two starting sequences are isomorphic if and only if $\dim F_1 = \dim F_2$. Two different injective spaces can play the roles of the ℓ_∞s if we proceed with a bit of additional care.

(d) Sequences $0 \longrightarrow A \longrightarrow L_1 \longrightarrow X \longrightarrow 0$ and $0 \longrightarrow B \longrightarrow L_1 \longrightarrow X \longrightarrow 0$ with A and B reflexive lead to finite-dimensional spaces F, G such that the sequences

$$\begin{array}{ccccccccc} 0 & \longrightarrow & A \times F & \longrightarrow & L_1 \times F & \longrightarrow & X & \longrightarrow & 0 \\ & & & & & & \| & & \\ 0 & \longrightarrow & B \times G & \longrightarrow & L_1 \times G & \longrightarrow & X & \longrightarrow & 0 \end{array}$$

are isomorphic, thus $A \times F \simeq B \times G$. Two different $\mathscr{L}_1$-spaces can play the role of L_1 without difficulty.

Assertion (f) was proved straight in Theorem 3.7.9, but it is also an easy victim of diagonal principles using that $\mathfrak{L}(L_p, \ker \rho) = 0$ and $\mathfrak{L}(L_p, \ker \pi) = 0$.

(g) is Proposition 7.2.16 and is *not* a victim of diagonal principles.

Two Automorphic Spaces

Since $\ell_2(I)$ is automorphic for any index set I, and since the Lindenstrauss-Rosenthal theorem says that c_0 is automorphic, it is easy to believe that $c_0(I)$ could also be automorphic. The remainder of this section is to prove that:

Theorem 7.2.3 *The space $c_0(I)$ is automorphic.*

We may assume I to be uncountable since for countable I, we already have the Lindestrauss–Rosenthal theorem 7.2.2 (b).

The following notation shall be used during the proof: if $J \subset I$, we write $c_0(J)$ for the subset of $c_0(I)$ whose elements have support in J; by P_J, we denote the natural projection of $c_0(I)$ onto $c_0(J)$ and for a subspace $X \subset c_0(I)$, $X_J = X \cap c_0(J)$. The heart of the proof is showing that given any subspace $X \subset c_0(I)$, there is a partition $\mathfrak{J}$ of I into countable subsets such that for all $J \in \mathfrak{J}$, there are subspaces $H^J \subset c_0(J)$ with $X = c_0(\mathfrak{J}, H^J)$. The rest is combinatorics, because that fact plainly implies that one actually has $H^J = X_J$ and thus $X = c_0(\mathfrak{J}, X_J)$; then show that if Y is another subspace of $c_0(I)$ such that $\dim (c_0(I)/Y) = \dim (c_0(I)/X)$ and $T\colon X \longrightarrow Y$ is an isomorphism, one can choose $\mathfrak{J}$ such that $Y = c_0(\mathfrak{J}, Y_J)$ and $T[X_J] = Y_J$ also; and then make a final combinatorial touch to arrange the partitions in such a way that $\dim(c_0(J)/X_J) = \dim(c_0(J)/Y_J)$ for all $J \in \mathfrak{J}$. That done, the operator T can be decomposed as $(T_J)_{J\in\mathfrak{J}}$, where T_J is an isomorphism between X_J and Y_J. Every T_J can be extended to an automorphism $\widehat{T}_J$ of $c_0(J)$ using the following 'uniform' version of the Lindenstrauss–Rosenthal theorem (whose proof is just the usual amalgamation argument): there is a function $g\colon [1, \infty) \longrightarrow [1, \infty)$

such that if $A, B \subset c_0$ are such that $\dim(c_0/A) = \dim(c_0/B) = \aleph_0$ and $u\colon A \longrightarrow B$ is an isomorphism such that $\|u\|, \|u^{-1}\| \le C$ then u extends to an automorphism U of c_0 such that $\|U\|, \|U^{-1}\| \le g(C)$. Finally, amalgamate all those $\widehat{T}_J$.

The first decomposition we need can be found in [236, Lemma 2]:

Lemma 7.2.4 *Let X be a closed subspace of $c_0(I)$. For every countable subset $I' \subset I$, there is a countable set J containing I' such that $P_J[X] \subset X$.*

It is clear that $P_J[X] \subset X$ implies $(\mathbf{1}-P_J)[X] \subset X$ and so $X = X_J \oplus X_{J^c}$, where $J^c = I \setminus J$. Notice that if $\dim(X) < |I|$ there is $J \subset I$ such that $X \cup Y \subset c_0(J)$ and $|J| = \dim(X)$. One can therefore assume that $\dim(X) = |I|$ and $\operatorname{supp}(X \cup Y) = I$.

Lemma 7.2.5 *For every countable subset $I' \subset I$, there is a countable set J containing I' such that $P_J[X] \subset X$ and $P_J[Y] \subset Y$.*

Proof Construct a sequence $(J_n)_{n\ge1}$ of countable subsets of I as follows: Use the preceding lemma to obtain a countable set $J_1 \supset I'$ for which $P_{J_1}[X] \subset X$. Let $J_2 \supset J_1$ be a countable set such that $P_{J_2}[Y] \subset Y$. Let $J_3 \supset J_2$ be such that $P_{J_3}[X] \subset X$. In this way we obtain an increasing sequence (J_n) of countable sets such that, for every n, $P_{J_n}[X] \subset X$ and $P_{J_{n+1}}[Y] \subset Y$. The set $J = \bigcup_{n\ge1} J_n$ has the required properties. □

Lemma 7.2.6 *For every countable set $I' \subset I$, there is a countable $J \supset I'$ such that $P_J[X] \subset X, P_J[Y] \subset Y$ and $T[X_J] = Y_J$.*

Proof This is a back-and-forth argument. Beginning with $J_1 = I'$, we construct an increasing sequence $(J_n)_{n\ge1}$ of countable subsets of I as follows. Assume J_n has been constructed. If n is even then pick a countable set $J_{n+1} \supset J_n$ as in the preceding lemma and such that $T[X_{J_n}] \subset Y_{J_{n+1}}$. If n is odd, pick $J_{n+1} \supset J_n$ such that $T^{-1}[Y_{J_n}] \subset X_{J_{n+1}}$ instead. The set $J = \bigcup_{n=1}^{\infty} J_n$ satisfies the conditions of Lemma 7.2.5 and, obviously, $T[X_J] \subset Y_J$: it quickly follows that $T[X_J] = Y_J$. □

Lemma 7.2.7 *There is a partition $\mathfrak{J}$ of I into infinite countable subsets such that, for every $J \in \mathfrak{J}$, one has $T[X_J] = Y_J$ and $X = c_0(\mathfrak{J}, X_J), Y = c_0(\mathfrak{J}, Y_J)$.*

Proof If the set of families of infinite countable disjoint subsets J of I satisfying the conclusions of the preceding lemma is ordered by inclusion, every chain has an upper bound. Thus, Zorn's lemma guarantees the existence of a maximal family $\mathfrak{J}$. If $I \setminus \bigcup \mathfrak{J}$ is empty, we are done. If it is countable, Lemma 7.2.4 and the maximality of the family yield a contradiction. If it is finite then add this finite set of points to the $J \in \mathfrak{J}$ you like the most. □

At this point the only obstacle to deducing Theorem 7.2.3 from Lindenstrauss–Rosenthal is that we cannot guarantee that $\dim(c_0(J)/X_J) = \dim(c_0(J)/Y_J)$ for any $J \in \mathcal{J}$. The following purely combinatorial argument shows how to group the pieces in the right way:

Lemma 7.2.8 *Let A be an infinite index set and B, C be arbitrary abstract sets. Let $(B_\alpha)_{\alpha\in A}$ and $(C_\alpha)_{\alpha\in A}$ be families of pairwise disjoint subsets of B and C, respectively, each of them countable (finite or infinite) or empty. If $\bigcup_{\alpha\in A} B_\alpha$ and $\bigcup_{\alpha\in A} C_\alpha$ have the same infinite cardinal then A can be decomposed into a disjoint union of countable sets, $A = \bigcup_{\gamma\in\Gamma} A_\gamma$, such that for any $\gamma \in \Gamma$, either both $\bigcup_{\alpha\in A_\gamma} B_\alpha$ and $\bigcup_{\alpha\in A_\gamma} C_\alpha$ are infinite or both are empty.*

Proof Set $A_{00} = \{\alpha \in A : B_\alpha = \emptyset = C_\alpha\}$, $A_{10} = \{\alpha \in A : B_\alpha \neq \emptyset = C_\alpha\}$, $A_{01} = \{\alpha \in A : B_\alpha = \emptyset \neq C_\alpha\}$ and $A_{11} = \{\alpha \in A : B_\alpha \neq \emptyset \neq C_\alpha\}$. Assume A_{11} is infinite. Take a decomposition $A_{11} = \bigcup_{i\in I} A_i$ into infinite countable sets. If $|A_{10}| > |A_{11}|$, the hypothesis forces $|A_{10}| = |A_{01}|$; let h be a bijection between those sets. If $A_{10} = \bigcup_{j\in J} A'_j$ and $A_{00} = \bigcup_{k\in K} A''_k$ are decompositions into infinite countable sets then $A = \left(\bigcup_i A_i\right) \cup \left(\bigcup_j (A'_j \cup h[A'_j])\right) \cup \left(\bigcup_k A''_k\right)$ is the desired decomposition. If, however, there exist injections $f : A_{10} \longrightarrow A_{11}$ and $g : A_{01} \longrightarrow A_{11}$ then the decomposition is $\bigcup_i \left(A_i \cup f^{-1}[A_i] \cup g^{-1}[A_i]\right) \cup \left(\bigcup_k A''_k\right)$. If A_{11}, A_{10}, A_{01} are all finite then the decomposition is $A = \left(\bigcup_k A''_k\right) \cup (A_{11} \cup A_{01} \cup A_{10})$. If, say, A_{01} is finite and A_{10} is infinite, the same decomposition of A works since now A_{10} must be countable. Finally, if A_{10}, A_{01} are both infinite then $|A_{10}| = |A_{01}|$; fix a bijection $h : A_{10} \longrightarrow A_{01}$ and an injection $f : A_{11} \longrightarrow A_{10}$ and the decomposition is $A = \bigcup_j \left(A_j \cup h[A_j] \cup f^{-1}[A_j]\right) \cup \bigcup_k A_k$. □

We are ready to conclude the proof. Pick two subspaces Y, X of $c_0(I)$. If they have the same finite codimension, the result is trivial; otherwise:

Lemma 7.2.9 *If $\dim(c_0(I)/X) = \dim(c_0(I)/Y)$ is infinite then I admits a partition $\mathcal{S}$ into countable subsets such that*

(a) *$T[X_S] = Y_S$ for every $S \in \mathcal{S}$,*
(b) *both $c_0(S)/X_S$ and $c_0(S)/Y_S$ are either infinite-dimensional or* 0,
(c) *$X = c_0(\mathcal{S}, X_S)$ and $Y = c_0(\mathcal{S}, Y_S)$.*

Proof Let $\mathcal{J}$ be a decomposition of I as in Lemma 7.2.7 so that $|\mathcal{J}| = |I|$. Write $\mathcal{J}$ as $(J_\alpha)_{\alpha\in A}$, where A is an index set, necessarily with $|A| = |I|$. For every $\alpha \in A$, the inclusion $X_{J_\alpha} \longrightarrow c_0(J_\alpha)$ induces a separable quotient $c_0(J_\alpha)/X_{J_\alpha}$ which could be zero. For each $\alpha \in A$, consider sets B_α and C_α of cardinality equal to $\dim(c_0(J_\alpha)/X_{J_\alpha})$ and $\dim(c_0(J_\alpha)/Y_{J_\alpha})$, respectively, such that the families $(B_\alpha)_{\alpha\in A}$ and $(C_\alpha)_{\alpha\in A}$ are formed by pairwise disjoint elements. Note that we can take B_α to be a subset of vectors in $c_0(I)/X$ that 'fall' in $c_0(J_\alpha)/X_{J_\alpha}$ in the obvious way; if $c_0(J_\alpha)/X_{J_\alpha} = 0$, we take $B_\alpha = \emptyset$. Similar considerations apply to C_α. The hypothesis yields $\left|\bigcup_{\alpha\in A} B_\alpha\right| = \dim(c_0(I)/X) = \dim(c_0(I)/Y) = \left|\bigcup_{\alpha\in A} C_\alpha\right| \geq \aleph_0$. We can apply Lemma 7.2.8 to obtain a decomposition $A = \bigcup_{\gamma\in\Gamma} A_\gamma$ such that, if we set $S_\gamma = \bigcup_{\alpha\in A_\gamma} J_\alpha$ for each $\gamma\in\Gamma$ then $\mathcal{S} = (S_\gamma)_{\gamma\in\Gamma}$ is the required partition: indeed, (a) and (c) follow easily from the obvious fact that $X_{S_\gamma} = c_0(A_\gamma, X_{J_\alpha})$ and $Y_{S_\gamma} = c_0(A_\gamma, Y_{J_\alpha})$ for all $\gamma\in\Gamma$, while (b) follows from the choice of the A_γs since

$$\dim\left(c_0(S_\gamma)/X_{S_\gamma}\right) = \left|\bigcup_{\alpha\in A_\gamma} B_\alpha\right| = \left|\bigcup_{\alpha\in A_\gamma} C_\alpha\right| = \dim\left(c_0(S_\gamma)/Y_{S_\gamma}\right). \qquad \square$$

The Automorphic Space Problem: a Dichotomy

The general form of the Lindenstrauss–Rosenthal conjecture has been called the automorphic space problem: *must an automorphic space be isomorphic to either $c_0(I)$ or $\ell_2(I)$?* The dichotomy for UFO spaces is somehow the best clue we currently have that the conjecture could be true.

Theorem 7.2.10 *A UFO must be either an $\mathscr{L}_\infty$-space or a near-Hilbert space with the Maurey extension property.*

Proof The result is based on the dichotomy 1.4.7. Assume X is a UFO containing ℓ_1^n uniformly. The ultrapower $X_{\mathcal{U}}$ via a free ultrafilter on $\mathbb{N}$ is a UFO and contains ℓ_1. So, by Lemma 7.1.4 (b) and (d), its bidual $(X_{\mathcal{U}})^{**}$ is extensible and contains ℓ_1.

Claim An extensible space E containing ℓ_1 must be separably injective.

Proof of the claim Let B be a separable space and $A \subset B$ a closed subspace. Pick a projective presentation of B/A and form the pushout diagram

$$\begin{array}{ccccccccc}
0 & \longrightarrow & \kappa & \xrightarrow{\;k\;} & \ell_1 & \longrightarrow & B/A & \longrightarrow & 0\\
 & & {\scriptstyle\phi}\downarrow & & \downarrow & & \| & & \\
0 & \longrightarrow & A & \xrightarrow{\;\imath\;} & B & \longrightarrow & B/A & \longrightarrow & 0
\end{array}$$

Let $\jmath\colon \ell_1 \longrightarrow E$ be an embedding, and let $\tau\colon A \longrightarrow E$ be an operator. Let $\widehat{\tau\phi}\colon E \longrightarrow E$ be an extension of $\tau\phi$ through $\jmath k$. Since $\widehat{\tau\phi}\jmath k = \tau\phi$, by the universal property of the pushout there exists an operator $\nu\colon B \longrightarrow E$ making the diagram

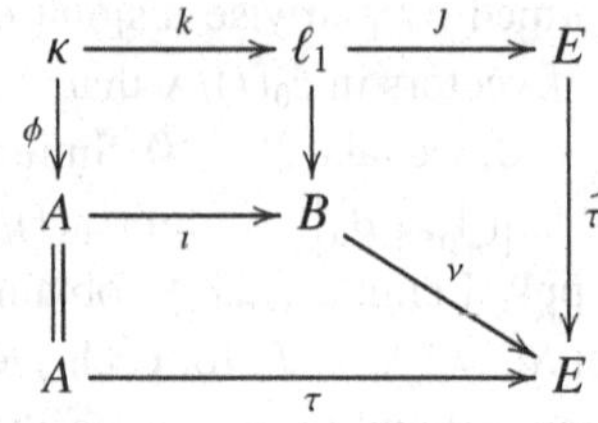

commute, which in particular means $\nu\imath = \tau$. □

This shows that the extensible space $(X_{\mathcal{U}})^{**}$ is separably injective, thus (Proposition 5.1.12) it must be an $\mathscr{L}_\infty$-space, as must $X_{\mathcal{U}}$ and thus X. This proves the first alternative of the dichotomy. If X contains ℓ_2^n uniformly complemented then the ultrapower $X_{\mathcal{U}}$ contains ℓ_2 complemented, as well as its bidual $(X_{\mathcal{U}})^{**}$ since the projection is a weakly compact operator. If $X \longrightarrow X_{\mathcal{U}}$ is the diagonal embedding and $Y \subset X$ is a subspace then, like the Mirror of Erised, the diagram

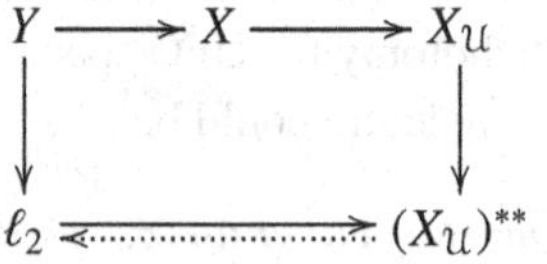

shows our desire that X has the Maurey extension property. In particular, copies of ℓ_2^n inside X are uniformly complemented. However, from the Maurey–Pisier theorem 1.4.10 we also get that X contains $\ell_{p(X)}^n$ and $\ell_{q(X)}^n$ uniformly. If $p(X) < 2$ then, by the result in [37] mentioned after Proposition 1.3.1, X contains a sequence of subspaces E_n uniformly isomorphic to ℓ_2^n, for which $\lim_n \lambda(E_n, X) = \infty$, yielding a contradiction. Thus, $p(X) = 2$. Assume $q(X) > 2$. If X has cotype $q(X)$ then [362, Proposition at 13.16] provides uniformly complemented copies of $\ell_{q(X)}^n$ in X which, in combination with the local argumentation behind Proposition 1.3.1, yields a contradiction with Proposition 7.1.4 (f). Otherwise, according to [359] (see also [363]), X contains $\ell_{q(X)}^n$ uniformly with $\lambda(\ell_{q(X)}^n, X) \leq C_\varepsilon n^\varepsilon$ for all $\varepsilon > 0$, and the usage of Proposition 7.1.4 (f) has to be subtler: in $\ell_q^n \oplus_\infty \ell_2^n$, pick the canonical bases $(e_i)_1^n$ of ℓ_q^n and $(h_i)_1^n$ of ℓ_2^n, and put $f_i = e_i + n^{(2-q)/2q} h_i$ for $i = 1, \ldots, n$. By the Hölder inequality, one has

$$\sum_{i=1}^n |a_i|^2 \leq \Big(\sum_{i=1}^n |a_i|^{2\frac{q}{2}}\Big)^{\frac{2}{q}} n^{1-\frac{2}{q}} \implies \Big\|\sum_{i=1}^n a_i h_i\Big\| \leq \Big\|\sum_{i=1}^n a_i e_i\Big\|\, n^{\frac{q-2}{2p}},$$

thus the subspace $[f_1, \ldots, f_n]$ is isometric to ℓ_q^n. We now need a particularisation of [412, Lemma 2A] in which all the coefficients are equal to obtain that any projection $P\colon \ell_q^n \oplus_\infty \ell_2^n \longrightarrow [f_1, \ldots, f_n]$ has norm at least $\frac{1}{2} n^{(q-2)/2q}$.

And, since $\ell_q^n \oplus_\infty \ell_2^n$ are uniformly complemented in ℓ_q^m for large m, one gets $\lambda(F_n, X) \geq 2^{-1} n^{(q-2)/2q}$ for certain uniformly isomorphic copies F_n of ℓ_q^n. □

Finitely Automorphic Spaces

As suggested in the introduction to this chapter, we consider the following property:

Definition 7.2.11 A quasi-Banach space is said to be finitely automorphic if every isomorphism between two finite-dimensional subspaces can be extended to an automorphism of the space.

To proceed with appropriate caution, let us consider two related properties:

Definition 7.2.12 A quasi-Banach space X is called transitive if, given $x, y \in X$, there is $T \in \mathfrak{L}(X)$ such that $Tx = y$. It is said to be strictly transitive if, for every linearly independent $x_1, \ldots, x_n \in X$ and arbitrary $y_1, \ldots, y_n \in X$, there is $T \in \mathfrak{L}(X)$ such that $Tx_i = y_i$ for all $1 \leq i \leq n$.

It is obvious that finitely automorphic spaces are transitive and 'almost' (when the second set of points is also linearly independent) strictly transitive. Banach spaces and, in general, quasi-Banach spaces with separating dual are finitely automorphic, as proved in the introduction. To go further, we need to think about $A(X) = \{T \in \mathfrak{L}(X) \colon TS = ST \text{ for every } S \in \mathfrak{L}(X)\}$.

Lemma 7.2.13 *Complex transitive quasi-Banach spaces are strictly transitive. A real transitive quasi-Banach space X such that $A(X) = \{\lambda \mathbf{1}_X \colon \lambda \in \mathbb{R}\}$ is strictly transitive.*

Proof If X is a complex quasi-Banach space then $A(X)$ is a field: if $S \in A(X)$ then $\ker S$ and $S[X]$ are invariant subspaces for all $T \in \mathfrak{L}(X)$. For a transitive X, this means that either $S = 0$ or S is an automorphism. Żelazko's extension of the Gelfand–Mazur theorem [283, Theorem 7.2] establishes that the only complex quasi-Banach algebra that is a field is $\mathbb{C}$.

Let us move to the real case. It will be sufficient to show that if $x_1, \ldots, x_n$ are linearly independent and $y \in X$ there exists $T \in \mathfrak{L}(X)$ such that $Tx_1 = \cdots = Tx_{n-1} = 0$ and $Tx_n = y$ which, since X is transitive, can be obtained whenever $Tx_n \neq 0$. The proof of this statement proceeds by induction on $n \geq 2$. For $n = 2$, pick independent x_1, x_2. If no such T exists then $Tx_1 = 0 \implies Tx_2 = 0$ for all $T \in \mathfrak{L}(X)$. Therefore, if $S, T \in \mathfrak{L}(X)$ then $Tx_2 - Sx_2 = Tx_1 - Sx_1$, and one can define a linear (not necessarily continuous) map $Dx = Tx_2$ for $Tx_1 = x$. This map D commutes with all operators: pick $x \in X$ and, by transitivity, S such that $Sx_1 = x$; since $Dx_1 = x_2$, one has $DTx = DTSx_1 = TSx_2 = TDSx_1 = TDx$. The map D is also continuous: for each x the evaluation map $\delta_x : \mathfrak{L}(X) \longrightarrow X$ is surjective, hence open, and thus there is $C > 0$ such that for each $y \in X$ there exists T such that $T(x) = y$ with $\|T\| \leq C\|y\|$. Pick a sequence (y_k) converging to zero in X and then a sequence (T_k) such that $T_k(x_1) = y_k$ with $\lim \|T_k\| = 0$. One has $\lim Dy_k = \lim DT_k x_1 = \lim T_k x_2 = 0$. Since D is continuous, the assumption for the real case yields that $D = d\mathbf{1}_X$ for some $d \in \mathbb{R}$, and the linear independence of x_1, x_2 is compromised. Now the induction step. Assume the statement is true for families up to n points. Either the statement also holds for families of $n + 1$ points or there exist $x_1, \ldots, x_n, x_{n+1} \in X$ such that $Tx_1 = \cdots = Tx_n = 0$ implies $Tx_{n+1} = 0$ for all $T \in \mathfrak{L}(X)$. Let T be such that $Tx_1 = \cdots = Tx_{n-1} = 0$ and Tx_n, Tx_{n+1} are linearly independent. There is an S such that $STx_n = 0$ (as well as $STx_j = 0$ for $j \leq n - 1$) and $STx_{n+1} \neq 0$, which is a contradiction. So Tx_n and Tx_{n+1} cannot be linearly independent. So here it goes, in a rather lorem ipsum form, a disquisition leading to the conclusion that there are scalars a, b, not simultaneously zero, such that $aTx_n + bTx_{n+1} = 0$ *for all* $T \in \mathfrak{L}(X)$. Let us show off one case: each T has an a_T such that $Tx_{n+1} = a_T x_n$, and the goal is to show that all the a_T coincide. If there exist operators S, T whose values at x_n are linearly independent, then $a_S S x_n + a_T T x_n = (S + T)x_{n+1} = a_{S+T}(S + T)x_n$ and thus $a_{T+S} = a_T + a_S$, which leads to $2a_T = a_{T+S} + a_{T-S}$, which leads to $2a_T = a_{2T}$, which cannot be since $a_{2T} = a_T$ – and the same for the other possibilities. The uniqueness of a, b means that $Tx_1 = \cdots = Tx_{n-1} = 0$ implies $T(ax_n + bx_{n+1}) = 0$, contradicting the induction hypothesis when applied to $x_1, \ldots, x_{n-1}, ax_n + bx_{n+1}$. □

Proposition 7.2.14 *If X is strictly transitive and isomorphic to its square then it is finitely automorphic.*

Proof Let $F \subset X$ be a finite-dimensional subspace. Let us show first that there is a projection P on X such that $\dim P[F] = \dim F$: pick a projection $P \colon X \longrightarrow X$ such that $P[X] \simeq X$, $(\mathbf{1}_X - P)[X] \simeq X$ and $\dim P[F]$ is maximal. Since $(\mathbf{1}_X - P)[X] \simeq X \times X$, if $\pi \colon X \times X \longrightarrow X$ is the canonical projection onto one factor space, $Q = \pi(\mathbf{1}_X - P)$ is a projection on X such that $PQ = QP = 0$ and $Q[X] \simeq X$ and $(\mathbf{1}_X - P - Q)[X] \simeq X$ since $(\mathbf{1}_X - P - Q) = \mathbf{1}_X - ((\mathbf{1}_X - P)\pi)(\mathbf{1}_X - P)$. Thus $(P + Q)[X] \simeq (\mathbf{1}_X - P - Q)[X] \simeq X$ and one must have $\dim(P + Q)[F] \leq \dim P[F]$. Since $Px = (P + Q)(x - Qx)$, it turns out that $P[F] \subset (P + Q)[F]$, hence $\dim(P + Q)[F] = \dim P[F]$. This implies that P is injective on $(P + Q)[F]$ since $P(P + Q) = P$. Therefore, $Pf = 0 \implies (P + Q)f = 0$ for $f \in F$. We repeat the argument with $(\mathbf{1}_X - Q)$, now *andante*: $(\mathbf{1}_X - Q)[X] \simeq X$, thus $\dim(\mathbf{1}_X - Q)[F] \leq \dim P[F]$, and since $Px = (\mathbf{1}_X - Q)(Px)$ we get $P[F] \subset (\mathbf{1}_X - Q)[F]$; still $P(\mathbf{1}_X - Q) = P$, and thus P is injective on $(\mathbf{1}_X - Q)[F]$. In conclusion, $Pf = 0 \implies (\mathbf{1}_X - Q)f = 0$. Both things combined yield $Pf = 0 \implies f = 0$ when $f \in F$, as we expected. Now we are ready: let

F, G be isomorphic finite-dimensional subspaces of X. Do as before and obtain projections P_F, P_G such that $\dim P_F[F] = \dim F = \dim G = \dim P_G[G]$. Let $\alpha\colon P_F[F] \longrightarrow P_G[G]$ be an isomorphism. Things have been boring so far, but they now improve: in the diagram

$$\begin{array}{ccccccccc}
0 & \longrightarrow & F & \longrightarrow & X & \longrightarrow & X/F & \longrightarrow & 0 \\
 & & \downarrow{\scriptstyle P_F} & & \downarrow{\scriptstyle (P_F, 1_X - P_F)} & & & & \\
0 & \longrightarrow & P_F[F] & \longrightarrow & X \times X & \longrightarrow & X/P_F[F] \times X & \longrightarrow & 0 \\
 & & \downarrow{\scriptstyle \alpha} & & & & & & \\
0 & \longrightarrow & P_G[G] & \longrightarrow & X \times X & \longrightarrow & X/P_G[G] \times X & \longrightarrow & 0 \\
 & & {\scriptstyle P_G}\uparrow & & \uparrow{\scriptstyle (P_G, 1_X - P_G)} & & & & \\
0 & \longrightarrow & G & \longrightarrow & X & \longrightarrow & X/G & \longrightarrow & 0
\end{array} \qquad (7.1)$$

the two upper and two lower sequences are isomorphic, as we have already proved, while strict transitivity makes the sequences

$$\begin{array}{ccccccccc}
0 & \longrightarrow & P_F[F] & \longrightarrow & X & \longrightarrow & X/P_F[F] & \longrightarrow & 0 \\
 & & {\scriptstyle \alpha}\downarrow & & & & & & \\
0 & \longrightarrow & P_G[G] & \longrightarrow & X & \longrightarrow & X/P_G[G] & \longrightarrow & 0
\end{array}$$

semi-equivalent. Thus the diagonal principles make the two central sequences in Diagram (7.1) isomorphic. □

We are ready to consider the case of L_p for $0 < p < 1$.

Lemma 7.2.15 *L_p is strictly transitive for* $0 < p < 1$.

Proof We first show that L_p is transitive. Assume f is positive, is normalised and has full support. Define $\phi(t) = \int_0^t f^p$. Then ϕ is an absolutely continuous homeomorphism of $[0, 1]$. Define $T \in \mathfrak{L}(L_p)$ by $Tg(t) = f(t)g(\phi(t))$. Then T is a *surjective isometry* and $T1_{[0,1]} = f$. Moral: if f, g have full support in L_p then there is an automorphism $T \in \mathfrak{L}(L_p)$ such that $g = Tf$. Now assume $f, g \neq 0$ are arbitrary, and consider the following composition:

$$L_p \xrightarrow{\text{restriction}} L_p(\operatorname{supp} f) \xrightarrow{\text{isomorphism}} L_p(\operatorname{supp} g) \xrightarrow{\text{extension}} L_p$$

where the first arrow takes h to its restriction, the second arrow is an isomorphism that exists by the 'full support' case and the third arrow extends a function by setting 0 out of $\operatorname{supp} g$. We conclude that complex L_p is strictly transitive. For real L_p, one still must show that $A(L_p) = \mathbb{R}$: for each $t \in (0, 1)$ and each $a \in L_\infty$, we consider the 'dilation' operator $D_t(f)(x) = f(tx)$ and the multiplication operator $M_a \in \mathfrak{L}(L_p)$ given by $M_a(f) = af$. We have $\|M_a\| = \|a\|_\infty$ and $\|D_t\| = t^{-1/p}$. If $S \in A(L_p)$ and $S(1) = \varphi$ then for each $f \in L_\infty$, we have $Sf = SM_f(1) = M_f S(1) = f\varphi$. Thus, $S = M_\varphi$. For $0 < t < 1$, we have $\varphi = S(1) = SD_t(1) = D_t S(1) = D_t(\varphi)$, hence $\varphi(x) = \varphi(tx)$ almost everywhere on x and for all $0 < t < 1$. But this also means that $\varphi(x) = \varphi(tx)$ almost everywhere on t and for all $0 < x < 1$ by Fubini's theorem, and thus φ is constant (almost everywhere). □

Since $L_p \simeq L_p \times L_p$, Lemma 7.2.15, Propositions 7.2.14 and 6.2.10 yield:

Proposition 7.2.16 *L_p and G_p are finitely automorphic for $p \in (0, 1)$.*

A space X with non-trivial dual X^* that does not separate points cannot be transitive: non-trivial minimal extensions of Banach spaces and the Lorentz function space $L_{1,\infty}$ are such spaces. There are also Orlicz function spaces with trivial dual that are not finitely automorphic [248, Theorem 5.4].

7.3 Isometries: Universal Disposition

Broadly speaking, automorphic spaces have the property that every isomorphism between subspaces can be extended to an automorphism of the space. But there are only two examples available: $c_0(I)$ and Hilbert spaces. It is then to be expected that if one attempts to consider spaces in which every isometry between subspaces can be extended to an isometry of the space then one will find that, apart from Hilbert spaces, there are none in sight. Thus, just relax.

Spaces of Universal Disposition

We have already encountered spaces of (almost) universal disposition treated with their due respect in Chapter 6 and spaces of separable universal disposition, which were obtained in 2.13.1. In general, given a class $\mathscr{A}$ of Banach spaces, a Banach space is said to be of universal disposition for $\mathscr{A}$ if, given $A, B \in \mathscr{A}$ and isometries $u: A \longrightarrow U$ and $v: A \longrightarrow B$, there is an isometry $w: B \longrightarrow U$ such that $u = wv$. Of special importance are the choices $\mathscr{A} = \mathscr{F}$ (finite-dimensional spaces), which lead to spaces of UD and $\mathscr{A} = \mathscr{S}$ (separable spaces), which lead to the spaces of SUD. How to obtain spaces of universal disposition for other classes? The Device (Chapter 2, Section 2.13) is especially well suited to the task. Let us begin with a sharpened definition.

Definition 7.3.1 Let θ and ζ be infinite cardinals, with $\theta \leq \zeta$. The space U is said to be of (θ, ζ)-universal disposition when, given isometries $u\colon A \longrightarrow U$ and $v\colon A \longrightarrow B$, where $\dim(A) < \theta, \dim(B) < \zeta$, there is an isometry $w\colon B \longrightarrow E$ such that $u = wv$.

For each Banach space X and cardinals $\theta \leq \zeta$ we now construct a special isometric embedding of X into a space of (θ, ζ)-universal disposition. Given a cardinal γ, let $\mathscr{S}_\gamma$ be a fixed *set* of Banach spaces containing an isometric copy of each Banach space of dimension less than γ and minimal with respect to that property. Let us put the Device to work with the following instructions:

- Take an ordinal μ with cofinality greater than any $\alpha < \theta$, say $\mu = 2^\theta$.
- Work in the category of Banach spaces and contractions. Construct an inductive system of Banach spaces $(X_\alpha)_{0 \leq \alpha \leq \mu}$ beginning with $X_0 = X$ as usual: assuming X_α is already defined for $\alpha < \beta$, take X_β to be the completion of $\bigcup_{\alpha<\beta} X_\alpha$ if β is a limit ordinal. Otherwise, write $\beta = \alpha + 1$ and take $\mathfrak{L}_\alpha$ to be the set of all X_α-valued isometries defined on elements of $\mathscr{S}_\theta$ and $\mathfrak{J}_\alpha$ to be the family of all isometries acting from spaces in $\mathscr{S}_\theta$ into spaces in $\mathscr{S}_\zeta$ and construct X_β as the pushout of $\mathfrak{J}_\alpha$ and $\mathfrak{L}_\alpha$.

Cooked with taste, the recipe will produce a space $\mathsf{F}^{\theta,\zeta}(X) = X_\mu$. We have:

Proposition 7.3.2 *$\mathsf{F}^{\theta,\zeta}(X)$ is a space of (θ, ζ)-universal disposition.*

Proof Let $u\colon A \longrightarrow X_\mu$ and $v\colon A \longrightarrow B$ be isometries, where $A \in \mathscr{S}_\theta$ and $B \in \mathscr{S}_\zeta$. Since the cofinality of μ is greater than any $\alpha < \theta$, the image of u must fall in some X_α for $\alpha < \mu$ and thus $u \in \mathfrak{L}_\alpha, v \in \mathfrak{J}_\alpha$. Therefore, both u and v are part of the amalgam of operators with respect to which we perform the pushout to obtain $X_{\alpha+1} = \mathrm{PO}$ so that there is an isometry $w\colon B \longrightarrow X_{\alpha+1} = \mathrm{PO}$ extending u. Composing this w with the inclusion $X_{\alpha+1} \longrightarrow X_\mu$ we obtain an isometry $B \longrightarrow X_\mu$ 'extending' u. □

Injectivity-like notions appearing in the literature (separable, universal separable, $\aleph$-injectivity, see [22]), can be unified as follows:

Definition 7.3.3 Let θ, ζ be infinite cardinals with $\theta \leq \zeta$. A Banach space E is said to be (θ, ζ)-injective if, for every Banach space B with $\dim(B) < \zeta$ and every subspace $A \subset B$ such that $\dim(A) < \theta$, every operator $\tau\colon A \longrightarrow E$ can be extended to an operator $T\colon B \longrightarrow E$. When, furthermore, there exists for every τ some extension T such that $\|T\| \leq \lambda\|\tau\|$, we say that E is (θ, ζ, λ)-injective.

Thus, $\aleph$-injectivity corresponds to $(\aleph, \aleph)$-injectivity, while universal $\aleph$-injectivity corresponds to $(\aleph, 2^\aleph)$-injectivity, since every Banach space with dimension $\aleph$ embeds into an injective space with dimension $2^\aleph$. In particular, separable injectivity is $(\aleph_1, \aleph_1)$-injectivity and universal separable injectivity is $(\aleph_1, \mathfrak{c})$-injectivity. From our point of view, the crucial fact is:

Proposition 7.3.4 *A Banach space is $(\theta, \zeta, 1)$-injective if and only if it is isometric to a 1-complemented subspace of a space of (θ, ζ)-universal disposition.*

Proof An obvious modification of Lemma 7.1.1(a) shows that Banach spaces of (θ, ζ)-universal disposition are $(\theta, \zeta, 1)$-injective, and the same is true for their 1-complemented subspaces. Conversely, let X be a $(\theta, \zeta, 1)$-injective space, and consider the isometry $X \longrightarrow \mathsf{F}^{\theta,\zeta}(X)$ as in Proposition 7.3.2. We shall show that X is 1-complemented in $\mathsf{F}^{\theta,\zeta}(X)$. Since X is $(\theta, \zeta, 1)$-injective, X-valued operators with domains in $\mathscr{S}_\theta$ extend with the same norm to superspaces in $\mathscr{S}_\zeta$. This occurs in particular for the operators forming $\bigoplus \mathfrak{L}_0$; see Diagram 2.42 in Section 2.13. Hence $\bigoplus \mathfrak{L}_0 \colon \ell_1(I_0, \operatorname{dom}(u)) \longrightarrow X_0$ has a norm-preserving extension through the isometry $\prod \mathfrak{J}_0 \colon \ell_1(I_0, \operatorname{dom}(u) = \operatorname{dom}(v)) \longrightarrow \ell_1(I_0, \operatorname{cod}(v))$. In other words, $X = X_0$ is 1-complemented in $\mathrm{PO} = X_1$. The argument can be transfinitely iterated. □

Universal disposition (i.e. $(\aleph_0, \aleph_0)$-universal disposition) does not imply separable injectivity (i.e. $(\aleph_1, \aleph_1)$-injectivity), as will be shown in 7.3.7. In its highly popular recognizable form, Proposition 7.3.4 reads:

Corollary 7.3.5 *A Banach space is 1-separably injective if and only if it is a 1-complemented subspace of a space of separable universal disposition.*

Time to have fun and face the unknown. Let us simplify the notation by writing $\mathsf{F}_0(X) = \mathsf{F}^{\theta,\zeta}(X)$ when $\theta = \zeta = \aleph_0$ (i.e. when only finite-dimensional spaces are involved) and writing $\mathsf{F}_1(X) = \mathsf{F}^{\theta,\zeta}(X)$ when $\theta = \zeta = \aleph_1$ (i.e. when only separable spaces are involved). In particular, we set $\mathsf{F}_0 = \mathsf{F}_0(\mathbb{K})$ and $\mathsf{F}_1 = \mathsf{F}_1(\mathbb{K})$. In all cases the ordinal μ that decides the 'length' of the iteration is to be fixed at ω_1. Some aspects of the classification of spaces of UD and SUD having dimension at most $\mathfrak{c}$ are known:

- [CH] All SUD spaces having dimension at most $\mathfrak{c}$ are isometric [22].
- $[\aleph_1 < \mathfrak{c}]$ and assuming that no Banach space of dimension $\mathfrak{c}$ is universal for all Banach spaces with dimension $\mathfrak{c}$, there are $\mathfrak{c}^+$ mutually non-isomorphic spaces of SUD having dimension $\mathfrak{c}$ [22, Proposition 3.18].
- $\mathsf{F}_0 \approx \mathsf{F}_0(X)$ for all separable X [22, Theorem 3.23 (2)].

One, moreover, has:

7.3.6 *The space F_1 is of SUD, hence 1-separably injective, and, consequently, it does not contain c_0 complemented.*

Proof One can find the 'consequently' part of the result among the exhaustive study of the properties of F_1 in [22, Chapter 3]: 1-separably injective spaces are Grothendieck spaces, and thus they cannot contain c_0 complemented. Note that 'our' F_1 is denoted by $\mathscr{S}^{\omega_1}$ there. □

7.3.7 *The space* F_0 *is not separably injective. Under* CH, *there is a space of UD in which all copies of* c_0 *are complemented.*

Proof We will first prove that if $\mathcal{L}$ is a Lindenstrauss space that is not separably injective then $\mathsf{F}_0(\mathcal{L})$ is neither separably injective. We adhere to the notation of Proposition 7.3.2 and show first that $\mathcal{L}$ is 1^+-complemented in all the intermediate pushout spaces $\mathcal{L}_\alpha$, with $\alpha < \omega_1$ appearing in the construction of $\mathsf{F}_0(\mathcal{L})$.

We already know that Lindenstrauss spaces are locally 1^+-injective, thus $\mathcal{L}$-valued operators defined on finite-dimensional spaces admit 1^+-extensions to finite-dimensional superspaces. Fix $\varepsilon > 0$ and a limit ordinal $\beta < \omega_1$. Pick a family $(\varepsilon_j)_{j<\beta}$ with $\varepsilon_j > 0$ such that $\prod_{j<\beta}(1+\varepsilon_j) < 1+\varepsilon$. All the elements in the composition $\mathbf{1}_{\mathcal{L}}(\oplus\mathfrak{L}_1)$ admit $(1+\varepsilon_1)$-extensions; thus, $\mathbf{1}_{\mathcal{L}}(\oplus\mathfrak{L}_1)$ extends to an operator $\ell_1(I_1, \operatorname{cod} u) \longrightarrow \mathcal{L}$ with norm at most $1+\varepsilon_1$, hence to an operator $\mathcal{L}_1 = \mathrm{PO} \longrightarrow \mathcal{L}$ with norm at most $1+\varepsilon_1$. This means that $\mathcal{L}$ is $(1+\varepsilon_1)$-complemented in $\mathcal{L}_1$ via some projection P_1. Repeat the argument with $P_j(\oplus\mathfrak{L}_1)$ to obtain a $(1+\varepsilon_{j+1})$-extension P_{j+1} of P_j to $\mathcal{L}_{j+1}$ for $j \geq 1$. If β is a successor ordinal, the argument is contained in the preceding explanation. Now try to believe that $\mathsf{F}_0(\mathcal{L})$ is separable injective. Let A be a subspace of a separable Banach space B, and let $\tau\colon A \longrightarrow \mathcal{L}$ be an operator. The composition of τ with the embedding $\mathcal{L} \longrightarrow \mathsf{F}_0(\mathcal{L})$ can be extended to an operator $T\colon B \longrightarrow \mathsf{F}_0(\mathcal{L})$ by separable injectivity. The uncountable cofinality of ω_1 means that T actually has its range contained in $\mathcal{L}_\alpha$ for some $\alpha < \omega_1$, thus composing with a projection $\mathcal{L}_\xi \longrightarrow \mathcal{L}$ provides an extension $B \longrightarrow \mathcal{L}$ of τ. In other words, $\mathcal{L}$ is separably injective, which it is not.

To prove the second assertion, we will construct a variant of F_0 which we shall imaginatively denote by F_0'. As usual, $\mathscr{F}$ denotes any set of finite-dimensional Banach spaces containing exactly one isometric copy of each finite-dimensional space. Let $\mathfrak{J}$ denote the set of isometries with domain and codomain in $\mathscr{F}$. Given a Banach space X, we denote by $\mathfrak{L}^X$ the set of all contractive operators $u\colon F \longrightarrow X$ with $F \in \mathscr{F}$. It is clear that $|\mathscr{F}| = |\mathfrak{J}| = \mathfrak{c}$, and the same is true for $\mathfrak{L}^X$ if X is separable. If $\{s_\beta\colon \beta < \omega_1\}$ is an ω_1-enumeration of the set S, we will write $S_\alpha = \{s_\beta\colon \beta \leq \alpha\}$ for $\alpha < \omega_1$. Assuming CH, we can fix an ω_1-enumeration $\mathfrak{J} = \{v_\beta : 0 \leq \beta < \omega_1\}$ of $\mathfrak{J}$, and the meaning of the sets $\mathfrak{J}_\alpha$ is clear. We first construct F_0' and then prove that it has the required properties. We will obtain F_0' as the union of an ω_1-sequence of separable spaces $(P_\alpha)_{\alpha<\omega_1}$.

Each P_α depends on the 'preceding' P_β and on certain enumerations of the sets of contractive operators from the spaces of $\mathscr{F}$ to P_β.

We take $P_0 = 0$. The set $\mathfrak{L}^{P_0}$ contains only the zero operator. Let $\beta < \omega_1$ fixed, and assume that for every $0 \leq \alpha < \beta$, the spaces P_α have already been obtained and each set $\mathfrak{L}^{P_\alpha}$ has been ω_1-enumerated. If β is a limit ordinal, we take P_β to be the completion of $\bigcup_{\alpha<\beta} P_\alpha$ and then fix an ω_1-enumeration of $\mathfrak{L}^{P_\beta}$. Otherwise, $\beta = \alpha + 1$. Consider the set

$$I_\alpha = \left\{ (u, v) \in \mathfrak{L}^{P_\alpha} \times \mathfrak{J} : v \in \mathfrak{J}_\alpha, u \in \bigcup_{\mu \leq \alpha} \mathfrak{L}_\alpha^{P_\mu}, \operatorname{dom}(u) = \operatorname{dom}(v) \right\}.$$

For $\nu < \eta$, operators in $\mathfrak{L}^{P_\nu}$ are understood as operators in $\mathfrak{L}^{P_\eta}$ after plain composition with the canonical isometry $P_\nu \longrightarrow P_\eta$. We consider now the sums $\ell_1(I_\alpha, \operatorname{dom}(u)) = \ell_1(I_\alpha, \operatorname{dom}(v))$ and $\ell_1(I_\alpha, \operatorname{cod}(v))$, the isometry $V_\alpha\colon \ell_1(I_\alpha, \operatorname{dom}(v)) \longrightarrow \ell_1(I_\alpha, \operatorname{cod}(v))$ given by $V_\alpha((x_{(u,v)})_{(u,v)\in I_\alpha}) = (v(x_{(u,v)})_{(u,v)\in I_\alpha})$ and the contraction $U_\alpha\colon \ell_1(I_\alpha, \operatorname{dom}(u)) \longrightarrow P_\alpha$ given by $U_\alpha((x_{(u,v)})_{(u,v)\in I_\alpha}) = (u(x_{(u,v)})_{(u,v)\in I_\alpha})$. Form the pushout

$$\begin{array}{ccc} \ell_1(I_\alpha, \operatorname{dom}(u)) = \ell_1(I_\alpha, \operatorname{dom}(v)) & \xrightarrow{V_\alpha} & \ell_1(I_\alpha, \operatorname{cod}(v)) \\ \Big\downarrow{\scriptstyle U_\alpha} & & \Big\downarrow \\ P_\alpha & \longrightarrow & \mathrm{PO} = P_{\alpha+1} \end{array}$$

and fix an enumeration of $\mathfrak{L}^{P_{\alpha+1}}$. This finishes the induction step.

The resulting $\mathsf{F}_0' = \bigcup_{\alpha<\omega_1} P_\alpha$ is a Banach space of UD: to prove it, let $v\colon F \longrightarrow G$ be an isometry between finite-dimensional spaces, and let $u\colon F \longrightarrow \mathsf{F}_0'$ be an isometry. Without loss of generality, we may assume $v \in \mathfrak{J}$ so that v is defined on some $F \in \mathscr{F}$, and thus we have $v = v_\alpha$ for some $\alpha < \omega_1$. The subspace $u[F]$ is necessarily contained in some P_μ, and consequently, $u \in \mathfrak{L}^{P_\mu}$. Let β be the ordinal corresponding to u in the enumeration of $\mathfrak{L}^{P_\mu}$. Letting $\gamma = \max\{\alpha, \mu, \beta\}$, it is clear that the pair (u, v) belongs to I_γ, and consequently, there is an isometry $\overline{u}\colon G \longrightarrow P_{\gamma+1}$ that 'extends' $u\colon F \longrightarrow u[F] \longrightarrow P_\mu \longrightarrow P_\gamma$, which easily yields the desired extension $G \longrightarrow \mathsf{F}_0'$.

Proving now that every copy of c_0 inside F_0' is complemented is just skimming the surface while surfing through life: let $\jmath\colon c_0 \longrightarrow \mathsf{F}_0'$ be an embedding and assume without loss of generality that $\|\jmath(x)\| \leq \|x\| \leq \lambda\|\jmath(x)\|$ for all $x \in c_0$ and a certain $\lambda \geq 1$. Since ω_1 has uncountable cofinality, there must be some $\alpha < \omega_1$ such that $\jmath\,[c_0] \subset P_\alpha$, which is *now* separable. By Sobczyk's theorem, $\jmath$ has a left-inverse $P\colon P_\alpha \longrightarrow c_0$ with $\|P\| \leq 2\lambda$. Since c_0-valued finite-rank operators can be extended with the same norm everywhere

(Lemma 8.0.1), the composition PU_α admits an equal norm extension to $\ell_1(I_\alpha, \mathrm{cod}(v))$, hence to $P_{\alpha+1}$ by the properties of the pushout. This process, after transfinite iteration, can only conclude with a 2λ-projection $\mathsf{F}_0' \longrightarrow c_0$ along $\jmath$. □

We do not know whether F_0' is a subspace of F_0 or even isometric to F_0. The same argumentation above shows that F_0' is not separably injective. We do not know if there is a single X such that $\mathsf{F}_0(X)$ is separably injective. Such an X should be non-separable and separably injective. Alas! this is exactly the X we will consider next. So we pause for a necessary *digression on polyhedral spaces*: rush to Section 10.2 if you need a reminder on anything about polyhedral spaces, including their definition. You will not be disappointed. The only thing we need now to know about the fine-tuning of polyhedrality is that finite-rank operators with values in *certain* Lindenstrauss polyhedral spaces admit 1-extensions. The standard name for the 'certain' required additional property is $(*)$, and [179] is a good place to learn about it. The spaces $C(\Delta_{\mathcal{M}})$ as well as preduals of ℓ_1 admit polyhedral renormings with property $(*)$. None of these facts are simple, by a long way: the first requires a combination of [318] and [89] and the second is a consequence of the deep analysis in [179], while the third is due to Fonf [177]. *End of the digression.* Why do we care about property $(*)$ now? Because asking $\mathcal{L}$ to be locally 1^+-injective is enough to get it complemented in the $\mathcal{L}_\alpha$ spaces that appear in the device for $\alpha < \omega_1$, as we just did. To get $\mathcal{L}$ complemented in $\mathcal{L}_{\omega_1}$, one needs that finite-dimensional $\mathcal{L}$-valued operators admit 1-extensions, which is what property $(*)$ guarantees. Thus, since $\mathsf{F}_0 = \mathsf{F}_0(X)$ for any separable X, F_0 contains complemented copies of all separable polyhedral Lindenstrauss spaces with property $(*)$, which include all preduals of ℓ_1. We return to the classification problem we were in:

7.3.8 *If $C(*)$ denotes a polyhedral renorming of $C(\Delta_{\mathcal{M}})$ with property $(*)$ then $\mathsf{F}_0(C(*))$ contains both complemented and uncomplemented copies of c_0.*

Proof Since $C(*) \simeq C(\Delta_{\mathcal{M}})$ contains an uncomplemented copy of c_0, the same is true of $\mathsf{F}_0(C(*))$. The existence of complemented copies of c_0 follows simply by observing that $C(*)$ is 1-complemented in $\mathsf{F}_0(C(*))$. □

In particular, $\mathsf{F}_0(C(*))$ cannot be isomorphic to either F_0' or F_1, although it might be to F_0. It has been conjectured [22, Section 6.6. Problem 16] that there is a continuum of mutually non-isomorphic spaces of universal disposition having dimension $\mathfrak{c}$.

7.4 Positions in Banach Spaces

Definition 7.4.1 Let Y, X be Banach spaces. A position of Y in X is a subspace $Y_0 \subset X$ of X isomorphic to Y such that X/Y_0 is infinite-dimensional. Two positions Y_0, Y_1 of Y in X are said to be equivalent if there is an automorphism β of X such that $\beta[Y_0] = Y_1$.

That two positions Y_0, Y_1 of Y in X are equivalent does not mean only that the subspaces Y_0, Y_1 are isomorphic (as they actually are!) but that the sequences

$$\begin{array}{ccccccccc} 0 & \longrightarrow & Y_0 & \longrightarrow & X & \longrightarrow & X/Y_0 & \longrightarrow & 0\\ & & \downarrow{\scriptstyle \alpha} & & \downarrow{\scriptstyle \beta} & & \downarrow{\scriptstyle \gamma} & & \\ 0 & \longrightarrow & Y_1 & \longrightarrow & X & \longrightarrow & X/Y_1 & \longrightarrow & 0 \end{array}$$

are isomorphic. This definition corresponds to Kalton's notion [274] of 'equivalent embeddings'. Our notion for the equivalence of embeddings is:

Definition 7.4.2 Two embeddings $\imath\colon Y \longrightarrow X$ and $\jmath\colon Y \longrightarrow X$ are said to be equivalent if there is an automorphism β of X such that $\beta\imath = \jmath$.

The translation into a diagram is easy: $\imath$ and $\jmath$ are equivalent if and only if there exist isomorphisms β, γ making the following diagram commute:

$$\begin{array}{ccccccccc} 0 & \longrightarrow & Y & \xrightarrow{\ \imath\ } & X & \longrightarrow & X/\imath[Y] & \longrightarrow & 0\\ & & \| & & \downarrow{\scriptstyle \beta} & & \downarrow{\scriptstyle \gamma} & & \\ 0 & \longrightarrow & Y & \xrightarrow{\ \jmath\ } & X & \longrightarrow & X/\jmath[Y] & \longrightarrow & 0 \end{array}$$

This definition corresponds to Kalton's notion of 'strongly equivalent' embeddings [274]. It is clear that equivalent embeddings $\imath, \jmath$ yield equivalent positions $\imath[Y], \jmath[Y]$ of Y, while the converse does not hold, as the example following Lemma 2.1.7 shows. The keystone in this vault is a simple observation:

7.4.3 *Complemented and uncomplemented positions are not equivalent.*

In how many different positions does a Banach space X admit another Y?

Pełczyński's Problem Regarding Positions in ℓ_1

While equivalence is the correct notion of equality for exact sequences (hence for embeddings) from the homological point of view, the notion of equivalent positions is somehow more accurate from the point of view of Banach spaces. Pełczyński formulated what is likely the nicest and deepest problem in the area: *In how many positions can one encounter ℓ_1 inside ℓ_1?* Clearly, there is a complemented position and, after Bourgain's example, a non-complemented one. Pełczyński's cunning question was, *Are there other 'different' copies of ℓ_1 inside ℓ_1?* The definition says that equivalent positions yield isomorphic quotients, while Proposition 2.7.3 goes the other way around: isomorphic quotients yield equivalent positions for the kernels. Therefore:

Lemma 7.4.4 *Two subspaces A, B of ℓ_1 are in equivalent positions if and only if ℓ_1/A and ℓ_1/B are isomorphic.*

How many non-isomorphic quotients of ℓ_1 by isomorphs of ℓ_1 are there? A reasonable approach is to look for properties of the quotient space that could distinguish two copies of ℓ_1 inside ℓ_1. The two known copies yield quotients that are both duals of subspaces of c_0, so they are Schur spaces. This is not an accident:

Proposition 7.4.5 *Every quotient of ℓ_1 by a subspace isomorphic to ℓ_1 has the Schur property.*

Proof Let $\jmath\colon \ell_1 \longrightarrow \ell_1$ be an embedding, and put $X = \ell_1/\jmath[\ell_1]$. The result is a combination of two facts to be proved in Chapter 8. The first fact is that every operator $\ell_1 \longrightarrow C(\Delta)$ can be extended to any separable superspace (Theorem 8.5.4). Thus, the lower sequence in the pushout diagram

$$\begin{array}{ccccccccc}
0 & \longrightarrow & \ell_1 & \xrightarrow{\jmath} & \ell_1 & \longrightarrow & X & \longrightarrow & 0 \\
 & & \downarrow & & \downarrow & & \| & & \\
0 & \longrightarrow & C(\Delta) & \longrightarrow & \cdot & \longrightarrow & X & \longrightarrow & 0
\end{array}$$

splits and then $\operatorname{Ext}_{\mathbf{B}}(X, C(\Delta)) = 0$. The second fact is Proposition 8.6.4, asserting that if $\operatorname{Ext}_{\mathbf{B}}(X, C(\Delta)) = 0$ then X must have the Schur property. □

Let us keep moving on along this line: Kalton asks in [271, Problem 2] whether such an X is always the dual of a subspace of c_0 and remarks that 'this problem asks for a global construction'. Observe that whilst it is not hard to find subspaces of ℓ_1 having two non-isomorphic positions, it is not as easy to find them in three or more positions. Indeed, any separable Banach space X not containing ℓ_1 complemented has $\kappa(X)$ in at least two non-isomorphic positions in ℓ_1: one gives X as quotient, and the other gives $X \times \ell_1$. Twisted sums make an inconspicuous entrance in this problem: it turns out that there are at least as many non-isomorphic positions for $\kappa(X)$ in ℓ_1 as non-isomorphic twisted sum Banach spaces $\ell_1 \oplus_\Omega X$ (and, probably, exactly as many), since:

Lemma 7.4.6 *For every Banach space X, one has $\kappa(\ell_1 \oplus_\Omega X) \simeq \kappa(X)$.*

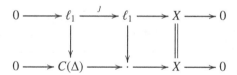

For the proof, just stare at that diagram appearing in Lemma 2.7.5,

$$
\begin{array}{ccccccccc}
 & & 0 & & 0 & & 0 & & \\
 & & \downarrow & & \downarrow & & \downarrow & & \\
0 & \longrightarrow & 0 & \longrightarrow & \kappa(\ell_1 \oplus_\Omega X) & \longrightarrow & \kappa(X) & \longrightarrow & 0 \\
 & & \downarrow & & \downarrow & & \downarrow & & \\
0 & \longrightarrow & \ell_1 & \longrightarrow & \ell_1 \times \ell_1 & \longrightarrow & \ell_1 & \longrightarrow & 0 \\
 & & \downarrow {\scriptstyle \mathbf{1}_{\ell_1}} & & \downarrow {\scriptstyle \mathbf{1}_{\ell_1} \times \overline{\pi}} & & \downarrow {\scriptstyle \pi} & & \\
0 & \longrightarrow & \ell_1 & \longrightarrow & \ell_1 \oplus_\Omega X & \longrightarrow & X & \longrightarrow & 0 \\
 & & \downarrow & & \downarrow & & \downarrow & & \\
 & & 0 & & 0 & & 0 & &
\end{array}
$$

until your eyes water. The peculiar case $X = c_0$ is considered in Note 7.5.2.

The Automorphy Index in Classical Spaces

The 'how many' question for equivalent embeddings can be cleanly formulated and solved.

Definition 7.4.7 The automorphy index $\mathfrak{a}(Y, X)$ of Y in X is the cardinal of the set of embeddings of Y into X having infinite-codimensional range, modulo equivalence of embeddings. The automorphy index of X is $\mathfrak{a}(X) = \sup_{Y \subset X} \mathfrak{a}(Y, X)$.

Thus, the automorphy index of Y in X measures how many non-equivalent embeddings $Y \longrightarrow X$ there are, while the automorphy index of X measures the maximum number of non-equivalent embeddings a space may have in X. It is clear that X is automorphic if and only if $\mathfrak{a}(X) = 1$. The index $\mathfrak{a}(Y, X)$ is an isomorphic invariant: if Y', X' are isomorphic to Y, X, respectively then $\mathfrak{a}(Y', X') = \mathfrak{a}(Y, X)$. Also, the automorphy index $\mathfrak{a}(Y, X)$ of Y and of all its finite codimensional subspaces are the same. Since the number of embeddings of a separable space into a separable superspace is at most $\mathfrak{c}$, one always has $\mathfrak{a}(Y, X) \leq \mathfrak{c}$ for separable X. Let us estimate the indices of some classical Banach spaces. The remainder of the section contains the proof of:

Theorem 7.4.8 *Let X be a separable Banach space.*

(a) $\mathfrak{a}(c_0, X) \in \{0, 1, 2, \aleph_0\}$.
(b) $\mathfrak{a}(X, \ell_p) \in \{0, \mathfrak{c}\}$ *for* $p \neq 2$.
(c) $\mathfrak{a}(X, L_p) \in \{0, \mathfrak{c}\}$ *for* $2 < p < \infty$ *and X non-Hilbert.*

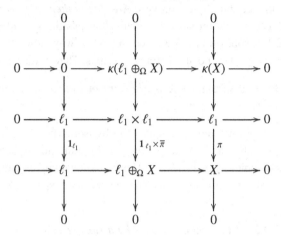

(d) $\mathfrak{a}(X, L_p) = 1$ *for some (all)* $2 < p < \infty$ *if and only if* X *is a Hilbert space.*
(e) $\mathfrak{a}(X, L_p) = \mathfrak{c}$ *for every subspace* $X \subset L_p$ *containing a copy of* ℓ_p *complemented in* L_p *for* $1 \leq p < 2$.
(f) $\mathfrak{a}(X, L_1) \in \{0, \mathfrak{c}\}$ *for every non-reflexive* X.
(g) $\mathfrak{a}(X, L_1) \geq 2$ *for every subspace* $X \subset L_1$ *containing* ℓ_2 *complemented.*
(h) $\mathfrak{a}(X, C(\Delta)) \in \{1, \mathfrak{c}\}$.

We prove (a). If X is separable and contains c_0, it must be complemented by Sobczyk's theorem. There are three possibilities:

(1) if every complement of every copy of c_0 in X contains c_0, $\mathfrak{a}(c_0, X) = 1$;
(2) if there is a copy of c_0 whose complement does not contain c_0 but is isomorphic to its hyperplanes, $\mathfrak{a}(c_0, X) = 2$;
(3) if there is a copy of c_0 whose complement is not isomorphic to its hyperplanes, $\mathfrak{a}(c_0, X) = \aleph_0$.

Proof (1) Since $X \simeq c_0 \times Z$ and $Z \simeq c_0 \times Z'$, $X \simeq c_0 \times c_0 \times Z' \simeq c_0 \times Z' \simeq Z$. Therefore, if $Y_1 \simeq Y_2 \simeq c_0$ are subspaces of X then $X \simeq Y_1 \oplus Z_1 \simeq Y_2 \oplus Z_2$ and $Z_1 \simeq Z_2 \simeq X$. One can thus extend any isomorphism between Y_1 and Y_2 to an automorphism in X. (2) The spaces c_0 and Z are totally incomparable now. It is clear that $\mathfrak{a}(c_0, X) \geq 2$: one position is a copy Y of c_0 in X whose complement is Z, and the second is any infinite codimensional copy of c_0 inside Y whose complement is $c_0 \times Z$. To prove that the index is 2, observe that we can assume without loss of generality that $V \cap Z = 0$ and $V + Z$ is closed for every subspace $V \simeq c_0$ of X. Let V be a subspace of X isomorphic to c_0. Since V is complemented in $X \simeq c_0 \oplus Z$ and c_0 and Z are incomparable, by the Edelstein–Wojtaszczyk decomposition (see 10.6.2), $V \simeq A \oplus B$ with A a complemented subspace of c_0, B a complemented subspace of Z, hence incomparable with A, and $A \oplus B$ isomorphic to c_0; therefore, B must be finite-dimensional. If Z is isomorphic to its hyperplanes, $Z \simeq Z/B$. Therefore $X/V \simeq (c_0/A) \oplus Z/B \simeq (c_0/A) \oplus Z$. If c_0/A is finite-dimensional then again, $X/V \simeq Z$, and V occupies the first position we mentioned above. If c_0/A is infinite-dimensional then $c_0/A \simeq c_0$, and $X/V \simeq c_0 \oplus Z$, so V occupies the second position we mentioned above. Finally, (3), if Z is not isomorphic to its hyperplanes then there are countably many different positions, according to the dimension of c_0/A. □

We pass to (b). We prove first that $\mathfrak{a}(\ell_p, \ell_p) = \mathfrak{c}$ for all $p \neq 2$:

Lemma 7.4.9 *For* $p \in [1, \infty)$ *different from 2, there exists a sequence* $(Y_i)_{i \geq 1}$ *of subspaces of* ℓ_p *uniformly isomorphic to* ℓ_p *and isomorphisms* $\tau_{ij} : Y_i \longrightarrow Y_j$, $i \neq j$ *that cannot be extended to operators on* ℓ_p.

Proof Let us construct two non-equivalent uncomplemented positions of ℓ_p in itself. Write $X = \ell_p$ in the form $X = X_1 \oplus_p X_2$, where X_1 and X_2 are isometric to ℓ_p. Denote the natural decomposition of X_1 into sum of n-dimensional subspaces by $X_1 = \overline{[\ell_p^n : n \in \mathbb{N}]}$. Let E_k be the aforementioned k-dimensional subspaces of ℓ_p^n, $n = n(k)$, which are c-isomorphic to ℓ_p^k and whose projection constants $\lambda(E_k, \ell_p^n) \to \infty$ as $k \to \infty$. Put $Y_1 = \overline{[E_k : k \in \mathbb{N}]}$. Of course, Y_1 is c-isomorphic to ℓ_p. Pick an increasing sequence $(i(k))_{k\in\mathbb{N}}$ of positive integers such that

$$\frac{\lambda(E_k, \ell_p^{n(k)})}{\lambda(E_{i(k)}, \ell_p^{n(i(k))})} \longrightarrow \infty \text{ as } k \longrightarrow \infty.$$

For every k, let F_k be $(k - i(k))$-dimensional subspaces spanned by the 'consecutive' unit vectors of X_2. Take $G_k = E_{i(k)} \times F_k$. Of course, the G_k are c-isomorphic to ℓ_p^k. Put $Y_2 = \overline{[G_k : k \in \mathbb{N}]}$; then Y_2 is c-isomorphic to ℓ_p. Let $\tau_{1,2} : Y_1 \longrightarrow Y_2$ be the natural linear operator which maps E_k c^2-isomorphically onto G_k. Then $\tau_{1,2}$ is a c^2-isomorphism. Since ℓ_p^n is 1-complemented in X, $\lambda(E_k, X) = \lambda(E_k, \ell_p^{n(k)})$; $\lambda(E_{i(k)}, X) = \lambda(E_{i(k)}, \ell_p^{n(i(k))})$ and $\lambda(G_k, X) = \lambda(E_{i(k)}, X)$. Therefore,

$$\frac{\lambda(E_k, X)}{\lambda(G_k, X)} = \frac{\lambda(E_k, X)}{\lambda(E_{i(k)}, X)} \frac{\lambda(E_k, \ell_p^{n(k)})}{\lambda(E_{i(k)}, \ell_p^{n(i(k))})} \longrightarrow \infty \text{ as } k \longrightarrow \infty.$$

By the arguments in the proof of Proposition 7.1.4 (f), $\tau_{1,2}$ cannot be extended to any bounded linear operator in X. Now write $X = X_1 \oplus_p X_2 \oplus_p X_3$ with X_1, X_2, X_3 isometric to ℓ_p. Let (E_k) and (G_k) be the subspaces from $X_1 \oplus_p X_2$ obtained above. Beginning from (G_k) and following the same construction, one can construct in X a sequence of subspaces H_k, c-isomorphic to ℓ_p^k, such that

$$\frac{\lambda(G_k, X)}{\lambda(H_k, X)} \longrightarrow \infty \text{ as } k \longrightarrow \infty \quad \left(\text{hence } \frac{\lambda(E_k, X)}{\lambda(H_k, X)} \longrightarrow \infty\right).$$

Put $Y_3 = \overline{[H_k : k \in \mathbb{N}]}$; then Y_3 is c-isomorphic to ℓ_p. Let $\tau_{2,3} : Y_2 \longrightarrow Y_3$ be the natural linear operator that maps c^2-isomorphically G_k onto H_k. Then $\tau_{2,3}$ is a c^2-isomorphism that, once again by Proposition 7.1.4 (f), cannot be extended to a bounded linear operator on X. The same arguments work for the natural linear operator $\tau_{1,3} : Y_1 \longrightarrow Y_3$ mapping E_k c^2-isomorphically onto H_k. Continue the construction inductively. □

We can now jump ahead to show that $\mathfrak{a}(Y, X) = \mathfrak{c}$ for any two infinite-dimensional subspaces of ℓ_p for $p \neq 2$. This follows from the following inequality:

Lemma 7.4.10 *If $V \subset Y \subset X \subset \ell_p$ and V is isomorphic to ℓ_p and complemented in X, $\mathfrak{a}(V, V) \leq \mathfrak{a}(Y, X)$.*

Proof Let I be a set of cardinal $|I| < \alpha(V, V)$, so that there are subspaces $(V_\gamma)_{\gamma \in I}$ of V isomorphic to V and isomorphisms $\tau_{\gamma\delta} \colon V_\gamma \longrightarrow V_\delta$, $\gamma < \delta$, that cannot be extended to operators on V, and let us show that $\mathfrak{a}(Y, X) \geq |I|$.

Let P be the projection of X onto V, and let $U = Y \cap \ker P$. Of course, each $U \times V_\gamma$ is isomorphic to Y. For $\gamma, \delta \in I$ define an isomorphism $\sigma_{\gamma\delta} \colon U \times V_\gamma \longrightarrow U \times V_\delta$, by $\sigma_{\gamma\delta}(u + v) = u + \tau_{\gamma\delta} v$. If $\sigma_{\gamma\delta}$ extends to an automorphism S of X then PS is an extension of $\tau_{\gamma\delta}$, which is impossible. □

To prove (c), we use the Kadec–Pełczyński dichotomy: when $2 < p < \infty$, every subspace of L_p is either isomorphic to ℓ_2 or contains a copy of ℓ_p that is complemented in L_p (see [241, Corollary 2] or [5, Corollary 6.4.9]); once the first case has been ruled out, apply the previous Lemma 7.4.10.

To prove (d), observe that $\mathfrak{a}(\ell_2, L_p) = 1$ since every copy of ℓ_2 inside L_p is complemented when $2 < p < \infty$; the rest is as in (a). More precisely, we can rewrite now (a) as:

(a′) *If X is a Banach space in which every copy of ℓ_2 is complemented then $\mathfrak{a}(\ell_2, X) \in \{1, 2, \aleph_0\}$. If, moreover, every copy of ℓ_2 contains another copy that is complemented in X then $\mathfrak{a}(\ell_2, X) = 1$.*

The implication $\mathfrak{a}(X, L_p) = 1 \implies X \simeq \ell_2$ is contained in (c). Assertion (e) follows from Lemma 7.4.10: X contains a complemented copy of ℓ_p and thus $\mathfrak{c} = \mathfrak{a}(\ell_p, \ell_p) \leq \mathfrak{a}(X, L_p)$.

Assertion (f) follows from another result of Kadec and Pełczyński [241, Theorem 6]: a non-reflexive subspace of L_1 contains a copy of ℓ_1 that is complemented in L_1. The estimate $\mathfrak{c} = \mathfrak{a}(\ell_1, \ell_1) \leq \mathfrak{a}(X, L_1)$ then follows from Lemma 7.4.10.

Finally, we prove (g) by showing that ℓ_2 has two non-equivalent positions in L_1. One of them is the subspace R spanned by the Rademacher sequence (r_n), and the other, which we denote by G, is spanned by a sequence (g_n) of independent standard Gaussian random variables on $[0, 1]$. The space R is isomorphic to ℓ_2 by Khintchine's inequality (see [153, pp. 9–14] or [334, p. 66]); the space G is isometric to ℓ_2 by the reproductive property of the normal distribution (see [391, p. 14]). Moreover, the distribution of $g \in G$ depends only on $\|g\|_1$.

Lemma 7.4.11 *If $T \in \mathfrak{L}(L_1)$ maps R to G then $T|_R$ is compact.*

Proof The key fact is that each endomorphism of L_1 arises as the difference of two positive operators and therefore preserves order-bounded sets; see [424, p. 232, Theorem 1.5(ii)]. Recall that a subset S of an ordered space A is said to be order-bounded if there are $a, b \in A$ such that $a \le s \le b$ for every $s \in S$. A typical order-bounded set in L_1 is the Rademacher sequence. Let us see that if $\gamma \in L_1$ is unbounded (not in L_∞) and γ_n are independent and have the same distribution as γ then $(\gamma_n)_{n\ge1}$ cannot be order bounded. Assume there is a measurable φ such that $|\gamma_n| \le \varphi$ for all n. Given $\varepsilon > 0$, there is $c > 0$ such that $\lambda(\{t \in [0,1] : \varphi(t) < c\}) > 1 - \varepsilon$, where λ denotes Lebesgue measure. However, since the γ_i are independent,

$$\lambda\Big\{t : \max_{i\le n} |\gamma_i(t)| > c\Big\} = 1 - \prod_{i\le n} \lambda\{t : |\gamma_i(t)| \le c\} \ge 1 - \big(\lambda\{t : |\gamma(t)| \le c\}\big)^n \longrightarrow 1,$$

as $n \to \infty$ and the sequence (γ_n) is not order bounded. The obvious consequence is that no operator on L_1 can send (a subsequence of) the Rademachers to (a subsequence of) the Gaussians g_n. Now, if $T|_R$ is not compact then the sequence Tr_n is weakly null but does not converge to zero in norm. Therefore, there is a $c > 0$, a subsequence $n(k)$ and a sequence $(\gamma_k)_{k\ge1}$ of (disjoint) blocks of $(g_n)_{n\ge1}$ such that $\|\gamma_k\| = c$ and $\|Tr_{n(k)} - \gamma_k\| \le 2^{-k}$ for all k. We know that the sequence $(\gamma_k)_k$ cannot be order bounded and neither can $Tr_{n(k)}$. Indeed, if we assume that there exists some $\varphi \in L_1$ such that $|Tr_{n(k)}| \le \varphi$ for all k then by letting $\eta_k = \gamma_k - Tr_{n(k)}$ and $\eta = \sum_{k=1}^{\infty} |\eta_k|$ (summation in L_1) we obtain $|\gamma_k| \le |\gamma_k - Tr_{n(k)}| + |Tr_{n(k)}| \le \eta + \varphi$, a contradiction. □

Let X be a subspace of L_1 containing a complemented copy of ℓ_2. Lemma 7.4.10 is now useless for the purpose of obtaining $\mathfrak{a}(X, L_1) = \mathfrak{c}$ since $\mathfrak{a}(\ell_2, \ell_2) = 1$; thus, a slightly more sophisticated version is necessary:

Lemma 7.4.12 *Let L be isomorphic to its square and let H be a complemented subspace of some subspace X of L. Let $H' \subset L$ be isomorphic to H and such that no isomorphism $H \longrightarrow H'$ can be extended to an automorphism in L. Then $\mathfrak{a}(X, L) > 1$.*

Proof Write $L = V \times V'$ with $V \simeq L \simeq V'$. Without loss of generality, we can assume that $X \subset V$ and get a subspace $H' \subset X' \subset V'$ with X' isomorphic to X

and H' complemented in X'. An isomorphism $\tau\colon X \longrightarrow X'$ that sends H to H' cannot be extended to an automorphism in L. □

To obtain (g), just set $L = L_1, H = \ell_2$.

A head-on attack on (h) will likely result in forlornness: since $C(\Delta) \simeq c_0(C(\Delta))$, one easily convinces oneself that two non-equivalent positions automatically yield $\mathfrak{c}$ non-equivalent positions, something we cannot in fact guarantee (we cannot prove that $\kappa(c_0)$ has $\mathfrak{c}$ positions in ℓ_1, can we?). Think calmly on that while we, apparently, take a detour. The conclusion of the proof of (h) will come right after Proposition 7.4.15.

Partially Automorphic Properties of $C(K)$-Spaces

A number of results attest to the 'partially automorphic' character of several classical Banach spaces: by Theorem 7.4.8 (d), Hilbert spaces occupy a unique position inside L_p for $p > 2$, while ℓ_∞, ℓ_∞/c_0 and ultrapowers of $\mathscr{L}_\infty$-spaces via countably incomplete ultrafilters are 'separably automorphic' [22, Section 2.6]. We can formalise this idea as follows:

Definition 7.4.13 A Banach space X is said to be Y-automorphic if $\mathfrak{a}(Y, X) = 1$. The space X is said to be $\mathscr{A}$-automorphic for a given class $\mathscr{A}$ if it is Y-automorphic for every Y in $\mathscr{A}$.

Thus, X is Y-automorphic if and only if any two embeddings $\imath, \jmath\colon Y \longrightarrow X$ with $\dim(X/\imath[Y]) = \dim(X/\jmath[Y])$ are equivalent. Theorem 7.4.8 (h) asserts that the situation for the space $C(\Delta)$ is as clear as blinding sunshine: if X is separable, $\mathfrak{a}(X, C(\Delta))$ is either 1 or $\mathfrak{c}$. The identification of which Banach spaces have $\mathfrak{a}(X, C(\Delta)) = 1$ and which have $\mathfrak{a}(X, C(\Delta)) = \mathfrak{c}$ is nowadays incomplete. Moving to the lip of inconsistency, we bring a few items forward from Chapter 8 in order to discuss this topic, in particular, the notion of a $\mathscr{C}$-extensible space (Section 8.5): *a separable Banach space X is $\mathscr{C}$-extensible when every $\mathscr{C}$-valued operator defined on X can be extended to any separable superspace.* We present two proofs for the following fact showing that $\mathscr{C}$-extensibility puts a severe restriction on embeddings:

Lemma 7.4.14 *If X is $\mathscr{C}$-extensible then for every embedding $u\colon X \longrightarrow C(\Delta)$, the space $C(\Delta)/u[X]$ has non-separable dual.*

First proof Let $0 \longrightarrow C(\Delta) \stackrel{\imath}{\longrightarrow} D(\Delta;\Delta_0) \stackrel{\jmath}{\longrightarrow} c_0 \longrightarrow 0$ be the original Foiaş–Singer sequence (2.5). Let us ease the notation by setting $D = D(\Delta;\Delta_0)$.

Assume the existence of an embedding $u\colon X \longrightarrow C(\Delta)$ for which $C(\Delta)/u[X]$ has separable dual and form the commutative diagram

$$\begin{array}{ccccccccc}
 & & 0 & & 0 & & & & \\
 & & \downarrow & & \downarrow & & & & \\
 & & X & = & X & & & & \\
 & & \downarrow{\scriptstyle u} & & \downarrow{\scriptstyle \imath u} & & & & \\
0 & \longrightarrow & C(\Delta) & \xrightarrow{\ \imath\ } & D & \xrightarrow{\ J\ } & c_0 & \longrightarrow & 0\\
 & & \downarrow & & \downarrow{\scriptstyle \rho} & & \| & & \\
0 & \longrightarrow & C(\Delta)/u[X] & \longrightarrow & D/\imath u[X] & \xrightarrow[\underline{J}]{} & c_0 & \longrightarrow & 0\\
 & & \downarrow & & \downarrow & & & & \\
 & & 0 & & 0 & & & &
\end{array}$$

If X is $\mathscr{C}$-extensible, the two vertical sequences are semi-equivalent, in which case the Parallel lines principle 2.11.5 shows that the two horizontal sequences are also semi-equivalent, so that there is an operator $\tau\colon D/\imath u[X] \longrightarrow C(\Delta)$ such that $J\tau = \underline{J}$. Since $C(\Delta)/u[X]$ has separable dual, $D/\imath u[X]$ also has separable dual, by an obvious 3-space argument that can be found in [102, Lemma 2.4.g]. Let (x_k^*) be a countable dense subset of the unit ball of $(D/\imath u[X])^*$ and let (V_n) be a countable base for the topology of Δ. The map $(x_1^*\rho, \dots, x_n^*\rho)\colon D \longrightarrow \mathbb{R}^n$ is compact, and thus, for each n, we can extract from the image of each sequence $(f_{d(m)})_{d(m)\in V_n}$ a convergent subsequence. This yields

$$\forall n\, \exists\, d_n, d_n' \in V_n\colon \forall k \in \{1, \dots, n\}\colon\ \langle x_k^*, \rho f_{d_n} - \rho f_{d_n'}\rangle \le 2^{-n}.$$

Since $J\tau(\rho f_{d_n} - \rho f_{d_n'}) = \underline{J}\ f_{d_n} - \underline{J}\ f_{d_n'} = e_{d_n} - e_{d_n'}$, the sequence $(\tau(\rho f_{d_n} - \rho f_{d_n'}))_n$ is a weakly convergent lifting of $(e_{d_n} - e_{d_n'})_n$. But there is a germ of a contradiction here, because arguing as in Lemma 2.2.3, we can find a decreasing subsequence $(V_{m_n})_n$ of neighbourhoods and signs $\varepsilon_n = \pm 1$ such that, allaying the reader's reading by writing $d(n) = d_{m_n}$ and $d'(n) = d'_{m_n}$, one has $\varepsilon_n(f_{d(n)}(t) - f_{d'(n)}(t)) \ge 1/2$ for all $t \in V_{m_{n+1}}$. Therefore, we can get by a compactness argument some x^* for which $|\langle x^*, f_{d(n)} - f_{d'(n)}\rangle| \ge 1/2$ for all n. □

Second proof Let $0 \longrightarrow C(\Delta) \longrightarrow \diamond \longrightarrow c_0 \longrightarrow 0$ be the modification of the Foiaş–Singer sequence that we will present in Corollary 9.3.5. It has the property that no subsequence of the canonical basis of c_0 admits a weakly Cauchy lifting. Assume the existence of a $\mathscr{C}$-extensible subspace $X \subset C(\Delta)$ such that $C(\Delta)/X$ has separable dual and form the commutative diagram

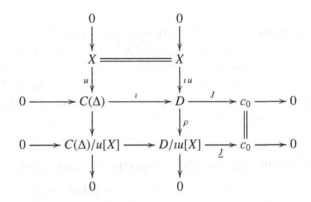

$$
\begin{array}{ccccccccc}
 & & 0 & & 0 & & & & \\
 & & \downarrow & & \downarrow & & & & \\
 & & X & = & X & & & & \\
 & & \downarrow & & \downarrow & & & & \\
0 & \longrightarrow & C(\Delta) & \longrightarrow & \diamond & \longrightarrow & c_0 & \longrightarrow & 0 \\
 & & \downarrow & & \downarrow & & \| & & \\
0 & \longrightarrow & C(\Delta)/X & \longrightarrow & \diamond/X & \longrightarrow & c_0 & \longrightarrow & 0 \\
 & & \downarrow & & \downarrow & & & & \\
 & & 0 & & 0 & & & &
\end{array}
$$

The space $C(\Delta)/X$ does not contain ℓ_1, so Lohman's lifting [339] yields a weakly Cauchy lifting in $\diamond/X$ of a subsequence of the canonical basis of c_0. Since X is $\mathscr{C}$-extensible, the two vertical sequences are semi-equivalent, in which case the Parallel lines principle 2.11.5 shows that the two horizontal sequences are also semi-equivalent, and therefore there is a weak Cauchy lifting in $\diamond$ of a subsequence of the canonical basis of c_0, and this is impossible. □

Proposition 7.4.15 *A separable Banach space X is $\mathscr{C}$-extensible if and only if $C(\Delta)$ is X-automorphic.*

Proof Milutin's theorem and the existence of the $\mathscr{C}$-trivial embedding $\delta_X \colon X \longrightarrow C(B_X^*)$ prove sufficiency. To prove necessity, assume that X is $\mathscr{C}$-extensible, so that any two sequences

$$
\begin{array}{ccccccccc}
0 & \longrightarrow & X & \xrightarrow{\jmath_1} & C(\Delta) & \xrightarrow{\pi_1} & Q_1 & \longrightarrow & 0 \\
 & & \| & & & & & & \\
0 & \longrightarrow & X & \xrightarrow{\jmath_2} & C(\Delta) & \xrightarrow{\pi_2} & Q_2 & \longrightarrow & 0
\end{array}
$$

are semi-equivalent. Applying the Diagonal principle 2.11.7, the sequences

$$
\begin{array}{ccccccccc}
0 & \longrightarrow & X & \longrightarrow & C(\Delta)\times C(\Delta) & \longrightarrow & Q_1 \times C(\Delta) & \longrightarrow & 0 \\
 & & \| & & & & & & \\
0 & \longrightarrow & X & \longrightarrow & C(\Delta)\times C(\Delta) & \longrightarrow & Q_2 \times C(\Delta) & \longrightarrow & 0
\end{array}
$$

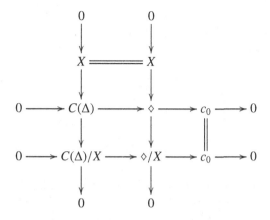

are isomorphic. From Lemma 7.4.14, we know that both Q_1 and Q_2 have non-separable dual, and Rosenthal's theorem [415] mentioned in 1.6.3(d) applies to make π_i ($i = 1, 2$) an isomorphism on some copy of $C(\Delta)$ inside $C(\Delta)$. By 1.6.4, this new copy contains another copy C_i of $C(\Delta)$ complemented in $C(\Delta)$ so that the sequences

$$\begin{array}{ccccccccc} 0 & \longrightarrow & X & \xrightarrow{\jmath_i} & C(\Delta) & \xrightarrow{\pi_i} & Q_i & \longrightarrow & 0 \\ & & \| & & & & & & \\ 0 & \longrightarrow & X & \xrightarrow{(\jmath_i,0)} & C(\Delta)\times C_i & \xrightarrow{\pi_i} & Q_i \times C_i & \longrightarrow & 0 \end{array}$$

are isomorphic for $i = 1, 2$. Thus, the embeddings $\jmath_1, \jmath_2$ are equivalent. □

Conclusion of the proof of Theorem 7.4.8 (h) If X has two positions in $C(\Delta)$, one of them is the canonical embedding $X \longrightarrow C(B_X^*)$ (considered as an embedding $\delta\colon X \longrightarrow C(\Delta)$ via Milutin's theorem) which is $\mathscr{C}$-trivial. Any other $\mathscr{C}$-trivial embedding $\jmath\colon X \longrightarrow C(\Delta)$ must be semi-equivalent to δ, and it is easy to convince oneself that it must also be equivalent. Proposition 7.4.15 takes no prisoners: since $C(\Delta)$ is not X-automorphic, X is not $\mathscr{C}$-extensible, so there is an embedding $\jmath$ not semi-equivalent to δ. Now apply Lemma 7.4.9 without missing a beat. □

The list of known spaces X for which $\mathfrak{a}(X, C(\Delta)) = 1$ can be found in Theorem 8.5.6. We now list those for which it is known that $\mathfrak{a}(X, C(\Delta)) = \mathfrak{c}$:

(a) all separable $\mathscr{C}$-spaces different from c_0;
(b) ℓ_p for $1 < p < \infty$;
(c) Banach spaces X such that $C(\Delta)/X$ does not contain ℓ_1;
(d) spaces containing a complemented copy of any of the previous spaces.

(a) follows from the fact that $C(\Delta)$ contains both a complemented and an uncomplemented copy of any separable $C(K)$ not isomorphic to c_0.
(b) is because ℓ_p is not $\mathscr{C}$-extensible for $1 < p < \infty$, Proposition 8.5.8.
(c) is contained in the proof of Proposition 7.4.15.
It could make sense to consider the largest class $\mathscr{A}$ of subspaces of X for which X is $\mathscr{A}$-automorphic, namely those Banach spaces Y such that $\mathfrak{a}(Y, X) = 1$. The symmetric question *given Y, characterise those Banach spaces X such that* $\mathfrak{a}(Y, X) = 1$ is tricky because enlarging a Banach space can produce conflicting results: if Y is an infinite-dimensional Banach space, there exist superspaces A, B for which $\mathfrak{a}(Y, A) = 1$ and $\mathfrak{a}(Y, B) > 1$ (see [121] for details). A few arguments can be presented supporting the idea that

$\mathscr{C}$-spaces enjoy a partially automorphic character. As we will say somewhere, the general non-separable case is, by the time of writing these lines, too difficult. However, Eberlein compacta offer a reasonably safe haven: *if K is an Eberlein compact then $C(K)$ is automorphic for every subspace of c_0*. This is a consequence of the fact that c_0-automorphic $C(K)$ spaces are also automorphic for every subspace of c_0, which is not as simple to prove as it sounds. A reading of [22, Section 2.6] can do only good. There are other results in the literature worth mentioning: if K is an Eberlein compact of finite height, the space $C(K)$ is separably automorphic and, in fact, automorphic for all possible subspaces of dimension less than $\aleph_\omega$ [22, Proposition 2.57]; and universally separably injective spaces are separably automorphic [22]. Those results are in some sense optimal since there exist separably injective spaces, such as $c_0 \times \ell_\infty$, which are not c_0-automorphic; there also exist non-Eberlein compacta of height 3, such as $\Delta_{\mathcal{M}}$, whose spaces of continuous functions are not c_0-automorphic because they contain complemented and uncomplemented copies of c_0 – and while Δ is Eberlein, the space $C(\Delta)$ is not separably automorphic. Beyond that, everything is twilight.

7.5 Notes and Remarks

7.5.1 Isomorphic but Different Twisted Sums

A delightful but scary consequence of Proposition 7.2.16 is that all quotients of L_p by an n-dimensional subspace are isomorphic. So, let us denote its isomorphism type by $L_p/(n)$ and observe that $L_p/(n+m) \simeq L_p/(n) \times L_p/(m)$ and thus $L_p/(n) \simeq (L_p/(1))^n$, since $L_p \simeq L_p \times L_p$. We have:

Proposition *Let $p \in (0,1)$ and $n, m \geq 1$.*

(a) *$\dim \operatorname{Ext}(L_p/(n), \mathbb{K}^m) = nm$.*
(b) *If $m \leq n$, there are exactly $nm+1$ twisted sums of $\mathbb{K}^m$ and $L_p/(n)$ up to isomorphism: n of each type $\mathbb{K}^{m-k} \times L_p/(n-k)$ for $1 \leq k \leq m$, the non-trivial ones, and the trivial one $\mathbb{K}^m \times L_p/(n)$.*
(c) *If $m > n$, there are exactly n^2+1 twisted sums of $\mathbb{K}^m$ and $L_p/(n)$ up to isomorphism: n of each type $\mathbb{K}^{m-n} \times \mathbb{K}^{n-k} \oplus L_p/(n-k)$ for $1 \leq k \leq n$, the non-trivial ones, and the trivial one $\mathbb{K}^m \times L_p/(n)$.*

Proof Assertion (a) is 4.2.6 (6), setting $E = \mathbb{K}^m$, $\dim(Y) = n$ and $Z = L_p$. Regarding the isomorphism type of those twisted sums, first consider (b), i.e

$m \leq n$. Observe that every exact sequence $0 \longrightarrow \mathbb{K}^m \longrightarrow X \longrightarrow L_p/(n) \longrightarrow 0$ arises as a pushout

$$\begin{array}{ccccccccc} 0 & \longrightarrow & \mathbb{K}^n & \longrightarrow & L_p & \longrightarrow & L_p/(n) & \longrightarrow & 0 \\ & & \downarrow{\scriptstyle f} & & \downarrow & & \| & & \\ 0 & \longrightarrow & \mathbb{K}^m & \longrightarrow & X & \longrightarrow & L_p/(n) & \longrightarrow & 0 \end{array}$$

and thus everything depends on the dimension k of the range $f[\mathbb{K}^n]$, which in this case can be $0 \leq k \leq m$. The diagram

$$\begin{array}{ccccccccc} & & 0 & & 0 & & & & \\ & & \downarrow & & \downarrow & & & & \\ & & \ker f & = & \ker f & & & & \\ & & \downarrow & & \downarrow & & & & \\ 0 & \longrightarrow & \mathbb{K}^n & \longrightarrow & L_p & \longrightarrow & L_p/(n) & \longrightarrow & 0 \\ & & \downarrow{\scriptstyle f} & & \downarrow & & \| & & \\ 0 & \longrightarrow & f[\mathbb{K}^m] & \longrightarrow & \mathrm{PO} & \longrightarrow & L_p/(n) & \longrightarrow & 0 \\ & & \downarrow & & \downarrow & & & & \\ & & 0 & & 0 & & & & \end{array}$$

shows that $\mathrm{PO} = L_p/\ker f = L_p/(n-k)$; therefore, $X \simeq \mathbb{K}^{m-k} \times \mathrm{PO} \simeq \mathbb{K}^{m-k} \times L_p/(n-k)$. The trivial case corresponds to $f = 0$. When $m > n$, case (c), the previous spaces must be multiplied by $\mathbb{K}^{m-n}$. □

In particular both $\mathrm{Ext}(L_p/(n), \mathbb{K})$ and $\mathrm{Ext}(L_p/(1), \mathbb{K}^n)$ have dimension n. However, while in the first case, we have $n+1$ mutually non-isomorphic twisted sums (including the trivial one), in the second case, all non-trivial twisted sums are isomorphic, and we have two non-isomorphic twisted sums only.

7.5.2 How Many Twisted Sums of Two Spaces Exist?

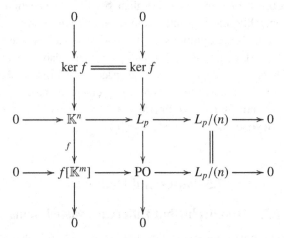

The question can be posed about every pair of spaces. Moreover, a 'how many' question is always open to interpretation: does one count as equal isomorphic spaces, equivalent exact sequences, isomorphic exact sequences, having the subspace in equivalent positions ... ? The nuances of the question when X is the quotient of L_p by a finite-dimensional subspace, $0 < p < 1$, and Y is finite-dimensional have been treated in the previous note. Ribe's remarkable paper [401] must be mentioned regarding the accounting of twisted sums: in addition to his counterexample to the 3-space problem for local convexity, embryonic forms of the transfer principle and the amalgamation technique to be treated in Chapter 9, it also packs in its barely five pages the noteworthy Theorem 2 asserting (translated) that *there is an uncountable family of extensions* $0 \longrightarrow \mathbb{R} \longrightarrow \cdot \longrightarrow \ell_1 \longrightarrow 0$ *such that none of them is a pullback of any of the others*, from where he deduces that the corresponding twisted sums are pairwise non-isomorphic. Cabello Sánchez used that same technique unawarely in [56] to answer a question asked in [108], showing that the counterexamples of Ribe and Kalton to the 3-space problem for local convexity are not isomorphic (see comments after Proposition 9.1.9).

Now we focus on the particular case $X = c_0$. The 'how many' question sets us now underneath the lintel:

Proposition $\ell_1 \oplus_\Omega c_0 \simeq \ell_1 \times c_0$ *if and only if* $\Omega \sim 0$.

Proof The proof consists in showing that a subspace Y of $\ell_1 \times c_0$ isomorphic to ℓ_1 and such that $(\ell_1 \times c_0)/Y$ is isomorphic to c_0 must be complemented. To prove this, let π be the canonical projection of $\ell_1 \times c_0$ onto ℓ_1. Since Y and c_0 are totally incomparable, any common subspace is finite-dimensional, and we can assume without loss of generality that $Y \cap \ker \pi = 0$. Moreover, $Y + \ker \pi$ is closed [410], as remarked in Diamond's lemma (note 2.15.2). Thus the restriction of π to Y is an embedding, and we get a commutative diagram

$$\begin{array}{ccccccccc} 0 & \longrightarrow & Y & \longrightarrow & \ell_1 \times c_0 & \longrightarrow & c_0 & \longrightarrow & 0 \\ & & \| & & \downarrow{\scriptstyle \pi} & & \downarrow{\scriptstyle \overline{\pi}} & & \\ 0 & \longrightarrow & Y & \xrightarrow{\pi|_Y} & \ell_1 & \longrightarrow & H & \longrightarrow & 0 \end{array}$$

where $\overline{\pi}$ is surjective, so H is isomorphic to a subspace of c_0. Since Y is isomorphic to ℓ_1, we conclude by observing that an exact sequence $0 \longrightarrow \ell_1 \longrightarrow \ell_1 \longrightarrow H \longrightarrow 0$ cannot exist when H is an infinite-dimensional subspace of c_0, which is clear since Proposition 7.4.5 states that H must have the Schur property. Thus, the only remaining possibility is that H is finite-dimensional, in which case, $\pi[Y]$ is complemented in ℓ_1 by a projection P, which makes Y complemented in $\ell_1 \times c_0$ by $(\pi|_Y)^{-1} P\pi$. □

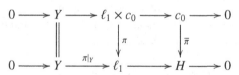

The argument also works for incomparable A, B provided B is incomparable with all quotients of A by subspaces isomorphic to A itself. Thus, understanding ℓ_∞ as c_0 here, one gets that $\ell_p \oplus_\Omega \ell_q \simeq \ell_p \times \ell_q$ *if and only if* $\Omega \sim 0$ *for different indices* $p, q \in [1, \infty]$. Thus, ℓ_1 admits $\kappa(c_0)$ in at least three non-equivalent positions: one having c_0 as quotient, another one having $\ell_1 \oplus c_0$ and a third one having a non-trivial sum $\ell_1 \oplus_\Omega c_0$. A fourth position is also available [126], but that does not give any inkling of what is to come. And what is to come is that one might easily conjecture that there are countably many non-isomorphic positions, but proving that is an entirely different story. Moreover, a new moving walls labyrinth opens here: the existence of a non-trivial twisted sum $\ell_1 \oplus_\Omega c_0$ is, well, an existence result – there is no known effective way to construct them or to distinguish their isomorphism types.

7.5.3 Moving towards the Automorphic Space Problem

The dichotomy for UFO spaces presented in Theorem 7.2.10 is a step towards the solution of the automorphic space problem. To move further, we must either find a way of using the automorphic property of the space instead of the (much weaker) UFO character or otherwise use additional properties to improve the dichotomy for UFO spaces. An example in this direction follows:

Proposition *A UFO space with unconditional basis is either* c_0 *or a reflexive and extensible near-Hilbert space with the MEP.*

Proof Since c_0 is the only $\mathscr{L}_\infty$-space with an unconditional basis, the first alternative is clear. To get the other, recall that the space cannot contain ℓ_1^n uniformly. Since spaces with unconditional basis contain either c_0 or ℓ_1 or are reflexive, the space must be reflexive and thus extensible. □

An extra turn of this screw can be given in terms of the so-called Boyd indices of a Banach space X with unconditional basis (e_n), defined as

$$\alpha(X) = \liminf_{n\to\infty} \frac{\log n}{\log \|\sum_1^n e_i\|}; \quad \beta(X) = \limsup_{n\to\infty} \frac{\log n}{\log \|\sum_1^n e_i\|}.$$

If X is a UFO with unconditional basis different from c_0 then $\alpha(X) = \beta(X) = 2$: since X does not contain ℓ_∞^n, one can apply [418] to obtain that ℓ_p is block finitely representable in (e_n) for every $\alpha(X) \le p \le \beta(X)$. So, by [367, Corollary 4.6], unless $\alpha(X) = \beta(X) = 2$, the space X is not extensible.

7.5.4 The Product of Spaces of (Almost) Universal Disposition

In this note the authors cleave to dictums of two Raymonds. First, Chandler: the contents look interesting at first glance, but after closer inspection, they look like something made up to be given a first glance. And second, Reddington: one of the keys of success is a clear and consistent understanding of one's own limitations.

Is the product of spaces of (almost) universal disposition isomorphic to a space of (almost) universal disposition? First difficulty: how to characterise the property of 'being isomorphic to a space of (almost) universal disposition'. Are partial results available? Not really. A few remarks follow. First of all, Gurariy space is isomorphic to its square, so the product of two separable spaces of AUD is isomorphic to a space of AUD. It follows that under [CH], the space F_1, being isometric to a countable ultrapower of G, is isomorphic to its square. A less direct consequence is that every space of AUD has an ultrapower isomorphic to its square; see [22, Proposition 4.29]. In the opposite direction, under the diamond principle (an axiom of set theory which implies CH), Shelah constructed a space $\mathscr{S}_\diamond$ of AUD and density $\aleph_1$ with the property that every operator $\mathscr{S}_\diamond \longrightarrow \mathscr{S}_\diamond$ has the form $\lambda\mathbf{1} + S$ for some scalar λ and some separable range operator S, which makes an isomorphism between $\mathscr{S}_\diamond$ and $\mathscr{S}_\diamond \times \mathscr{S}_\diamond$ impossible and leaves open the question of whether the latter space can be renormed to be of AUD. Regarding 3-space problems, we have:

Proposition *AUD and UD are not 3-space properties. Under* CH, *SUD is not a 3-space property.*

Proof The proof for AUD depends on three facts: that an AUD space must be Lindenstrauss and therefore enjoy property (V); that property (V) passes to complemented subspaces and on the existence of twisted sums of separable Lindenstrauss spaces that fail property (V); such as the sequence $0 \longrightarrow C(\omega^\omega) \longrightarrow \diamond \longrightarrow c_0 \longrightarrow 0$, in (2.6). Applying Corollary 6.4.7 and multiplying as appropriate by G, we get an exact sequence $0 \longrightarrow \mathsf{G} \longrightarrow \mathsf{G} \times \diamond \longrightarrow \mathsf{G} \longrightarrow 0$ in which the middle space does not have property (V) and cannot therefore be isomorphic to a space of AUD. The proof for UD follows from Fact 3 in Section 7.3: if L is a polyhedral Lindenstrauss space with property $(*)$ then $\mathsf{F}_0(L)$ contains a complemented copy of L. Apply this to both c_0 and $C(\omega^\omega)$. Multiplying the same sequence in which $\diamond$ appears on the left and on the right, we obtain an exact sequence

$$0 \longrightarrow \mathsf{F}_0(C(\omega^\omega)) \longrightarrow \diamond \times \spadesuit \longrightarrow \mathsf{F}_0(c_0) \longrightarrow 0$$

in which $\diamond \times \spadesuit$ cannot have property (V). We need a different strategy to show that SUD is not a 3-space property, since spaces of SUD do not contain complemented separable subspaces. A whirlwind quest through the pages of this book will let us know from Proposition 10.5.3 that ultrapowers of $\mathscr{L}_\infty$-spaces are universally separably injective (USI, Definition 10.5.2) and from Theorem 10.5.6 that, under CH, USI is not a 3-space property. More precisely, if $\mathcal{U}$ is a free ultrafilter on $\mathbb{N}$, there exists an exact sequence $0 \longrightarrow C(\Delta)_\mathcal{U} \longrightarrow \diamond \longrightarrow \ell_\infty \longrightarrow 0$ in which the space $\diamond$ is not USI. Now, the SUD space $\mathsf{G}_\mathcal{U}$ contains complemented copies of $C(\Delta)_\mathcal{U}$ and ℓ_∞. Proceeding as before, one gets an exact sequence $0 \longrightarrow \mathsf{G}_\mathcal{U} \longrightarrow \cdot \longrightarrow \mathsf{G}_\mathcal{U} \longrightarrow 0$ in which the middle space fails to be USI. Finally, use an obvious modification of Proposition 2.13.2 to get that spaces of SUD are USI. □

Sources

The notion of extendability is from Moreno and Plichko [367], and the automorphic term is from [109]. The Lindenstrauss–Rosenthal theorem appeared in [332], and the unified proof through the diagonal principles for all results here presented is from [109]. The L_1-version is essentially from Kislyakov [295]; the ℓ_p-version, $0 < p < 1$, is from Kalton [255]; and the L_p-versions are from Kalton and Peck [281]. UFO spaces were introduced in [121]. All results in the UFO section belong to that paper, except (e) of Proposition 7.1.4, which was taken from [99]. That $c_0(I)$ is automorphic is in [367], as well as all results about extensible spaces. Johnson and Zippin had proved in [236] that $c_0(I)$ was extensible, while Zippin [26, Remark 5.8] pointed to the decomposition Lemma 7.2.4 as an instrument for proving the automorphic character of $c_0(I)$, and the authors of [367] say that Castillo and Johnson also suggested that possibility in private communications. The material on transitivity is from [283, Chapter 7] and [281]. The proof of Lemma 7.2.15 is something of a classic, already used by Pełczyński–Rolewicz to establish that L_p is almost isotropic; see [408, Lemma 4.6.2]. The notions of (θ, ζ)-universal disposition and (θ, ζ)-injectivity were introduced in [117] with the declared purpose of obtaining Proposition 7.3.4, which is a unifying generalisation of Kubiś [308, Theorem 3.22]: *Assume* CH. *Let V be the unique Banach space of density $\aleph_1$ that is of SUD. A Banach space of density $\aleph_1$ or less is isometric to a* 1-*complemented subspace of V if and only if it is* 1-*separably injective.* The new examples of spaces of universal disposition are from [127]. The results about positions in ℓ_1 are from [126]. The automorphy index was introduced in

[366]; Theorem 7.4.8 is from [121]. Lemma 7.4.11 was whispered to Castillo by Kalton and to Plichko by Pełczyński. The proof works for $0 \leq p < 1$ as well and shows that $\mathfrak{a}(\ell_2, L_p) = \mathfrak{c}$ since each $T \in \mathfrak{L}(L_p)$ is the difference of two positive operators; see [283, Chapter 8] for explicit representations of $\mathfrak{L}(L_p)$ when $0 \leq p < 1$. It is not known whether L_p is ℓ_p-automorphic for any $0 < p < 1$. Most of the results on automorphisms of L_p for $0 < p < 1$ displayed in the chapter have versions for $p = 0$; moreover, Peck and Starbird proved in [375] that L_0 is $\mathbb{K}^{\mathbb{N}}$-automorphic. Lemma 7.4.9 is from [121] and was thoroughly stretched to its limits in [14, Theorem 2.7], where positions in non-UFO spaces were studied from the point of view of descriptive set theory. The beautiful Proposition 7.4.15 is a strange mix: it is undoubtedly from Kalton [274], who somehow considers it a 'more precise version of a theorem of Castillo and Moreno' [109]; the proof here presented made its appearance in [121, Proposition 4.8] as 'following an idea taken from Kalton [274]'. Section 7.5.3 is from [121], and Section 7.5.4 is from [106].

8

Extension of $C(K)$-Valued Operators

Homological and operator extension methods live in symbiosis, and the theory of extension of $\mathscr{C}$-valued operators shares their habitat. However, there are a number of techniques for obtaining extensions of $\mathscr{C}$-valued operators that are so firmly anchored to the existence of an underlying compact space that they cannot be translated to more general extension problems. Those techniques depend, one way or another, on the following facts:

- The norm of $C(K)$ is related to the order structure: $\|f\| \leq \lambda$ if and only if $-\lambda \leq f(s) \leq \lambda$ for every $s \in K$.
- An operator $\tau : X \longrightarrow C(K)$ is a family of functionals on X parametrised by K. More precisely, if $\varphi \colon K \longrightarrow X^*$ is a weak*-continuous map, the formula $(\tau x)(s) = \langle \varphi(s), x \rangle$ defines an operator, and all $C(K)$-valued operators on X arise in this way.
- If g and h are functions on K such that g is upper semicontinuous, h is lower semicontinuous and $g \leq h$, then there is $f \in C(K)$ such that $g \leq f \leq h$.

Each extension problem involves a class of operators to be extended and, more or less implicitly, an often unnamed embedding. The following definitions emphasise the role of the different elements.

$\mathscr{C}$-trivial embedding. An embedding $\jmath$ is said to be $\mathscr{C}$-trivial if every $\mathscr{C}$-valued operator τ admits an extension through $\jmath$. It will be called $(\lambda, \mathscr{C})$-trivial if every $\mathscr{C}$-valued operator admits a λ-extension.

$\mathscr{C}$-extension property. A space X has the $\mathscr{C}$-extension property if all embeddings into X are $\mathscr{C}$-trivial. If all such embeddings are $(\lambda, \mathscr{C})$-trivial then X is said to have the $(\lambda, \mathscr{C})$-extension property. A space with the $(1^+, \mathscr{C})$-extension property is said to have the almost isometric $\mathscr{C}$-extension property.

$\mathscr{C}$-extensible. A separable space X is said to be $\mathscr{C}$-extensible if all embeddings $\jmath\colon X \longrightarrow C(\Delta)$ are $\mathscr{C}$-trivial.

Throughout this chapter, we will only consider real separable Banach spaces: the general non-separable case is, at the time of writing these lines, just too difficult, even if we do make progress in the non-separable world, as recounted in the story of the CCKY problem in Section 8.7. This chapter, devoted to the single topic of extending $\mathscr{C}$-valued operators has, notwithstanding that and for very good reasons, a whirlpool organisation. Let us explain why. Section 8.1 presents the *global* approach to the extension of operators: Zippin's characterisation of $\mathscr{C}$-trivial embeddings by means of weak*-continuous selectors and a few remarkable applications. Section 8.2 is devoted to the Lindenstrauss–Pełczyński theorem, one of the cornerstones of the theory. Two very different proofs for this important result are presented: the first one combines homological techniques with the global approach, while the second is Lindenstrauss–Pełczyński's original proof. The analysis of their proof is indispensable for understanding Kalton's imaginative, not once but twice, inventions that lead to the so-called (L^*) and m_1-type properties and to a decent list of $\mathscr{C}$-extensible spaces that subsumes all previously known cases. Kalton did not stop there: he further produced a no less impressive list of non-$\mathscr{C}$-extensible spaces. Kalton's approach to the $\mathscr{C}$-extension property was primarily designed to deal with Lipschitz maps. Accordingly, in Section 8.3 we present those points of the non-linear theory that are necessary to develop the linear theory. Kalton's subtle analysis crystallises into an asymptotic property of the norm, property (L^*), which implies the almost isometric $\mathscr{C}$-extension property and is enjoyed by most classical sequence spaces … after suitable renormings. The list can be found in Section 8.4. The techniques in Section 8.5 are for the most part independent of those in the rest of the chapter, although the results are not. Section 8.6 contemplates different aspects of Zippin's problem about the extension of $\mathscr{C}$-valued operators from subspaces of ℓ_1, a seemingly offline question that is, however, central for this book: Zippin's problem is to determine which separable Banach spaces X satisfy $\operatorname{Ext}(X, C(K)) = 0$, where Ext is taken, here and for the rest of the chapter, in the category of Banach spaces. The question admits an interesting gradation in terms of the topological complexity of K, and so the chapter continues with a detailed analysis of the class of Banach spaces X for which $\operatorname{Ext}(X, C(\Delta)) = 0$ and the larger class of those for which $\operatorname{Ext}(X, C(\omega^\omega)) = 0$. In Section 8.7, we report the complete solution of the problem of whether $\operatorname{Ext}(C(K), c_0) \neq 0$ for all non-metrisable compacta K. The preparations for this travel conclude with a simple and usually overlooked result:

Lemma 8.0.1 *Finite-rank c_0-valued operators admit compact 1-extensions.*

Proof We first consider the extension from finite-dimensional subspaces. Let F be a finite-dimensional subspace of a Banach space X, and let $\tau\colon F \longrightarrow c_0$ be an operator. Represent τ by a pointwise null sequence (τ_n) in F^*. For each n, let T_n be a Hahn–Banach extension of τ_n to X. The sequence (T_n) defines a 1-extension $T\colon X \longrightarrow \ell_\infty$ of τ. But since F is finite-dimensional, the pointwise null sequence (τ_n) is actually norm null, as is (T_n), which makes T compact and c_0-valued. The general case follows from this, but the reader is left to discover why: given a finite-rank operator $Y \longrightarrow c_0$ with range F defined on a subspace $Y \subset X$, consider the pushout diagram

$$\begin{array}{ccc} Y & \longrightarrow & X \\ \downarrow & & \downarrow \\ F & \longrightarrow & \mathrm{PO} \\ \downarrow & & \\ c_0 & & \end{array}$$

By the argumentation above, the inclusion $F \longrightarrow c_0$ admits a compact 1-extension to PO. And that is all. □

Our previous exposure to local injectivity in Chapter 5 now lets us indulge ourselves in saying:

Lemma 8.0.2 *Every finite-rank operator taking values in a Lindenstrauss space admits* 1^+*-extensions to any superspace.*

8.1 Zippin Selectors

We focus here on determining when a given operator $\tau\colon Y \longrightarrow C(K)$ has an extension through an embedding $\jmath\colon Y \longrightarrow X$. We look at the situation from the perspective provided by the adjunction in Note 4.6.1(d): the operator $\tau\colon Y \longrightarrow C(K)$ is associated with a weak*-continuous map $\varphi\colon K \longrightarrow Y^*$ by the formula $\langle \varphi(k), y\rangle = (\tau(y))(k)$ and, if the operator $T\colon X \longrightarrow C(K)$ is associated with $\Phi\colon K \longrightarrow X^*$, then T extends τ through $\jmath$ if and only if Φ lifts φ through $\jmath^*$:

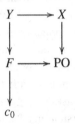

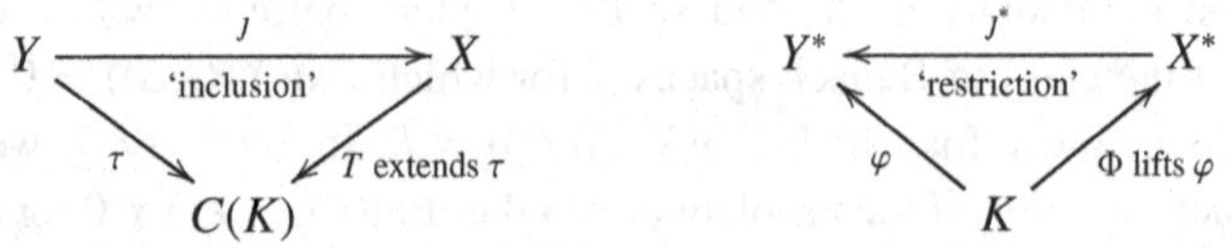

If we assume that $\|\tau\| \le 1$ so that $\varphi[K] \subset B^*_Y$, it is clear that an overwhelmingly sufficient condition for τ to admit an extension is the existence of a weak*-continuous mapping $\omega\colon B_Y^* \longrightarrow X^*$ such that $\jmath^* \circ \omega$ is the identity on B_Y^*. It was Zippin who introduced this 'global approach' idea into the business of extending $\mathscr{C}$-valued operators.

Definition 8.1.1 Let $\jmath\colon Y \longrightarrow X$ be an embedding between Banach spaces. A Zippin selector for $\jmath$ is a weak*-continuous map $\omega\colon B_Y^* \longrightarrow X^*$ such that $\jmath^* \circ \omega$ is the identity on B_Y^*. If $\|\omega(y^*)\| \leq \lambda$ for every $y^* \in B_Y^*$ as well then we call it a λ-Zippin selector.

Since weak*-compact sets are bounded, we see that each Zippin selector can be labelled with some λ. The following proof shows that the canonical isometry $\delta\colon Y \longrightarrow C(B_Y^*)$, which was shown to be the best embedding into a $\mathscr{C}$-space regarding complementation in 2.12.2, is the most difficult operator to extend.

Proposition 8.1.2 *Let $\jmath\colon Y \longrightarrow X$ be an embedding and $\lambda \geq 1$. The following are equivalent:*

(i) *$\jmath$ is $(\lambda, \mathscr{C})$-trivial.*
(ii) *$\delta\colon Y \longrightarrow C(B_Y^*)$ admits a λ-extension to X.*
(iii) *$\jmath$ admits a λ-Zippin selector.*

Proof The implication (i) $\Longrightarrow$ (ii) is trivial. To prove (ii) $\Longrightarrow$ (iii), let D be an operator making commutative the diagram

$$\begin{array}{ccc} Y & \xrightarrow{\ \jmath\ } & X \\ & {}_{\delta}\searrow \quad \swarrow_{D} & \\ & C(B_Y^*) & \end{array}$$

Since $D\jmath = \delta$ then $\delta^* = \jmath^* D^*$, and thus the restriction of D^* to B_Y^* is a $\|D\|$-Zippin selector for $\jmath$. Note that $D^*|_{B_Y^*}\colon B_Y^* \longrightarrow X^*$ is the weak*-continuous mate of D in the adjunction, while the mate of the inclusion $B_Y^* \longrightarrow Y^*$ is $\delta\colon Y \longrightarrow C(B_Y^*)$.

The implication (iii) $\Longrightarrow$ (i) has been basically proved above, but we do it again anyway. Assume $\omega\colon B_Y^* \longrightarrow X^*$ is a λ-Zippin selector for $\jmath$, and let $\tau\colon Y \longrightarrow C(K)$ be an operator with $\|\tau\| = 1$. Define $T\colon X \longrightarrow C(K)$ by $Tx(k) = \langle \omega(\tau^*\delta_k), x\rangle$ for $x \in X$ and $k \in K$. Then $\|T\| \leq \lambda$ and $T\jmath y = \tau y$ since $T\jmath y(k) = \langle \omega(\tau^*\delta_k), \jmath y\rangle = \langle \jmath^*(\omega(\tau^*\delta_k)), y\rangle = \langle \tau^*\delta_k), y\rangle = \langle \delta_k), \tau y\rangle = (\tau y)(k)$. □

Having built an understanding in which we are reasonably confident of what it means to admit a Zippin selector, let us present some natural non-trivial examples of $\mathscr{C}$-trivial embeddings. A warning for the reader about these results: some of them will become obsolete by the end of the chapter.

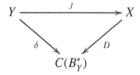

Proposition 8.1.3 *The following embeddings admit Zippin selectors:*

(a) *The inclusion of any subspace of ℓ_p into ℓ_p for $p \in (1, \infty)$.*
(b) *The natural embedding of ℓ_p into the Kalton–Peck space $\ell_p(\varphi)$ when $\varphi \in \mathrm{Lip}_0(\mathbb{R})$ and $p \in (1, \infty)$.*
(c) *The inclusion of any finite-dimensional subspace whose unit ball is a polyhedron into any Banach space.*
(d) *The inclusion of any finite-dimensional subspace into a uniformly smooth Banach space.*

Proof We shall obtain homogeneous 1-Zippin selectors in all cases except (b). Note that in cases (a) and (b) the weak* and weak topologies coincide and that the required selectors in (c) and (d) have to be continuous in the norm topology. The proof of (a) depends on the unbeatable behaviour of the duality map on ℓ_p. For a Banach space X with strictly convex dual, the duality map (also called the support map) $J\colon X \longrightarrow X^*$ takes each $x \in X$ to the only $x^* \in X^*$ such that $\|x\| = \|x^*\|$ and $\langle x^*, x\rangle = \|x\|\,\|x^*\|$. When $X = \ell_p$ for $1 < p < \infty$, we can identify the dual of ℓ_p with ℓ_q, where q is the conjugate exponent of p and the duality map $J_p\colon \ell_p \longrightarrow \ell_q$ is given by the clean formula $J_p(x) = \|x\|^{2-p}\,\mathrm{sgn}(x)\,|x|^{p-1}$ and turns out to be weak (= weak*) sequentially continuous. Let Y be a subspace of ℓ_p. Since ℓ_q is strictly convex, for each $y^* \in Y^*$, there is a unique $x^* \in \ell_q$ extending y^* with $\|x^*\| = \|y^*\|$. Let $\omega\colon Y^* \longrightarrow \ell_q$ be the resulting map. We show that the restriction $\omega\colon B_Y^* \longrightarrow B_{\ell_p}^*$ is continuous in the weak topology. Indeed, the map ω is the composition of three maps,

$$\begin{array}{ccc} Y^* & \xrightarrow{\ \omega\ } & \ell_q \\ {\scriptstyle J^{-1}}\big\downarrow & & \big\uparrow{\scriptstyle J_p} \\ Y & \xrightarrow{\ \imath\ } & \ell_p \end{array}$$

where $J\colon Y \longrightarrow Y^*$ is the duality map of Y and $\imath$ is plain inclusion. We know that J_p is weakly sequentially continuous, as is the inclusion $\imath$. To see that $J^{-1}\colon B_{Y^*} \longrightarrow B_Y$ is weak* to weak continuous, just observe that $J\colon B_Y \longrightarrow B_{Y^*}$ is one to one and weak to weak* continuous since $J(y) = J_p(y)|_Y$ and B_Y is weakly compact, while B_{Y^*} is weak*-compact.

To prove (b), we revisit the duality of Kalton–Peck spaces $\ell_p(\varphi)$ in Section 3.8 and Kalton–Peck maps (3.27) now garbed in their full-grown centralizer clothes (Section 3.12): consider the mappings $\ell_p \longrightarrow \ell_\infty$ given by

$$\mathsf{KP}_{p,\varphi}(x) = x\,\varphi\left(p \log \frac{\|x\|_p}{|x|}\right), \qquad \mathsf{kp}_{p,\varphi}(x) = x\,\varphi(-p\log|x|),$$

and form the space $\ell_p(\varphi) = \{(y, x) \in \ell_\infty \times \ell_p : \|y - \mathsf{KP}_{p,\varphi}x\|_p + \|x\|_p < \infty\}$ endowed with the obvious quasinorm. Consider the canonical embedding

$\jmath\colon \ell_p \longrightarrow \ell_p(\varphi)$ given by $\jmath(y) = (y,0)$ and let us try to find a Zippin selector $\omega\colon \ell_p^* \longrightarrow \ell_p(\varphi)^*$ for $\jmath$. Identify $\ell_p(\varphi)^*$ with $\ell_q(-\varphi)$ as in Proposition 3.8.5 through the duality pairing $\langle (x^*,y^*),(y,x)\rangle = \langle x^*,x\rangle + \langle y^*,y\rangle$. Since $\jmath^*(x^*,y^*) = y^*$, we set the Zippin selector $\omega : \ell_q \longrightarrow \ell_q(\varphi)$ to be the map

$$\omega(y^*) = (-\mathsf{kp}_{q,\varphi}(y^*), y^*),$$

which takes values in the right space and is bounded, since, by estimate (3.7),

$$\begin{aligned}\|\omega(y^*)\| &= \left\|(-\mathsf{kp}_{q,\varphi}(y^*), y^*)\right\|_{-\mathsf{KP}_{q,\varphi}} \\ &= \|\mathsf{KP}_{q,\varphi}(y^*) - \mathsf{kp}_{q,\varphi}(y^*)\| + \|y^*\| \\ &\le q\,\mathrm{Lip}(\varphi)\|y^*\|_q \log \|y^*\|_q.\end{aligned}$$

Finally, ω is weak*-continuous on bounded sets because a sequence converges in the weak (= weak*) topology of $\ell_q(-\varphi)$ if and only if it is bounded and pointwise convergent, and $\mathsf{kp}_{q,\varphi}$ preserves pointwise convergence (the value of each coordinate of $\mathsf{kp}_{q,\varphi}(x)$ depends on the corresponding coordinate of x in a continuous way).

To prove (c), let $\imath\colon F \subset X$ be a finite-dimensional subspace of the Banach space X, and let $\imath^*\colon X^* \longrightarrow F^*$ be the restriction map. If B_F is a polyhedron (i.e., the convex hull of a finite set) then so is B_F^*. Let $(S_i)_{1\le i\le k}$ be a *triangulation* of B_F^*, which means that

- each S_i is a simplex of the same dimension as F^* and $B_F^* = \bigcup_{i=1}^k S_i$,
- the intersection of any pair of them is a (possibly empty) common face.

Let V be the set of all vertices of the triangulation, and observe that not every vertex has to be in the boundary of the ball. For each $v \in V$, let $x_v^* \in X^*$ be a norm-preserving extension of v. Now, each $y^* \in B_F^*$ has a unique convex representation as $y^* = \sum_{v\in V} c_v(y^*)v$, and the coordinate functions $y^* \longmapsto c_v(y^*)$ are all continuous. The map $\omega\colon B_F^* \longrightarrow X^*$ given by

$$\omega(y^*) = \sum_{v\in V} c_v(y^*)x_v^*$$

is a 1-Zippin selector. Note that ω is piecewise linear but not necessarily homogeneous. One can obtain a homogeneous version of ω by just taking $\tilde{\omega}(0) = 0$ and

$$\tilde{\omega}(y^*) = \frac{\|y^*\|}{2}\left(\omega\left(\frac{y^*}{\|y^*\|}\right) - \omega\left(\frac{-y^*}{\|y^*\|}\right)\right).$$

This map is again weak* continuous since F^* is finite-dimensional.

(d) Let F be a finite-dimensional subspace of X. Then, for every $n \in \mathbb{N}$, there is a Zippin selector $\omega_n\colon B_F^* \longrightarrow (1+\frac{1}{n})B_X^*$. A weak*-accumulation point of $\omega_n(y^*)$ must be an extension of y^*, which in a uniformly smooth space is unique. So it makes sense to define $\omega(y^*)$ as the weak*-accumulation point of $\omega_n(x^*)$, and this yields a selector $\omega\colon B^*_F \longrightarrow B^*_X$. Again by uniform smoothness, for every normalised $y^* \in F^*$, we have $\omega(y^*) = \lim_n \omega_n(x^*)$ in the norm topology of X^*. Hence ω is weak*-continuous on the unit sphere of F^*. □

8.2 The Lindenstrauss–Pełczyński Theorem

We now present the Lindenstrauss–Pełczyński theorem, which is perhaps the first significant result about the extension of $\mathscr{C}$-valued operators. We shall provide two (actually three) proofs for this important result. The first has a homological flavour but ultimately depends on the simple structure of the subspaces and quotients of c_0 and comes without any explicit bound on the norm of the extension. Here it is.

Theorem 8.2.1 *Every $\mathscr{C}$-valued operator defined on a subspace of c_0 can be extended to c_0.*

Proof The proof proceeds in several steps.

Step 1 We test our abilities when the subspace and the embedding have the simplest conceivable form. For each integer k, let $\jmath_k\colon A_k \longrightarrow \ell_\infty^{n(k)}$ be an isometry and let $\jmath\colon c_0(\mathbb{N}, A_k) \longrightarrow c_0(\mathbb{N}, \ell_\infty^{n(k)})$ be the isometry given by $\jmath((a_k)_k) = (\jmath_k(a_k))_k$. For each $k \in \mathbb{N}$, let $\omega_k\colon B^*_{A_k} \longrightarrow \ell_1^{n(k)}$ be the homogeneous 1-Zippin selector constructed in the proof of Proposition 8.1.3(c). Paste together all the maps ω_k into one map $\omega\colon B_{\ell_1(\mathbb{N},A_k^*)} \longrightarrow B_{\ell_1(\mathbb{N},\ell_1^{n(k)})}$ by setting $\omega((x_k^*)_k) = (\omega_k(x_k^*))_k$. This is clearly a selector for $\jmath$, and it turns out to be weak*-continuous: observe that its domain is metrisable and that a bounded sequence in the space $\ell_1(\mathbb{N}, A_k)$ is weak*-null if and only if the norms of its projections into each A_k are convergent to 0. This shows that the embedding $\jmath$ is $\mathscr{C}$-trivial.

Step 2 We keep the subspace isomorphic to some $c_0(\mathbb{N}, A_k)$, with A_k finite-dimensional, but now consider any possible embedding $\imath\colon c_0(\mathbb{N}, A_k) \longrightarrow c_0$. Let us pick an almost isometric embedding $\jmath\colon c_0(\mathbb{N}, A_k) \longrightarrow c_0(\mathbb{N}, \ell_\infty^{n(k)}) = c_0$ once again. Replacing c_0 by its square if necessary, we can assume that both $\jmath$ and $\imath$ have infinite codimensional ranges, and so the automorphic character of c_0 comes to the rescue by providing an automorphism u that intertwines $\jmath$ and $\imath$ as in the commutative diagram

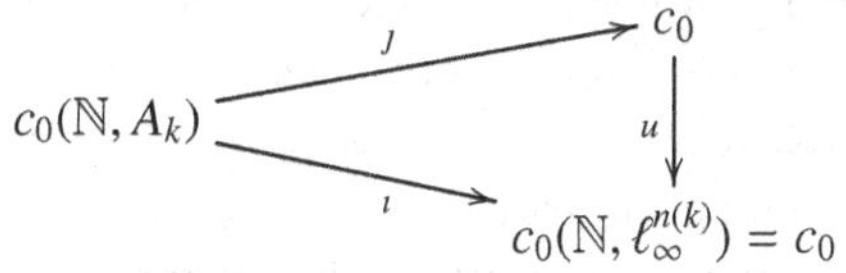

Needless to say, an operator defined on $c_0(\mathbb{N}, A_k)$ can be extended through $\imath$ if and only if it can be extended through $\jmath$.

Step 3 Finally, if Y is any arbitrary closed subspace of c_0, we use the Johnson–Rosenthal–Zippin result in Proposition 5.3.1 to decompose Y as a twisted sum $0 \longrightarrow c_0(\mathbb{N}, A_k) \longrightarrow Y \longrightarrow c_0(\mathbb{N}, B_k) \longrightarrow 0$ in which A_k and B_k are finite-dimensional. Now we draw the pushout diagram:

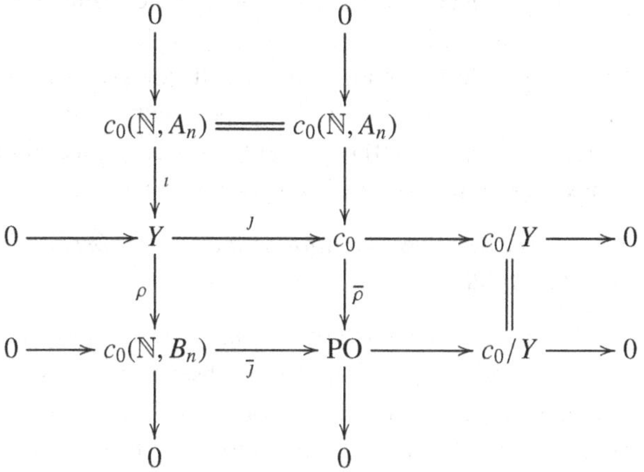

Let $\tau\colon Y \longrightarrow E$ be a $\mathscr{C}$-valued operator. Its restriction $\tau\imath$ can be extended to c_0. Let $T_1 : c_0 \longrightarrow E$ be any extension of $\tau\imath$. The difference $\tau - T_1\jmath$ vanishes on the image of $\imath$, and thus it factors through the quotient ρ in the form $\tau - T_1\jmath = \tau_2\rho$ for some operator $\tau_2\colon c_0(\mathbb{N}, B_k) \longrightarrow E$. Since every quotient of c_0 is isomorphic to a subspace of c_0, τ_2 admits an extension, say T_2, to PO. The operator $T = T_1 + T_2\overline{\rho}$ is an extension of τ since, according to our records, for every $y \in Y$, we have $T_2(\overline{\rho}(y)) = T_2(\overline{\jmath}(\rho(y)) = \tau_2(\rho(y)) = \tau(y) - T_1(y)$. □

The statement just proved implies the existence of a constant C such that every $\mathscr{C}$-valued operator defined on any subspace of c_0 has a C-extension to c_0: this follows from an obvious amalgamation argument, taking into account that if (H_n) is a sequence of subspaces of c_0 then so is $c_0(\mathbb{N}, H_n)$, and that if (E_n) is a sequence of $\mathscr{C}$-spaces then so is $\ell_\infty(\mathbb{N}, E_n)$. However, there is no easy way to follow the track of C throughout the proof since one has no control over the parameters of the left vertical sequence in the preceding diagram nor over the norm of the automorphism in the second step of the proof. In [465, Proposition 3], Zippin provides a complete proof for the following: *every $\mathscr{C}$-valued operator defined on a subspace of c_0 admits a 4^+-extension.* The proof uses his global approach, very much in the spirit of the previous proof, but without the good (homological) parts.

We will stop burdening the reader with these worries and next recover the original proof supplied by Lindenstrauss and Pełczyński [330], which provides sharp bounds for the norm of the extension. The ripples from this stone thrown in the $\mathscr{C}$-extension pond will spread out through the entire chapter.

Let K be a compact space, and suppose we are given two bounded (but not necessarily continuous) functions $f, g\colon K \longrightarrow \mathbb{R}$ such that $f \leq g$. Under what circumstances can we separate them by a continuous $h\colon K \longrightarrow \mathbb{R}$ in the sense that $f(s) \leq h(s) \leq g(s)$ for all $s \in K$? The Hahn–Tong theorem (see the proof of Lemma 2.2.2) states that this is the case if f is upper-semicontinuous and g is lower-semicontinuous. It quickly follows that such an $h \in C(K)$ exists if and only if the upper semicontinuous envelope $f^{\mathrm{usc}}(s) = \min\left(f(s), \limsup_{t\to s} f(t)\right)$ of f and the lower semicontinuous envelope $g_{\mathrm{lsc}}(s) = \max\left(f(s), \liminf_{t\to s} g(t)\right)$ of g satisfy $f^{\mathrm{usc}} \leq g_{\mathrm{lsc}}$. Enough said:

Theorem 8.2.2 *Every $\mathscr{C}$-valued operator defined on a subspace of c_0 admits a 1^+-extension to the whole c_0.*

Proof Let $\tau\colon H \longrightarrow C(K)$ be an operator, where H is a subspace of c_0 and K is a compact space. Since every separable subset of $C(K)$ is contained in a separable subalgebra and these are $\mathscr{C}$-spaces, we may assume that K is metrisable and also that $\|\tau\| = 1$. It is enough to prove that τ can be extended to an operator T on $H + [x]$ having norm at most λ for each $x \in c_0 \setminus H$ and $\lambda > 1$. Searching for an admissible value $h = T(x)$ means showing that there exists $h \in C(K)$ such that $\|h - \tau(y)\| \leq \lambda\|x - y\|$ for all $y \in H$. Using the order structure of $C(K)$, this is equivalent to

$$\tau(y) - \lambda\|y - x\|1_K \leq h \leq \tau(y) + \lambda\|y - x\|1_K. \tag{8.1}$$

Letting

$$F = \bigvee_{y\in H} \left(\tau(y) - \lambda\|x - y\|1_K\right) \quad \text{and} \quad G = \bigwedge_{y\in H} \left(\tau(y) + \lambda\|x - y\|1_K\right),$$

we see that $F \leq G$ and that h fits in (8.1) if and only if $F \leq h \leq G$. The proof will be complete if we show that $F^{\mathrm{usc}} \leq G_{\mathrm{lsc}}$. Here is where c_0 comes into play. If for some $s \in K$ we have $F^{\mathrm{usc}}(s) > G_{\mathrm{lsc}}(s)$ then there are sequences $(s_n), (t_n)$ in K converging to s for which

$$\lim_n F(s_n) > \lim_n G(t_n).$$

From the definition of F and G, it can easily be deduced that there exist $y_n, z_n \in H$ such that

$$\lim_n \left(\tau(y_n)(s_n) - \lambda\|x - y_n\|\right) > \lim_n \left(\tau(z_n)(t_n) + \lambda\|x - z_n\|\right). \tag{8.2}$$

Now we consider the adjoint $\tau^* : C(K)^* \longrightarrow H^*$. For each n, $\tau^*(\delta_{s_n})$ is a linear functional on H that admits an equal norm extension to c_0, which we will call $u_n \in \ell_1$. Similarly, for each n, let $v_n \in \ell_1$ be a norm-preserving extension of $\tau^*(\delta_{t_n})$ to c_0. Note that $\|u_n\|, \|v_n\| \le 1$. Passing to subsequences if necessary, we can assume that both $(u_n)_n$ and $(v_n)_n$ are weak*-convergent in ℓ_1. If we denote the corresponding limits by u and v, then $u|_H = v|_H = \tau^*(\delta_s)$ since τ^* is continuous in the weak* topologies and $\delta_{t_n}, \delta_{s_n}$ converge to δ_s in the weak* topology of $C(K)^*$. On the other hand, if $(w_n)_n$ is weak*-null in ℓ_1, it is almost obvious that for every $w \in \ell_1$ we have $\lim_n (\|w_n + w\| - \|w_n\| - \|w\|) = 0$. In particular,

$$\lim_{n\to\infty} \left(\|u_n\| - \|u_n - u\| - \|u\|\right) = \lim_{n\to\infty} \left(\|v_n\| - \|v_n - v\| - \|v\|\right) = 0. \tag{8.3}$$

If we restrict these functionals to H then the norms of u_n and v_n do not vary (they are norm-preserving extensions of $\tau^*\delta_{s_n}$ and $\tau^*\delta_{t_n}$), while the norms of the other functionals in (8.3) only decrease. Letting $r = \left\| u|_H \right\| = \left\| v|_H \right\| = \left\| \tau^*(\delta_s) \right\|$, we have $\limsup_n \|u_n - u\| \le 1 - r$ and also $\limsup_n \|v_n - v\| \le 1 - r$, both with respect to the ℓ_1 norm. Since $\tau(y_n)(s_n) = \langle \tau^* s_n, y_n \rangle = \langle u_n, y_n \rangle$ and $\tau(z_n)(t_n) = \langle \tau^* t_n, z_n \rangle = \langle v_n, z_n \rangle$, we can rewrite (8.3) as

$$L = \lim_n \left(\langle v_n, z_n \rangle - \langle u_n, y_n \rangle + \lambda(\|x - z_n\| + \|x - y_n\|)\right) < 0. \tag{8.4}$$

But $\lim_n \langle u_n - u, z \rangle = \lim_n \langle v_n - v, z \rangle = 0$ for every $z \in c_0$, so

$$\begin{aligned}
L &= \lim_n \left(\langle v_n, z_n \rangle - \langle u_n, y_n \rangle + \lambda(\|x - z_n\| + \|x - y_n\|)\right) \\
&= \lim_n \left(\langle v_n - v, z_n \rangle - \langle u_n - u, y_n \rangle + \langle \tau^*(\delta_s), z_n - y_n \rangle + \lambda(\|x - z_n\| + \|x - y_n\|)\right) \\
&= \lim_n \left(\langle v_n - v, z_n - x \rangle - \langle u_n - u, y_n - x \rangle + \langle \tau^*\delta_s, z_n - y_n \rangle + \lambda(\|x - z_n\| + \|x - y_n\|)\right) \\
&= \lim_n (\alpha_n + \beta_n + \gamma_n),
\end{aligned}$$

with the obvious choices

$$\begin{aligned}
\alpha_n &= \langle v_n - v, z_n - x \rangle + (\lambda - r)\|z_n - x\|, \\
\beta_n &= -\langle u_n - u, y_n - x \rangle + (\lambda - r)\|y_n - x\|, \\
\gamma_n &= \langle \tau^*(\delta_s), z_n - y_n \rangle + r(\|x - z_n\| + \|x - y_n\|).
\end{aligned}$$

Now, α_n and β_n are non-negative for large n, while $\gamma_n \ge 0$ for all n. This contradicts (8.4), which concludes the proof. □

We cannot usually achieve 1-extensions, as the following example shows.

8.2.3 *There is a compact operator from an hyperplane of c_0 into c which has no norm-preserving extension to c_0.*

Proof Fix $0 < \lambda < \frac{1}{2}$ and define $f = \sum_{n=1}^{\infty} \lambda^{n-1}(e_{2n-1} + e_{2n})$. Clearly, $f \in \ell_1$. Treating ℓ_1 as the dual of c_0 in the obvious way, we set $H = \ker f$. Let

$$g = \sum_{n=1}^{\infty} \lambda^{n-1} e_{2n} - \sum_{n=2}^{\infty} \lambda^{n-1} e_{2n-1},$$

and for $n = 2, 3, ...,$ we set

$$g_n = \sum_{j=1}^{n} \lambda^{j-1} e_{2j} - \sum_{j=2}^{n} \lambda^{j-1} e_{2j-1},$$

$$h_n = \sum_{j=1}^{n} \lambda^{j-1} e_{2j} - \sum_{j=2}^{n} \lambda^{j-1} e_{2j-1} + \lambda^{n-1}(e_{2n-1} + e_{2n+2}).$$

It is clear that $g = \lim g_n = \lim h_n$ in the norm of ℓ_1. In particular, the convergence is maintained in H^*. We put

$$G_n = \frac{g_n|_H}{\|g_n|_H\|}, \qquad H_n = \frac{h_n|_H}{\|h_n|_H\|}, \qquad G = \frac{g|_H}{\|g|_H\|}.$$

The set $K = \{G\} \bigcup \{G_n, H_n : n \in \mathbb{N}\}$ is a norm-closed subset of H^* homeomorphic to $\alpha\mathbb{N}$, the one point compactification of $\mathbb{N}$, and the inclusion of K into the unit ball of H^* is *norm* continuous so it induces a *compact* operator $\tau : H \longrightarrow C(K)$. By the considerations made at the beginning of the section, everything we have to show is that there is no weak*-continuous map $\phi : K \longrightarrow B_{\ell_1}$ that satisfies $\phi(k)|_H = k$ for all $k \in K$. To show that, it is enough to show that all G_n and H_n have unique Hahn–Banach extensions g_n and $h_n - f$, respectively. Since $\lim g_n = g$ and $\lim h_n - f = g - f$, the proof is essentially done. To fill in the details, observe that any extension of G_n must have the form $g_n + tf$. Since

$$\|g_n + tf\| = |t| \left(1 + \frac{2\lambda^n}{1 - \lambda}\right) + |1 - t| \frac{1 - \lambda^n}{1 - \lambda} + \lambda |1 + t| \frac{1 - \lambda^{n-1}}{1 - \lambda}$$

and the minimum of this quantity occurs only at $t = 0$, it follows that the unique norm-preserving extension of G_n is g_n. Also, any extension of H_n must have the form $h_n + tf$. Since

$$\|h_n + tf\| = |t| \left(1 + \frac{2\lambda^{n+1}}{1 - \lambda}\right) + |1 - t| \frac{1 - \lambda^n}{1 - \lambda} + \lambda |1 + t| \frac{1 - \lambda^{n-1}}{1 - \lambda} + 2\lambda^{n-1} |1 - \lambda t|$$

and the minimum of this quantity occurs only at $t = 1$, it follows that the unique Hahn–Banach extension of H_n is $h_n - f$. □

8.3 Kalton's Approach to the $\mathscr{C}$-Extension Property

Kalton's approach to the extension of $\mathscr{C}$-valued operators is rooted in his studies of Lipschitz maps in Banach spaces. Lipschitz maps have appeared in Banach space theory since its inception, sporadically at first but occupying an increasingly central role. The authoritative book by Benyamini and Lindenstrauss [41] is responsible to a great extent for this turnaround. In the series of papers [271; 272; 273; 274], Kalton revisits the topic and, as he always did, sets new standards. Of course, the linear and non-linear theories are different, and thus we will present here only the facts of the non-linear theory that are indispensable for understanding the linear part. The interested reader can behold a more complete picture in Section 8.8.3. The following observation helps avoid possible misunderstandings concerning the role of the space c in the ensuing exposition.

Proposition 8.3.1 *There is a 2-Lipschitz retraction of ℓ_∞ onto c.*

Proof Given $x \in \ell_\infty$, set

$$x^- = \liminf_n x(n), \quad x^- = \liminf_n x(n), \quad m(x) = \frac{x^- + x^+}{2}, \quad r(x) = \frac{x^+ - x^-}{2},$$

and define $\tilde{x}\colon \mathbb{N} \longrightarrow \mathbb{R}$ by

$$\tilde{x}(n) = \begin{cases} m(x) & \text{if } x(n) \in [x^-, x^+], \\ x(n) - r(x) & \text{if } x(n) > x^+, \\ x(n) + r(x) & \text{if } x(n) < x^-. \end{cases}$$

It is clear that $\tilde{x} \in c$ for all $x \in \ell_\infty$ and also that $\tilde{x} = x$ when $x \in c$, so the mapping $x \longrightarrow \tilde{x}$ extends the identity of c. To prove that this map is 2-Lipschitz, one has to check that $|\tilde{x}(n) - \tilde{y}(n)| \leq 2\|x - y\|$ for all n. This is very easy, taking into account that $|m(x) - m(y)|, |r(x) - r(y) \leq \|x - y\|$ and reasoning on a case-by-case basis. □

However, no Lipschitz retraction of ℓ_∞ onto c can be linear at all because c is not injective in the category of Banach spaces (even if it is injective in 'the Lipschitz category of Banach spaces'). Is separability the problem? Not on its own. In fact, c is injective in both the Lipschitz and the linear categories of separable Banach spaces, but the constants are different: c is 3-separably injective as a Banach space, while it is 2-injective in the Lipschitz category. So, even in this setting, the previous proposition highlights a subtle difference between the extension of Lipchitz maps and the extension of linear operators.

Corollary 8.3.2 *Let Y be a subset of a metric space X. Every Lipschitz map $\tau\colon Y \longrightarrow c$ admits a 2-extension to X.*

Proof Extend a contraction τ to a contraction $X \longrightarrow \ell_\infty$ and then compose with the map provided by Proposition 8.3.1. □

Extension of $\mathscr{C}$-Valued Lipchitz Maps for Dummies

For a first contact with the problems we face, let us cheat a bit about the simplest conceivable extension problem for non-linear Lipschitz $\mathscr{C}$-valued maps by picking arguably the simplest infinite compactum there is: the one-point compactification $\alpha\mathbb{N}$ of $\mathbb{N}$ and its associated space of convergent sequences $c = C(\alpha\mathbb{N})$. Let Y be a subset of a metric space X, and let $\tau\colon Y \longrightarrow c$ be a fixed contraction, which is not assumed to be linear for obvious reasons. Given $x \in X\backslash Y$ and $\lambda \geq 1$, under what conditions can τ be extended to a λ-Lipschitz map $Y \cup \{x\} \longrightarrow c$? As we already know by Corollary 8.3.2, this can always be achieved when $\lambda \geq 2$. Of course, what has to be done is to assign an admissible value $\xi \in c$ to x such that $\|\xi - \tau(y)\| \leq \lambda d(x, y)$ for every $y \in Y$. This means, taking advantage of the order structure of c, that ξ must satisfy

$$\tau(y) - \lambda d(x, y)1_{\mathbb{N}} \leq \xi \leq \tau(y) + \lambda d(x, y)1_{\mathbb{N}}. \tag{8.5}$$

We define two bounded functions τ^-, τ^+ on $\mathbb{N}$ by

$$\tau^- = \bigvee_{y\in Y} \tau(y) - \lambda d(x, y)1_{\mathbb{N}}, \qquad \tau^+ = \bigwedge_{z\in Y} \tau(z) + \lambda d(x, z)1_{\mathbb{N}},$$

where the supremum and the infimum are defined pointwise on $\mathbb{N}$. It is clear that $\tau^- \leq \tau^+$ and also that $\xi \in \ell_\infty$ satisfy (8.5) if and only if $\tau^- \leq \xi \leq \tau^+$. Now, the point is that some $\xi \in c$ exists separating τ^- from τ^+ if and only if

$$\limsup_{n\to\infty} \tau^-(n) \leq \liminf_{n\to\infty} \tau^+(n),$$

and this condition is clearly equivalent to:

- For every $\varepsilon > 0$, there is $N \in \mathbb{N}$ such that for every $n, k > N$ and every $y, z \in Y$, one has $|\tau(y)(n) - \tau(z)(m)| \leq \lambda(d(y, x) + d(z, x)) + \varepsilon$.

Corollary 8.3.2 says that this condition is satisfied by all Lipschitz maps on all metric spaces when $\lambda = 2$. It is up to the reader whether to decide whether or not this was unforeseen.

Extension of c-Valued Lipschitz Maps

Let's keep up our steady trotting pace with a slight modification to the previous problem: we now want to fix a λ and extend all contractions $Y \longrightarrow c$ to one more point while keeping the Lipschitz constant of the extension bounded by λ. This will be the key step in the extension of $\mathscr{C}$-valued Lipschitz maps.

Lemma 8.3.3 *Let Y be a separable subset of a metric space X, $x \in X \setminus Y$ and $\lambda \geq 1$. The following are equivalent:*

(i) *Every Lipschitz map $\tau\colon Y \longrightarrow c$ admits a λ-extension to $Y \cup \{x\}$.*
(ii) *Given sequences $(y_n), (z_n)$ in Y and $\varepsilon > 0$, there is $u \in Y$ such that*

$$d(u, y_n) + d(u, z_n) \leq \lambda(d(x, y_n) + d(x, z_n)) + \varepsilon \tag{8.6}$$

for infinitely many n.
(iii) *Every $\mathscr{C}$-valued Lipschitz map on Y admits a λ-extension to $Y \cup \{x\}$.*

Proof We will work with contractions. To prove the implication (i) $\Longrightarrow$ (ii), let us fix a countable dense subset $(e_n)_{n\geq 1}$ of Y. The following condition is clearly stronger than (ii):

(†) Given $\varepsilon > 0$, there is $n \in \mathbb{N}$ such that for any $y, z \in Y$ we have

$$\min_{1\leq j\leq n} \left(d(e_j, y) + d(e_j, z)\right) \leq \lambda\,(d(x, y) + d(x, z)) + \varepsilon.$$

Now we establish the implication (i) $\Longrightarrow$ (†). If (†) fails, we can construct sequences (y_n) and (z_n) in Y such that, for $j < n$, we have

$$\begin{aligned} \lambda\,(d(x, y_n) + d(x, z_n)) + \varepsilon &< d(y_j, y_n) + d(y_j, z_n),\\ \lambda\,(d(x, y_n) + d(x, z_n)) + \varepsilon &< d(z_j, y_n) + d(y_j, z_n),\\ \lambda\,(d(x, y_n) + d(x, z_n)) + \varepsilon &< d(e_j, y_n) + d(e_j, z_n). \end{aligned} \tag{8.7}$$

Given $e \in Y$, the sequence $(d(e, y_n) - d(x, y_n))$ is bounded by $d(e, x)$, and thus, since Y is separable, by merciless diagonalisation, we can assume that $\lim_n d(e, y_n) - d(x, y_n)$ exists for all $e \in Y$. We define now a sequence of Lipschitz maps $f_n\colon Y \longrightarrow \mathbb{R}$. First the odd terms:

$$f_{2n-1}(y) = d(y, y_n) - d(x, y_n),$$

which have $\mathrm{Lip}(f_{2n-1}) = 1$. To define the even terms, we first define new Lipschitz maps $\varphi_n\colon \{y_k\}_{k=1}^n \cup \{z_k\}_{k=1}^n \cup \{e_k\}_{k=1}^n \longrightarrow \mathbb{R}$ by

$$\begin{cases} \varphi_n(y_j) = f_{2n-1}(y_j), & j < n,\\ \varphi_n(z_j) = f_{2n-1}(z_j), & j < n,\\ \varphi_n(e_j) = f_{2n-1}(e_j), & j \leq n, \end{cases}$$

and

$$\begin{cases} \varphi_n(y_n) = f_{2n-1}(y_n) + \lambda(d(y_n, x) + d(z_n, x)) - d(y_n, z_n) + \varepsilon,\\ \varphi_n(z_n) = f_{2n-1}(z_n) + \lambda\,(d(y_n, x) + d(z_n, x)) - d(y_n, z_n) + \varepsilon. \end{cases}$$

Claim $\operatorname{Lip}(\varphi_n) \le 1$.

Proof of the claim It suffices to estimate $\varphi_n(w)-\varphi_n(w')$ when $w \in \{y_n, z_n\}$ and $w' \in \{y_1, \dots, y_{n-1}, z_1, \dots, z_{n-1}, e_1, \dots, e_n\}$. Because if so,

$$\varphi_n(w) - \varphi_n(w') \ge f_{2n-1}(w) - f_{2n-1}(w') \ge -d(w, w'),$$

and the claim is proved using the inequalities (8.7) to get

$$\varphi_n(w) - \varphi_n(w') \le d(w, y_n) + d(z_n, w') - d(y_n, z_n) \le d(w, w'). \qquad \square$$

The even term f_{2n} can be any contractive extension of φ_n to Y. Now let $\tau\colon Y \longrightarrow \ell_\infty$ be the contractive map given by $\tau(y) = (f_n(y))_n$. Note that $\tau(e_j) \in c$ for every j, and since the e_j generate a dense subspace in Y, it turns out that τ takes values in c. We prove that τ cannot be extended to a λ-Lipschitz mapping $\tilde{\tau}\colon Y \cup \{x\} \longrightarrow c$. Suppose that $\tilde{\tau} = (g_n)_{n\le 1}$ is such an extension. Then it follows from

$$\begin{cases} g_{2n}(x) \ge f_{2n}(z_n) - \lambda d(x, z_n) \\ g_{2n-1}(x) \le f_{2n-1}(y_n) + \lambda d(x, y_n) \end{cases}$$

that $g_{2n}(x) - g_{2n-1}(x) \ge \varepsilon$, which flagrantly contradicts our assumption that $\tilde{\tau}$ takes values in c.

Let us prove the much easier implication (ii) $\Longrightarrow$ (iii). Let $\tau : Y \longrightarrow C(K)$ be a contractive map that we want to extend to a λ-Lipschitz map on $Y \cup \{x\}$. This amounts to finding $f \in C(K)$ such that $\|f - \tau(y)\|_\infty \le \lambda d(x, y)$ for all $y \in Y$; that is,

$$\tau(y) - \lambda d(x, y)1_K \le f \le \tau(y) - \lambda d(x, y)1_K. \tag{8.8}$$

Define $F, G \in \ell_\infty(K)$ by

$$F = \bigvee_{y\in Y} \tau(y) - \lambda d(x, y)1_K, \qquad G = \bigwedge_{z\in Y} \tau(z) + \lambda d(x, z)1_K.$$

Clearly, $F \le G$, and f satisfies (8.8) if and only if $F \le f \le G$. Thus the proof will be complete if we show for each $s \in K$ that

$$\limsup_{t\to s} F(t) \le \liminf_{t\to s} G(t).$$

Assume on the contrary that $\limsup_{t\to s} F(t) > \liminf_{t\to s} G(t)$ for some $s \in K$. Then there is $\varepsilon > 0$ and a pair of sequences $(s_n), (t_n)$ converging to s such that $F(s_n) > G(t_n) + 2\varepsilon$, and then we may choose sequences $(y_n), (z_n)$ in Y such that

$$(\tau(y_n))(s_n) - \lambda d(x, y_n) > \tau(z_n)(t_n) + \lambda d(x, z_n) + 2\varepsilon.$$

Now we apply (ii) with these $(y_n), (z_n)$ and ε to find $u \in Y$ such that

$$d(u, y_n) + d(u, z_n) \le \lambda(d(x, y_n) + d(x, z_n)) + \varepsilon$$

for infinitely many n. Let us take a look at the function $g = \tau(u)$. We have $\tau(y_n) \leq g + d(u, y_n)$ and $\tau(z_n) \geq g - d(u, z_n)$. For those n we thus obtain

$$\begin{aligned} g(s_n) - g(s_n) &\geq \tau(y_n)(s_n) - d(u, y_n) - \tau(z_n)(t_n) - d(u, z_n) \\ &> \lambda(d(x, y_n) + d(x, z_n)) + 2\varepsilon - d(u, y_n) - d(u, z_n) \geq \varepsilon, \end{aligned}$$

which contradicts the continuity of g at s. □

Back to Linear Operators

Let us see if the connections we have sown between linear and non-linear extensions bring forth a harvest by returning to the main topic of the chapter. First of all, we isolate the metric configuration supporting Lemma 8.3.3:

Definition 8.3.4 Let Y be a subset of a metric space X and $\lambda \geq 1$. We say the pair (Y, X) satisfies condition $\Sigma_1(\lambda)$ if, given sequences $(y_n), (z_n)$ in Y, $x \in X$ and $\varepsilon > 0$, there exists $u \in Y$ such that

$$d(u, y_n) + d(u, z_n) \leq \lambda(d(x, y_n) + d(x, z_n)) + \varepsilon$$

for infinitely many n.

Thus, the real content of Lemma 8.3.3 is that if Y is a subset of a separable Banach space X then the pair (Y, X) satisfies $\Sigma_1(\lambda)$ if and only if $\mathscr{C}$-valued (or c-valued) Lipschitz maps defined on Y admit λ-extensions to one further point in X. A simple remark provides the necessary irrigation for our seeds.

Lemma 8.3.5 *Let H be a closed hyperplane of a Banach space X, and let $\tau\colon H \longrightarrow E$ be a contractive operator. Let $x \in X \setminus H$ and $\lambda \geq 1$. If τ has a λ-Lipschitz extension to $H \cup \{x\}$ then it has a λ-linear extension $T\colon X \longrightarrow E$.*

Proof If $\tilde{\tau}\colon H \cup \{x\} \longrightarrow E$ is λ-Lipschitz extension, pick $\xi = \tilde{\tau}(x)$ and set $T(y + tx) = \tau(y) + t\xi$. This T is a linear extension of τ and $\|T\| \leq \lambda$ since

$$\|T(y + tx)\| = |t|\, \|\tau(y/t) + \xi\| = |t|\, \|\xi - \tau(-y/t)\| \leq |t|\, \lambda\, \|x + y/t\| = \lambda \|y + tx\|. \quad \square$$

Time to provide some restorative shadow confirming that linear extension properties are much more demanding than their Lipschitz counterparts.

Theorem 8.3.6 *Let E be a closed subspace of a separable Banach space X and let $\lambda > 1$. If every operator $\tau\colon E \longrightarrow c$ admits a λ-extension then (E, X) has property $\Sigma_1(\lambda)$.*

Proof If the pair (Y, X) does not satisfy $\Sigma_1(\lambda)$ then there are two sequences $(y_n), (z_n)$ in Y, a point $x \in X \setminus Y$ and an $\varepsilon > 0$ such that for every $u \in Y$ the set of n for which

$$\|u - y_n\| + \|u - z_n\| < \lambda(\|x - y_n\| + \|x - z_n\|) + 2\varepsilon$$

is finite. This implies that for each compact subset $K \subset Y$ there is an $n(K)$ such that for all $u \in K$ and $n \geq n(K)$ we have

$$\|u - y_n\| + \|u - z_n\| > \lambda(\|x - y_n\| + \|x - z_n\|) + \varepsilon.$$

Since Y is separable, there is an increasing sequence of compact subsets of Y containing the origin, say (K_m), whose union is dense in Y. It then follows that we can choose a subsequence $\mathbb{M} = \{n(1), n(2), \ldots\}$ such that when $u \in K_m$,

$$\|u - y_{n(m)}\| + \|u - z_{n(m)}\| > \lambda(\|x - y_{n(m)}\| + \|x - z_{n(m)}\|) + \varepsilon. \tag{8.9}$$

Next, observe that if A is a Banach space and V is a compact, convex subset of A such that $\|v - a\| > c$ for some $a \in A$ and $c \geq 0$ and all $v \in V$, then there exists a functional $a^* \in A^*$ with $\|a^*\| \leq 1$ such that $\langle a^*, v - a\rangle \geq c$ for all $v \in V$. To see this, just apply the Hahn–Banach theorem to the effect of separating the compact convex set $V - a$ from the closed ball of radius c centered at the origin. If we now interpret $\|u - y_{n(m)}\| + \|u - z_{n(m)}\|$ as the norm of the difference $(u, u) - (y_{n(m)}, z_{n(m)})$ in the space $X \oplus_1 X$, since $(X \oplus_1 X)^* = X \oplus_\infty X$, from (8.9), we get the existence of $y_m^*, z_m^* \in X^*$ with $\|y_m^*\|, \|z_m^*\| \leq 1$ such that when $u \in K_m$,

$$\langle y_m^*, u - y_{n(m)}\rangle + \langle z_m^*, u - z_{n(m)}\rangle \geq \lambda(\|x - y_{n(m)}\| + \|x - z_{n(m)}\|) + \varepsilon. \tag{8.10}$$

Hence, for $u \in K_m$,

$$\begin{aligned}\langle y_m^* + z_m^*, u\rangle &\geq \langle y_m^*, y_{n(m)}\rangle + \langle z_m^*, z_{n(m)}\rangle + \lambda(\|x - y_{n(m)}\| + \|x - z_{n(m)}\|) + \varepsilon \\ &\geq \langle y_m^*, y_{n(m)}\rangle + \langle z_m^*, z_{n(m)}\rangle + \lambda(\langle y_m^*, x - y_{n(m)}\rangle + \langle z_m^*, x - z_{n(m)}\rangle) + \varepsilon \\ &\geq \langle y_m^*, x\rangle + \langle z_m^*, x\rangle + \varepsilon. \end{aligned} \tag{8.11}$$

At this point, we pass to a further subsequence $\mathbb{L}$ of $\mathbb{M}$ such that $(y_m^*)_{m\in\mathbb{L}}$ and $(z_m^*)_{m\in\mathbb{L}}$ are weak*-convergent. Take limits in the weak* topology to set

$$y^* = \lim_{m\in\mathbb{L}} y_m^* \qquad \text{and} \qquad z^* = \lim_{m\in\mathbb{L}} z_m^*.$$

Using (8.11), we get that for $u \in \bigcup_m K_m$,

$$\langle y^*, u\rangle + \langle z^*, u\rangle \geq \langle y^*, x\rangle + \langle z^*, x\rangle + \varepsilon,$$

and since (K_m) has dense union in Y, we conclude that $\langle y^* + z^*, u\rangle \geq \langle y^* + z^*, x\rangle$ for every $u \in Y$. Since Y is a linear subspace of X, this implies that $(y^* + z^*)|_Y = 0$. We define a contractive operator $\tau \colon X \longrightarrow \ell_\infty(\{1, -1\} \times \mathbb{L})$ by

$$\tau(z) = (\langle y_m^*, z\rangle - \langle z_m^*, z\rangle)_{m\in\mathbb{L}}.$$

Then τ maps Y into $c(\{1,-1\}\times\mathbb{L})$, and the hypothesis provides an extension $T\colon X\longrightarrow c(\{1,-1\}\times\mathbb{L})$, with $\|T\|\leq\lambda$. Let us write T in the form

$$T(z)=(\langle\tilde{y}^*_m,z\rangle-\langle\tilde{z}^*_m,z\rangle)_{m\in\mathbb{L}}.$$

For each n, we have $\|T(x-y_n)\|+\|T(x-z_n)\|\leq\lambda(\|x-y_n\|+\|x-z_n\|)$. Hence, for any $n\in\mathbb{N}, m\in\mathbb{L}$ and any choice of signs, we also have

$$\pm\langle\tilde{y}^*_m,x-y_n\rangle\pm\langle\tilde{z}^*_m,x-z_n\rangle\leq\lambda(\|x-y_n\|+\|x-z_n\|). \tag{8.12}$$

But since every value of T is a convergent sequence, we have that $(\tilde{y}^*_m)_m$ and $(\tilde{z}^*_m)_m$ converge in the weak* topology of X^*. If we let

$$\tilde{y}^*=\lim_{m\in\mathbb{L}}\tilde{y}^*_m \qquad\text{and}\qquad \tilde{z}^*=\lim_{m\in\mathbb{L}}\tilde{z}^*_m,$$

then $\tilde{y}^*+\tilde{z}^*=0, \tilde{y}^*|_Y=y^*|_Y, \tilde{z}^*|_Y=z^*|_Y$. In particular,

$$\lim_{m\in\mathbb{L}}\langle\tilde{y}^*_m+\tilde{z}^*_m,x\rangle=\langle\tilde{y}^*+\tilde{z}^*,x\rangle=0.$$

Thus, if we take both signs in (8.12) to be negative and set $n=n(m)$, we obtain

$$\begin{aligned}\limsup_{m\in\mathbb{L}}\ (\langle\tilde{y}^*_m,y_{n(m)}\rangle+\langle\tilde{z}^*_m,z_{n(m)}\rangle)&=\limsup_{m\in\mathbb{L}}\ (\langle y^*_m,y_{n(m)}\rangle+\langle z^*_m,z_{n(m)}\rangle)\\&\leq\lambda(\|x-y_{n(m)}\|+\|x-z_{n(m)}\|),\end{aligned}$$

which contradicts (8.10) when $u=0$. □

Types and the Almost Isometric $\mathscr{C}$-Extension Property

While the $(\lambda,\mathscr{C})$-extension property seems elusive, the almost isometric $\mathscr{C}$-extension property for separable spaces can be readily characterised. It is all a matter of showing that when Lipschitz maps extend to Lipschitz maps, operators extend to operators. This, which was very simple for extensions to one more dimension when done in Lemma 8.3.5, is considerably harder in the general situation. The passage can be smoothed using the language of types. Usually, a *type* on a Banach space X is defined to be a function $\sigma\colon X\longrightarrow[0,\infty)$ having the form $\sigma(x)=\|x-a\|$ for some a belonging to some ultrapower of X. However, in separable spaces, we can adopt the following equivalent formulation:

Definition 8.3.7 A type on a separable Banach space X is a function $\sigma:X\longrightarrow[0,\infty)$ having the form

$$\sigma(x)=\lim_{n\to\infty}\|x+x_n\|$$

for some bounded sequence $(x_n)_n$ of X. We will say in that case that σ is defined by (x_n) and when $(x_n)\subset Y\subset X$, σ is said to be supported on Y.

It is a matter of an elementary Ramsey-like argument that every bounded sequence $(x_n)_{n\in\mathbb{N}}$ in a separable space X contains a subsequence $(x_n)_{n\in\mathbb{M}}$ such that $\lim_{n\in\mathbb{M}} \|x + x_n\|$ exists for all $x \in X$, thus defining a type on X. Condition $\Sigma_1(\lambda)$ can be reformulated as:

Lemma 8.3.8 *Let X be a separable Banach space, Y a subset of X and $\lambda \geq 1$. If (Y, X) satisfies condition $\Sigma_1(\lambda)$ then, for every pair of types σ, β supported on Y, we have*

$$\inf_{u\in Y} (\sigma(u) + \beta(u)) \leq \lambda \inf_{x\in X} (\sigma(x) + \beta(x)). \tag{8.13}$$

The converse is true if $\lambda > 1$.

Proof We first remark that the reason for not having an equivalence in the case $\lambda = 1$ is that, while sequences defining types have to be bounded, the same is not the case for the sequences appearing in condition $\Sigma_1(\lambda)$. For $\lambda > 1$, the inequality required in condition $\Sigma_1(\lambda)$ is automatic when (y_n) or (z_n) is unbounded, and we can thus work with bounded sequences only. Now assume that (Y, X) satisfies $\Sigma_1(\lambda)$ and let σ and β be types on X supported on Y. Then there exist sequences $(y_n), (z_n)$ of Y such that

$$\sigma(x) = \lim_{n\to\infty} \|x + y_n\| \qquad \text{and} \qquad \beta(x) = \lim_{n\to\infty} \|x + z_n\|$$

for all $x \in X$. Fix $\varepsilon > 0$ and pick $x_0 \in X$ such that

$$\sigma(x_0) + \beta(x_0) < \inf_{x\in X} (\sigma(x) + \beta(x)) + \varepsilon.$$

Applying condition $\Sigma_1(\lambda)$ to $(y_n), (z_n), x_0$ and ε, we obtain a point $u \in Y$ such that $\sigma(u) + \beta(u) \leq \lambda(\sigma(x_0) + \beta(x_0)) + \varepsilon$, which is enough. As for the other part, suppose $\lambda > 1$, and let $(y_n), (z_n)$ be bounded sequences of Y. Then there is an increasing function $n : \mathbb{N} \longrightarrow \mathbb{N}$ such that the functions

$$\sigma(x) = \lim_{k\to\infty} \|x + y_{n(k)}\| \qquad \text{and} \qquad \beta(x) = \lim_{k\to\infty} \|x + z_{n(k)}\|$$

are correctly defined types on X. Now, if (8.13) holds then for every $x \in X$ and every $\varepsilon > 0$, there is $u \in Y$ and $N \in \mathbb{N}$ such that for all $k \geq N$ we have

$$\|u - y_{n(k)}\| + \|u - z_{n(k)}\| \leq \lambda(\|x - y_{n(k)}\| + \|x - z_{n(k)}\|) + \varepsilon. \qquad \square$$

Lemma 8.3.9 *Let X be a separable Banach space and $\lambda > 1$. If the pair (H, X) satisfies $\Sigma_1(\lambda)$ for every hyperplane H of X then so does (Y, X) for every closed subspace $Y \subset X$.*

Proof If (Y, X) fails to have property $\Sigma_1(\lambda)$ then there exist types σ, τ supported on Y, $\varepsilon > 0$ and $x_\varepsilon \in X$ such that for all $y \in Y$ we have

$$\sigma(y) + \tau(y) > \lambda(\sigma(x_\varepsilon) + \tau(x_\varepsilon)) + 2\varepsilon.$$

Let $D = \{x \in X : \sigma(x) + \tau(x) < \lambda(\sigma(x_\varepsilon) + \tau(x_\varepsilon)) + \varepsilon\}$. Note that $x_\varepsilon \in D$. Since $D + \varepsilon B_X$ does not meet Y we can arrange the geometric form of the Hahn–Banach theorem to obtain $x^* \in X^*$ such that $Y \subset \ker x^*$ and $\langle x^*, x\rangle > 0$ for all $x \in D$. Thus, $(\ker x^*, X)$ also fails to have $\Sigma_1(\lambda)$. □

The following result should be compared with the theorem in Section 8.8.3.

Theorem 8.3.10 *Let X be a separable Banach space. The following are equivalent:*

(i) *For every hyperplane $H \subset X$ and every $\lambda > 1$, every operator $\tau: H \longrightarrow c$ admits a λ-extension.*
(ii) *For every hyperplane $H \subset X$ and every $\lambda > 1$, the pair (H, X) has property $\Sigma_1(\lambda)$.*
(iii) *For every subspace $Y \subset X$ and every $\lambda > 1$, the pair (Y, X) has property $\Sigma_1(\lambda)$.*
(iv) *X has the almost isometric $\mathscr{C}$-extension property.*

Proof The proof is an assembly of previous results. The implications (iv)$\Longrightarrow$(i) and (iii)$\Longrightarrow$(ii) are trivial. The implication (ii)$\Longrightarrow$(iii) is the content of Lemma 8.3.9, and Theorem 8.3.6 provides both (iv)$\Longrightarrow$(iii) and (i)$\Longrightarrow$(ii). Thus, it suffices to show that (iii)$\Longrightarrow$(iv), which is an easy consequence of Lemma 8.3.3 inlaid with Lemma 8.3.5. To see why, let $\tau: Y \longrightarrow E$ be a $\mathscr{C}$-valued operator, $\lambda > 1$ and let (λ_n) be a sequence with $\lambda_n > 1$ for every n and such that $\prod_{n\geq 1} \lambda_n \leq \lambda$. Let $(Y_n)_{n\geq 0}$ be an increasing sequence of subspaces of X such that $Y_0 = Y$, each Y_n has codimension 1 (or 0) in Y_{n+1} and $\bigcup_n Y_n$ is dense in X. Assume that τ has been extended to a linear operator $\tau_n: Y \longrightarrow E$, with $\|\tau_n\| \leq \prod_{1\leq k\leq n} \lambda_k$. Since (Y_n, X) has property $\Sigma(\lambda_{n+1})$, Lemma 8.3.3 provides an extension $\tilde{\tau}_n: Y_{n+1} \longrightarrow E$ with Lipschitz constant at most $\prod_{1\leq k\leq n+1} \lambda_k$. Applying Lemma 8.3.5, we get a linear extension, say $\tau_{n+1}: Y_{n+1} \longrightarrow E$, with $\|\tau_{n+1}\| \leq \prod_{1\leq k\leq n+1} \lambda_k$. Continuing in this way, we arrive at a linear extension on $\bigcup_n Y_n$ bounded by λ, and we are done. □

Kalton's First Reading and Property (L^*)

No matter if one is aware or not, part of c_0's dowry in the proof of the Lindenstrauss–Pełczyński theorem is that weak*-null sequences $(x_n)_{n\geq 1}$ in ℓ_1 behave this way: for every $x \in \ell_1$, we have

$$\lim_{n\to\infty} (\|x - x_n\| - \|x_n\| - \|x\|) = 0. \tag{8.14}$$

Kalton reads this condition two times. The first reading [273] is in the language of types: if σ is a weak*-null type on ℓ_1 (that is, one realised by a weak*-null sequence) then one has $\sigma(x) = \sigma(0) + \|x\|$, and therefore c_0 (not ℓ_1!) has the following property:

Definition 8.3.11 A Banach space X has property (L^*) if any two weak*-null types on X^* that agree at the origin are equal.

Thus, $f \in X^* \longmapsto \|f\|$ is to be the only type on X^* vanishing at 0, and so the idea could cross one's mind that $f \longmapsto c+\|f\|$ is the only type taking the value c at 0. This is false, but only because we cannot guarantee that $c + \|\cdot\|$ is a type (types do not form a vector space). Property (L^*) can be stated equivalently with inequalities, namely, if σ and τ are weak*-null types on X^* and $\sigma(0) \leq \tau(0)$ then $\sigma \leq \tau$. The proof is trivial. This property is not as innocent as it seems: if X has property (L^*) then all types σ on X^* are symmetric (i.e. $\sigma(f) = \sigma(-f)$ since $\sigma(f) = \sigma(0) + \|f\|$) and therefore all bidual types (i.e. all types having the form $\sigma(f) = \|f + g\|$ for some g in the bidual of X^*) are symmetric. Maurey shows [357] that this implies that X^* contains ℓ_1.

Proposition 8.3.12 *Every* $\mathscr{C}$*-valued operator defined on a subspace of a separable space with property* (L^*) *admits a* 1^+*-extension to the whole space.*

Proof Let X be a separable space with property (L^*). Given two weak*-null types σ, τ on X^* and $u^*, v^* \in X^*$, there exists $0 \leq \theta \leq 1$ such that

$$\max\{\sigma(w^*), \tau(w^*)\} \leq \max\{\sigma(u^*), \tau(v^*)\} \tag{8.15}$$

whenever $w^* = (1-\theta)u^* + \theta v^*$. Indeed, assume without loss of generality $\|u^*\| \leq \|v^*\|$. Property (L^*) and $\theta = 1$ yield $\sigma(v^*) \leq \tau(v^*)$. To conclude, use Theorem 8.3.10 after the next lemma. □

Lemma 8.3.13 *If inequality* (8.15) *holds, every c-valued operator defined on a hyperplane of X admits a* 1^+*-extension to X.*

Proof With the same notation as in Lemma 1.8.4, if $K_M = \limsup_M K_n$ then all we need to show is that $\bigcap_M K_M \neq \emptyset$. Since H is an hyperplane of X, it suffices to prove that $K_M \cap K_N \neq \emptyset$ for any two infinite subsets $M, N \subset \mathbb{N}$. Indeed, let $C = \{M \colon K_M \text{ is convex}\}$; by Helly's theorem, $\bigcap_{M\in C} K_M \neq \emptyset$. But the second assertion of Lemma 1.8.4 implies that also $\bigcap_M K_M \neq \emptyset$. For every $n \in \mathbb{N}$, pick $f_n \in K_n$ with $\|f_n\| = 1$. The sequence $(f_n)_{n\in M}$ contains a subsequence $(f_n)_{n\in M_1}$ that is weak*-convergent to some point, say, x^*. Analogously, the sequence $(f_n)_{n\in N}$ contains a subsequence $(f_n)_{n\in N_1}$ that is weak*-convergent to some point, say, y^*. By an argument 'à la Brunel-Sucheston', there is no loss of generality in assuming that for every $z^* \in [x^*, y^*]$, the limits $\lim_{M_1} \|z^* + f_n - x^*\|$ and $\lim_{N_1} \|z^* + f_n - x^*\|$ exist. By condition (8.15), there exists some $u^* = (1-\theta)x^* + \theta y^*$ such that

$$\max\left\{\lim_{M_1} \|u^* + f_n - x^*\|, \lim_{N_1} \|u^* + f_n - x^*\|\right\} \leq 1.$$

Since $y^* - x^* \in H^\perp$, it follows that $u^* - x^* \in H^\perp$, and therefore $u^* + f_n - x^* \in K_n$ for large $n \in M_1$. Similarly, $u^* + f_n - y^* \in K_n$ for large $n \in N_1$. Taking weak*-limits, we have $u^* \in K_{M_1} \cap K_{N_1}$. □

Since c_0 has property (L^*), we recover the Lindenstrauss–Pełczyński theorem 8.2.1 from Proposition 8.3.12 with sharp bounds. Besides, it is nearly obvious that the ℓ_p-spaces have property (L^*): if $\sigma\colon \ell_q \longrightarrow \mathbb{R}$ is a weak* (= weak) null type and $f \in \ell_q$ then

$$\sigma(f) = (\sigma(0)^q + \|f\|^q)^{1/q}. \tag{8.16}$$

In this way, Proposition 8.3.12 implies that isometries $Y \longrightarrow \ell_p$ are $(1^+, \mathscr{C})$-trivial (compare with Proposition 8.1.3 (a)).

Proposition 8.3.14 *Let X be a Banach space with property (L^*) whose dual is separable. Every $\mathscr{C}$-valued operator defined on a weak*-closed subspace of X^* admits a 1^+-extension.*

Proof The goal is to show that property (L^*) implies that for every weak*-closed subspace E of X^*, the pair (E, X^*) has all properties $\Sigma_1(\lambda)$ for $\lambda > 1$. In view of Lemma 8.3.8, it suffices to see that if σ, τ are types on X^*, supported on E, then

$$\inf_{u^*\in E} (\sigma(u^*) + \tau(u^*)) = \inf_{x^*\in X^*} (\sigma(x^*) + \tau(x^*)). \tag{8.17}$$

Observe that since all weak*-null types $\sigma\colon X^* \longrightarrow \mathbb{R}$ are symmetric (i.e. they are even maps), given $u^*, v^* \in X^*$, we have

$$\sigma\left(\frac{u^* - v^*}{2}\right) + \sigma\left(\frac{v^* - u^*}{2}\right) \leq \sigma(u^*) + \sigma(v^*).$$

Let us show that 'the same' happens with any two weak*-null types σ, τ, namely that given points u^*, v^*, there exists $\theta \in [0, 1]$ such that

$$\sigma(\theta(u^* - v^*)) + \tau((1 - \theta)(v^* - u^*)) \leq \sigma(u^*) + \tau(v^*). \tag{8.18}$$

Indeed, taking $\theta \in [0, 1]$ such that $(1 - \theta)\sigma(0) = \theta\tau(0)$, we have

$$\begin{aligned}\sigma\left(\theta(u^* - v^*)\right) &= \lim \|\theta u^* - \theta v^* + x_n^*\| \\ &\leq \lim \|\theta u^* + \theta x_n^*\| + \lim \| - \theta v^* + (1 - \theta)x_n^*\| \\ &= \lim \|\theta u^* + \theta x_n^*\| + \lim \|\theta v^* + \theta y_n^*\| \\ &= \theta\left(\sigma(u^*) + \tau(v^*)\right).\end{aligned}$$

Analogously, $\tau((1-\theta)(v^*-u^*)) \leq (1-\theta)(\sigma(u^*)+\tau(v^*))$. We finally show that (8.18) implies (8.17). Let E be a weak*-closed subspace of X^* and $\sigma, \tau\colon X^* \longrightarrow \mathbb{R}$ arbitrary types supported on E. Since the unit ball of E is weak*-compact, for some $e^*, f^* \in E$, we have that $\sigma_0(x^*) = \sigma(x^*+e^*)$ and $\tau_0(x^*) = \tau(x^*+f^*)$ are weak*-null types on X^*. For arbitrary $x^* \in X^*$, taking $u^* = x^* - e^*$ and $v^* = x^* - f^*$, it is possible by (8.18) to find some $\theta \in [0,1]$ such that

$$\sigma_0(\theta(f^*-e^*)) + \tau_0((1-\theta)(e^*-f^*)) \leq \sigma_0(x^*-e^*) + \tau_0(x^*-f^*).$$

Put $w^* = (1-\theta)e^* + \theta f^*)$ to conclude that $\sigma(w^*)+\tau(w^*) \leq \sigma(x^*)+\tau(x^*)$. □

Even if our next assertion does not mean much at this moment, we want to remark that the Johnson–Zippin theorem 8.6.2 can be derived from here. The interest of this remark is to observe that Proposition 8.3.12 unifies most classical results on the extension of $\mathscr{C}$-valued operators. But since we did not arrive thus far merely to reprove oldies, the next section will contain different, and quite spectacular, applications.

8.4 Sequence Spaces with the $\mathscr{C}$-Extension Property

Many familiar sequence spaces, including most Orlicz sequence spaces and modular spaces, can be renormed to have property (L^*). This furnishes us with a significant number of spaces having the $\mathscr{C}$-extension property, something quite remarkable when one considers how difficult it is to establish the $\mathscr{C}$-extension property for those spaces in their native norms, and more remarkable yet when we take into account that we will actually use not property (L^*) but instead a close relative:

Definition 8.4.1 A Banach space has property (L) if any two weakly null types that agree at the origin are equal.

As the reader can imagine, some work is necessary to relate properties (L) and (L^*) since, at first glance, the latter depends on the behaviour of weak*-null sequences of X^*, while the former depends on the behaviour of weakly null sequences in X. All the connection we need is provided by:

Proposition 8.4.2 *A Banach space with a* 1*-unconditional shrinking basis and property* (L) *also has property* (L^*).

The shrinking property of the basis is necessary: the space ℓ_1 obviously has property (L) and clearly fails (L^*). Before beginning the proof, let us isolate the key fact linking weak*-null types on X^* and weakly null types on X:

Lemma 8.4.3 *Assume X has a* 1*-unconditional, shrinking basis. Suppose that $(x_n)_{n\geq 1}$ is weak*-null in X^* and that $u^* \in X^*$ is such that* $\lim_n \|u^* + x_n^*\|$ *exists. Then there is an infinite $\mathbb{M} \subset \mathbb{N}$, a weakly null sequence $(x_m)_{m\in\mathbb{M}}$ in the unit ball of X and a point $u \in X$ such that $\|u + x_m\| = 1$ for all $m \in \mathbb{M}$ and*

$$\langle u^*, u\rangle + \lim_{m\in\mathbb{M}} \langle x_m^*, x_m\rangle = \lim_{n\in\mathbb{N}} \|u^* + x_n^*\|. \tag{8.19}$$

Proof of the lemma Let $(e_n)_{n\geq 1}$ be the basis of X which we consider normalised. The hypotheses guarantee that the coordinate functionals (e_n^*) constitute a basis of X^*. Thus, we may assume that $(x_n^*)_n$ is a block sequence and also that u^* is finitely supported, say $u^* = \sum_{1\leq i\leq k} u_i e_i^*$. In particular, x_n^* and u^* are 'disjoint' for $n > k$. Now, using the 1-unconditional property of the basis, for each n, we can pick a normalised $v_n \in X$ such that

$$\langle u^* + x_n^*, v_n\rangle = \|u^* + x_n^*\|, \qquad \operatorname{supp}(v_n) \subset \operatorname{supp}(u^* + x_n^*).$$

Let $P = \sum_{1\leq i\leq k} e_i^* \otimes e_i$ be the projection of X onto the first k coordinates, and let $\mathbb{M} \subset \mathbb{N}$ be an infinite subset for which the limit $u = \lim_{n\in\mathbb{M}} P(v_n)$ exists in the norm topology of X. Now, set $x_n = v_n - u$. Clearly, $\|u + x_n\| = 1$ for all n, while $(x_n)_{n\in\mathbb{M}}$ is weakly null, with $\|x_n\| \leq 1$ for all n. The proof concludes with

$$\|u^* + x_n^*\| = \langle u^* + x_n^*, v_n\rangle = \langle u^* + x_n^*, u + x_n\rangle = \langle u^*, u\rangle + \langle u^*, x_n\rangle + \langle x_n^*, u\rangle + \langle x_n^*, x_n\rangle.$$

□

Proof of Proposition 8.4.2 It suffices to check that if (x_n^*) and (y_n^*) are normalised weak*-null sequences of X^* then

$$\lim_{n\to\infty} \|u^* + x_n^*\| \leq \lim_{n\to\infty} \|u^* + y_n^*\|,$$

as long as both limits exist. Let us apply the just proved lemma to (x_n^*) and u^* to get an infinite subset $\mathbb{M} \subset \mathbb{N}$, a weak*-null sequence $(x_n)_{n\in\mathbb{M}}$ and a $u \in X$, such that $\|u + x_n\| = 1$ for all $n \in \mathbb{M}$ and

$$\langle u^*, u\rangle + \lim_{n\in\mathbb{M}} \langle x_n^*, x_n\rangle = \lim_{n\in\mathbb{N}} \|u^* + x_n^*\|.$$

Applying the lemma a second time to $(y_n^*)_{n\in\mathbb{M}}$ and $0 \in Y^*$, we obtain an infinite $\mathbb{L} \subset \mathbb{M}$ and a normalised weakly null sequence $(y_n)_{n\in\mathbb{L}}$ such that $\lim_{n\in\mathbb{L}} \langle y_n^*, y_n\rangle = 1$. Since $(\|x_n\| y_n)_{n\in\mathbb{L}}$ is weakly null, the property (L) of X yields $\lim_{n\in\mathbb{L}} \big\| u + \|x_n\| y_n \big\| = 1$. Thus,

$$\begin{aligned}
\lim_{n\in\mathbb{L}} \|u^* + y_n^*\| &\geq \limsup_{n\in\mathbb{L}} \langle u^* + y_n^*, u + \|x_n\| y_n\rangle \geq \langle u^*, u\rangle + \limsup_{n\in\mathbb{L}} \|x_n\| \\
&\geq \langle u^*, u\rangle + \lim_{n\in\mathbb{L}} \|x_n\| \geq \langle u^*, u\rangle + \lim_{n\in\mathbb{L}} \langle x_n^*, x_n\rangle \\
&= \lim_{n\in\mathbb{L}} \|u^* + x_n^*\|.
\end{aligned}$$

□

We are ready to display Banach spaces with shrinking basis and property (L). Start with a sequence of finite-dimensional spaces $(V_k)_{k\geq 1}$. For each k, let N_k be a norm on $\mathbb{R} \times V_k$ (not V_k!) having the following properties:

★ $N_k(t, x) \geq \max(N_k(t, 0), N_k(0, x))$ for every $t \in \mathbb{R}$ and every $x \in V_k$,
★ $N_k(1, 0) = 1$ and $N_k(-t, x) = N_k(t, x)$ for every $(t, x) \in \mathbb{R} \times V_k$.

Next, we define an upper triangular infinite matrix of seminorms $(N_{m,n})_{m\leq n}$ on the space $c_{00}(V_k)$ as follows:

- For (n, n) on the diagonal, we set $N_{n,n}(v) = N_n(0, v_n)$, where $v = (v_k)_{k\geq 1}$.
- Then, if (m, n) is above the diagonal, that is, if $m < n$, we inductively define $N_{m,n}(v) = N_m(N_{m+1,n}(v), v_m)$ until reaching the diagonal from above.

We define a genuine norm on $c_{00}(V_k)$ by taking $\|v\|_L = \sup_{m\leq n} N_{m,n}(v)$. Let Λ_L be the completion of $c_{00}(V_k)$ with respect to $\|\cdot\|_L$. It is clear that each Λ_L can be regarded as the space of all infinite sequences $v = (v_k)_{k\geq 1}$ such that

- $\|v\|_L = \sup_k \|(v_1, \ldots, v_k, 0, \ldots)\|_L < \infty$,
- $\lim_{k\to\infty} \sup_{n>k} \|(0, \ldots, 0, v_k, \ldots, v_n, 0, \ldots)\|_L = 0$.

The time is now ripe for

Proposition 8.4.4 *The space Λ_L has property (L).*

Proof The proof is almost trivial after realising how the norm of Λ_L works. We must see that if $(y^n)_{n\geq 1}, (z^n)_{n\geq 1}$ are weakly null sequences in Λ such that the limits

$$\sigma(x) = \lim_n \|x + y^n\|_L \qquad \text{and} \qquad \rho(x) = \lim_n \|x + z^n\|_L$$

exist for every $x \in \Lambda_L$ and $\sigma(0) = \rho(0)$, then $\sigma(x) = \rho(x)$ for all $x \in \Lambda_L$. After a moment's reflection, we see that it suffices to prove our statements for finitely supported x assuming that y^n and z^n are finitely supported and that the sequences $(y^n)_{n\geq 1}, (z^n)_{n\geq 1}$ have the property that for each k there is m such that $y_i^n = z_i^n = 0$ for $i \leq k$ and $n \geq m$. Besides, we may and do assume that $\|y^n\|_L = \|z^n\|_L = \sigma(0) = \rho(0)$ for all n. In this way, property (L) falls victim to the following property of the norm of Λ_L: if x, y, z are finitely supported vectors such that

- there is an m such that $x_i = 0$ for $i > m$, while $y_i = z_i = 0$ for $i \leq m$, and
- $\|y\|_L = \|z\|_L$,

then $\|x + y\|_L = \|x + z\|_L$. Note that the crucial hypothesis is that when the supports of y and z 'start', x is already null. To prove this, note that one has $N_{m,n}(v) \geq N_{m+1,n}(v)$ by the very definition of the norms $N_{m,n}$ when $m < n$, so $\|v\|_L = \sup_n N_{1,n}(v)$. On the other hand, $N_{1,n+1}(v) \geq N_{1,n}(v)$, so actually,

$$\|v\|_L = \lim_{n\to\infty} N_{1,n}(v).$$

But if $v_{n+1} = 0$ then $N_{1,n+1}(v) = N_{1,n}(v)$, and so, for $v = (v_1, \dots, v_k, 0, \dots)$,

$$\|v\|_L = N_{1,k}(v) = N_1(N_2(\dots N_{k-1}(N_k(0, v_k), v_{k-1}) \dots, v_2), v_1). \tag{8.20}$$

Now, if k is so large that we can compute the L-norm of the points $x, y, z, x + y, x + z$ using $N_{1,k}$, then

$$\|x + y\|_L = N_1(N_2(\dots N_m \overbrace{(N_{m+1}(\dots N_k(0, y_k), \dots y_{m+1})}^{\|y\|_L}, x_m) \dots, x_2), x_1),$$
$$\|x + z\|_L = N_1(N_2(\dots N_m \underbrace{(N_{m+1}(\dots N_k(0, z_k), \dots z_{m+1})}_{\|z\|_L}, x_m) \dots, x_2), x_1),$$

hence if $\|y\|_L = \|z\|_L$ then the proof is complete by showing

$$\begin{aligned}\|x + y\|_L &= N_1(N_2(\dots N_m(\|y\|_L, x_m) \dots, x_2), x_1)\\ &= N_1(N_2(\dots N_m(\|z\|_L, x_m) \dots, x_2), x_1)\\ &= \|x + z\|_L.\end{aligned}$$

□

To identify the spaces Λ_L as modular sequence spaces, one can consider the family of functions $\Phi_k \colon V_k \longrightarrow \mathbb{R}_+$ given by $\Phi_k(x) = N_k(1, x) - 1$. It is clear that each Φ_k is a Young function and so it makes sense to consider the corresponding modular space

$$h((\Phi_k)_k) = \left\{ v \in \prod_{k\geq 1} V_k : \sum_{k=1}^{\infty} \Phi_k(tv_k) < \infty \text{ for all } t > 0 \right\}$$

(see Section 1.8.2) with the Luxemburg norm

$$\|v\|_{(\Phi_k)_k} = \inf\left\{ t > 0 : \sum_{k=1}^{\infty} \Phi_k(v_k/t) \leq 1 \right\}.$$

Proposition 8.4.5 *One has $\Lambda_L = h((\Phi_k)_k)$, with equivalence of norms.*

Proof We first prove that $h((\Phi_k)_k)$ contains Λ_L, and that the inclusion is bounded. Assume $v \in \Lambda_L$, with $\|v\|_L \leq 1$. Then, for $1 \leq k < n$, using that

$N_{k+1,n}(v) \leq 1$ and the convexity of Φ_k, we have

$$\begin{aligned} N_{k,n}(v) &= N_k(N_{k+1,n}(v), v_k)) \\ &= N_{k+1,n}(v) N_k\left(1, \frac{v_k}{N_{k+1,n}(v)}\right) \\ &= N_{k+1,n}(v)\left(1 + \Phi_k\left(\frac{v_k}{N_{k+1,n}(v)}\right)\right) \\ &= N_{k+1,n}(v) + N_{k+1,n}(v)\,\Phi_k\left(\frac{v_k}{N_{k+1,n}(v)}\right) \\ &\geq N_{k+1,n}(v) + \Phi_k(v_k). \end{aligned} \tag{8.21}$$

Thus, for each $n \geq 1$,

$$1 \geq N_{1,n}(v) = \sum_{k=1}^{n-1} (N_{k,n}(v) - N_{k+1,n}(v)) \geq \sum_{k=1}^{n-1} \Phi_k(v_k).$$

It quickly follows that $v \in h((\Phi_k))$, with norm at most 1. To prove the other inclusion, first suppose that $v = (v_k)$ is finitely non-zero and that $\sum_k \Phi_k(v_k) < 1$, and let us see that $\|v\|_L \leq 2e$. If $\|v\|_L \leq 2$, there is nothing to prove. Note that $N_{n,n}(v) = N_n(0, v_n) \leq N_n(1, v_n) = 1 + \Phi_n(v_n) \leq 2$, thus if $\|v\|_L > 2$ then $N_{m,n}(v) > 2$ for some $m < n$. Let r be the smallest index such that $N_{r,n}(v) \leq 2$. Clearly, $m < r \leq n$. We have

$$N_{r-1,n}(v) = N_{r-1}(N_{r,n}(v), v_{r-1}) \leq N_{r-1}(2, v_{r-1}) \leq 2(1 + \Phi_{r-1}(v_{r-1}/2)),$$

and then, for any $m < j < r$,

$$N_{j-1,n}(v) = N_{j-1}(N_{j,n}(v), v_{j-1}) \leq N_{j,n}(v)(1 + \Phi_{j-1}(v_{j-1}/2)).$$

Hence,

$$N_{m,n}(v) \leq 2 \prod_{k=m}^{r-1} (1 + \Phi_k(v_k)) \leq 2e,$$

as required. Now assume $v \in h((\Phi_k))$, that is, $\sum_{k \geq 1} \Phi_k(tv_k) < \infty$ for all $t > 0$. Then, for every $\varepsilon > 0$, there is an r such that $\sum_{r<k} \Phi_k(2ev_k/\varepsilon) < \infty$. Thus, for $r < k < n$ and $v \in \Lambda_L$, we have $\|(0, \ldots, 0, v_k, \ldots, v_n, 0, \ldots)\|_L < \varepsilon$. □

What have we obtained? That all modular sequence spaces and all Fenchel–Orlicz spaces fall within the range of application of Proposition 8.4.5! We first give the proof for modular spaces to then provide a description of the argument for Fenchel–Orlicz spaces, which is long and winding, although not terribly difficult.

Corollary 8.4.6 *Let $\Phi_k : V_k \longrightarrow \mathbb{R}_+$ be a sequence of Young functions, where each V_k is a finite-dimensional space. Assume that for each k, there is a norm N_k on $\mathbb{R} \times V_k$ such that, for every $x \in V_k$,*

- *$N_k(s, x) \leq N_k(t, x)$ for $|s| \leq |t|$,*
- *$N_k(1, x) = 1 + \Phi_k(x)$.*

Then $\Lambda_L((N_k)) = h((\Phi_k))$, with equivalence of norms.

Proof This is a direct consequence of Proposition 8.4.5, since the norms N_k clearly have the properties marked with ★ at the begining of this section. □

Theorem 8.4.7 *Every modular sequence space $h((\phi_k)_k)$ has an equivalent norm with property (L). If $h((\phi_k)_k)$ has separable dual then this norm has property (L^*) as well.*

Proof We can assume that $\phi_k(1) = 1$ for every k. This implies that the right derivative of each ϕ_k at $\frac{1}{2}$ is at most 2. Then we only need to replace the sequence (ϕ_k) with an equivalent sequence (φ_k) where each φ_k has the form $\varphi_k(t) = N_k(1, t) - 1$, where each N_k a is norm on $\mathbb{R}^2$ satisfying the hypotheses of Corollary 8.4.6. We will choose φ_k to be convex, with $(\varphi_k(t)+1)/t$ decreasing for $t > 0$, say

$$\varphi_k(t) = \begin{cases} \phi_k(t) & \text{for } 0 \leq t \leq \frac{1}{2} \\ \phi_k(\frac{1}{2}) + 2(t - \frac{1}{2}) & \text{for } t \geq \frac{1}{2}. \end{cases}$$

Now we define the required norm on $\mathbb{R}^2$ by

$$N_k(s, t) = \begin{cases} |s|(1 + \phi_k(|t|/|s|)) & \text{for } s \neq 0 \\ 2|t|, & \text{for } s = 0. \end{cases}$$

This concludes the proof of the first part. The second statement is implied by Proposition 8.4.2 and James' classical result [216] asserting that a Banach space with unconditional basis has separable dual if and only if it does not contain ℓ_1 and if and only if the given (or any other) basis is shrinking. □

Proposition 8.4.8 *Every modular sequence space not containing ℓ_1 has the $\mathscr{C}$-extension property.*

Note that *degenerate* Young functions have not been excluded so that c_0 can be considered as the modular space generated by the function $\varphi(t) = 2(t - \frac{1}{2})$ for $t \geq \frac{1}{2}$ and $\varphi(t) = 0$ for $0 \leq t \leq \frac{1}{2}$. Thus, finite products $X_1 \times \cdots \times X_k$, where each X_i is either c_0 or ℓ_p, with $1 < p < \infty$, can be renormed to enjoy the almost isometric $\mathscr{C}$-extension property. Observe that there is no direct proof for this fact because, in general, neither do the spaces $\ell_p \oplus_s \ell_r$ have property

(L^*), nor is the duality map weakly continuous (even when $p = s = 2, r = 4$). We conclude with the case of Fenchel–Orlicz spaces, thus providing a shallow description of the fourth section of [12]:

Proposition 8.4.9 *Every Fenchel–Orlicz space not containing ℓ_1 and built on a non-degenerate Young function has the $\mathscr{C}$-extension property.*

These include the spaces $\ell_p(\varphi)$ for $p > 1$, by Theorem 10.8.1. The key point is that each non-degenerate Young function $\Phi\colon \mathbb{R}^n \longrightarrow \mathbb{R}^+$ is equivalent near zero to another Young function Ψ that arises as $\Psi(x) = N(1, x) - 1$, where N is a norm on $\mathbb{R} \times \mathbb{R}^n$ satisfying the first condition of Corollary 8.4.6. The proof, which requires considerable skill in convex geometry, goes as follows. Let B be the Euclidean ball of $\mathbb{R}^n$. Starting with a Young function $\Phi\colon \mathbb{R}^n \longrightarrow \mathbb{R}^+$ vanishing only at zero, one constructs an even, convex function $\Gamma\colon B \longrightarrow \mathbb{R}^+$ vanishing only at zero, which is C^1 away from zero and equivalent to Φ on B. The lion's share of the proof consists of showing that Γ can be extended from a neighbourhood of zero to a Young function $\Psi\colon \mathbb{R}^n \longrightarrow \mathbb{R}^+$ such that

- for each $x \in \mathbb{R}^n$, the map $t \longmapsto (1 + \Psi(tx))/t$ is decreasing on $\mathbb{R}^+$,
- the function $x \longmapsto \lim_{t\to\infty}(1 + \Psi(tx))/t$ is a norm on $\mathbb{R}^n$.

The function Ψ, which is again equivalent to Φ near zero, is finally used to define a norm N on $\mathbb{R} \times \mathbb{R}^n$ through the formula

$$N(s, x) = \begin{cases} |s|\,(1 + \Psi\,(x/|s|))\,, & \text{if } s \neq 0,\\ \lim_{t\to 0} |t|\,(1 + \Psi\,(x/|t|))\,, & \text{otherwise.} \end{cases}$$

From this, it is clear that $N(s, x) \leq N(t, x)$ for $|s| \leq |t|$ and also that $N(1, x) = 1 + \Psi(x)$ for all $x \in \mathbb{R}^n$. Corollary 8.4.6 then implies that $h(\Phi) = h(\Psi) = \Lambda_L(N)$. Whether the result remains true for families of Young functions is unclear because the size of the neighbourhood of zero in the second step seems to depend on the given Young function.

8.5 $\mathscr{C}$-Extensible Spaces

Since, because of Lemma 1.6.2, $C(\Delta)$ contains 1-complemented copies of all separable $\mathscr{C}$-spaces as well as isometric copies of all separable Banach spaces, $\mathscr{C}$-extendibility is a game played in $C(\Delta)$: X is $\mathscr{C}$-extensible if and only if every operator $\tau\colon X \longrightarrow C(\Delta)$ extends to $C(\Delta)$ through whatever embedding $X \longrightarrow C(\Delta)$. As we know, every separable Banach space X admits some $\mathscr{C}$-trivial embedding into $C(\Delta)$: the composition of the canonical embedding

$\delta\colon X \longrightarrow C(B_X^*)$ with an isomorphism between $C(B_X^*)$ and $C(\Delta)$. To be $\mathscr{C}$-extensible means that all embeddings into $C(\Delta)$ are $\mathscr{C}$-trivial. Sobczyk's theorem records c_0 as the first $\mathscr{C}$-extensible space, while the Lindentrauss–Pełczyński theorem files all its subspaces in the list of $\mathscr{C}$-extensible spaces. The following result is, formally at least, a generalisation of both. A crucial step in its proof is the use of a *homogeneous* Zippin selector. Upgrading Zippin selectors in this manner is possible for separable Banach spaces (see Note 8.8.1) but far from trivial.

Lemma 8.5.1 *If E is $\mathscr{C}$-extensible then so is $c_0(E)$.*

Proof Assume $c_0(E)$ has been embedded as a subspace of a separable Y, and let $\pi_n\colon c_0(E) \longrightarrow E$ be the projection onto the nth-coordinate; when necessary, we will write $E_n = \pi_n[c_0(E)]$. Let (F_k) be an increasing sequence of finite-dimensional subspaces of Y spanning a dense subspace and such that $F_1 = 0$, $F_k \cap c_0(E) = 0$. Let $W_k = F_k + c_0(E)$.

Claim For each k, there is $n_0(k)$ such that for all $n \geq n_0(k)$, the operator $T_n\colon W_k \longrightarrow E$ defined as $f + z \longmapsto \pi_n(z)$ has norm at most $2 + \varepsilon$.

Proof of the claim If not, for some k, there are sequences $(f_n) \subset F_k$, $(z_n) \subset c_0(E)$ and $(m(n)) \subset \mathbb{N}$ such that $\|f_n + z_n\| < 1$ but $\pi_{m(n)}(z_n) > 2 + \varepsilon$. Since $F_k \cap c_0(E) = 0$, the sequence (f_n) is bounded and it must contain a convergent subsequence, which, after relabelling, is itself. But this is in contradiction with the fact that $\|z_m - z_n\| > 2 + \varepsilon$ for large m, which means $\|f_m - f_n\| > \varepsilon$. $\quad\square$

Let $T_n\colon F_{k(n)} + c_0(E) \longrightarrow E$ be the operators $T_n(f + z) = \pi_n(z)$ with $\|T_n\| \leq 2 + \varepsilon$. Consider the pushout diagrams

$$
\begin{array}{ccccccccc}
0 & \longrightarrow & W_{k(n)} & \longrightarrow & Y & \longrightarrow & Z & \longrightarrow & 0 \\
& & \downarrow{\scriptstyle T_n} & & \downarrow{\scriptstyle \tau_n} & & \| & & \\
0 & \longrightarrow & E & \xrightarrow{\ \imath_n\ } & \mathrm{PO}_n & \longrightarrow & Z & \longrightarrow & 0
\end{array}
$$

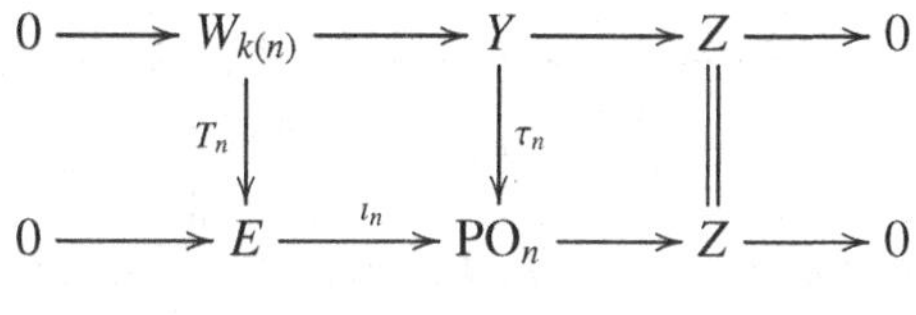

The open mapping theorem implies that every embedding $\jmath\colon E \longrightarrow C(\Delta)$ comes with a constant

$$\lambda_\jmath = \inf\big\{\lambda\colon \forall \tau \in \mathfrak{L}(E, C(\Delta))\ \exists\, T \in \mathfrak{L}(C(\Delta))\ \text{such that } \tau = T\jmath, \|T\| \leq \lambda\|\tau\|\big\}.$$

Showing that $\sup_\jmath \lambda_\jmath < \infty$ is a bit trickier. Let $\jmath_n$ be a sequence of 'uniform' embeddings, each with constant λ_n. Form their (multiple) pushout PO, and let $\imath\colon E \longrightarrow \mathrm{PO}$ be the resulting embedding. Since PO is separable, pick 'the canonical' embedding $\delta\colon \mathrm{PO} \longrightarrow C(\Delta)$ and then the embedding $\jmath = \delta\imath$. It is clear that $\lambda_\jmath \geq \sup_n \lambda_n$, which shows the assertion. Returning to the proof, we have obtained a constant λ such that the lower sequences in the preceding diagram $(\lambda, \mathscr{C})$-split and therefore all of them admit λ-Zippin selectors. For the time being, we will take for granted the existence of a homogeneous λ-Zippin selector ψ_n for $\imath_n$. With its aid, we can define a homogeneous map $\phi_n \colon c_0(E)^* \longrightarrow Y^*$ in the form $\phi_n(z^*) = \tau_n^* \psi_n(z^*|_{E_n})$. This map is weak*-continuous on bounded sets and satisfies the estimate $\|\phi_n(z^*)\| \leq (2+\varepsilon)\lambda\|\pi_n^*(z^*)\|$ since $\|\tau_n\| \leq 1$ (by the usual properties of pushouts). Moreover, if $z \in c_0(E)$ and $z^* \in c_0(E)^*$ then

$$\langle \phi_n(z^*), z\rangle = \langle \tau_n^* \psi_n(z^*|_{E_n}), z\rangle = \langle \psi_n(z^*|_{E_n}), \tau_n z\rangle = \langle \psi_n(z^*|_{E_n}), \imath_n \pi_n z\rangle = \langle z^*, \pi_n(z)\rangle.$$

Now define $\omega \colon c_0(E)^* \longrightarrow Y^*$ by $\omega(z^*) = \sum \phi_n(z^*)$; this has $\|\omega\| \leq (2+\varepsilon)\lambda$ and is a selector since $\omega(z^*)(z) = z^*$. Moreover, it is weak*-continuous on bounded sets: if $f \in \bigcup F_k$ then eventually, $T_n f = 0$. Thus, $z^* \longmapsto \omega(z^*)(f)$ is weak*-continuous on bounded sets, as well as ω. □

Now, the typical 3-space result:

Lemma 8.5.2 *$\mathscr{C}$-extensibility is a 3-space property.*

Proof Let X be a Banach space with a $\mathscr{C}$-extensible subspace Y such that X/Y is $\mathscr{C}$-extensible. Let $\tau \colon X \longrightarrow E$ be a $\mathscr{C}$-valued operator and $\jmath \colon X \longrightarrow U$ an embedding in which U is separable. Consider the commutative diagram

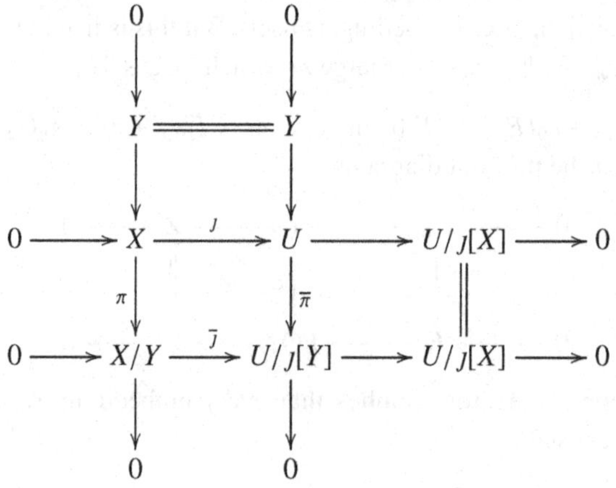

We sympathise with any reader believing this is groundhog day, since this diagram is formally identical to those that already appeared in the proofs of Theorem 8.2.1 and Lemma 2.14.3 (central diagram; the hypotheses now are about sequences 2 and 4 and the thesis about sequence 1); moreover, we will proceed exactly as we did then: we first extend $\tau|_Y$ to an operator $T_1 \colon U \longrightarrow E$ such that $T_1(\jmath(y)) = \tau(y)$ for $y \in Y$. As the difference $\tau - T_1\jmath$ vanishes on Y, there is a $\tau_2\colon X/Y \longrightarrow E$ such that $\tau - T_1 \jmath = \tau_2 \pi$. If $T_2\colon U/\jmath[Y] \longrightarrow E$ is an extension of τ_2 through $\bar{\jmath}$ then $T = T_1 + T_2 \pi$ is the required extension of τ. □

Kalton's Second Reading and the Secret Life of ℓ_1

We now show that ℓ_1 is also $\mathscr{C}$-extensible. To understand why, let us recall that the key fact (8.14) about weak*-null sequences of ℓ_1 that appears during the proof of the Lindenstrauss–Pełczyński theorem was transformed in Kalton's hands into property (L^*). In that same paper, Kalton returned to the crime scene and read it again in a new way: let $\sigma\colon \ell_1 \longrightarrow \mathbb{R}_+$ be *any* type defined by a bounded sequence (x_n). The sequence can be assumed to be weak*-convergent to some point $u \in \ell_1$, hence $(x_n - u)$ is weak*-null, and we have

$$\lim_{n\to\infty} (\|x + u - x_n\| - \|u - x_n\| - \|x\|) = 0$$

for every $x \in \ell_1$. Replacing x by $x - u$ yields

$$\lim_{n\to\infty} (\|x - x_n\| - \|u - x_n\| - \|x - u\|) = 0,$$

equivalently, $\sigma(x) - \sigma(u) = \|x - u\|$. This peculiarity of ℓ_1 deserves a name:

Definition 8.5.3 A Banach space X has the m_1-type property if, for every type σ on X, there exists $u \in X$ such that for all $x \in X$, we have

$$\sigma(x) = \|x - u\| + \sigma(u).$$

Of course, ℓ_1 has the m_1-type property. It is not the first time we have encountered something similar: recall from (8.16) that ℓ_p spaces, $1 < p < \infty$, have the analogous property that *any* type $\sigma\colon \ell_p \longrightarrow \mathbb{R}_+$ has the form $\sigma(x) = (\|x-u\|^p + \sigma(u)^p)^{1/p}$. The exact value of p is essential because (compare with Proposition 8.5.8):

Theorem 8.5.4 *A separable Banach space with the m_1-type property is $\mathscr{C}$-extensible. In particular, ℓ_1 is $\mathscr{C}$-extensible.*

Proof We shall proceed in two steps. The first step is a construction that allows one to replace a given nasty enlargement by a more pleasant one. The trick for doing that is as follows. Let Z be a Banach space with a subspace X. An X-seminorm on Z is a convex, symmetric function $\varphi\colon Z \longrightarrow [0, \infty)$ such that

- $\varphi(z) \le \|z\|$ for $z \in Z$,
- $\varphi(x) = \|x\|$ for $x \in X$.

Observe that a minimal X-seminorm (with respect to the pointwise order) is actually a seminorm on Z, because if $0 < t < 1$ then $\phi(y) = t^{-1}\varphi(ty)$ defines another X-seminorm with $\phi \leq \varphi$. Hence $\phi = \varphi$ and since $t\phi(tx) = \varphi(ttx) = \phi(ttx)$, it follows that $t\phi(x) = \phi(tx)$ for $|t| < 1$. For $t > 1$, just set $\phi(tx) = \phi(t^{-1}ttx) = t^{-1}\phi(ttx)$ to get $t\phi(tx) = \phi(ttx)$ and we are done again. So, $\varphi(tx) = |t|\varphi(x)$ for t real, and it is in fact a seminorm. Let φ be any minimal X-seminorm on Z and let Z_φ be the completion of $Z/\ker\varphi$ with respect to the induced norm. It is clear that there is a contractive operator $Z \longrightarrow Z_\varphi$ whose restriction to X is an isometry:

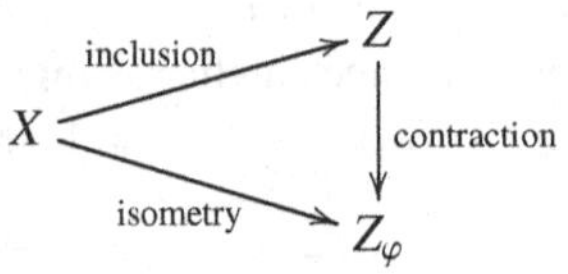

Thus, it suffices to show that $\mathscr{C}$-valued operators on X admit 1^+-extensions to Z_φ. Before going any further, let us remark that Z_φ has the property that the only X-seminorm on Z_φ is the norm itself. Clearly, Z_φ is separable. Let $(Z_n)_{n\geq 0}$ be an increasing sequence of subspaces of Z_φ whose union is dense, with $Z_0 = X$ and $\dim Z_{n+1}/Z_n = 1$ for $n \geq 1$. Now, the plan is to prove that each $\mathscr{C}$-valued operator on Z_n has a 1^+-extension to Z_{n+1}. To do this, it suffices to show that for every $n \geq 0$, the pair (Z_n, Z_φ) has property $\Sigma_1(\lambda)$ for all $\lambda > 1$, and then to apply Lemma 8.3.3 followed by Lemma 8.3.5. We get the idea, right? So the proof will be complete after showing that for each $n \geq 0$, the pair (Z_n, Z_φ) satisfies the inequality (8.13) of Lemma 8.3.8 for $\lambda = 1$. Fix $n \geq 0$ and let σ be any type on Z_φ that is supported on Z_n. Since Z_n/X is finite-dimensional, there is a $u_0 \in Z_n$ such that $\sigma(z) = \sigma'(z - u_0)$ for some type σ' supported on X. Since X has the m_1-type property, there exists $u_1 \in X$ such that $\sigma'(x) - \sigma(u_1) = \|x - u_1\|$ for all $x \in X$. Letting $u = u_0 + u_1$, define $\sigma_0 \colon Z_\varphi \longrightarrow [0, \infty)$ by $\sigma_0(z) = \sigma(z + u)$. Then, for $x \in X$, we have $\sigma_0(x) = \|x\| + \sigma(u)$ and, in general,

$$\sigma_0(z) \leq \sigma_0(0) + \|z\| = \sigma(u) + \|z\|. \tag{8.22}$$

Consider the function

$$\psi(z) = \frac{\sigma_0(z) + \sigma_0(-z) - 2\sigma(u)}{2}.$$

This is an X-seminorm on Z_φ: that it is convex and symmetric is obvious; similarly that $\psi(0) = 0$, and since $\mathrm{Lip}(\psi) = 1$, we have $\psi(z) \leq \|z\|$ and, clearly, $\psi(x) = \|x\|$ for all $x \in X$. But every X-seminorm on Z_φ is actually the norm, and so $\psi(z) = \|z\|$ for all $z \in Z_\varphi$. Taking (8.22) into account, it quickly follows that $\sigma_0(z) - \sigma(u) = \|z\|$ for all $z \in Z_\varphi$, which can be written as $\sigma(z) = \|z - u\| + \sigma(u)$.

Finally, if β is another type on Z_φ that is supported on Z_n, then one can find $v \in Z_n$ such that $\beta(z) = \|z - v\| + \beta(v)$. Hence,

$$\begin{aligned}\sigma(u) + \beta(u) &= \sigma(u) + \beta(v) + \|u - v\| \\ &\leq \sigma(u) + \beta(v) + \|z - u\| + \|z - v\| \\ &= \sigma(u) + \beta(v) + \sigma(z) - \sigma(u) + \beta(z) - \beta(v) \\ &= \sigma(z) + \beta(z).\end{aligned}$$

□

Speechless? What has actually been proved is

Corollary 8.5.5 *Every $\mathscr{C}$-valued operator defined on a separable space with the m_1-type property admits a 1^+-extension to every separable superspace.*

The 3-space argument given in Lemma 8.5.2 immediately yields that weak*-closed subspaces of ℓ_1 are $\mathscr{C}$-extensible because spaces of the form $\ell_1(\mathbb{N}, F_n)$ with each F_n finite-dimensional have the m_1-type property and thus they must be $\mathscr{C}$-extensible, while each weak*-closed subspace of ℓ_1 is a twisted sum of two subspaces of that form (Proposition 5.3.1). This argument does not, however, provide an estimate for the norm of the extending operator; on the other hand, that is unnecessary since Theorem 8.5.4 and Proposition 8.3.14 combine to yield 1^+ in all cases! It is perhaps about time to present the carnival parade of all $\mathscr{C}$-extensible spaces currently known:

Theorem 8.5.6 *The following Banach spaces are $\mathscr{C}$-extensible:*

- *c_0 and all its subspaces,*
- *ℓ_1 and all its weak*-closed subspaces,*
- *all spaces with property (L*) as well as their duals,*
- *every space with the m_1-type property,*
- *$c_0(E)$ for every E in the list,*
- *twisted sums of two spaces in the list.*

What about ℓ_p?

It is difficult not to believe that ℓ_p spaces are $\mathscr{C}$-extensible too; after all, it is true for the extreme values $p = 1, \infty$, no matter how we interpret the case ∞: either as c_0 or, bending the rules, as ℓ_∞. So, what could go wrong in the middle? Well, what goes wrong is that ℓ_p is not $\mathscr{C}$-extensible for $1 < p < \infty$.

Proposition 8.5.7 *Let X be a separable Banach space containing ℓ_p, where $p \in (1, \infty)$. If every $\mathscr{C}$-valued operator on ℓ_p can be extended to X then X*

can be given an equivalent norm $|\cdot|$ such that for every weakly null type σ supported on ℓ_p and all $x \in X$, we have

$$(|x|^p + \sigma(0)^p)^{1/p} \leq \sigma(x). \tag{8.23}$$

Proof If the embedding $\ell_p \longrightarrow X$ is $\mathscr{C}$-trivial then there is a homogeneous Zippin selector $\omega\colon \ell_q \longrightarrow X^*$, where q is the conjugate exponent of p and ℓ_q is treated as the dual space of ℓ_p. We consider the following seminorm on X:

$$|x|_0 = \sup\left\{\langle x, \omega(u^*)\rangle : \ \|u^*\|_q \leq 1\right\}.$$

It is clear that $|u|_0 = \|u\|$ for $u \in \ell_p$ and that $|x|_0 \leq \|\omega\|\,\|x\|$, where, as usual, $\|\omega\| = \sup\{\|\omega(u^*)\|\colon \|u^*\|_q \leq 1\}$. Assume (u_n) is a weakly null sequence in ℓ_p. We want to see that

$$\liminf_{n\to\infty} (|x|_0^p + \|u_n\|^p)^{1/p} \leq \limsup_{n\to\infty} |x + u_n|_0 \tag{8.24}$$

for each $x \in X$. Pick normalised $u_n^* \in \ell_q$ such that $\langle u_n^*, u_n\rangle = \|u_n\|$. Fix $x \in X$ and $\varepsilon > 0$. Take a normalised $u^* \in \ell_q$ such that $\langle x, \omega(u^*)\rangle = |x|_0$. We have

$$\langle x + u_n, \omega(u^* + \varepsilon u_n^*)\rangle \leq \|u^* + \varepsilon u_n^*\|\, |x + u_n|_0,$$

hence

$$\liminf_{n\to\infty} \langle x + u_n, \omega(u^* + \varepsilon u_n^*)\rangle \leq \limsup_{n\to\infty} \|u^* + \varepsilon u_n^*\|\, |x + u_n|_0. \tag{8.25}$$

For the terms on the right, we have $\lim \|u^*+\varepsilon u_n^*\| = (1+\varepsilon^q)^{1/q}$ as we are working in ℓ_q. The terms on the left can be decomposed as $\langle x, \omega(u^* + \varepsilon u_n^*)\rangle + \langle u_n, \omega(u^* + \varepsilon u_n^*)\rangle$ and thus $\lim_n \langle x, \omega(u^* + \varepsilon u_n^*)\rangle = \langle x, \omega(u^*)\rangle = |x|_0$ since ω is weak*-continuous and u_n^* is weak*-null, while $\langle u_n, \omega(u^* + \varepsilon u_n^*)\rangle = u^*(u_n) + \varepsilon\|u_n\|$ because ω is a selector. Thus (8.25) implies

$$|x|_0 + \varepsilon \liminf_{n\to\infty} \|u_n\| \leq (1 + \varepsilon^q)^{1/q} \limsup_{n\to\infty} |x + u_n|_0,$$

and Hölder's inequality yields (8.24). Now, we introduce a true renorming on X as follows:

$$|x| = \inf\{\|x - v\| + |x - v|_0 + \|v\| : v \in \ell_p\}.$$

Note that $\|x\| \leq |x| \leq (1 + \|\omega\|)\|x\|$ for all $x \in X$ and that $|x| = \|x\| = \|x\|_p$ for $x \in \ell_p$. This norm has 'the same' property as $|\cdot|_0$, namely that if (u_n) is weakly null in ℓ_p then

$$\liminf_{n\to\infty} (|x|^p + \|u_n\|^p)^{1/p} \leq \limsup_{n\to\infty} |x + u_n|_0, \tag{8.26}$$

which is just a restatement of (8.23). Fix (u_n) and x. Passing to a subsequence if necessary, we may assume that both $|x + u_n|$ and $\|u_n\|$ converge. Choose a sequence (v_n) of points in ℓ_p such that

$$\lim_n |x + u_n| = \lim_n \left(\|x + u_n - v_n\| + |x + u_n - v_n|_0 + \|v_n\|\right).$$

The sequence (v_n) is bounded in ℓ_p. Hence, passing to further subsequences, we may assume it is weakly convergent to, say $v \in \ell_p$ so that

$$\lim_n \|v_n\| = \lim_n \left(\|v - v_n\|^p + \|v\|^p\right)^{1/p}.$$

The relevant property of $|\cdot|_0$ now intervenes: assuming that all limits exist, we have

$$\lim_n \left(|x - v|_0^p + \|u_n + v - v_n\|^p\right)^{1/p} \leq \lim_n |y - v + v - v_n + u_n|_0. \tag{8.27}$$

In conclusion, assuming again that all limits exist, we have

$$\begin{aligned}
&\lim_n |x + u_n| \\
&= \lim_n \left(\|x + u_n - v_n\| + |x + u_n - v_n|_0 + \|v_n\|\right) \\
&\geq \lim_n \left(\|x + u_n - v_n\| + \left(|x - v|_0^p + \|u_n + v - v_n\|^p\right)^{1/p} + \|v_n\|\right) \\
&\geq \lim_n \left(\|x + u_n - v_n\| + \left(|x - v|_0^p + \|u_n + v - v_n\|^p\right)^{1/p} + \left(\|v - v_n\|^p + \|v\|^p\right)^{1/p}\right) \\
&\geq \lim_n \left(\left(\|x - v\| + |x - v|_0 + \|v\|\right)^p + \|u_n\|^p\right)^{1/p} \\
&\geq \lim_n \left(|x|^p + \|u_n\|^p\right)^{1/p}.
\end{aligned}$$

□

Proposition 8.5.8 *The space ℓ_p is not $\mathscr{C}$-extensible for $1 < p < \infty$.*

Proof The idea is to construct a separable enlargement $X(p)$ of ℓ_p for which no renorming satisfies (8.23). Let $\mathcal{D}$ be the dyadic tree, whose elements are all finite sequences $a = (s_1, \ldots, s_m)$ of zeros and ones, including the empty sequence, denoted by $\emptyset$. The number m is called the *depth* of a. The depth of $\emptyset$ is zero. Given $a = (s_1, \ldots, s_m)$ and $b = (t_1, \ldots, t_n)$, we write $a \preceq b$ if $m \leq n$ and $s_i = t_i$ for $1 \leq i \leq m$. If $a \preceq b$ and $a \neq b$, are write $a \prec b$. A *segment* of $\mathcal{D}$ is a finite subset of the form $\{a_1, \ldots, a_k\}$ with $a_1 \prec a_2 \prec \cdots \prec a_k$. The collection of all segments of $\mathcal{D}$ will be denoted S. Let $c_{00}(\mathcal{D})$ be the space of all finitely supported functions $\xi\colon \mathcal{D} \longrightarrow \mathbb{K}$. As usual, we use e_a to denote the function that takes the value 1 at a, and zero otherwise. We consider the natural bilinear pairing on $c_{00}(\mathcal{D})$ given by

$$\langle \xi, \eta \rangle = \sum_{a \in \mathcal{D}} \xi(a)\eta(a).$$

We now introduce two mutually dual norms on $c_{00}(\mathcal{D})$. The first is

$$\|\xi\|_* = \max\left(\|\xi\|_{\ell_q(\mathcal{D})}, \sup_{\beta \in S} \sum_{a \in \beta} |\xi(a)|\right), \tag{8.28}$$

where $q^{-1} + p^{-1} = 1$. This norm will play a supporting role in the construction. The lead role is played by

$$\|\eta\| = \sup\{\langle \xi, \eta \rangle : \|\xi\|_* \leq 1\}.$$

Let $X(p)$ be the completion of $c_{00}(\mathfrak{D})$ with respect to $\|\cdot\|$. Although these norms may remind us of that of the James tree space, they are unconditional in the sense that if $\eta, \xi \in c_{00}(\mathfrak{D})$ are such that $|\xi| \leq \eta$, then $\|\xi\| \leq \|\eta\|$ and $\|\xi\|_* \leq \|\eta\|_*$. Clearly, if $\beta = \{a_1, \ldots, a_k\}$ is a segment, then $\|1_\beta\|_* = k$ and $\|1_\beta\| = 1$. Our immediate task is to find a copy of ℓ_p inside $X(p)$. To this end, given $a \in \mathfrak{D}$, let $a0, a1$ be the two successors of a and set $u_a = e_{a0} + e_{a1}$; that is,

$$u_a(b) = \begin{cases} 1 & \text{if } b \text{ is a successor of } a \\ 0 & \text{otherwise.} \end{cases}$$

We now prove that the sequence $(u_a)_{a\in\mathfrak{D}}$ spans a subspace isomorphic to ℓ_p in $X(p)$. More precisely, we shall show that, for $\eta \in c_{00}(\mathfrak{D})$, we actually have

$$\frac{\|\eta\|_{\ell_p(\mathfrak{D})}}{C_p} \leq \Big\| \sum_{a\in\mathfrak{D}} \eta(a) u_a \Big\|_X \leq 2^{1/p} \|\eta\|_{\ell_p(\mathfrak{D})}, \tag{8.29}$$

where $C_p > 0$ is a constant depending only on p. The right inequality is obvious from (8.28). The left inequality follows from the following:

Claim For every $\eta \in c_{00}(\mathfrak{D})$, there is $\xi \in c_{00}(\mathfrak{D})$ such that $\langle \xi, u_a \rangle = \eta(a)$ and $\|\xi\|_* \leq C_p \|\eta\|_{\ell_q(\mathfrak{D})}$, where $C_p = \max\,(1, (2^p - 2)^{-1/p})$.

Proof of the claim It suffices to consider the case in which $\eta, \xi \geq 0$. For $n \geq 0$, let $\mathfrak{D}_n$ be the subset of those points of $\mathfrak{D}$ whose depth is at most n. For fixed p, let C_n be the best constant such that, whenever $\eta \geq 0$ is supported on $\mathfrak{D}_n$, there is $\xi \geq 0$ satisfying $\langle \xi, u_\alpha \rangle = \eta(a)$ and $\|\xi\|_* \leq C_n \|\eta\|_{\ell_q(\mathfrak{D})}$. Note that $C_0 = 1$. We will estimate C_n in terms of C_{n-1}. Let us split $\mathfrak{D}_n$ into three disjoint subsets $\mathfrak{D}_n = \{\emptyset\} \cup 0\mathfrak{D}_{n-1} \cup 1\mathfrak{D}_{n-1}$, where, as should be obvious, $k\mathfrak{D}_{n-1} = \{(a_1, \ldots, a_n) : a_1 = k\}$, for $k = 0, 1$. If η is supported on $\mathfrak{D}_n$ and we call $\eta^{(k)} = \eta 1_{k\mathfrak{D}_{n-1}}$ for $k = 0, 1$ then $\eta = \eta(\emptyset) e_\emptyset + \eta^{(0)} + \eta^{(1)}$. In view of the obvious symmetries of the norms $\|\cdot\|_*$ and $\|\cdot\|$, and the definition of C_{n-1}, for $k = 0, 1$, we can find $\xi^{(k)} \geq 0$, supported on $k\mathfrak{D}_{n-1}$, such that

$$\|\xi^{(k)}\|_* \leq C_{n-1} \|\eta^{(k)}\|_{\ell_{p^*}(\mathfrak{D})} \quad \text{and} \quad \langle \xi^{(k)}, u_a \rangle = \eta(a) \quad \text{for} \quad a \in k\mathfrak{D}_{n-1}.$$

Note that for every $a \in \mathfrak{D}_n$ different from $\emptyset$, we do have $\langle \xi^{(0)} + \xi^{(1)}, u_a \rangle = \eta(a)$. Thus, we try to construct ξ by just tuning $\xi^{(0)} + \xi^{(1)}$, keeping $\xi(\emptyset) = 0$. To do this, for $k = 0, 1$, let $\beta^{(k)}$ be the segment starting at (k) such that

$$\sum_{a\in\beta^{(k)}} \xi^{(k)}(a) = \max_{\beta\in S} \sum_{a\in\beta} \xi^{(k)}(a).$$

Assume, without loss of generality, that $\sum_{a\in\beta^{(0)}} \xi^{(0)}(a) \le \sum_{a\in\beta^{(1)}} \xi^{(1)}(a)$. Thus,

- if $\eta(\emptyset) \le \sum_{a\in\beta^{(1)}} \xi^{(1)}(a) - \sum_{a\in\beta^{(0)}} \xi^{(0)}(a)$, set $\xi = \eta(\emptyset)e_{(0)} + \xi^{(0)} + \xi^{(1)}$;
- otherwise, set

$$\xi = \eta(\emptyset)\frac{e_{(0)} + e_{(1)}}{2} + \left(\sum_{a\in\beta^{(1)}} \xi^{(1)}(a) - \sum_{a\in\beta^{(0)}} \xi^{(0)}(a)\right)\frac{e_{(0)} - e_{(1)}}{2} + \xi^{(0)} + \xi^{(1)}.$$

In either case, we have

$$\begin{aligned}\|\xi\|_{\ell_q(\mathcal{D})} &\le \left(\eta(\emptyset)^q + C_{n-1}\|\eta^{(0)}\|^q_{\ell_q(\mathcal{D})} + C_{n-1}\|\eta^{(1)}\|^q_{\ell_q(\mathcal{D})}\right)^{1/q}\\ &\le \max(1, C_{n-1})\,\|\eta\|_{\ell_q(\mathcal{D})}\end{aligned}$$

since, in the first case, $\max_{\beta\in S}\sum_{a\in\beta}\xi(a) \le \sum_{a\in\beta^{(1)}}\xi^{(1)}(a) \le C_{n-1}\|\eta^{(0)}\|_{\ell_q(\mathcal{D})}$; and, in the second case,

$$\begin{aligned}\max_{\beta\in S}\sum_{a\in\beta}\xi(a) &= \frac{\eta(\emptyset) + \sum_{a\in\beta^{(0)}}\xi^{(0)}(a) + \sum_{a\in\beta^{(1)}}\xi^{(1)}(a)}{2}\\ &\le \frac{\eta(\emptyset) + C_{n-1}\|\eta^{(0)}\|_{\ell_q(\mathcal{D})} + C_{n-1}\|\eta^{(1)}\|_{\ell_q(\mathcal{D})}}{2}\\ &\le \frac{(1 + 2C^p_{n-1})^{1/p}}{2}\,\|\eta\|_{\ell_q(\mathcal{D})}\,.\end{aligned}$$

Hence $C_n \le \max\left(1, C_{n-1}, \frac{1}{2}(1 + 2C^p_{n-1})^{1/p}\right)$. It follows that for $2^p \ge 3$, we get $C_n \le 1$, while if $2^p < 3$ then $C_n \le (2^p - 2)^{-1/p}$. □

The left estimate in (8.29) is now easy. Pick a finitely supported η, which we assume positive without loss of generality. Let ξ be the output of the claim when the input is η^{p-1} so that

$$\begin{aligned}&\langle \xi, u_a\rangle = \eta(a)^{p-1},\\ &\|\xi\|_* \le C_p\|\eta^{p-1}\|_{\ell_q(\mathcal{D})} = C_p\|\eta\|^{p/q}_{\ell_p(\mathcal{D})}.\end{aligned}$$

For the last equality, note that $q(p-1)/p = 1$. Thus,

$$\left\|\sum_{a\in\mathcal{D}}\eta(a)u_a\right\| \ge \frac{\langle \xi, \sum_a \eta(a)u_a\rangle}{\|\xi\|_*} \ge \frac{\sum_a \eta^p(a)}{C_p\|\eta\|^{p/q}_{\ell_p(\mathcal{D})}} \ge \frac{\|\eta\|^p_{\ell_p(\mathcal{D})}}{C_p\|\eta\|^{p/q}_{\ell_p(\mathcal{D})}} = \frac{\|\eta\|_{\ell_p(\mathcal{D})}}{C_p}.$$

Up to here, we have the proof that $X(p)$ is a separable enlargement of ℓ_p. The remainder of the proof is to establish that the inclusion of ℓ_p into $X(p)$ is not

$\mathscr{C}$-trivial. This is done by exploiting the criterion in Proposition 8.5.7 against the sequence $(\frac{1}{2}u_a)_{a\in\mathfrak{D}}$: if every $\mathscr{C}$-valued operator on ℓ_p can be extended to $X(p)$, some equivalent norm $|\cdot|$ on $X(p)$ should satisfy

$$(|\eta|^p + 2c^p)^{1/p} \leq \liminf_a \left|\eta + \tfrac{1}{2}u_a\right| \tag{8.30}$$

for some constant $c > 0$ and every $\eta \in X(p)$. One can actually take $c = |u_\emptyset|/2^{1+1/p}$. In particular, for each finitely supported η and every $a \in \mathfrak{D}$, there is $a^* > a$ such that

$$(|\eta|^p + c^p)^{1/p} < \left|\eta + \tfrac{1}{2}u_{a^*}\right|. \tag{8.31}$$

No norm on $c_{00}(\mathfrak{D})$ having this property can be equivalent to the norm of $X(p)$; let us see why. If (8.31) holds then we can find, for every n, a segment β (of cardinality n, if we want) such that $|1_\beta| \geq cn^{1/p}$, where $1_\beta = \sum_{a\in\beta} e_a$. On the other hand, $\|1_\beta\| = 1$ for every segment β: it is trivially true for $n = 1$, so we can assume that $\beta = \{a_1, \ldots, a_n\}$ is a segment such that $|1_\beta| \geq cn^{1/p}$. Letting $\eta = 1_\beta$ in (8.31), we can find $a^* > a_n$ such that

$$c(n+1)^{1/p} \leq (|1_\beta|^p + c^p)^{1/p} < \left|1_\beta + \tfrac{1}{2}u_{a^*}\right|.$$

Since $u_{a^*} = e_{a^*0} + e_{a^*1}$, either

$$c(n+1)^{1/p} < \left|1_\beta + e_{a^*0}\right| \qquad \text{or} \qquad c(n+1)^{1/p} < \left|1_\beta + e_{a^*1}\right|.$$

In the first case, the new segment is $\{a_1, \ldots, a_n, a^*0\}$; in the second case, it is $\{a_1, \ldots, a_n, a^*1\}$. □

We now want to obtain a superreflexive version of $X(p)$. The idea is to obtain an intermediate space between $X(p)$ and $\ell_p(\mathfrak{D})$. A comfortable way to do so would be by using the *complex* interpolation method, which requires a temporary lift of the ban on complex scalars. But the alleged simplification that this way of acting brings is mostly a delusion, resulting in the reader's disappointment. We prefer instead to avoid that tactical move and just explain the bare facts as they are:

Corollary 8.5.9 *For each $1 < p < \infty$, there is an embedding of ℓ_p into a superreflexive space that is not $\mathscr{C}$-trivial.*

Proof Let X_0 and X_1 be two Banach lattices of functions defined on the same set S. For $0 < \theta < 1$, we define the intermediate space $X_\theta = X_0^{1-\theta}X_1^\theta$ of functions $f\colon S \longrightarrow \mathbb{R}$ that admit a decomposition $|f| = g^{1-\theta}h^\theta$ for non-negative $g \in X_0$ and $h \in X_1$, endowed with the norm

$$\|f\|_\theta = \inf \|g\|_0^{1-\theta}\|h\|_1^\theta,$$

where the infimum is taken over all g, h in the decomposition above. The space X_θ is again a Banach lattice, and it is uniformly convex if either X_0 or X_1 is: this we know because a nice computation performed by Cwikel and Reisner in [143, Theorem 1 (i)] shows that the modulus of convexity of X_θ obeys an estimate of the form $\delta_{X_\theta}(\varepsilon) \geq c_1\delta_{X_1}(c_2\varepsilon^{1/\theta})$ for some $c_1, c_2 > 0$ and all $0 < \varepsilon < 2$. That said, let us specialise to the case $X_0 = X(p)$ and $X_1 = \ell_p(\mathfrak{D})$. We claim that any of the spaces $X_\theta = X(p)^{1-\theta}\ell_p(\mathfrak{D})^\theta$ for $0 < \theta < 1$ provides the required example as a consequence of the following facts:

- X_θ is uniformly convex because $\ell_p(\mathfrak{D})$ is.
- The sequence $(u_a)_{a\in\mathfrak{D}}$ is equivalent to the unit basis of ℓ_p in X_θ since this happens both in $\ell_p(\mathfrak{D})$ and in $X(p)$, by the very definition of the norm of X_θ.
- If β is a segment of length n, then the norm of 1_β in $\ell_p(\mathfrak{D})$ is $n^{1/p}$, while its norm in $X(p)$ is 1. Therefore, the norm of 1_β in X_θ is at most $n^{\theta/p}$.

The embedding of ℓ_p into X_θ provided by the sequence (u_a) cannot be $\mathscr{C}$-trivial since it was established during the proof of Proposition 8.5.8 that for any norm $|\cdot|$ on $c_{00}(\mathfrak{D})$ satisfying the condition (8.30) there is a constant $c > 0$ such that, for every n, there is a segment β of length n such that $|1_\beta| \geq cn^{1/p}$. This contradicts $\|1_\beta\|_\theta$ being dominated by $n^{\theta/p}$. □

8.6 The Dark Side of the Johnson–Zippin Theorem

The Johnson–Zippin theorem has its origin in the papers [464; 465]. At the end of [464] Zippin poses three problems; one of them, Problem 2, reads: *Is it true that every $\mathscr{C}$-valued operator defined on any subspace of ℓ_1 admits a λ-extension for some $\lambda \geq 2$?* Why has $\lambda < 2$ been ruled out? Because Zippin shows in [465] that if E is the kernel of the sum functional on ℓ_1, the inclusion $E \longrightarrow \ell_1$, which obviously admits a 2-Zippin selector, does not admit a λ-Zippin selector for any $\lambda < 2$. That is startling. Problem 2 is explicitly implicit in [466, pp. 1732–1735], has a negative solution as we all know (don't we?) and mutates along the way into

8.6.1 Zippin's problem Characterise the subspaces E of ℓ_1 such that every $C(K)$-valued operator defined on E can be extended to ℓ_1.

We already know that $\mathscr{C}$-valued operators defined on weak*-closed subspaces of ℓ_1 admit extensions. The following theorem asserts the same for $\mathscr{L}_\infty$-valued operators. Just use Proposition 5.2.9 (b) to get

8.6.2 The Johnson–Zippin theorem *$\mathscr{L}_\infty$-valued operators defined on any weak*-closed subspace of ℓ_1 can be extended to ℓ_1.*

Proof $\text{Ext}(H^*, \mathscr{L}_\infty) = 0$ for any subspace $H \subset c_0$. □

The original Johnson–Zippin theorem [237] yields an estimate of 3^+ for the norm of the extending operator. Under the additional assumption that the quotient space has the AP, they reduce the estimate to an optimal 1^+. The question of whether the apparent duality between the Lindenstrauss–Pełczyński and the Johnson–Zippin theorems is a side effect of the structure of subspaces of c_0 or, as some of the authors think, a real duality that deserves careful consideration.

In doing so, one must take into account that the sequence $0 \longrightarrow \ell_2 \longrightarrow C(B^*_{\ell_2}) \longrightarrow \diamond \longrightarrow 0$ is $\mathscr{C}$-trivial (with the meaning that the embedding is $\mathscr{C}$-trivial) but has non-$\mathscr{C}$-trivial dual or bidual sequences: indeed, the dual sequence has the form $0 \longrightarrow \diamond^* \longrightarrow L_1(\mu) \longrightarrow \ell_2 \longrightarrow 0$ and is not $\mathscr{C}$-trivial because it is semi-equivalent to any projective presentation $0 \longrightarrow \kappa(l_2) \longrightarrow \ell_1 \longrightarrow \ell_2 \longrightarrow 0$ since $\kappa(\ell_2)$ is an ultrasummand by Lemma 10.4.1. Thus, either both sequences are $\mathscr{C}$-trivial or neither of them is. But a $\mathscr{C}$-trivial projective presentation of ℓ_2 means $\text{Ext}(\ell_2, C(\Delta)) = 0$, which is simply false for many reasons, among which is the next Proposition 8.6.4. The bidual sequence cannot be $\mathscr{C}$-trivial because the natural embedding $\delta\colon \ell_2 \longrightarrow C(B^*_{\ell_2})$ cannot extend to an operator $C(B^*_{\ell_2})^{**} \longrightarrow C(B^*_{\ell_2})$: the former space is injective, thus it enjoys the Dunford–Pettis and Grothendieck properties. The latter makes every operator from it into a separable space weakly compact; the former makes it completely continuous. Thus, its restriction to ℓ_2 should be compact and cannot be δ. Similar reasoning shows that while any sequence $0 \longrightarrow H \longrightarrow c_0 \longrightarrow c_0/H \longrightarrow 0$ is $\mathscr{C}$-trivial, as is its dual sequence, its bidual sequence cannot be $\mathscr{C}$-trivial. Since weak*-closed subspaces of ℓ_1 are $\mathscr{C}$-extensible, the 'converse Johnson–Zippin theorem' would be to decide whether weak*-closed subspaces are *the only* $\mathscr{C}$-extensible subspaces of ℓ_1. But even this formulation remains fishy: does one mean *subspaces isomorphic to weak*-closed subspaces*? The additional information that $\mathscr{C}$-valued operators on weak*-closed subspaces admit almost isometric extensions paves the way for a true converse:

Proposition 8.6.3 *A subspace $E \subset \ell_1$ is weak*-closed if and only if every $\mathscr{C}$-valued operator defined on E admits 1^+-extensions to ℓ_1.*

Proof The 'only if' part was proved in Proposition 8.3.14, so we prove the 'if' part. Let (y_n) be a sequence in a closed subspace $E \subset \ell_1$. Assume that the sequence is weak*-convergent to some point $y \in \ell_1$. By passing to a subsequence, we may assume that (y_n) defines a type σ on ℓ_1, which, by the m_1-type property, must have the form $\sigma(x) = \|x - y\| + \|y\|$. Since the pair (E, ℓ_1) has all properties $\Sigma_1(\lambda)$ for $\lambda > 1$, one has $\inf_{u \in E} \sigma(u) \leq \lambda\sigma(y)$, hence

$\inf_{u\in E} \sigma(u) \le \sigma(y)$. These two things together yield $\inf_{u\in E} \|u-y\| + \|y\| \le \sigma(y)$, which necessarily means $y \in E$. □

Subspaces of ℓ_1 that are weak*-closed with respect to other dual pairings are not necessarily $\mathscr{C}$-extensible. To show that, we anticipate from Proposition 8.6.4 that there are heavy restrictions on a subspace E of ℓ_1 in order for it to be $\mathscr{C}$-extensible: ℓ_1/E has to be at least a Schur space. Thus, given a non-Schur separable space X and a quotient map $Q\colon \ell_1 \longrightarrow X$, there exists an operator $\ker Q \longrightarrow C(\Delta)$ that cannot be extended to ℓ_1. Now pick an isometric embedding of Schreier's space S into $C(\omega^\omega)$. The kernel of the quotient map $\ell_1 \longrightarrow \mathsf{S}^*$ is obviously weak*-closed when ℓ_1 is treated as the dual of $C(\omega^\omega)$. But S^* is not Schur since S does not have the Dunford–Pettis property [102].

When Is $\mathrm{Ext}(X, C(\Delta)) = 0$?

A reformulation of Zippin's problem is: *Characterise the Banach spaces X such that* $\mathrm{Ext}(X, C(\Delta)) = 0$. Actually, it is not known whether there exists a single 'non-trivial' example of a Banach space X for which $\mathrm{Ext}(X, C(\Delta)) = 0$. What do we mean by a non-trivial example? Well, something like 'X does not have the form $\ell_1(F_n)$ for finite-dimensional F_n, or is not a twisted sum of such spaces ... or it is not too close to ℓ_1 ... somehow'.

Proposition 8.6.4 *If* $\mathrm{Ext}(X, C(\Delta)) = 0$ *then X has the Schur property.*

Proof Let X be a separable Banach space. If X is not Schur then it contains a weakly null normalised basic sequence $(x_n)_n$. By considering the basis expansion, we obtain a map $\tau_0\colon \overline{[x_n : n \in \mathbb{N}]} \longrightarrow c_0$ such that $\tau_0(x_n) = e_n$. Since c_0 is separably injective, τ_0 admits a 2-extension $\tau\colon X \longrightarrow c_0$. Pick the Foiaş–Singer sequence (2.5) and draw the diagram

$$\begin{array}{ccccccccc} 0 & \longrightarrow & C(\Delta) & \longrightarrow & D & \longrightarrow & c_0 & \longrightarrow & 0 \\ & & \| & & \uparrow{\scriptstyle \underline{\tau}} & & \uparrow{\scriptstyle \tau} & & \\ 0 & \longrightarrow & C(\Delta) & \longrightarrow & \mathrm{PB} & \longrightarrow & X & \longrightarrow & 0 \end{array}$$

The pullback sequence cannot split because a linear continuous section $s\colon X \longrightarrow \mathrm{PB}$ would produce a weakly null lifting $(\underline{\tau} s(x_n))_n$ of (e_n), which is impossible by Lemma 2.2.4. □

Actually, [267, Theorem 5.1] shows that X must have a stronger version of the Schur property, but we will not continue on that path. The paper [267] provides, in addition to the previous one, the best available answer to Zippin's problem: if X is a separable space with a UFDD and $\mathrm{Ext}(X, C(\Delta)) = 0$ then

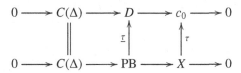

X is isomorphic to the dual of a subspace of c_0 (see the next section). This, in turn, provides the best evidence we have for a positive answer to the following conjecture [271, Problem 1]: *let X be a separable Banach space.* $\operatorname{Ext}(X, C(\Delta)) = 0$ *if and only if X is isomorphic to the dual of a subspace of* c_0. Equivalently, a subspace of ℓ_1 is $\mathscr{C}$-extensible if and only if it occupies a weak*-closed *position*, that is, if there is an automorphism $\tau \colon \ell_1 \longrightarrow \ell_1$ such that $\tau[E]$ is weak*-closed. A different approach to Zippin's problem was undertaken in [113] by characterising when $\operatorname{Ext}(X, C(K)) = 0$ via properties of a metric projection $m \colon Q^{(1)}(X, \mathbb{R}) \longrightarrow \mathsf{L}(X, \mathbb{R})$. The crucial point in all this is that while $\operatorname{Ext}(X, C(\Delta)) = 0$ imposes tight conditions on X, $\operatorname{Ext}(X, c) = 0$ imposes none when X is separable. What about intermediate cases? The simplest $\mathscr{C}$-spaces intermediate between c and $C(\Delta)$, or $C[0,1]$, are the spaces $C(\alpha)$ for countable ordinals α. Since $C(\alpha)$ is a complemented subspace of $C(\Delta)$, it is clear that $\operatorname{Ext}(X, C(\Delta)) = 0 \Longrightarrow \operatorname{Ext}(X, C(\alpha)) = 0$. It is then natural to look for properties of X, weaker than Schur although probably of the same type, that might characterise $\operatorname{Ext}(X, C(\alpha)) = 0$. Brunel and Sucheston [55] introduced the notion of a *spreading model*, which we will use only as a tool, using Ramsey-style arguments to observe that each normalised sequence in a Banach space has a subsequence $(x_n)_{n\geq 1}$ such that the limit $\lim_{\substack{n_1 \longrightarrow \infty \\ n_1 < \cdots < n_k}} \left\| \sum_{i=1}^k \lambda_i x_{n_i} \right\|$ exists for every finite sequence of scalars $\lambda_1, \ldots, \lambda_k$. This limit defines a norm on the space of finitely supported sequences if and only if the subsequence $(x_n)_n$ is not convergent [34, I. 1. Proposition 2]. The spreading model generated by the sequence (x_n) is the completion of that space.

Proposition 8.6.5 *Let X be a separable Banach space. If* $\operatorname{Ext}(X, C(\omega^\omega)) = 0$ *then, for some $\mu > 0$, every weakly null normalised sequence (x_n) and every $k \in \mathbb{N}$, there are integers $n_1 < n_2 < \cdots < n_k$ and signs $\varepsilon_k = \pm 1$ such that $\| \sum_j^k \varepsilon_j \lambda_j x_{n_j} \| \geq \mu \sum_{j=1}^k \lambda_j$ for all choices of positive scalars λ_j. Therefore, every spreading model of X generated by a normalised weakly null sequence is isomorphic to ℓ_1.*

Proof Consider the embedding of ω^ω into $[0,1]$ as in Lemma 2.2.6, form the sequence $0 \longrightarrow C(\omega^\omega) \longrightarrow D(\omega^\omega; (\omega^\omega)') \longrightarrow c_0((\omega^\omega)') \longrightarrow 0$ and observe that since $(\omega^\omega)'$ is countable, we can identify the quotient space with c_0.

Let (x_n) be a basic sequence in X. Proceeding as in the proof of Proposition 8.6.4, we obtain an operator $\tau \colon X \longrightarrow c_0$ such that $\tau(x_n) = e_n$, where (e_n) is the unit basis of c_0, and we can form the pullback diagram

$$\begin{array}{ccccccccc}
0 & \longrightarrow & C(\omega^\omega) & \longrightarrow & D(\omega^\omega; (\omega^\omega)') & \xrightarrow{J} & c_0 & \longrightarrow & 0 \\
 & & \| & & \uparrow{\scriptstyle \underline{\tau}} & & \uparrow{\scriptstyle \tau} & & \\
0 & \longrightarrow & C(\omega^\omega) & \longrightarrow & \mathrm{PB} & \longrightarrow & X & \longrightarrow & 0
\end{array}$$

Since $\operatorname{Ext}(X, C(\omega^\omega)) = 0$, the pullback sequence splits and so τ admits a bounded linear lifting $L: X \longrightarrow D(\omega^\omega; (\omega^\omega)')$. Place the functions $L(x_n)$ in Lemma 2.2.7 to get, for every n, integers $n \le n_1 < \cdots < n_k$ and signs $\varepsilon_i = \pm 1$ such that for all positive λ_i, we have

$$\|L\| \left\| \sum_{1\le i\le k} \varepsilon_i \lambda_i x_{n_i} \right\| \ge \left\| \sum_{1\le i\le k} \varepsilon_i \lambda_i L(x_{n_i}) \right\| \ge (1-\delta) \sum_{1\le i\le k} \lambda_i,$$

which proves the first part. A result of Beauzamy [34, I. 5. Proposition 1] implies that the canonical basis of the spreading model constructed over a normalised weakly null sequence is 1-unconditional, and this means that the spreading model constructed over the sequence (x_n) is isomorphic to ℓ_1. □

Ok, enough striking; we go to the ground and pound.

When Is $\operatorname{Ext}(X, C(\omega^\omega)) = 0$?

Let X be a separable Banach space. For every $N \in \mathbb{N}$, the space $C(\omega^N)$ is isomorphic to c_0, hence it is separably injective and $\operatorname{Ext}(X, C(\omega^N)) = 0$. On the other hand, the space ω^ω can be represented as the one-point compactification of the disjoint union $\bigcup_N \omega^N$, hence $C_0\big(\bigcup_N \omega^N\big)$ is a hyperplane of $C(\omega^\omega)$ and

$$C(\omega^\omega) \simeq C_0\left(\textstyle\bigcup_N \omega^N\right) = c_0\left(\mathbb{N}, C(\omega^N)\right).$$

Thus, Corollary 5.2.6 yields $\operatorname{Ext}(X, C(\omega^\omega)) = 0$ if and only if $\operatorname{Ext}(X, C(\omega^N)) = 0$ uniformly on N, and the existence of non-trivial elements of $\operatorname{Ext}(X, C(\omega^\omega))$ therefore depends on quantitative aspects of the diagram

$$\begin{array}{ccccccccc}
0 & \longrightarrow & C(\omega^N) & \longrightarrow & \ell_\infty(\omega^N) & \longrightarrow & Q_N & \longrightarrow & 0\\
 & & \| & & \uparrow & \nwarrow & \uparrow & & \\
0 & \longrightarrow & C(\omega^N) & \longrightarrow & \cdot & \longrightarrow & X & \longrightarrow & 0
\end{array}$$

The upper sequence is the 'obvious' injective presentation of $C(\omega^N)$, and we have written $Q_N = \ell_\infty(\omega^N)/C(\omega^N)$ for the sake of simplicity. The lower sequence splits, and therefore every operator $X \longrightarrow Q_N$ admits a lifting to $\ell_\infty(\omega^N)$, as the dotted arrow reminds us.

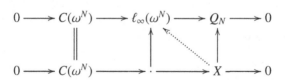

We will focus on the ratio between the norms of the operators $X \longrightarrow Q_N$ and the norms of their liftings to $\ell_\infty(\omega^N)$. Since there is not much we can say about the structure of Q_N, we will proceed the other way a round (taking advantage, let us say it once more, of the fact that *all* such operators have liftings): we start with an operator $\tau: X \longrightarrow \ell_\infty(\omega^N)$ and obtain, after composition with the quotient map, an operator $\tau': X \longrightarrow Q_N$. This operator might have a much smaller norm, and we then study the norms of its liftings to $\ell_\infty(\omega^N)$. All of them have the form $\tau - \tau''$, with $\tau'' \in \mathfrak{L}(X, C(\omega^N))$.

If K is a compact space, operators $\tau\colon X \longrightarrow \ell_\infty(K)$ and (not necessarily continuous) bounded mappings $\varphi : K \longrightarrow X^*$ correspond one to each other in the obvious way's $(\tau x)(a) = \langle \delta_a, \tau x\rangle = \langle \varphi(a), x\rangle$. Identification of a with the corresponding evaluation functional δ_a on $C(K)$ means that $\varphi(a) = \tau^*(\delta_a)$ can be interpreted as 'φ is the restriction of τ^* to K', and we can just write $\varphi(a) = \tau^*(a)$.

It is thus clear that $\|\tau\| = \sup_{a\in K} \|\tau^*(a)\|$. If τ' is the composition of τ with the quotient map $\ell_\infty \longrightarrow \ell_\infty/C(K)$ then $\|\tau'\| = \sup_{\|x\|\leq 1} \operatorname{dist}(\tau(x), C(K))$, and Lemma 2.2.2 immediately yields

$$2\|\tau'\| = \sup_{\|x\|\leq 1} \operatorname{osc}_K\langle \tau^*(\cdot), x\rangle.$$

We will informally refer to the right-hand member of the preceding equation as *the oscillation* of τ. The operator τ takes values in $C(K)$ if and only if it has oscillation 0; equivalently, τ^* is weak*-continuous on K (notice that the canonical inclusion $K \longrightarrow \ell_\infty(K)^*$ is not weak*-continuous).

Returning to ω^N and ω^ω, it is true that, using the constants $K^{(1)}[\cdot,\cdot]$ of (3.20), we have $\operatorname{Ext}(X, C(\omega^\omega)) = 0$ if and only if $\sup_{N\in\mathbb{N}} K^{(1)}[X, C(\omega^N)] < \infty$, but this does not fit very well with our current approach via operators. The following parameters provide more 'computable' forms for $K^{(1)}[X, C(\omega^N)]$:

8.6.6 Let $\pi_N(X)$ (resp. $\sigma_N(X)$) be the smallest constant such that if

$$0 \longrightarrow C(\omega^N) \xrightarrow{\ \jmath\ } Z \xrightarrow{\ \rho\ } X \longrightarrow 0 \tag{8.32}$$

is an isometrically exact sequence of Banach spaces and $\varepsilon > 0$ then there is a linear projection P through $\jmath$ with $\|P\| \leq \pi_N(X) + \varepsilon$ (resp. a linear section S of the quotient map with $\|S\| \leq \sigma_N(X) + \varepsilon$).

It is clear that the sequences $(\pi_N(X))_N, (\sigma_N(X))_N$ and $(K^{(1)}[X, C(\omega^N)])_N$ are equivalent: actually $|\pi_N(X) - \sigma_N(X)| \leq 1$ for all N by the discussion preceding Definition 2.1.6, while the argument in Corollary 3.3.8 implies that $\sigma_N(X) \leq 4K^{(1)}[X, C(\omega^N)]$ and the proof of Proposition 3.6.7 shows that $K^{(1)}[X, C(\omega^N)] \leq 2\sigma_N(X)$. Any other estimates the reader may find are welcome.

Given a separable Banach space X, we set

$$\rho_N(X) = \sup \operatorname{dist}(\tau, \mathfrak{L}(X, C(\omega^\omega)), \tag{8.33}$$

where the supremum is taken over all bounded operators $\tau : X \longrightarrow \ell_\infty(\omega^N)$ satisfying $d(\tau x, C(\omega^N)) \leq \|x\|$ for all $x \in X$. A brief reflection suffices to realise that $\rho_N(X)$ is the infimum of those constants ϱ such that every operator

$X \longrightarrow Q_N$ admits a ϱ-lifting to $\ell_\infty(\omega^N)$. We must be Gandalf on this: you should not pass from here without assimilating this fact.

The sequences $(\rho_N(X))_N$ and $(\pi_N(X))_N$ are equivalent too:

Lemma 8.6.7 *For any Banach space X, one has* $|\rho_N(X) - \pi_N(X)| \le 1$.

Proof Consider an isometrically exact sequence as in (8.32), let $I\colon Z \longrightarrow \ell_\infty(\omega^N)$ be an extension of the identity of $C(\omega^N)$ with $\|I\| = 1$ and form the commutative diagram

$$\begin{array}{ccccccccc} 0 & \longrightarrow & C(\omega^N) & \xrightarrow{\jmath} & Z & \xrightarrow{\rho} & X & \longrightarrow & 0 \\ & & \| & & \downarrow{\scriptstyle I} & & \downarrow{\scriptstyle I'} & & \\ 0 & \longrightarrow & C(\omega^N) & \longrightarrow & \ell_\infty(\omega^N) & \longrightarrow & Q_N & \longrightarrow & 0 \end{array}$$

in which $\|I'\| \le 1$. If $L\colon X \longrightarrow \ell_\infty(\omega^N)$ is a lifting of I' with $\|L\| \le \rho_N(X) + \varepsilon$, then $I - L\rho$ is a projection along $\jmath$ of norm at most $1 + \rho_N(X) + \varepsilon$. Hence $\pi_N(X) \le \rho_N(X) + 1$. The other inequality is clear after the following remark, which is nothing but *the* pullback trick in disguise:

Claim Let $\tau\colon X \longrightarrow \ell_\infty(\omega^N)$ be a linear map such that $\operatorname{dist}(\tau(x), C(\omega^N) \le \|x\|$ for all $x \in X$. For every $\varepsilon > 0$, there is a linear map $\tau''\colon X \longrightarrow C(\omega^N)$ such that $\|\tau - \tau''\| \le \sigma_N(X) + \varepsilon$. Moreover, if τ is bounded then τ'' is bounded.

Indeed, let $Z[\tau]$ be the space $C(\omega^N) \times X$ normed by $\|(f, x)\| = \max(\|f - \tau(x)\|, \|x\|)$ and consider the sequence $0 \longrightarrow C(\omega^N) \longrightarrow Z[\tau] \longrightarrow X \longrightarrow 0$ with embedding $f \longmapsto (f, 0)$ and quotient map $(f, x) \longmapsto x$. The sequence is isometrically exact by the hypothesis on τ: if $\|x\| < 1$ then there is f such that $\|f - \tau(x)\| < 1$ and therefore the point (f, x) is in the open unit ball of $Z[\tau]$. By definition, there is a section $S : X \longrightarrow Z[\tau]$ of the quotient map with $\|S\| \le \sigma_N(X) + \varepsilon$. This S must have the form $S(x) = (\tau''(x), x)$ for some linear map $\tau''\colon X \longrightarrow C(\omega^N)$, and since $\|S(x)\| = \|(\tau''(x), x)\| = \max(\|\tau''(x) - \tau(x)\|, \|x\|)$, it follows that $\|\tau'' - \tau\| \le \sigma_N(X) + \varepsilon$. □

Consequently:

Proposition 8.6.8 *Let X be a separable Banach space.* $\operatorname{Ext}(X, C(\omega^\omega)) = 0$ *if and only if* $\sup_N \rho_N(X) < \infty$.

Amir [8; 9] and Baker [28] proved that given an isometrically exact sequence $0 \longrightarrow C(\omega^N) \longrightarrow Z \longrightarrow X \longrightarrow 0$ with X separable, there is always linear projection with norm at most $2N + 1$. Moreover, if $X = C(\omega^{N-1})$ then for every $\varepsilon > 0$, there is an exact sequence as above such that any projection has norm at least $2N + 1 - \varepsilon$. Corollary 2.2.8 yields $\sigma_N(c_0) \ge N$ (hence $\pi_N(c_0) \ge N - 1$ and

$\rho_N(c_0) \geq N-2$ by Lemma 8.6.7), while a more precise estimate [73, Theorem 3.5] yields $\pi_N(c_0) = 2N+1$, as in the Amir–Baker result, and thus $\rho_N(c_0) \geq 2N$. Let us continue with the general case.

When dealing with ordinals or with trees, as we are soon to do, we must not be stark and lose our heads (with unnecessary details): recall that despite its ordinal pedigree, the space ω^N was declared in Section 1.6 to be the only countable compact whose Nth derived set is a singleton, no matter which peculiar representation of it we choose.

To simplify the analysis of the oscillation of the involved operators, we choose the following disguised $\sigma_N(2^{\mathbb{N}})$ form: ω^N is the set whose points are subsets of $\mathbb{N}$ with at most N elements (including the empty set) with the topology inherited from $2^{\mathbb{N}}$. We write the elements of ω^N in increasing order: $a = (n_1, \ldots, n_k)$ with $n_1 < n_2 < \cdots < n_k$. We define an order on ω^N as follows: if $b = (m_1, \ldots, m_l)$ then $a \leq b$ means that $k \leq l$ and $n_j = m_j$ for $1 \leq j \leq k$. In particular, $\emptyset \leq b$ for all $b \in \omega^N$. Observe that this is a mere partial order, making our particular ω^N a tree, but not order isomorphic to any ordinal. Given a as above, $a^- = (n_1, \ldots, n_{k-1})$ and $a^+ = \{b\colon a \leq b\colon |b| = |a|+1\} = \{(n_1, \ldots, n_k, m)\colon m > n_k\}$. And in case of doubt, $\lim_{b\in a^+} b = a$ and $\lim a = \emptyset$ as $\min a \to \infty$.

Definition 8.6.9 A map $\Upsilon\colon \omega^N \longrightarrow X^*$ is a weak*-null tree map if $\Upsilon(\emptyset) = 0$ and $\lim_{b\in a^+} \Upsilon(b) = 0$ in the weak*-topology for all $|a| < N$.

Definition 8.6.10 An operator $\tau_\Upsilon\colon X \longrightarrow \ell_\infty(\omega^N)$ is said to be tree generated by a map $\Upsilon\colon \omega^N \longrightarrow X^*$ if

$$\langle \tau_\Upsilon x, a\rangle = \left\langle \sum_{b\leq a} \Upsilon(b), x \right\rangle.$$

Observe that all linear maps $\tau\colon X \longrightarrow C_0(\omega^N\backslash\{\emptyset\})$ are tree generated by weak*-null tree maps: set $\Upsilon(\emptyset) = 0$ and $\langle \Upsilon(a), x\rangle = \langle \tau x, a\rangle - \langle \tau x, a^-\rangle$. If τ_γ is tree generated, then since $\|\tau_\Upsilon x\|$ is essentially attained at the isolated points $|a| = N$ of ω^N, we have $\|\tau_\Upsilon\| = \sup_{|a|=N} \left\|\sum_{b\leq a} \Upsilon(b)\right\|$.

Given a separable Banach space X, we introduce the parameter

$$\alpha_N(X) = \sup_{\Upsilon} \inf_{|a|=N} \left\| \sum_{b\leq a} \Upsilon(b) \right\|, \tag{8.34}$$

where the supremum is taken over all weak*-null tree maps $\Upsilon\colon \omega^N \longrightarrow B_{X^*}$.

Lemma 8.6.11 *If X is a Banach space with a monotone shrinking basis then*

$$\rho_{2N}(X) \leq 4\alpha_N(X) \leq 4\rho_N(X).$$

Proof Let $(e_j)_j$ be the basis so that $x = \sum_j \langle e_j^*, x\rangle e_j$ for all $x \in X$, and let $P_n(x) = \sum_{j=1}^n \langle e_j^*, x\rangle e_j$ be the canonical projection. Since the basis is (monotone and) shrinking, (P_n^*) is a sequence of (contractive) operators on X^* that is pointwise convergent to the identity. We prove the first inequality. Let $\tau\colon X \longrightarrow \ell_\infty(\omega^{2N})$ be a linear operator with oscillation at most 2. Our plan is to find a linear $\ell\colon X \longrightarrow C(\omega^{2N})$ such that $\|\tau - \ell\| \leq 4\alpha_N(X)$; or, what is the same, $\|\tau^*(a) - \ell^*(a)\| \leq 4\alpha_N(X)$ for all a. Fix $\varepsilon > 0$. A compactness argument used in combination with the definition of oscillation yields for each finite-dimensional subspace $F \subset X$ and a^- an index $\psi(F, a) \geq a^-$ such that for all $a^- \leq b, c \leq \psi(F, a)$, we have $\|\,(\tau^*(b) - \tau^*(c))\,|_F\| = \sup_{x\in B_F} |\langle \tau^*(b) - \tau^*(c), x\rangle| \leq 2 + \varepsilon$. Choose the spaces $P_n[X]$ as F and let $\nu(a) = \max\{n\colon \psi(P_n[X], a^-) \geq a\}$. Thus, for $b \leq a$, we have

$$\|P_{\nu(a)}^*\,(\tau^*(b) - \tau^*(a^-))\,\| \leq 2 + \varepsilon. \tag{8.35}$$

Fix λ. We proceed by reverse induction to define what is a λ-acceptable set $\{f_1, \ldots, f_k\} \subset B_X^*$ of cardinality $0 \leq k \leq N$. The set $\{f_1, \ldots, f_N\} \subset B_X^*$ is λ-acceptable if $\|f_1 + \cdots + f_N\| \leq \lambda$. For $k \leq N - 1$, the set $\{f_1, \ldots, f_k\} \subset B_X^*$ is *λ-acceptable* if there is a weak*-neighbourhood V of 0 such that for each $f \in V \cap B_X^*$, the set $\{f_1, \ldots, f_k, f\}$ is λ-acceptable. Our interest in this notion is that if $\lambda > \alpha_N(X)$ then $\emptyset$ is λ-acceptable. A collection of $k \leq N$ block subspaces $\{G_1, \ldots, G_k\}$ is *λ-good* if, for some $\mu < \lambda$, every set $\{x_1^*, \ldots, x_k^*\}$ with $x_j^* \in G_j$ is μ-acceptable. At this moment, the magical function $g\colon \mathbb{N} \longrightarrow \mathbb{N}$ appears.

Claim There is a function $g\colon \mathbb{N} \longrightarrow \mathbb{N}$ such that if $k < N$ and $\{G_1, \ldots, G_k\}$ is a λ-good family of block subspaces of $P_n^*[X^*]$ then for any block subspace $G_{k+1} \subset (\mathbf{1} - P_{g(n)})^*[X^*]$, the collection $\{G_1, \ldots, G_k, G_{k+1}\}$ is λ-good.

Proof of the claim The family of block subspaces of $P_n^*[X^*]$ is finite; therefore, there is $\mu < \lambda$ such that every λ-good collection $\{G_1, \ldots, G_k\}$ of block subspaces is μ-good. Pick $\varepsilon > 0$ such that $\mu + N/\varepsilon < \lambda$, and choose in each block subspace G an ε-net for B_G. This produces a finite collection $\mathscr{G}$ of μ-acceptable sets so that whenever $\{G_1, \ldots, G_k\}$ is a λ-good collection of block subspaces of $P_n^*[X^*]$ and $g_j \in B_{G_j}$, there then is $\{f_1, \ldots, f_k\} \in \mathscr{G}$ with $\|g_j - f_j\| \leq \varepsilon$ for all $1 \leq j \leq k$. Find $g(n)$ such that for every $x^* \in P_{g(n)}^*[X^*] \cap B_X^*$ and every $\{f_1, \ldots, f_k\} \in \mathscr{G}$, the set $\{f_1, \ldots, f_k, x^*\}$ is μ-acceptable. A perturbation argument yields that for every λ-good family $\{G, \ldots, G_k\}$ of block subspaces of $P_n[X^*]$ with $k < N$ and any block subspace $G \subset \mathbf{1} - P_{g(n)}^*[X]$, the collection $\{G_1, \ldots, G_k, G\}$ is $(\mu + N/\varepsilon)$-good, hence λ-good. □

For the rest of the proof, set $\lambda = \alpha_N + \varepsilon$ and set g to be the corresponding function. Define $\varphi\colon \omega^{2N} \longrightarrow \mathbb{N}$ by

$$\varphi(a) = \begin{cases} g(\emptyset), & \text{if } a = \emptyset, \\ \varphi(a^-), & \text{if } \nu(a) < g(\varphi(a^-)), \\ \nu(a), & \text{if } \nu(a) \geq g(\varphi(a^-)). \end{cases}$$

To get $\ell\colon X \longrightarrow C(\omega^{2N})$, we define a weak*-continuous map $\ell^*\colon \omega^{2N} \longrightarrow X^*$ by setting $\ell^*(\emptyset) = \tau^*(\emptyset)$ and

$$\ell^*(a) = \sum_{\emptyset < b \leq a} \left(P_{\varphi(b)} - P_{\varphi(b^-)}\right)^* \tau^*(b^-) + \left(\mathbf{1} - P_{\varphi(a)}\right)^* \tau^*(a)$$

for $a \neq \emptyset$. The map ℓ^* is weak*-continuous because if $c \in a^+$,

$$\begin{aligned} \ell^*(c) - \ell^*(a) &= \left(P_{\varphi(c)} - P_{\varphi(a)}\right)^* \tau^*(a) + \left(\mathbf{1} - P_{\varphi(c)}\right)^* \tau^*(c) - \left(\mathbf{1} - P_{\varphi(a)}\right)^* \tau^*(a) \\ &= \left(\mathbf{1} - P_{\varphi(c)}\right)^* \tau^*(c) - \left(\mathbf{1} - P_{\varphi(c)}\right)^* \tau^*(a) \\ &= \left(\mathbf{1} - P_{\varphi(c)}\right)^* (\tau^*(c) - \tau^*(a)), \end{aligned}$$

and thus

$$\begin{aligned} \lim_{c \in a^+} \langle \ell^*(c) - \ell^*(a), x \rangle &= \lim_{c \in a^+} \left\langle \left(\mathbf{1} - P_{\varphi(c)}\right)^* (\tau^*(c) - \tau^*(a)), x \right\rangle \\ &= \lim_{c \in a^+} \left\langle \tau^*(c) - \tau^*(a), (\mathbf{1} - P_{\varphi(c)})x \right\rangle = 0. \end{aligned}$$

Finally, we need to estimate $\|\tau - \ell\| = \sup_{a \in \omega^{2N}} \|\tau^*(a) - \ell^*(a)\|$. Given $a = (n_1, \ldots, n_k) \in \omega^{2N}$, set $m_0 = \varphi(\emptyset)$ and $m_j = \varphi(n_1, \ldots, n_j)$. If we look carefully at the family $(P^*_{m_1} - P^*_{m_0})[X^*], (P^*_{m_2} - P^*_{m_1})[X^*], \ldots, (P^*_{m_k} - P^*_{m_{k-1}})[X^*]$ then we detect that all the even and all the odd elements form one of those λ-good families in the claim. In other words, if $x^*_j \in (P^*_{m_j} - P^*_{m_{j-1}})[X^*]$ are chosen with $\|x^*_j\| \leq 1$ then $\|\sum_{j=1}^k x^*_j\| \leq 2\lambda$. We thus have

$$\begin{aligned} \tau^*(a) - \ell^*(a) &= \tau^*(a) - \sum_{\emptyset < b \leq a} \left(P_{\varphi(b)} - P_{\varphi(b^-)}\right)^* \tau^*(b^-) - \left(\mathbf{1} - P_{\varphi(a)}\right)^* \tau^*(a) \\ &= \sum_{\emptyset < b \leq a} P^*_{\varphi(b)} \tau^*(a) + \left(\mathbf{1} - P_{\varphi(a)}\right)^* \tau^*(a) \\ &\quad - \sum_{\emptyset < b \leq a} \left(P_{\varphi(b)} - P_{\varphi(b^-)}\right)^* \tau^*(b^-) - \left(\mathbf{1} - P_{\varphi(a)}\right)^* \tau^*(a) \\ &= \sum_{\emptyset < b \leq a} \left(P_{\varphi(b)} - P_{\varphi(b^-)}\right)^* \tau^*(a) - \sum_{\emptyset < b \leq a} \left(P_{\varphi(b)} - P_{\varphi(b^-)}\right)^* \tau^*(b^-) \\ &= \sum_{\emptyset < b \leq a} \left(P_{\varphi(b)} - P_{\varphi(b^-)}\right)^* (\tau^*(a) - \tau^*(b^-)). \end{aligned}$$

From (8.35), we get $\left\|(P_{\varphi(b)} - P_{\varphi(b^-)})^* (\tau^*(a) - \tau^*(b^-))\right\| \leq 2 + \varepsilon$, hence

$$\|\tau^*(a) - \ell^*(a)\| = \left\| \sum_{\emptyset < b \leq a} \left(P_{\varphi(b)} - P_{\varphi(b^-)}\right)^* (\tau^*(a) - \tau^*(b^-)) \right\| \leq (4 + \varepsilon)\alpha_N(X).$$

We pass to prove the second inequality $\alpha_N \le \rho_N$. Let us inform the reader that $\mathcal{A} \subset \omega^N$ is a subtree if $a \in \mathcal{A}$ implies $a^- \in \mathcal{A}$.

Lemma 8.6.12 *Let $\Upsilon : \omega^N \longrightarrow X^*$ be a weak*-null tree map. There is a subtree $\mathcal{A} \subset \omega^N$ that is order isomorphic to ω^N such that $\lim_{\max a \to \infty} \Upsilon(a) = 0$ in the weak* topology for all $a \in \mathcal{A}$.*

Proof Let (V_n) be a countable base of weak*-neighbourhoods of 0 such that $V_{n+1} + V_{n+1} \subset V_n$ for all n. The subtree we need is $\mathcal{A} = \{a \in \omega^N : \text{ if } \emptyset < b \le a \text{ then } \Upsilon(b) \in V_{\max b}\}$. □

Let $\Upsilon : \omega^N \longrightarrow B_X^*$ be a weak*-null tree map such that $\alpha_N(X) \le \left\|\sum_{b\le a} \Upsilon(b)\right\| + \varepsilon$ for all $|a| = N$. Let $\sigma : \mathbb{N} \longrightarrow \mathbb{N}$ be any surjective map such that for each $k \in \mathbb{N}$, the set $\sigma^{-1}(k)$ is infinite. Say, $1, 2, 1, 2, 3, 1, 2, 3, 4, \ldots$. We will set $\sigma\{n_1, \ldots, n_k\} = \{\sigma(n_1), \ldots, \sigma(n_k)\}$. To make this σ a map $\omega^N \longrightarrow \omega^N$ in the representation we use, it is only necessary to work with sets $\{n_1, \ldots, n_k\}$ such that $\sigma(n_j) > \sigma(n_{j-1})$ whenever $n_j > n_{j-1}$, and so we will do to keep things tidy. We then consider the operator $\tau_{\Upsilon\sigma}$ tree generated by $\Upsilon\sigma$ (which is not a weak*-null tree map) for which $\tau^*_{\Upsilon\sigma}(a) = \sum_{\emptyset<b\le a} \Upsilon\sigma(b)$. The operator $\tau_{\Upsilon\sigma}$ has oscillation at most 1 at every point x: indeed, if $d \ge c \in a^+$ then

$$\limsup_{d\to a} \left|\langle \tau^*_{\Upsilon\sigma}(d) - \tau^*_{\Upsilon\sigma}(a), x\rangle\right| = \limsup_{d\to a} \left|\left\langle \tau^*_{\Upsilon}(\sigma c) + \sum_{c<b\le d} \tau^*_{\Upsilon}(\sigma b), x\right\rangle\right|$$

$$\le \left|\langle \tau^*_{\Upsilon}(\sigma c), x\rangle\right| + \left|\left\langle \lim_{d\to a} \sum_{c<b\le d} \tau^*_{\Upsilon}(\sigma b), x\right\rangle\right| \le 1 + 0$$

by the additional property of Lemma 8.6.12 that we assume Υ enjoys. Therefore, by hypothesis, there exists a weak*-continuous map $\ell^* : \omega^N \longrightarrow X^*$ such that $\|\tau^*_{\Upsilon\sigma}(a) - \ell^*(a)\| \le \frac{1}{2}\rho_N(X)$ for all a. Now we just need to find some $|a| = N$ for which $\left\|\sum_{b\le a} \Upsilon\sigma(b)\right\|$ can be properly bounded by a multiple of $\rho_N(X)$. Begin with $\emptyset$ and find m_1 such that $\left\|(\mathbf{1} - P_{m_1})^*(\Upsilon\sigma(\emptyset) - \ell^*(\emptyset))\right\| \le \varepsilon$, which is possible since the FDD is shrinking; then pick $|c| = 1$ such that $\|P^*_{m_1}(\Upsilon(c))\| \le \varepsilon$, which is possible since Υ is weak*-null. Now pick $n_1 > m_1$ such that $\|(\mathbf{1} - P_{n_1})^*(\Upsilon(c))\| \le \varepsilon$. Among the infinitely many $|b| = 1$ with $\sigma(b) = c$, pick one a_1 such that $\|P^*_{n_1}(\ell^*(a_1) - \ell^*(\emptyset))\| \le \varepsilon$. This a_1 is our choice. Repeat the same construction inductively N times until obtaining some $|a_N| = N$, and keep track of the pairs (m_j, n_j):

$$\begin{aligned}
&\left\|\sum_{k=1}^{N} \Upsilon\sigma(a_k)\right\| \\
&\leq \left\|\sum_{k=1}^{N}(P_{n_k} - P_{m_k})^*(\Upsilon\sigma(a_k))\right\| + 2N\varepsilon \\
&\leq \left\|\sum_{k=1}^{N}((P_{n_k} - P_{m_k})^*(\Upsilon\sigma(a_k)) + (P_{m_k} - P_{n_{k-1}})^*(\ell^*\sigma(a_k) - \ell^*\sigma(a_{k-1}))\right\| + 4N\varepsilon \\
&\leq \left\|\sum_{k=1}^{N}(\Upsilon\sigma(a_k) + \ell^*\sigma(a_k) - \ell^*\sigma(a_{k-1}))\right\| + 6N\varepsilon \\
&\leq \|\Upsilon\sigma(a_N) - \ell^*(a_N) + \ell^*(\emptyset) - \Upsilon\sigma(\emptyset)\| + 6N\varepsilon \\
&\leq \rho_N(X) + 2N\varepsilon.
\end{aligned}$$

□

The conclusion we get is that a Banach space X with a shrinking basis satisfies $\mathrm{Ext}(X, C(\omega^\omega)) = 0$ if and only if $\sup_N \alpha_N(X) < \infty$.

Let E be a Banach space. A tree map $\Upsilon\colon \omega^N \longrightarrow E$ is said to be weakly null if, for every $a \in \omega^N$, we have $\Upsilon(b) \longrightarrow 0 \longrightarrow$ weakly as $b \in a^+$. Given $\sigma > 0$, we define $N(E, \sigma)$ to be the least integer N such that there exists a weakly null tree map $\Upsilon\colon \omega^{N+1} \longrightarrow E$ such that $\|\Upsilon(a)\| \leq \sigma$ for all a and $\|\sum_{b \leq a} \Upsilon(b)\| > 1$ for $|a| = N$. We put $N(E, \sigma) = \infty$ if no such integer exists. We say that E has a summable Szlenk index if there is $\sigma > 0$ such that $N(E, \sigma) = \infty$. This is not the 'original' definition, which is much funnier, but an equivalent formulation; see the equivalence between (i) and (ii) in [191, Theorem 4.10]. We see that $N(E, \sigma) = N$ means (dividing by σ) that

$$\sup_{\Upsilon} \inf_{|a|=N} \left\|\sum_{b \leq a} \Upsilon(b)\right\| \leq \frac{1}{\sigma}, \tag{8.36}$$

where the supremum runs *now* over the weakly null tree maps $\Upsilon\colon \omega^N \longrightarrow B_E$. By taking $E = X^*$, a comparison between (8.36) and (8.34) shows that $\sup_N \alpha_N(X) \leq \lambda \Longrightarrow N(X^*, 1/\lambda) = \infty$ (and therefore X^* has a summable Szlenk index). If X is moreover reflexive, the converse is also true. We have:

Proposition 8.6.13 *Let X be a Banach space with a shrinking basis.*

(a) *If* $\mathrm{Ext}(X, C(\omega^\omega)) = 0$ *then X^* has a summable Szlenk index.*
(b) *If the basis is unconditional and* $\mathrm{Ext}(X, C(\omega^\omega)) = 0$ *then X is reflexive.*
(c) *If X is reflexive,* $\mathrm{Ext}(X, C(\omega^\omega)) = 0$ *if and only if X^* has a summable Szlenk index.*

Proof By classical results of James, X^* is separable. Part (a) has already been proved: $\mathrm{Ext}(X, C(\omega^\omega)) = 0 \implies \sup_N \rho_N(X) < \infty$ (Proposition 8.6.8) $\implies \sup_N \alpha_N(X) < \infty$ (Lemma 8.6.11) $\implies X^*$ has a summable Szlenk index.

(b) X cannot contain c_0 (otherwise, since it is separable, it would contain it complemented, and then $\mathrm{Ext}(X, C(\omega^\omega)) \neq 0$). Thus, the basis is boundedly complete, and, being unconditional, X must be reflexive [334, Theorem. 1.b.5].

(c) If X is reflexive, the weak and weak* topologies of X^* coincide, and the implications in the proof of (a) are all reversible. □

So far, the hunt for spaces X such that $\mathrm{Ext}\,(X, C(\omega^\omega)) = 0$ but at the same time $\mathrm{Ext}(X, C(\Delta)) \neq 0$ required us to look at non-Schur spaces whose spreading models are all ℓ_1. We see now that we can restrict our hunt further to spaces whose duals have a summable Szlenk index.

The property of having a summable Szlenk index goes back to [296] and can be considered a sophisticated way of saying that the space is close to being a subspace of c_0. Indeed, spaces uniformly homeomorphic to subspaces of c_0 have a summable Szlenk index [191]. But, fortunately, there are more: in [191, Remark p. 3911], the authors claim that the original Tsirelson space T^* has a summable Szlenk index, as is proved in [296, p. 196]. Furthermore, since T is reflexive, its basis is shrinking. In other (our) words,

8.6.14 $\mathrm{Ext}\,(\mathsf{T}, C(\omega^\omega)) = 0$, *while* $\mathrm{Ext}\,(\mathsf{T}, C(\Delta)) \neq 0$.

There is no special difficulty adapting Lemma 8.6.11 to cover the case of spaces with shrinking FDD (see [73] for details): firstly, transform the FDD into a bi-monotone FDD with a renorming, then prove that $\rho_{2N} \leq 4\alpha_N$ and finally obtain $\alpha_N \leq \rho_N$ when the space has a (monotone) shrinking FDD. The interest all this has for us is it allows us to obtain a real, though modest, improvement of Proposition 8.6.13:

Proposition 8.6.15 *Let X be a separable reflexive Banach space.*

(a) *If X^* has a summable Szlenk index then* $\mathrm{Ext}(X, C(\omega^\omega)) = 0$.
(b) *If X has a FDD and* $\mathrm{Ext}(X, C(\omega^\omega)) = 0$ *then X^* has a summable Szlenk index.*

Proof (a) We appeal to the full force of the Johnson–Rosenthal decomposition 5.3.1 to represent X in the form $0 \longrightarrow A \longrightarrow X \longrightarrow B \longrightarrow 0$ with both A, B having shrinking FDD. Since having a summable Szlenk index passes to subspaces and quotients, if X^* has a summable Szlenk index then so have A^* and B^*, hence $\mathrm{Ext}(A^*, C(\omega^\omega)) = 0, \mathrm{Ext}(B^*, C(\omega^\omega)) = 0$ and thus also $\mathrm{Ext}(X^*, C(\omega^\omega)) = 0$.

(b) Since X is reflexive, the FDD must be shrinking. Thus, $\operatorname{Ext}(X, C(\omega^\omega)) = 0$ implies that both $\rho_N(X)$ and $\alpha_N(X)$ are uniformly bounded so that X^* has a summable Szlenk index. □

8.7 The Astounding Story behind the CCKY Problem

The paper [73] ended with a comment on a problem that had been ricocheting around our heads since [79, final problem], which we will call the CCKY problem: *show that if K is a non-metrisable compact then* $\operatorname{Ext}(C(K), c_0) \neq 0$. Since $\operatorname{Ext}(c_0(\aleph_1), c_0) \neq 0$ by everything said in Section 2.2, while $\operatorname{Ext}(\ell_\infty, c_0) \neq 0$ by the construction in 2.12.9, it is clear that $\operatorname{Ext}(X, c_0) \neq 0$ for every Banach space containing a complemented copy of either $c_0(\aleph_1)$ or ℓ_∞. Compacta K for which $C(K)$ does not contain $c_0(\aleph_1)$ are characterised by the *countable chain condition* (*ccc* in short): every family of pairwise disjoint open sets is countable; see [413, Theorem 4.5]. No characterisation is currently known for compacta K such that $C(K)$ contains / does not contain ℓ_∞. Thus, the list of compacta K for which the CCKY problem was known *by then* to have an affirmative answer is:

(a) Eberlein compacta,
(b) Valdivia compacta failing the *ccc*,
(c) $C(K)$ contains ℓ_∞,
(d) $C(K)$ admits a continuous injection into $C(\mathbb{N}^*)$ but not into ℓ_∞,
(e) K is an ordinal space.

Assertion (a) follows from the following general statement [103, Theorem 3.4]:

Lemma 8.7.1 *Every non-separable Banach space admits a non-WCG extension by c_0.*

Proof We assume that X is WCG, since the result is trivial otherwise. According to [400, pp. 336–337], the space X admits a Markuševič basis $(x_\gamma, f_\gamma)_{\gamma\in\Gamma}$, which means a biorthogonal system in $X \times X^*$ for which $(x_\gamma)_{\gamma\in\Gamma}$ separates the points of X^* and $(f_\gamma)_{\gamma\in\Gamma}$ separates the points of X. One may assume without loss of generality that $(f_\gamma)_{\gamma\in\Gamma}$ is bounded. Thus, the map $T\colon X \longrightarrow c_0(\Gamma)$ defined by $Tx = (f_\gamma(x))_{\gamma\in\Gamma}$ is easily checked to be an operator with dense range. Consider any non-trivial sequence $0 \longrightarrow c_0 \longrightarrow X \longrightarrow c_0(\Gamma) \longrightarrow 0$ in which X is not WCG, as in Section 2.2. The pullback space in the diagram

$$\begin{array}{ccccccccc} 0 & \longrightarrow & c_0 & \longrightarrow & X & \longrightarrow & c_0(\Gamma) & \longrightarrow & 0 \\ & & \| & & \uparrow{\scriptstyle \underline{T}} & & \uparrow{\scriptstyle T} & & \\ 0 & \longrightarrow & c_0 & \longrightarrow & \mathrm{PB} & \longrightarrow & X & \longrightarrow & 0 \end{array}$$

cannot be WCG because $\underline{T}$ has dense range by Lemma 2.1.8. □

Thus, the lower sequence cannot split because X is WCG, so (a) is true.

We prove (b): compacta failing *ccc* contain $c_0(\aleph_1)$. Use [16, Theorem 1.2] – *if* $c_0(\Gamma) \subset C(K)$ *for a Valdivia compact* K *then there is a subset* $J \subset \Gamma$ *such that* $|J| = |\Gamma|$ *and* $c_0(J)$ *is complemented in* $C(K)$ – to get a complemented copy of $c_0(\aleph_1)$ inside $C(K)$, which is enough.

Assertion (c) is clear since the copy of ℓ_∞ is necessarily complemented.

To get (d), form the pullback diagram

$$\begin{array}{ccccccccc} 0 & \longrightarrow & c_0 & \longrightarrow & \ell_\infty & \longrightarrow & C(\mathbb{N}^*) & \longrightarrow & 0 \\ & & \| & & \uparrow & & \uparrow{\scriptstyle \imath} & & \\ 0 & \longrightarrow & c_0 & \longrightarrow & \mathrm{PB} & \longrightarrow & C(K) & \longrightarrow & 0 \end{array}$$

in which $\imath$ is the continuous injection claimed in the hypothesis. Instances of (d) appear when K contains a dense set of weight at most $\aleph_1$ but $C(K)$ is not a subspace of ℓ_∞, by Parovičenko's first theorem mentioned in Section 1.6, example 6, or else $C(K)$ spaces with non-weak*-separable dual, but admitting continuous injections into $C(\mathbb{N}^*)$, such as $C[0, \omega_1]$.

The uncountable ordinal cases of (e) can be reduced to $[0, \omega_1]$.

Now the story goes that [73, p. 4539–4540] claims that Corson compacta can also play the role of Eberlein compacta just using 'similar arguments'. However, Correa and Tausk noticed that a 'similar argument' cannot work (an explanation of why can be found in [93, p. 115]), amended the situation in [136, Theorem 3.1] and obtained a result of general interest in passing:

Proposition 8.7.2 $\operatorname{Ext}(X, c_0) \neq 0$ *for every Banach space* X *admitting a biorthogonal system* $(x_{n,\gamma}, f_{n,\gamma})_{n\in\omega,\gamma\in\mathfrak{c}}$ *such that* $(f_{n,\gamma}(x)) \in c_0(\omega \times \mathfrak{c})$ *and*

$$\sup \left\| \sum_{i=1}^{k} x_{n_i,\gamma_i} \right\| < \infty,$$

where the sup *is taken over all finite sets of* $\omega \times \mathfrak{c}$.

Proof The idea is to obtain an operator $\tau \colon X \longrightarrow \ell_\infty/c_0$ that cannot be lifted to ℓ_∞, which is clearly enough to conclude that the lower pullback sequence

$$\begin{array}{ccccccccc} 0 & \longrightarrow & c_0 & \longrightarrow & \ell_\infty & \longrightarrow & \ell_\infty/c_0 & \longrightarrow & 0 \\ & & \| & & \uparrow{\scriptstyle \underline{\tau}} & & \uparrow{\scriptstyle \tau} & & \\ 0 & \longrightarrow & c_0 & \longrightarrow & \mathrm{PB} & \longrightarrow & X & \longrightarrow & 0 \end{array}$$

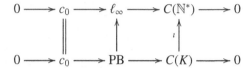

does not split. The operator τ will appear as a composition $\tau = uv$ with $v \colon X \longrightarrow c_0(\omega \times \mathfrak{c})$ and $u \colon c_0(\omega \times \mathfrak{c}) \longrightarrow \ell_\infty/c_0$, where v is obviously $v(x) = (f_{n,\gamma}(x))$, while u is $u(e_{n,\gamma}) = 1_{A_{n,\gamma}} + c_0$ for $(A_{n,\gamma})_{n\in\omega,\gamma\in\mathfrak{c}}$, an almost disjoint family of subsets of $\mathbb{N}$ with the following property: for every family $(B_{n,\gamma})_{n\in\omega,\gamma\in\mathfrak{c}}$ with $B_{n,\gamma} \subset A_{n,\gamma}$ cofinite, we have $\sup_{m\in\mathbb{N}} \left|\{n \in \omega\colon m \in \bigcup_{\gamma\in\mathfrak{c}} B_{n,\gamma}\}\right| < \infty$. The existence of such a family is shown in [136, Lemma 2.1]. Assume that τ admits a lifting T. The set $B_{n,\gamma} = \{m \in A_{n,\gamma}\colon \delta_m T(x_{n,\gamma}) \geq \frac{1}{2}\}$ is cofinite in $A_{n,\gamma}$, so for each $k \geq 1$, there exists $p \in \omega$, $n_1, \ldots, n_k \in \mathbb{N}$ pairwise distinct and $\gamma_1, \ldots, \gamma_k \in \mathfrak{c}$ such that $p \in B_{n_i,\gamma_i}$ for $i = 1, \ldots, k$. Therefore, we get a contradiction with

$$\frac{k}{2} \leq \delta_p\left(\sum_{i=1}^{k} x_{n_i,\gamma_i}\right) \leq \sup\left\|\sum_{i=1}^{k} x_{n_i,\gamma_i}\right\| \|T\|. \qquad \square$$

A simple way to satisfy the conditions of Proposition 8.7.2 in a $C(K)$ space is to have a bounded weak*-null biorthogonal system $(x_{n,\gamma}, f_{n,\gamma})_{n\in\omega,\gamma\in\mathfrak{c}}$ (a biorthogonal system $(x_i, f_i)_{i\in I}$ is bounded if $\sup_i (\|x_i\|, \|f_i\|) < \infty$ and weak*-null if $(f_i(x)) \in c_0(I)$ for all $x \in X$) such that $x_{n,\alpha}x_{m,\beta} = 0$ provided $(n, \alpha) \neq (m, \beta)$ in $\omega \times \mathfrak{c}$. This situation occurs under the following condition, which is satisfied by Corson compacta [136, Lemmata 2.8 and 3.2] under CH: there is a sequence $(F_n)_{n\in\omega}$ of closed subsets of K and a bounded biorthogonal weak*-null system $(x_{n,\gamma}, f_{n,\gamma})_{n\in\omega,\gamma\in\mathfrak{c}}$ in $C(K)$ such that $F_n \cap \overline{\bigcup_{n\neq m} F_m} = \emptyset$ and $\operatorname{supp} f_{n,\gamma} \subset F_n$ for all $n \in \omega$ and all $\gamma \in \mathfrak{c}$. Therefore,

Proposition 8.7.3 [CH] *If K is a non-metrisable Corson compact then*

$$\operatorname{Ext}(C(K), c_0) \neq 0.$$

In the meantime, one more offspring emerged. During the final stages of the writing of [22], Avilés asked: if a $C(K)$ space is itself a twisted sum of c_0 and $c_0(I)$ and c_0, does it admit a non-trivial twisted sum with c_0?

The intended purpose of [93] was to answer with a consistent yes. In a sense, the paper proves more than announced: any space that is a twisted sum of two $c_0(I)$ can be twisted against c_0 under CH; the same is true for the newly obtained twisted sums, and for the new twisted sums and so on. The proof waddles between homology and cardinal set theory, with not much left to $C(K)$-spaces. Its formulation in $C(K)$ terms could, however, be as follows:

Proposition 8.7.4 [CH] *If K is a non-metrisable compact space of finite height then* $\operatorname{Ext}(C(K), c_0) \neq 0$.

The proof goes as follows. Since K must be infinite, $K' \neq \emptyset$. If $K^{(2)} = \emptyset$ then K' is finite, $C(K)$ is isomorphic to a finite product $c_0(I_1) \times \cdots \times c_0(I_n)$ and the assertion is true. So our first serious concern is with compacta K having $K^{(2)} \neq \emptyset$ and $K^{(3)} = \emptyset$. The natural exact sequence $0 \longrightarrow C_0(K\backslash K') \longrightarrow C(K) \longrightarrow C(K') \longrightarrow 0$ becomes $0 \longrightarrow c_0(I) \longrightarrow C(K) \longrightarrow c_0(J) \longrightarrow 0$. Now, if J is countable, the sequence splits, and $C(K)$ is isomorphic to $c_0(\mathfrak{c})$, so the conclusion follows. Otherwise, the following result applies:

Lemma 8.7.5 *If X fits into an exact sequence $0 \longrightarrow c_0(I) \longrightarrow X \longrightarrow c_0(\mathfrak{c}) \longrightarrow 0$ then* $\operatorname{Ext}(X, c_0) \neq 0$.

Proof If $|I| \leq \mathfrak{c}$, there is a pushout diagram

$$\begin{array}{ccccccccc} 0 & \longrightarrow & c_0(\mathfrak{c}) & \longrightarrow & X_0 & \longrightarrow & c_0(\mathfrak{c}) & \longrightarrow & 0 \\ & & \downarrow{\scriptstyle \imath} & & \downarrow & & \| & & \\ 0 & \longrightarrow & c_0(I) & \longrightarrow & X & \longrightarrow & c_0(\mathfrak{c}) & \longrightarrow & 0 \end{array} \tag{8.37}$$

and the same if $|I| \geq \mathfrak{c}$ using Lemma 3.9.4. Apply homology with target c_0 to the lower exact sequence in (8.37) to get an exact sequence

$$\mathfrak{L}(c_0(I), c_0) \xrightarrow[\text{morphism}]{\text{connecting}} \operatorname{Ext}(c_0(\mathfrak{c}), c_0) \longrightarrow \operatorname{Ext}(X, c_0).$$

If $\operatorname{Ext}(X, c_0) = 0$, the connecting morphism is surjective, which by composition with the map $\imath$ in (8.37) in turn yields a surjective map $\mathfrak{L}(c_0(\mathfrak{c}), c_0) \longrightarrow \operatorname{Ext}(c_0(\mathfrak{c}), c_0)$, something that cardinal arithmetics will show to be impossible. Indeed, $\left|\mathfrak{L}(c_0(\mathfrak{c}), c_0)\right| < \left|\operatorname{Ext}(c_0(\mathfrak{c}), c_0)\right|$. To prove this, fix a Banach space Y and observe that $|\mathfrak{L}(Y, c_0)| \leq |\mathfrak{L}(\ell_1, Y^*)|$: since $\mathfrak{L}(\ell_1, Y^*)$ is the set of bounded sequences of Y^*, there are $|(Y^*)^{\mathbb{N}}|$ countable subsets of Y^* and each of them admits $\mathfrak{c}$ bounded sequences, we get

$$|\mathfrak{L}(Y, c_0)| \leq \left|\mathbb{R} \times (|Y^*|^{\aleph_0})^{\aleph_0}\right|.$$

Therefore, if $|Y^*| \leq \mathfrak{c}$, as is the case when $Y = c_0(\mathfrak{c})$, one gets $|\mathfrak{L}(Y, c_0)| \leq \mathfrak{c}^{\aleph_0} = \mathfrak{c}$. On the other hand, Marciszewski and Pol show in [355, Proposition 7.4] (see Proposition 8.7.18 below) that there exist $2^{\mathfrak{c}}$ non-equivalent exact sequences $0 \longrightarrow c_0 \longrightarrow \cdot \longrightarrow c_0(\mathfrak{c}) \longrightarrow 0$; i.e. $|\operatorname{Ext}(c_0(\mathfrak{c}), c_0)| \geq 2^{\mathfrak{c}}$. □

It is easy now to believe, even if this sounds like the shoulder on which mathematicians come to cry after making a gaffe, that the ideas above can be inductively continued. Full details are in [93].

The point that is relevant in this tale is that no CH has been used so far, and the reader has our word that it was not used for the rest of the proof in [93]. So the question is unavoidable: does that mean that we have proved that any non-metrisable finite height compact K is such that $\operatorname{Ext}(C(K), c_0) \neq 0$? No. The reason is cardinal arithmetics. What has been proved is that any non-metrisable compact K with height at most 3 and such that $|K'| \geq \mathfrak{c}$ has $\operatorname{Ext}(C(K), c_0) \neq 0$;

at the end of the day, the proof worked because $\mathfrak{c}^{\aleph_0} < 2^{\mathfrak{c}}$. But the inequality $\aleph^{\aleph_0} < 2^{\aleph}$ does not necessarily hold for all cardinals. Indeed, see [221, Theorem 5.15]: assuming GCH, if $\aleph$ has cofinality greater than $\aleph_0$ then $\aleph^{\aleph_0} = \aleph$, but $\aleph^{\aleph_0} = 2^{\aleph}$ if $\aleph$ has cofinality $\aleph_0$. In particular, $\aleph_1^{\aleph_0} < 2^{\aleph_1}$ does not necessarily hold. On the other hand, even the Marciszewski–Pol argument could not work for $\aleph_1$ without CH: if $\mathfrak{c} < 2^{\aleph_1}$ then the same proof as in [355] yields that there are $2^{\aleph_1}$ different exact sequences $0 \longrightarrow c_0 \longrightarrow X \longrightarrow c_0(\aleph_1) \longrightarrow 0$. However, if $2^{\aleph_1} = \mathfrak{c}$ then the method in [355] does not decide. Summing up, the proof works when $\mathfrak{c}$ is the first step, something that happens under CH. In fact, Proposition 8.7.4 can be formulated as a theorem in ZFC [25, Theorem 6.2]:

8.7.6 $\operatorname{Ext}(C(K), c_0) \neq 0$ *for every compact space of finite height and weight at least* $\mathfrak{c}$.

In this scenario, Marciszewski and Plebanek [354] show that *that* first step cannot actually be done for $\aleph_1 < \mathfrak{c}$ under Martin's axiom. Let us briefly sketch, but we are just a hunchback digging on a hill of gold, how their ideas go. The key property behind the Marciszewski–Plebanek construction is the following:

Definition 8.7.7 Let K be a compact space. A countable discrete extension of K is a compact space L containing (a homeomorphic copy of) K whose complement is countable and discrete. A compact space K has the $*$-extension property if, whenever L is a countable discrete extension of $B^*_{C(K)}$ (not K!), the canonical embedding $\delta\colon C(K) \longrightarrow C(B^*_{C(K)})$ lifts to $C(L)$ through the restriction map.

The kernel of the restriction map arising from any countable discrete extension is isometric to c_0. We have:

Proposition 8.7.8 *If K has the $*$-extension property then* $\operatorname{Ext}(C(K), c_0) = 0$.

Proof We show that every exact sequence $0 \longrightarrow c_0 \xrightarrow{\imath} X \xrightarrow{\rho} C(K) \longrightarrow 0$ actually fits into a commutative diagram

$$\begin{array}{ccccccccc} 0 & \longrightarrow & c_0 & \longrightarrow & C(L) & \xrightarrow[\text{restriction}]{r} & C(B^*_{C(K)}) & \longrightarrow & 0 \\ & & \| & & \uparrow & & \uparrow{\scriptstyle\delta} & & \\ 0 & \longrightarrow & c_0 & \longrightarrow & X & \longrightarrow & C(K) & \longrightarrow & 0 \end{array} \tag{8.38}$$

Construction of L: there is no loss of generality assuming that $\imath\colon c_0 \longrightarrow X$ is an isometric embedding and $\rho\colon X \longrightarrow C(K)$ is an isometric quotient. Then let

x_n^* be Hahn–Banach extensions of the coordinate functionals of c_0 through $\imath$ so that

$$M = \rho^*[B^*_{C(K)}] \cup \{x_n^* : n \in \mathbb{N}\},$$

equipped with the weak*-topology is a countable and discrete extension of $\rho^*[B^*_{C(K)}]$ since all accumulation points of $\{x_n^* \colon n \in \mathbb{N}\}$ vanish on c_0. Now we form a countable discrete extension of $B^*_{C(K)}$ taking

$$L = B^*_{C(K)} \cup \{x_n^* \colon n \in \mathbb{N}\},$$

topologised to be homeomorphic with M via the obvious bijection $h\colon L \longrightarrow M$ given by $h(\mu) = \rho^*(\mu)$ and $h(x_n^*) = x_n^*$. This yields Diagram (8.38), whose lower sequence must split. □

The lower sequence in (8.38) splits if and only if there is an extension operator $E\colon C(K) \longrightarrow C(L)$, i.e. a lifting of δ. Since we can assume without loss of generality that X is the pullback space, $f \longmapsto (Ef, f)$ is a linear continuous selection for the lower sequence and thus $(g, f) \longmapsto (g - Ef, 0)$ defines a projection $X \longrightarrow c_0$. This means that whenever $g \in C(L)$ is such that $rg = \delta f$, $g - Ef \in c_0$, namely $\lim_n (g(n) - Ef(n)) = 0$. Since $Ef(n) = \langle E^*(\delta_n), f\rangle$, it turns out that a lifting E exists if and only if there is a bounded sequence $(\nu_n) \in C(K)^*$ such that

$$\lim_{n\to\infty} (g(n) - \nu_n(f)) = \lim_{n\to\infty} (Ef(n) - \nu_n(f)) = 0 \tag{8.39}$$

for every $f \in C(K)$ and every $g \in C(L)$ such that $rg = \delta f$.

The next natural step towards the solution of the CCKY problem is to find a way to arrive at a bounded sequence (ν_n) as above. So, the authors place the action in the duality between Boolean algebras $\mathfrak{A}$ and their Stone compacta $\mathrm{ult}(\mathfrak{A})$ as described in Note 4.6.1. Concatenation of the functors Boolean algebras $\rightsquigarrow$ Stone compacta $\rightsquigarrow$ $C(K)$-spaces yields the correspondence $a \rightsquigarrow a^\circ \rightsquigarrow a^{\circ\circ}$, taking Boolean homomorphisms to continuous functions and then to operators. It is clear that when an arrow at a 'lower level' (Boole, compact) has a (left, right) inverse then the same is true for the induced arrows at 'higher levels' (compact, Banach), but not the converse. In particular, if $a\colon \mathfrak{A} \longrightarrow \mathfrak{B}$ and $b\colon \mathfrak{B} \longrightarrow \mathfrak{A}$ are Boolean morphisms such that $ab = \mathbf{1}_{\mathfrak{B}}$ then $a^{\circ\circ}b^{\circ\circ} = \mathbf{1}_{C(\mathrm{ult}(\mathfrak{B}))}$. An especially important case is that of a Boolean algebra $\mathfrak{A} \subset \mathcal{P}(\mathbb{N})$ containing the finite subsets: if $\rho\colon \mathfrak{A} \longrightarrow \mathfrak{A}/\mathrm{fin}(\mathbb{N})$ is the natural quotient map, then $\rho^{\circ\circ} = r$ is the restriction operator $C(\mathrm{ult}(\mathfrak{A})) \longrightarrow C(\mathrm{ult}(\mathfrak{A}/\mathrm{fin}(\mathbb{N})))$. It turns out then that if s is a right inverse for ρ, $s^{\circ\circ}$ is an extension operator for r.

We will set $M(\mathfrak{A}) = C(\mathrm{ult}(\mathfrak{A}))^*$ to both simplify notation and stress the fact that elements $C(\mathrm{ult}(\mathfrak{A}))^*$ and bounded additive functions $\mu\colon \mathfrak{A} \longrightarrow \mathbb{R}$ can be identified via $\mu(A) = \langle \mu, 1_A \rangle$. It will also simplify the notation to write $M_1(\mathfrak{A})$ for the unit ball of $M(\mathfrak{A})$. Given a subalgebra $\mathfrak{B} \subset \mathfrak{A}$, we define a seminorm for bounded functions $\varphi\colon \mathfrak{A} \longrightarrow \mathbb{R}$ by

$$\| - \varphi\|_{\mathfrak{B}} = \sup_{B \in \mathfrak{B}} |\varphi(B)|,$$

which induces, via the identification above,

$$\mathrm{dist}_{\mathfrak{B}}(\varphi, M_1(\mathfrak{A})) = \inf_{\mu \in M_1(\mathfrak{A})} \|\varphi - \mu\|_{\mathfrak{B}}.$$

Definition 8.7.9 A Boolean algebra $\mathfrak{A}$ has the approximation property ($\mathfrak{a.p.}$) if, given any sequence (f_n) of functions $f_n\colon \mathfrak{A} \longrightarrow [-1, 1]$ such that for any finite subalgebra $\mathfrak{B} \subset \mathfrak{A}$,

$$\lim_n \mathrm{dist}_{\mathfrak{B}}(f_n, M_1(\mathfrak{A})) = 0,$$

there is a bounded sequence $(\nu_n) \subset M(\mathfrak{A})$ such that $\lim(f_n(a) - \nu_n(a)) = 0$ for all $a \in \mathfrak{A}$.

The approximation property of the Boolean algebra $\mathfrak{A}$ is crying out: the associated Stone compact $\mathrm{ult}(\mathfrak{A})$ enjoys the $*$-extension property!

Lemma 8.7.10 *If K is a totally disconnected space whose Boolean algebra of clopen sets has the $\mathfrak{a.p.}$ then K has the $*$-extension property.*

Proof Let $\mathfrak{A} = \mathfrak{cl}(K)$ so that $K = \mathrm{ult}(\mathfrak{A})$. For each $A \in \mathfrak{A}$, consider the evaluation map $M_1(\mathfrak{A}) \longrightarrow [-1, 1]$ that sends μ to $\mu(A)$. Given a countably discrete extension L of $M_1(\mathfrak{A})$, obtain a continuous extension $\theta_A\colon L \longrightarrow [-1, 1]$, and then form the sequence of functions $f_n\colon \mathfrak{A} \longrightarrow [-1, 1]$ given by $f_n(A) = \theta_A(n)$. Since L is a countably discrete extension of $M_1(\mathfrak{A})$, all accumulation points of the elements of the countable discrete addition are in $M_1(\mathfrak{A})$, thus $\lim_n \mathrm{dist}_{\mathfrak{B}}(f_n, M_1(\mathfrak{A})) = 0$ precisely because $\mathfrak{B}$ is finite. So, there is a bounded sequence (ν_n) such that $\lim(\nu_n(A) - f_n(A)) = 0$ for every $A \in \mathfrak{A}$; that is,

$$\lim_{n\to\infty} (\nu_n(A) - \theta_A(n)) = 0,$$

which implies (8.39) for functions $f = 1_A$ with $A \in \mathfrak{A}$: if $rg = 1_A$ then necessarily $\lim(g(n) - \theta_A(n)) = 0$ and thus

$$\lim (g(n) - \nu_n(1_A)) = \lim (g(n) - \theta_A(n) + \theta_A(n) - \nu_n(A)) = 0.$$

Finally, if (8.39) holds for all functions 1_A with $A \in \mathfrak{A}$ then it holds for all functions $f \in C(\mathrm{ult}(\mathfrak{A}))$. □

What else has to be done? Oh, yes, to construct a Boolean algebra $\mathfrak{A}$ with the $\mathfrak{a.p.}$! And this is where Martin's axiom, which is a statement about certain partially ordered sets, has a leading role. A strong antichain (downwards) in a partially ordered set P is a subset in which no two elements have a common lower bound. P is said to satisfy the countable chain condition (*ccc*) if every strong antichain is countable. A subset $D \subset P$ is *dense* if, for every $p \in P$, there is $d \in D$ such that $d \leq p$; a subset $F \subset P$ is said to be a *filter* (on P) if it is directed ($\forall f, g \in F\ \exists\, h \in F$: $f, g \leq h$) and downwards closed (if $g \leq f$ and $f \in F$ then $g \in F$). Consider the following statement for a cardinal $\aleph_0 \leq \aleph \leq \mathfrak{c}$:

8.7.11 $\mathsf{MA}(\aleph)$ Given a partially ordered set P satisfying the *ccc*, for every collection $\mathcal{D}$ of dense subsets of P such that $|\mathcal{D}| \leq \aleph$, there is a filter on P meeting all the elements of $\mathcal{D}$.

Assertion $\mathsf{MA}(\aleph_0)$ is a theorem in ZFC (the Rasiowa–Sikorski lemma), while $\mathsf{MA}(\mathfrak{c})$ is false. Martin's axiom MA is the statement that $\mathsf{MA}(\aleph)$ holds for all $\aleph < \mathfrak{c}$. As Levy mentions [321, p. 280], MA is a formal consequence of CH that, instead of necessarily denying the existence of cardinals between $\aleph_0$ and $\mathfrak{c}$, asserts that, if they exist, they behave like $\aleph_0$; moreover, it adds no information about the value of $\mathfrak{c}$, and it is therefore consistent with any reasonable specification of which value for $\mathfrak{c}$ one assumes. We only need Martin's axiom at the first level, namely $\mathsf{MA}(\aleph_1)$:

Lemma 8.7.12 [$\mathsf{MA}(\aleph_1)$] *Let $\mathcal{M}$ be an almost disjoint family of subsets of $\mathbb{N}$ of size $\aleph_1$. The Boolean algebra $\mathfrak{A}_{\mathcal{M}}$ generated by the sets in $\mathcal{M}$ and the finite sets has the* $\mathfrak{a.p.}$

The proof is by no means simple: Marciszewski and Plebanek introduce [354, Definition 4.6] the complicated notion of the local extension property of order r (LEP(r)) for Boolean algebras, which turns out to be the fulcrum on which the lever of MA is placed.

And with this we arrive at the awful truth: the space that Marciszewski and Plebanek show as impossible to twist against c_0 is the same space of continuous functions on $\Delta_{\mathcal{M}}$ we know so well from 2.2.10 and which was the first one we twisted against c_0 in Lemma 8.7.5 to answer Avilés question: the Johnson–Lindenstrauss space JL_∞ of Diagram (2.37) – the reason is that the Stone compact corresponding to $\mathfrak{A}_{\mathcal{M}}$ is $\Delta_{\mathcal{M}}$. We have:

Theorem 8.7.13 *Given an almost disjoint family $\mathcal{M}$ of size $\aleph_1$,*

- [CH] $\mathrm{Ext}(C(\Delta_{\mathcal{M}}), c_0) \neq 0$.
- [$\mathsf{MA}(\aleph_1)$] $\mathrm{Ext}(C(\Delta_{\mathcal{M}}), c_0) = 0$.

Funny, isn't it? Thus, the CCKY problem cannot be solved in ZFC! And more nails have been hammered into the CCKY problem's coffin: [25; 134; 135; 136; 137; 354]. So, all's well that ends well? Well, Avilés, Marciszewski and Plebanek decided this was the ideal place to be an apple tree [25]:

Theorem 8.7.14 [CH] *If K is a non-metrisable compact,* $\operatorname{Ext}(C(K), c_0) \neq 0$.

The proof contains a stab at the heart of [93] in the form of an incredibly clever counting lesson plus some general results of independent interest:

Proposition 8.7.15 *If X is a Banach space of density $\aleph_1$ such that $|X^*| < 2^{\aleph_1}$ then* $\operatorname{Ext}(X, c_0) \neq 0$.

This result depends only on the following general estimate [25, Lemma 4]: *if K is a compact space of weight $\aleph_1$ then $|C(\mathbb{N}^*, K)| \geq 2^{\aleph_1}$*. The proof then goes smoothly once the implication ✓ is established in the following chain:

$$\dim(X) \leq \aleph_1 \Rightarrow \operatorname{weight}(B_X^*) \leq \aleph_1 \Rightarrow |C(\mathbb{N}^*, B_X^*)| \geq 2^{\aleph_1} \overset{\checkmark}{\Rightarrow} \operatorname{Ext}(X, c_0) \neq 0.$$

This implication is simultaneously impossible to figure out and easy to see in light of the following definition:

Definition 8.7.16 Let K be a compact subspace of another topological space H. We say that a countable discrete extension L of K can be realised in H if there is a homeomorphic embedding $L \longrightarrow H$ making a commutative diagram

$$\begin{array}{ccc} K & \longrightarrow & L \\ & \searrow \quad \swarrow & \\ & H & \end{array}$$

Avilés, Marciszewski and Plebanek [25, Theorem 3.1 and Corollary 4.3] then show the following:

8.7.17 *Let X be an infinite-dimensional Banach space. The following are equivalent:*

(i) $\operatorname{Ext}(X, c_0) = 0$.
(ii) *Every countable discrete extension of B_X^* can be realised in* $(X^*, \text{weak*})$.
(iii) *Every weak*-continuous function $\mathbb{N}^* \longrightarrow X^*$ extends to a weak*-continuous function $\beta\mathbb{N} \longrightarrow X^*$.*

It is then clear that implication ✓ holds: $\operatorname{Ext}(X, c_0) = 0$ is impossible since otherwise, using (iii),

$$2^{\aleph_1} \leq |C(\mathbb{N}^*, B_X^*)| \leq |X^*|^{\aleph_0} = |X^*| < 2^{\aleph_1}.$$

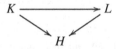

We continue to sketch the proof of Theorem 8.7.14. The first task is to settle the case in which K has weight $\aleph_1 = \mathfrak{c}$. This requires the counting lesson mentioned above but also the pièce de résistance [25, Corollary 5.6]: [CH] *if* $\operatorname{Ext}(C(K), c_0) = 0$ *then there is a finite set* $F \subset K$ *such that for every closed subspace* $L \subset K$ *of weight* $\mathfrak{c}$ $L \setminus F$ *is locally metrisable*. Having the case $\mathfrak{c}$ in hand, the authors appeal to Juhász's theorem [239]: [CH] *if* K *has weight greater than* $\mathfrak{c}$ *then there is a compact subspace of* K *of weight* $\mathfrak{c}$. This, and a lot of know-how, will result in the final contradiction: if $\operatorname{Ext}(C(K), c_0) = 0$ then K must be metrisable. Ok this is the end of the line (see Note 8.8.5 if it is not immediately clear why).

Or maybe not. Koszmider posed [298] five problems on the spaces $C(\Delta_{\mathcal{M}})$:

(1) Is there a separable Banach space X not isomorphic to c_0 whose only non-trivial decompositions are of the form $c_0 \times X$?
(2) [ZFC] Are there almost disjoint families $\mathcal{M}$ such that every decomposition of $C(\Delta_{\mathcal{M}})$ into two factors has one separable factor?
(3) [MA] Is it true that if $|\mathcal{M}| = |\mathcal{N}| < \mathfrak{c}$, then $C(\Delta_{\mathcal{M}}) \simeq C(\Delta_{\mathcal{N}})$?
(4) [MA] Is it true that if $|\mathcal{M}| < \mathfrak{c}$ then $C(\Delta_{\mathcal{M}})$ is isomorphic to its square?
(5) [ZFC] Are there two almost disjoint families $\mathcal{M}$ and $\mathcal{N}$ of the same cardinality such that $C(\Delta_{\mathcal{M}})$ and $C(\Delta_{\mathcal{N}})$ are not isomorphic?

These problems are relevant to our discussion. Koszmider himself gives a partial solution to (2) by showing that under either CH or MA, there is a family $\mathcal{M}$, constructed ad hoc, such that if $C(\Delta_{\mathcal{M}}) = A \oplus B$ with A and B infinite-dimensional, then $A \simeq c_0$ and $B \simeq C(\Delta_{\mathcal{M}})$, or vice versa. Argyros and Raifkotsalis [20] showed the existence of separable spaces $\mathsf{AR}(p)$ for $1 \leq p < \infty$ such that if $\mathsf{AR}(p) = A \oplus B$ then $A \simeq \ell_p$ and $B \simeq \mathsf{AR}(p)$ (or vice versa). Problem (5) is about how many different $C(\Delta_{\mathcal{M}})$ spaces exist: it was solved by Marciszewski and Pol [355], who showed that there exist $2^{\mathfrak{c}}$ non-isomorphic spaces with $|\mathcal{M}| = \mathfrak{c}$; we already used this during the proof of Proposition 8.7.4. The proof in [355], or at least a large share of its three lines, could well be considered implicit. An explicit proof has been produced in [74]. Question (3) is the same question (5), but under MA plus $|\mathcal{M}| < \mathfrak{c}$. Koszmider's questions have surprising answers:

Proposition 8.7.18

(a) *There exist* $2^{\mathfrak{c}}$ *almost disjoint families* $\mathcal{M}$ *of size* $\mathfrak{c}$ *such that the Banach spaces* $C(\Delta_{\mathcal{M}})$ *are pairwise non-isomorphic.*
(b) *Under* $\mathsf{MA}(\aleph)$*, all the spaces* $C(\Delta_{\mathcal{M}})$ *with* $|\mathcal{M}| = \aleph$ *are isomorphic and isomorphic to their squares.*

Proof Part (a) is the result of Marciszewski and Pol [355] already mentioned. Part (b) is a formal consequence of Theorem 8.7.13: Let $\mathcal{M}, \mathcal{N}$ be almost disjoint families of size $\aleph < \mathfrak{c}$. Since $\operatorname{Ext}(C(\Delta_{\mathcal{M}}), c_0) = 0$ and $\operatorname{Ext}(C(\Delta_{\mathcal{N}}), c_0) = 0$, the two exact sequences in the diagram

$$\begin{array}{ccccccccc} 0 & \longrightarrow & c_0 & \longrightarrow & C(\Delta_{\mathcal{M}}) & \longrightarrow & c_0(\aleph_1) & \longrightarrow & 0 \\ & & & & & & \| & & \\ 0 & \longrightarrow & c_0 & \longrightarrow & C(\Delta_{\mathcal{N}}) & \longrightarrow & c_0(\aleph_1) & \longrightarrow & 0 \end{array}$$

are semi-equivalent; thus, the diagonal principle in Theorem 2.11.6 yields

$$C(\Delta_{\mathcal{M}}) \simeq c_0 \times C(\Delta_{\mathcal{M}}) \simeq c_0 \times C(\Delta_{\mathcal{N}}) \simeq C(\Delta_{\mathcal{N}})$$

since it is plain that $C(\Delta_{\mathcal{M}}) \simeq c_0 \times C(\Delta_{\mathcal{M}})$. The last assertion is clear since for every $\aleph_0 \le \aleph \le \mathfrak{c}$, there exist families $\mathcal{M}$ of size $\aleph$ such that $C(\Delta_{\mathcal{M}})$ is isomorphic to its square. □

We can polish the idea behind the argument above as follows:

Lemma 8.7.19 *Given two exact sequences*

$$\begin{array}{ccccccccc} 0 & \longrightarrow & c_0 & \longrightarrow & Z & \longrightarrow & c_0(\aleph) & \longrightarrow & 0 \\ & & & & & & \| & & \\ 0 & \longrightarrow & c_0 & \longrightarrow & Z' & \longrightarrow & c_0(\aleph) & \longrightarrow & 0 \end{array}$$

if $\operatorname{Ext}(Z, c_0) = 0$ *and* $\operatorname{Ext}(Z', c_0) = 0$ *then* $Z \simeq Z'$.

Proof Working as above, one gets $Z \times c_0 \simeq Z' \times c_0$, and we just need to check that both Z, Z' contain c_0 complemented, which is obvious since both quotient maps are invertible on every separable subspace of $c_0(\aleph)$. □

8.8 Notes and Remarks

8.8.1 Homogeneous Zippin Selectors

One might wonder which additional properties a Zippin selector could enjoy and also which classes of Banach spaces could play the role of $\mathscr{C}$-spaces throughout this chapter. Both questions are somehow connected since there are simple correspondences between certain types of Lindenstrauss spaces and certain types of Zippin selectors.

Consider homogeneous Zippin selectors (i.e. $\omega(\lambda y^*) = \lambda(y^*)$ for $|\lambda| \le 1$) and positively homogeneous Zippin selectors (only for $0 \le \lambda \le 1$). They correspond to two well-known types of Lindenstrauss spaces. A $\mathscr{G}$-space is a Banach space X for which there exists a compact space K and a set of triples $\{k_\alpha^1, k_\alpha^2, \lambda_\alpha\}_{\alpha\in A}$ with $k_\alpha^1, k_\alpha^2 \in K$ and $\lambda_\alpha \in \mathbb{K}$ such that $X = \{f \in C(K)\colon f(k_\alpha^1) = \lambda_\alpha f(k_\alpha^2) \quad \forall\alpha \in A\}$. An $\mathscr{M}$-space is a $\mathscr{G}$-space with $\lambda_\alpha \geq 0$ for all α, equivalently, a sublattice of a $\mathscr{C}$-space.

Lemma *An embedding is $\mathscr{G}$-trivial (resp. $\mathscr{M}$-trivial) if and only if it admits a homogeneous (resp. positively homogeneous) Zippin selector.*

Proof We prove the case of $\mathscr{G}$-spaces, and the other is similar. Form the $\mathscr{G}$-space $G(B_Y^*) = \{f \in C(B_Y^*)\colon f(\lambda y^*) = \lambda f(y^*),\ \forall|\lambda| \leq 1,\ \forall y^* \in B_Y^*\}$ and observe that the natural embedding $\delta\colon Y \longrightarrow G(B_Y^*)$ has the universal property that every $\mathscr{G}$-valued operator on Y factors through δ. Now, let $\jmath\colon Y \longrightarrow X$ be an embedding. There is a correspondence between homogeneous Zippin selectors for $\jmath$ and extensions D of δ through $\jmath$ as in the diagram

$$\begin{array}{ccc} Y & \xrightarrow{\ \jmath\ } & X \\ & {\scriptstyle\delta}\searrow \quad \swarrow{\scriptstyle D} & \\ & G(B_Y^*) & \end{array} \tag{8.40}$$

given by $D(x)(y^*) = \langle\omega(y^*), x\rangle$. The universal property of δ yields, then, the extension of any $\mathscr{G}$-valued operator defined on Y. □

The following result was instrumental in the proofs of Propositions 8.5.1 and 8.5.7.

Proposition *Let Y be a separable Banach space and let $\jmath\colon Y \longrightarrow X$ be an embedding. If $\jmath$ admits a Zippin selector then it admits a homogeneous Zippin selector.*

Proof Everything rests on Benyamini's magical result [38] that separable $\mathscr{G}$-spaces are actually isomorphic to $\mathscr{C}$-spaces. However, to control the bound of the homogeneous selector, we need to go inside [38, Proof of the Theorem] where it is shown that if G is a separable $\mathscr{G}$-space then there is a metric compactum K and an operator $u\colon G \longrightarrow C(K)$ whose range is 1-complemented and such that for all $g \in G$, one has $\frac{1}{2}\|g\| \leq \|u(g)\| \leq \frac{3}{2}\|g\|$. Now, if $\jmath$ admits a λ-Zippin selector then $\jmath$ is $\mathscr{C}$-trivial, and there is a commutative diagram as (8.40) with $\|D\| \leq 3\lambda$ that yields a homogeneous 3λ-Zippin selector. □

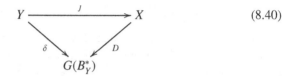

Therefore, most of the material of this chapter for $\mathscr{C}$-spaces could be adapted for $\mathscr{G}$-spaces just multiplying by 3 here and there. Does this deliver all answers with it? For that, you're going to need a bigger boat: Kalton shows in [274, Proposition 4.5] that, for $1 < p < \infty$, the canonical embedding $\delta\colon \ell_p \longrightarrow C(B_{\ell_p}^*)$, which obviously admits a 1-Zippin selector (and therefore a homogeneous 3-Zippin selector), does not admit a homogeneous λ-Zippin selector for any $\lambda < (1 + (q-1)q^{-p})^{1/q}$, where $q^{-1} + p^{-1} = 1$. Moreover, in the non-separable case, the existence of a Zippin selector does not guarantee at all the existence of a homogeneous one: in [39], Benyamini constructs a non-separable $\mathscr{M}$-space M that is not complemented in any $\mathscr{C}$-space. The embedding $M \longrightarrow C(B^*_M)$ is $\mathscr{C}$-trivial but not $\mathscr{M}$-trivial and therefore cannot be $\mathscr{G}$-trivial. Moreover, as we already know, an embedding can be $\mathscr{C}$-trivial without being $\mathscr{L}_\infty$-trivial: in fact, the identity on the Gurariy space G cannot be extended through the canonical embedding $\mathsf{G} \longrightarrow C(B^*_{\mathsf{G}})$ since G is not complemented in any $\mathscr{C}$-space [22, Theorem 3.34]. It is an open question whether there exists a global approach to the extension problem for arbitrary Lindenstrauss-valued operators.

8.8.2 Lindenstrauss-Valued Extension Results

We will show in Section 10.6 that $\mathscr{C}$-valued extension results do not automatically pass to $\mathscr{L}_\infty$-valued results, except the Johnson–Zippin theorem 8.6.2, of course. Let us focus here on results for operators with values in Lindenstrauss spaces. The proof of the Lindenstrauss–Pełczyński theorem presented in Theorem 8.2.2 is a slightly edited version of the original proof in [330]. In that paper, the authors suggest that a version for Lindenstrauss spaces should also hold and give some hints about how to proceed: by using a generalisation of Edwards' separation theorem for Lindenstrauss spaces [319, Theorem 2.1] instead of the naive Hahn–Tong insertion trick. The version of the separation theorem that best suits our needs seems to be the following due to Olsen; see [371, Theorem 4.1] or [316, Theorem 1 on p. 220]:

Lemma *A Banach space $\mathcal{L}$ is a Lindenstrauss space if and only if, for every lower semicontinuous, concave function $G\colon B^*_{\mathcal{L}} \longrightarrow \mathbb{R}$ such that $G(e^*) + G(-e^*) \geq 0$ for all $e^* \in B^*_{\mathcal{L}}$, there exist $\xi \in \mathcal{L}$ such that $\langle \xi, e^* \rangle \leq G(e^*)$ for every $e^* \in B^*_{\mathcal{L}}$.*

We are ready to prove:

Theorem *Every operator from a subspace of c_0 to a Lindenstrauss space has a 1^+-extension to c_0.*

Proof Let $\mathcal{L}$ be a Lindenstrauss space. The proof treads in the footprints of that of Theorem 8.2.2, with $B^*_{\mathcal{L}}$ in the role of the underlying compact space. Lindenstrauss spaces are exactly the $\mathscr{L}_{\infty,1^+}$-spaces, and so each separable subspace of $\mathcal{L}$ is contained into a separable Lindenstrauss subspace. Thus we may assume that $\mathcal{L}$ is separable and that $\|\tau\| = 1$. We prove that, for each $\lambda > 1$ and for each $x \in c_0 \setminus H$, τ can be extended to an operator on $H + [x]$ having norm at most λ. This amounts to showing that there exists $\xi \in \mathcal{L}$ such that $\|\xi - \tau y\| \le \lambda \|y - x\|$ for all $y \in H$. This ξ has to satisfy

$$\langle e^*, \tau y\rangle - \lambda\|y - x\| \le \langle e^*, \xi\rangle \le \langle e^*, \tau y\rangle + \lambda\|y - x\| \tag{8.41}$$

for all $e^* \in B^*_{\mathcal{L}}$ and all $y \in H$. We define two functions on $B^*_{\mathcal{L}}$ as follows:

$$G(e^*) = \inf_{y\in H}\Big(\langle e^*, \tau y\rangle + \lambda\|y - x\|\Big),$$

$$F(e^*) = \sup_{y\in H}\Big(\langle e^*, \tau y\rangle - \lambda\|y - x\|\Big).$$

It is clear that $F(e^*) = -G(-e^*)$ and so ξ satisfies (8.41) if and only if $\langle e^*, \xi\rangle \le G(e^*)$ for every $e^* \in B^*_{\mathcal{L}}$. Keeping an eye on Lemma 8.8.2, notice that G is concave since it is a pointwise infimum of affine functions. Form its lower semicontinuous envelope G_{lsc}, which is also concave. Thus, to apply Olsen's lemma, we only have to see that $G_{\mathrm{lsc}}(e^*) + G_{\mathrm{lsc}}(-e^*) \ge 0$ for every e^* in the ball of $\mathcal{L}^*$. Assuming, on the contrary, that there exists some $e^* \in B^*_{\mathcal{L}}$ such that $-G_{\mathrm{lsc}}(-e^*) > G_{\mathrm{lsc}}(e^*)$, then there also exist sequences $(s_n), (t_n)$ converging to e^* in $B^*_{\mathcal{L}}$ such that $\lim_n F(s_n) > \lim_n G(t_n)$ and, therefore, there exist points $y_n, z_n \in H$ such that

$$\lim_n\Big(\langle s_n, \tau(y_n)\rangle - \lambda\|x - y_n\|\Big) > \lim_n\Big(\langle t_n, \tau(z_n)\rangle + \lambda\|x - z_n\|\Big).$$

Now, switch to the part of the proof of Theorem 8.2.2 starting at (8.2), with e^* replacing δ_s, until reaching a contradiction. □

The $\mathscr{C}$-extensibility of ℓ_1 also passes without modification to Lindenstrauss-space extensibility:

Proposition *Every operator from ℓ_1 to a Lindenstrauss space admits 1^+ extensions to any separable superspace.*

Proof Let $\mathcal{L}$ be a Lindenstrauss space, and let $\tau\colon \ell_1 \longrightarrow \mathcal{L}$ be an operator. We can assume that $\mathcal{L}$ is separable by the same argument as in the previous proof. Johnson and Zippin [234] proved that there is an isometric quotient map $Q\colon C(\Delta) \longrightarrow \mathcal{L}$. Pick a 1^+-lifting t of τ through Q and then a 1^+-extension T of t. The operator QT is a 1^+-extension of τ. □

8.8.3 The Last Stroke on the Extension of $\mathscr{C}$-Valued Lipschitz Maps

As we have already mentioned, the material contained in Section 8.3 originates in Kalton's studies about the extension of $\mathscr{C}$-valued Lipschitz maps, which we have avoided in its full generality as much as possible thus far. Oh well, in for a penny, in for a pound: it is time to explain why the non-linear context yields even better results. Suppose we are given a metric space X, not necessarily normed, and a Lipschitz map $\tau\colon Y \longrightarrow C(K)$, where Y is a subset of X and K is a metrisable compact space. If we want to extend τ to X to one more point $x \in X \setminus Y$, we typically generate two bounded functions $g, h\colon K \longrightarrow \mathbb{R}$, depending on x, in such a way that g is upper semicontinuous, h is lower semicontinuous and $g \leq h$ in order to then use the Hahn–Tong sandwich theorem to insert a continuous f between g and h. We choose this f as the value of the extension of τ at x. If, moreover, we could almost preserve the Lipschitz constant of the extension then we could proceed through an enumeration of a dense subset of $X \setminus Y$ to get a global extension to X, as in Theorem 8.3.10. Otherwise, when no almost-isometric preservation of the Lipschitz constant can be done, one instead needs to control the Lipschitz constant of the extension by means of the distances between the semicontinuous functions entering into the Hahn–Tong theorem. All this is treated by Kalton in a rather unexpected way. Before proceeding, let us make our lives easier by recalling that, in our current separable setting, all problems about extensions of $\mathscr{C}$-valued maps can be reduced to the case where the underlying compactum is the Cantor set, as Lemma 1.6.2 clearly explains. Now, consider SUB(Δ), the subset of those pairs $(g, h) \in \ell_\infty(\Delta) \times \ell_\infty(\Delta)$ such that g is upper semicontinuous, h is lower semicontinuous and $g \leq h$. We measure distances in SUB(Δ) just using the restriction of the norm:

$$d((g_1, h_1); (g_2, h_2)) = \max\left(\|g_1 - g_2\|_\infty, \|h_1 - h_2\|_\infty\right).$$

We have (we skip the proof):

Proposition *There is a contraction $\theta\colon$ SUB(Δ) $\longrightarrow C(\Delta)$ such that, for every $(g, h) \in$ SUB(Δ), we have $g \leq \theta(g, h) \leq h$.*

The task now at hand is to give a metric characterisation of those pairs (Y, X) for which every $\mathscr{C}$-valued contraction on Y has a λ-Lipschitz extension to X (for fixed λ!). It is clear from Lemma 8.3.3 that (Y, X) satisfies condition $\Sigma_1(\lambda)$ if and only if every contraction $Y \longrightarrow c$ admits a λ-Lipschitz extension to one more point, no matter which point one chooses. It seems like magic that this suffices to get a global, coherent extension with the same bound, but it does:

Theorem *If $Y \subset X$ is a separable subset of a metric space and $\lambda \geq 1$, the following are equivalent:*

(i) *Every $\mathscr{C}$-valued Lipschitz map on Y admits a λ-extension to X.*
(ii) *The pair (Y, X) verifies condition $\Sigma_1(\lambda)$.*

Proof The implication (i) $\Longrightarrow$ (ii) is contained in Lemma 8.3.3. To prove the converse, it suffices to consider the case in which the target space is $C(\Delta)$. So, let $\tau\colon Y \longrightarrow C(\Delta)$ be a contraction. We define two λ-Lipschitz maps $\tau^-, \tau^+\colon X \longrightarrow \ell_\infty(\Delta)$ by means of

$$\tau^-(x) = \bigvee_{y\in Y} (\tau(y) - \lambda d(y,x)1_\Delta),$$
$$\tau^+(x) = \bigwedge_{z\in Y} (\tau(z) + \lambda d(z,x)1_\Delta),$$

where the order refers to $\ell_\infty(\Delta)$, a complete lattice. Clearly, $\tau^-(x) \le \tau^+(x)$ for every $x \in X$, and $\tau^-(y) = \tau^+(y) = \tau(y)$ for $y \in Y$. Now, with an eye on Proposition 8.8.3, we define $G, H\colon X \longrightarrow \ell_\infty(\Delta)$ by $G(x) = \tau^-(x)^{\mathrm{usc}}$ and $H(x) = \tau^+(x)_{\mathrm{lsc}}$. It is easy to check that these are again λ-Lipschitz and that $G(y) = H(y) = \tau(y)$ for $y \in Y$. The core of the argument is contained in the following:

Claim For every $x \in X \setminus Y$, one has $G(x) \le H(x)$.

Proof of the claim Following the proof of Lemma 8.3.3, assume that there are $x \in X \setminus Y$, $s \in \Delta$ and $\varepsilon > 0$ such that $G(x)(s) > H(x)(s) + 2\varepsilon$. Then there are sequences $(s_n), (t_n)$ in Δ, both converging to s, such that $\tau^-(x)(s_n) > \tau^+(x)(t_n) + 2\varepsilon$. For each n, we may select y_n and z_n in Y such that $\tau(y_n)(s_n) - \lambda d(y_n, x) > \tau(z_n)(t_n) + \lambda d(z_n, x) + 2\varepsilon$. Applying condition $\Sigma_1(\lambda)$ to the sequences $(y_n), (z_n)$ and ε, we get $u \in Y$ such that $d(u, y_n) + d(u, z_n) \le \lambda(d(x, y_n) + d(x, z_n)) + \varepsilon$ for infinitely many ns. Let us set $\eta = \tau(u)$ and see what happens. We have $\tau(y_n) \le \eta + d(u, y_n)$ and $\tau(z_n) \ge \eta - d(u, z_n)$ and, in particular,

$$\tau(y_n)(s_n) \le \eta(s_n) + d(u, y_n) \qquad \text{and} \qquad \tau(z_n)(t_n) \ge \eta(t_n) - d(u, z_n).$$

Combining these, we obtain that for infinitely many n,

$$\begin{aligned}\eta(s_n) - \eta(t_n) &\ge \tau(y_n)(s_n) - d(u, y_n) - \tau(z_n)(t_n) - d(u, z_n)\\ &> \lambda(d(x, y_n) + d(x, z_n)) + 2\varepsilon - d(u, y_n) - d(u, z_n) \ge \varepsilon,\end{aligned}$$

which contradicts the continuity of η at s. □

Thus, the λ-Lipschitz map $x \in X \longmapsto (G(x), H(x)) \in \ell_\infty(\Delta) \times \ell_\infty(\Delta)$ actually takes values in $\mathrm{SUB}(\Delta)$. Composing with the contraction $\theta\colon \mathrm{SUB}(\Delta) \longrightarrow C(\Delta)$ provided by Proposition 8.8.3, we obtain the required λ-Lipschitz extension of τ, concluding the proof. □

This provides a complete characterisation of the separable subsets Y of a metric space X for which all $\mathscr{C}$-valued Lipschitz maps admit λ-extensions. Of course, it includes the case in which X is a Banach space. Compare to Theorem 8.3.10. A version of Lindenstrauss' classic comes as a bonus:

Corollary *Every separable $\mathscr{C}$-space is an absolute 3-Lipschitz retract.*

8.8.4 Property (M) and M-Ideals

Properties (L) and (L^*) are the ugly mates of properties (M) and (M^*) introduced by Kalton in his study of M-ideals [264]. A Banach space X is said to have property (M) if every weakly null type on X is a function of the norm, and X is said to have property (M^*) if every weak*-null type on X^* is a function of the norm. It is clear that c_0 and the spaces ℓ_p for $1 < p < \infty$ have properties (M) and (M^*). The Yellow Book [209] incorporates most of the discoveries of [264], although not the ultimate connections between properties (M^*), (M) and M-ideals of compact operators, which came later [370; 287; 322]. Namely

- A Banach space X has property (M^*) if and only if it has the metric compact approximation property (obvious meaning) and $\mathfrak{K}(X)$ is an M-ideal in $\mathfrak{L}(X)$ – see [264, Theorem 2.4] and [287] for separable X and [370; 322] for the general case.
- Property (M^*) implies (M) for separable spaces [264, Proposition 2.5], and Property (M) implies (M^*) for separable spaces containing no copy of ℓ_1 [287, Theorem 2.6].

The paper [287] contains a bunch of examples (and counterexamples) concerning these properties too. Regarding the topic of this chapter, separable spaces having property (M^*) have the almost isometric $\mathscr{C}$-extension property [273, Theorem 7.5]. One could actually develop Section 8.4 focusing on properties (M) and (M^*) instead of (L) and (L^*). In some sense, this is what Kalton did in [264, Section 4] and [273, Section 3]. We decided to present the results in their L-version only because the M-version of Proposition 8.4.2 (which is a particular case of [273, Proposition 3.4]) is much harder to prove.

8.8.5 Set Theoretic Axioms and Twisted Sum Affairs

Cardinal axiomatics have made an essential irruption in homological affairs. We list the places where they played or will play a role, if not for better, at least for good:

- The CCKY-problem has different solutions under CH or MA.
- Under CH, there are $2^{\aleph_1}$ non-isomorphic spaces $C(\Delta_{\mathcal{M}})$ for $|\mathcal{M}| = \aleph_1$. All of them are isomorphic under MA.

- Under CH, every 1-separably injective Banach space contains ℓ_∞, while under MA + $\aleph_2 = \mathfrak{c}$, there exists a 1-separably injective $\mathscr{C}$-space that does not contain ℓ_∞ [24] .
- Under CH, universal separable injectivity is not a 3-space property (Theorem 10.5.6 (c)).
- Under CH, there exist non-trivial sequences $0 \longrightarrow C(\mathbb{N}^*) \longrightarrow C(\mathbb{N}^*) \longrightarrow C(\mathbb{N}^*) \longrightarrow 0$ (Theorem 10.5.6 (b)). In particular, $\operatorname{Ext}(C(\mathbb{N}^*), C(\mathbb{N}^*)) \neq 0$ [23]. It is apparently unknown whether the same holds in ZFC.
- Under CH, there is just one space of separable universal disposition and dimension $\mathfrak{c}$, the Kubiś space F_1. Under different axiomatics (see after Corollary 7.3.5), there is a continuum of mutually non-isomorphic spaces of that type.

Sources

The material in this chapter is classical, old and new, all at once. It is classical because some parts can be already found in Semadeni's book [430]; it is as old as the 1970s [330] and as new as 2007 because a large part of it develops Kalton's extraordinary series of four papers [271; 272; 273; 274]. Weak*-continuous selectors are among the extremely nice contributions of Zippin [463; 464; 465], who coined the term *global approach*. The four 'elementary' examples in Proposition 8.1.3 have different origins. Example (a), embeddings into ℓ_p, is from Zippin, although the simple proof we present was the idea of Yost and appeared in [76]. Example (b), the embedding $\ell_p \longrightarrow \ell_p(\varphi)$, is a particular case of the $\mathscr{C}$-extensibility of the spaces $\ell_p(\varphi)$, although the argument presented is original. Example (c) and its proof are a reworking of some parts of Kalman's paper [245]; however, the use of triangularisations to prove Kalman's theorem was suggested to us by Francisco Santos. The Lindenstrauss–Pełczyński theorem is one of the beautiful contributions in [330]. As mentioned in the text, Zippin [464; 465] produced his own proof; the homologically flavoured proof in Theorem 8.2.1 is from [76], and the Lindenstrauss-valued extension à la Lindenstrauss–Pełczyński in Section 8.8.2 is from [129]. The result remained isolated until Johnson and Zippin [236] obtained an extension to subspaces of $c_0(I)$ and later in [237] to weak*-closed subspaces of ℓ_1, which is the Johnson–Zippin theorem. It involves $\mathscr{L}_\infty$-valued operators and is, in fact, a characterisation of them [118, Proposition 3.1]: *X is an $\mathscr{L}_\infty$-space if and only if every X-valued operator defined on a weak*-closed subspace of ℓ_1 can be extended to ℓ_1*. The Ext-version of this result is Proposition 5.2.9 (b). The paper [76] contains the first example showing that $\mathscr{L}_\infty$-spaces cannot replace $\mathscr{C}$-spaces in the Lindenstrauss–Pełczyński theorem and answers Zippin's problem 6.15 in [466]. The existence of such 'rare' $\mathscr{L}_\infty$-spaces sparked the theory of Lindenstrauss–Pełczyński spaces, to be developed in Section 10.6. The nice unexpected example in 8.2.3 has been taken from [236]. The concept of a type on a Banach space was introduced by Krivine and Maurey [307]. That ℓ_2 is not extensible and all that comes with it was Kalton's response (see [271, Section 4]) to the final comments in [109]. In this paper, the connection between the $\mathscr{C}$-extensibility property of X and the X-automorphic character of $C[0,1]$ was established (Proposition 7.4.15), something that Kalton reformulates as [271, Theorem 4.1] or else as [274, Proposition 2.5]. The connection between the two properties was established while studying [237, Problem 4.2]: *if E is a reflexive subspace of X, does every $\mathscr{C}$-valued operator on E extend to X?* After that initial impetus, Kalton's imagination ran free. He considered the question of when $\operatorname{Ext}(X, C[0,1]) = 0$ in [267], obtaining Proposition 8.6.4 and much more, and then again [113] following a different approach. The problem was revisited in [73] considering also the case $C(\omega^\omega)$. The reader is warned that the reading of [73] is difficult at some points, and quite difficult for the rest of the time; it is advisable to have [191] at hand. In [73], the parameter α_N defined in (8.34) is rather associated with X^*, or with some of its subspaces, and the role of X is to define the weak* topology in X^*. In the text, we have treated α_N as a constant associated with X that is computed on its dual. Very recently, Causey, Fovelle and Lancien have shown that having a summable Szlenk index is a 3-space property [131]. Zippin's problem 8.6.1 still remains open. All adventurous episodes of the astounding story have been told during Section 8.7. But there remain outrageous possibilities: Proposition 8.7.2 has been generalised in [25, Proposition 8.2] to:

Proposition *Let X be a Banach space and let $c_n, d_n > 0$ be two sequences such that $\lim n\, d_n\, c_n^{-1} = \infty$. Suppose that, for every n, there exist $\Phi_1, \ldots, \Phi_n \subset B_X^*$ and $\Psi_1, \ldots, \Psi_n \subset B_X$ such that*

- $\left\|\sum_{i=1}^n x_i\right\| \leq c_n$ *for any choice of vectors $x_i \in \Psi_i$;*
- *for every i and $x^* \in \Phi_i$, there is $x \in \Psi_i$ such that $x^*(x) > d_n$;*
- *the sets Φ_F are pairwise disjoint and $|\Phi_i| \geq \mathfrak{c}$;*
- *$\Phi_1 \cup \cdots \cup \Phi_n$ is discrete, and its weak*-closure is its one-point compactification.*

Then there is a non-trivial twisted sum of c_0 and X.

What the proposition says, in its somewhat technical formulation, is that biorthogonality is not essential in Proposition 8.7.2. The assumption CH on Proposition 8.7.3 can be relaxed to MA; see [25, Corollary 4.2]. The problem of whether the assertion holds in ZFC is, however, open. What the whole story teaches us is that it is by now clear that 'there are nonmetrisable compacta K for which the question of whether $\mathrm{Ext}(C(K), c_0) \neq 0$ is undecidable within the usual axioms of set theory' [25]. Which compacta those are and for which ones there is a plain answer in ZFC is an entire world of research. Many more results about good-natured compacta not mentioned in the astounding tale can be found in [25; 134; 135; 136; 137; 354]. A few additional steps towards 8.7.6 are worth mentioning: Correa [135] proves it under MA and, in private communications long before [25] was available, she informed us that the result was true in ZFC for metrisable height 3 compacta. Property LEP(r) has been reconsidered in [137] for finite height compacta. The results in Sections 8.8.1 and 8.8.2 are from [129], and those in Section 8.8.3 are from [272] and [273].

9

Singular Exact Sequences

Exact sequences whose quotient map is a strictly singular operator are strange and somewhat cryptic objects. They have been called *singular* sequences; accordingly, a quasilinear map will be called *singular* if it generates a singular sequence. These objects are not only non-trivial but in a sense some of the most non-trivial objects there are. Singular sequences have been studied per se in several papers [56; 62; 78; 108; 111; 98; 445] and are used as a tool to obtain counterexamples in many others [280; 73; 265]. The list of folklore examples is not too long:

- (General fact) If Z, X are totally incomparable, every operator $Z \longrightarrow X$ is strictly singular, hence every exact sequence $0 \longrightarrow \cdot \longrightarrow Z \longrightarrow X \longrightarrow 0$ is singular. In particular, the exact sequences $0 \longrightarrow \kappa(X) \longrightarrow \ell_p(I) \longrightarrow X \longrightarrow 0$ are singular for every p-Banach space X not containing ℓ_p.
- (H.I. folklore) Every exact sequence in which the middle space is H.I. is singular.
- (Obvious fact) No singular p-linear map can be defined on a p-Banach space with a copy of ℓ_p. In particular, $\mathrm{Ext}_{\mathbf{B}}(X, Y)$ has no singular element when X contains ℓ_1. More particularly still, there is no singular element in $\mathrm{Ext}(C(\Delta), Y)$ when Y is a Banach space.

In this chapter we will add a few more examples, some impressive, some unexpected:

- The Kalton–Peck map KP_X is singular on many sequence spaces X, in particular, on $X = \ell_p$ for every $0 < p < \infty$.
- The Kalton–Peck centralizers KP_φ are not singular on L_p.
- For every separable Banach space X not containing ℓ_1 there exist singular sequences $0 \longrightarrow C(\Delta) \longrightarrow \cdot \longrightarrow X \longrightarrow 0$. Additional but reasonable requirements on X will provide singular twistings against $C(\omega^\omega)$.

- There exists a singular sequence $0 \longrightarrow \mathbb{R} \longrightarrow \cdot \longrightarrow \ell_1 \longrightarrow 0$.

For the rest of this chapter we will study singular sequences, different methods of construction and examples. A unusually spicy 'Notes and Remarks' gumbo concludes the chapter.

9.1 Basic Properties and Techniques

It is obvious that if the quotient map of an exact sequence is strictly singular then the same is true for any equivalent or even isomorphic sequence.

Definition 9.1.1 An exact sequence $0 \longrightarrow Y \longrightarrow Z \longrightarrow X \longrightarrow 0$ is singular if the quotient map is a strictly singular operator. A quasilinear map $\Omega\colon X \longrightarrow Y$ is said to be singular if the exact sequence $0 \longrightarrow Y \longrightarrow Y \oplus_\Omega X \longrightarrow X \longrightarrow 0$ is singular.

Observe that an operator $T\colon Z \longrightarrow X$ fails to be strictly singular if and only if there is an infinite-dimensional subspace $H \subset X$ and an operator $L\colon H \longrightarrow Z$ such that TL is the inclusion of H into X. Thus, an exact sequence $0 \longrightarrow Y \longrightarrow Z \longrightarrow X \longrightarrow 0$ is singular if and only if, whenever $H \subset X$ is an infinite-dimensional subspace, the pullback sequence (here $\mathrm{PB} = \rho^{-1}[H]$)

$$\begin{array}{ccccccccc} 0 & \longrightarrow & Y & \longrightarrow & Z & \xrightarrow{\rho} & X & \longrightarrow & 0 \\ & & \| & & \uparrow & & \uparrow{\scriptstyle\text{inclusion}} & & \\ 0 & \longrightarrow & Y & \longrightarrow & \mathrm{PB} & \longrightarrow & H & \longrightarrow & 0 \end{array}$$

is non-trivial. The quasilinear description of pullbacks immediately yields:

Proposition 9.1.2 *A quasilinear map $\Omega\colon X \longrightarrow Y$ is singular if and only if its restriction to every infinite-dimensional subspace of X is not trivial.*

In this way, the definition of singular quasilinear maps is in complete analogy with that of strictly singular operators, at least if one imaginatively understands this as an operator that is not trivial (i.e., an isomorphism) on any infinite-dimensional subspace. The complete analogy is not that complete: 0 is a strictly singular operator, while 0 is not a singular sequence. That fact will cause some troubles. The following equivalent formulations of singularity will be useful:

Proposition 9.1.3 *Each of the following assertions is equivalent to the singularity of the sequence $0 \longrightarrow Y \xrightarrow{\jmath} Z \xrightarrow{\rho} X \longrightarrow 0$:*

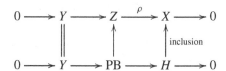

(i) *Whenever one has a pushout diagram*

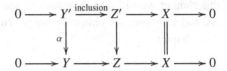

$$\begin{array}{ccccccccc} 0 & \longrightarrow & Y' & \xrightarrow{\text{inclusion}} & Z' & \longrightarrow & X & \longrightarrow & 0 \\ & & \Big\downarrow{\scriptstyle\alpha} & & \Big\downarrow & & \Big\| & & \\ 0 & \longrightarrow & Y & \longrightarrow & Z & \longrightarrow & X & \longrightarrow & 0 \end{array}$$

the operator α cannot be extended to any subspace of Z' containing Y' in which Y' has infinite codimension.

(ii) *Whenever one has a pullback diagram*

$$\begin{array}{ccccccccc} 0 & \longrightarrow & Y & \longrightarrow & Z' & \longrightarrow & X' & \longrightarrow & 0 \\ & & \Big\| & & \Big\uparrow & & \Big\uparrow{\scriptstyle\gamma} & & \\ 0 & \longrightarrow & Y & \longrightarrow & Z & \longrightarrow & X & \longrightarrow & 0 \end{array}$$

the operator γ has a finite-dimensional kernel.

Proof **Equivalence with (i).** If the sequence is not strictly singular then there is an infinite-dimensional $H \subset X$ where the quotient map is invertible. Consider the 'obvious' pushout diagram

$$\begin{array}{ccccccccc} 0 & \longrightarrow & \jmath[Y] & \xrightarrow{\text{inclusion}} & Z & \longrightarrow & X & \longrightarrow & 0 \\ & & \Big\downarrow{\scriptstyle\jmath^{-1}} & & \Big\downarrow & & \Big\| & & \\ 0 & \longrightarrow & Y & \longrightarrow & Z & \longrightarrow & X & \longrightarrow & 0 \end{array}$$

and note that $\jmath^{-1}$ extends to $\rho^{-1}[H]$. Conversely, if α has an extension of some $Z'' \subset Z'$ containing Y' with $\dim(Z''/Y') = \infty$ then the lower row in the pullback-and-then-pushout diagram

$$\begin{array}{ccccccccc} 0 & \longrightarrow & Y' & \xrightarrow{\text{inclusion}} & Z' & \longrightarrow & X & \longrightarrow & 0 \\ & & \Big\| & & \Big\uparrow & & \Big\uparrow{\scriptstyle\text{obvious embedding}} & & \\ 0 & \longrightarrow & Y' & \xrightarrow{\text{inclusion}} & Z'' & \longrightarrow & Z''/Y' & \longrightarrow & 0 \\ & & \Big\downarrow{\scriptstyle\alpha} & & \Big\downarrow & & \Big\| & & \\ 0 & \longrightarrow & Y & \longrightarrow & \mathrm{PO} & \longrightarrow & Z''/Y' & \longrightarrow & 0 \end{array}$$

splits and is the same (Section 2.10) as the one obtained by first taking the pushout and then the pullback, namely

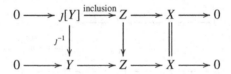

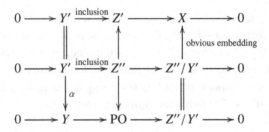

$$\begin{array}{ccccccccc}
0 & \longrightarrow & Y' & \xrightarrow{\text{inclusion}} & Z' & \longrightarrow & X & \longrightarrow & 0 \\
 & & \downarrow{\scriptstyle\alpha} & & \downarrow & & \| & & \\
0 & \longrightarrow & Y & \xrightarrow{\jmath} & Z & \xrightarrow{\rho} & X & \longrightarrow & 0 \\
 & & \| & & \uparrow & & \uparrow{\scriptstyle\text{obvious embedding}} & & \\
0 & \longrightarrow & Y & \longrightarrow & \mathrm{PB} & \longrightarrow & Z''/Y' & \longrightarrow & 0
\end{array}$$

Thus ρ is invertible on the image of Z''/Y' under the obvious embedding.
Equivalence with (ii). If the sequence is not singular and the quotient map is invertible on $H \subset X$ then the corresponding pullback sequence

$$\begin{array}{ccccccccc}
0 & \longrightarrow & Y & \longrightarrow & Z & \longrightarrow & X & \longrightarrow & 0 \\
 & & \| & & \uparrow & & \uparrow{\scriptstyle\text{inclusion}} & & \\
0 & \longrightarrow & Y & \longrightarrow & \mathrm{PB} & \longrightarrow & H & \longrightarrow & 0
\end{array}$$

is trivial. The homology sequence (4.5) yields a pullback diagram

$$\begin{array}{ccccccccc}
0 & \longrightarrow & Y & \longrightarrow & Z' & \longrightarrow & X/H & \longrightarrow & 0 \\
 & & \| & & \uparrow & & \uparrow{\scriptstyle\text{quotient map}} & & \\
0 & \longrightarrow & Y & \longrightarrow & Z & \longrightarrow & X & \longrightarrow & 0
\end{array}$$

which contradicts (ii). The other implication is obvious since, given a diagram as in (ii), the quotient map $Z \longrightarrow X$ is always invertible on $\ker \gamma$. □

We are dragged now to Banach space results since Criterion 0.0.2, on which these results are based, ultimately requires the existence of basic sequences (see Note 9.4.4).

9.1.4 *An exact sequence of Banach spaces* $0 \longrightarrow Y \xrightarrow{\imath} Z \xrightarrow{\rho} X \longrightarrow 0$ *is singular if and only if for every infinite-dimensional subspace* $Z' \subset Z$ *there exist infinite-dimensional subspaces* $Y_1 \subset Y$ *and* $Z_1 \subset Z'$ *and a nuclear operator* $\nu\colon Y_1 \longrightarrow Z_1$ *such that* $\imath + \nu$ *is an isomorphism between* Y_1 *and* Z_1.

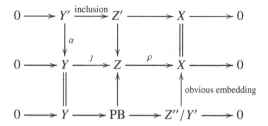

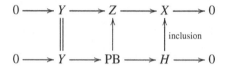

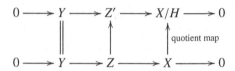

Proof We may assume that $Y = \ker \rho$ and $\imath$ is the inclusion map. The condition is obviously sufficient (even if we replace nuclear by compact; even for quasi-Banach spaces): if $k\colon Y_1 \longrightarrow Z_1$ is a compact operator such that $\imath + k$ is an isomorphism then $\rho(\imath + k) = \rho k$ means that the restriction of ρ to $(\imath + k)[Y_1]$ cannot be an isomorphism. For the other implication, we use 0.0.2 to get, for every $\varepsilon > 0$ and every subspace $Z' \subset Z$, a subspace B spanned by a basic sequence $(b_n)_{n\geq 1}$ such that $\|\rho|_B\| \leq \varepsilon$. A careful diagonalisation yields a basic sequence (z_n) of blocks of (b_n) such that $\|\rho(z_n)\| \leq 2^{-n}$, and this generates a basic sequence $(y_n) \subset Y$ equivalent to (z_n), thus $y_n \longmapsto y_n + (z_n - y_n)$ establishes an isomorphism $\overline{[y_n : n \in \mathbb{N}]} \longrightarrow \overline{[z_n : n \in \mathbb{N}]}$ of the form $\imath + \nu$ with ν nuclear. □

And from that one gets:

Lemma 9.1.5 *Given a commutative diagram in which the rows are exact sequences of Banach spaces*

$$\begin{array}{ccccccccc} 0 & \longrightarrow & A & \stackrel{\imath}{\longrightarrow} & B & \stackrel{\rho}{\longrightarrow} & C & \longrightarrow & 0 \\ & & \Big\downarrow{\scriptstyle\tau} & & \Big\downarrow{\scriptstyle\overline{\tau}} & & \Big\| & & \\ 0 & \longrightarrow & D & \longrightarrow & E & \longrightarrow & C & \longrightarrow & 0 \end{array}$$

$\overline{\tau}$ is strictly singular if and only if both ρ and τ are.

Proof The necessity is obvious, so let us prove the sufficiency. Fix $\varepsilon > 0$. If ρ is strictly singular, the characterisation 9.1.4 yields that given $B_1 \subset B$ there are subspaces $B_2 \subset B_1$, $A_2 \subset A$ and a compact operator $K : A_2 \longrightarrow B_2$ such that $\imath + K : A_2 \longrightarrow B_2$ is an embedding. Since τ is strictly singular, there must be a further subspace $A_3 \subset A_2$ such that $\|\tau|_{A_3}\| < \varepsilon$ by the characterisation 0.0.2. There is no loss of generality in also assuming that $\|K|_{A_3}\| < \varepsilon$, and thus one gets $\|\overline{\tau}|_{(\imath+K)(A_3)}\| = \|\tau|_{A_3} + \overline{\tau}K|_{A_3}\| < (1 + \|\overline{\tau}\|)\varepsilon$. □

Lemma 9.1.6 *Consider a commutative diagram*

$$\begin{array}{ccccccccc} & & 0 & & 0 & & & & \\ & & \Big\downarrow & & \Big\downarrow & & & & \\ 0 & \longrightarrow & Y & \stackrel{\omega}{\longrightarrow} & Z & \stackrel{\rho_\omega}{\longrightarrow} & X & \longrightarrow & 0 \\ & & \Big\downarrow{\scriptstyle\phi} & & \Big\downarrow{\scriptstyle\psi} & & \Big\| & & \\ 0 & \longrightarrow & W & \stackrel{\lambda}{\longrightarrow} & \mathrm{PO} & \stackrel{\rho_\lambda}{\longrightarrow} & X & \longrightarrow & 0 \\ & & \Big\downarrow{\scriptstyle\rho_\phi} & & \Big\downarrow{\scriptstyle\rho_\psi} & & & & \\ & & W' & = & W' & & & & \\ & & \Big\downarrow & & \Big\downarrow & & & & \\ & & 0 & & 0 & & & & \end{array}$$

in which both the rows and the columns are exact sequences of Banach spaces.

(a) *If ρ_λ and ρ_ϕ are strictly singular then ρ_ψ is strictly singular.*

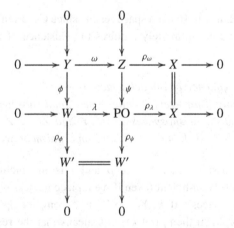

(b) *If ρ_ω and ρ_ϕ are strictly singular and Z and Z' are totally incomparable then also ρ_λ and ρ_ψ are strictly singular.*

Proof (a) Let V be a subspace of PO. Since ρ_λ is strictly singular, some subspace $V_1 \subset V$ is the isomorphic image $(\lambda + \nu_1)[W_1]$ of some $W_1 \subset W$ with ν_1 nuclear; since ρ_ϕ is strictly singular, some subspace $V_2 \subset V_1$ is the isomorphic image $(\phi + \nu_2)[Y_2]$ of some subspace $\jmath\colon Y_2 \longrightarrow Y$, with ν_2 nuclear. Therefore the subspace $(\lambda + \nu_1)(\jmath + \nu_2)[Y_2] = (\psi\omega + \lambda\nu_2 + \nu_1\phi + \nu_1\nu_2)[Y_2]$ of PO is the isomorphic image of $Y_2 \subset Y$. By extension, there exists a nuclear operator $\nu\colon X \longrightarrow$ PO such that $\lambda\nu_2 + \nu_1\phi + \nu_1\nu_2 = \nu\omega$. Hence the subspace $(\psi\omega + \lambda\nu_2 + \nu_1\phi + \nu_1\nu_2)[Y_2] = (\psi\omega + \nu\omega)[Y_2] = (\psi + \nu)\omega[Y_2]$ of PO is the isomorphic image of $\omega[Y_2] \subset X$ through an embedding $\psi + \nu$.

(b) Let V be a subspace of PO. If it contains a subspace V_1 where both ρ_λ and ρ_ψ are isomorphisms, then Z and W' are not incomparable. If both ρ_λ, ρ_ψ are singular on every subspace of V then they are singular. So, assume there is a subspace V_1 of V where ρ_λ is singular and ρ_ψ is an isomorphism. Then there is a further subspace V_2 of V_1 that is the isomorphic image $(\lambda + \nu_1)[W_2]$ of some $W_2 \subset W$ with ν_1 nuclear; since ρ_ψ is strictly singular, some subspace $V_3 \subset V_2$ is the isomorphic image $(\phi + \nu_2)[Y_3]$ of some $Y_3 \subset Y$, with ν_2 nuclear. This makes $\rho_\psi(\lambda + \nu_1)(\phi + \nu_2)[Y_3] = \rho_\psi(\psi\omega + \nu')[Y_3] = \rho_\psi\nu'[Y_3]$, hence $\rho_\psi\nu'$ is an isomorphism, which is impossible since ν' is nuclear. □

There is a remarkable connection between H.I. spaces and singularity:

Lemma 9.1.7 *Let* $0 \longrightarrow Y \xrightarrow{\imath} Z \xrightarrow{\rho} X \longrightarrow 0$ *be an exact sequence of Banach spaces in which Y is H.I. Then Z is H.I. if and only if ρ is strictly singular.*

Proof If ρ is not strictly singular, there is a subspace Z_1 of Z such that $\rho|_{Z_1}$ is an isomorphism. Therefore, $Y \oplus Z_1$ is a subspace of Z and thus Z cannot be H.I. Now assume that ρ is strictly singular and fix two infinite-dimensional subspaces Z_1 and Z_2 of Z. Since $\rho|_{Z_1}$ is singular, for any n, one can find a normalised vector x_1 in Z_1 such that $\operatorname{dist}(x_1, Y) < 2^{-n}$. Using this, and fixing $0 < \varepsilon < 1$, it is easy to find a subspace Y_1 of Y and an embedding $\imath_1 : Y_1 \longrightarrow Z_1$ such that $\|\imath_1 - \imath|_{Y_1}\| \le \varepsilon$. The same holds for some subspace Y_2 of Y and some embedding $\imath_2\colon Y_2 \longrightarrow Z_2$. Since Y is H.I., there are normalised vectors y_1 in Y_1 and y_2 in Y_2 such that $\|y_1 - y_2\| \le \varepsilon$. Then $x_1 = \imath_1 y_1$ and $x_2 = \imath_2 y_2$ are vectors of Z_1 and Z_2, respectively, of norm at least $1/2$, for which $\|x_1 - x_2\| \le \|x_1 - y_1\| + \|y_1 - y_2\| + \|y_2 - x_2\| \le 3\varepsilon$. Thus Z_1 and Z_2 do not form a direct sum, and Z is H.I. □

Perhaps we should mention that there is no known method for twisting an H.I. space with itself to obtain an H.I. twisted sum. In general, it is an open problem to determine, out of trivialities, when $\operatorname{Ext}(X, X)$ contains singular elements. We conclude this introductory section by presenting a couple of *balance principles* for singular sequences which the reader can savour as a sip of dandelion wine. The version for Banach spaces requires us to work with the minimal cardinal $\mathfrak{n}(X)$ of a norming subset of X.

Proposition 9.1.8 *If X, Y are Banach spaces with $|X| > \mathfrak{c}^{\mathfrak{n}(Y)}$ then* $\operatorname{Ext}_{\mathbf{B}}(X, Y)$ *contains no singular element.*

Proof Let $0 \longrightarrow Y \longrightarrow Z \longrightarrow X \longrightarrow 0$ be an exact sequence. Let $N \subset B_{Y^*}$ be a minimal norming set for Y, $\imath\colon Y \longrightarrow \ell_\infty(N)$ be an isometry, $\jmath\colon Z \longrightarrow \ell_\infty(N)$ be an extension and $\bar{\jmath}$ be the induced operator. The pullback diagram

$$\begin{array}{ccccccccc} 0 & \longrightarrow & Y & \xrightarrow{\ \imath\ } & \ell_\infty(N) & \longrightarrow & \ell_\infty(N)/\imath[Y] & \longrightarrow & 0 \\ & & \Big\| & & \uparrow{\scriptstyle \jmath} & & \uparrow{\scriptstyle \bar{\jmath}} & & \\ 0 & \longrightarrow & Y & \longrightarrow & Z & \longrightarrow & X & \longrightarrow & 0 \end{array}$$

implies that if the lower sequence is singular then $\ker \bar{\jmath}$ must be finite-dimensional (Proposition 9.1.3 (ii)). A computation of sizes yields a contradiction:

$$\mathfrak{c}^{\mathfrak{n}(Y)} < |X| \leq |\ell_\infty(N)| = |[-1, 1]^N| = \mathfrak{c}^{|N|} = \mathfrak{c}^{\mathfrak{n}(Y)}. \qquad \square$$

The taste of the quasi-Banach version will improve after Note 9.4.4:

Proposition 9.1.9 *No singular sequence $0 \longrightarrow \mathbb{K} \longrightarrow \cdot \longrightarrow \ell_1(I) \longrightarrow 0$ is possible when $|I| > \mathfrak{c}$.*

Proof For $p < 1$, one has $|\operatorname{Ext}(\ell_1, \mathbb{K})| = |\operatorname{Ext}^{\mathrm{proj}}_{\mathbf{pB}}(\ell_1, \mathbb{K})| \leq |\mathfrak{L}(\kappa_p(X), \mathbb{K})| \leq \mathfrak{c}$. Now let $\Omega\colon \ell_1(I) \longrightarrow \mathbb{K}$ be a quasilinear map, where $|I| > \mathfrak{c}$. Partition I into $|I|$ countable sets $I = \bigcup_\alpha I_\alpha$, and let $\sigma_\alpha\colon \ell_1 \longrightarrow \ell_1(I_\alpha)$ be the obvious isomorphism obtained from an enumeration of I_α. Since $|I| > |\operatorname{Ext}(\ell_1, \mathbb{K})|$, there necessarily exist $\alpha \neq \beta$ such that $\Omega \circ \sigma_\alpha \sim \Omega \circ \sigma_\beta$, hence Ω is trivial on $\overline{(\sigma_\alpha - \sigma_\beta)[\ell_1]} = \overline{[\sigma_\alpha(e_n) - \sigma_\beta(e_n)]} \approx \ell_1$. $\square$

One actually has $\dim \operatorname{Ext}(\ell_1, \mathbb{K}) \geq \mathfrak{c}$ since the existence of at least $\mathfrak{c}$ non-projectively equivalent elements in $\operatorname{Ext}(\ell_1, \mathbb{K})$ follows from $\operatorname{Ext}(\ell_1, \mathbb{K}) \neq 0$ and $\ell_1(\ell_1) = \ell_1$. Moreover, a dexterous swirling reveals that if $\varrho_\varphi\colon \ell_1^0 \longrightarrow \mathbb{R}$ are the Ribe-like quasilinear maps generated by a Lipschitz map φ as in Corollary 4.5.9 then, whenever θ, ϕ are Lipschitz functions such that $\liminf_{t\to\infty} \theta(t)/\phi(t) = 0$, the exact sequence $0 \longrightarrow \mathbb{R} \longrightarrow Z(\varrho_\phi) \longrightarrow \ell_1 \longrightarrow 0$ cannot be obtained via pullback from $0 \longrightarrow \mathbb{R} \longrightarrow Z(\varrho_\theta) \longrightarrow \ell_1 \longrightarrow 0$. This implies in particular that $Z(\varrho_\theta)$ and $Z(\varrho_\phi)$ cannot be isomorphic [56, Theorem 2].

9.2 Singular Quasilinear Maps

A sequence of blocks of a Schauder basis in a quasi-Banach space will be called a *block sequence*, and the subspace spanned by it will be referred to as a *block subspace.*

Lemma 9.2.1 *Let Z be a quasi-Banach space and X be a quasi-Banach space with a Schauder basis. A quotient map $\rho\colon Z \longrightarrow X$ is either strictly singular or invertible on some infinite-dimensional block subspace of X.*

Proof There is no loss of generality in assuming that Z is p-normed for some $p \in (0, 1]$ and ρ is an isometric quotient, so that X is also p-normed. Assume ρ is not strictly singular. Then there is an infinite-dimensional subspace $H \subset X$ and an operator $s\colon H \longrightarrow Z$ such that ρs is the inclusion of H in X. A standard procedure allows one to find a normalised sequence (h_n) in H with $h_n(i) = 0$ for $i < n$ ([81, Proof of Corollary 5.3] or [334, Proof of Proposition 1.a.11]), and an even more standard gliding hump argument yields a subsequence (we don't relabel) and a block sequence (u_n) in X such that $\|h_n - u_n\| < \varepsilon^n$. Since (u_n) is a basic sequence in X, by eliminating finitely many terms, we have that (h_n) is also a basic sequence, by the basic sequence criterion. Let U and H' be the subspaces of X spanned by (u_n) and (h_n), respectively. The operator $v\colon U \longrightarrow H$ defined by $v(u_n) = h_n$ is an isomorphism: it suffices to check that given a sequence (t_n), which can be assumed to be bounded, the series $\sum_n t_n h_n$ converges if and only if the series $\sum_n t_n u_n$ does as well. This is due to the fact that $\sum_n t_n h_n - \sum_n t_n u_n = \sum_n t_n(h_n - u_n)$, and this last series converges since

$$\sum_n \|t_n(h_n - u_n)\|_X^p \le \sum_n |t_n|^p \|h_n - u_n\|_X^p \le \|(t_n)\|_\infty^p \frac{\varepsilon^p}{1-\varepsilon^p}.$$

Next we construct an operator $S : U \longrightarrow Z$ such that $\rho S = \mathbf{1}_U$. For each n, choose z_n in Z such that $\rho(z_n) = u_n$ and $\|z_n - s(h_n)\| < \varepsilon^n$ and define $S : U^0 \longrightarrow Z$ as follows: given $u = \sum_n t_n u_n$, set $S(u) = \sum_n t_n z_n$. To compute $\|S(u)\|$, take $h = \sum_n t_n u_n = v(u)$ to obtain

$$\|S u\|^p \le \|s(h)\|^p + \|S(u) - s(h)\|^p \le \|s\|^p \|h\|^p + \left\|\sum_n t_n(z_n - s(h_n))\right\|^p$$

$$\le \|s\|^p \|v\|^p \|u\|^p + \sum_n |t_n|^p \varepsilon^{pn} \le \left(\|s\|^p \|v\|^p + \frac{\varepsilon^p}{1-\varepsilon^p}\right) \|u\|^p. \qquad \square$$

And this is how we find what we were looking for:

9.2.2 Transfer principle *Let X be a quasi-Banach space with a Schauder basis, let X^0 be the subspace of finite linear combinations of the basis and let*

$\Omega\colon X^0 \longrightarrow Y$ be a quasilinear map. Then either Ω is singular or there is a block subspace of X^0 on which Ω is trivial.

9.3 Amalgamation Techniques

The idea behind amalgamation techniques is to pass, under certain conditions, from the existence of a non-trivial sequence $0 \longrightarrow Y \longrightarrow \cdot \longrightarrow X \longrightarrow 0$ to the existence of a singular sequence $0 \longrightarrow \ell_\infty(I, Y) \longrightarrow \cdot \longrightarrow X \longrightarrow 0$ for I large enough. That done, there is the further problem of shrinking the range from $\ell_\infty(I, Y)$ down to Y. This can be handled, under additional conditions.

Lemma 9.3.1 *Let X, Y be p-Banach spaces. If there exists a non-trivial p-linear map $\Omega\colon X \longrightarrow Y$ then for every $C \geq 1$, there exists an index set I and a p-linear map $\Omega_\infty\colon X \longrightarrow \ell_\infty(I, Y)$ whose restriction to any subspace C-isomorphic to X is non-trivial.*

Proof Let I be the set formed by all the infinite-dimensional C-complemented subspaces of X that are C-isomorphic to X. Given $E \in I$, let $\gamma_E\colon E \longrightarrow X$ be a C-isomorphism, and let $P_E\colon X \longrightarrow E$ be a C-projection. Given a non-trivial p-linear map $\Omega\colon X \longrightarrow Y$, we define $\Lambda\colon X \longrightarrow Y^I$ by $\Lambda(x) = (\Omega \circ \gamma_E P_E(x))_{E\in I}$. This map is p-linear from X to $\ell_\infty(I, Y)$, with $Q^{(p)}(\Lambda) \leq C^2 Q^{(p)}(\Omega)$, since for finite sums, one has

$$\sup_{E\in I}\left\|\Omega \circ \gamma_E P_E\left(\sum x_i\right) - \sum_i \Omega \circ \gamma_E P_E(x_i)\right\| \leq C^2 Q^{(p)}(\Omega)\left(\sum_i \|x_i\|^p\right)^{1/p},$$

which moreover guarantees that the Cauchy differences of Λ fall in $\ell_\infty(I, Y)$. Therefore there exists a linear map $L\colon X \longrightarrow Y^I$ such that $\Omega_\infty = \Lambda - L\colon X \longrightarrow \ell_\infty(I, Y)$ is a p-linear map. This Ω_∞ cannot be trivial on any C-copy of X (which, in particular, means that it is not trivial). For assume otherwise that there is some $E \in I$ such that the restriction $\Omega_\infty|_E$ is trivial. Let $\pi_E\colon \ell_\infty(I, Y) \longrightarrow Y$ be the evaluation functional at the coordinate E. One has

$$\pi_E \circ \Omega_\infty|_E = \Omega \circ \gamma_E + \pi_E L|_E,$$

This map $E \longrightarrow Y$ is trivial if and only if $\Omega \circ \gamma_E$ is trivial, which occurs if and only if Ω is trivial since γ_E is an isomorphism. □

Since every quasi-Banach space is p-Banach for some p (see 1.1.1) and every quasilinear map is r-linear for some r (Corollary 3.7.7), the result remains true replacing 'p-Banach' and 'p-linear' by, respectively, 'quasi-Banach' and 'quasilinear'. Moreover, the special structure of $C(\Delta)$ allows one to reduce the range of Ω_∞.

Corollary 9.3.2 *If there exists a non-trivial quasilinear map $X \longrightarrow C(\Delta)$ then there exists, for each $C > 1$, a quasilinear map $X \longrightarrow C(\Delta)$ that is not trivial on any subspace C-isomorphic to X. If X is p-Banach then 'quasilinear' can be replaced by 'p-linear'.*

Proof By Lemma 3.9.4, the map $\Omega_\infty \colon X \longrightarrow \ell_\infty(I, C(\Delta))$ of Lemma 9.3.1 admits a version having separable range, and since every separable subspace of $\ell_\infty(I, C(\Delta))$ is contained in a copy of $C(\Delta)$, that version provides a quasilinear map $X \longrightarrow C(\Delta)$. This new map cannot be trivial on any C-isomorphic copy of X since it is a 'version' of Ω_∞. □

Seasons can turn on a dime, and when combined with Proposition 1.2.5, these barely useful results yield:

Proposition 9.3.3 *Let X be a minimal Banach space and let Y be a CK)-space. If there exists a non-trivial quasilinear map $\Omega \colon X \longrightarrow Y$ then there exists a singular quasilinear map $\Omega_\infty \colon X \longrightarrow C(\Delta)$. If Y is finite-dimensional then there exists a singular quasilinear map $\Omega_\infty : X \longrightarrow \ell_\infty$.*

Proof Proposition 1.2.5 shows that minimality implies C-minimality for some $C > 1$, thus the map in Lemma 9.3.1 becomes singular $X \longrightarrow \ell_\infty(I, Y)$. Since a minimal space must be separable, by Lemma 3.9.4, one can assume without loss of generality that Y is separable too, thus the reduction of the range to either $C(\Delta)$ or ℓ_∞ (depending on the hypothesis on Y) can be performed without any trouble. □

Two especially artful ways of using this proposition are:

Corollary 9.3.4 *There exist singular quasilinear maps $\ell_1 \longrightarrow \ell_\infty$ and $\ell_1 \longrightarrow C(\Delta)$.*

Corollary 9.3.5 *There exist exact sequences $0 \longrightarrow C(\Delta) \longrightarrow \cdot \longrightarrow c_0 \longrightarrow 0$ in which no subsequence of the canonical basis of c_0 admits a weakly Cauchy lifting.*

Proof For each subsequence $M = (n_m)_m$ of $\mathbb{N}$, let $\tau_M \colon M \longrightarrow \mathbb{N}$ be the canonical counting $\tau_M(n_m) = m$ and let $P_M : c_0(\mathbb{N}) \longrightarrow c_0(M)$ be the canonical projection. Consider the Foiaş–Singer sequence (2.5) $0 \longrightarrow C(\Delta) \longrightarrow D(\Delta; \Delta_0) \longrightarrow c_0 \longrightarrow 0$ and recall from Lemma 2.2.4 that since Δ_0 is dense in Δ, the canonical basis of c_0 does not admit a weakly Cauchy lifting. Let Ω be a 1-linear map associated to the Foiaş–Singer sequence and form the map $\Omega_\infty \colon c_0 \longrightarrow \ell_\infty(\mathscr{P}_\infty(\mathbb{N}), C(\Delta))$ given by

$$\Omega_\infty(x)(M) = \Omega \circ \tau_M P_M(x)$$

up to a linear map. If some $(e_m)_{m\in M}$ admits a weakly Cauchy lifting then the canonical basis of $c_0(M)$ admits a weakly Cauchy lifting too, but $\pi_M \circ \Omega_\infty|_{c_0(M)} = \Omega \circ \tau_M$ is 'isomorphic' to Ω, which makes that impossible. Now perform the standard reduction of the range to get a map $c_0 \longrightarrow C(\Delta)$. □

Definition 9.3.6 An element $\Upsilon \in \operatorname{Ext}(X, \diamond)$ is said to be initial for $E \subset \operatorname{Ext}(X, Y)$ if every element of E is a pushout of it. If, moreover, $\Upsilon \in E$ then we say that E has an initial element.

If $\operatorname{Ext}(X, Y)$ has an initial element and a singular element then the initial element is itself singular. With this in mind, we present a variation of the amalgamation technique.

Lemma 9.3.7 *Given p-Banach spaces X, Y, there is an index set I and a p-linear map $\Upsilon_X : X \longrightarrow \ell_\infty(I, Y)$ that is initial for* $\operatorname{Ext}_{p\mathbf{B}}(X, Y)$.

Proof Let $0 \longrightarrow \kappa(X) \longrightarrow Z(\Lambda) \longrightarrow X \longrightarrow 0$ be a projective presentation for X in $p\mathbf{B}$. Let I be the unit ball of $\mathfrak{L}(\kappa(X), Y)$, and form the operator $\Phi : \kappa(X) \longrightarrow \ell_\infty(I, Y)$ defined by $\Phi(k)(\tau) = \tau(k)$. Let us check that $\Upsilon_X = \Phi \circ \Lambda$ is initial. Since Λ is projective, given any p-linear map $\Omega : X \longrightarrow Y$, there is an operator ϕ such that $\phi \circ \Lambda \sim \Omega$. Since $\phi/\|\phi\|$ factorises through Φ in the form $\phi/\|\phi\| = \pi_{\phi/\|\phi\|}\Phi$, where $\pi_\eta : \ell_\infty(I, Y) \longrightarrow Y$ is the natural projection onto the η-coordinate then $\Omega \sim \phi \circ \Lambda = \|\phi\|\pi_{\phi/\|\phi\|}\Phi \circ \Lambda = \|\phi\|\pi_{\phi/\|\phi\|} \circ \Upsilon_X$, and Ω is a pushout of Υ_X □

The same reduction of range performed for Ω_∞ can be done for Υ_X:

Corollary 9.3.8 *For every separable quasi-Banach space X, there is an initial element in* $\operatorname{Ext}(X, C(\Delta))$. *If X is p-Banach then there is an initial element in* $\operatorname{Ext}_{p\mathbf{B}}(X, C(\Delta))$.

We are ready to present a tangible example of a singular sequence:

Theorem 9.3.9 *Let X be a Banach space that does not contain ℓ_1. Every initial element of* $\operatorname{Ext}_{\mathbf{B}}(X, C(\Delta))$ *is singular.*

Proof Let Υ_X be the initial element of $\operatorname{Ext}_{\mathbf{B}}(X, C(\Delta))$ constructed in Lemma 9.3.7 and whose range has been reduced according to Corollary 9.3.8. Let E be an infinite-dimensional closed subspace of X. Since E cannot contain ℓ_1, it contains a weakly null basic sequence $(x_n)_n$ by [334, p. 5, Remark]. Let $E_0 = \overline{[x_n : n \in N]}$ be the subspace it spans and let $\jmath : E_0 \longrightarrow X$ be the canonical inclusion. By considering the basis expansion, it is easy to construct a linear continuous map $\tau_0 : E_0 \longrightarrow c_0$ such that $\tau_0(x_n) = e_n$. Since c_0 is separably injective, the operator τ_0 extends to an operator $\tau\colon X \longrightarrow c_0$. Pick the Foiaş–Singer sequence Ω and form the pullback diagram

$$\begin{array}{ccccccccc}
0 & \longrightarrow & C(\Delta) & \longrightarrow & Z(\Omega) & \longrightarrow & c_0 & \longrightarrow & 0\\
 & & \| & & \uparrow{\scriptstyle \underline{\tau}} & & \uparrow{\scriptstyle \tau} & & \\
0 & \longrightarrow & C(\Delta) & \longrightarrow & Z(\Omega\circ\tau) & \longrightarrow & X & \longrightarrow & 0\\
 & & \| & & \uparrow & & \uparrow{\scriptstyle \jmath} & & \\
0 & \longrightarrow & C(\Delta) & \longrightarrow & Z(\Omega\circ\tau\jmath) & \longrightarrow & E_0 & \longrightarrow & 0
\end{array}$$

The lower sequence $\Omega\circ\tau\jmath$ cannot split since otherwise there would be a linear continuous lifting $T\colon E_0 \longrightarrow Z(\Omega)$ of $\tau\jmath = \tau_0$ which would provide a weakly 1-summable lifting of (e_n), which is impossible by Lemma 2.2.4. The proof of Lemma 9.3.7 yields $\Omega\circ\tau \sim \phi\circ\Upsilon_X$ for some operator ϕ. Hence $\Omega\circ\tau\jmath \sim \phi\circ\Upsilon_X\circ\jmath$ does not split, thus $\Upsilon_X\circ\jmath$ cannot split either, which shows that Υ_X is singular. □

Singularity (or Not) of Kalton–Peck Maps

In this section, X is a sequence space and $\mathsf{KP}_X\colon X^0 \longrightarrow X$ is the Kalton–Peck map

$$\mathsf{KP}_X(x) = \sum_n x_n \log\frac{\|x\|}{|x_n|} e_n.$$

If $(u_n)_{n\geq 1}$ is a normalised sequence of disjointly supported elements of X, its closed linear span U is also a sequence subspace with basis (u_n), and we can consider *its own* Kalton–Peck map

$$\mathsf{KP}_U(u) = \sum_n \lambda_n \log\frac{\|u\|}{|\lambda_n|} u_n$$

defined for $u = \sum_n \lambda_n u_n$ in U^0, the subspace of all finite linear combinations of the basis (u_n). The next result uncovers the surprising similarities between the two maps.

Lemma 9.3.10 *There is a linear map $L\colon U^0 \longrightarrow X$ such that $\mathsf{KP}_X|_{U^0} = \mathsf{KP}_U + L$.*

Proof We define a linear map $L\colon U^0 \longrightarrow X$ by $L(u_n) = -u_n \log|u_n|$ and linearly on the rest. Let $u = \sum_n \lambda_n u_n$ be normalised in U^0. We have

$$\begin{aligned}\mathsf{KP}_X(u) &= -u\log|u| = -\sum_n \lambda_n u_n \log(|\lambda_n||u_n|)\\ &= -\left(\sum_n \lambda_n u_n \log|\lambda_n|\right) - \left(\sum_n \lambda_n u_n \log|u_n|\right) = \mathsf{KP}_U(u) + L(u),\end{aligned}$$

and the result follows from homogeneity. □

What the lemma asserts is that whenever U is a block subspace of X there is a commutative diagram

$$\begin{array}{ccccccccc} 0 & \longrightarrow & X & \longrightarrow & X \oplus_{\mathsf{KP}_X} X & \longrightarrow & X & \longrightarrow & 0\\ & & \uparrow{\scriptstyle \text{inclusion}} & & \uparrow{\scriptstyle T_U} & & \uparrow{\scriptstyle \text{inclusion}} & & \\ 0 & \longrightarrow & U & \longrightarrow & U \oplus_{\mathsf{KP}_U} U & \longrightarrow & U & \longrightarrow & 0 \end{array} \tag{9.1}$$

in which the three vertical maps are isometries: the inclusions obviously are, while $T_U(v, u) = (v + L(u), u)$ for $u \in U^0$ and thus

$$\|(v + Lu, u)\|_{\mathsf{KP}_X} = \|y + Lu - \mathsf{KP}_X(u)\| + \|u\| = \|y - \mathsf{KP}_U(u)\| + \|u\| = \|(v, u)\|_{\mathsf{KP}_U}.$$

The goal now is to show that KP_X is singular on a wide range of sequence spaces, in particular the *complementably block-saturated* sequence spaces, i.e. those spaces X with a Schauder basis in which every infinite-dimensional block subspace contains a block subspace that is complemented in X.

Theorem 9.3.11 *The map KP_X is singular when X is either*

(a) *a complementably block-saturated sequence space not containing c_0,*
(b) *ℓ_p for $0 < p < \infty$,*
(c) *L_p for $p \in [2, \infty)$ when regarded as a sequence space through the Haar system.*

Proof (a) If not, the transfer principle 9.2.2 implies that KP_X must be trivial on some block subspace U which we may assume to be complemented in X by a bounded projection $P\colon X \longrightarrow U$. Hence there is a linear map $\ell\colon U^0 \longrightarrow X$ with $\mathsf{KP}_X - \ell$ bounded. Let $L\colon U^0 \longrightarrow X$ be the linear map given by Lemma 9.3.10 such that $\mathsf{KP}_X = \mathsf{KP}_U + L$. Since $\mathsf{KP}_U + L - \ell$ is bounded, so is the composition $P \circ (\mathsf{KP}_U + L - \ell) = \mathsf{KP}_U + P \circ (L - \ell)$. Since $P \circ (L - \ell)\colon U \longrightarrow X$ is linear, this means that $\mathsf{KP}_U\colon U \longrightarrow U$ is trivial, contradicting Proposition 3.3.5.

(b) Every normalised block sequence in ℓ_p is equivalent to the unit basis of ℓ_p itself. Thus, if U is a block subspace of ℓ_p then $\ell_p \oplus_{\mathsf{KP}} U$ cannot be isomorphic to the product $\ell_p \times U$ by Proposition 3.2.7.

(c) Let $p \in [2, \infty)$. By the Kadec–Pełczyński dichotomy [241], every normalised weakly null sequence in L_p contains a subsequence equivalent to the unit basis of ℓ_r, where r is either p or 2. Hence it suffices to check that $\mathsf{KP}: U_0 \longrightarrow L_p$ is not trivial when U is a block subspace spanned by a sequence equivalent to the unit basis of ℓ_r for $r = p, 2$. Clearly, $L_p \oplus_{\mathsf{KP}|_U} U$ has a subspace isomorphic to Z_r. But Z_r does not embed into L_p: for Z_2, this is because L_p has type 2 when $2 \leq p$, and thus every copy of ℓ_2 is complemented, and for Z_p, it is because it fails to have cotype p. □

Complementably minimal sequence spaces provide examples of complementably block-saturated spaces, although there are many others: Tsirelson's space T and its dual T^* [85, Corollary 9 and Remark (2) in p. 93] or vector sums $\ell_p(\ell_q)$ for $p, q \in [1, \infty)$. The product of two complementably block-saturated spaces is also complementably block saturated.

On function spaces, however, things are different:

Proposition 9.3.12 KP_φ *is not singular on* L_p *for* $0 < p < \infty$.

Proof The proof is written assuming real scalars. The adaptation to complex scalars requires only minor changes. We switch to random mode, exploiting the otherwise obvious fact that the distribution of $\mathsf{KP}_\varphi f$ depends only on that of f. The idea is to find a subspace G of L_p where all functions with the same quasinorm have the same distribution. If $\mathsf{KP}_\varphi(g) \in L_p$ for some non-zero $g \in G$, then KP_φ has to be bounded from G to L_p. The following estimate shows that Gaussian variables work for this purpose.

Claim Let g be a standard Gaussian variable on a probability space (S, P), and let $\varphi \in \mathrm{Lip}(\mathbb{R})$, not necessarily vanishing at zero. Then, for every $c > 0$ and every $0 < p < \infty$, the random variable $g\,\varphi(\log(c/|g|))$ belongs to $L_p(S, P)$.

Proof of the claim For every $c > 0$ and each finite p, we have

$$\mathbb{E}\left[\left|g\,\varphi(\log(c/|g|))\right|^p\right] = \frac{1}{\sqrt{2\pi}} \int_{-\infty}^{\infty} \left|t\,\varphi(\log(c/|t|))\right|^p e^{-t^2/2} dt < \infty. \qquad \square$$

Now, fix p and $\varphi \in \mathrm{Lip}(\mathbb{R})$ and set

$$\alpha = \left(\mathbb{E}\left[|g|^p\right]\right)^{1/p}, \qquad \beta = \left(\mathbb{E}\left[\left|g\,\varphi(\log(\alpha/|g|))\right|^p\right]\right)^{1/p},$$

where g is a Gaussian variable. Let $(g_n)_{n\geq 1}$ be a sequence of independent Gaussian variables in L_p, and let G be the subspace they span. If $(c_i)_{i\leq n}$ are scalars, then $g = \sum_{i\leq n} c_i g_i$ is normal with mean zero and standard deviation $\sigma = \left(\sum_{i\leq n} c_i^2\right)^{1/2}$, and so $\|g\|_p = \alpha\,\sigma$. Since

$$\mathsf{KP}_\varphi\left(\frac{g}{\sigma}\right) = \frac{g}{\sigma}\varphi\left(\log\left(\frac{\alpha}{g/\sigma}\right)\right)$$

and g/σ is Gaussian, we have

$$\|\mathsf{KP}_\varphi(g/\sigma)\|_p = \beta \quad\Longrightarrow\quad \|\mathsf{KP}_\varphi(g)\|_p = \beta\sigma = \frac{\beta}{\alpha}\|g\|_p.$$

Thus, the 'obvious' operator $g \in G \longmapsto (0, g) \in L_p(\varphi)$ is a section of the quotient map $L_p(\varphi) \longrightarrow L_p$ on G. □

It is an open problem posed in [62] whether there exists a singular quasilinear map $\Omega\colon L_p \longrightarrow L_p$ for $0 < p < 2$. The behaviour of the KP maps looks so different in sequence and function spaces because ... it is actually the same! In all cases, KP maps are disjointly singular, but while disjointly singular maps are singular on sequence spaces, they are not necessarily singular on function spaces. Note 9.4.2 will dig into this.

Singular Twisted Sums with $C(\omega^\omega)$

In principle, the reduction of the range of either Ω_∞ (Lemma 9.3.1) or Υ_X (Lemma 9.3.7) performed when the target space is $C(\Delta)$ and leading to Corollaries 9.3.2 and 9.3.8, cannot be done for other $\mathscr{C}$-spaces. For instance, singular sequences $0 \longrightarrow c_0 \longrightarrow \cdot \longrightarrow c_0 \longrightarrow 0$ do not exist even if singular sequences $0 \longrightarrow C(\Delta) \longrightarrow \cdot \longrightarrow c_0 \longrightarrow 0$ exist. What occurs for other $\mathscr{C}$-spaces between c_0 and $C(\Delta)$? As we mentioned after Proposition 8.6.4, these are the spaces $C(\alpha)$ for countable ordinal α, among which the simplest non-trivial example is $C(\omega^\omega)$. We have already seen that the cases of $C(\Delta)$ and $C(\omega^\omega)$ are different regarding the mere existence of non-trivial twistings: $\operatorname{Ext}(X, C(\Delta)) \neq 0$ whenever X does not have the Schur property (Proposition 8.6.4), while the additional condition that some spreading model of X is not isomorphic to ℓ_1 is necessary to get $\operatorname{Ext}(X, C(\omega^\omega)) \neq 0$ (Proposition 8.6.5). It is to be expected that something similar occurs regarding singularity: while the existence of a singular map $\Omega\colon X \longrightarrow C(\Delta)$ requires only the minimum input (X does not contain ℓ_1, Theorem 9.3.9), the existence of a singular map $\Omega\colon X \longrightarrow C(\omega^\omega)$ has to be taken by assault, and that is precisely the purpose of this section.

Our battering rams are the equivalence of the parameters $\rho_N(\cdot)$ and $\alpha_N(\cdot)$ (see equations (8.33) and (8.34)) plus the fact that $\operatorname{Ext}(X, C(\omega^\omega)) \neq 0$ if and only if $\rho_N(X) \longrightarrow \infty$, something that, when X has a shrinking UFDD, happens if and only if $\alpha_N(X) \longrightarrow \infty$. The existence of operators $T_N : X \longrightarrow \ell_\infty(\omega^N)$ with small oscillation whose distance to $\mathfrak{L}(X, C(\omega^N))$ tends to infinity is thus enough to guarantee $\operatorname{Ext}(X, C(\omega^\omega)) \neq 0$ but not the existence of a singular element, and the example $T_N \oplus 0\colon c_0 \times c_0 \longrightarrow \ell_\infty(\omega^N)$ shows why. To overrun the walls of singularity, we need an operator which is 'everywhere like T_N'.

Definition 9.3.13 A map $\curlywedge : \omega^N \longrightarrow X^*$ is a dense tree map if

- $\curlywedge(\emptyset) = 0$ and $\| \curlywedge (a)\| \leq 1$ for all $a \in \omega^N$,
- for each $|a| < N$, there is a weak*-neighbourhood V_a of 0 such that $\{\curlywedge(b) : b \in a^+\}$ is weak*-dense in $V_a \cap B_X^*$,
- if $b_n \to a$ and $|b_n| \geq |a| + 2$ then $\curlywedge(b_n) \to 0$ weak*.

The operator $\tau_\curlywedge : X \longrightarrow \ell_\infty(\omega^N)$ tree generated by a dense tree map $\curlywedge$ will be called a *dense-type operator*.

Dense tree maps exist: let (V_n) be a countable base of weak* neighbourhoods of 0 in X^* such that $V_{n+1} + V_{n+1} \subset V_n$. For each a with $|a| < N$, choose a countable weak* dense subset (d_j) of $V_{\max a} \cap B_X^*$, and if $a = (a_1, \ldots, a_n)$ and $(a_1, \ldots, a_n, m) \in a^+$ then set $\curlywedge(a_1, \ldots, a_n, m) = d_m$. Our assault ladder is:

Lemma 9.3.14 *If X has UFDD and $\tau_\curlywedge$ is a dense-type operator then $\rho_{2N}(X) \leq 4\|\tau_\curlywedge - L\|^+$ for every linear $L : X \longrightarrow C(\omega^N)$.*

Proof Fix $\varepsilon > 0$. For some weak*-null tree map $\Upsilon : \omega^N \longrightarrow B_X^*$, $|a| = N$, $x \in B_X^*$ and $a \in \omega^N$, we have

$$\rho_{2N}(X) \leq 4\alpha_N(X) \leq 4\|\sum_{b \leq a} \Upsilon(b)\| + \varepsilon \leq 4\left\langle \sum_{b \leq a} \Upsilon(b), x\right\rangle + 2\varepsilon.$$

'Spread' Υ using an onto map $\sigma : \mathbb{N} \longrightarrow \mathbb{N}$ such that each $\sigma^{-1}(n)$ is infinite, say $1, 2, 1, 2, 3, 1, 2, 3, 4 \ldots$ as in the second part of the proof of Lemma 8.6.11, forming $\Upsilon\sigma$. It turns out that $\Upsilon\sigma(a) = \Upsilon\sigma(a')$ for a countable cofinal set of a'.

Claim There is $\psi : \omega^N \longrightarrow \omega^N$ such that $\tau_{\Upsilon\sigma} - \tau_{\curlywedge\psi} : X \longrightarrow \ell_\infty(\omega^N)$ is weak*-continuous and $(\Upsilon\sigma - \curlywedge\psi)(\emptyset) = 0$.

Proof of the claim We build the map ψ inductively. First $\psi(\emptyset) = \emptyset$. Once $\psi(a)$ has been defined and $|a| < N$ then set $\psi(b) \in \psi(a)^+$ for each $b \in a^+$ such that ψ is injective (impossible to miss) and $\lim_{b \in a^+} \Upsilon\sigma(b) - \curlywedge\psi(b) = 0$ in the weak*-topology. We want to check that $\lim_{b \in a^+} \tau^*_{\Upsilon\sigma}(b) - \tau^*_{\curlywedge\psi}(b) = 0$ in the weak*-topology which, once again, is similar to what we did during the second part of the proof of Lemma 8.6.11: if $b \geq a$ then

$$(\tau^*_{\Upsilon\sigma}(b) - \tau^*_{\Upsilon\sigma}(a)) - \left(\tau^*_{\curlywedge\psi}(b) - \tau^*_{\curlywedge\psi}(a)\right) = \sum_{a < c \leq b} (\Upsilon\sigma(c) - \curlywedge\psi(c)).$$

Thus, if $b_n \longrightarrow a$, and d_n is chosen so that $b_n \leq d_n \leq a$ and $|d_n| = |a| + 1$ then

$$\lim_{n \to \infty} \sum_{d_n < c < b} (\Upsilon\sigma(c) - \curlywedge\psi(c)) = 0 = \lim (\Upsilon\sigma(d_n) - \curlywedge\psi(d_n))$$

in the weak*-topology because of the assumptions on both maps. □

Let $L\colon X \longrightarrow C(\omega^N)$ be a linear map with $L^*(\emptyset) = 0$. The map $\tau_{\Upsilon\sigma} - \tau_{\curlywedge\psi} - L\psi$ is also weak* continuous. Given a and x, pick a' large enough so that $|(\tau_{\Upsilon\sigma} - \tau_{\curlywedge\psi} - L\psi))(x)(a')| \leq \varepsilon$ and $\tau_{\Upsilon\sigma}(x)(a) = \tau_{\Upsilon\sigma}(x)(a')$ to get

$$\begin{aligned}
\left\langle \sum\nolimits_{b\leq a} \Upsilon(b), x\right\rangle &= |\tau_{\Upsilon}(x)(a)| = |\tau_{\Upsilon\sigma}(x)(a)| \\
&= |\tau_{\Upsilon\sigma}(x)(a) - (\tau_{\Upsilon\sigma} - \tau_{\curlywedge\psi} - L\psi)(x)(\emptyset)| \\
&\leq |\tau_{\Upsilon\sigma}(x)(a') - (\tau_{\Upsilon\sigma} - \tau_{\curlywedge\psi} - L\psi)(x)(a')| + \varepsilon \\
&= |(\tau_{\curlywedge\psi} - L\psi)(x)(a')| + \varepsilon \\
&\leq \|(\tau_{\curlywedge\psi} - L\psi)(x)\| + \varepsilon \\
&\leq \|\tau_{\curlywedge\psi} - L\psi\| + \varepsilon \\
&\leq \|\tau_{\curlywedge} - L\| + \varepsilon.
\end{aligned}$$

□

Our mole inside the castle informs us that when $\tau_{\curlywedge} : X \longrightarrow \ell_\infty(\omega^N)$ is a dense-type operator then its restriction to any subspace $E \subset X$ is of dense type on E just because, calling $\jmath_E\colon E \longrightarrow X$ the inclusion, $\jmath_E^*\curlywedge\colon \omega^N \longrightarrow E^*$ is still a dense tree map. With these tools in hand, have fun storming the castle!

Proposition 9.3.15 *Let X be a separable Banach space with a monotone shrinking UFDD such that* $\operatorname{Ext}(X, C(\omega^\omega)) \neq 0$ *and such that every closed infinite-dimensional subspace contains a further subspace E with monotone shrinking UFDD such that* $\operatorname{Ext}(E, C(\omega^\omega)) \neq 0$. *Then there is a singular element in* $\operatorname{Ext}(X, C(\omega^\omega))$.

Proof For each N, pick a dense-type operator $T_N\colon X \longrightarrow \ell_\infty(\omega^N)$ and form the space $Z[T_N]$ as in the final part of the proof of Lemma 8.6.7, with its associated isometrically exact sequence $0 \longrightarrow C(\omega^N) \longrightarrow Z[T_N] \longrightarrow X \longrightarrow 0$. Pick a projective presentation of X and form the pushout diagram

$$\begin{array}{ccccccccc}
0 & \longrightarrow & \kappa(X) & \longrightarrow & \ell_1 & \xrightarrow{Q} & X & \longrightarrow & 0 \\
 & & \downarrow{\scriptstyle \phi_N} & & \downarrow{\scriptstyle \overline{\phi_N}} & & \| & & \\
0 & \longrightarrow & C(\omega^N) & \longrightarrow & Z[T_N] & \longrightarrow & X & \longrightarrow & 0
\end{array}$$

with some operator ϕ_N that we assume is normalised. Let (F_N) be an increasing sequence of finite-dimensional subspaces of $\kappa(X)$ such that $\bigcup F_N$ is dense in $\kappa(X)$. Since $C(\omega^N)$ is an $\mathscr{L}_{\infty,1+}$-space, there are finite-rank projections P_N on $C(\omega^N)$ whose range includes $\phi_N(F_N)$ and with $\|P_N\| \leq 1 + \varepsilon$. The sequence of operators $\psi_N = \phi_N - P_N\phi_N$ is uniformly bounded ($\|\psi_N\| \leq 2 + \varepsilon$) and satisfies $\lim_N \|\psi_N x\| = 0$ for all $x \in \kappa(X)$. This allows us to define a norm $2 + \varepsilon$ operator $\psi\colon \kappa(X) \longrightarrow c_0(C(\omega^N))$ and form a pushout diagram

$$\begin{array}{ccccccccc}
0 & \longrightarrow & \kappa(X) & \longrightarrow & \ell_1 & \longrightarrow & X & \longrightarrow & 0\\
 & & \downarrow{\scriptstyle \psi} & & \downarrow{\scriptstyle \overline{\psi}} & & \| & & \\
0 & \longrightarrow & c_0(\mathbb{N}, C(\omega^N)) & \longrightarrow & \mathrm{PO} & \xrightarrow{\pi} & X & \longrightarrow & 0
\end{array}$$

We claim that π is strictly singular. Otherwise, there is a subspace $E \subset X$ such that the lower pullback sequence

$$\begin{array}{ccccccccc}
0 & \longrightarrow & c_0(\mathbb{N}, C(\omega^N)) & \longrightarrow & \mathrm{PO} & \xrightarrow{\pi} & X & \longrightarrow & 0\\
 & & \| & & \uparrow & & \uparrow{\scriptstyle J_E} & & \\
0 & \longrightarrow & c_0(\mathbb{N}, C(\omega^N)) & \longrightarrow & \mathrm{PB(PO)} & \longrightarrow & E & \longrightarrow & 0
\end{array}$$

splits. Since pullback and pushout commute, the lower sequence in the diagram

$$\begin{array}{ccccccccc}
0 & \longrightarrow & \kappa(X) & \longrightarrow & \ell_1 & \xrightarrow{Q} & X & \longrightarrow & 0\\
 & & \| & & \uparrow & & \uparrow{\scriptstyle J_E} & & \\
0 & \longrightarrow & \kappa(X) & \longrightarrow & \mathrm{PB} & \longrightarrow & E & \longrightarrow & 0\\
 & & {\scriptstyle \psi}\downarrow & & \downarrow & & \| & & \\
0 & \longrightarrow & c_0(\mathbb{N}, C(\omega^N)) & \longrightarrow & \mathrm{PB(PO)} & \longrightarrow & E & \longrightarrow & 0
\end{array}$$

splits. Thus, $\psi = (\psi_N)$ can be extended to an operator $\Psi\colon \mathrm{PB} \longrightarrow c_0(\mathbb{N}, C(\omega^N))$, which provides a uniformly bounded (by $\|\Psi\|$) sequence $(\Psi_N)_{N\in\mathbb{N}}$ of extensions of the operators $(\psi_N)_{N\in\mathbb{N}}$. But, like all $\mathscr{C}$-valued finite-rank operators, $P_N\phi_N\colon \kappa(X) \longrightarrow C(\omega^N)$ also admits 1^+-extensions (wherever). Therefore, $\phi_N = \psi_N + P_N\phi_N$ also admits $(\|\Psi\| + 1)^+$-extensions $\Phi_N\colon \mathrm{PB} \longrightarrow C(\omega^N)$. Combine the upper pullback diagram with the projective presentation to get

$$\begin{array}{ccccccccc}
0 & \longrightarrow & \kappa(X) & \longrightarrow & \mathrm{PB} & \xrightarrow{\underline{Q}} & E & \longrightarrow & 0\\
 & & \| & & \downarrow{\scriptstyle \underline{J_E}} & & \downarrow{\scriptstyle J_E} & & \\
0 & \longrightarrow & \kappa(X) & \longrightarrow & \ell_1 & \xrightarrow{Q} & X & \longrightarrow & 0\\
 & & {\scriptstyle \phi_N}\downarrow & & \downarrow{\scriptstyle \overline{\phi_N}} & & \| & & \\
0 & \longrightarrow & C(\omega^N) & \longrightarrow & Z[T_N] & \xrightarrow{q_N} & X & \longrightarrow & 0
\end{array}$$

Since $\Phi_N - \overline{\phi_N}\underline{J}_E = 0$, there is an operator $L\colon E \longrightarrow Z[T_N]$ such that $L\underline{Q} = \Phi_N - \overline{\phi_N}\underline{J}_E$ and $\|L\| = \|\Phi_N - \overline{\phi_N}\underline{J}_E\| \le \|\Psi\| + 2 + \varepsilon$. If one writes $Le = (L_N e, e)$

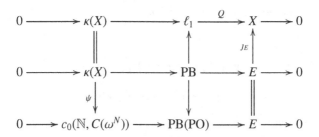

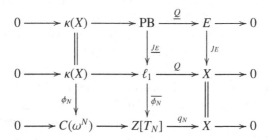

then $\|L_N e - T_N e\| \leq \|(L_N e, e)\|_{T_N} \leq (\|\Psi\| + 2 + \varepsilon)\|e\|$. Since T_{NJE} is still a dense-type operator, we have

$$\rho_{2N}(E) \leq \|L_N - T_{NJE}\| \leq \|\Psi\| + 2 + \varepsilon. \qquad \square$$

Let us secure the control of the terrain: since the list of spaces X for which $\operatorname{Ext}(X, C(\omega^\omega)) \neq 0$ includes ℓ_p for $1 < p < \infty$ or c_0, it is natural to think that ℓ_p-saturated spaces will admit a singular twisting against $C(\omega^\omega)$. Proposition 9.3.15 *almost* says that, except for the 'shrinking UFDD' assumption. Thus, by virtue of either Proposition 8.4.8 or Proposition 8.4.9, we have:

Corollary 9.3.16 *There is a singular element in* $\operatorname{Ext}(X, C(\omega^\omega))$ *when*

- *X is a c_0-saturated space with shrinking UFDD,*
- *X is an ℓ_p-saturated modular sequence space, $p \in (1, \infty)$,*
- *X is an ℓ_p-saturated Fenchel–Orlicz space, $p \in (1, \infty)$,*
- *X is a twisted sum of the above enjoying the $\mathscr{C}$-extension property.*

9.4 Notes and Remarks

9.4.1 Super Singularity

One way to reinforce singularity is by considering exact sequences such that every ultrapower is singular. Let us call them super singular, a notion introduced by Plichko [395]. An operator T will be called super strictly singular (SSS) if, for every $c > 0$, there exists $n \in \mathbb{N}$ such that every $E \subset X$ of dimension n (or greater) contains a point x such that $\|Tx\| \leq c\|x\|$. An operator is SSS if and only if every ultrapower is SSS [128, Lemma 1.1] and thus, in particular, an SSS operator has strictly singular bidual [395, Corollary]. It is useful to introduce a parameter to measure super singularity of quasilinear maps: the *modulus of super singularity* of a quasilinear map $\Omega\colon X \longrightarrow Y$ is defined as $\mathfrak{s}_\Omega(n) = \inf \operatorname{dist}(\Omega|_E, \mathfrak{L}(E, Y))$, where the infimum is taken over all n-dimensional subspaces E of X. It turns out that Ω is super singular $\iff \lim_n \mathfrak{s}_\Omega(n) = \infty \iff$ the quotient map of Ω is SSS. Examples? Any exact sequence $0 \longrightarrow K \longrightarrow \mathscr{L}_\infty \longrightarrow X \longrightarrow 0$ in which X does not contain ℓ_∞^n uniformly is super singular since any operator $\mathscr{L}_\infty \longrightarrow X$ satisfying those conditions is weakly compact. Another example is any exact sequence $0 \longrightarrow \kappa \longrightarrow \mathscr{L}_1 \longrightarrow X \longrightarrow 0$ when X does not contain ℓ_1^n uniformly [395, Proposition 6]. In particular, projective presentations of B-convex spaces are super singular, what makes surprising that:

Lemma *No super singular quasilinear map between B-convex spaces exists.*

Proof Assume $0 \longrightarrow Y \longrightarrow Z \stackrel{\rho}{\longrightarrow} X \longrightarrow 0$ is an allegedly super singular quasilinear map between B-convex spaces X, Y. Then Z must be B-convex since B-convexity is a 3-space property [102]. On the other hand, Plichko shows in [395, Theorem 3] that SSS operators acting between B-convex spaces have SSS adjoint. Since ρ^* is an embedding, ρ cannot be SSS. □

In particular, there are ultrapowers of KP_{ℓ_p} that are not singular, a result that matches with the fact that KP_{L_p} is not singular. Explicit constructions are available.

9.4.2 Disjoint Singularity

We said it before and repeat it now: the behaviour of KP maps is different in sequence and function spaces because it is actually the same. And this 'the same' is *disjoint singularity*:

Definition A quasilinear map defined on a Banach lattice is said to be disjointly singular if its restriction to every infinite-dimensional subspace generated by a disjointly supported sequence is non-trivial.

It turns out that KP_{ℓ_p} is disjointly singular, just like KP_{L_p}, with the difference that disjoint singularity on a space with unconditional basis implies singularity, something that does not occur, in general, on function spaces on non-atomic measure spaces, aside from some ad hoc results: for example, the Kadec–Pełczyński alternative immediately yields that a quasilinear map on L_p, $2 \leq p < \infty$, that is both disjointly singular and ℓ_2-singular, must be singular. The theory in [98] was created with the purpose of obtaining singular quasilinear maps, but what it actually does is obtain disjointly singular maps. Let us show how different the 'standard' and the 'disjoint' approaches can be. Observe first that one can define super disjoint singularity without too many problems: a quasilinear map is super disjointly singular if every ultrapower is disjointly singular. This is not so simple to do in terms of the quotient map of an exact sequence: after all, $Y \oplus_\Omega X$ is not (necessarily) a lattice. Anyway, if one defines the *modulus of super disjoint singularity* of a quasilinear map $\Omega\colon X \longrightarrow Y$ as $\mathfrak{ds}_\Omega(n) = \inf \operatorname{dist}(\Omega|_{E_n}, \mathfrak{L}(E_n, Y))$, where the infimum is taken over all n-dimensional subspaces E_n of X generated by disjointly supported vectors, one then has that Ω is super disjointly singular if and only if $\lim \mathfrak{ds}_\Omega(n) = \infty$. And now, the surprises:

(a) KP is super disjointly singular on L_p for $0 < p < \infty$.
(b) There exist disjointly singular quasilinear maps on $C[0, 1]$.
(c) There exist disjointly singular not super disjointly singular maps on ℓ_2.

To prove (a), consider a subspace $E = [u_1, \ldots, u_n]$, with u_i pairwise disjoint in L_p. We prove that the Banach–Mazur distance between $L_p \oplus_{\mathsf{KP}} E$ and the direct sum is at least $c_p \log n$, where $c_p > 0$ is independent of E. We do the case $0 < p \leq 2$; the case $p > 2$ is similar. Without loss of generality, assume $\|u_i\| = 1$ for all i, so that the quasinorm of $u = \sum_{i \leq n} u_i$ is $n^{1/p}$. Clearly, for each $t \in [0, 1]$, one has $|\sum_{i \leq n} r_i(t) u_i| = |u|$, so

$$\mathsf{KP}\left(\sum_{i \leq n} r_i(t) u_i\right) = \sum_{i \leq n} r_i(t) u_i \log \frac{n^{1/p}}{|u_i|}, \text{ while } \sum_{i \leq n} r_i(t) \mathsf{KP}\, u_i = \sum_{i \leq n} r_i(t) u_i \log \frac{1}{|u_i|}.$$

Thus (see Lemma 3.11.1) the type p-constant of $L_p \oplus_{\mathsf{KP}} E$ is at least $p^{-1} \log n$, and the result follows.

To prove assertion (b), observe that it is enough to construct a quasilinear map on $C[0, 1]$ that is not trivial on any copy of c_0, something that can be done using the amalgamation technique in Section 9.3. Let us provide a few hints for the construction: pick a non-trivial quasilinear map $\Omega \colon c_0 \longrightarrow C[0, 1]$; set I to be the set of all 2-isomorphic copies γ of c_0 inside $C[0, 1]$, which are necessarily 4-complemented in $C[0, 1]$ via some projection π_γ; and let $\alpha_\gamma \colon \gamma \longrightarrow c_0$ be a 2-isomorphism and construct the quasilinear map $\Omega_\infty \colon C[0, 1] \longrightarrow \ell_\infty(I, C[0, 1])$ given by $\Omega_\infty(f)(\gamma) = \omega(\alpha_\gamma \pi_\gamma(f))$, which is c_0-singular and admits an equivalent version having separable range that can be considered contained in $C[0, 1]$. An example for (c) is the quasilinear map $\Omega(x) = \left(x^k \log\left(\frac{\|x^k\|_2}{\|x\|}\right)\right)_k$ on $\ell_2(\mathbb{N}, \ell_2^k)$. It is disjointly singular (this requires proof) but not super disjointly singular since $\Omega(x) = 0$ for every $x \in \ell_2^k$ and every $k \in \mathbb{N}$.

Other notions from the quasilinear world can be reformulated in disjointland: a quasilinear map Ω defined on a Banach lattice is said to be *disjointly trivial* if it is trivial on any subspace generated by a sequence of disjointly supported elements; it is *locally disjointly trivial* if there exists $C > 0$ such that for any finite-dimensional subspace F generated by disjointly supported vectors, there exists a linear map L_F such that $\|\Omega|_F - L_F\| \leq C$. We should not underestimate the power of these versions, as we will shortly discover.

Proposition *Let Ω be a quasilinear map defined on a function space and with values on a quasi-Banach space. Consider the following assertions:*

(i) *Ω is trivial.*

(ii) *Ω is disjointly trivial.*

(iii) *Ω is locally disjointly trivial.*

(iv) *Ω is locally trivial.*

Then (i) $\Rightarrow$ (ii) $\Rightarrow$ (iii) $\Leftrightarrow$ (iv). *Moreover, if the codomain space is an ultrasummand then all assertions are equivalent.*

The reader is addressed to [77; 97] for the proof and further implications.

9.4.3 Cosingularity

Strict singularity admits a dual notion, strict cosingularity, also introduced by Kato. An operator $\tau: E \longrightarrow F$ is said to be *strictly cosingular* if $\rho\tau$ is never a quotient operator for any quotient operator ρ with infinite-dimensional range. Observe that an isomorphic embedding $\jmath: Y \longrightarrow Z$ is strictly cosingular if and only if no (infinite-dimensional) quotient map $Y \longrightarrow E$ can be extended through $\jmath$ to a quotient map $X \longrightarrow E$. Analogously:

Definition A quasilinear map Ω is said to be cosingular if the embedding in the exact sequence it generates is strictly cosingular.

In other words, Ω is cosingular if and only if whenever ρ is a quotient map with infinite-dimensional range, the composition $\rho \circ \Omega$ is not trivial. Cosingular maps enjoy their own characterisation and balance principle. Proposition 9.1.3 can be dualised: (i) is slightly awkward (see [111] for details), while (ii) becomes *If* $\Omega \sim \psi \circ \Phi$, *the operator,* ψ *has finite-dimensional kernel.* We also have a

Balance principle for strictly cosingularity *Let* Y, X *be quasi-Banach spaces. If* $\dim(X) < \dim(Y)$ *then no quasilinear map* $\Omega: X \longrightarrow Y$ *is cosingular.*

The result is valid for general quasi-Banach spaces since its proof does not require injective presentations as the singular case did. A careful application of Lemma 3.9.4 suffices. We conclude with the observation that since an operator with singular adjoint must be cosingular, KP is cosingular.

9.4.4 The Basic Sequence Problem

Bumping up against Proposition 9.1.9 on one side and limited by finite-dimensional truisms on the other, Kalton [265] obtained by simple witchcraft one of the most extremal examples of singular exact sequence there is:

Kalton's singular sequence *There exists a singular sequence*

$$0 \longrightarrow \mathbb{R} \longrightarrow \mathsf{W} \longrightarrow \ell_1 \longrightarrow 0 \tag{9.2}$$

This result is astonishing no matter how one looks at it. Moreover, there is no simple way to describe it. Maurey gives a crystal-clear exposition in [358], where he explains the mysteries of the construction one by one; yet, the reader is warned, there are at least a couple of places where one encounters 'and here are necessary a number of quite complicated calculations to be done'. Let us present some additional information about the meaning of this construction. As Maurey remarks, W can be considered an improved quasi-Banach version of the Gowers–Maurey [197] Banach space without an unconditional basis, with an additional feature that is impossible to obtain (like sequence (9.2)) in a Banach space: *the space* W *does not admit a basic sequence*. And it is so because there is a point in W, the vector u that generates the remarkable one-dimensional subspace $[u] = \mathbb{R}$ in Diagram (9.2), that by singularity must belong to the closure of any infinite-dimensional subspace of W. This makes the existence of a basic sequence (e_n) for W impossible, since for every infinite subset $M \subset \mathbb{N}$, one has $u \in \overline{[(e_n)_{n\in M}]}$. The observation is shocking since one of the first results in Banach space theory, known to Banach himself [30], is that every Banach space contains a basic sequence. In particular, W contains no infinite-dimensional Banach subspace. Another side effect of the existence of W is that the classical criterion 0.0.2 for strict singularity fails in the quasi-Banach ambient: if Q is the quotient map in sequence (9.2) and $E \subset$ W is *any* infinite-dimensional subspace then $\|Q|_E\| = 1$ because E contains $[u]$, and thus Riesz's lemma provides $p \in E$ with $\|p\| = 1$ such that $\mathrm{dist}(p, [u]) \geq 1$. And with it, Lemma 9.1.5 is lost too: just set $\overline{\tau} = \mathbf{1}_{\mathsf{W}}$. It can be proved that W is indecomposable [265] .

A space without proper closed infinite-dimensional subspaces has been called *atomic*. It is an open question, the so-called atomic space problem, whether atomic spaces exist. The space W is still far from atomic since one would need a space in which all lines behave like $[u]$ does in W. Of course, Banach spaces cannot be atomic since linear continuous (non-zero) functionals exists and their kernels exist too. Reese [399] obtained an 'almost atomic' space $\mathscr{R}$, i.e. a space with a sequence of finite-dimensional subspaces F_n with $\dim F_n > n$ so that if $x_n \in F_n$ is any sequence which is non-zero infinitely often then $\overline{[x_n : n \in \mathbb{N}]} = \mathscr{R}$. However, the space $\mathscr{R}$ is only an F-space and not a quasi-Banach space (it is not locally bounded). In a rare moment of intimacy, Kalton says that 'we suspect, however, that an atomic quasi-Banach space will eventually be found'.

Sources

The definition and essential facts about strictly singular operators are from Kato [290], while the definition and essential facts about strictly singular quasilinear maps, including the balance principles, are from [108; 111]. The quasi-Banach balance principle in Proposition 9.1.9 is from [56]. The transfer principle appears as an embryo in Ribe's seminal paper [401], as well as the amalgamation technique, but the construction of a singular sequence $0 \longrightarrow \ell_\infty \longrightarrow \cdot \longrightarrow \ell_1 \longrightarrow 0$ is explicit in Theorem 3A there. It was nonetheless independently rediscovered in [108] in a quasilinear form and reappeared in [78], this time with a linear proof. Amalgamation techniques were introduced once again in [108] and reappeared in [73], where the initial sequences also made their appearance. That the Kalton–Peck map is singular on ℓ_p was known to Kalton and Peck [280] for $1 < p < \infty$; its singularity for $p = 1$ was obtained in [108] with a proof à la Roberts [403]. The unifying proof for all $0 < p < \infty$ is from [78]. That KP is not singular in L_p was shown in [445]. A more general proof that no centralizer is singular on L_p was obtained in [62] and generalised in [97]. That KP is not super singular on ℓ_p is from [128], while the argument that no quasilinear map between B-convex Banach spaces can be super singular was obtained in [96]. The singular sequences involving $C[0, 1]$ and $C(\omega^\omega)$ appeared in [73] and in a draft form in [108]. Argyros and Haydon [18, Theorem 10.2] produce a singular element in $\operatorname{Ext}(c_0, C(\omega^\omega))$ in their own way. Disjoint singularity made its appearance in a lead role in [98], where it was proved that centralizers generated from a complex interpolation process between function spaces have a strong tendency to be disjointly singular.

10

Back to Banach Space Theory

Supper's ready. After the display of quasilinear and homological techniques we have presented so far, we are ready to return to the place the journey started: classical Banach space theory. Much in the spirit, we hope, of Eliot, 'We shall not cease from exploration, and the end of all our exploring will be to arrive where we started and know the place for the first time'. But the twist is that we can now provide solutions for, or at least a better understanding of, a number of open problems. Among the topics covered, the reader will encounter vector-valued forms of Sobczyk's theorem, isomorphically polyhedral $\mathscr{L}_\infty$-spaces, Lipschitz and uniformly homeomorphic $\mathscr{L}_\infty$-spaces, properties of kernels of quotient operators from $\mathscr{L}_1$-spaces, sophisticated 3-space problems, the extension of $\mathscr{L}_\infty$-valued operators, Kadec spaces, Kalton–Peck spaces and, at last, the space Z_2. All these topics can be easily considered as part of classical Banach space theory, even if the techniques we employ involve most of the machinery developed throughout the book.

10.1 Vector-Valued Versions of Sobczyk's Theorem

We are ready to tackle the question of what exactly a 'vector-valued Sobczyk's theorem' should mean. Probably the answer should be a result along one of the following lines. Let X be a separable quasi-Banach space and let (E_n) be quasi-Banach spaces.

(u) If $\operatorname{Ext}(X, E_n) = 0$ uniformly on n then $\operatorname{Ext}(X, c_0(\mathbb{N}, E_n)) = 0$.

(s) Let Y be a subspace of a quasi-Banach space Z. Let $\tau\colon Y \longrightarrow c_0(\mathbb{N}, E_n)$ be an operator. If there is a λ such that each component $\pi_n\tau\colon Y \longrightarrow E_n$ admits a λ-extension to Z then τ admits an $f(\lambda)$-extension to Z for some function f.

(w) Variations of these, maybe under additional hypotheses on the spaces involved.

Here (u) stands for 'uniform', (s) for 'single' and (w) for 'who knows'. We already know from 5.2.5 and 2.14.8 that (u) and (s) hold for Banach spaces X that have the BAP; we also know from the Pełczyński–Lusky sequence that (s) is false when X fails the BAP. Let us show that (u) is extremely false for p-Banach spaces if X fails the BAP.

10.1.1 A p-Banach space oddity *For each $p \in (0,1)$ there is a non-trivial exact sequence of p-Banach spaces* $0 \longrightarrow c_0(\mathbb{N}, \ell_p^n) \longrightarrow \cdot \longrightarrow L_p \longrightarrow 0$.

Proof Write $L_p = \overline{\bigcup_n F_n}$ with $F_n \approx \ell_p^n$ and $F_n \subset F_{n+1}$. The Pełczyński–Lusky sequence $0 \longrightarrow c_0(\mathbb{N}, \ell_p^n) \longrightarrow c(\mathbb{N}, F_n) \longrightarrow L_p \longrightarrow 0$ does not split since actually there is no non-zero operator $L_p \longrightarrow c(\mathbb{N}, F_n)$. □

Therefore, even if $\operatorname{Ext}_{p\mathbf{B}}(L_p, c_0) = 0$ and $\operatorname{Ext}_{p\mathbf{B}}(L_p, \ell_p^n) = 0$ uniformly on n (because $\operatorname{Ext}_{p\mathbf{B}}(L_p, \ell_p) = 0$), still $\operatorname{Ext}_{p\mathbf{B}}(L_p, c_0(\ell_p)) \neq 0$. We do not know if something similar can occur in a Banach space ambient. Thus, the role of the BAP seems to be central in this matter, even if the true truth underneath is that a vector-valued Sobczyk's theorem, whatever it is, requires a balance between properties of the spaces E_n and properties of the space X. And this balance even determines which operators (one, all, some, ...) can be extended. The equilibrium can be achieved in different ways: in the classical scalar case of Sobczyk's theorem, in which all the spaces $E_n = \mathbb{K}$ are 1-injective, the '$\operatorname{Ext}(X, E_n) = 0$ uniformly on n' condition requires nothing from X beyond its separability. On the other hand, asking nothing on the E_n side requires asking for the BAP on X, or at least something near to that. All versions known or presented so far fit into this schema one way or another. Thus, let us enter into the non-locally-convex zone, where M-ideals do not dare to thread. If E_n is a sequence of quasi-Banach spaces then $\ell_\infty(\mathbb{N}, E_n)$ or $c_0(\mathbb{N}, E_n)$ is a quasi-Banach space if and only if the moduli of concavity of the spaces E_n are uniformly bounded, in which case they can be given uniformly equivalent r-norms for some $r \in (0,1]$. If $\Phi\colon X \longrightarrow c_0(\mathbb{N}, E_n)$ is quasilinear, its components $\Phi_n\colon X \longrightarrow E_n$ form a sequence of quasilinear maps such that $\lim \|\Phi_n x\| = 0$ for all $x \in X$ and $Q(\Phi_n) \leq Q(\Phi)$.

Lemma 10.1.2 *Let $\Phi_n\colon X \longrightarrow Y$ be a sequence of p-linear maps vanishing on a fixed Hamel basis of X. If* $\sup_n Q^{(p)}(\Phi_n) < \infty$ *and Φ_n is pointwise null, then* $\lim_n Q^{(p)}(\Phi_n) = 0$. *If, moreover, X is finite-dimensional,* $\lim_n \|\Phi_n\| = 0$.

Proof The first assertion is part of the completeness Theorem 3.6.3. The second part is straightforward after that. □

Proposition 10.1.3 *Let E_n and X be p-Banach spaces and let $\Phi: X \longrightarrow c_0(\mathbb{N}, E_n)$ be a p-linear map whose components $\Phi_k: X \longrightarrow E_k$ are μ-trivial. If X is separable and has the λ-AP, then Φ is trivial. More precisely, for each $\varepsilon > 0$, there is a dense subspace $X_\varepsilon \subset X$ and a linear map $\Lambda: X_\varepsilon \longrightarrow c_0(\mathbb{N}, E_n)$ such that $\|\Phi - \Lambda\| < \mu(1 + \lambda^p)^{1/p} + \varepsilon$ on X_ε.*

Proof Fix $\varepsilon \in (0, 1)$, a chain of finite-dimensional subspaces (X_k) whose union is dense in X and operators $B_k \in \mathfrak{F}(X)$ such that $B_k|_{X_k} = \mathbf{1}_{X_k}, B_k[X] = X_{k+1}$ and $\|B_k\| < \lambda + \varepsilon$ for all k. Set $X_\varepsilon = \bigcup X_n$. Fix a Hamel basis $\mathscr{H}$ of X that contains a basis for each of the X_k and assume without loss of generality that Φ vanishes on $\mathscr{H}$, so that each Φ_n vanishes on $\mathscr{H}$. For each j, choose a natural number $N(j)$ such that $\|\Phi_n|_{X_{j+1}}\| \leq \varepsilon^j$ for $n \geq N(j)$, with $N(j+1) > N(j)$ for all j. We are ready to construct a linear map $\Lambda: X \longrightarrow c_0(\mathbb{N}, E_n)$ at a finite distance from Φ as follows: since Φ_n is μ-trivial, we pick a linear map L_n such that $\|\Phi_n - L_n\| \leq \mu$ and set

$$\Lambda(x)(n) = \begin{cases} L_n(x) & \text{for } n < N(1), \\ (L_n - L_n B_j)(x) & \text{for } N(j) \leq n < N(j+1). \end{cases}$$

It is clear that $\Lambda(x) \in c_0(\mathbb{N}, E_n)$ for $x \in \bigcup_{n\geq 1} X_n$. Indeed, if $x \in X_k$, then for $n \geq N(k)$, we have $\Lambda(x)(n) = 0$ since $B_j(x) = x$ for $j \geq k$. The argument concludes by taking into account that for $n \geq N(j)$ we have

$$\|L_n|_{X_{j+1}}\|^p \leq \|\Phi_n|_{X_{j+1}} - L_n|_{X_{j+1}}\|^p + \|\Phi_n|_{X_{j+1}}\|^p \leq \mu^p + \varepsilon^{jp}.$$

Finally, let us estimate $\|\Phi - \Lambda\| = \sup_n \|\Phi_n - \Lambda_n\|$, where $\Lambda_n(x) = \Lambda(x)(n)$. For $n < N(1)$, we have $\|\Phi_n - \Lambda_n\| = \|\Phi_n - L_n\| \leq \mu$, while for $N(j) \leq n < N(j+1)$, the number $\|\Phi_n - \Lambda_n\|^p$ is at most

$$\|\Phi_n - L_n + L_n B_j\|^p \leq \|\Phi_n - L_n\|^p + \|L_n B_j\|^p \leq \mu^p + (\mu^p + \varepsilon^{jp})(\lambda + \varepsilon)^p. \quad \square$$

We translate this into an extension result

Proposition 10.1.4 *Let Z and E_n be p-Banach spaces. Let Y be subspace of Z such that Z/Y is separable and has the λ-AP. If $\tau: Y \longrightarrow c_0(\mathbb{N}, E_n)$ is an operator such that every component $\tau_n: Y \longrightarrow E_n$ admits a μ-extension to Z then, for every $\varepsilon > 0$, τ admits a $(\lambda^p + \mu^p 2^{1-1/p}\lambda^p)^{1/p} 2^{1/p} + \varepsilon$ -extension to Z.*

Proof Put $X = Z/Y$, take $\varepsilon > 0$ and use Corollary 3.3.8 to obtain a quasilinear map $\Phi: X \longrightarrow Y$ and an isomorphism $u: Y \oplus_\Phi X \longrightarrow Z$ with $\|u\| < 2^{1/p-1} + \varepsilon$ and $\|u^{-1}\| < 2^{1/p} + \varepsilon$. Now, each τ_n admits μ-extensions to Z, hence they admit

$(\mu 2^{1/p-1} + \varepsilon)$-extensions to $Y \oplus_\Phi X$ and thus, by Lemma 3.5.4, each $\tau_n \circ \Phi$ is $(\mu 2^{1/p-1}+\varepsilon)$-trivial. From here, Lemma 3.5.4 yields that $\tau\circ\Phi \colon X \longrightarrow c_0(\mathbb{N}, E_n)$ is $(\mu 2^{1/p-1} + \varepsilon)((1 + \lambda^p)^{1/p} + \varepsilon)$-trivial on a certain dense subspace $X_\varepsilon \subset X$. Lemma 3.5.4 once again says that τ admits a $(\mu 2^{1/p-1} + \varepsilon)((1 + \lambda^p)^{1/p} + \varepsilon)$-extension to $Y \oplus_\Phi X_\varepsilon$ (which is dense in $Y \oplus_\Phi X$) and so a $(\mu 2^{1/p-1} + \varepsilon)((1 + \lambda^p)^{1/p} + \varepsilon)(2^{2/p-1} + \varepsilon)$-extension to the corresponding subspace of Z. Since p-norms are continuous, the proof is done. □

10.2 Polyhedral $\mathscr{L}_\infty$-Spaces

A Banach space is said to be *polyhedral* if the unit ball of every finite-dimensional subspace is a polyhedron. Since this a geometrical notion – c_0 is polyhedral, while c is not – we will consider the isomorphism version: a space is said to be *isomorphically polyhedral* if it can be renormed to be polyhedral. The space $C(\alpha)$ and all of its subspaces are isomorphically polyhedral for every ordinal α. The wicked ways of polyhedral spaces were exhausted by Fonf in results that are 'most enjoyable to encounter, to lecture on, and to write about. Plainly speaking, they are too clever by half' (Diestel [152, p. 172]). Among them, a fundamental result [178, Theorem 6.21]: infinite-dimensional polyhedral spaces are c_0-saturated (closed infinite-dimensional subspaces contain a copy of c_0). Polyhedral spaces are baffling objects: there exist polyhedral spaces that admit ℓ_p, $1 < p < \infty$, as a quotient [185]; other polyhedral spaces are even more exotic [341]. Polyhedral $\mathscr{L}_\infty$ or Lindenstrauss spaces are baffling too. A clean presentation of the connections between polyhedral *and* Lindenstrauss spaces is in [178, Section 6]. We have:

- There are polyhedral spaces that are not $\mathscr{L}_\infty$-spaces, such as the Schreier space, or any subspace of c_0 not isomorphic to c_0, since subspaces of $c_0(I)$ are $\mathscr{L}_\infty$-spaces if and only if they are isomorphic to some $c_0(J)$ by 1.6.3 (b).
- There are Lindenstrauss spaces with no polyhedral renorming: $C(\Delta)$ is one.
- A Banach space whose dual is isometric to ℓ_1 has a polyhedral renorming for which the dual space is still isometric to ℓ_1 [177]. This result it optimal in view of the next two items.
- The Bourgain–Delbaen (second) space is not isomorphically polyhedral since it does not contain c_0, while its dual is isomorphic to ℓ_1.
- Isometric preduals of $\ell_1(I)$ with I uncountable are not necessarily isomorphically polyhedral. All known counterexamples are $C(K)$-spaces with K

scattered (so that $C(K) = \ell_1(K)$) that for some reason have no polyhedral renorming. This was first established for Kunen's compactum K: under [CH], the space $C(\mathsf{K})$ has the rare property that every uncountable set of elements contains one that belongs to the closed convex hull of the others. This property was used by Jiménez and Moreno [222] to show that every renorming of $C(\mathsf{K})$ has a countable boundary, which forbids equivalent polyhedral renormings. Later on, it was proved that spaces of continuous functions on some tree spaces do the same in [ZFC]; see [179].

The following counterexample was requested by Fonf and appears, along with variations, in [120]. Somehow it shows that some conjectures one might come up with are false:

Proposition 10.2.1 *There exist separable polyhedral $\mathscr{L}_\infty$-spaces that lack Pełczyński's property (V); in particular, they are not isomorphic to any Lindenstrauss space.*

Proof Recall from Lemma 9.3.14 that for each N there is an operator $T_N\colon c_0 \longrightarrow \ell_\infty(\omega^N)$ with $\operatorname{dist}(T_N(x), C(\omega^N)) \leq \|x\|$ for all $x \in c_0$ and such that $\|T_N - L\| \geq \rho_{2N}(c_0)$ for every linear map $L\colon c_0 \longrightarrow C(\omega^N)$. Taking into account the comment after Proposition 8.6.8, we can simply assume $\|T_N - L\| \geq N$. We shall use the spaces $Z[T_N]$ introduced during the proof of Proposition 9.3.15 and the isometrically exact sequences

$$0 \longrightarrow C(\omega^N) \longrightarrow Z[T_N] \xrightarrow{Q_N} c_0 \longrightarrow 0 \qquad (10.1)$$

with inclusion $f \longmapsto (f, 0)$ and quotient map $(f, x) \longmapsto x$. The twisted sum space $Z[T_N]$ is isomorphic in this occasion to c_0, thus it has a polyhedral norm. The main result in [146] asserts that in a separable isomorphically polyhedral space the polyhedral norms are dense; thus, let $\|\cdot\|_N$ be a polyhedral norm in $d_{T_N}(C(\omega^N, c_0))$ that is 2-equivalent to $\|\cdot\|_{T_N}$. The sequences (10.1) split but they are increasingly far from trivial by the choice of T_N; thus, their c_0-sum cannot split and neither can the c_0-sum (here $Q = (Q_N)$):

$$0 \longrightarrow c_0(\mathbb{N}, C(\omega^N)) \longrightarrow c_0(\mathbb{N}, Z[T_N]) \xrightarrow{Q} c_0(c_0) \longrightarrow 0$$

The space $c_0(\mathbb{N}, Z[T_N])$ is polyhedral since any c_0-sum of polyhedral spaces is polyhedral [208]. Define the 'diagonal' operator $\tau\colon c_0 \longrightarrow c_0(c_0)$ by $\tau(x) = (N^{-1/2}x)_N$ and form the pullback diagram

$$\begin{array}{ccccccccc}
0 & \longrightarrow & c_0(\mathbb{N}, C(\omega^N)) & \longrightarrow & c_0(\mathbb{N}, Z[T_N]) & \xrightarrow{Q} & c_0(c_0) & \longrightarrow & 0\\
 & & \| & & \uparrow{\scriptstyle \underline{\tau}} & & \uparrow{\scriptstyle \tau} & & \\
0 & \longrightarrow & c_0(\mathbb{N}, C(\omega^N)) & \longrightarrow & \mathrm{PB} & \xrightarrow{\underline{Q}} & c_0(c_0) & \longrightarrow & 0
\end{array}$$

The plan is to show that $\underline{Q}$ is strictly singular, which prevents PB from having property (V), and thus prevents it from being Lindenstrauss under any equivalent norm. Assume that $\underline{Q}$ is not strictly singular, and find an infinite-dimensional subspace $E \subset c_0$ and a lifting $T\colon E \longrightarrow \mathrm{PB}$ such that $\underline{Q}T = \tau|_E$. By the c_0 standard saturation and distortion properties of $c_0 = c_0(c_0)$, there is no loss of generality assuming that E is a 2-isomorphic copy of c_0. Since $Q\underline{\tau}T = \tau|_E$, we get $Q_N\underline{\tau}T(e) = N^{-1/2}e$ for all $e \in E$, which in particular means that $\underline{\tau}T$ has the form $(L_N e, N^{-1/2}e)_N$ where $L_N\colon E \longrightarrow C(\omega^N)$ is a certain linear map. Therefore, there is a constant M such that $\|(L_N e, N^{-1/2}e)\| \le M\|e\|$, which means $\|L_N e - T_N N^{-1/2}e\| \le M\|e\|$, yielding a contradiction:

$$2N \le \rho_{2N}(c_0) \le 2\rho_{2N}(E) \le 2\|N^{1/2}L_N - T_N\| \le 2MN^{1/2}.$$

To conclude, the space PB is a subspace of $c_0(\mathbb{N}, Z[T_N]) \oplus_\infty c_0$, and thus it is polyhedral. □

10.3 Lipschitz and Uniformly Homeomorphic $\mathscr{L}_\infty$-Spaces

Non-linear geometry of Banach spaces has seen spectacular advances since the Benyamini–Lindenstrauss book [42]; [275; 192] are good surveys on the matter. The part we are interested in here is the collision between three ways of classifying Banach spaces: isomorphic, Lipschitz and uniformly homeomorphic. Our entrance is:

Lemma 10.3.1 *Let* $0 \longrightarrow A \longrightarrow B \xrightarrow{\pi} C \longrightarrow 0$ *be an exact sequence of Banach spaces. If* π *admits a Lipschitz / uniformly continuous section then* $A \times C$ *and* B *are Lipschitz / uniformly homeomorphic.*

Proof Let $s\colon C \longrightarrow B$ be a Lipschitz / uniformly continuous section of π. The Lipschitz / uniformly continuous map $f\colon A \times C \longrightarrow B$ given by $f(a, c) = a + s(c)$ has Lipschitz / uniformly continuous inverse $f^{-1}(b) = (b - s\pi(b), \pi(b))$. □

Thus, if one is able to find good reasons to prevent $B \simeq A \times C$, the game is over. Let us say that a Banach space is determined by its Lipschitz

/ uniform structure if it is linearly isomorphic to every Banach space to which it is Lipschitz / uniformly homeomorphic. Next we consider the Lipschitz and uniform structure of $\mathscr{L}_\infty$-spaces. To tackle the problem, it would help us, as always, to know the behaviour of Lipschitz / uniform structures under homological manipulations:

Lemma 10.3.2 *If (the quotient map of) a short exact sequence has a Lipschitz / uniformly continuous section then so does any pushout or pullback sequence.*

Proof Indeed, given a pushout diagram

$$\begin{array}{ccccccccc} 0 & \longrightarrow & A & \longrightarrow & B & \xrightarrow{\pi} & C & \longrightarrow & 0 \\ & & {\scriptstyle\alpha}\downarrow & & \downarrow{\scriptstyle\beta} & & \| & & \\ 0 & \longrightarrow & A' & \longrightarrow & B' & \xrightarrow{\overline{\pi}} & C & \longrightarrow & 0 \end{array}$$

if s is a Lipschitz / uniformly continuous section of π then $\beta \circ s$ is a Lipschitz / uniformly continuous section of $\overline{\pi}$. In a pullback diagram

$$\begin{array}{ccccccccc} 0 & \longrightarrow & A & \longrightarrow & B & \xrightarrow{\pi} & C & \longrightarrow & 0 \\ & & \| & & {\scriptstyle\underline{\gamma}}\uparrow & & \uparrow{\scriptstyle\gamma} & & \\ 0 & \longrightarrow & A & \longrightarrow & \mathrm{PB} & \xrightarrow{\underline{\pi}} & C' & \longrightarrow & 0 \end{array}$$

the map $(s\gamma, \mathbf{1}_{C'})$ is a Lipschitz / uniformly continuous section of $\underline{\pi}$. □

The pullback part is due to Kalton [276] and the pushout part is due to Suárez [446]. We are ready for (counter) examples and surprises. We know from Proposition 8.3.1 that the sequence $0 \longrightarrow c_0 \longrightarrow \ell_\infty \longrightarrow \ell_\infty/c_0 \longrightarrow 0$ admits a Lipschitz projection (retraction sounds better in this context), which raises the question of whether a Lipschitz or at least a uniformly continuous section can be found. Kalton shows in [276] that the quotient map $\ell_\infty \longrightarrow \ell_\infty/c_0$ has no uniformly continuous section, which is simultaneously unexpected and surprising. Unexpected because it makes Lemma 2.1.7, the basic fuel for most homological arguments, contain empty calories in the non-linear world: the existence of Lipschitz retraction and Lipschitz section are no longer equivalent! And surprising because Aharoni and Lindenstrauss [1] had already shown:

Lemma 10.3.3 *The Nakamura–Kakutani sequences admit Lipschitz sections.*

Proof We show that the quotient map $\pi\colon C_0(\wedge_{\mathcal{M}}) \longrightarrow c_0(\mathcal{M})$ admits a Lipschitz selector on the finitely supported elements of $c_0(\mathcal{M})$. Write

$$x = \sum_{n=1}^{N} a_n e_{\gamma_n} - \sum_{n=1}^{M} b_m e_{\mu_m}$$

with $a_1 \geq a_2 \geq \cdots \geq 0$ and $b_1 \geq b_2 \geq \cdots \geq 0$. This is a representation of x as a difference $x = x^+ - x^-$ of two disjointly supported positive elements. Set

$$\gamma_n^* = \gamma_n \setminus \bigcup_{1 \leq i < n} \gamma_i \qquad \mu_m^* = \mu_m \setminus \bigcup_{1 \leq j < m} \mu_j$$

for $1 \leq n \leq N$ and $1 \leq m \leq M$, and define

$$s(x) = \sum_{n=1}^{N} a_n 1_{\gamma_n^*} - \sum_{n=1}^{M} b_m 1_{\mu_m^*}.$$

Clearly, $\pi(s(x)) = x$ and s is Lipschitz because

$$s(x)(k) = \operatorname{dist}(x^+, [e_\gamma : k \notin \gamma]) - \operatorname{dist}(x^-, [e_\gamma : k \notin \gamma]). \qquad \square$$

Thus, $C(\Delta_{\mathcal{M}})$ is Lipschitz but not linearly homeomorphic to $c_0(\mathcal{M}) = c_0 \times c_0(\mathcal{M})$. The same is true for the spaces JL_p and CC obtained via the pullback diagrams (2.38) and (2.39) thanks to Lemma 10.3.2:

Corollary 10.3.4 *The Johnson–Lindenstrauss space JL_p is Lipschitz homeomorphic but not linearly homeomorphic to $c_0 \times \ell_p(\mathfrak{c})$. The space CC is Lipschitz homeomorphic but not linearly homeomorphic to $c_0 \times \ell_\infty$.*

Therefore, there is no doubt that the linear structure of non-separable $\mathscr{L}_\infty$- or even $\mathscr{C}$-spaces is not determined by their Lipschitz structure. Aharoni and Lindenstrauss [1] asked whether a similar result holds in the separable setting. A partial answer was provided by Johnson, Lindenstrauss and Schechtman [229, Corollary 3.2]: if a $\mathscr{C}$-space is uniformly homeomorphic to c_0 then it is linearly homeomorphic to c_0. This result raises the question [229, Problem (d)] of whether every separable $\mathscr{L}_\infty$-space is determined by its uniform structure. Again, the answer is no, as shown by Suárez in [446]. Let X be a Banach space. Let $\omega\colon \mathbb{R}^+ \longrightarrow \mathbb{R}^+$ be the function $\omega(t) = \sqrt{t}$ if $t \leq 1$ and $\omega(t) = 1 + \frac{1}{2}(t-1)$ if $t \geq 1$. It is clear that ω is concave, hence subadditive, and therefore, the function $d(x, y) = \omega(\|x - y\|)$ is a metric on X. Let $X(\omega)$ be the metric space obtained by endowing X with d. The formal identity $X(\omega) \longrightarrow X$ is a 2-Lipschitz map with uniformly continuous inverse. Form the exact sequence involving the Lipschitz-free space of Section 4.6.1 (e):

$$0 \longrightarrow \ker\beta \longrightarrow \mathcal{F}(X(\omega)) \xrightarrow{\ \beta\ } X \longrightarrow 0$$

The quotient map β admits a uniformly continuous section, namely the composition of the identity $X \longrightarrow X(\omega)$ with the natural Lipschitz map $\delta\colon X(\omega) \longrightarrow \mathcal{F}(X(\omega))$. The space $\mathcal{F}(X(\omega))$ is Schur [270, Theorem 4.6], and thus it cannot be linearly homeomorphic to $\ker\beta \times X$ when X fails the

Schur property, even if they *are* uniformly homeomorphic. To obtain an $\mathscr{L}_\infty$ counterexample, pick $X = c_0$ and form the diagram

$$
\begin{array}{ccccccccc}
 & & 0 & & 0 & & & & \\
 & & \downarrow & & \downarrow & & & & \\
0 & \longrightarrow & \ker\beta & \longrightarrow & \mathcal{F}(c_0(\omega)) & \xrightarrow{\beta} & c_0 & \longrightarrow & 0\\
 & & \downarrow & & \downarrow & & \| & & \\
0 & \longrightarrow & \mathscr{L}_\infty^{\mathrm{BP}}(\ker\beta) & \longrightarrow & \mathrm{PO} & \xrightarrow{\overline{\beta}} & c_0 & \longrightarrow & 0\\
 & & \downarrow & & \downarrow & & & & \\
 & & S & = & S & & & & \\
 & & \downarrow & & \downarrow & & & & \\
 & & 0 & & 0 & & & &
\end{array} \tag{10.2}
$$

Since β admits a uniformly continuous selection, $\overline{\beta}$ does as well, and thus PO is a separable $\mathscr{L}_\infty$-space that is uniformly homeomorphic to $\mathscr{L}_\infty^{\mathrm{BP}}(\ker\beta) \times c_0$. By the 3-space property of Schur spaces, PO must be Schur, thus it cannot be isomorphic to $\mathscr{L}_\infty^{\mathrm{BP}}(\ker\beta) \times c_0$. In conclusion:

Proposition 10.3.5 *There are separable $\mathscr{L}_\infty$-spaces that are uniformly homeomorphic but not linearly homeomorphic.*

10.4 Properties of Kernels of Quotient Maps on $\mathscr{L}_1$ Spaces

Throughout this section, $Q\colon L_1(\mu) \longrightarrow X$ will denote a quotient map, and we consider the sequence

$$0 \longrightarrow \ker Q \longrightarrow L_1(\mu) \xrightarrow{\;Q\;} X \longrightarrow 0$$

Our aim is to connect properties of X with those of $\ker Q$. We have already treated some aspects of the problem: both the BAP and the UAP pass from X to $\ker Q$; see Proposition 5.3.4 and its Corollary 5.3.12. Most questions about kernels of quotient maps on $L_1(\mu)$ are simply too difficult; we will see good examples in this and the following two sections. Bourgain's paper [48] concludes with the following remarks: 'Let Δ be the Cantor group and define E as the subspace of $L_1(\Delta)$ generated by the Walsh functions w_S for

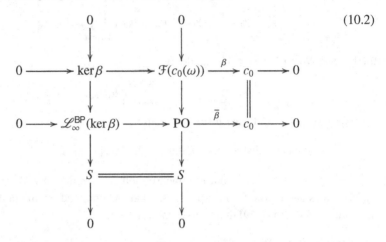

$|S| \geq 2$. Obviously E is uncomplemented. What about the following: Is E an $\mathscr{L}_1$-space? Is E isomorphic to $L_1(\Delta)$?' Bourgain attributes these questions to Pisier. To understand what is being asked, write $\Delta = \{\pm 1\}^{\mathbb{N}}$ and consider the characters $\chi_n \colon \Delta \longrightarrow \{\pm 1\}$ given by $\chi_n(x) = x_n$. Note that (χ_n) is the Rademacher sequence in disguise. Then $E = \{f \in L_1(\Delta)\colon \int_\Delta f\chi_n = 0 \,\forall n \in \mathbb{N}\}$, where the integral is taken with respect to the Haar measure on Δ. The space E is the kernel of the operator $\chi\colon L_1(\Delta) \longrightarrow \ell_\infty$ given by $\chi(f) = \left(\int_\Delta f\chi_n\right)_{n\geq 1}$. By the Riemann–Lebesgue lemma, χ takes values in c_0, and basic harmonic folklore implies that χ is onto c_0. Thus, the questions of Bourgain and Pisier refer to the kernel of a well-behaved quotient map $L_1 \longrightarrow c_0$. As Johnson [224] says; 'at any rate, both of them as well as Kisliakov, Zippin, Schechtman, and I thought about them around that time'. Again following Johnson, the underlying problem seems to be whether there is a 'natural', uncomplemented $\mathscr{L}_1$-subspace of an $L_1(\mu)$-space, in particular an invariant one in $L_1(G)$, where G is a locally compact Abelian group. The narration continues with Johnson mentioning that 'while lecturing on their results in 1995 and 1996, the authors of [284] asked whether such an E could have local unconditional structure'. To show how ugly the subspace E can be, we will first show that it is not an ultrasummand (so it is not an $L_1(\mu)$-space) and its bidual is not complemented in a Banach lattice (so it is not even an $\mathscr{L}_1$-space).

When Is ker Q an Ultrasummand?

Our use of the Radon–Nikodým property (RNP) is merely pragmatic. Anyway, everything there is to know about it can be found in [155, Chapter III]. What we need is to know that a Banach space X has the RNP if, for every finite measure μ, every operator $T\colon L_1(\mu) \longrightarrow X$ is *representable*, that is, has the form $Tf = \int fg\,d\mu$ for some $g \in L_\infty(\mu, X)$. This implies that for any measure λ, finite or not, every operator $L_1(\lambda) \longrightarrow X$ factorises through some $\ell_1(I)$. Clearly, the RNP is inherited by subspaces. Separable dual spaces and subspaces of spaces with RNP have RNP, whereas c_0 and L_1 do not.

Lemma 10.4.1 *If X is an ultrasummand with the RNP then the kernel of every quotient map $L_1(\mu) \longrightarrow X$ is an ultrasummand.*

Proof The second and third rows in the following diagram are equivalent (see Section 2.10; also compare with Diagrams (2.28) and (2.29)):

$$\begin{array}{ccccccccc}
0 & \longrightarrow & K & \longrightarrow & L_1(\mu) & \xrightarrow{Q} & X & \longrightarrow & 0 \qquad (\mathrm{z})\\
 & & \downarrow{\scriptstyle \delta_K} & & \downarrow & & \| & & \\
0 & \longrightarrow & K^{**} & \longrightarrow & \mathrm{PO} & \longrightarrow & X & \longrightarrow & 0\\
 & & \| & & & & \| & & \\
0 & \longrightarrow & K^{**} & \longrightarrow & \mathrm{PB} & \xrightarrow{\underline{Q^{**}}} & X & \longrightarrow & 0\\
 & & \| & & \downarrow{\scriptstyle \underline{\delta_X}} & & \downarrow{\scriptstyle \delta_X} & & \\
0 & \longrightarrow & K^{**} & \longrightarrow & L_1(\mu)^{**} & \xrightarrow{Q} & X & \longrightarrow & 0 \qquad (\mathrm{z}^{**})
\end{array}$$

The existence of a projection $P\colon X^{**} \longrightarrow X$ and the fact that $Q^{**}\underline{\delta_X} = \delta_X\underline{Q^{**}}$ imply $PQ^{**}\underline{\delta_X} = P\delta_X\underline{Q^{**}} = \underline{Q^{**}}$; thus, since X has the RNP and $L_1(\mu)^{**}$ is isometric to some $L_1(\lambda)$-space, $\underline{Q^{**}}$ factorises through some $\ell_1(I)$. Therefore $[\mathrm{z}\ \underline{Q^{**}}] = 0$. Moreover, quite obviously, $[\mathrm{z}^{**}\ \delta_X Q] = 0$. The diagonal principle implies that $K \times \mathrm{PB} \simeq K^{**} \times L_1(\mu)$, thus K is an ultrasummand. □

One cannot replace $L_1(\mu)$ by an arbitrary $\mathscr{L}_1$-space: after all, not every $\mathscr{L}_1$-space is an ultrasummand. Replacing L_1 with ℓ_1 allows us to present a local version of the result above:

Proposition 10.4.2 *If X is a Banach space without the RNP then $\kappa(X)$ is not an ultrasummand.*

Proof Since X lacks the RNP, there is a finite measure μ and an operator $\phi\colon L_1(\mu) \longrightarrow X$ that is not representable. Take a projective presentation of X and form the pullback with ϕ:

$$\begin{array}{ccccccccc}
0 & \longrightarrow & \kappa(X) & \longrightarrow & \ell_1(I) & \xrightarrow{\pi} & X & \longrightarrow & 0\\
 & & \| & & \uparrow & & \uparrow{\scriptstyle \phi} & & \\
0 & \longrightarrow & \kappa(X) & \longrightarrow & \mathrm{PB} & \longrightarrow & L_1(\mu) & \longrightarrow & 0
\end{array}$$

The lower sequence does not split since ϕ cannot be lifted to $\ell_1(I)$ and Lindenstrauss' lifting tells us that $\kappa(X)$ is not an ultrasummand. □

From this it immediately follows:

Corollary 10.4.3 *If X contains c_0 then $\kappa(X)$ is not complemented in its bidual.*

We supply an alternative proof for this corollary. Let X be a space containing c_0 and consider the situation

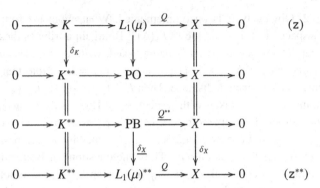

$$\begin{array}{ccccccccc}
0 & \longrightarrow & \kappa(X) & \longrightarrow & \ell_1(I) & \longrightarrow & X & \longrightarrow & 0 \\
 & & \| & & \uparrow & & \uparrow{\scriptstyle J} & & \\
0 & \longrightarrow & \kappa(X) & \longrightarrow & \mathrm{PB} & \longrightarrow & c_0 & \longrightarrow & 0 \\
 & & \| & & \uparrow & & \uparrow{\scriptstyle \imath} & & \\
0 & \longrightarrow & \kappa(X) & \longrightarrow & \mathrm{PB}' & \longrightarrow & \ell_2 & \longrightarrow & 0
\end{array}$$

in which $J\colon c_0 \longrightarrow X$ is the embedding and $\imath\colon \ell_2 \longrightarrow c_0$ is the natural inclusion. An appeal to the surprising sequence (5.1) implies that $\kappa(X)$ cannot be an ultrasummand: otherwise the lower pullback sequence would split and $J\imath$ could be lifted to an operator $\ell_2 \longrightarrow \ell_1(I)$, which would be compact, something that $J\imath$ is not. □

Now, we pass to the general situation [284, Proposition 2.2]:

Proposition 10.4.4 *If μ is a finite measure, X is a Banach space containing c_0 and $Q\colon L_1(\mu) \longrightarrow X$ is a quotient map then* $\ker Q$ *is not an ultrasummand.*

Proof Since μ is finite, the space X must be WCG, and thus all copies of c_0 it contains must be complemented. The hard work of [284] is a delicate analysis, which we will omit, showing the existence of a copy $J\colon c_0 \longrightarrow X$ that is complemented via a projection $P\colon X \longrightarrow c_0$ such that PQ is representable. Once this is done, the rest is easy: since c_0 lacks the RNP, there is an operator $\tau\colon L_1 \longrightarrow c_0$ that is not representable. This means that the composition $J\tau$ cannot be lifted to an operator $T\colon L_1 \longrightarrow L_1(\mu)$ since otherwise $\tau = PJ\tau = PQT$ would be representable. By Lindenstrauss' lifting, if $J\tau$ cannot be lifted then $\ker Q$ cannot be complemented in its bidual. □

If the measure is not σ-finite then the kernel can be an ultrasummand: consider a quotient map $Q\colon L_1(\mu) \longrightarrow c_0$ and the double adjoint $Q^{**}\colon L_1(\mu)^{**} \longrightarrow \ell_\infty$. Then $L_1(\mu)^{**} = L_1(\lambda)$ for some (very large) λ, and $\ker Q^{**} = (\ker Q)^{**}$ is an ultrasummand.

When Is $\ker Q$ an $\mathscr{L}_1$-Space?

The results so far suggest that $\ker Q$ 'tends not to be an $L_1(\mu)$ space'. Could it be at least an $\mathscr{L}_1$-space? The answer to this question is a resounding no.

Proposition 10.4.5 *Let X be a Banach space containing ℓ_∞^n uniformly, and let $Q\colon \mathscr{L}_1 \longrightarrow X$ be a quotient map. Then* $\ker Q$ *is not an $\mathscr{L}_1$-space.*

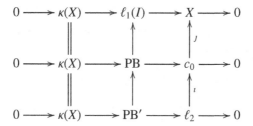

Proof By Proposition 5.2.22 there is a non-trivial sequence of Banach spaces $0 \longrightarrow \ell_2 \longrightarrow E \longrightarrow X \longrightarrow 0$. By Lindenstrauss' lifting, Q lifts to an operator $\widehat{Q}\colon \mathscr{L}_1 \longrightarrow E$, and one gets the diagram

$$\begin{array}{ccccccccc} 0 & \longrightarrow & \ker Q & \longrightarrow & \mathscr{L}_1 & \xrightarrow{Q} & X & \longrightarrow & 0 \\ & & \Big\downarrow{\scriptstyle \widehat{Q}|_{\ker Q}} & & \Big\downarrow{\scriptstyle \widehat{Q}} & & \Big\| & & \\ 0 & \longrightarrow & \ell_2 & \longrightarrow & E & \longrightarrow & X & \longrightarrow & 0 \end{array}$$

The restriction $\widehat{Q}|_{\ker Q}$ cannot extend to $\mathscr{L}_1$ since the lower sequence is not trivial, which means that $\widehat{Q}|_{\ker Q}$ is not 2-summing, and thus $\ker Q$ cannot be an $\mathscr{L}_1$-space. □

When Does $\ker Q$ Have l.u.st.?

A Banach space X has local unconditional structure (l.u.st.) if there is a constant Λ such that for every finite-dimensional subspace $E \subset X$, there is another finite-dimensional subspace $F \subset X$ containing E and admitting a Λ-unconditional basis.

Lemma 10.4.6 *A Banach space X with l.u.st. embeds as a locally complemented subspace of a Banach lattice L which is (crudely) finitely representable in X. If X is separable then L can be taken separable.*

Proof Let $\mathfrak{U}$ be an ultrafilter refining the order filter on $\mathscr{F}(X)$. For every $E \in \mathscr{F}(X)$, we select $F_E \in \mathscr{F}(X)$ containing E with a normalised Λ-unconditional basis, and we renorm it such that the new constant of the basis is 1. If we denote this renorming of F_E by $\tilde{E}$, then it is clear that $\|x\| \le \|x\|_{\tilde{E}} \le \Lambda\|x\|$ for every $x \in F$. Consider the operators

$$X \xrightarrow{\text{`inclusion'}} [E]_{\mathfrak{U}} \xrightarrow{\text{inclusion}} [F_E]_{\mathfrak{U}} \xrightarrow{\text{identity}} [\tilde{E}]_{\mathfrak{U}} \xrightarrow{\text{inclusion}} X_{\mathfrak{U}}$$

Here, the first arrow sends each $x \in X$ into the class of $(x1_E(x))_E$, and the others are the obvious arrows. The ultraproduct $[\tilde{E}]_{\mathfrak{U}}$ is a Banach lattice where X sits locally complemented because the composition of all the arrows is the diagonal embedding of X into $X_{\mathfrak{U}}$. Also, $[\tilde{E}]_{\mathfrak{U}}$ is finitely representable in X because it embeds into $X_{\mathfrak{U}}$. Finally, if X is separable, then so is $L(X)$, the (closed) sublattice generated by X in $[\tilde{E}]_{\mathfrak{U}}$, where X is again locally complemented. □

Thus we get:

Proposition 10.4.7 *If $Q : L_1 \longrightarrow c_0$ is a quotient map then* $\ker Q$ *does not have l.u.s.t.*

Proof Assume then that $\ker Q$ has l.u.st. and let $L(\ker Q)$ be a separable lattice as in the lemma. Form the pushout diagram

$$\begin{array}{ccccccccc}
 & & 0 & & 0 & & & & \\
 & & \downarrow & & \downarrow & & & & \\
0 & \longrightarrow & \ker Q & \longrightarrow & L_1 & \xrightarrow{Q} & c_0 & \longrightarrow & 0 \\
 & & \downarrow & & \downarrow & & \| & & \\
0 & \longrightarrow & L(\ker Q) & \longrightarrow & \mathrm{PO} & \longrightarrow & c_0 & \longrightarrow & 0 \\
 & & \downarrow & & \downarrow & & & & \\
 & & V & = & V & & & & \\
 & & \downarrow & & \downarrow & & & & \\
 & & 0 & & 0 & & & &
\end{array}$$

The local splitting of the left column has two immediate consequences. First is that V is (crudely) finitely representable in $L(\ker Q)$, hence also in $\ker Q$, and so also in L_1. Second is that the middle column, which is a pushout of the left column, also locally splits. This entails that PO is (crudely) finitely representable in $V \times L_1$, hence also in L_1. Using [329], we obtain that PO embeds into L_1, which was our goal. Now form the pushout diagram

$$\begin{array}{ccccccccc}
 & & 0 & & 0 & & & & \\
 & & \downarrow & & \downarrow & & & & \\
 & & L(\ker Q) & = & L(\ker Q) & & & & \\
 & & \downarrow & & \downarrow & & & & \\
0 & \longrightarrow & \mathrm{PO} & \longrightarrow & L_1 & \longrightarrow & M & \longrightarrow & 0 \\
 & & \downarrow & & \downarrow & & \| & & \\
0 & \longrightarrow & c_0 & \longrightarrow & N & \longrightarrow & M & \longrightarrow & 0 \\
 & & \downarrow & & \downarrow & & & & \\
 & & 0 & & 0 & & & &
\end{array}$$

and observe that N contains c_0. Proposition 10.4.4 yields that $L(\ker Q)$ cannot be an ultrasummand, something that it actually *is*: $L(\ker Q)$ does not contain c_0, and any Banach lattice that does not contain c_0 is an ultrasummand [335, Theorem 1.c.4]. □

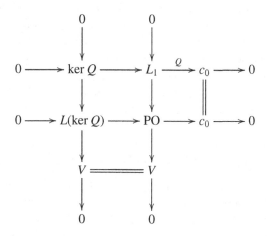

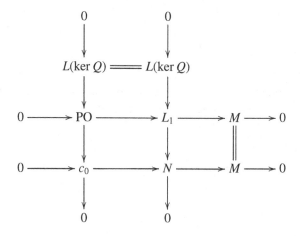

In the Appendix of [224] Johnson manages to prove that if L embeds into a Banach lattice that does not contain ℓ_∞^n uniformly and $Q: L \longrightarrow X$ is a quotient map whose kernel has Gordon–Lewis local unconditional structure (GL-l.u.st.), then X does not contain ℓ_∞^n uniformly. Even if this is just a result of [174], let us take as a definition that a Banach space Y has GL-l.u.st. if Y^{**} is complemented in a Banach lattice or, equivalently, if Y is locally complemented in a Banach lattice. See [153, p. 348] for a proof (bearing in mind Johnson's warning [224]: 'but keep in mind that in [153] GL-l.u.st. is called l.u.st. while l.u.st. is called DPR-l.u.st').

When Does ker Q Have the Dunford–Pettis Property?

Obtaining spaces with the DPP is always a challenge. Let us consider now the question of when kernels of quotient maps on $\mathscr{L}_1$-spaces have the DPP. The classical Lohman's lifting yields that if Z has the DPP and Y is a subspace of Z that does not contain ℓ_1 then Z/Y has the DPP [151]. Thus, quotients of an $\mathscr{L}_\infty$-space by a reflexive subspace have the DPP, as well as all their higher duals [294], and consequently, the same occurs with the kernels of quotient maps from an $\mathscr{L}_1$-space onto a reflexive space. If we, however, relax the conditions and simply ask for this quotient to be a separable dual then $(\ker Q)^*$ could fail the DPP, as the following example shows. The kernel in the sequence $0 \longrightarrow \ell_1(\mathbb{N}, \ell_2^n) \longrightarrow \ell_1(\mathbb{N}, \ell_1^{2^n}) \longrightarrow \ell_1(\mathbb{N}, \ell_1^{2^n}/\ell_2^n) \longrightarrow 0$ is a Schur space and thus it enjoys the DPP, but its dual $\ell_\infty(\mathbb{N}, \ell_2^n)$ does not because it contains complemented copies of ℓ_2: just take a free ultrafilter on $\mathbb{N}$ and lift (the identity of) any infinite-dimensional separable subspace of $[\ell_2^n]_{\mathcal{U}}$ to $\ell_\infty(\mathbb{N}, \ell_2^n)$. However, ker Q itself has the DPP:

Proposition 10.4.8 *Let U be an ultrasummand with the RNP and let $Q: \mathscr{L}_1 \longrightarrow U$ be a quotient map. Then* ker Q *has the DPP.*

Proof Let us observe the commutative diagram:

$$\begin{array}{ccccccccc}
0 & \longrightarrow & \kappa(U) & \longrightarrow & \ell_1(I) & \longrightarrow & U & \longrightarrow & 0 \\
 & & \downarrow & & \downarrow & & \| & & \\
0 & \longrightarrow & \ker Q & \longrightarrow & \mathscr{L}_1 & \xrightarrow{Q} & U & \longrightarrow & 0
\end{array}$$

The space $\kappa(U)$ is an ultrasummand by Lemma 10.4.1. Thus, $\operatorname{Ext}(\mathscr{L}_1, \kappa(U)) = 0$ by Lindenstrauss' lifting. The Diagonal principle then yields $\kappa(U) \times \mathscr{L}_1 \simeq \ker Q \times \ell_1(I)$. The Schur space $\kappa(U)$ has the DPP, and thus $\ker Q \times \ell_1(I)$ also has the DPP, as well as ker Q. □

Just make a sideways step to spaces far from 'ultrasummands with the RNP' and you will encounter a problem treated in [284]: does the kernel of a quotient map $Q: L_1(\mu) \longrightarrow c_0(I)$ have the DPP? In particular, does the kernel of a quotient map $Q: L_1(\mu) \longrightarrow c_0$ have the DPP? Does the kernel of a quotient map $Q: X \longrightarrow c_0$ have the DPP when X has the DPP (a question from Pełczyński)? Regarding these problems, Kalton and Pełczyński observe that if S is a Sidon set in a locally compact abelian group G then the 'truncated' Fourier transform $Q_S : L_1(G) \longrightarrow c_0(S)$ given by $Q_S(f) = (\hat{f}(\gamma))_{\gamma\in S}$ is onto (this can be taken as the definition of a Sidon set, if one wants) and prove, swallowed with a good draught of hard analysis, that $\ker Q_S$ does have the DPP. Of course that $\kappa(c_0)$ has the DPP (it is Schur), but the hard questions are whether either $\kappa(c_0)^* = \ell_\infty/\ell_1$ [100, Question 1 (c)] and [109, Problem B] or $\kappa(c_0)^{**}$ [109, below Problem B] has the DPP. Both questions are treated next.

When Does $\mathcal{L}_\infty/X$ Have the Dunford–Pettis Property?

Related to questions of when $\ker Q$ has the DPP, and somewhat dual, are questions of when quotients of $\mathcal{L}_\infty$-spaces have the DPP. It is clear that the DPP is stable under products and passes to complemented subspaces. In fact, it passes to locally complemented subspaces because weakly compact operators extend to weakly compact operators from locally complemented subspaces; see Proposition 5.1.9. The DPP is not a 3-space property, even in locally trivial sequences: there exist Schur spaces S and non-trivial sequences $0 \longrightarrow S \longrightarrow \ell_1 \times \ell_2 \longrightarrow L_1 \longrightarrow 0$ (go to Proposition 2.12.5 and set $X = \ell_2$ and L_1 in the place of $C(K)$). However, if X does not contain ℓ_1 and X^{**}/X has the DPP then X^{**} also has the DPP [101]. A Banach space is called Asplund if all its separable subspaces have separable duals; equivalently, if its dual has the RNP. An Asplund space cannot contain ℓ_1, and thus every quotient of an $\mathcal{L}_\infty$-space by an Asplund subspace has the DPP. Moreover:

Proposition 10.4.9 *The dual of every quotient of an $\mathcal{L}_\infty$ space by an Asplund space has the DPP.*

Proof If A is an Asplund space, A^* has the RNP, and Proposition 10.4.8 applies to the dual sequence $0 \longrightarrow (\mathcal{L}_\infty/A)^* \longrightarrow \mathcal{L}_1 \longrightarrow A^* \longrightarrow 0$. □

The bidual of $\mathcal{L}_\infty/A$ can, however, fail the DPP, as the sequence $0 \longrightarrow c_0(\mathbb{N}, K_n) \longrightarrow c_0(\mathbb{N}, \ell_\infty^{2^n}) \longrightarrow c_0(\mathbb{N}, \ell_2^n) \longrightarrow 0$ shows. Probably the least Asplund space in sight is ℓ_1, and thus deciding whether $\ell_\infty/\ell_1 = \kappa(c_0)^*$ has DPP would round off the situation. Recall that the Lindenstrauss–Rosenthal theorem makes the space ℓ_∞/ℓ_1 well defined and, since $C(\Delta)$ is ℓ_1-automorphic (Proposition 7.4.15 plus Theorem 8.5.4), so is $C(\Delta)/\ell_1$. We have:

10.4.10 *ℓ_∞/ℓ_1 has the DPP $\iff$ $C(\Delta)/\ell_1$ has the DPP.*

Proof $\Rightarrow$ is a consequence of the following general fact: if E is a subspace of an $\mathscr{L}_\infty$-space $\mathscr{L}'_\infty$, and in turn a subspace of another $\mathscr{L}_\infty$-space $\mathscr{L}_\infty$ such that $\mathscr{L}_\infty/E$ has the DPP, then $\mathscr{L}'_\infty/E$ also has the DPP. Just have a look at the diagram

$$\begin{array}{ccccccccc}
 & & 0 & & 0 & & & & \\
 & & \downarrow & & \downarrow & & & & \\
 & & E & = & E & & & & \\
 & & \downarrow & & \downarrow & & & & \\
0 & \longrightarrow & \mathscr{L}'_\infty & \longrightarrow & \mathscr{L}_\infty & \longrightarrow & \mathscr{L}_\infty/\mathscr{L}'_\infty & \longrightarrow & 0 \\
 & & \downarrow & & \downarrow & & \| & & \\
0 & \longrightarrow & \mathscr{L}'_\infty/E & \longrightarrow & \mathscr{L}_\infty/E & \longrightarrow & \mathscr{L}_\infty/\mathscr{L}'_\infty & \longrightarrow & 0 \\
 & & \downarrow & & \downarrow & & & & \\
 & & 0 & & 0 & & & &
\end{array}$$

Since the middle row splits locally, the same is true of the lower sequence, and thus the DPP passes from $\mathscr{L}_\infty/E$ to $\mathscr{L}'_\infty/E$.

To prove $\Leftarrow$ it is clearly enough to show that if $C(\Delta)/\ell_1$ had the DPP, every separable subspace of ℓ_∞/ℓ_1 would be contained in some subspace with the DPP. Let $E \subset \ell_\infty/\ell_1$ be separable and take a separable $F \subset \ell_\infty$ such that $\pi[F] \supset E$ (here π is the quotient map). Now, every separable subspace $E \subset \ell_\infty$ is contained in a copy G of $C(\Delta)$ inside ℓ_∞. The rest is easy: pick $G \subset \ell_\infty$ isomorphic to $C(\Delta)$ and containing both F and the relevant copy of ℓ_1. Then $\pi[G]$ contains E and is isomorphic to $C(\Delta)/\ell_1$. □

The space ℓ_∞ can be replaced in the proposition by any injective space. The general problem of which quotients of an $\mathscr{L}_\infty$-space have the DPP is wide open, but the previous discussion suggests another question [109, Problem B and Conjecture C]: If X, Y are isomorphic subspaces of $C(\Delta)$, is it true that $C(\Delta)/X$ has the DPP if and only if $C(\Delta)/Y$ has the DPP?

10.5 3-Space Problems

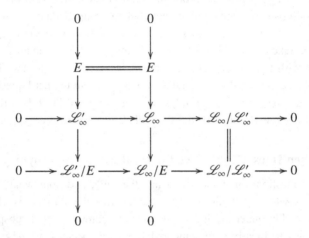

Three-space problems that only require a direct application of basic homological techniques were treated in Section 2.12. Here we will consider 3-space problems that require either more sophisticated applications of the basic techniques or more sophisticated tools altogether.

Pełczyński's and Rosenthal's Property (V)

The 3-space problem for Pełczyński's property (V) was solved in the negative using an involved construction of Ghoussoub and Johnson that leads to a strictly singular surjection onto c_0 whose kernel has property (V). The construction can be found in [102, Section 6.9]. Other clean examples follow from Corollary 9.3.8 or Proposition 9.3.15, in which singular sequences $0 \longrightarrow C(K) \longrightarrow \cdot \longrightarrow c_0 \longrightarrow 0$ appear for either $K = \Delta$ or $K = \omega^\omega$. The singularity of such sequences implies that the middle space cannot have property (V). The role of c_0 cannot be reversed:

Proposition 10.5.1 *Let* $0 \longrightarrow c_0(I) \xrightarrow{\imath} Z \xrightarrow{\rho} X \longrightarrow 0$ *be an exact sequence of Banach spaces. If X has Pełczyński's property (V) then so does Z.*

Proof Let $\phi\colon Z \longrightarrow E$ be an operator. If the restriction $\phi\imath$ is an isomorphism on some copy of c_0 then the so is ϕ. Otherwise, $\phi\imath$ is weakly compact and admits a weakly compact extension $\varphi\colon Z \longrightarrow E$. As $\phi - \varphi$ vanishes on $\ker\rho$, we have $\phi - \varphi = \psi\rho$ for some $\psi \in \mathfrak{L}(X, E)$. If ψ is weakly compact then so is $\phi = \varphi + \psi\rho$. Otherwise, there is a subspace of X isomorphic to c_0 on which ψ is an isomorphism. Let $\jmath\colon c_0 \longrightarrow X$ be the corresponding embedding

$$\begin{array}{ccccccccc} 0 & \longrightarrow & c_0(I) & \longrightarrow & Z & \longrightarrow & X & \longrightarrow & 0 \\ & & & & & & \uparrow{\scriptstyle \jmath} & & \\ & & & & & & c_0 & & \end{array}$$

If $J\colon c_0 \longrightarrow Z$ is a lifting of $\jmath$, the existence of which is clear from $\operatorname{Ext}_{\mathbf{B}}(c_0, c_0(I)) = 0$, the composition ϕJ cannot be strictly singular since $\phi J = (\varphi + \psi\rho)J = \varphi J + \psi\jmath$, where φJ is strictly singular and $\psi\jmath$ is an isomorphism onto its range. □

A counterexample can also be supplied for the 3-space problem for Rosenthal's property (V): use Proposition 5.2.20 to get a non-trivial element of $\operatorname{Ext}(\ell_\infty, \ell_2)$, and then Lemma 9.3.1 to obtain an exact sequence $0 \longrightarrow \ell_\infty(I, \ell_2) \longrightarrow \diamond \longrightarrow \ell_\infty \longrightarrow 0$ whose quotient map is not an isomorphism on any copy of ℓ_∞, so that $\diamond$ fails Rosenthal's property (V). However, $\ell_\infty(I, \ell_2)$ has property (V) because Rosenthal's property (V) obviously passes to quotients and $\ell_\infty(I, \ell_2)$ is a quotient of $\ell_\infty(I, \ell_\infty) \approx \ell_\infty(I)$.

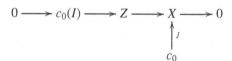

Universal Separable Injectivity

Injectivity and separable injectivity are classical notions that describe the behaviour of ℓ_∞ and c_0, respectively. Universal separable injectivity, defined next, reflects the behaviour of ℓ_∞/c_0. A thorough study of all these variations of injectivity can be found in [22].

Definition 10.5.2 A Banach space U is said to be universally separably injective (USI) if every operator $\tau\colon X \longrightarrow U$ with separable range can be extended anywhere.

Replacing 'separable range' by 'separable domain' does not affect the definition. Important examples of USI spaces are provided by the following result:

Proposition 10.5.3 *The following Banach spaces are USI:*

(a) $\ell_\infty/c_0 = C(\mathbb{N}^*)$ *and, more generally, the quotient of any USI space by a separably injective subspace.*

(b) *All ultraproducts of families of $\mathscr{L}_{\infty,\lambda}$-spaces following countably incomplete ultrafilters.*

Proof (a) Assume Z is injective and $Y \subset Z$ is separably injective. Let X be a separable Banach space and $\tau\colon X \longrightarrow Z/Y$ be an operator. It is clear that τ lifts to Z: just consider the diagram

$$\begin{array}{ccccccccc} 0 & \longrightarrow & Y & \longrightarrow & Z & \xrightarrow{\pi} & Z/Y & \longrightarrow & 0 \\ & & & & & & \uparrow{\scriptstyle\tau} & & \\ & & & & & & X & & \end{array}$$

Let $L\colon X \longrightarrow Z$ be a lifting of τ and $T \in \mathfrak{L}(X,Z)$ be an extension of L. Then $\pi T\colon X \longrightarrow Z/Y$ is the required extension of τ.

(b) Let us check that $[X_i]_{\mathcal{U}}$ is USI if X_i are $\mathscr{L}_{\infty,\lambda}$-spaces and $\mathcal{U}$ is countably incomplete. It clearly suffices to show that every separable subspace S of $[X_i]_{\mathcal{U}}$ is contained in a USI subspace. Let $(E^n)_{n\geq 1}$ be a chain of finite-dimensional subspaces whose union is dense in S. For each n we may take a lifting L^n of $\mathbf{1}_{E_n}$ to $\ell_\infty(I, X_i)$ with norm at most $1 + 1/n$ (use Theorem 2.14.5 or just do it by hand). Write $L^n = (L^n_i)_{i\in I}$ and put $E^n_i = L^n_i[E^n]$. Since X_i is an $\mathscr{L}_{\infty,\lambda}$-space, one can pick a finite-dimensional $F^n_i \subset X_i$ containing E^n_i and λ-isomorphic to some ℓ^k_∞. Now use the hypothesis on $\mathcal{U}$ to 'diagonalise': pick $n\colon I \longrightarrow \mathbb{N}$ such that

$n(i) \to \infty$ along $\mathcal{U}$ and consider the family $(F_i^{n(i)})_{i\in I}$. We have a commutative diagram

$$\begin{array}{ccccccccc} 0 & \longrightarrow & c_0^{\mathcal{U}}(I, X_i) & \longrightarrow & \ell_\infty(I, X_i) & \longrightarrow & [X_i]_{\mathcal{U}} & \longrightarrow & 0 \\ & & \uparrow & & \uparrow & & \uparrow & & \\ 0 & \longrightarrow & c_0^{\mathcal{U}}(I, F_i^{n(i)}) & \longrightarrow & \ell_\infty(I, F_i^{n(i)}) & \longrightarrow & [F_i^{n(i)}]_{\mathcal{U}} & \longrightarrow & 0 \end{array}$$

in which the vertical arrows are plain inclusions and $[F_i^{n(i)}]_{\mathcal{U}}$ contains S. But since $c_0^{\mathcal{U}}(I, F_i^{n(i)})$ is an M-ideal in $\ell_\infty(I, F_i^{n(i)}) \simeq \ell_\infty(I, \ell_\infty^{k(i)})$, which is injective, and $[F_i^{n(i)}]_{\mathcal{U}}$ obviously has the BAP, we conclude that $[F_i^{n(i)}]_{\mathcal{U}}$ is USI in much the same way as before. □

Injectivity and separable injectivity are 3-space properties. Universal separable injectivity is not, at least under CH, even if it admits the following clean characterisation: a Banach space U is USI if and only if every separable subspace $S \subset U$ is contained in another subspace $V \subset U$ isomorphic to ℓ_∞; see [22, Definition 2.25 and Theorem 2.26]. Although it might throw the reader for a loop, we begin our treatment of the 3-space problem for USI spaces with the following:

Proposition 10.5.4 *No ultrapower of the Foiaş–Singer sequence splits.*

We adhere to the notation of Section 2.2. Take the sequence

$$0 \longrightarrow C(\Delta) \xrightarrow{\text{inclusion}} D \xrightarrow{J} c_0(\Delta_0) \longrightarrow 0$$

and let $\mathcal{U}$ be an ultrafilter on I and form the ultrapower sequence

$$0 \longrightarrow C(\Delta)_{\mathcal{U}} \xrightarrow{\text{inclusion}} D_{\mathcal{U}} \xrightarrow{J_{\mathcal{U}}} c_0(\Delta_0)_{\mathcal{U}} \longrightarrow 0 \tag{10.3}$$

To show that this sequence does not split, we will prove that the quotient space contains a copy of $c_0(\mathbb{N}^{\mathcal{U}})$ that cannot be lifted to $D_{\mathcal{U}}$. To see this, observe that if (q_i) is a family of points of Δ_0 indexed by I, then the class of (e_{q_i}) in the ultrapower $c_0(\Delta_0)_{\mathcal{U}}$ depends only on the class of (q_i) in the set-theoretic ultrapower $\Delta_0^{\mathcal{U}}$. Thus, given $q \in \Delta_0^{\mathcal{U}}$, let us write $e_q = [(e_{q_i})]$, where $\langle (q_i) \rangle = q$. Clearly, if $q^1, \ldots, q^n$ are different points of $\Delta_0^{\mathcal{U}}$, then

$$\left\| \sum_{k=1}^{n} \lambda_k e_{q^k} \right\|_{c_0(\Delta_0)_{\mathcal{U}}} = \max_{1 \le k \le n} |\lambda_k|.$$

In this way, we may consider $c_0(\Delta_0^{\mathcal{U}})$ as a closed subspace of $c_0(\Delta_0)_{\mathcal{U}}$. We now prove that the pullback sequence

$$0 \longrightarrow C(\Delta)_{\mathcal{U}} \xrightarrow{\text{inclusion}} J_{\mathcal{U}}^{-1}[c_0(\Delta_0^{\mathcal{U}})] \xrightarrow{J_{\mathcal{U}}} c_0(\Delta_0^{\mathcal{U}}) \longrightarrow 0 \tag{10.4}$$

does not split. The proof consists of 'interpreting' what we did in Lemma 2.2.3 in the ultrapower structure, so let us give heartfelt homage to Larry Tesler:

Lemma 10.5.5 *Let (f_q) be any family in $D_{\mathcal{U}}$ such that $J_{\mathcal{U}}(f_q) = e_q$ for every $q \in \Delta_0^{\mathcal{U}}$. Then, given $\lambda^1, \dots, \lambda^n \in \mathbb{R}; q^1, \dots, q^n \in \Delta_0^{\mathcal{U}}$ and $\varepsilon > 0$, there exist $q \in \Delta_0^{\mathcal{U}} \setminus \{q^1, \dots, q^n\}$ and $\lambda = \pm 1$ such that*

$$\left\|\lambda f_q + \sum_{k=1}^{n} \lambda^k f_{q^k}\right\|_{D_{\mathcal{U}}} \geq 1 + \left\|\sum_{k=1}^{n} \lambda^k f_{q^k}\right\|_{D_{\mathcal{U}}} - \varepsilon.$$

Proof Notice that if $f \in D_{\mathcal{U}}$ and $t \in \Delta^{\mathcal{U}}$, then the 'value of f at t' is given by $f(t) = \lim_{\mathcal{U}(i)} f_i(t_i)$, where $[f_i]$ is any representative of f and (t_i) is any representative of t. Clearly, $\|f\|_{D_{\mathcal{U}}} = \sup_{q \in \Delta_0^{\mathcal{U}}} |f(q)|$. Also, note that for each $q \in \Delta_0^{\mathcal{U}}$, we can define the 'right limit' $f(q^+) = \lim_{\mathcal{U}(i)} f_i(q_i^+)$, where (f_i) is any representative of f and (q_i) is any representative of q. Now, assume that there is $q \in \Delta_0^{\mathcal{U}} \setminus \{q^1, \dots, q^n\}$ such that

$$\sum_{k=1}^{n} \lambda^k f_{q^k}(q) > \left\|\sum_{k=1}^{n} \lambda^k f_{q^k}\right\|_{D_{\mathcal{U}}} - \varepsilon.$$

Clearly, $f_{q^k}(q^+) = f_{q^k}(q)$ for $1 \leq k \leq n$. Now, since $J_{\mathcal{U}}(f_q) = e_q$, if $f_q(q) \geq -1$, $f_q(q^+) \geq 1$ and

$$\left\|f_q + \sum_{k=1}^{n} \lambda^k f_{q^k}\right\|_{D_{\mathcal{U}}} \geq \left(f_q + \sum_{k=1}^{n} \lambda^k f_{q^k}\right)(q^+) \geq 1 + \left\|\sum_{k=1}^{n} \lambda^k f_{q^k}\right\|_{D_{\mathcal{U}}} - \varepsilon.$$

And if $f_q(q) < -1$ then

$$\left\|-f_q + \sum_{k=1}^{n} \lambda^k f_{q^k}\right\|_{D_{\mathcal{U}}} \geq \left(-f_q(q) + \sum_{k=1}^{n} \lambda^k f_{q^k}(q)\right) > 1 + \left\|\sum_{k=1}^{n} \lambda^k f_{q^k}\right\|_{D_{\mathcal{U}}} - \varepsilon.$$

This is enough to conclude the argument. □

Theorem 10.5.6 *Let $\mathcal{U}$ be a countably incomplete ultrafilter.*

(a) $\mathrm{Ext}(X, C(\Delta)_{\mathcal{U}}) \neq 0$ *for* $X = C(\Delta)_{\mathcal{U}}, (c_0)_{\mathcal{U}}, c_0(\mathbb{N}^{\mathcal{U}})$.
(b) [CH] $\mathrm{Ext}(X, C(\mathbb{N}^*)) \neq 0$ *for* $X = c_0(\aleph_1), C(\mathbb{N}^*), \ell_\infty$.
(c) [CH] *Universal separable injectivity is not a* 3*-space property.*

Proof (a) The exact sequence (10.4) is not trivial, so $\mathrm{Ext}(c_0(\Delta_0^{\mathcal{U}}), C(\Delta)_{\mathcal{U}}) \neq 0$ and, of course, $c_0(\Delta_0^{\mathcal{U}}) \approx c_0(\mathbb{N}^{\mathcal{U}})$. Since (10.4) is a pullback of (10.3), one also has $\mathrm{Ext}(c_0(\Delta_0)_{\mathcal{U}}), C(\Delta)_{\mathcal{U}}) \neq 0$. As c_0 is complemented in $C(\Delta)$, $C(\Delta)_{\mathcal{U}}$ also contains a complemented copy of $(c_0)_{\mathcal{U}}$, and the result follows.

(b) Let us explain the role of CH here. If $\mathcal{U}$ is a free ultrafilter on $\mathbb{N}$, then the ultrapower algebra $C(\Delta)_{\mathcal{U}}$ is isometric to a $C(K)$, where K is a compactum having the properties that characterise $\mathbb{N}^*$ in Section 1.6 (6): it is a totally disconnected F-space without isolated points of weight $\mathfrak{c}$ and such that non-empty G_δ subsets have non-empty interior; see [22, Proposition 4.12] for a proof. Thus, by Parovičenko's theorem, under CH it follows that K is homeomorphic to $\mathbb{N}^*$ and the case $X = c_0(\aleph_1)$ of (b) follows from (a). The case $X = \ell_\infty$ follows from $\operatorname{Ext}(C(\mathbb{N}^*), C(\mathbb{N}^*)) \neq 0$. Indeed, starting with a non-trivial self-extension of $C(\mathbb{N}^*)$, the middle sequence in the pullback diagram

$$\begin{array}{ccccccccc}
 & & & & 0 & & 0 & & \\
 & & & & \uparrow & & \uparrow & & \\
0 & \longrightarrow & C(\mathbb{N}^*) & \longrightarrow & \diamond & \xrightarrow{\rho} & C(\mathbb{N}^*) & \longrightarrow & 0 \\
 & & \| & & \uparrow & & \uparrow{\scriptstyle \pi} & & \\
0 & \longrightarrow & C(\mathbb{N}^*) & \longrightarrow & \mathrm{PB} & \longrightarrow & \ell_\infty & \longrightarrow & 0 \\
 & & & & \uparrow & & \uparrow & & \\
 & & & & c_0 & = & c_0 & & \\
 & & & & \uparrow & & \uparrow & & \\
 & & & & 0 & & 0 & &
\end{array} \tag{10.5}$$

cannot split since otherwise π would admit a lifting $L\colon \ell_\infty \longrightarrow \diamond$ and one would have a pushout diagram

$$\begin{array}{ccccccccc}
0 & \longrightarrow & c_0 & \longrightarrow & \ell_\infty & \xrightarrow{\pi} & C(\mathbb{N}^*) & \longrightarrow & 0 \\
 & & \downarrow{\scriptstyle L|_{c_0}} & & \downarrow{\scriptstyle L} & & \| & & \\
0 & \longrightarrow & C(\mathbb{N}^*) & \xrightarrow{\jmath} & \diamond & \xrightarrow{\rho} & C(\mathbb{N}^*) & \longrightarrow & 0
\end{array}$$

which that cannot be: the USI property of $C(\mathbb{N}^*)$ would then allow to extend $L|_{c_0}$, and the lower sequence should split.

(c) Let $\mathcal{U}$ be a free ultrafilter on $\mathbb{N}$, consider the ultrapower sequence (10.3), and multiply it by a complement C of $(c_0)_\mathcal{U}$ in $C(\Delta)_\mathcal{U}$ so as to obtain a sequence $0 \longrightarrow C(\Delta)_\mathcal{U} \longrightarrow D_\mathcal{U} \times C \longrightarrow C(\Delta)_\mathcal{U} \longrightarrow 0$ which, under CH, takes the form

$$0 \longrightarrow C(\Delta)_\mathcal{U} \longrightarrow \diamond \xrightarrow{\rho} C(\mathbb{N}^*) \longrightarrow 0$$

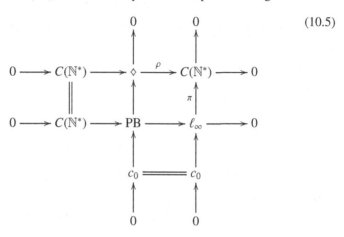

Keep in mind that the quotient map ρ is not invertible on a certain copy of $c_0(\mathfrak{c})$ inside $C(\Delta)_\mathcal{U}$. Place the sequence as the upper row in Diagram (10.5). We claim that the pullback space PB fails to be USI because the inclusion $c_0 \longrightarrow \mathrm{PB}$ does not extend to ℓ_∞. If it did, the two vertical sequences in that diagram would be semi-equivalent, and then the Parallel lines principle 2.11.5 would make the two horizontal sequences in that same diagram semi-equivalent too, which results in a commutative diagram

$$\begin{array}{ccccccccc}
0 & \longrightarrow & C(\Delta)_{\mathfrak{U}} & \longrightarrow & \mathrm{PB} & \longrightarrow & \ell_\infty & \longrightarrow & 0 \\
 & & \| & & \uparrow & & \uparrow & & \\
0 & \longrightarrow & C(\Delta)_{\mathfrak{U}} & \longrightarrow & \diamond & \xrightarrow{\rho} & C(\mathbb{N}^*) & \longrightarrow & 0
\end{array}$$

But this is impossible: as we said, ρ is not invertible on a certain copy of $c_0(\mathfrak{c})$, and thus the new pullback sequence

$$\begin{array}{ccccccccc}
0 & \longrightarrow & C(\Delta)_{\mathfrak{U}} & \longrightarrow & \mathrm{PB} & \longrightarrow & \ell_\infty & \longrightarrow & 0 \\
 & & \| & & \uparrow & & \uparrow & & \\
0 & \longrightarrow & C(\Delta)_{\mathfrak{U}} & \longrightarrow & \diamond & \xrightarrow{\rho} & C(\mathbb{N}^*) & \longrightarrow & 0 \\
 & & \| & & \uparrow & & \uparrow & & \\
0 & \longrightarrow & C(\Delta)_{\mathfrak{U}} & \longrightarrow & \mathrm{PPB} & \longrightarrow & c_0(\mathfrak{c}) & \longrightarrow & 0
\end{array}$$

does not split. But since every operator $c_0(\mathfrak{c}) \longrightarrow \ell_\infty$ has separable range and $C(\Delta)_{\mathfrak{U}}$ is separably injective, the operator lifts to PB, and the lower pullback sequence splits. □

Proposition 10.5.4 and Theorem 10.5.6 are taken from [23]. A previous analysis of the 3-space problem for universal separable injectivity can be found in [22, Section 6.2], where the interested reader will find a more systematic study of the injectivity properties of $\mathscr{C}$-spaces and ultraproducts. The assertion $\mathrm{Ext}(C(\mathbb{N}^*), C(\mathbb{N}^*)) \neq 0$ provides an improvement on the assertion '$C(\mathbb{N}^*)$ contains an uncomplemented copy of itself' that can be found in [121, Proposition 5.3]. The paper [64] contains some results on 'abstract' sequences of Foiaş–Singer type, explains why the quotient space tends to be $c_0(\Gamma)$ and shows, among other things, that $\mathrm{Ext}\,(c_0(\mathfrak{c}), C(\mathbb{N}^*)) \neq 0$ in the Cohen standard model, a model for set theory in which the first thing we do is renege on CH.

Vogt's Duality Problem in Focus

Recall Vogt's problem from Proposition 2.12.3: *must an exact sequence* $0 \longrightarrow A^* \longrightarrow Z \longrightarrow B^* \longrightarrow 0$ *be the dual sequence of another exact sequence? Must* Z *be a dual space?* A counterexample was already presented in Proposition 2.12.3.

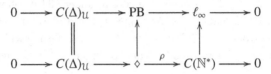

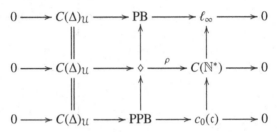

Now we will present an exposition of duality issues in Banach spaces that is lush with detail and buoyed by simpatico and that produces, in the end, an optimal concrete counterexample to Vogt's problem. The shimmering details of a quasilinear version of what follows can be found in [65]. Let us begin by elucidating exactly what a dual space and dual exact sequence are.

Lemma 10.5.7 *Let P and D be Banach spaces. The following are equivalent:*

(i) *D is isomorphic to the dual of P.*

(ii) *There is an embedding $\jmath\colon P \longrightarrow D^*$ such that $\jmath^*\delta_D\colon D \longrightarrow D^{**} \longrightarrow P^*$ is an isomorphism.*

Proof Only the implication (i) $\implies$ (ii) needs a proof. Assume that we have an isomorphism $\phi\colon D \longrightarrow P^*$. Then $\phi^*\colon P^{**} \longrightarrow D^*$ is an isomorphism, $\phi^*\delta_P\colon P \longrightarrow D^*$ is an embedding and D is the dual of $\phi^*\delta_P[P]$ through the restriction of the duality between D^* and D: $\langle\phi(d), p\rangle = \langle\phi^*(\delta_P(p)), d\rangle$. □

If P and D are as in (i), we say that P is an isomorphic predual of D; the advantage of (ii) is that one can always find a copy of each isomorphic predual of D in D^* acting through the restriction of the duality between D^* and D. Simply out of curiosity, Dixmier a long time ago characterised the corresponding subspaces of D^* as those closed subspaces that are total over D and minimal with respect to the property of being total over D. From now on, when referring to a predual of D, we tacitly assume it lies in D^*. A Banach space can have many preduals, some even isometric and others not even isomorphic: if K is a metrisable scattered compactum, then $C(K)^* = \ell_1(K) \approx \ell_1$. Furthermore, both ℓ_1 and L_1 are preduals of ℓ_∞. Each exact sequence of Banach spaces $0 \longrightarrow A \xrightarrow{\imath} B \xrightarrow{\pi} C \longrightarrow 0$ has an *adjoint* sequence, namely

$$0 \longrightarrow C^* \xrightarrow{\pi^*} B^* \xrightarrow{\imath^*} A^* \longrightarrow 0$$

The fact that Banach spaces can have many different preduals compels us to make the following definition:

Definition 10.5.8 An exact sequence is said to be a dual sequence if it is isomorphic to an adjoint sequence.

In other words, the sequence (z) in the next diagram is a dual sequence if there exists some exact sequence $0 \longrightarrow A \xrightarrow{\imath} B \xrightarrow{\pi} C \longrightarrow 0$ and a commutative diagram

$$\begin{array}{ccccccccc}
0 & \longrightarrow & C^* & \xrightarrow{\pi^*} & B^* & \xrightarrow{\imath^*} & A^* & \longrightarrow & 0\\
 & & \uparrow{\scriptstyle\gamma} & & \uparrow{\scriptstyle\beta} & & \uparrow{\scriptstyle\alpha} & & \\
0 & \longrightarrow & Y & \xrightarrow{\jmath} & Z & \xrightarrow{\rho} & X & \longrightarrow & 0
\end{array} \qquad \text{(z)}$$

in which α, β, γ are isomorphisms. This means, in particular, that A, B, C are preduals of X, Z, Y, respectively, with embeddings given by $\alpha^*\delta_A, \beta^*\delta_B, \gamma^*\delta_C$. Every dual sequence is itself the adjoint of a sequence formed with suitably chosen subspaces of the duals of the spaces occurring in it. Indeed, taking adjoints in the previous diagram one gets the commutative diagram

$$\begin{array}{ccccccccc}
0 & \longrightarrow & A & \xrightarrow{\imath} & B & \xrightarrow{\pi} & C & \longrightarrow & 0\\
 & & \downarrow{\scriptstyle\delta_A} & & \downarrow{\scriptstyle\delta_B} & & \downarrow{\scriptstyle\delta_C} & & \\
0 & \longrightarrow & A^{**} & \xrightarrow{\imath^{**}} & B^{**} & \xrightarrow{\pi^{**}} & C^{**} & \longrightarrow & 0\\
 & & \downarrow{\scriptstyle\alpha^*} & & \downarrow{\scriptstyle\beta^*} & & \downarrow{\scriptstyle\gamma^*} & & \\
0 & \longrightarrow & X^* & \xrightarrow{\rho^*} & Z^* & \xrightarrow{\jmath^*} & Y^* & \longrightarrow & 0
\end{array}$$

Set $X_* = \alpha^*\delta_A[A]$, $Z_* = \beta^*\delta_B[B]$ and $Y_* = \gamma^*\delta_C[C]$. Then $(X_*)^* = X, (Z_*)^* = Z$ and $(Y_*)^* = Y$ under the obvious dualities. Moreover, $\rho^*[X_*] \subset Z_*$ and $\jmath^*[Z_*] = Y_*$, so (z) is the adjoint of the exact sequence

$$0 \longrightarrow X_* \xrightarrow{\rho^*|_{X_*}} Z_* \xrightarrow{\jmath^*|_{Z_*}} Y_* \longrightarrow 0$$

We now characterise dual sequences:

Proposition 10.5.9 *An exact sequence* $0 \longrightarrow Y \xrightarrow{\jmath} Z \xrightarrow{\rho} X \longrightarrow 0$ *is a dual sequence if and only if there are preduals* $X_* \subset X^*$ *of* X *and* $Y_* \subset Y^*$ *of* Y, *an exact sequence* $0 \longrightarrow X_* \xrightarrow{\imath} V \xrightarrow{\pi} Y_* \longrightarrow 0$ *and a commutative diagram*

$$\begin{array}{ccccccccc}
0 & \longrightarrow & X_* & \xrightarrow{\imath} & V & \xrightarrow{\pi} & Y_* & \longrightarrow & 0\\
 & & \downarrow{\scriptstyle\text{inclusion}} & & \downarrow{\scriptstyle v} & & \downarrow{\scriptstyle\text{inclusion}} & & \\
0 & \longrightarrow & X^* & \xrightarrow{\rho^*} & Z^* & \xrightarrow{\jmath^*} & Y^* & \longrightarrow & 0
\end{array} \qquad (10.6)$$

Proof The 'only if' part is obvious from the definition. As for the 'if', taking adjoints in the hypothesised diagram and splicing with it the starting sequence, we obtain

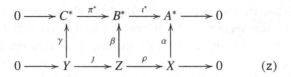

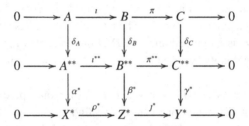

$$\begin{array}{ccccccccc}
0 & \longrightarrow & Y^*_* & \xrightarrow{\pi^*} & V^* & \xrightarrow{\imath^*} & X^*_* & \longrightarrow & 0\\
 & & \uparrow{\scriptstyle \text{restriction}} & & \uparrow{\scriptstyle v^*} & & \uparrow{\scriptstyle \text{restriction}} & & \\
0 & \longrightarrow & Y^{**} & \xrightarrow{\rho^*} & Z^{**} & \xrightarrow{\jmath^*} & X^{**} & \longrightarrow & 0\\
 & & \uparrow{\scriptstyle \delta_Y} & & \uparrow{\scriptstyle \delta_Z} & & \uparrow{\scriptstyle \delta_X} & & \\
0 & \longrightarrow & Y & \xrightarrow{\jmath} & Z & \xrightarrow{\rho} & X & \longrightarrow & 0
\end{array}$$

By the 3-lemma, $v^*\delta_Z$ is an isomorphism: this shows that the starting sequence is isomorphic to the adjoint of $0 \longrightarrow X_* \xrightarrow{\imath} V \xrightarrow{\pi} Y_* \longrightarrow 0$ and also that $v[V]$ is a predual of Z. □

The following two propositions yield especially remarkable examples of dual sequences:

Proposition 10.5.10 *If* $\mathrm{Ext}(X, Y^{**}/Y) = 0$ *then every exact sequence* $0 \longrightarrow X^* \longrightarrow Z \longrightarrow Y^* \longrightarrow 0$ *is the adjoint of a sequence* $0 \longrightarrow Y \longrightarrow Z_* \longrightarrow X \longrightarrow 0$ *for some* $Z_* \subset Z^*$, *which is necessarily a predual of* Z.

Proof Taking adjoints and forming the pullback we obtain the diagram

$$\begin{array}{ccccccccc}
0 & \longrightarrow & Y^{**} & \longrightarrow & Z^* & \longrightarrow & X^{**} & \longrightarrow & 0\\
 & & \| & & \uparrow & & \uparrow{\scriptstyle \delta_X} & & \\
0 & \longrightarrow & Y^{**} & \longrightarrow & \mathrm{PB} & \longrightarrow & X & \longrightarrow & 0
\end{array}$$

The hypothesis implies that the lower sequence becomes trivial after forming the pushout with the quotient map $Y^{**} \longrightarrow Y^{**}/Y$. Therefore, there is a pushout diagram

$$\begin{array}{ccccccccc}
0 & \longrightarrow & Y^{**} & \longrightarrow & \mathrm{PB} = \mathrm{PO} & \longrightarrow & X & \longrightarrow & 0\\
 & & \uparrow{\scriptstyle \delta_Y} & & \uparrow & & \| & & \\
0 & \longrightarrow & Y & \longrightarrow & Z_* & \longrightarrow & X & \longrightarrow & 0
\end{array}$$

Now assemble the two diagrams and apply the preceding proposition. □

An unrefined version of this result appears in [150, proposition 3]. The following statement is in some sense dual to it.

Proposition 10.5.11 $0 \longrightarrow Y \longrightarrow Z \longrightarrow X \longrightarrow 0$ *is a dual sequence if and only if there is a predual* Z_* *of* Z *for which* Y *is weak*-closed in* Z.

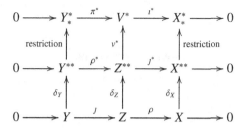

Proof If Y is weak*-closed then $Y = (Z_*/Y_\perp)^*$, where $Y_\perp = \{f \in Z_* : \langle f, y\rangle = 0 \,\forall y \in Y\}$ by the bipolar theorem, thus $0 \longrightarrow Y \longrightarrow Z \longrightarrow X \longrightarrow 0$ is the adjoint of $0 \longrightarrow Y_\perp \longrightarrow Z_* \longrightarrow Z/Y_\perp \longrightarrow 0$. The other implication is obvious. □

We are ready for the promised counterexample.

10.5.12 A counterexample to Vogt's duality problem *There is a separable dual W^* and an exact sequence $0 \longrightarrow \ell_2 \longrightarrow P \longrightarrow W^* \longrightarrow 0$ that is not a dual sequence. In particular, the space P is not isomorphic to a dual space.*

Proof The basic idea is simple: we start with a Banach space W complemented in its bidual and write $W^{**} = W \oplus A$, where A is a complement of W in W^{**}. Now we take a non-trivial extension $0 \longrightarrow A \longrightarrow E \longrightarrow R \longrightarrow 0$ where R is reflexive. Multiplying on the left by W, we obtain

$$\begin{array}{ccccccccc} 0 & \longrightarrow & A & \longrightarrow & E & \longrightarrow & R & \longrightarrow & 0 & \qquad (\mathtt{e}) \\ & & \downarrow{\scriptstyle \imath_A} & & \downarrow & & \| & & & \\ 0 & \longrightarrow & W^{**} = W \oplus A & \longrightarrow & W \oplus E & \longrightarrow & R & \longrightarrow & 0 & \qquad (\imath_A \mathtt{e}) \end{array}$$

By Proposition 10.5.10, $(\imath_A \mathtt{e})$ is the adjoint of a sequence

$$0 \longrightarrow R^* \longrightarrow P \longrightarrow W^* \longrightarrow 0 \qquad (\mathtt{p})$$

This sequence cannot be the adjoint of a sequence $0 \longrightarrow W \longrightarrow F \longrightarrow R \longrightarrow 0$ because, if it were, one could form the following diagram:

$$\begin{array}{ccccccccc} 0 & \longrightarrow & W & \longrightarrow & F & \longrightarrow & R & \longrightarrow & 0 & \qquad (\mathtt{f}) \\ & & \downarrow{\scriptstyle \imath_W} & & \downarrow{\scriptstyle v} & & \| & & & \\ 0 & \longrightarrow & W^{**} = W \oplus A & \longrightarrow & W \oplus E & \longrightarrow & R & \longrightarrow & 0 & \qquad (\imath_A \mathtt{e}) \\ & & \downarrow{\scriptstyle \pi_A} & & \downarrow & & \| & & & \\ 0 & \longrightarrow & A & \longrightarrow & \mathrm{PO} & \longrightarrow & R & \longrightarrow & 0 & \end{array}$$

But this cannot be unless $(\mathtt{e})$ is trivial: on one hand, the class of the lower row is $[\pi_A \imath_W \mathtt{f}] = 0$, since $\pi_A \imath_W = 0$ while, on the other hand, it is $[\pi_A \imath_A \mathtt{e}] = [\mathtt{e}]$, since $\pi_A \imath_A = \mathbf{1}_A$. This shows that sequence $(\mathtt{p})$ cannot be the adjoint of a sequence in which the subspace is W. To complete the proof, we must prove the same for all possible preduals of W^*, and the idea is to choose W such that any other predual of W^* is in 'essentially the same position' as W in W^{**}. Let W be a separable Banach space such that $W^{**}/W \simeq \ell_1$. Such a W exists by a result of Lindenstrauss [5, Section 15.1]: every separable Banach space U can be represented as $U \simeq W^{**}/W$ for some separable space W. As ℓ_1 is projective, we can write $W^{**} = W \oplus A$, where $A \simeq \ell_1$ is a fixed subspace of W^{**}. Brown and Ito proved in [54] that if $V \subset W^{**}$ is another predual of W^*

then there is a decomposition $W^{**} = V \oplus B$, where $B \cap A$ has finite codimension in A. Now we can start with a non-trivial element of $\mathrm{Ext}(\ell_2, \ell_1)$ to be used as the sequence (e) and whose existence is guaranteed by Proposition 5.2.20, and conclude the proof as follows: if (p) is a dual sequence there is an exact sequence $0 \longrightarrow V \longrightarrow G \longrightarrow R^* \longrightarrow 0$ where $V \subset W^{**}$ is a predual of W^* and a commutative diagram

$$\begin{array}{ccccccccc}
0 & \longrightarrow & V & \longrightarrow & G & \longrightarrow & R & \longrightarrow & 0 \\
 & & \downarrow{\scriptstyle \imath_V} & & \downarrow & & \| & & \\
0 & \longrightarrow & W^{**} = V \oplus B & \longrightarrow & W \oplus E & \longrightarrow & R & \longrightarrow & 0 \\
 & & \downarrow{\scriptstyle \pi_B} & & \downarrow & & \| & & \\
0 & \longrightarrow & B & \longrightarrow & \mathrm{PO} & \longrightarrow & R & \longrightarrow & 0
\end{array}
\qquad
\begin{array}{c}
(\mathsf{g}) \\ \\ (\imath_A \mathsf{e}) \\ \\ (\pi_B \imath_A \mathsf{e})
\end{array}$$

It follows that $[\pi_B \imath_A \mathsf{e}] = [\pi_B \imath_B \mathsf{g}] = 0$. Since $\pi_A \imath_A - \pi_B \imath_A$ has finite rank, $[\mathsf{e}] = 0$, which completes the proof of the first part. To conclude, the space P is not isomorphic to any dual space: otherwise, R^* would be weak*-closed in P and Proposition 10.5.11 would imply that (p) is a dual sequence. □

We can modify the counterexample $0 \longrightarrow \ell_2 \longrightarrow P \longrightarrow W^* \longrightarrow 0$ to obtain one whose quotient space is a bidual: take a separable Banach space V such that $V^{**}/V = c_0$ and set $W = V^*$ in the construction above. The final subtleties in the previous proof cannot be avoided. If J is James quasireflexive space such that $\mathsf{J}^{**}/\mathsf{J}$ has dimension 1 then $\ell_2(\mathsf{J})^{**} = \ell_2(\mathsf{J}) \oplus H$, where H is a separable Hilbert space. Take the Kalton–Peck sequence $0 \longrightarrow \ell_2 \longrightarrow Z_2 \longrightarrow \ell_2 \longrightarrow 0$ and multiply it by $\ell_2(\mathsf{J})$ to get the non-trivial sequence $0 \longrightarrow \ell_2(\mathsf{J}) \times \ell_2 \longrightarrow \ell_2(\mathsf{J}) \times Z_2 \longrightarrow \ell_2 \longrightarrow 0$; this sequence can be identified with one of the form $0 \longrightarrow \ell_2(\mathsf{J})^{**} \longrightarrow \diamond \longrightarrow \ell_2(\mathsf{J})^* \longrightarrow 0$, which must be the transpose of some sequence $0 \longrightarrow \ell_2^* \longrightarrow \diamond_* \longrightarrow \ell_2(\mathsf{J})^* \longrightarrow 0$. This sequence cannot be the adjoint of one of the form $0 \longrightarrow \ell_2(\mathsf{J}) \longrightarrow \diamond_{**} \longrightarrow \ell_2 \longrightarrow 0$ by the same reasoning as in the first part of the proof of the counterexample. However, the sequence is the adjoint of a sequence in which the subspace is another predual of $\ell_2(\mathsf{J})^*$ inside $\ell_2(\mathsf{J})^{**}$ because any hyperplane of J^{**} is a predual of J^*.

10.6 Extension of $\mathscr{L}_\infty$-Valued Operators

Do the $\mathscr{C}$-valued extension results of Chapter 8 remain valid for $\mathscr{L}_\infty$-valued operators? Some obviously do, such as the Johnson–Zippin theorem 8.6.2. This question for the Lindenstrauss–Pełczyński theorem is posed by Zippin as Problem 6.15 in [466], and the answer is a strong no:

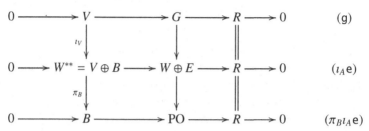

10.6.1 Example *Let H be a subspace of c_0 such that $c_0/H \not\simeq c_0$. The Bourgain–Pisier embedding $\imath\colon H \longrightarrow \mathscr{L}_\infty^{\mathsf{BP}}(H)$ cannot be extended to c_0.*

Proof Observe the two exact sequences

$$\begin{array}{ccccccccc} 0 & \longrightarrow & H & \stackrel{\jmath}{\longrightarrow} & c_0 & \longrightarrow & c_0/H & \longrightarrow & 0 \\ & & \| & & & & & & \\ 0 & \longrightarrow & H & \stackrel{\imath}{\longrightarrow} & \mathscr{L}_\infty^{\mathsf{BP}}(H) & \longrightarrow & S & \longrightarrow & 0 \end{array}$$

Since $\mathscr{L}_\infty^{\mathsf{BP}}(H)$ is separable, Sobczyk's theorem provides an extension of $\jmath$ through $\imath$. If $\imath$ would also extend through $\jmath$, the two sequences would be semi - equivalent, and then the diagonal principles yield $\mathscr{L}_\infty^{\mathsf{BP}}(H) \times c_0/H \simeq c_0 \times S$. In particular, c_0/H is a complemented subspace of $c_0 \times S$. Since S and c_0 are totally incomparable by the Schur property of S, we can apply the

10.6.2 Edelstein–Wojtaszczyk decomposition *Let X and Y be Banach spaces such that every operator from Y into X is strictly singular. Let P be a projection of $X \times Y$ onto an infinite-dimensional subspace E. Then there exists an automorphism τ_0 of $X \oplus Y$ and complemented subspaces $X_0 \subset X$ and $Y_0 \subset Y$ such that $\tau_0[E] = X_0 \times Y_0$*

(see [334, 2.c.13]) to obtain that c_0/H is isomorphic to some $A \times B$ with A complemented in c_0 and B complemented in S. Since c_0/H is a subspace of c_0, the space B must be finite-dimensional, hence $c_0/H \simeq c_0$, against the hypothesis. □

The Kalton extendability Theorem 8.5.4 also is not valid for ℓ_1:

10.6.3 Example *Let $\jmath\colon \ell_1 \longrightarrow C(\Delta)$ be any embedding. The Bourgain–Pisier embedding $\imath\colon \ell_1 \longrightarrow \mathscr{L}_\infty^{\mathsf{BP}}(\ell_1)$ cannot be extended to $C(\Delta)$.*

Proof An extension of $\imath$ through $\jmath$ yields a pullback diagram

$$\begin{array}{ccccccccc} 0 & \longrightarrow & \ell_1 & \stackrel{\jmath}{\longrightarrow} & C(\Delta) & \longrightarrow & B & \longrightarrow & 0 \\ & & \| & & \downarrow{\scriptstyle J} & & \downarrow & & \\ 0 & \longrightarrow & \ell_1 & \stackrel{\imath}{\longrightarrow} & \mathscr{L}_\infty^{\mathsf{BP}}(\ell_1) & \longrightarrow & S & \longrightarrow & 0 \end{array}$$

Since $\mathscr{L}_\infty^{\mathsf{BP}}(\ell_1)$ is a Schur space, it does not contain c_0, and therefore J must be weakly compact. But, then, its restriction $J|_{\ell_1} = \imath$ should be compact. □

On the other hand, a live agenda item in Lindenstrauss' memoir [323] is:

Proposition 10.6.4 *Compact $\mathscr{L}_{\infty,\lambda}$-valued operators admit compact λ-extensions anywhere.*

This can be proved by approximation since finite-rank operators have finite-rank λ-extensions. However, even weakly compact $\mathscr{C}$-valued operators do not necessarily admit extensions (weakly compact or otherwise), while weakly compact operators defined *on* $\mathscr{L}_\infty$-spaces admit weakly compact extensions. Thus, there are juicy classes of operators for which $\mathscr{L}_\infty$-valued operator extensions exist, even though Lindenstrauss–Pelczynski's and Kalton's theorems do not hold. Ok, worse things happen at sea.

Lindenstrauss–Pełczyński Spaces

Definition 10.6.5 A Banach space E is said to be a Lindenstrauss–Pełczyński space (LP) if all operators from subspaces of c_0 into E can be extended to c_0. When every operator $\tau\colon H \longrightarrow E$ admits a λ-extension, we shall say that E is an LP_λ space.

Each LP space is clearly an LP_λ space for some λ. As c_0 contains almost isometric copies of every finite-dimensional space, we see each LP_λ space is λ^+-locally injective and hence an $\mathscr{L}_\infty$-space. It therefore makes sense to ask; which $\mathscr{L}_\infty$-spaces are LP spaces?

Proposition 10.6.6 *The Banach spaces in* (a)–(e) *are* LP *spaces:*

(a) *$\mathscr{L}_\infty$-spaces not containing c_0,*
(b) *complemented subspaces of Lindenstrauss spaces,*
(c) *separably injective space,*
(d) *every quotient of an* LP *space by a separably injective subspace,*
(e) *the c_0-sum (in particular, the product) of* LP_λ *spaces.*
(f) *To be an* LP *space is not a* 3*-space property.*

Proof In what follows, H is always a subspace of c_0. To prove (a), observe that when a Banach space X contains no copy of c_0, every operator $H \longrightarrow X$ must be compact, and thus Proposition 10.6.4 applies.

Assertion (b) follows from the theorem in Section 8.8.2, and (c) is obvious.

To prove (d), let E be a separably injective space, let $0 \longrightarrow E \longrightarrow \mathsf{LP} \stackrel{\rho}{\longrightarrow} X \longrightarrow 0$ be an exact sequence and let $\tau\colon H \longrightarrow X$ be an operator. Since $\operatorname{Ext}(H, E) = 0$, τ can be lifted through ρ to an operator $H \longrightarrow \mathsf{LP}$ which, in turn, can be extended to an operator $T\colon c_0 \longrightarrow \mathsf{LP}$. The operator $\rho T\colon c_0 \longrightarrow X$ is the desired extension of τ.

(e) Keep in a handful of quietness the balance we mentioned after 10.1.1, the observation that one can easily consider all E_n equal; say, E, and the fact that since LP_λ spaces are $\mathscr{L}_{\infty,\mu}$-spaces for some μ, $\ell_\infty(E)$ is an $\mathscr{L}_{\infty,\mu}$-space, as well as their quotient, who must, therefore, have the BAP. Pick X a quotient of c_0 (with or without the BAP – as it could well be the case: see the space Z_∞ in [448, p. 276]). Then, apply Corollary 2.14.7 setting $Z = c_0$, $Z/Y = X$, $J = c_0(E)$ and $A = \ell_\infty(E)$.

(f) Combine a singular sequence $0 \longrightarrow C(K) \longrightarrow \Diamond \longrightarrow c_0 \longrightarrow 0$ with Bourgain's $0 \longrightarrow \mathcal{B} \longrightarrow c_0 \longrightarrow c_0 \longrightarrow 0$ in a pullback diagram

$$\begin{array}{ccccccccc}
 & & & & 0 & & 0 & & \\
 & & & & \uparrow & & \uparrow & & \\
0 & \longrightarrow & C(K) & \longrightarrow & \Diamond & \xrightarrow{\rho} & c_0 & \longrightarrow & 0 \\
 & & \| & & \uparrow & & \uparrow{\scriptstyle \pi} & & \\
0 & \longrightarrow & C(K) & \longrightarrow & \mathrm{PB} & \longrightarrow & c_0 & \longrightarrow & 0 \\
 & & & & \uparrow{\scriptstyle \jmath} & & \uparrow{\scriptstyle \imath} & & \\
 & & & & \mathcal{B} & = & \mathcal{B} & & \\
 & & & & \uparrow & & \uparrow & & \\
 & & & & 0 & & 0 & &
\end{array}$$

and show that PB cannot be an LP space. If $\jmath$ could be extended to c_0 through $\imath$, there would be an operator $\bar{\jmath}\colon c_0 \longrightarrow \Diamond$ yielding a commutative diagram

$$\begin{array}{ccccccccc}
0 & \longrightarrow & \mathcal{B} & \longrightarrow & c_0 & \xrightarrow{\pi} & c_0 & \longrightarrow & 0 \\
 & & \| & & \uparrow & & \uparrow{\scriptstyle \rho} & & \\
0 & \longrightarrow & \mathcal{B} & \longrightarrow & \mathrm{PB} & \longrightarrow & \Diamond & \longrightarrow & 0 \\
 & & \| & & \uparrow & & \uparrow{\scriptstyle \bar{\jmath}} & & \\
0 & \longrightarrow & \mathcal{B} & \longrightarrow & c_0 & \xrightarrow{\pi} & c_0 & \longrightarrow & 0
\end{array}$$

Since ρ is strictly singular, $\rho\bar{\jmath}$ is also strictly singular and therefore compact. It follows from Lemma 4.3.3 that the diagram above is impossible. □

It is clear how to use (b) to obtain concrete examples of LP spaces, while (c) and (d) can even be used to obtain non-separable LP spaces. On the other hand, at least three types of $\mathscr{L}_\infty$-spaces that do not contain c_0 appear in the literature:

- the Bourgain–Pisier space $\mathscr{L}_\infty^{\mathsf{BP}}(X)$ when X does not contain c_0, since not containing c_0 is a 3-space property [102, Theorem 3.2.e];
- the Bourgain–Delbaen isomorphic preduals of ℓ_1 without copies of c_0 [51];
- H.I. $\mathscr{L}_\infty$-spaces [17].

LP spaces enjoy additional properties: cheating a bit, they are those $\mathscr{L}_\infty$-spaces having all subspaces of c_0 placed in a unique position. That is not completely correct since the LP space $c_0 \times \ell_\infty$ contains complemented and uncomplemented copies of c_0 and thus it cannot be c_0-automorphic. Put precisely:

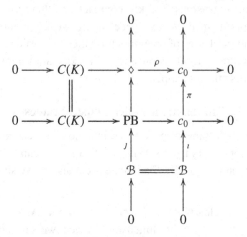

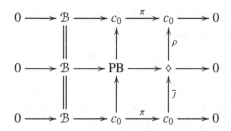

Proposition 10.6.7

(a) *A Banach space that contains c_0 and is automorphic for all subspaces of c_0 is an* LP *space.*

(b) *A separable $\mathscr{L}_\infty$-space is an* LP *space if and only if it is automorphic for all subspaces H of c_0.*

Proof (a) Let X be automorphic for all subspaces of c_0 and assume that there is an embedding $\jmath\colon c_0 \longrightarrow X$. Assume there is a subspace $H \subset c_0$ and an operator $T\colon H \longrightarrow X$ that cannot be extended to c_0. It turns out that, for small $\varepsilon > 0$, the operator $\jmath|_H + \varepsilon T\colon H \longrightarrow X$ is an embedding that cannot be extended to an operator $R \in \mathfrak{L}(X)$ through $\jmath|_H$; if, otherwise, $R\jmath|_H = \jmath|_H + \varepsilon T$ then $\varepsilon^{-1}(R\jmath - \jmath)$ would be an extension of T.

(b) The 'if' part is contained in (a) for spaces that contain c_0 and in Proposition 10.6.6(a) for those that do not. Let us show the other implication. Let X be a separable LP space. If X does not contain c_0 then the result is (vacuously) true. So let $\imath\colon H \longrightarrow X$ be an embedding where H is a subspace of c_0 and let $\jmath\colon H \longrightarrow c_0$ be the inclusion map. We can assume that $\imath$ has infinite-dimensional cokernel and that H is uncomplemented in c_0. Otherwise, the result follows directly from Sobczyk's theorem. The extension $J\colon c_0 \longrightarrow X$ of $\imath$ that exists because X is an LP space yields the commutative diagram

$$
\begin{array}{ccccccccc}
0 & \longrightarrow & H & \xrightarrow{\ \jmath\ } & c_0 & \xrightarrow{\ \rho\ } & c_0/H & \longrightarrow & 0 \\
 & & \| & & \downarrow{\scriptstyle J} & & \downarrow{\scriptstyle J'} & & \\
0 & \longrightarrow & H & \xrightarrow{\ \imath\ } & X & \xrightarrow{\ \pi\ } & X/\imath[H] & \longrightarrow & 0
\end{array}
\tag{10.7}
$$

which, in combination with Sobczyk's theorem, makes those two sequence semi-equivalent. The diagonal principles yield that the sequences

$$
\begin{array}{ccccccccc}
0 & \longrightarrow & H & \xrightarrow{(\imath,0)} & X \times c_0 & \longrightarrow & X/\imath[H] \times c_0 & \longrightarrow & 0 \\
 & & \| & & & & & & \\
0 & \longrightarrow & H & \xrightarrow{(\jmath,0)} & c_0 \times X & \longrightarrow & c_0/H \times X & \longrightarrow & 0
\end{array}
\tag{10.8}
$$

are isomorphic. We now show that the operator $\pi J = J'\rho$ is not weakly compact. Otherwise, it would be compact, and thus J' would also be compact. Since X is separable, $\imath$ can be extended to c_0, which yields a commutative diagram

$$
\begin{array}{ccccccccc}
0 & \longrightarrow & H & \xrightarrow{\ \imath\ } & X & \xrightarrow{\ \pi\ } & X/\imath[H] & \longrightarrow & 0 \\
 & & \| & & \downarrow{\scriptstyle I} & & \downarrow{\scriptstyle I'} & & \\
0 & \longrightarrow & H & \xrightarrow{\ \jmath\ } & c_0 & \xrightarrow{\ \rho\ } & c_0/H & \longrightarrow & 0
\end{array}
\tag{10.9}
$$

Putting Diagrams (10.7) and (10.9) together, we get

$$\begin{array}{ccccccccc} 0 & \longrightarrow & H & \xrightarrow{J} & c_0 & \longrightarrow & c_0/H & \longrightarrow & 0 \\ & & \| & & \big\downarrow{\scriptstyle IJ} & & \big\downarrow{\scriptstyle I'J'} & & \\ 0 & \longrightarrow & H & \xrightarrow{J} & c_0 & \xrightarrow{\rho} & c_0/H & \longrightarrow & 0 \end{array}$$

in which $I'J'$ is compact. Lemma 4.3.3 shows that this is impossible. Since $\mathscr{C}$-spaces have property (V) and πJ is not weakly compact, it must be an isomorphism on a subspace isomorphic to c_0, as well as π. This last copy of c_0 on which π is an isomorphism will necessarily be complemented in both $X/\imath[H]$ and X, which means that the sequences

$$\begin{array}{ccccccccc} 0 & \longrightarrow & H & \xrightarrow{\imath} & X & \longrightarrow & X/\imath[H] & \longrightarrow & 0 \\ & & \| & & & & & & \\ 0 & \longrightarrow & H & \xrightarrow{(\imath,0)} & X\times c_0 & \longrightarrow & X/\imath[H]\times c_0 & \longrightarrow & 0 \end{array} \tag{10.10}$$

are isomorphic. Matching (10.8) with (10.10), we get that the sequences

$$\begin{array}{ccccccccc} 0 & \longrightarrow & H & \xrightarrow{\imath} & X & \longrightarrow & X/\imath[H] & \longrightarrow & 0 \\ & & \| & & & & & & \\ 0 & \longrightarrow & H & \xrightarrow{(\jmath,0)} & c_0\times H & \longrightarrow & c_0/H\times X & \longrightarrow & 0 \end{array}$$

must be isomorphic too. Since the same is true starting with a different embedding $H \longrightarrow X$ with infinite-dimensional cokernel, the proof of (b) is done. □

$\mathscr{L}_\infty$-Envelopes

The natural embedding $X \longrightarrow C(B_X^*)$ enjoys the universal property that every $\mathscr{C}$-valued operator on X admits a 1-extension to $C(B_X^*)$. In other words, it is a $\mathscr{C}$-envelope in the following sense:

Definition 10.6.8 Let $\mathscr{A}$ be a class of Banach spaces. An $\mathscr{A}$-envelope of X is a space A in $\mathscr{A}$, together with an isometry $\delta\colon X \longrightarrow A$ such that, for every B in $\mathscr{A}$, every operator $\tau\colon X \longrightarrow B$ admits a 1-extension through δ.

Do similar envelopes exist for other classes of $\mathscr{L}_\infty$-spaces? For instance, any embedding of a Banach space X into a λ-injective space yields a λ-injective envelope of X, and Proposition 7.3.2, which yields the construction of an (α,β)-universal disposition envelope, can be easily modified to yield an (α,β)-injective envelope. We now focus on the construction of Lindenstrauss envelopes of separable Banach spaces. Since G contains 1-complemented copies of all separable Lindenstrauss spaces, every separable Banach space X has an isometry $\delta\colon X \longrightarrow \mathsf{G}$ that acts as a Lindenstrauss envelope. This can be arranged from the 'Fraïssé' construction of G simply by taking a chain of finite-dimensional subspaces $(X_n)_{n\geq 1}$ whose union is dense in X and considering a Fraïssé class of isometries containing the inclusions $X_n \longrightarrow X_m$ for $n \geq m$. Let us take a different approach based on the Bourgain–Pisier construction and which will presumably lead to a smaller envelope. Note that since $\mathscr{L}_\infty^{\mathsf{BP}}(X)/X$ is Schur, $\mathscr{L}_\infty^{\mathsf{BP}}(X)$ cannot be a Lindenstrauss space. On the other hand, as was shown in [129] and will be proved here soon, Lindenstrauss-valued operators on X admit 1^+-extensions to $\mathscr{L}_\infty^{\mathsf{BP}}(X)$, so we are close to the goal.

Proposition 10.6.9 *Every separable Banach space admits a Lindenstrauss envelope.*

Proof A drawing might help the reader understand the extension schema we will follow:

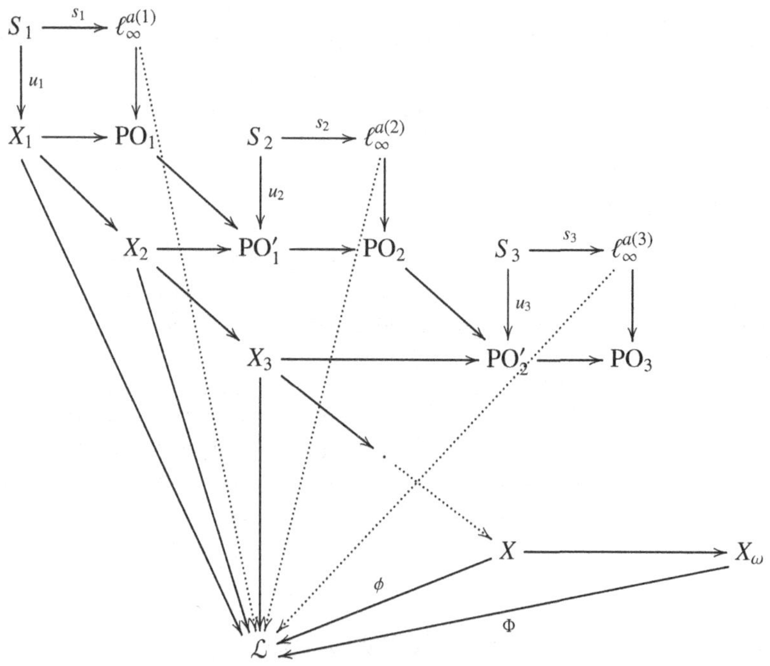

Keep the construction of $\mathscr{L}_\infty^{\mathsf{BP}}(X)$ in mind, especially Diagram (2.43). A subtle change is needed to make the resulting space Lindenstrauss: instead of fixing the parameters λ, η, choose sequences $(\lambda_n)_{n\geq 1}$ and $(\mu_n)_{n\geq 1}$ with

$\lambda_n^{-1} < \eta_n < 1$ and $\lim \lambda_n = 1$, and use λ_n, η_n at step n. The relevant property of the sequence of 'almost isometries' $(u_n)_{n\geq 1}$ is the estimate $\lambda_n^{-1}\|s\| \leq \|u_n(s)\| \leq \eta_n\|s\|$, for $s \in S_n$. And thus, PO_n is λ_n-isomorphic to $\ell_\infty^{a(n)}$, and this makes X_ω a Lindenstrauss space. The isometry $X \longrightarrow X_\omega$ remains unchanged. We prove the extension property: let $\mathcal{L}$ be a Lindenstrauss space and $\phi\colon X \longrightarrow \mathcal{L}$ be a norm 1 operator. The composition ϕu_1 is a finite-rank operator with norm at most $\eta_1 < 1$, and since $\mathcal{L}$ is locally 1^+-injective, there is a contractive extension $\ell_\infty^{a(1)} \longrightarrow \mathcal{L}$. Now use the pushout properties to get a contractive extension $\Phi_1 : \mathrm{PO}_1 \longrightarrow \mathcal{L}$; applying the universal property of PO_1' to this Φ_1 and $\phi|_{X_2} : X_2 \longrightarrow \mathcal{L}$ yields a new contractive extension $\phi_1 : \mathrm{PO}_1' \longrightarrow \mathcal{L}$, and the process can be iterated, now using $\phi_1 u_2$. The desired extension of ϕ is then defined locally by $\Phi(x) = \Phi_n(x)$ if $x \in \mathrm{PO}_n$. □

The original Bourgain–Pisier space $\mathscr{L}_\infty^{\mathsf{BP}}(X)$ corresponds to the choice $\lambda_n = \lambda$ and $\eta_n = \eta$, thus

Corollary 10.6.10 *Every Lindenstrauss-valued operator defined on a separable Banach space X admits a 1^+-extension to $\mathscr{L}_\infty^{\mathsf{BP}}(X)$.*

To deal with $\mathscr{L}_{\infty,\lambda}$-valued operators, we have to make a few tricky adjustments to simultaneously get a bigger $\mathscr{L}_{\infty,\lambda}$-superspace and the equal norm extension of $\mathscr{L}_{\infty,\lambda}$-valued operators on X.

Proposition 10.6.11 *For fixed $1 < \lambda < \infty$, every separable Banach space admits an $\mathscr{L}_{\infty,\lambda}$-envelope.*

Proof Set $\eta = \lambda^{-1}$ and simplify everything by choosing isometric embeddings $S_n \longrightarrow C(\Delta)$. All the rest goes as before. The resulting space X_ω is a separable $\mathscr{L}_{\infty,\lambda}$-space since it is the inductive limit of the spaces PO_n, all of them λ-isomorphic to $C(\Delta)$. □

More than 9/10 of the authors of this book are convinced that any reader arriving at this point will be able to construct an LP_λ-envelope and other similar envelopes without difficulties. A different matter is how to construct $\mathscr{L}_\infty$-envelopes. But, seriously: *Do $\mathscr{L}_\infty$-envelopes exist*?

10.7 Kadec Spaces

The p-Kadec space treated in Chapter 6 is the unique separable p-Banach space of AUCD with a skeleton. Proposition 6.3.13 provides different separable p-Banach spaces with property [Ͽ]. Moreover, Note 6.5.1 explained that any attempt to set $\varepsilon = 0$ and obtain a space of UCD (defined right below) directly expels it to the density of the continuum outside. Let us explore that outside.

Definition 10.7.1 A p-Banach space U is of universal complemented disposition (UCD) if, for all 1-pairs $u\colon\ E \rightleftarrows U$ and $v\colon\ E \rightleftarrows F$, where F is a finite-dimensional p-normed space, there exists a 1-pair $w\colon\ F \rightleftarrows U$ such that $u = w \circ v$.

The funny thing is that unearthing spaces of UCD is simpler than it was in the *almost* case. We just need a less rambunctious use of the Device:

Proposition 10.7.2 *Every p-Banach space can be isometrically embedded as a 1-complemented subspace of a p-Banach space of UCD.*

Proof Let's prepare a `recipe` with the following ingredients:

- the ordinal ω_1 and your favourite p-Banach space X;
- a set $\mathscr{F}^{(p)}$ containing exactly one isometric copy of each finite-dimensional p-Banach space;
- we work in the category of p-Banach spaces and 1-pairs and will construct an inductive system $(X_\alpha)_{0\leq\alpha<\omega_1}$ starting with $X_0 = X$;
- assuming all X_α have been defined for $\alpha < \beta$, if β is a limit ordinal then $X_\beta = \overline{\bigcup_{\alpha<\beta} X_\alpha}$ being $\imath_{\alpha,\beta}\colon\ X_\alpha \rightleftarrows X_\beta$ the obvious pairs;
- to obtain $X_{\alpha+1}$, let $\mathfrak{J}$ be the set of 1-pairs between the spaces in $\mathscr{F}^{(p)}$ and let $\mathfrak{L}_\alpha$ be the set of 1-pairs $u = \langle u^\flat, u^\sharp\rangle$ with domain in $\mathscr{F}^{(p)}$ and codomain X_α, excluding – this caution is crucial – those for which $u^\flat = \imath_{\eta,\alpha} v^\flat$ for some 1-pair v with codomain in X_η for some $\eta < \alpha$. Then consider the set $I_\alpha = \{(u, v) \in \mathfrak{L}_\alpha \times \mathfrak{J}\colon \operatorname{dom} u = \operatorname{dom} v\}$ and apply Lemma 6.3.15 to obtain a p-Banach space $X_{\alpha+1}$ and a 1-pair $\imath_{\alpha,\alpha+1}\colon\ X_\alpha \rightleftarrows X_{\alpha+1}$ with the corresponding pushout property. Furthermore, for every 1-pairs $v = \langle v^\flat, v^\sharp\rangle\colon\ E \rightleftarrows F$ in $\mathfrak{J}$ and $u = \langle u^\flat, u^\sharp\rangle\colon\ E \rightleftarrows X_\alpha$, regardless of whether u has been excluded, the construction miraculously provides a 1-pair $w\colon\ F \rightleftarrows X_{\alpha+1}$ such that $\imath_{\alpha,\alpha+1} \circ u = w \circ v$.

Let us prove that the space X_{ω_1} is of UCD. Consider a 1-pair $v \in \mathfrak{J}$ with domain E and codomain F and any 1-pair $u\colon\ E \rightleftarrows X_{\omega_1}$. Pick $\alpha < \omega_1$ such that $u^\flat[E]$ is contained in X_α. For each $\beta > \alpha$, the pair $\langle \imath_{\alpha,\beta} u^\flat,\ u^\sharp|_{X_\beta}\rangle$: $E \rightleftarrows X_\beta$ is one of the elements excluded from $\mathfrak{L}_\beta$, so the extending (through v) 1-pair $w_\beta\colon\ F \rightleftarrows X_{\beta+1}$, which exists, has $w_\beta^\flat = \imath_{\alpha,\beta+1} u^\flat$. Denoting the canonical pair by $\imath_\alpha\colon\ X_\alpha \rightleftarrows X_{\omega_1}$, we are ready to construct the desired 1-pair $w\colon\ F \rightleftarrows X_{\omega_1}$, that extends u. To do this, set $w^\flat = \imath_\alpha^\flat u^\flat$ and $w^\sharp x = \lim_{\mathfrak{U}} w_\alpha^\sharp \imath_{\alpha+1}^\sharp x$ using some ultrafilter refining the order filter on ω_1. □

There is no control on the size of the output in this construction because even with a 1-dimensional seed $X_0 = \mathbb{K}$, the dimension of the first space X_1 in the chain is already $\mathfrak{c}$. Skeleton issues mark the difference here. An *ω-skeleton* of a p-Banach space Y is a continuous ω_1-chain $(Y_\alpha)_{\alpha<\omega_1}$ of separable subspaces in which each Y_α is 1-complemented in $Y_{\alpha+1}$ and whose union is dense in Y. Here, *continuous* means that $Y_\beta = \overline{\bigcup_{\alpha<\beta} Y_\alpha}$ for all limit ordinals $\beta < \omega_1$. Kubiś proved [309, Lemma 6.1] that if $(Y_\alpha)_{\alpha<\omega_1}$ is an ω-skeleton of a Banach space Y (the proof works for p-Banach spaces as well) then Y admits a PRI $(P_\alpha)_{\omega\leq\alpha\leq\omega_1}$ such that $P_\alpha[Y] = Y_\alpha$.

No serious trouble is caused by our using ω_1 as index set for the skeleton and $[\omega, \omega_1)$ for the indices of the PRI: assume, if you like, that $Y_\alpha = Y_0$ for $0 \leq \alpha \leq \omega$.

Proposition 10.7.3 [CH] *A p-Banach space with ω-skeleton can be isometrically embedded as a* 1*-complemented subspace of a p-Banach space of UCD with ω-skeleton.*

Proof We need to perform a scrupulously ordered version of the previous recipe. By doing so, we construct the directed system of 1-pairs $X_\alpha \rightleftarrows X_\beta$ for $0 \leq \alpha \leq \beta < \omega_1$ with X_β separable for all $\beta < \omega_1$ in such a way that its limit X_{ω_1} is of UCD and contains a 1-complemented copy of Y. Assume Y has an ω-skeleton $(Y_\alpha)_{0\leq\alpha<\omega_1}$. We start with $X_0 = Y_0$ (not with Y!). As in the previous recipe, fix the set $\mathfrak{J}$ of 1-pairs with domain and codomain in $\mathscr{F}^{(p)}$ and let $\mathfrak{L}_0$ be the set of 1-pairs with domain in $\mathscr{F}^{(p)}$ and codomain X_0. Since, under CH, a set of size $\mathfrak{c}$ can be written as an increasing union of ω_1 countable sets and $|\mathfrak{L}_0 \times \mathfrak{J}| = \mathfrak{c}$, form $I_0 = \{(u, v) \in \mathfrak{L}_0 \times \mathfrak{J}\colon \operatorname{dom} u = \operatorname{dom} v\}$ and write $I_0 = \bigcup_{\alpha<\omega_1} I_0^\alpha$ with each I_0^α countable. Apply Lemma 6.3.15 using only I_0^1 to obtain a separable space X_1' and a 1-pair $Y_0 \rightleftarrows X_1'$. Forming the pushout

$$\begin{array}{ccc} Y_0 & \longrightarrow & X_1' \\ \downarrow & & \downarrow \\ Y_1 & \longrightarrow & X_1 \end{array}$$

produces our first separable space X_1 and 1-pair $\jmath_1\colon\ Y_1 \rightleftarrows X_1$. Now assume that the separable space X_β has already been obtained. Let $\mathfrak{L}_\beta$ be the set of 1-pairs with domain in $\mathscr{F}^{(p)}$ and codomain X_β, from which we exclude those 1-pairs u for which $u^\flat = \imath_{\eta,\beta} v^\flat$, for some $\eta < \beta$ and some 1-pair v, which have already been used in the construction of X_β. Let $I_\beta = \{(u, v) \in \mathfrak{L}_\beta \times \mathfrak{J} \colon \operatorname{dom} u = \operatorname{dom} v\}$ and write $I_\beta = \bigcup_{\alpha<\omega_1} I_\beta^\alpha$ with each I_β^α countable. Apply Lemma 6.3.15 using only I_β^β to obtain a separable space $X'_{\beta+1}$ and a 1-pair $Y_\beta \rightleftarrows X'_{\beta+1}$. Then form a new pushout

$$\begin{array}{ccc} Y_\beta & \longrightarrow & X'_{\beta+1} \\ \downarrow & & \downarrow \\ Y_{\beta+1} & \longrightarrow & X_{\beta+1} \end{array}$$

to obtain the 1-pair $\jmath_{\beta+1} \colon Y_{\beta+1} \rightleftarrows X_{\beta+1}$ with $X_{\alpha+1}$ separable. Set $X_\beta = \overline{\bigcup_{\alpha<\beta} X_\alpha}$ when β is a limit ordinal. The space X_{ω_1} has an ω-skeleton formed by $(X_\alpha)_{\alpha<\omega_1}$ and is of UCD, by a proof similar to that of Proposition 10.7.2. The space X_{ω_1} contains an isometric 1-complemented copy of $\bigcup_{\alpha<\omega_1} Y_\alpha$ with isometry $\jmath^\flat y = \imath_\alpha^\flat \jmath_\alpha^\flat y$ for $y \in Y_\alpha$ and projection $\jmath^\sharp x = \jmath_\alpha^\sharp \imath_\alpha^\sharp x$ for $x \in X_\alpha$. □

When the resulting space X_{ω_1} has ω-skeleton or the BAP then the same is true for Y and in this way we obtain the existence of spaces of UCD with or without ω-skeleton and without the BAP. To obtain spaces of UCD with the BAP, start with a space Y with the BAP and proceed methodically to show that each X_α can be obtained with a skeleton. Indeed, if X_α has a skeleton then so does $X_{\alpha+1}$: simply observe that the required countable pushout can also be formed via a countable number of pushouts performed with a finite number of operators each. Finally, if each X_α has the BAP then so does X_{ω_1} since any of its finite-dimensional subspaces is contained in some X_α. Finding different kinds of UCD spaces is an open problem: for instance, it is anybody's guess whether an ultrapower of a space of AUCD is of UCD (or has property $[\supset]$).

10.8 The Kalton–Peck Spaces

$\ell_p(\varphi)$ as a Fenchel–Orlicz Space

The Fenchel–Orlicz spaces (Section 1.8.2) are exactly those modular (hence Banach) spaces built over a Young funcion. Our purpose here is to show that $\ell_p(\varphi)$ is the Fenchel–Orlicz space associated with a certain function on $\mathbb{K}^2$. The general argument contains an elementary proof that the spaces $\ell_p(\varphi)$ are Banach spaces for $p > 1$.

Theorem 10.8.1 *For every $1 < p < \infty$ and every $\varphi \in \mathrm{Lip}_0(\mathbb{R}^+)$, there is a Young function $\Phi \colon \mathbb{K}^2 \longrightarrow \mathbb{R}^+$ such that $\ell_p(\varphi) = h(\Phi)$, with equivalent quasinorms.*

Proof The natural candidate to be the Young function for the space

$$\ell_p(\varphi) = \left\{ (y, x) \in \mathbb{K}^{\mathbb{N}} \times \mathbb{K}^{\mathbb{N}} : \sum_{n=1}^{\infty} \left| y(n) - x(n)\,\varphi\left(\log \frac{\|x\|}{|x(n)|}\right)\right|^p + |x(n)|^p < \infty \right\}$$

is $\Phi_p(y, x) = \big|y - x\,\varphi(-\log|x|)\big|^p + |x|^p$, which, unfortunately, is not convex. Let us circumvent this difficulty. A function $\phi\colon V \longrightarrow \mathbb{R}$ defined on a convex subset V of a linear space is said to be *quasiconvex* if there is some $C > 0$ such that, for all $v_1, v_2 \in V$ and all $t \in [0, 1]$, one has $\phi(tv_1+(1-t)v_2) \le C(t\phi(v_1)+(1-t)\phi(v_2))$.

Claim The function $\Phi_p\colon \mathbb{K}^2 \longrightarrow \mathbb{R}^+$ above is quasiconvex for $1 < p < \infty$.

Proof of the claim The proof is based on Lemma 3.2.5, the convexity of the function $x \longmapsto |x|^p$ and the following simple inequalities:

- $|x + y|^p \le 2^{1-1/p}(|x|^p + |y|^p)$, for every $x, y \in \mathbb{C}$,
- $|t \log t|^p \le t$ for $t \in (0, 1]$.

So, for $i = 1, 2$, choose $(y_i, x_i) \in \mathbb{K}^2$ and $t_i \in [0, 1]$ such that $t_1 + t_2 = 1$. Then,

$$\Phi_p\Big(\sum t_i(y_i, x_i)\Big) = \overbrace{\left|\sum_i t_i y_i - \Big(\sum_i t_i x_i\Big)\varphi\Big(-\log\Big|\sum_i t_i x_i\Big|\Big)\right|^p}^{\text{(a)}} + \overbrace{\left|\sum_i t_i x_i\right|^p}^{\text{(b)}},$$

$$\sum t_i \Phi_p(y_i, x_i) = \underbrace{\sum_i t_i\big|y_i - x_i\varphi(-\log|x_i|)\big|^p}_{\text{(c)}} + \underbrace{\sum_i t_i|x_i|^p}_{\text{(d)}},$$

We have (b) $\le$ (d), by convexity. Let us focus on (a). It will be helpful to recall the function $\omega_\varphi(x) = x\varphi(-\log|x|)$ from Section 3.2:

$$\begin{aligned}\text{(a)} &= \left|\sum_i t_i y_i - \omega_\varphi\Big(\sum_i t_i x_i\Big)\right|^p \\ &\le 2\left(\left|\sum_i (t_i y_i - \omega_\varphi(t_i x_i))\right|^p + \left|\omega_\varphi\Big(\sum_i t_i x_i\Big) - \sum_i \omega_\varphi(t_i x_i)\right|^p\right).\end{aligned}$$

Thanks to Lemma 3.2.5, the second summand can be bounded by

$$\left|\omega_\varphi\Big(\sum_i t_i x_i\Big) - \sum_i \omega_\varphi(t_i x_i)\right|^p \le \left(\frac{2\,\mathrm{Lip}(\varphi)}{e}\right)^p \left(\sum_i |t_i x_i|\right)^p \le C \sum_i t_i|x_i|^p.$$

As for the first part, we treat each summand separately as

$$\big|t_i y_i - \omega_\varphi(t_i x_i)\big|^p \le 2\left(\big|t_i y_i - t_i\omega_\varphi(x_i)\big|^p + \big|t_i\omega_\varphi(x_i) - \omega_\varphi(t_i x_i)\big|^p\right).$$

Its first chunk is dominated by (c), while its second chunk satisfies

$$\left|t_i\omega_\varphi(x_i) - \omega_\varphi(t_ix_i)\right|^p = \left|t_ix_i\varphi(-\log|x_i|) - t_ix_i\varphi(-\log|t_ix_i|)\right|^p$$
$$\leq \left|t_ix_i \operatorname{Lip}(\varphi)\log t_i\right|^p \leq \operatorname{Lip}(\varphi)t_i|x_i|^p. \qquad \square$$

Quasiconvex functions cannot be used as Young functions in principle, but Φ_p is equivalent to a convex function, in the same way that the usual quasinorm of $\ell_p(\varphi)$ is not a norm while being equivalent to a norm. Not every quasiconvex function is equivalent to a convex function: after all, every quasinorm is quasiconvex. The situation is more favourable in finite dimensions:

Lemma 10.8.2 *Every quasiconvex function defined on a finite-dimensional convex set is equivalent to some convex function.*

Proof This is a very standard proof based on Carathéodory's theorem [164, p. 34]: every point v of the convex hull of a subset W of a real k-dimensional space can be written as $v = \sum_{1\leq i\leq k+1} t_iv_i$, where $v_i \in W, t_i \in [0,1]$ and $\sum_{1\leq i\leq k+1} t_i = 1$. To be fussy, if W is connected, k points suffice. Let $\phi\colon V \longrightarrow \mathbb{R}^+$ be any non-negative function defined on a convex set. The greatest convex minorant of ϕ is the function $\operatorname{gcm}(\phi)\colon V \longrightarrow \mathbb{R}^+$ given by

$$\operatorname{gcm}(\phi)(v) = \inf\left\{\sum_{i=1}^m t_i\phi(v_i) : m \in \mathbb{N}, v = \sum_{i=1}^m t_iv_i, 1 = \sum_{i=1}^m t_i, t_i \in [0,1]\right\}.$$

It is clear that $\operatorname{gcm}(\phi)$ is convex, that $\operatorname{gcm}(\phi) \leq \phi$ and that $\operatorname{gcm}(\phi)(v)$ equals the infimum of those $t \in \mathbb{R}^+$ for which the point (v,t) belongs to the convex hull of the set $W = \{(u,\phi(u)) \in V \times \mathbb{R}^+ : u \in V\}$. If V is n-dimensional then W has dimension at most $n+1$, and every point (v,t) in the convex hull of W can be written as a convex combination of $n+2$ or fewer points in W so that $v = \sum_{i=1}^m t_iv_i$ and $t = \sum_{i=1}^m t_i\phi(v_i)$ with $m \leq n+2$ and, in the end,

$$\operatorname{gcm}(\phi)(v) = \inf\left\{\sum_{i=1}^m t_i\phi(v_i) : m \leq n+2, v = \sum_{i=1}^m t_iv_i, 1 = \sum_{i=1}^m t_i, t_i \in [0,1]\right\}.$$

Finally, if ϕ is quasiconvex then, for each $m \geq 2$, there is c_m such that for every $v_1,\ldots,v_m \in V$ and every $t_i \in [0,1]$ satisfying $\sum_{i=1}^m t_i = 1$, we have

$$\phi\left(\sum_{i=1}^m t_iv_i\right) \leq c_m \sum_{i=1}^m t_i\phi(v_i).$$

Hence $\operatorname{gcm}(\phi) \leq \phi \leq c_{n+2}\operatorname{gcm}(\phi)$. That is, ϕ is equivalent to $\operatorname{gcm}(\phi)$. $\square$

Let V be a finite-dimensional vector space, and let $\Phi\colon V \longrightarrow \mathbb{R}^+$ be a quasiconvex and even function such that $\Phi(0) = 0$ and $\lim_{t\to\infty}\Phi(tx) = \infty$ for all non-zero $x \in V$. Then $\operatorname{gcm}(\Phi)$ is a Young function. With a slight abuse of notation, we can consider the space

$$h(\Phi) = \left\{ v \in V^{\mathbb{N}} : \sum_{k=1}^{\infty} \Phi(tv(k)) < \infty \text{ for all } t > 0 \right\},$$

which agrees with the straight modular space $h(\mathrm{gcm}(\Phi))$, and the functional

$$\|v\|_{\Phi} = \inf\left\{ t > 0 : \sum_{k=1}^{\infty} \Phi(v(k)/t) \leq 1 \right\},$$

which is a quasinorm equivalent to the Luxemburg norm on $h(\mathrm{gcm}(\Phi))$. It being obvious after Claim 1 that these considerations apply to Φ_p, we are ready to conclude the proof that $\ell_p(\varphi) = h(\Phi_p)$ with equivalent quasinorms. This is done with the aid of the Kalton–Peck map $\mathsf{KP}_\varphi(x) = x\varphi\left(\frac{\|x\|}{|x|}\right)$ and its non-homogeneous version $\mathsf{kp}_\varphi \colon \ell_p \longrightarrow \mathbb{K}^{\mathbb{N}}$ defined as $\mathsf{kp}_\varphi(x) = x\varphi(-\log|x|)$ in Proposition 3.12.5 and the inequality $\|\mathsf{kp}_\varphi(x) - \mathsf{KP}_\varphi(x)\| \leq \mathrm{Lip}(\varphi)\big|\|x\|\log\|x\|\big|$ obtained there. Letting

$$|(y,x)|_{\mathsf{kp}_\varphi} = \|y - \mathsf{kp}_\varphi(x)\|^p + \|x\|^p = \sum_{k=1}^{\infty} \Phi_p(y(k), x(k)),$$

it should be obvious that $\ell_p(\varphi) = \{(y,x) \in \mathbb{K}^{\mathbb{N}} \times \mathbb{K}^{\mathbb{N}} \colon |(y,x)|_{\mathsf{kp}_\varphi} < \infty\}$ and also that $|(\cdot,\cdot)|_{\mathsf{kp}_\varphi}$ is 'coarsely equivalent' to $\|(\cdot,\cdot)\|_{\mathsf{KP}_\varphi}$ in the sense that

$$|(y,x)|_{\mathsf{kp}_\varphi} \leq f(\|(y,x)\|_{\mathsf{KP}_\varphi}) \qquad \text{and} \qquad \|(y,x)\|_{\mathsf{KP}} \leq g(|(y,x)|_{\mathsf{kp}_\varphi}),$$

for $f(t) = 2\,\mathrm{Lip}(\varphi)\max(t^p, |t\log t|^p)$ and $g(s) = 2\,\mathrm{Lip}(\varphi)\max(s^{1/p}, |s^{1/p}\log s^{1/p}|)$. In particular,

$$0 < r = \inf\{\|(y,x)\|_{\mathsf{KP}_\varphi} : |(y,x)|_{\mathsf{kp}_\varphi} \geq 1\} \leq \sup\{\|(y,x)\|_{\mathsf{KP}} : |(y,x)|_{\mathsf{kp}_\varphi} \leq 1\} = R < \infty.$$

Assuming $\|(y,x)\|_{\Phi_p} < 1$, by the very definition, there is an $s > 1$ such that

$$\sum_{k\geq 1} \Phi_p(s(y(k),x(k))) = |(sy,sx)|_{\mathsf{kp}_\varphi} \leq 1 \implies \|(y,x)\|_{\mathsf{KP}} \leq \|(sy,sx)\|_{\mathsf{KP}_\varphi} \leq R.$$

The other inclusion is also easy: if $\|(y,x)\|_{\mathsf{KP}} < r$ then $|(y,x)|_{\mathsf{kp}_\varphi} \leq 1$, and so $\|(y,x)\|_{\Phi_p} \leq 1$. □

Orlicz Subspaces of $\ell_p(\varphi)$

Let $0 < p < \infty$. The subspace $D = \{x \in \mathbb{K}^{\mathbb{N}} \colon (0,x) \in \ell_p(\varphi)\}$ of $\ell_p(\varphi)$ will be called the domain of the centralizer $\mathsf{KP}_\varphi \colon \ell_p \longrightarrow \mathbb{K}^{\mathbb{N}}$ because $D = \{x \in \ell_p \colon \mathsf{KP}_\varphi(x) \in \ell_p\}$. It is clear from Lemma 3.12.4 that D is an unconditional sequence space: if $x \in D$ and $a \in \ell_\infty$, then $ax \in D$ and

$$\|ax\|_D = \|(0,ax)\|_{\mathsf{KP}_\varphi} \leq C\|a\|_\infty\|(0,x)\|_{\mathsf{KP}_\varphi} = C\|a\|_\infty\|x\|_D,$$

where C is a constant depending only on $\mathrm{Lip}(\varphi)$ and p. If we put $\phi_p(t) = \Phi_p(0,t) = |t\varphi(-\log|t|)|^p + |t|^p$ then ϕ_p is an Orlicz function, and it should be obvious from the proof of the preceding claim that

- $D = \ell_{\phi_p}$ and the quasinorm of D is equivalent to the Luxemburg quasinorm of ℓ_{ϕ_p} given by $\|x\|_{\phi_p} = \inf\{r > 0\colon \sum_k \phi_p(x(k)/r) \le 1\}$;
- if $p > 1$ then ϕ_p is equivalent to a convex Orlicz function, hence to $\mathrm{gcm}(\phi_p)$, and so $D = \ell_{\phi_p} = h(\mathrm{gcm}(\phi_p))$. This also follows from the fact that $\phi_p(t)/t$ is increasing near zero along with classical material on Orlicz functions.

Lemma 10.8.3 *Let $(v_n)_{n\ge1}$ be a normalised block sequence in ℓ_p. If $w_n = (\mathsf{KP}_\varphi(v_n), v_n)$ then*

(a) *$(w_n)_{n\ge1}$ is an unconditional basic sequence in $\ell_p(\varphi)$;*
(b) *if $\varphi'(t)$ decreases to zero as $t \longrightarrow \infty$, then $(w_n)_{n\ge1}$ has a subsequence equivalent to the unit basis of either ℓ_{ϕ_p} or ℓ_p;*
(c) *if φ is the identity on $\mathbb{R}^+$, then (w_n) is equivalent to the unit basis of ℓ_{ϕ_p}.*

Proof Part (a) is clear from the centralizer property of the Kalton–Peck maps. Let $(c_n)_{n\ge1}$ be a sequence of scalars for which the series $\sum_n c_n w_n$ converges in $\ell_p(\varphi)$, and assume that $|d_n| \le |c_n|$ for all n. We set

$$w = \sum_n c_n w_n, \qquad v = \sum_n c_n v_n, \qquad u = \sum_n c_n \mathsf{KP}_\varphi(v_n),$$

$$\tilde{w} = \sum_n d_n w_n, \qquad \tilde{v} = \sum_n d_n v_n, \qquad \tilde{u} = \sum_n d_n \mathsf{KP}_\varphi(v_n),$$

where the series defining w converges in $\ell_p(\varphi)$, those of v and $\tilde{v}$ converge in ℓ_p and the other three are just pointwise sums. We must show that $\tilde{w} \in \ell_p(\varphi)$ and $\|\tilde{w}\| \le C\|w\|$ for some C independent on (c_n). Define $a\colon \mathbb{N} \longrightarrow \mathbb{K}$ by taking $a(k) = d_n(k)/c_n(k)$ for $k \in \operatorname{supp} v_n$ and filling with zeros. Then $\|a\|_\infty \le 1$ and

$$w = (u, v), \qquad \tilde{w} = (\tilde{u}, \tilde{v}), \qquad \tilde{u} = au, \qquad \tilde{v} = av,$$

hence $\tilde{w} \in \ell_p(\varphi)$ with $\|\tilde{w}\|_{\mathsf{KP}} \le \max(\Delta C(\mathsf{KP}_\varphi), 1)\|w\|_{\mathsf{KP}}$ by Lemma 3.12.4, where Δ is the modulus of concavity of ℓ_p. The proof of (b) is simpler after realising that the 'coordinate functionals' $(y, x) \longmapsto y(k)$ and $(y, x) \longmapsto x(k)$ are (uniformly) bounded on $\ell_p(\varphi)$, which follows trivially from the centralizer property of KP_φ. Indeed, for each $k \in \mathbb{N}$ and all $(y, x) \in \ell_p(\varphi)$, we have

$$\begin{aligned} |y(k)| + |x(k)| &= \|e_k y - \mathsf{KP}(e_k x)\| + \|e_k x\| \\ &= \|(e_k y, e_k x)\|_{\mathsf{KP}_\varphi} \le \max(\Delta C(\mathsf{KP}_\varphi), 1)\|(y, x)\|_{\mathsf{KP}_\varphi}. \end{aligned}$$

The immediate consequence is that convergence in $\ell_p(\varphi)$ implies convergence in every coordinate of $\mathbb{K}^{\mathbb{N}} \times \mathbb{K}^{\mathbb{N}}$, which provides a manageable criterion for the convergence of a series of the form $\sum_{n\ge1} t_n w_n$.

Claim The series $\sum_{n\geq 1} t_n w_n$ converges in $\ell_p(\varphi)$ if and only if the pointwise sum $w = (\sum_{n=1}^{\infty} t_n \mathsf{KP}_\varphi(v_n), \sum_{n=1}^{\infty} t_n v_n)$ belongs to $\ell_p(\varphi)$.

Proof of the claim The 'only if' direction is clear. To prove the converse, assume $w \in \ell_p(\varphi)$ and let us prove that the 'remainder'

$$r_N = w - \sum_{n\leq N} (t_n \mathsf{KP}_\varphi(v_n), t_n v_n) = \left(\sum_{n>N} t_n \mathsf{KP}_\varphi(v_n), \sum_{n>N} t_n v_n \right)$$

converges to zero in $\ell_p(\varphi)$. While this leads to serious difficulties if we insist on using the quasinorm of $\ell_p(\varphi)$, it becomes transparent if we use the 'equivalent' modular $|(y, x)|_{\mathsf{kp}_\varphi} = \|y - \mathsf{kp}_\varphi(x)\|^p + \|x\|^p$ instead, taking advantage of the fact that kp_φ and $\|\cdot\|^p$ are additive on disjoint families. Now,

$$|w|_{\mathsf{kp}_\varphi} = \left\| \sum_{n=1}^{\infty} t_n v_n \left(\varphi\left(\log \frac{1}{|v_n|}\right) - \varphi\left(\log \frac{1}{|t_n||v_n|}\right)\right)\right\|^p + \left\| \sum_{n=1}^{\infty} t_n v_n \right\|^p$$

$$= \sum_{n=1}^{\infty} |t_n|^p \left\| v_n \left(\varphi\left(\log \frac{1}{|v_n|}\right) - \varphi\left(\log \frac{1}{|t_n||v_n|}\right)\right)\right\|^p + \sum_{n=1}^{\infty} |t_n|^p,$$

$$|r_N|_{\mathsf{kp}_\varphi} = \sum_{n>N} |t_n|^p \left\| v_n \left(\varphi\left(\log \frac{1}{|v_n|}\right) - \varphi\left(\log \frac{1}{|t_n||v_n|}\right)\right)\right\|^p + \sum_{n>N} |t_n|^p.$$

So, certainly, $|r_N|_{\mathsf{kp}} \to 0$ as $N \to \infty$ if $|w|_{\mathsf{kp}} < \infty$. □

In more computable terms, the series $\sum_{n\geq 1} t_n w_n$ converges if and only if the numerical series $\sum_n |t_n|^p$ and

$$\sum_{n=1}^{\infty} \sum_{k\in \operatorname{supp} v_n} |t_n|^p |v_n(k)|^p \underbrace{\left| \varphi\left(\log \frac{1}{|v_n(k)|}\right) - \varphi\left(\log \frac{1}{|t_n||v_n(k)|}\right)\right|^p}_{(\star)} \tag{10.11}$$

are convergent. Now, if $\varphi'(t)$ is decreasing, with limit 0, then φ is increasing and concave, and for every n and k,

$$(\star) \leq \left| \varphi(0) - \varphi\left(\log \frac{1}{|t_n|}\right)\right| = \left| \varphi\left(\log \frac{1}{|t_n|}\right)\right|.$$

It follows that if $(t_n) \in \ell_{\phi_p}$ then $\sum_n t_n w_n \in \ell_p(\varphi)$ and

$$\left| \sum_n t_n w_n \right|_{\mathsf{kp}} \leq \sum_n \phi_p(t_n).$$

We now need to distinguish two cases, depending on the behaviour of the norms $\|v_n\|_\infty$.

• If $\|v_n\|_\infty \geq \varepsilon$ for some $\varepsilon > 0$ and all n, then (w_n) is equivalent to the unit basis of ℓ_{ϕ_p}.

Indeed, selecting for each n some $k \in \operatorname{supp} v_n$ such that $|v_n(k)| \geq \varepsilon$, we get

$$|w|_{\mathsf{KP}} \geq \sum_{n=1}^{\infty} |t_n|^p + \sum_{n=1}^{\infty} |t_n|^p |v_n(k)|^p \left| \varphi\left(\log \frac{1}{|v_n(k)|}\right) - \varphi\left(\log \frac{1}{|t_n||v_n(k)|}\right) \right|^p$$

$$\geq \sum_{n=1}^{\infty} |t_n|^p + \varepsilon^p \sum_{n=1}^{\infty} |t_n|^p \left| \varphi\left(\log \frac{1}{|t_n|}\right) \right|^p \geq \min(1, \varepsilon^p) \sum_{n=1}^{\infty} \phi_p(t_n).$$

• If, otherwise, $\liminf_n \|v_n\|_\infty = 0$, then we may assume, passing to a subsequence if necessary, that $|\varphi'(t)| \leq 2^{-n}$ for $t \geq \log(1/\|v_n\|_\infty)$. In this case, (w_n) is equivalent to the unit basis of ℓ_p. Assume (t_n) is a sequence in the unit ball of ℓ_p and consider the pointwise sum $\sum_n t_n w_n$.

Let us estimate ($\star$). By the mean value theorem, for each n and every $k \in \operatorname{supp} v_n$, there is $t \in (-\log|v_n(k)|, -\log|t_n||v_n(k)|)$ such that

$$\left| \varphi\left(\log \frac{1}{|v_n(k)|}\right) - \varphi\left(\log \frac{1}{|t_n||v_n(k)|}\right) \right| = |\varphi'(t)| \log \frac{1}{|t_n|} \leq 2^{-n} \log \frac{1}{|t_n|}$$

since $-\log|v_n(k)| \geq -\log|v_n|_\infty$. Recalling once again that $|t \log t| \leq e^{-1}$ for $0 \leq t \leq 1$, we have

$$\left| \sum_n t_n w_n \right|_{\mathsf{KP}} \leq \sum_n |t_n|^p + \sum_{n=1}^{\infty} \sum_{k \in \operatorname{supp} v_n} |t_n|^p |v_n(k)|^p 2^{-pn} \left| \log|t_n| \right|^p$$

$$\leq 1 + \sum_{n=1}^{\infty} \sum_{k \in \operatorname{supp} v_n} e^{-p} 2^{-pn} |v_n(k)|^p = 1 + \frac{e^{-p} 2^{-p}}{1 - 2^{-p}}. \qquad \square$$

Theorem 10.8.4 *Suppose that either $\varphi'(t)$ decreases to zero or φ is the identity on $\mathbb{R}^+$. Then every normalised pointwise null basic sequence in $\ell_p(\varphi)$ has a subsequence equivalent to the unit basis of either ℓ_p or ℓ_{ϕ_p}.*

Proof Let $w_n = (u_n, v_n)$ be such a sequence. If $\|v_n\| \to 0$, then $\tilde{w}_n = w_n - (\mathsf{KP}_\varphi v_n, v_n)$ belongs to $\imath[\ell_p]$ and $\|\tilde{w}_n - w_n\| \to 0$. It follows from the customary argument on perturbation of bases that (w_n) has a subsequence equivalent to the unit basis of ℓ_p. If (v_n) is not null, we may directly assume that $\|v_n\| \geq \varepsilon$ for some $\varepsilon > 0$ and all n. The hypothesis implies that (v_n) is pointwise null in ℓ_p and, passing to a subsequence, we may assume that there is a block basic sequence $(x_n)_{n\geq 1}$ in ℓ_p such that $\|v_n - x_n\| \leq 2^{-n}$. By applying Lemma 10.8.3 and relabelling, we can assume that $z_n = (\mathsf{KP}_\varphi x_n, x_n)$ is equivalent to one of the two bases under consideration. Now we distinguish two cases:

• If $\liminf_n \|w_n - z_n\| = 0$, then (w_n) and (z_n) have equivalent subsequences, and we are done.

• Otherwise, assume that (a subsequence of) $(w_n - z_n)$ is a basic sequence. But $w_n - z_n = (u_n - \mathsf{KP}_\varphi(x_n), v_n - x_n)$, and since $\|v_n - x_n\| \longrightarrow 0$ in ℓ_p, the first paragraph of the proof shows that $(w_n - z_n)$ has a subsequence equivalent to the unit basis of ℓ_p. It turns out that (w_n) and (z_n) are equivalent bases: if $\sum_n t_n w_n$ converges then $\sum_n t_n v_n$ and $\sum_n t_n x_n$ converge, hence $\sum_n |t_n|^p < \infty$ and $\sum_n t_n(w_n - z_n)$ converge, and so does $\sum_n t_n z_n$. The argument is reversible. □

Corollary 10.8.5 *Given* $0 < r \leq 1$, *define* $\varphi_r(t) = \min(t, t^r)$. *Then, for each fixed* $0 < p < \infty$, *the spaces* $\ell_p(\varphi_r)$ *are mutually non-isomorphic.*

Proof In fact, none of the spaces $\ell_p(\varphi_r)$ can be embedded into any other. Assume $T \colon \ell_p(\varphi_r) \longrightarrow \ell_p(\varphi_s)$ is an embedding. Then $f_n = T(0, e_n)$ is a basic sequence in $\ell_p(\varphi_s)$. If (f_n) is pointwise null then some subsequence would be equivalent to the unit basis of ℓ_p or to the unit basis of the Orlicz space associated with the function $t^p(1 + \log^{sp}(1/t))$, which cannot be, since (e_n) is equivalent to the unit basis of the Orlicz space associated with the function $t^p(1 + \log^{rp}(1/t))$. Otherwise, some subsequence of (f_n), which we do not relabel, is pointwise convergent in $\mathbb{K}^{\mathbb{N}} \times \mathbb{K}^{\mathbb{N}}$, and the preceding argument applies to $(e_{2n-1} - e_{2n})_{n\geq 1}$. □

Many other examples can be created, for instance, by introducing a second parameter and considering the family of functions $\varphi(t) = t^r \log^s(1/t)$ for $0 \leq r < 1, 0 < s < \infty$, and so on.

Z_p Obtained by Complex Interpolation

We briefly describe the complex interpolation method for pairs, following [43; 278]. Let $\mathbb{S}$ denote the open strip $\{z \in \mathbb{C} \colon 0 < \mathrm{Re}(z) < 1\}$ in the complex plane, and let $\overline{\mathbb{S}}$ be its closure. A pair (X_0, X_1) of complex Banach spaces will form an admissible or interpolation pairs if there exist injective operators $X_0 \longrightarrow \Sigma$ and $X_1 \longrightarrow \Sigma$ into some Banach space Σ. We will identify both X_0, X_1 with their continuous images in Σ without further mention. The Calderón space $\mathcal{H} = \mathcal{H}(X_0, X_1)$ is the space of continuous bounded functions $G \colon \overline{\mathbb{S}} \longrightarrow \Sigma$ that are holomorphic on $\mathbb{S}$ and satisfy the following boundary condition:

- for $k = 0, 1$, $G(k + it) \in X_k$ for each $t \in \mathbb{R}$ and $\sup_t \|G(k + it)\|_{X_k} < \infty$.

The space $\mathcal{H}$ is complete under the norm $\|G\| = \sup\{\|G(k + it)\|_{X_k} \colon k = 0, 1; t \in \mathbb{R}\}$. The evaluation map $\delta_z \colon \mathcal{H} \longrightarrow \Sigma$ is continuous for all $z \in \overline{\mathbb{S}}$. Given $\theta \in (0, 1)$, one defines the *interpolation* space as

$$X_\theta = \{x \in \Sigma \colon x = G(\theta) \text{ for some } G \in \mathcal{H}\}$$

endowed with the norm $\|x\|_{X_\theta} = \inf\{\|G\| : x = G(\theta), G \in \mathcal{H}\}$. This space is isometric to the quotient $\mathcal{H}/\ker\delta_\theta$, hence it is a Banach space. Now, if $z \in \mathbb{S}$ then the map $\delta'_\theta \colon \mathcal{H} \longrightarrow \Sigma$ given by evaluation of the derivative at θ, being the pointwise limit of a sequence of operators, is also bounded by the uniform boundedness principle. Thc connection between complex interpolation and twisted sums is provided by the following lemma:

Lemma 10.8.6 $\delta'_\theta \colon \ker\delta_\theta \longrightarrow X_\theta$ *is bounded and onto for* $0 < \theta < 1$.

Proof The crucial property of $\mathcal{H}$ in this bussiness is that if $\phi \colon \mathbb{S} \longrightarrow \mathbb{D}$ is a conformal equivalence vanishing at θ then $\ker\delta_\theta = \phi\cdot\mathcal{H}$, in the sense that every $F \in \mathcal{H}$ vanishing at θ has a factorisation $F = \phi\cdot G$, with $G \in \mathcal{H}$ and $\|G\| = \|F\|$. A conformal equivalence vanishing at θ is

$$\phi(z) = \frac{\exp(i\pi z) - \exp(i\pi\theta)}{\exp(i\pi z) - \exp(-i\pi\theta)}$$

(any other has the form $u\phi$ with $|u| = 1$). Now, if $F \in \ker\delta_\theta$, writing $F = \phi\cdot G$, we have $F' = \phi' G + \phi G'$, and so $\delta'_\theta(F) = \phi'(\theta)\delta_\theta(G)$, hence $\|\delta'_\theta \colon \ker\delta_\theta \longrightarrow X_\theta\| \leq |\phi'(\theta)|$. That δ'_θ maps $\ker\delta_\theta$ onto X_θ is also clear: if $x \in X_\theta$, then $x = G(\theta)$ for some $G \in \mathcal{H}$, and x is the derivative of $\phi(\theta)^{-1}\phi\cdot G$ at θ. □

Thus, for each $\theta \in (0, 1)$, there is a pushout diagram

$$\begin{array}{ccccccccc} 0 & \longrightarrow & \ker\delta_\theta & \longrightarrow & \mathcal{H}(X_0, X_1) & \xrightarrow{\delta_\theta} & X_\theta & \longrightarrow & 0 \\ & & \downarrow{\scriptstyle \delta'_\theta} & & \downarrow{\scriptstyle \overline{\delta'_\theta}} & & \| & & \\ 0 & \longrightarrow & X_\theta & \xrightarrow{\jmath} & \mathrm{PO} & \xrightarrow{\rho} & X_\theta & \longrightarrow & 0 \end{array} \tag{10.12}$$

where the lower row is a self-extension of X_θ. The twisted sum space PO, which is a quotient of $\mathcal{H}$, admits a nice description as a certain subspace of $\Sigma\times\Sigma$, as we now see. We call $dX_\theta = \{(f'(\theta), f(\theta)) \colon f \in \mathcal{H}\}$ equipped with the quotient norm $\|(y, x)\|_{dX_\theta} = \inf\{\|F\| \colon y = F'(\theta), x = F(\theta))\}$ the *derived* space. Let $Q \colon \mathcal{H} \longrightarrow dX_\theta$ be the natural quotient map. To prove that $dX_\theta \simeq \mathrm{PO}$, we show that the pushout sequence in (10.12) is equivalent to

$$0 \longrightarrow X_\theta \xrightarrow{\imath} dX_\theta \xrightarrow{\pi} X_\theta \longrightarrow 0 \tag{10.13}$$

with $\imath(y) = (y, 0)$ and $\pi(y, x) = x$. The operator π is correctly defined and maps the (open) unit ball of dX_θ onto that of X_θ. The kernel of π consists of those points $(y, 0) \in \Sigma\times\Sigma$ where y is the value at θ of the derivative of some function in $\mathcal{H}$ vanishing at θ; the previous lemma not only tells us that this is exactly X_θ

but also that $\|(y,0)\|_{dX_\theta} = |\phi'(\theta)|\,\|y\|_{X_\theta}$. Thus, $\imath$ is continuous, and the sequence is exact. We show that there is a commutative diagram

$$\begin{array}{ccccccccc} 0 & \longrightarrow & X_\theta & \xrightarrow{\jmath} & \mathrm{PO} & \xrightarrow{\rho} & X_\theta & \longrightarrow & 0 \\ & & \| & & \downarrow{\scriptstyle\gamma} & & \| & & \\ 0 & \longrightarrow & X_\theta & \xrightarrow{\imath} & dX_\theta & \xrightarrow{\pi} & X_\theta & \longrightarrow & 0 \end{array} \tag{10.14}$$

The operator γ is defined by the universal property of the pushout applied to the commutative square

$$\begin{array}{ccc} \ker\delta_\theta & \longrightarrow & \mathcal{H} \\ {\scriptstyle\delta'_\theta}\downarrow & & \downarrow{\scriptstyle Q} \\ X_\theta & \xrightarrow{\imath} & dX_\theta \end{array}$$

which produces the unique operator $\gamma\colon \mathrm{PO} \longrightarrow dX_\theta$ such that $\gamma\jmath = \imath$ and $\gamma\overline{\delta'_\theta} = Q$. This makes the left square of (10.14) commutative. The commutativity of the right square is also clear: $\rho\colon \mathrm{PO} \longrightarrow X_\theta$ is the only operator satisfying $\rho\jmath = 0$ and $\delta_\theta = \rho\overline{\delta'_\theta}$, and $\pi\gamma$ does the same. Of course, the sequence (10.13) can be described by a quasilinear map. To see which one, fix $\varepsilon > 0$ and, for each $x \in X_\theta$, (homogeneously) select $F_x \in \mathcal{H}(X_0, X_1)$ such that $x = F_x(\theta)$ and $\|F_x\| \le (1+\varepsilon)\|x\|_{X_\theta}$. Define $\Omega\colon X_\theta \longrightarrow \Sigma$ by $\Omega(x) = F'_x(\theta)$. With the notation of Section 3.12, we have:

Lemma 10.8.7 *Ω is quasilinear from X_θ to X_θ and $dX_\theta = X_\theta \oplus_\Omega X_\theta$, with equivalent quasinorms.*

Proof Pick $x, y \in X_\theta$ and let $F_x, F_y, F_{x+y} \in \mathcal{H}(X_0, X_1)$ be the corresponding extremals. One has $\Omega(x+y) - \Omega(x) - \Omega(y) = \delta'_\theta(F_{x+y} - F_x - F_y) \in X_\theta$, and since $F_{x+y} - F_x - F_y \in \ker\delta_\theta$, Lemma 10.8.6 applies to yield $\Omega(x+y) - \Omega(x) - \Omega(y) \in X_\theta$. Moreover,

$$\begin{aligned} \|\Omega(x+y) - \Omega(x) - \Omega(y)\|_{X_\theta} &= \|\delta'_\theta(F_{x+y} - F_x - F_y)\|_{X_\theta} \\ &\le \|\delta'_\theta : \ker\delta_\theta \longrightarrow X\|\big(\|F_{x+y}\| + \|F_x\| + \|F_y\|\big) \\ &\le \|\delta'_\theta\|(1+\varepsilon)\big(\|x+y\|_{X_\theta} + \|x\|_{X_\theta} + \|y\|_{X_\theta}\big) \\ &\le 2(1+\varepsilon)\|\delta'_\theta\|\big(\|x\|_{X_\theta} + \|y\|_{X_\theta}\big). \end{aligned}$$

We now check that $dX_\theta = X_\theta \oplus_\Omega X_\theta$ with equivalent norms. First note that $f'(\theta) - \Omega(f(\theta)) \in X_\theta$ for every $f \in \mathcal{H}$. Indeed, since $f - F_{f(\theta)} \in \ker\delta_\theta$, we have $f'(\theta) - \Omega(f(\theta)) = f'(\theta) - F'_{f(\theta)}(\theta)) = (f - F_{f(\theta)})'(\theta)) \in X_\theta$. One thus obtains the containment $dX_\theta \subset X_\theta \oplus_\Omega X_\theta$. Conversely, if $y - \Omega(x) \in X_\theta$ then $y - \Omega(x) = g'(\theta)$ for some $g \in \ker\delta_\theta$ since $\delta'_\theta\colon \ker\delta_\theta \longrightarrow X_\theta$ is onto. Thus,

$y = \Omega(x) + g'(\theta) = (F_x + g)'(\theta)$, and therefore $(y, x) = ((F_x + g)'(\theta), (F_x + g)(\theta)$. To prove the equivalence of norms, pick $(y, x) \in X_\theta \oplus_\Omega X_\theta$ so that x and $y - \Omega(x)$ belong to X_θ. Let F and G be the corresponding extremals:

$$x = F(\theta), \quad \|F\| \leq (1+\varepsilon)\|x\|_{X_\theta}, \quad y - \Omega(x) = G(\theta), \quad \|G\| \leq (1+\varepsilon)\|y - \Omega(x)\|_{X_\theta}.$$

If $\phi\colon \mathbb{S} \longrightarrow \mathbb{D}$ is a conformal map such that $\phi(\theta) = 0$ and we define $H(z) = \phi(\theta)^{-1}\phi(z)G(z) + F(z)$, then $H \in \mathcal{H}(X_0, X_1)$, with

$$\|H\| \leq |\phi(\theta)|^{-1}\|G\| + \|F\| \leq \max\left(|\phi(\theta)|^{-1}, 1\right)(1 + \varepsilon)\|(y, x)\|_\Omega,$$

and one has $H(\theta) = F(\theta) = x$ and $H'(\theta) = G(\theta) + F'(\theta) = y - \Omega(x) + \Omega(x) = y$. To prove the other inclusion, take $H \in \mathcal{H}(X_0, X_1)$ and set $(y, x) = (H'(\theta), H(\theta))$. Then $x \in X_\theta$ and $\|x\|_{X_\theta} \leq \|H\|$. Besides $\Omega(x) = F_x'(\theta)$, with $\|F_x\| \leq (1 + \varepsilon)\|x\|_{X_\theta}$. Hence $y - \Omega(x) = H'(\theta) - F_x'(\theta) = \delta_\theta'(H - F) \in X_\theta$ since $(H - F)(\theta) = 0$. Also,

$$\begin{aligned}\|y - \Omega(x)\|_{X_\theta} &\leq \|\delta_\theta'\|\|H - F\| \leq \|\delta_\theta'\|\left(\|H\| + \|F\|\right)\\ &\leq \|\delta_\theta'\|\left(\|H\| + (1 + \varepsilon)\|x\|_{X_\theta}\right) \leq (2 + \varepsilon)\|\delta_\theta'\|\|H\|,\end{aligned}$$

hence $\|(y, x)\|_\Omega = \|y - \Omega(x)\|_{X_\theta} + \|x\|_{X_\theta} \leq (2 + \varepsilon)\|\delta_\theta'\|\|H\| + \|H\|$. □

And now, and this is the main event of the evening, the Kalton–Peck spaces appear as the derived spaces associated with the pair (ℓ_1, ℓ_∞). It is no exaggeration to say that complex interpolation theory is founded on the fact that if we interpolate the pair (ℓ_p, ℓ_q) by the complex method, with both spaces sitting in ℓ_∞ and $1 \leq p, q \leq \infty$, then $(\ell_p, \ell_q)_\theta = \ell_r$, where $r^{-1} = (1-\theta)p^{-1} + \theta q^{-1}$. Thus, for every $x \in \ell_r$ and every $\varepsilon > 0$, there is an $F_x \in \mathcal{H}(\ell_p, \ell_q)$ such that $F_x(\theta) = x$ with $\|F_x\| \leq (1 + \varepsilon)\|x\|_r$. If $q < \infty$ then

$$F_x(z) = x\left(\frac{|x|}{\|x\|_r}\right)^{(r/q - r/p)(z-\theta)}$$

works even for $\varepsilon = 0$. If $q = \infty$, the same extremal can be used when x has finite support, but not in general, because F_x may be discontinuous on the right border of the strip. In any case, we get

$$\Omega(x) = \delta_\theta' F_x = \left(\frac{r}{q} - \frac{r}{p}\right) x \log \frac{|x|}{\|x\|_r} = \left(\frac{r}{p} - \frac{r}{q}\right) \mathsf{KP}(x).$$

Amazing, isn't it? This is the form in which Rochberg and Weiss [405] rediscovered the Kalton–Peck spaces. Many important features of the spaces Z_p can only be properly understood after realising that they arise as derived spaces in an interpolation schema.

10.9 The Properties of Z_2 Explained by Itself

The Kalton–Peck space Z_2 is *the* archetypal twisted Hilbert space, the archetypal twisted sum in fact. And, if twisted Hilbert spaces are the King's Landing of the theory of twisted sums, the space Z_2 sits on the Iron Throne. To study it, let us first formulate properties of general twisted Hilbert spaces before we pass to the specifics of Z_2. Twisted Hilbert spaces of course enjoy all the 3-space properties that Hilbert spaces enjoy. There is no need to make a complete list of such properties; we will just mention a few especially important ones:

⊛ *Twisted Hilbert spaces are ℓ_2-saturated.* That is, every closed infinite-dimensional subspace contains a copy of ℓ_2.

⊛ *Twisted Hilbert spaces are superreflexive.*

⊛ *Twisted Hilbert spaces are near-Hilbert*, i.e. they have type $2-\varepsilon$ and cotype $2+\varepsilon$ for all $\varepsilon > 0$ by Corollary 3.11.4. Near-Hilbert spaces were isolated by Szankowski [447] while studying Banach spaces all of whose subspaces have the approximation property (they must be near-Hilbert). On the other hand, it follows from the Maurey–Pisier Theorem 1.4.10 that near-Hilbert spaces have ℓ_2^n finitely represented in them. Near-Hilbert spaces are meaningful in the twisted context for at least two reasons: (a) the dichotomy theorem 7.2.10 for automorphic spaces – UFO spaces are either $\mathscr{L}_\infty$ or near-Hilbert – and (b) twisted Hilbert spaces are near-Hilbert.

⊛ *Non-trivial twisted Hilbert spaces do not have type* 2 *or cotype* 2 since type 2 spaces cannot contain uncomplemented Hilbert subspaces. Cotype 2 superreflexive spaces have type 2 duals, so the dual space (and thus the starting space) must be Hilbert. In the case of Z_2, this was explicitly shown during the proof of Proposition 3.2.7.

⊛ *Non-trivial twisted Hilbert spaces do not have unconditional bases.* The proof for Z_2 is in [280]. The state-of-the-art general proof is a complicated and not well-understood result of Kalton [268], see also [392], asserting that a twisted Hilbert space with unconditional basis must be Hilbert. It is also remarkable in this context that Z_2 admits an (obvious) unconditional 2-dimensional decomposition. It is not known whether the result remains valid for other p; that is, does there exist a non-trivial twisted sum of ℓ_p with unconditional basis? It can be proven that non-trivial twisted sums of ℓ_p spaces obtained from centralizers cannot have an unconditional basis [86, Theorem 3.9], but it is unknown whether non-trivial exact sequences $0 \longrightarrow \ell_p \longrightarrow \ell_p \longrightarrow \ell_p \longrightarrow 0$ exist for $p \neq 1, 2, \infty$.

⊛ *Twisted Hilbert spaces with additional properties.* The original Enflo–Lindenstrauss–Pisier example [167] has the form $\ell_2(\mathbb{N}, F_n)$, with F_n finite-dimensional. Twisted Hilbert spaces of the form $\ell_2(\mathbb{N}, F_n)$ enjoy property $\mathcal{W}_2$

[122]: they are reflexive and weakly null sequences admit weakly 2-summable subsequences. Quotient operators $X \longrightarrow \ell_2$ defined on a $\mathcal{W}_2$ space X are 'strictly non-singular', with the meaning that every infinite-dimensional subspace of ℓ_2 contains a further infinite-dimensional subspace on which the quotient map is invertible (after all, weakly 2-summable sequences are exactly the linear continuous images of the canonical basis of ℓ_2). Any twisted Hilbert space with property $\mathcal{W}_2$ thus contains complemented copies of ℓ_2.

The standard quasinorm of Z_2 is $\|(y, x)\| = \|y - \mathsf{KP}x\| + \|x\|$, which is equivalent to a norm because B-convex Banach spaces are $\mathscr{K}$-spaces (Corollary 3.11.3). We have:

★ *Z_2 is isomorphic to its dual.* To be honest, we don't know if this is a property of all twisted Hilbert spaces since all Kalton–Peck spaces $\ell_2(\varphi)$ and actually all twisted Hilbert spaces generated by centralizers have it [63]. This tree will grow in the Fiddler's Green of another book.

★ *The space Z_2 is 'self-similar'.* The feature of the Kalton–Peck maps KP_X formally described before Lemma 9.3.10, that they 'look the same everywhere', seems to be a property peculiar to Z_p. In fact, two-thirds of the authors of this book conjecture that it characterises Z_p. In the particular case $X = \ell_2$, Diagram (9.1) becomes

$$\begin{array}{ccccccccc} 0 & \longrightarrow & \ell_2 & \longrightarrow & Z_2 & \longrightarrow & \ell_2 & \longrightarrow & 0\\ & & \| & & \uparrow{\scriptstyle T_U} & & \uparrow & & \\ 0 & \longrightarrow & U & \longrightarrow & U\oplus_{\mathsf{KP}_U} U & \longrightarrow & U & \longrightarrow & 0\end{array} \qquad (10.15)$$

The middle operator is $T_U(u_n, 0) = (u_n, 0)$ and $T_U(0, u_n) = (\mathsf{KP}u_n, u_n)$ and is an isometry, as the proof of Lemma 9.3.10 clearly shows. Thus, its range is an isometric copy of Z_2. We show that it is complemented in Z_2. Let $D\colon Z_2 \longrightarrow Z_2^*$ be the isomorphism $D(y, x)(y', x') = \langle y, x'\rangle - \langle x, y'\rangle$ provided at Corollary 3.8.6. Let $D_U\colon U\oplus_{\mathsf{KP}_U} U \longrightarrow (U\oplus_{\mathsf{KP}_U} U)^*$ be the corresponding isomorphism whose action is determined by $\langle D_U(u_i, u_j), (u_k, u_l)\rangle = \delta_{il} - \delta_{jk}$. The diagram

$$\begin{array}{ccc} U\oplus_{\mathsf{KP}_U} U & \xrightarrow{\;T_U\;} & Z_2\\ {\scriptstyle D_U}\downarrow & & \downarrow{\scriptstyle D}\\ (U\oplus_{\mathsf{KP}_U} U)^* & \xleftarrow{\;T_U^*\;} & Z_2^* \end{array}$$

is commutative since, for arbitrary $i, j, k, l \in \mathbb{N}$, we have

$$\begin{aligned}(T_U^* D T_U(u_i, u_j))(u_k, u_l) &= \langle D T_U(u_i, u_j), T_U(u_k, u_l)\rangle \\ &= \langle D(u_i + \mathsf{KP}u_j, u_j), (u_k + \mathsf{KP}u_l, u_l)\rangle \\ &= \langle u_i + \mathsf{KP}u_j, u_l\rangle - \langle u_j, u_k + \mathsf{KP}u_l\rangle \\ &= \delta_{il} - \delta_{jk}.\end{aligned}$$

It follows that $D_U^{-1} T_U^* D$ a projection onto the range of T_U.

★ *Operators on* Z_2. The space Z_2 enjoys a surprising '(V)-like' property:

Proposition 10.9.1 *Every operator defined on* Z_2 *is either strictly singular or an isomorphism on a complemented copy of* Z_2.

Proof Since KP is singular, Lemma 9.1.5 yields that an operator $\tau\colon Z_2 \longrightarrow X$ is strictly singular if and only if its restriction to the canonical copy of ℓ_2 also is. Thus, if τ is not strictly singular then there is a block subspace U of ℓ_2 where the restriction of τ is an isomorphism. Replacing Z_2 by the range of the operator T_U, we can continue with the proof by assuming that U is the whole of ℓ_2, so that the restriction $\tau|_{\ell_2}$ is an embedding, say $\|\tau(y, 0)\| \geq \|y\|$ for all $y \in \ell_2$. Let us stare for few seconds at the pushout diagram:

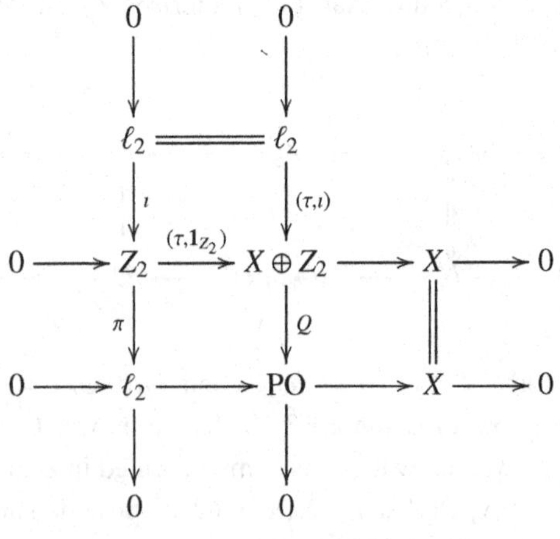

- The composition $Q(\tau, \mathbf{1}_{Z_2})$ is strictly singular since it factors through π.
- $Q(\tau, \mathbf{1}_{Z_2}) = Q(\tau, 0) + Q(0, \mathbf{1}_{Z_2})$.
- $Q(0, \mathbf{1}_{Z_2})$ is an embedding since

$$\begin{aligned}\|Q(0, z)\| &= \inf_{y \in \ell_2} \|(0, z) - (\tau, \imath)(y)\| \\ &= \inf_{y \in \ell_2} \|(-\tau y, z - y)\| \\ &= \inf_{y \in \ell_2} \{\|\tau(y, 0)\| + \|z - y\|\} \geq \|y\| + \|z\| - \|y\| = \|z\|.\end{aligned}$$

Hence, $Q(\tau, 0)$, being the difference (or sum) between a strictly singular operator and an embedding, has to have closed range and finite-dimensional kernel [334, Proposition 2.c.10]. Therefore it must be an isomorphism on some finite codimensional subspace of Z_2, and the same is true for τ. All subspaces of Z_2 with even codimension are isomorphic to Z_2, and thus we are done. □

Thus, it is not only the quotient map $Z_2 \longrightarrow \ell_2$ that is strictly singular: every operator $Z_2 \longrightarrow \ell_2$ is strictly singular. In particular:

Corollary 10.9.2 *Z_2 does not contain complemented copies of ℓ_2.*

Old Ideas and Popular Problems

- In [266], Kalton used a version of Z_2 to obtain quite natural examples of non-isomorphic complex spaces whose underlying real spaces are isomorphic, thus providing alternative solutions to a problem first solved by Bourgain [50]. Recall that if X is a complex Banach space then the conjugate $\overline{X}$ is the same space X with the 'new' multiplication $cx = \overline{c}x$ for $c \in \mathbb{C}, x \in X$. The spaces X and $\overline{X}$ are always isomorphic as *real* spaces by means of the identity. Given $\alpha \in \mathbb{R}$, Kalton considers the Lipschitz map $\varphi_\alpha(t) = t^{1+i\alpha}$ and then the twisted Hilbert space $Z_2(\alpha)$ generated by the Kalton–Peck map induced by φ_α. It is easy to see that the conjugate of $Z_2(\alpha)$ is $Z_2(-\alpha)$; in particular, the underlying *real* spaces are isomorphic. In a real tour de force, Kalton shows that *if $\alpha \neq \beta$ then the complex spaces $Z_2(\alpha)$ and $Z_2(\beta)$ are not isomorphic*. In particular the complex space $Z_2(\alpha)$ is not isomorphic to $Z_2(-\alpha)$ if $\alpha \neq 0$. The proof is technically demanding. Fortunately, Benyamini and Lindenstrauss [42, Theorem 16.17] come to our aid: yes, their proof is still complicated, but it is far simpler than Kalton's original.
- There are non-separable versions of Z_2 used in [171, p. 576] to show that admitting an injection into Hilbert spaces is not a 3-space property.
- **Is Z_2 prime?** This is unknown [228; 254]. What is known [254] is that complemented subspaces of Z_2 that are isomorphic to their square are isomorphic to Z_2. Perhaps Z_2 fails to be prime for the simplest of reasons.
- **Is Z_2 isomorphic to its hyperplanes?** This is a still open problem that Kalton was very fond of [228]. The common belief is that it is not [228; 254], which would make of Z_2 the first natural Banach space that is not isomorphic to its hyperplanes. The first example was due to Gowers [195]. Codimension 2 subspaces of Z_2 are obviously isomorphic to Z_2.
- **Do hyperplanes of Z_2 admit a complex structure?** Connected to the previous question is Ferenczi's observation behind [96] that, since Z_2 admits

a complex structure, if we could prove that hyperplanes of Z_2 do not admit complex structures then it would follow that Z_2 is not isomorphic to its hyperplanes.

- **New problems.** If, ultimately, the origin of the problems in this book is the question 'is being Hilbert a 3-space property?' then a new problem could be suggested now: 'is being twisted Hilbert a 3-space property?' Unlike the Hilbert question, this one has its origins well documented: it is due to D. Yost, appears posed in [102], is considered in [58] and is negatively solved in [72]. One could write a book about all that.

Bibliography

[1] I. Aharoni, J. Lindenstrauss, *Uniform equivalence between Banach spaces*, Bull. Amer. Math. Soc. 84 (1978) 281–283.

[2] P. Aiena, M. González, M. Chō, *The perturbation classes problem for closed operators*, Filomat 31 (2017) 621–627.

[3] M. Aigner, G. M. Ziegler, *Proofs from THE BOOK*. Springer, 1999.

[4] G. P. Akilov, *Necessary conditions for the extension of linear operators*, Doklady Acad. Sci. URSS 59 (1948) 417-418.

[5] F. Albiac, N. J. Kalton, *Topics in Banach Space Theory (2nd Ed.)*, GTM 233, Springer, 2016.

[6] A. B. Aleksandrov, *Essays on non locally convex Hardy classes,* in *Complex Analysis and Spectral Theory*. Edited by V. P. Havin, N. K. Nikol'skii. Lecture Notes in Math. 864. Springer, 1981, pp. 1–89.

[7] D. E. Alspach, S. Argyros, *Complexity of weakly null sequences*, Dissertationes Math. 321 (1992).

[8] D. Amir, *Continuous function spaces with the separable projection property*. Bull. Res. Council Israel Sect. F 10F (1962) 163–164.

[9] D. Amir, *Projections onto continuous function spaces*, Proc. Amer. Math. Soc. 15 (1964) 396–402.

[10] D. Amir, J. Lindenstrauss, *The structure of weakly compact sets in Banach Spaces*, Ann. Math. 88 (1968) 35–46.

[11] T. Ando, *Closed range theorems for convex sets and linear liftings*, Pacific J. Math. 44 (1973) 393–410.

[12] G. Androulakis, C. D. Cazacu, N. J. Kalton, *Twisted sums, Fenchel-Orlicz spaces and property* (M). Houston J. Math. 24 (1998) 105–126.

[13] G. Androulakis, T. Schlumprecht, *The Banach space S is complementably minimal and subsequentially prime*, Studia Math. 156 (2003) 227–242.

[14] R. Anisca, V. Ferenczi, Y. Moreno, *On the classification of positions and complex structures in Banach spaces*, J. Funct. Anal. 272 (2017) 3845–3868.

[15] S. A. Argyros, A. D. Arvanitakis, *A characterization of regular averaging operators and its consequences*, Studia Math. 151 (2002) 207–226.

[16] S. A. Argyros, J. M. F. Castillo, A. S. Granero, M. Jiménez, J. P. Moreno, *Complementation and embeddings of $c_0(I)$ in Banach spaces,* Proc. London Math. Soc. 85 (2002) 742–768.

[17] S. A. Argyros, R. Haydon, *A hereditarily indecomposable $\mathscr{L}_\infty$-space that solves the scalar-plus-compact problem,* Acta Math. 206 (2011) 1–54.

[18] S. A. Argyros, R. Haydon, *Bourgain-Delbaen $\mathscr{L}_\infty$-spaces, the scalar-plus-compact property and related problems*, Proceedings of the International Congress of Mathematicians Rio de Janeiro 2018, Vol III. Edited by B. Sirakov, P. Ney de Souza, M. Viana. World Scientific, 2018, pp. 1495–1531.

[19] S. A. Argyros, S. Mercourakis, S. Negrepontis, *Functional-analytic properties of Corson compact spaces*, Studia Math. 89 (1988) 197–229.

[20] S. A. Argyros, T. Raikoftsalis, *Banach spaces with a unique nontrivial decomposition*, Proc. Amer. Math. Soc. 136 (2008) 3611–3620.

[21] M. A. Ariño, M. A. Canela, *Complemented subspaces and the Hahn-Banach extension property in ℓ_p* $(0 < p < 1)$, Glasgow Math. J. 28 (1986) 115–120.

[22] A. Avilés, F. Cabello Sánchez, J. M. F. Castillo, M. González, Y. Moreno, *Separably injective Banach spaces*, Lecture Notes in Math. 2132, Springer, 2016.

[23] A. Avilés, F. Cabello Sánchez, J. M. F. Castillo, M. González, Y. Moreno, *Corrigendum to "On separably injective Banach spaces [Adv. Math. 234 (2013) 192-216]"*; Adv. Math. 318 (2017) 737–747.

[24] A. Avilés, P. Koszmider, *A* 1*-separably injective space that does not contain ℓ_∞*, Bull. London Math. Soc. 50 (2018) 249–260.

[25] A. Avilés, W. Marciszewski, G. Plebanek, *Twisted sums of c_0 and $C(K)$-spaces: a solution to the CCKY problem*, Adv. Math. 369 (2020) 107–168.

[26] A. Avilés, Y. Moreno, *Automorphisms in spaces of continuous functions on Valdivia compacta*, Topology Appl. 155 (2008) 2027–2030.

[27] S. Awodey, *Category Theory* (2nd Ed.), Oxford Logic Guides 52. Oxford University Press, 2010.

[28] J. W. Baker, *Projection constants for $C(S)$ spaces with the separable projection property*. Proc. Amer. Math. Soc. 41 (1973) 201–204.

[29] K. Ball, *An Elementary Introduction to Modern Convex Geometry,* in *Flavors of Geometry*. Edited by Silvio Levy. MSRI Lecture Notes 31, Cambridge University Press, 1997, pp. 1–58.

[30] S. Banach, *Théorie des Operations linéaires* (French), Monografie Matematyczne 1, Inst. Mat. Polskiej Akad. Nauk, Warszawa 1932; freely available at the Polish Digital Mathematical Library on http://pldml.icm.edu.pl. Reprinted by Chelsea, 1955 and Éditions Jacques Gabay, 1993.

[31] P. Bankston. *A survey of ultraproduct constructions in general topology,* Top. Atlas 8 (1993) 1–32.

[32] D. Bartošová, J. López-Abad, M. Lupini, B. Mbombo, *The Ramsey property for Banach spaces and Choquet simplices*, J. Eur. Math. Soc. 24 (2022) 1353–1388.

[33] J. Bastero, *ℓ_q-subspaces of stable p-Banach spaces,* $0 < p \leq 1$, Arch. Math. (Basel) 40 (1983) 538–544.

[34] B. Beauzamy, J.-T. Lapresté, *Modèles étalés des espaces de Banach*, Travaux en Cours. Hermann, 1984.

[35] E. Behrends, *On Rosenthal's ℓ_1 theorem,* Arch. Math. 62 (1994) 345–348.

[36] M. Bell, W. Marciszewski, *On scattered Eberlein compact spaces*, Israel J. Math. 158 (2007) 217–224.

[37] G. Bennett, L. E. Dor, V. Goodman, W. B. Johnson, C. M. Newman, *On uncomplemented subspaces of L_p, $1 < p < 2$*, Israel J. Math. 26 (1977) 178–187.

[38] Y. Benyamini, *Separable G-spaces are isomorphic to C(K)-spaces*, Israel J. Math. 14 (1973) 287–293.

[39] Y. Benyamini, *An M-space which is not isomorphic to a C(K)-space*, Israel J. Math. 28 (1977) 98–102.

[40] Y. Benyamini, *An extension theorem for separable Banach spaces*, Israel J. Math. 29 (1978) 24–30.

[41] Y. Benyamini, J. Lindenstrauss, *A predual of ℓ_1 which is not isomorphic to a C(K)-space*, Israel J. Math. 13 (1972) 246–259.

[42] Y. Benyamini, J. Lindenstrauss, *Geometric Nonlinear Functional Analysis, Vol. 1.*, Amer. Math. Soc., 1999.

[43] J. Bergh, J. Löfstrom, *Interpolation Spaces*, Grund. der math. Wissenschaften. 223. Springer, 1976.

[44] C. Bessaga, A. Pełczyński, *Spaces of continuous functions IV.* Studia Math. 19 (1960) 53–62.

[45] A. Błaszyk, A. Szymański, *Concerning Parovičenko's theorem*, Bull. Acad. Polon. Sci. Math. 28 (1980) 311–314.

[46] J. Blatter, E. W. Cheney, *Minimal projections on hyperplanes in sequence spaces*, Ann. Mat. Pura Appl. 101 (1974) 215–227.

[47] N. Bourbaki, *Espaces Vectoriels Topologiques*, Masson, 1981.

[48] J. Bourgain, *A counterexample to a complementation problem*, Compo. Math. 43 (1981) 133–144.

[49] J. Bourgain, *On the Dunford-Pettis property*, Proc. Amer. Math. Soc. 81 (1981) 265–272.

[50] J. Bourgain, *Real isomorphic complex Banach spaces need not be complex isomorphic*, Proc. Amer. Math. Soc. 96 (1986) 221–226.

[51] J. Bourgain, F. Delbaen, *A class of special $\mathscr{L}_\infty$-spaces*, Acta Math. 145 (1980) 155–176.

[52] J. Bourgain, G. Pisier, *A construction of $\mathscr{L}_\infty$-spaces and related Banach spaces*, Bol. Soc. Bras. Mat. 14 (1983) 109–123.

[53] K. S. Brown, *Cohomology of Groups*, GTM 87. Springer, 1982.

[54] L. Brown, T. Ito, *Some non-quasi-reflexive spaces having unique isomorphic preduals*, Israel J. Math. 20 (1975) 321–325.

[55] A. Brunel, L. Sucheston, *On B-convex Banach spaces*, Math. Systems Theory 7 (1974) 294–299.

[56] F. Cabello Sánchez, *Contribution to the classification of minimal extensions*, Nonlinear Anal.- TMA 58 (2004) 259–269.

[57] F. Cabello Sánchez, *Maximal symmetric norms on Banach spaces*, Math. Proc. Royal Irish Acad. 98A (2), (1998) 121–130.

[58] F. Cabello Sánchez, *Twisted Hilbert spaces*, Bull. Austral. Math. Soc. 59 (1999) 177–180.

[59] F. Cabello Sánchez, *A simple proof that super-reflexive spaces are K-spaces*, Proc. Amer. Math. Soc. 132 (2004) 697–698.

[60] F. Cabello Sánchez, *Quasi-additive mappings*, J. Math. Anal. Appl. 290 (2004) 263–270.

[61] F. Cabello Sánchez, *Yet another proof of Sobczyk's theorem*, in *Methods in Banach Space Theory*. Edited by J. M. F. Castillo, W. B. Johnson. London Math. Soc. LN 337. Cambridge University Press, 2006, pp. 133–138.

[62] F. Cabello Sánchez, *There is no strictly singular centralizer on* L_p, Proc. Amer. Math. Soc. 142 (2014) 949–955.

[63] F. Cabello Sánchez, *Nonlinear centralizers in homology*, Math. Ann. 358 (2014) 779–798.

[64] F. Cabello Sánchez, A. Avilés, P. Borodulin-Nadzieja, D. Chodounský, O. Guzmán, *Splitting chains, tunnels and twisted sums*, Israel J. Math. 241 (2021) 955–989.

[65] F. Cabello Sánchez, J. M. F. Castillo, *Duality and twisted sums of Banach spaces*, J. Funct. Anal. 175 (2000) 1–16.

[66] F. Cabello Sánchez, J. M. F. Castillo, *Banach space techniques underpinning a theory for nearly additive mappings*. Dissertationes Math. 404 (2002).

[67] F. Cabello Sánchez, J. M. F. Castillo, *Uniform boundedness and twisted sums of Banach spaces*, Houston J. Math. 30 (2004) 523–536.

[68] F. Cabello Sánchez, J. M. F. Castillo, *The long homology sequence in quasi-Banach spaces, with applications*, Positivity 8 (2004) 379–394.

[69] F. Cabello Sánchez, J. M. F. Castillo, *Stability constants and the homology of quasi-Banach spaces*, Israel J. Math. 198 (2013) 347–370.

[70] F. Cabello Sánchez, J. M. F. Castillo, W. H. G. Corrêa, V. Ferenczi, R. García, *On the* Ext^2*-problem in Hilbert spaces*, J. Funct. Anal. 280 (2021) 108863.

[71] F. Cabello Sánchez, J. M. F. Castillo, R. García, *Homological dimensions of Banach spaces*, Mat. Sbornik. 212 (2021) 531–550.

[72] F. Cabello Sánchez, J. M. F. Castillo, N. J. Kalton, *Complex interpolation and twisted twisted Hilbert spaces*, Pacific J. Math. 276 (2015) 287–307.

[73] F. Cabello Sánchez, J. M. F. Castillo, N. J. Kalton, D. T. Yost, *Twisted sums with* $C(K)$ *spaces*, Trans. Amer. Math. Soc. 355 (2003) 4523–4541.

[74] F. Cabello Sánchez, J. M. F. Castillo, W. Marciszewski, G. Plebanek, A. Salguero, *Sailing over three problems of Koszmider*, J. Funct. Anal. 279 (2020) 108571.

[75] F. Cabello Sánchez, J. M. F. Castillo, Y. Moreno, *On the bounded approximation property on subspaces of* ℓ_p *when* $0 < p < 1$ *and related issues*, Forum Math. 14-08 (2019) 1–24.

[76] F. Cabello Sánchez, J. M. F. Castillo, Y. Moreno, D. Yost, *Extension of* $\mathscr{L}_\infty$*-spaces under a twisted light*, in *Functional Analysis and Its Applications*. Edited by V. Kadets, W. Żelazko. North-Holland Math. Stud. 197. Elsevier, 2004, pp. 59 –70.

[77] F. Cabello Sánchez, J. M. F. Castillo, A. Salguero, *The behaviour of quasilinear maps on* $C(K)$*-spaces*, J. Math. Anal. Appl. 475 (2019) 1714–1719.

[78] F. Cabello Sánchez, J. M. F. Castillo, J. Suárez, *On strictly singular exact sequences*, Nonlinear Anal. -TMA 75 (2012) 3313–3321.

[79] F. Cabello Sánchez, J. M. F. Castillo, D. Yost, *Sobczyk's theorems from A to B*, Extracta Math. 15 (2000) 391–420.

[80] F. Cabello Sánchez, J. Garbulińska-Węgrzyn, W. Kubiś, *Quasi-Banach spaces of almost universal disposition*, J. Funct. Anal. 267 (2014) 744–771.

[81] N. L. Carothers, *A Short Course on Banach Space Theory*, London Math. Soc. Student Texts 64. Cambridge University Press, 2005.

[82] M. J. Carro, J. Cerdà, F. Soria, *Commutators and interpolation methods*, Ark. Math. 33 (1995) 199–216.

[83] P. G. Casazza, *Approximation properties*, in *Handbook of the Geometry of Banach Spaces*, Vol. I. Edited by W. B. Johnson, J. Lindenstrauss. Elsevier, 2001, pp. 271–316.

[84] P. G. Casazza, *Some questions arising from the homogeneous Banach space problem*, in *Banach Spaces*. Edited by B.-L. Lin, W. B. Johnson. Contemporary Mathematics 144 (1993) 35–52.

[85] P. G. Casazza, W. B. Johnson, L. Tzafriri, *On Tsirelson's space*, Israel J. Math 47 (1984) 81–98.

[86] P. G. Casazza, N. J. Kalton, *Unconditional bases and unconditional finite-dimensional decompositions in Banach spaces*, Israel J. Math. 95 (1996) 349–373.

[87] P. G. Casazza, N. J. Kalton, D. Kutzarova, M. Mastylo, *Complex interpolation and complementably minimal spaces*, Lecture Notes in Pure and Applied Math., 175. Marcel Dekker, 1996, pp. 135–143.

[88] P. G. Casazza, T. J. Shura, *Tsirelson's Space*. Lecture Notes in Math. 1363. Springer, 1989.

[89] E. Casini, E. Miglierina, Ł. Piasecki, L. Veselý, *Rethinking polyhedrality for Lindenstrauss spaces*, Israel J. Math. 216 (2016) 355–369.

[90] J. M. F. Castillo, *Banach spaces, à la recherche du temps perdu*, Extracta Math. 15 (2000) 291–334.

[91] J. M. F. Castillo, *Wheeling around Sobczyk's theorem*, in *General Topology and Banach Spaces*. Edited by A. Plichko, T. Banakh. NOVA, 2001, pp. 103–110.

[92] J. M. F. Castillo, *The hitchhiker guide to categorical Banach space theory. Part I*, Extracta Math. 25 (2010) 103–149.

[93] J. M. F. Castillo, *Nonseparable $C(K)$-spaces can be twisted when K is a finite height compact*, Topology Appl. 198 (2016) 107–116.

[94] J. M. F. Castillo, *Simple twist of K.* in *Nigel J. Kalton Selecta*, Vol. 2, Edited by F. Gesztesy, G. Godefroy, L. Grafakos, I. Verbitsky. Contemporary Mathematicians. Birkhäuser, 2016, pp. 251–254.

[95] J. M. F. Castillo, *The hitchhiker guide to categorical Banach space theory, Part II*, Extracta Math. 37 (2022) 1–56.

[96] J. M. F. Castillo, W. Cuellar, V. Ferenczi, Y. Moreno, *Complex structures on twisted Hilbert spaces*, Israel J. Math. 222 (2017) 787–814.

[97] J. M. F. Castillo, W. Cuellar, V. Ferenczi, Y. Moreno, *On disjointly singular centralizers*, Israel J. Math., (2022) https://doi.org/10.1007/s11856-022-2347-x

[98] J. M. F. Castillo, V. Ferenczi, M. González, *Singular exact sequences generated by complex interpolation*, Trans. Amer. Math. Soc. 369 (2017) 4671–4708.

[99] J. M. F. Castillo, V. Ferenczi, Y. Moreno, *On uniformly finitely extensible Banach spaces*, J. Math. Anal. Appl. 410 (2014) 670–686.

[100] J. M. F. Castillo, M. González, *On the Dunford-Pettis property in Banach spaces*, Acta Univ. Carolinae Math. 35 (1994) 5–12.

[101] J. M. F. Castillo, M. González, *New results on the Dunford-Pettis property,* Bull. London Math. Soc. 27 (1995) 599–605.

[102] J. M. F. Castillo, M. González, *Three-space problems in Banach space theory*, Lecture Notes in Math. 1667, Springer, 1997.

[103] J. M. F. Castillo, M. González, A. Plichko, D. Yost, *Twisted properties of Banach spaces*, Math. Scand. 89 (2001) 217–244.

[104] J. M. F. Castillo, M. González, F. Sánchez, *M-ideals of Schreier type and the Dunford-Pettis property*, in *Non-Associative Algebra and Its Applications*. Edited by S. González, vol. 303, Kluwer Acad. Press, 1994, pp. 80–85.

[105] J. M. F. Castillo, M. González, F. Sánchez, *Oscillation of weakly null and Banach-Saks sequences*, Bolletino de'll U.M.I. 11-A (1997) 685–695.

[106] J. M. F. Castillo, M. González, M. A. Simões, *Universal disposition is not a 3-space property*, Filomat 33 (2019) 3203–3208.

[107] J. M. F. Castillo, Y. Moreno, *On isomorphically equivalent extensions of quasi-Banach spaces*, in *Recent Progress in Functional Analysis*. Edited by K. D. Bierstedt, J. Bonet, M. Maestre, J. Schmets. North-Holland Math. Stud. 187. Elsevier, 2000, pp. 263–272.

[108] J. M. F. Castillo, Y. Moreno, *Strictly singular quasi-linear maps*, Nonlinear Anal. - TMA. 49 (2002) 897–904.

[109] J. M. F. Castillo, Y. Moreno, *On the Lindenstrauss-Rosenthal theorem*, Israel J. Math. 140 (2004) 253–270.

[110] J. M. F. Castillo, Y. Moreno, *The category of exact sequences between Banach spaces*, in *Banach Space Methods*, Proceedings of the V Conference in Banach spaces, Cáceres, 2004. Edited by J. M. F. Castillo, W. B. Johnson. London Math. Society Lecture Notes 337, Cambridge University Press, 2006, pp. 139–158.

[111] J. M. F. Castillo, Y. Moreno, *Singular and cosingular quasi-linear maps*, Arch. Math. 88 (2007) 123–132.

[112] J. M. F. Castillo, Y. Moreno, *Twisted dualities in Banach space theory*, in *Banach Spaces and Their Applications in Analysis*, in Honor of Nigel Kalton's 60th Birthday. Edited by B. Randrianantoanina, N. Randrianantoanina. Walter de Gruyter, 2007, pp. 59–76.

[113] J. M. F. Castillo, Y. Moreno, *Extensions by spaces of continuous functions*, Proc. Amer Math. Soc. 136 (2008) 2417–2424.

[114] J. M. F. Castillo, Y. Moreno, *Sobczyk's theorem and the bounded approximation property in Banach spaces*, Studia Math. 201 (2010) 1–19.

[115] J. M. F. Castillo, Y. Moreno, *On the bounded approximation property in Banach spaces*, Israel J. Math. 198 (2013) 243–259.

[116] J. M. F. Castillo, Y. Moreno, *Banach spaces of almost universal complemented disposition*, Q. J. Math. Oxford 71 (2020) 139–174.

[117] J. M. F. Castillo, Y. Moreno, M. A. Simões, *1-complemented subspaces of Banach spaces of universal disposition*, New York J. Math. 24 (2018) 251–260.

[118] J. M. F. Castillo, Y. Moreno, J. Suárez de la Fuente, *On Lindenstrauss-Pełczyński spaces*, Studia Math. 174 (2006) 213–231.

[119] J. M. F. Castillo, Y. Moreno, J. Suárez de la Fuente, *On the structure of Lindenstrauss-Pełczyński spaces*, Studia Math. 194 (2009) 105–115.

[120] J. M. F. Castillo, P. L. Papini, *On isomorphically polyhedral $\mathscr{L}_\infty$-spaces*, J. Funct. Anal. 270 (2016) 2336–2342.

[121] J. M. F. Castillo, A. Plichko, *Banach spaces in various positions*, J. Funct. Anal. 259 (2010) 2098–2138.

[122] J. M. F. Castillo, F. Sánchez, *Weakly p-compact, p-Banach-Saks and superreflexive Banach spaces*, J. Math. Anal. Appl. 185 (1994) 256–261.

[123] J. M. F. Castillo, F. Sánchez, *Remarks about the range of a vector measure*, Glasgow Math. J. 36 (1994) 157–161.

[124] J. M. F. Castillo, M. A. Simões, *On the three-space problem for the Dunford-Pettis property*, Bull. Austral. Math. Soc. 60 (1999) 487–493.

[125] J. M. F. Castillo, M. A. Simões, *Property (V) still fails the 3-space property*, Extracta Math. 27 (2012) 5–11.

[126] J. M. F. Castillo, M. A. Simões, *Positions in ℓ_1*, Banach J. Math. 9 (2015) 395–404.

[127] J. M. F. Castillo, M. A. Simões, *On Banach spaces of universal disposition*, New York J. Math. 22 (2016) 605–613.

[128] J. M. F. Castillo, M. A. Simões, J. Suárez de la Fuente, *On a question of Pełczyński about strictly singular operators*, Bull. Polish Acad. Sci. 60 (2012) 21–25.

[129] J. M. F. Castillo, J. Suárez de la Fuente, *Extension of operators into Lindenstrauss spaces*, Israel J. Math. 169 (2009) 1–27.

[130] J. M. F. Castillo, J. Suárez de la Fuente, *On $\mathscr{L}_\infty$-envelopes of Banach spaces*, J. Math. Anal. Appl. 394 (2012) 152–158.

[131] R. M. Causey, A. Fovelle, G. Lancien, *Asymptotic smoothness in Banach spaces, three space properties and applications*, preprint (2021), arXiv:2110.06710.

[132] K. Ciesielski, R. Pol, *A weakly Lindelöf function space $C(K)$ without any continuous injection into $c_0(T)$*, Bull. Polish Acad. Sci. Math. 32 (1984) 681–688.

[133] W. W. Comfort, S. Negrepontis, *The theory of ultrafilters*. Grund. der math. Wissenschaften 211. Springer, 1974.

[134] C. Correa, *Additional set-theoretic assumptions and twisted sums of Banach spaces*, in *Logic Around the World: On the Occasion of 5th Annual Conference of the Iranian Association for Logic*. Edited by M. Pourmahdian, A. Sadegh Daghighi. Amirkabir Publisher, arXiv:1801.10439.

[135] C. Correa, *Nontrivial twisted sums for finite height space under Martin's axiom*, Fund. Math. 248 (2020) 195–204.

[136] C. Correa, D. V. Tausk, *Nontrivial twisted sums of c_0 and $C(K)$*, J. Funct. Anal. 270 (2016) 842–853.

[137] C. Correa, D. V. Tausk, *Local extension property for finite height spaces*, Fund. Math. 245 (2019) 149–165.

[138] M.-D. Choi, E. G. Effros, *Lifting problems and the cohomology of C*-algebras*, Can. J. Math. 29 (1977) 1092–1111.

[139] M. Cwikel, N. J. Kalton, M. Milman, R. Rochberg, *A unified theory of commutator estimates for a class of interpolation methods*, Adv. Math. 169 (2002), no. 2, 241–312.

[140] M. Cwikel, M. Milman, R. Rochberg, *An overview of Nigel Kalton's work on interpolation and related topics*, in *Nigel J. Kalton Selecta*, Vol. 2. Edited by F. Gesztesy, G. Godefroy, L. Grafakos, I. Verbitsky. Contemporary Mathematicians. Birkhäuser, 2016, pp. 507–515.

[141] M. Cwikel, M. Milman, R. Rochberg, *Nigel Kalton's work on differentials in complex interpolation*, in *Nigel J. Kalton Selecta*, Vol. 2. Edited by

F. Gesztesy, G. Godefroy, L. Grafakos, I. Verbitsky. Contemporary Mathematicians. Birkhäuser, 2016, pp. 569–578.

[142] M. Cwikel, M. Milman, R. Rochberg, *Nigel Kalton and the interpolation theory of commutators*, in *Nigel J. Kalton Selecta*, Vol. 2. Edited by F. Gesztesy, G. Godefroy, L. Grafakos, I. Verbitsky. Contemporary Mathematicians. Birkhäuser, 2016, pp. 652–664.

[143] M. Cwikel, S. Reisner, *Interpolation of uniformly convex Banach spaces*, Proc. Amer. Math. Soc. 84 (1982) 555–559.

[144] M. M. Day, *The spaces L_p with* $0 < p < 1$, Bull. Amer. Math. Soc. 46 (1940) 816–823.

[145] D. W. Dean, *The equation $L(E, X^{**}) = L(E, X)^{**}$ and the principle of local reflexivity*, Proc. Amer. Math. Soc. 40 (1973) 146–148.

[146] R. Deville, V. Fonf, P. Hájek, *Analytic and polyhedral approximation of convex bodies in separable polyhedral Banach spaces*, Israel J. Math. 105 (1998) 139–154.

[147] R. Deville, G. Godefroy, *Some applications of projective resolutions of identity*, Proc. London Math. Soc. 67 (1993) 183–199.

[148] R. Deville, G. Godefroy, V. Zizler, *Smoothness and Renormings in Banach Spaces*, Monographs and Surveys in Pure and Applied Math. 64. Pitman, 1993.

[149] S. Dierolf, *Über Vererbbarkeitseigenschaften in topologischen Vektorräumen*, Dissertation, Ludwing-Maximilians-Universität, München, 1973.

[150] J. C. Díaz, S. Dierolf, P. Domański, C. Fernández, *On the three space problem for dual Fréchet spaces*, Bull. Polish Acad. Sci. 40 (1992) 221–224.

[151] J. Diestel, *A survey of results related to the Dunford-Pettis property*, Contemp. Math. 2 (1980) 15–60.

[152] J. Diestel, *Sequences and Series in Banach Spaces*, GTM 92. Springer, 1992.

[153] J. Diestel, H. Jarchow, A. Tonge, *Absolutely Summing Operators*, Cambridge Studies in Advanced Math. 43. Cambridge University Press, 1995.

[154] J. Diestel, J. J. Uhl Jr., *The Radon-Nikodym theorem for Banach space valued measures*, Rocky Mtn. Math. J. 6 (1976) 1–46.

[155] J. Diestel, J. J. Uhl Jr., *Vector Measures*, Math. Surveys 15, Amer. Math. Soc. 1977.

[156] J. Dieudonné, *Natural homomorphisms in Banach spaces*, Proc. Amer Math. Soc. 1 (1950) 54–59.

[157] S. Dilworth, Y.-P. Hsu, *On a property of Kadec-Klee type for quasi-normed unitary spaces*, Far East J. Math. Sci., Special Volume, Part II (1996), 183–194.

[158] S. Z. Ditor, *On a lemma of Milutin concerning averaging operators in continuous function spaces*, Trans. Amer. Math. Soc. 149 (1970) 443–452.

[159] S. Z. Ditor, *Averaging operators in C(S) and lower semicontinuous sections of continuous maps*, Trans. Amer. Math. Soc. 175 (1973) 195–208.

[160] P. Domański, *Local convexity of twisted sums*, in *Proceedings of the 12th Winter School on Abstract Analysis, Section of Analysis*. Edited by Z. Frolík. Circolo Matematico di Palermo, Palermo, 1984. Rendiconti del Circolo Matematico di Palermo, Serie II, Supplemento No. 5., pp. 13–31.

[161] P. Domański, *On the splitting of twisted sums and the three space problem for local convexity*, Studia Math. 82 (1985) 155–189.

[162] P. Dodos, *Banach Spaces and Descriptive Set Theory: Selected Topics*, Lecture Notes in Math. 1993. Springer, 2010.

[163] P. L. Duren, *Theory of H^p spaces*. Pure and Appl. Math. 38. Academic, 1970.

[164] H. G. Eggleston, *Convexity*, Cambridge Tracts in Mathematics and Mathematical Physics 47. Cambridge University Press, 1958.

[165] S. Eilenberg, S. MacLane, *General theory of natural equivalences*, Trans. Amer. Math. Soc. 58 (1945) 231–294.

[166] P. H. Enflo, *Comments on the paper "The endomorphisms of L_p, $(0 \leq p \leq 1)$" by N.J. Kalton, Indiana Univ. Math. J. 27 (1978) 353–381*, in *Nigel J. Kalton Selecta*, Vol. 2. Edited by F. Gesztesy, G. Godefroy, L. Grafakos, I. Verbitsky. Contemporary Mathematicians. Birkhäuser, 2016, pp. 69–70.

[167] P. Enflo, J. Lindenstrauss, G. Pisier, *On the "three-space" problem*, Math. Scand. 36 (1975) 199–210.

[168] H. Fakhoury, *Sélections linéaires associées au théorème de Hahn-Banach*, J. Funct. Anal. 11 (1972) 436–452.

[169] V. Ferenczi, *A uniformly convex hereditarily indecomposable Banach space*, Israel J. Math. 102 (1997) 199–225.

[170] V. Ferenczi, J. López-Abad, B. Mbombo, S. Todorcevic, *Amalgamation and Ramsey properties of L_p spaces*, Adv. Math. 369 (2020) 107–190.

[171] J. Ferrer, *A note on zeroes of real polynomials in $C(K)$-spaces*, Proc. Amer. Math. Soc. 137 (2009) 573–577.

[172] T. Figiel, W. B. Johnson, *The dual form of the approximation property for a Banach space and a subspace*, Studia Math. 231 (2015) 287–292.

[173] T. Figiel, W. B. Johnson, A. Pełczyński, *Some approximation properties of Banach spaces and Banach lattices*, Israel J. Math. 183 (2011) 199–232.

[174] T. Figiel, W. B. Johnson, L. Tzafriri, *On Banach lattices and spaces having local unconditional structure with applications to Lorentz function spaces*, J. Approx. Theory 13 (1975) 395–412.

[175] C. Finol, M. Wójtowicz, *The structure of nonseparable Banach spaces with uncountable unconditional bases*, RACSAM 99 (2005) 15–22.

[176] C. Foiaş, I. Singer, *On bases in $C([0,1])$ and $L^1([0,1])$*, Rev. Roumaine Math. Pures Appl. 10 (1965), 931–960.

[177] V. P. Fonf, *Massiveness of the set of extreme points of the dual ball of a Banach spaces and polyhedral spaces*, Funct. Anal. Appl. 12 (1978) 237–239.

[178] V. P. Fonf, J. Lindenstrauss, R. R. Phelps, *Infinite dimensional convexity*, in *Handbook on the Geometry of Banach Spaces*, Vol. I. Edited by W. B. Johnson, J. Lindenstrauss. Elsevier, 2001, pp. 599–670.

[179] V. P. Fonf, A. J. Pallares, R. J. Smith, S. Troyanski, *Polyhedral norms on non-separable Banach spaces*, J. Funct. Anal. 255 (2008) 449–470.

[180] L. Frerick, D. Sieg, *Exact categories in functional analysis*, Script, 2010. Available at: www.mathematik.uni-trier.de:8080/abteilung/analysis/HomAlg.pdf.

[181] D. H. Fremlin, *Comments on the paper "Uniformly exhaustive submeasures and nearly additive set functions" by N. J. Kalton, J. W. Roberts, Trans. Amer. Math. Soc. 278 (1983) 803-816*, in *Nigel J. Kalton Selecta*, Vol. 1. Edited by F. Gesztesy, G. Godefroy, L. Grafakos, I. Verbitsky. Contemporary Mathematicians. Birkhäuser, 2016, pp. 86–89.

[182] E. Galego, A. Plichko, *On Banach spaces containing complemented and uncomplemented subspaces isomorphic to c_0*, Extracta Math. 18 (2003) 315–319.

[183] J. Garbulińska-Węgrzyn, *Isometric uniqueness of a complementably universal Banach space for Schauder decompositions*, Banach J. Math. Anal. 8 (2014) 211–220.

[184] J. Garbulińska, W. Kubiś, *Remarks on Gurarĭ spaces*, Extracta Math. 26 (2011) 235–269.

[185] I. Gasparis, *New examples of c_0-saturated Banach spaces*, Math. Ann. 344 (2009) 491–500.

[186] S. I. Gelfand, Yu. I. Manin, *Methods of Homological Algebra.* (2nd Ed.). Springer Monographs in Math. Springer, 2010.

[187] N. Ghoussoub, E. Saab, *On the weak Radon-Nikodým property*, Proc. Amer. Math. Soc. 81 (1981) 81–84.

[188] G. Godefroy, *The Banach space c_0*, Extracta Math. 16 (2001) 1–25.

[189] G. Godefroy, N. J. Kalton, *Lipschitz-free Banach spaces*, Studia Math. 159 (2003) 121–141.

[190] G. Godefroy, N. J. Kalton, G. Lancien, *Subspaces of $c_0(\mathbb{N})$ and Lipschitz isomorphisms*, Geom. Funct. Anal. 10 (2000) 798–820.

[191] G. Godefroy, N. J. Kalton, G. Lancien, *Szlenk index and uniform homeomorphisms*, Trans. Amer. Math. Soc. 353 (2001) 3895–3918.

[192] G. Godefroy, G. Lancien, V. Zizler, *The non linear geometry of Banach spaces after Nigel Kalton*, Rocky Mtn. J. Math. 44 (2014) 1529–1583.

[193] G. Godefroy, P. Saphar, *Three-space problems for the approximation properties*, Proc. Amer. Math. Soc. 105 (1989) 70–75.

[194] Y. Gordon, D. R. Lewis, *Absolutely summing operators and local unconditional structures*, Acta Math. 133 (1974) 27–48.

[195] T. Gowers, *A solution to the Schröder Bernstein problem for Banach spaces*, Bull. London Math. Soc. 28 (1996) 297–304

[196] T. Gowers, *An infinite Ramsey theorem and some Banach-space dichotomies*, Ann. Math. 156 (2002) 797–833.

[197] W. T. Gowers, B. Maurey, *The unconditional basic sequence problem*, J. Amer. Math. Soc. 6 (1993) 851–874.

[198] W. T. Gowers, B. Maurey, *Banach spaces with small spaces of operators*, Math. Ann. 307 (1997) 543–568.

[199] A. S. Granero, *On complemented subspaces of $c_0(I)$*, Atti Sem. Mat. Fis. Univ. Modena 46 (1998) 35–36.

[200] A. Grothendieck, *Sur les applications linéaires faiblement compactes d'espaces de type $C(K)$*, Can. J. Math. 5 (1953) 129–173.

[201] A. Grothendieck, *Une caractérisation vectorielle-métrique des espaces L^1*, Can. J. Math. 7 (1955) 552–561.

[202] B. Grünbaum, *Some applications of expansion constants*, Pacific J. Math. 10 (1960) 193–201.

[203] V. I. Gurariy, *Space of universal disposition, isotropic spaces and the Mazur rotations of Banach spaces*, Sib. Mat. J. 7 (1966) 1002–1013.

[204] V. I. Gurariy, M. I. Kadec, V. I. Macaev, *On Banach-Mazur distance between certain Minkowski spaces*, Bull. Acad. Polon. Sci. Sér. Sci. Math. Astronom. Phys. 13 (1965) 719–722.

[205] J. Hagler, C. Stegall, *Banach spaces whose duals contain complemented subspaces isomorphic to* $C[0,1]$, J. Funct. Anal. 13 (1973) 233–251.

[206] H. Hahn, *Über halbstetige und unstetige Funktionen*, Sitz. Akad. Wiss. Wien IIa 126 (1917) 91–110.

[207] P. Hájek, V. Montesinos Santalucía, J. Vanderwerff, V. Zizler, *Biorthogonal Systems in Banach Spaces*, CMS Books in Mathematics 26. Springer, 2008.

[208] A. B. Hansen, N. J. Nielsen, *On isomorphic classification of polyhedral preduals of* L_1, Preprint Series Aarhus University 24, 1973/74.

[209] P. Harmand, D. Werner, W. Werner, *M-ideals in Banach Spaces and Banach Algebras.* Lecture Notes in Math. 1547, Springer, 1993.

[210] R. G. Haydon. *A nonreflexive Grothendieck space that does not contain* ℓ_∞, Israel J. Math. 40 (1981), 65–73.

[211] S. Heinrich, *Ultraproducts in Banach space theory*, J. Reine Angew. Math. 313 (1980) 72–104.

[212] C. W. Henson, J. Iovino, *Ultraproducts in Analysis*, in *Analysis and Logic*. Edited by C. W. Henson, J. Iovino, A. S. Kechris, E. Odell. London Math. Soc. Lecture Notes 262, Cambridge University Press, 2002, pp. 1–114.

[213] P. J. Hilton, U. Stammbach, *A course in homological algebra*, GTM 4. Springer, 1971.

[214] W. Hodges, *Model Theory*. Encyclopedia of Mathematics and Its Applications 42, Cambridge University Press, 1993.

[215] V. Indumathi, S. Lalithambigai, *A new proof of proximinality for M-ideals*, Proc. Amer. Math. Soc. 135 (2007) no. 4, 1159–1162.

[216] R. C. James, *Bases and reflexivity of Banach spaces*, Ann. Math. 52 (1950) 518–527.

[217] R. C. James, *A non-reflexive Banach Space isometric with its second conjugate space*, Proc. Nat. Acad. Sci. U. S. A. 37 (1951) 174–177.

[218] R. C. James, *Separable conjugate spaces*, Pacific J. Math. 10 (1960) 563–571.

[219] G. J. O. Jameson, *Topology and Normed Spaces.* Chapman and Hall Math. Series. Chapman and Hall, 1974.

[220] F. B. H. Jamjoon, H. M. Jebreen, D. T. Yost, *Colocality and twisted sums of Banach spaces*, J. Math. Anal. Appl. 323 (2006) 864–875.

[221] T. Jech, *Set Theory*. Perspectives in Mathematical Logic. Springer, 1997.

[222] M. Jiménez Sevilla, J. P. Moreno, *Renorming Banach spaces with the Mazur intersection property,* J. Funct. Anal. 144 (1997) 486–504.

[223] W. B. Johnson, *On finite dimensional subspaces of Banach spaces with local unconditional structure*, Studia Math. 51 (1974) 225–240.

[224] W. B. Johnson, *Extensions of* c_0, Positivity 1 (1997) 55–74.

[225] W. B. Johnson, J. Lindenstrauss, *Some remarks on weakly compactly generated Banach spaces*, Israel J. Math. 17 (1974) 219–230.

[226] W. B. Johnson, J. Lindenstrauss, *Examples of* $\mathscr{L}_1$ *spaces*, Ark. Mat. 18 (1980) 101–106.

[227] W. B. Johnson, J. Lindenstrauss, *Basic concepts in the geometry of Banach spaces*, in *Handbook of the Geometry of Banach Spaces*, Vol. 1. Edited by W. B. Johnson, J. Lindenstrauss. Elsevier, 2001, pp. 1–84.

[228] W. B. Johnson, J. Lindenstrauss, G. Schechtman. *On the relation between several notions of unconditional structure*, Israel J. Math. 37 (1980) 120–129.

[229] W. B. Johnson, J. Lindenstrauss, G. Schechtman, *Banach spaces determined by their uniform structures*, Geom. Funct. Anal. 6 (1996) 430–470.

[230] W. B. Johnson, T. Oikhberg, *Separable lifting property and extensions of local reflexivity*, Illinois J. Math. 45 (2001) 123–137.

[231] W. B. Johnson, H. P. Rosenthal, M. Zippin, *On bases, finite dimensional decompositions and weaker structures in Banach spaces*, Israel J. Math. 9 (1971) 488–506 .

[232] W. B. Johnson, H. P. Rosenthal, *On w*-basic sequences and their applications to the study of Banach spaces*, Studia Math. 43 (1972) 77–92.

[233] W. B. Johnson, A. Szankowski, *Complementably universal Banach spaces*, Studia Math. 58 (1976) 91–97.

[234] W. B. Johnson, M. Zippin, *Separable L_1 preduals are quotients of $C(\Delta)$*, Israel J. Math. 16 (1973) 198–202.

[235] W. B. Johnson, M. Zippin, *Subspaces and quotient spaces of $(\sum G_n)_{\ell_p}$ and $(\sum G_n)_{c_0}$*, Israel J. Math. 17 (1974) 50–55.

[236] W. B. Johnson, M. Zippin, *Extension of operators from subspaces of $c_0(\Gamma)$ into $C(K)$ spaces*, Proc. Amer. Math. Soc. 107 (1989) 751–754.

[237] W. B. Johnson, M. Zippin, *Extension of operators from weak*-closed subspaces of ℓ_1 into $C(K)$ spaces*, Studia Math. 117 (1995) 43–55.

[238] I. Juhász, *Cardinal Functions in Topology*. Mathematical Centre Tract 34. Mathematisch Centrum, 1971.

[239] I. Juhász, *On the weight-spectrum of a compact space*, Israel J. Math. 81 (1993) 369–379.

[240] M. I. Kadec, *On complementably universal Banach spaces*, Studia Math. 40 (1971) 85–89.

[241] M. I. Kadec, A. Pełczyński, *Bases, lacunary sequences and complemented subspaces in the spaces L_p*, Studia Math. 21 (1962) 161–176.

[242] M. I. Kadec, M. G. Snobar, *Certain functionals on the Minkowski compactum,* Math. Zametki 10 (1971) 453–457 (Russian); English transl. Math. Notes 10 (1971) 694–696.

[243] O. Kalenda, *Valdivia compact spaces in topology and Banach space theory*, Extracta Math. 15 (2000) 1–85.

[244] O. Kalenda, W. Kubiś, *Complementation in spaces of continuous functions on compact lines*, J. Math. Anal. Appl. 386 (2012) 241–257.

[245] J. A. Kalman, *Continuity and convexity of projections and barycentric coordinates in convex polyhedra*, Pacific J. Math. 11 (1961) 1017–1022.

[246] N. J. Kalton, *Basic sequences in F-spaces and their applications*, Proc. Edinburgh Math. Soc. 19 (1974) 151–167.

[247] N. J. Kalton, *Universal spaces and universal bases in metric linear spaces*, Studia Math. 61 (1977) 161–191.

[248] N. J. Kalton, *Transitivity and quotients of Orlicz spaces*, Comment. Math. (Special issue in honor of the 75th birthday of W. Orlicz) (1978) 159–172.

[249] N. J. Kalton, *Compact and strictly singular operators on Orlicz spaces*, Israel J. Math. 26 (1977) 126–136.

[250] N. J. Kalton, *The endomorphisms of L_p, $0 \leq p \leq 1$*, Indiana Univ. Math. J. 27 (1978) 353–381.

[251] N. J. Kalton, *The three-space problem for locally bounded F-spaces*, Compositio Math. 37 (1978) 243–276.

[252] N. J. Kalton, *Convexity, type and the three-space problem*, Studia Math. 69 (1981) 247–287.

[253] N. J. Kalton, *Isomorphisms between L_p-function spaces when $p < 1$*, J. Funct. Anal. 42 (1981) 299–337.

[254] N. J. Kalton, *The space Z_2 viewed as a symplectic Banach space*, in *Proc. Research Workshop on Banach Space Theory (1981), Univ. of Iowa*. Edited by Bor-Luh Lin. University of Iowa, 1982, pp. 97–111.

[255] N. J. Kalton, *Locally complemented subspaces and $\mathscr{L}_p$ spaces for $p < 1$*, Math. Nachr. 115 (1984) 71–97.

[256] N. J. Kalton, *The metric linear spaces L_p for $0 < p < 1$*, Contemporary Math. 52 (1986) 55–69

[257] N. J. Kalton, *The Maharam problem*, Séminaire d'Initiation à l'Analyse 1988/89, Exp. No. 18, 13 pp., Publ. Math. Univ. Pierre et Marie Curie, 94, Univ. Paris VI, 1991.

[258] N. J. Kalton, *Banach envelopes of nonlocally convex spaces*, Can. J. Math. 38 (1986) 65–86.

[259] N. J. Kalton, *Nonlinear commutators in interpolation theory*, Mem. Amer. Math. Soc. 73 (1988).

[260] N. J. Kalton, *Trace-class operators and commutators*, J. Funct. Anal. 86 (1989) 41–74.

[261] N. J. Kalton, *A remark on bases in quotients of ℓ_p when $0 < p < 1$*, Note Mat. 11 (1991) 231–236.

[262] N. J. Kalton, *The atomic space problem and related problems for F-spaces*, Proc. Orlicz Memorial Conference, University of Mississippi, 1991.

[263] N. J. Kalton, *Differentials of complex interpolation processes for Köthe function spaces*, Trans. Amer. Math. Soc. 333 (1992) 479–529.

[264] N. J. Kalton, *M-ideals of compact operators*, Illinois J. Math. 37 (1993) 147–169.

[265] N. J. Kalton, *The basic sequence problem*, Studia Math. 116 (1995) 167–187.

[266] N. J. Kalton, *An elementary example of a Banach space not isomorphic to its complex conjugate*, Can. Math. Bull. 38 (1995) 218–222.

[267] N. J. Kalton, *On subspaces of c_0 and extension of operators into $C(K)$-spaces*, Q. J. Oxford 52 (2001) 313–328.

[268] N. J. Kalton, *Twisted Hilbert spaces and unconditional structure*, J. Inst. Math. Jussieu 2 (2003) 401–408.

[269] N. J. Kalton, *Quasi-Banach spaces*, in *Handbook of the Geometry of Banach Spaces*, Vol. 2. Edited by W. B. Johnson, J. Lindenstrauss. Elsevier, 2003, pp. 1099–1130.

[270] N. J. Kalton, *Spaces of Lipschitz and Hölder functions and their applications*, Collectanea Math. 55 (2004) 171–217.

[271] N. J. Kalton, *Extension problems for $C(K)$-spaces and twisted sums*, in *Methods in Banach Space Theory*. Edited by J. M. F. Castillo, W. B. Johnson, London Math. Soc. Lecture Notes 337. Cambridge University Press, 2006, pp. 159–168.

[272] N. J. Kalton, *Extension of Lipschitz maps into $C(K)$-spaces*, Israel J. Math. 162 (2007) 275–315.

[273] N. J. Kalton, *Extension of linear operators and Lipschitz maps into $C(K)$-spaces*, New York J. Math. 13 (2007) 317–381.

[274] N. J. Kalton, *Automorphisms of $C(K)$ spaces and extension of linear operators*, Illinois J. Math. 52 (2008) 279–317.

[275] N. J. Kalton, *The nonlinear geometry of Banach spaces*, Rev. Mat. Univ. Complutense 21 (2008) 7–60.

[276] N. J. Kalton, *Lipschitz and uniform embeddings into ℓ_∞*, Fund. Math. 212 (2011) 53–69.

[277] N. J. Kalton, *Examples of uniformly homeomorphic Banach spaces*, Israel J. Math. 194 (2013) 151–182.

[278] N. J. Kalton, S. J. Montgomery-Smith, *Interpolation of Banach Spaces,* in *Handbook of the Geometry of Banach Spaces*, Vol. 2. Edited by W. B. Johnson, J. Lindenstrauss. Elsevier, 2003, pp. 1131–1175.

[279] N. J. Kalton, M. I. Ostrovskii, *Distances between Banach spaces*, Forum Math. 11 (1999) 17–48.

[280] N. J. Kalton, N. T. Peck, *Twisted sums of sequence spaces and the three-space problem*, Trans. Amer. Math. Soc. 255 (1979) 1–30.

[281] N. J. Kalton, N. T. Peck, *Quotients of $L_p(0,1)$ for $0 \le p < 1$*, Studia Math. 64 (1979) 65–75.

[282] N. J. Kalton, N. T. Peck, *A remark on a problem of Klee*, Colloq. Math. 71 (1996) 1–5.

[283] N. J. Kalton, N. T. Peck, J. W. Roberts, *An F-space Sampler*, London Math. Soc. Lecture Notes 89. Cambridge University Press, 1984.

[284] N. J. Kalton, A. Pełczyński, *Kernels of surjections from $\mathscr{L}_1$-spaces with an application to Sidon sets,* Math. Ann. 309 (1997) 135–158.

[285] N. J. Kalton, J. W. Roberts, *Uniformly exhaustive submeasures and nearly additive set functions*, Trans. Amer. Math. Soc. 278 (1983) 803–816.

[286] N. J. Kalton, J. H. Shapiro, *Bases and basic sequences in F-spaces*, Studia Math. 56 (1976) 47–61.

[287] N. J. Kalton, D. Werner, *Property (M), M-ideals, and almost isometric structure of Banach spaces*, J. Reine Angew. Math. 461 (1995) 137–178.

[288] N. J. Kalton, P. Wojtaszczyk, *On nonatomic Banach lattices and Hardy spaces*, Proc. Amer. Math. Soc. 120 (1994) 731–741.

[289] A. Kamiśka, *Comments on the paper "Banach envelopes of non-locally convex spaces" by N. J. Kalton, Can. J. Math. 38 (1986) 65–86*, in *Nigel J. Kalton Selecta*, Vol. 1. Edited by F. Gesztesy, G. Godefroy, L. Grafakos, I. Verbitsky. Contemporary Mathematicians. Birkhäuser, 2016, pp. 114–117.

[290] T. Kato, *Perturbation Theory for Linear Operators*, Grund. der math. Wissenschaften 132. Springer, 1980.

[291] M. Katětov, *On real-valued functions in topological spaces*, Fund. Math. 38 (1951) 85–91; Fund. Math. 40 (1953) 203–205 (correction).

[292] Y. Katznelson, *An Introduction to Harmonic Analysis*, Dover, 1976.

[293] A. S. Kechris, V. Pestov, S. Todorcevic, *Fraïsé limits, Ramsey theory, and topological dynamics of automorphism groups*, Geom. Funct. Anal. 15 (2005), no. 1, 106–189.

[294] S. V. Kisliakov, *Spaces with "small" annihiliators,* J. Soviet Math. 16 (1981) 1181–1184.

[295] S. V. Kislyakov, *Isomorphisms and projections for quotient-spaces of $\mathscr{L}_1$-spaces by reflexive subspaces*, J. Soviet Math. 34 (1986) 2074–2080.

[296] H. Knaust, E. Odell, T. Schlumprecht, *On asymptotic structure, the Szlenk index and UKK properties in Banach spaces*, Positivity 3 (1999) 173–199.

[297] P. Koszmider, *Banach spaces of continuous functions with few operators*, Math. Ann. 330 (2004) 151–183.

[298] P. Koszmider, *On decomposition of Banach spaces of continuous functions on Mrówka's spaces*, Proc. Amer. Math. Soc. 133 (2005) 2137–2146.

[299] P. Koszmider, *On large indecomposable Banach spaces*, J. Funct. Anal. 264 (2013) 1779–1805.

[300] P. Koszmider, C. Rodríguez-Porras, *On automorphisms of the Banach space* ℓ_∞/c_0, Fund. Math. 235 (2016) 49–99.

[301] P. Koszmider, S. Shelah, M. Świętek, *There is no bound on sizes of indecomposable Banach spaces*, Adv. Math. 323 (2018) 745–783.

[302] P. Koszmider, P. Zieliński, *Complementation and decompositions in some weakly Lindenlöf Banach spaces*, J. Math. Ann. Appl. 376 (2011) 329–341.

[303] G. Köthe, *Topological vector spaces I*, Grund. der math. Wissenschaften 159. Springer, 1969.

[304] G. Köthe, *Topological vector spaces II*, Grund. der math. Wissenschaften 237. Springer, 1979.

[305] E. Kreyszig, *Introductory Functional Analysis with Applications*. New York: Wiley Classics Library, 1978.

[306] J. L. Krivine, *Sous-espaces de dimension finie des espaces de Banach réticulés*, Ann. Math. 104 (1976) 1–29.

[307] J. L. Krivine, B. Maurey, *Espaces de Banach stables*, Israel J. Math. 39 (1981) 273–295.

[308] W. Kubiś, *Fraïssé sequences – a category-theoretic approach to universal homogeneous structures*, Ann. Pure Appl. Logic 165 (2014) 1755–1811.

[309] W. Kubiś, *Linearly ordered compacta and Banach spaces with a projectional resolution of the identity*, Topology Appl. 154 (2007) 749–757.

[310] W. Kubiś, S. Solecki, *A proof of uniqueness of Gurarĭ space*, Israel J. Math. 115 (2013) 449–456.

[311] J. Kupka, *A short proof and generalization of a measure theoretic disjointization lemma*, Proc. Amer. Math. Soc. 45 (1974) 70–72.

[312] P. A. Kuchment, *Reconstruction of a continuous representation with respect to a subrepresentation and a factor representation*, Funk. Anal i Priloz. 10 (1976) 79–80; english transl. Functional Anal. Appl. 10 (1976) 67–68.

[313] P. A. Kuchment, *Three-representation problem in Banach spaces*. Complex Anal. Oper. Theory 15 (2021), paper 34.

[314] K. Kuratowski, A. Mostowski, *Set Theory*. P.W.N., 1968.

[315] S. Kwapien, *Isomorphic characterizations of inner product spaces by orthogonal series with vector valued coefficients*, Studia Math. 44 (1972) 583–595.

[316] H. E. Lacey, *The Isometric Theory of Classical Banach Spaces*, Grund. der math. Wissenschaften 208. Springer, 1974.

[317] G. Lancien, *On uniformly convex and uniformly Kadec-Klee renormings*, Serdica Math. J. 21 (1995) 1–18.

[318] A. J. Lazar, *Polyhedral Banach spaces and extensions of compact operators*, Israel J. Math. 7 (1970) 357–364.

[319] A. J. Lazar, J. Lindenstrauss, *Banach spaces whose duals are L_1 spaces and their representing matrices*, Acta Math. 126 (1971) 165–193.

[320] D. H. Leung, *Some stability properties of c_0-saturated spaces*, Math. Proc. Cambridge Philos. Soc. 118 (1995) 287–301.

[321] A. Levy, *Basic Set Theory*, Perspectives in Math. Logic. Springer, 1979.

[322] Å. Lima, *Property (wM^*) and the unconditional metric compact approximation property*, Studia Math. 113 (1995) 249–263.

[323] J. Lindenstrauss, *On the Extension of Compact Operators*, Mem. Amer. Math. Soc. 48 (1964).

[324] J. Lindenstrauss, *On a certain subspace of ℓ_1*, Bull. Polish Acad. Sci. 12 (1964) 539–542.

[325] J. Lindenstrauss, *On nonlinear projections in Banach spaces*, Michigan Math. J. 11 (1964) 263–287.

[326] J. Lindenstrauss, *On complemented subspaces of m*, Israel J. Math. 5 (1967) 153–156.

[327] J. Lindenstrauss, *A remark on $\mathscr{L}_1$-spaces*, Israel J. Math. 8 (1970) 80–82.

[328] J. Lindenstrauss, *On James's paper "separable conjugate spaces"*, Israel J. Math. 9 (1971) 279–284.

[329] J. Lindenstrauss, A. Pełczyński, *Absolutely summing operators in $\mathscr{L}_p$ spaces and their applications*, Studia Math. 29 (1968), 275–326.

[330] J. Lindenstrauss, A. Pełczyński, *Contributions to the theory of the classical Banach spaces*, J. Funct. Anal. 8 (1971) 225–249.

[331] J. Lindenstrauss, H. P. Rosenthal, *The $\mathscr{L}_p$-spaces*, Israel J. Math. 7 (1969) 325–349.

[332] J. Lindenstrauss, H. P. Rosenthal, *Automorphisms in c_0, ℓ_1 and m*, Israel J. Math. 9 (1969) 227–239.

[333] J. Lindenstrauss, L. Tzafriri, *On the complemented subspace problem*, Israel J. Math. 9 (1971) 263–269.

[334] J. Lindenstrauss, L. Tzafriri, *Classical Banach spaces I, sequence spaces*, Ergeb. der Math. Grenzgebiete 92. Springer, 1977.

[335] J. Lindenstrauss, L. Tzafriri, *Classical Banach spaces II*, Ergeb. der Math. Grenzgebiete 97. Springer, 1979.

[336] J. Lindenstrauss, D. E. Wulbert, *On the classification of the Banach spaces whose duals are L_1 spaces*, J. Funct. Anal. 4 (1969) 332–349.

[337] Z. Lipecki, MR 85f:28006 (this is the review of [285]).

[338] R. H. Lohman, *Isomorphisms of c_0*, Can. Math. Bull. 14 (1971) 571–572.

[339] R. H. Lohman, *A note on Banach spaces containing ℓ_1*, Can. Math. Bull. 19 (1976) 365–367.

[340] J. López-Abad, *A Bourgain-Pisier construction for general Banach spaces*, J. Funct. Anal. 265 (2013) 1423–1441.

[341] J. López-Abad, S. Todorcevic, *Generic Banach spaces and generic simplexes*, J. Funct. Anal. 261 (2011) 300–386.

[342] M. Lupini, *Fraïssé limits in functional analysis*, Adv. Math. 338 (2018) 93–174.

[343] W. Lusky, *The Gurariĭ spaces are unique*, Arch. Math. 27 (1976) 627–635.

[344] W. Lusky, *Separable Lindenstrauss spaces,* in *Functional Analysis: Surveys and Recent Results, Proceedings of the Paderborn Conference on Functional Analysis.* Edited by K. D. Bierstedt, B. Fuchssteiner, North-Holland Math. Studies 27. Elsevier, 1977, pp. 15–28.
[345] W. Lusky, *Some consequences of Rudin's paper "L_p-isometries and equimeasurability"*, Indiana Univ. Math. J. 27 (1978) 859–866.
[346] W. Lusky, *A note on rotation in separable Banach spaces*, Studia Math. 65 (1979) 239–242.
[347] W. Lusky, *A note on Banach spaces containing c_0 or C_∞*, J. Funct. Anal. 62 (1985) 1–7.
[348] W. Lusky, *Three-space problems and basis extensions*, Israel J. Math. 107 (1988) 17–27.
[349] W. Lusky, *Three-space problems and bounded approximation property*, Studia Math. 159 (2003) 417–434.
[350] S. Mac Lane, *Categories for the Working Mathematician*, GTM 5. Springer, 1971.
[351] S. Mac Lane, *Homology*, Grund. der math. Wissenschaften 114. Springer, 1975.
[352] W. Marciszewski. *A function space $C(K)$ not weakly homeomorphic to $C(K) \times C(K)$*, Studia Math. 88 (1988) 129–137.
[353] W. Marciszewski. *On Banach spaces $C(K)$ isomorphic to $c_0(\Gamma)$*, Studia Math. 156 (2003) 295–302.
[354] W. Marciszewski, G. Plebanek, *Extension operators and twisted sums of c_0 and $C(K)$ spaces*, J. Funct. Anal. 274 (2018) 1491–1529.
[355] W. Marciszewski, R. Pol, *On Banach spaces whose norm-open sets are F_σ sets in the weak topology*, J. Math. Anal. Appl. 350 (2009) 708–722.
[356] V. Mascioni, *Topics in the theory of complemented subspaces in Banach spaces*, Expo. Math. 7 (1989) 3–47.
[357] B. Maurey, *Types and ℓ_1 subspaces*, Longhorn Notes, Univ. Texas, Funct. Anal. Seminar 1982/83, 123–137.
[358] B. Maurey, *Comments on the paper "The basic sequence problem" by N. J. Kalton, Studia Math. 116 (1995), 167–187*, in *Nigel J. Kalton Selecta*, Vol. 1. Edited by F. Gesztesy, G. Godefroy, L. Grafakos, I. Verbitsky. Contemporary Mathematicians. Birkhäuser, 2016, pp. 141–146.
[359] B. Maurey, G. Pisier, *Series de variables aletoires vectorielles indépendantes et propriétés géométriques des espaces de Banach*, Studia Math. 58 (1976) 45–90.
[360] S. Mazurkiewicz, W. Sierpiński, *Contribution à la topologie des ensembles dénombrables*, Fund. Math. 1 (1920) 17–27.
[361] E. Michael, A. Pełczyński, *Separable Banach spaces which admit ℓ_∞^n approximations*, Israel J. Math. 4 (1966) 189–198.
[362] V. Milman, G. Schechtman, *Asymptotic Theory of Finite Dimensional Normed Spaces*, Lecture Notes in Math. 1200. Springer, 1986.
[363] V. Milman, M. Sharir, *A new proof of the Maurey-Pisier theorem*, Israel J. Math. 33 (1979) 73–87.
[364] A. A. Milyutin, *Isomorphism of the spaces of continuous functions over compact sets of the cardinality of the continuum.* Teor. Funkcii Funkcional. Anal. i Priložen. (Kharkov) 2 (1966), 150–156 (Russian).
[365] Y. Moreno, *Theory of z-linear maps*, PhD thesis, Univ. Extremadura, 2003.

[366] Y. Moreno, *The diagonal functors*, Appl. Cat. Structures 16 (2008) 617–627.
[367] Y. Moreno, A. Plichko, *On automorphic Banach spaces*, Israel J. Math. 169 (2009) 29–45.
[368] F. J. Murray, *On complementary manifolds and projections in spaces L_p and ℓ_p*, Trans. Amer. Math. Soc. 41 (1937) 138–152.
[369] M. Nakamura, S. Kakutani, *Banach limits and the Čech compactification of an uncountable discrete set*, Proc. Imp. Acad. Japan 19 (1943) 224–229.
[370] E. Oja, *A note on M-ideals of compact operators*, Acta Comment. Univ. Tartu. 960 (1993) 75–92.
[371] H. G. Olsen. *Edwards' separation theorem for complex Lindenstrauss spaces with application to selection and embedding theorems*, Math. Scandinavica 38 (1976) 97–105.
[372] A. Ortyński, *On complemented subspaces of $\ell^p(\Gamma)$ for $0 < p \leq 1$*, Bull. Acad. Pol. Sci. 26 (1978) 31–34.
[373] I. I. Parovičenko, *On a universal bicompactum of weight $\aleph$*, Dokl. Akad. Nauk SSSR 150 (1963) 36–39.
[374] J. R. Partington, *Subspaces of certain Banach sequence spaces*, Bull. London Math. Soc. 13 (1981) 162–166.
[375] N. T. Peck, T. Starbird, *L_0 is ω-transitive*, Proc. Amer. Math. Soc. 83 (1981), no. 4, 700–704.
[376] A. Pełczyński, *Projections in certain Banach spaces*, Studia Math. 19 (1960) 209–228.
[377] A. Pełczyński, *Linear extensions, linear averagings, and their applications to linear topological classification of spaces of continuous functions,* Dissertationes Math. 58 (1968).
[378] A. Pełczyński, *Banach spaces on which every unconditionally converging operator is weakly compact*, Bull. Polish Acad. Sci. 10 (1962) 641–648.
[379] A. Pełczyński, *Uncomplemented function algebras with separable annihilators*, Duke Math. J. 33 (1966), 605–612.
[380] A. Pełczyński, *On $C(S)$-subspaces of separable spaces*, Studia Math. 31 (1968) 513–522.
[381] A. Pełczyński, *Universal bases*, Studia Math. 32 (1969) 247–268.
[382] A. Pełczyński, *Any separable Banach space with the bounded approximation property is a complemented subspace of a Banach space with a basis*, Studia Math. 40 (1971) 239–242.
[383] A. Pełczyński, *Selected problems on the structure of complemented subspaces of Banach spaces*, in *Methods in Banach Space Theory*. Edited by J. M. F. Castillo, W. B. Johnson, London Math. Soc. Lecture Notes 337. Cambridge University Press, 2006, pp. 341–354.
[384] A. Pełczyński, P. Wojtaszczyk, *Banach spaces with finite dimensional expansions of identity and universal bases of finite dimensional spaces*. Studia Math. 40 (1971) 91–108.
[385] V. Pestov, *Dynamics of infinite-dimensional groups*, Univ. Lecture Series, 40. American Math. Soc., 2006.
[386] R. S. Phillips, *On linear transformations*, Trans. Amer. Math. Soc. 48 (1940) 516–541.
[387] A. Pietsch, *History of Banach Spaces and Linear Operators*. Birkhäuser 2007.

[388] G. Pisier, *Le problème des 3 espaces: un contre-exemple de J. Lindenstrauss*, (French) Séminaire Maurey-Schwartz 1974–1975: Espaces L_p, applications radonifiantes et géométrie des espaces de Banach, Exp. No. XVII, 10 pp. Centre Math., École Polytech., 1975.

[389] G. Pisier, *Counterexamples to a conjecture of Grothendieck*, Acta. Math. 151 (1983) 181–208.

[390] G. Pisier, *Holomorphic semi-groups and the geometry of Banach spaces*, Ann. Math. 115 (1982) 375–392.

[391] G. Pisier, *The Volume of Convex Bodies and Banach Space Geometry*, Cambridge Tracts in Math. 94. Cambridge University Press, 1989.

[392] G. Pisier, *Comments on the paper "Twisted Hilbert spaces and unconditional structure" by N. J. Kalton, J. Inst. Math. Jussieu 2 (2003), 401–408*, in *Nigel J. Kalton Selecta*, Vol. 2. Edited by F. Gesztesy, G. Godefroy, L. Grafakos, I. Verbitsky. Contemporary Mathematicians. Birkhäuser, 2016, pp. 424–427.

[393] G. Plebanek, A. Salguero Alarcón, *The complemented subspace problem for $C(K)$-spaces: a counterexample*, preprint 2022, arXiv:2111.13860v2.

[394] A. Plichko, *Examples of n-Sobzcyk spaces*, in *General Topology and Banach Spaces*. Edited by A. Plichko, T. Banakh. NOVA, 2001, pp. 111–113.

[395] A. Plichko, *Superstrictly singular and superstrictly cosingular operators*, in *Functional Analysis and Its Applications*. Edited by V. Kadets, W. Żelazko. North-Holland Math. Stud. 197. Elsevier, 2004, pp. 239–255.

[396] A. Plichko, D. Yost, *Complemented and uncomplemented subspaces of Banach spaces*, Extracta Math. 15 (2000) 335–371.

[397] N. Popa, *On complemented subspaces of ℓ_p*, $0 < p < 1$, Rev. Roum. Math. Pures Appl. 26 (1981) 287–299.

[398] B. Randrianantoanina, *On isometric stability of complemented subspaces of L_p*, Israel J. Math. 113 (1999) 45–60.

[399] M. L. Reese, *Almost-atomic spaces*, Illinois J. Math. 36 (1992) 316–324.

[400] J. Reif, *A note on Markusevic bases in weakly compactly generated Banach spaces*, Comment. Math. Univ. Carolin. 15 (1974) 335–340.

[401] M. Ribe, *Examples for the nonlocally convex three space problem*, Proc. Amer. Math. Soc. 73 (1979) 351–355.

[402] E. Riehl, *Category Theory in Context*, Aurora Mod. Math Orig. Dover, 2016.

[403] J. W. Roberts, *A non-locally convex F-space with the Hahn-Banach extension property*, in *Banach spaces of Analytic Functions*. Edited by J. Baker, C. Cleaver, J. Diestel. Lecture Notes in Math. 604. Springer, 1977, pp. 76–81.

[404] J. W. Roberts, *Every locally bounded space with trivial dual is the quotient of a rigid space*, Illinois J. Math. 45 (2001) 1119–1144.

[405] R. Rochberg, G. Weiss, *Derivatives of analytic families of Banach spaces*, Ann. Math. 118 (1983) 315–347.

[406] B. Rodríguez-Salinas, *On the complemented subspaces of $c_0(I)$ and $\ell_p(I)$ for* $1 < p < \infty$, Atti. Sem. Mat. Fis. Univ. Modena 42 (1994) 399–402.

[407] W. Roelcke, *Einige Permanenzeigenschaften bei topologischen Gruppen und topologischen Vektorräumen*. Vortrag Funkt. Oberwolfach, 1972.

[408] S. Rolewicz, *Metric Linear Spaces* (2nd Ed.). PWN and D. Reidel, 1984.

[409] H. P. Rosenthal, *Projections onto Translation Invariant Subspaces of $L_p(G)$*, Mem. Amer. Math. Soc. 63 (1966)

[410] H. P. Rosenthal, *On totally incomparable Banach spaces*, J. Funct. Anal. 4 (1969) 167–175.

[411] H. P. Rosenthal, *On relatively disjoint families of measures, with some applications to Banach space theory*, Studia Math. 37 (1970) 13–36.

[412] H. P. Rosenthal, *On the subspaces of L_p ($p > 2$) spanned by sequences of independent random variables*, Israel J. Math. 8 (1970) 273–303.

[413] H. P. Rosenthal, *On injective Banach spaces and the spaces $L^\infty(\mu)$ for finite measures μ*, Acta Math. 124 (1970) 205–248.

[414] H. P. Rosenthal, *The Banach spaces $C(K)$ and $L^p(\mu)$*, Bull. Amer. Math. Soc. 81 (1975) 763–781.

[415] H. P. Rosenthal, *On factors of $C[0, 1]$ with non-separable dual*, Israel J. Math. 13 (1972) 361–378.

[416] H. P. Rosenthal, *The complete separable extension property*, J. Operator Theory 43 (2000) 329–374.

[417] H. P. Rosenthal, *On the subspaces of L^p ($p > 2$) spanned by sequence of independent random variables*, Israel J. Math. 8 (1970) 273–303.

[418] H. P. Rosenthal, *On a theorem of J.L. Krivine concerning block finite representability of l^p in general Banach spaces*, J. Funct. Anal. 28 (1978) 197–225.

[419] H. P. Rosenthal, *Some aspects of the subspace structure of infinite-dimensional Banach spaces*, in *Approximation Theory and Functional Analysis*. Edited by C. Chuy. Academic Press, 1991, pp. 151–176.

[420] H. P. Rosenthal, *The Banach spaces $C(K)$*, in *Handbook of the Geometry of Banach Spaces*, Vol. 2. Edited by W. B. Johnson, J. Lindenstrauss. Elsevier, 2003, pp. 1547–1602.

[421] W. Rudin, *Functional Analysis,* Int. Series in Pure and Appl. Math. McGraw-Hill, 1991.

[422] K.-S. Saito, M. Kato, Y. Takahashi, *Absolute norms on $\mathbb{C}^{n,1}$*, J. Math. Anal. Appl. 252 (2000) 879–905.

[423] M. Scott Osborne, *Basic Homological Algebra*, GTM 196. Springer, 2000.

[424] H. H. Schaefer, *Banach Lattices and Positive Operators*, Grund. der math. Wissenschaften 215. Springer, 1974.

[425] R. L. Schilling, *Measures, Integrals and Martingales.* Cambridge University Press, 2005.

[426] T. Schlumprecht, *A complementably minimal space not containing c_0 or ℓ_p*, unpublished.

[427] M. Schreiber, *Quelques remarques sur les caractérisations des espaces L_p*, $0 \leq p < 1$, Ann. Inst. Henri Poincaré 8 (1972) 83–92.

[428] J. Schreier, *Ein Gegenbeispiel zur Theorie der schwachen Konvergenz*, Studia Math. 2 (1930), 58–62.

[429] Z. Semadeni, *The Banach Mazur functor and related functors*, Comment. Math. Prace Math. 14 (1970) 173–182.

[430] Z. Semadeni, *Banach Spaces of Continuous Functions, Vol. 1.* PWN Warszawa, 1971.

[431] Z. Semadeni, H. Zidenberg, *Inductive limits in the category of Banach spaces*, Bull. Acad. Sci. Pol. 13 (1965) 579–583.

[432] S. Shelah, *Uncountable constructions for B.A.e.c. groups and Banach spaces*, Israel J. Math. 51 (1985) 273–297.

[433] W. Sierpiński, *Cardinal and Ordinal Numbers*. PWN Warszawa, 1965.

[434] B. Sims, *"Ultra"-techniques in Banach Space Theory*. Queen's Papers in Pure and Appl. Math. 60. Kingston, ON, 1982.

[435] B. Sims, *A support map characterization of the Opial conditions*, Proc. Centre Math. Anal. Austral. Nat. Univ. 9 (1985) 259–264.

[436] V. A. Smirnov, C. Khuen, *On the functor* Ext *in the category of linear topological spaces*, Math. USSR Izv. 36 (1991) 199–210.

[437] V. A. Smirnov, V. A. Sheikhman, *Continuation of homogeneous functionals with a given convexity*, Mat. Zametki 50 (1991) 90–96. English transl. Math. Notes 50 (1991) 1157–1161.

[438] A. Sobczyk, *Projections in Minkowski and Banach spaces*, Duke Math. J. 8 (1941) 78–106.

[439] A. Sobczyk, *Projection of the space m on its subspace* c_0, Bull. Amer. Math. Soc. 47 (1941) 938–947.

[440] A. Sobczyk, *On the extension of linear transformations*, Trans. Amer. Math. Soc. 55 (1944) 153–169.

[441] C. Stegall, *Duals of certain spaces with the Dunford-Pettis property*, Not. Amer. Math. Soc. 19 (1972) A-799.

[442] C. Stegall, *Banach spaces whose duals contain* $\ell_1(\Gamma)$ *with applications to the study of dual* $L_1(\mu)$*-spaces*, Trans. Amer. Math. Soc. 176 (1973) 463–477.

[443] W. J. Stiles, *On properties of subspaces of* ℓ_p, $0 < p < 1$, Trans. Amer. Math. Soc. 149 (1970) 405–415.

[444] W. J. Stiles, *Some properties of* ℓ_p, $0 < p < 1$, Studia Math. 42 (1972) 109–119.

[445] J. Suárez de la Fuente, *The Kalton centralizer on* L_p *is not strictly singular*, Proc. Amer. Math. Soc. 141 (2013), no. 10, 3447–3451.

[446] J. Suárez de la Fuente, *On the uniform structure of separable* $\mathscr{L}_\infty$ *spaces*, J. Funct. Anal. 266 (2014) 1050–1067.

[447] A. Szankowski, *Subspaces without the approximation property*, Israel J. Math. 30 (1978) 123–129.

[448] A. Szankowski, *Three-space problems for the approximation property*, J. Eur. Math. Soc. 11 (2009) 273–282.

[449] T. Szarvas, *Uniform* $L_p(w)$*-spaces*, Illinois J. Math. 45 (2001) 1145–1160.

[450] H. Tong, *Some characterizations of normal and perfectly normal spaces*, Duke Math. J. 19 (1952) 289–292.

[451] W. Veech, *A short proof of Sobczyk theorem*, Proc. Amer. Math. Soc. 28 (1971) 627–628.

[452] D. Vogt, *Lectures on projective spectra of (DF) spaces,* Lectures held in the Functional Analysis Seminar, Dusseldorf/Wuppertal, 1987, unpublished.

[453] R. C. Walker, *The Stone-Čech Compactification,* Ergeb. Math. Grenzgebiete 83. Springer, 1974.

[454] H. M. Wark, *A class of primary Banach spaces*, J. Math. Anal. Appl. 326 (2007) 1427–1436.

[455] N. Weaver, *Lipschitz Algebras*. World Scientific, 1999.

[456] J. H. Williamson, *Compact linear operators in linear topological spaces*, J. London Math. Soc. 29 (1954) 129–256.

[457] P. Wojtaszczyk, *Some remarks on the Gurarij space,* Studia Math. 41 (1972) 207–210.

[458] P. Wojtaszczyk, *Banach Spaces for Analysts*, Cambridge Studies in Advanced Mathematics 25. Cambridge University Press, 1991.

[459] K.-W. Yang, *A note on reflexive Banach spaces*, Proc. Amer. Math. Soc. 18 (1967) 859–861.

[460] D. Yost, *The Johnson-Lindenstrauss space*, Extracta Math. 12 (1997) 185–192.

[461] D. Yost, *A different Johnson-Lindenstrauss space*, New Zealand J. Math. 36 (2007) 1–3.

[462] M. Zippin, *The separable extension problem*, Israel J. Math. 26 (1977) 372–387.

[463] M. Zippin, *The embedding of Banach spaces into spaces with structure*, Illinois J. Math. 34 (1990) 586–606.

[464] M. Zippin, *A Global Approach to Certain Operator Extension Problems*, Lecture Notes in Math. 1470. Springer, 1990.

[465] M. Zippin, *Applications of Michael's continuous selection theorem to operator extension problems*, Proc. Amer. Math. Soc. 127 (1999) 1371–1378.

[466] M. Zippin, *Extension of bounded linear operators*, in *Handbook of the Geometry of Banach Spaces*, Vol 2. Edited by W. B. Johnson, J. Lindenstrauss. Elsevier, 2003, pp. 1703–1741.

PS. This book grew in many places, and the second author has been happy in a few: La casa de muchos colores, Institute of the Polish Academy of Sciences (Warszawa), Residenza Galaxy Studi Superiori (Bologna), la casa de mi madre, apartamentos Escalo (Formentera), Hotel Golden Tower (Sao Paulo), la cabaña de Teresa y Alma (Formentera), Villa Adriana (Naxos), Oliaros House (Antiparos), Qubus Hotel (Lodz), Intercontinental Hotel (Warszawa), Casa la Danza (Formentera), Orea (Santorini), Apollon's Village (Anafi), Argyris and George's house at Psathi (Ios), Vrahos Boutique Hotel (Folegandros), MarNik Village (Milos), the Blue Café and *Αρμενακι* tavern, where I felt like a bird on a wire, the BIRS center at Banff, la casa del Tigre along an extraordinary June, Don's place (Chios), Annoula & Stavros Golden Sun hotel (Patmos), Aronis Studios – but don't order lamb for dinner – (Naxos), Porto Vila (Santorini), Stella (Naxos), Nostos (Santorini), Maganas (Astipalea), Bambi's house (Kos), Banach Center (Bedlewo), Hotel Grace (Shin-Yokohama), Lino's house (Bologna), Atlas Hotel (Lviv), Astoria Apartments (Bologna), Es Pins (Formentera).

Index

Printed in the United States
by Baker & Taylor Publisher Services